Edwin I. Stein

FUNDAMENTALS OF MATHEMATICS

Allyn and Bacon, Inc.

Boston Rockleigh, N.J. Atlanta
Dallas Belmont, Calif.

Seventh Edition

Cover design: Christopher Valente
Cover photo: Jerry Kresch

ISBN: 0–205–06895–2

Printed in the United States of America

2 3 4 5 6 7 8 9 88 87 86 85 84 83 82 81 80

Preface

Fundamentals of Mathematics is a comprehensive basal textbook in general mathematics for the junior and senior high schools. It may be used as a one-year or two-year textbook. It contains in a single volume a wealth of material in all the basic topics of mathematics and includes computational practice and related enrichment materials.

The preceding edition of *Fundamentals of Mathematics* was especially noted for its wide range of subject areas and topics of both traditional and contemporary mathematics. This new edition retains all these features and content but now also includes an expanded study of the metric system of measurement and applications, shortened names for large whole numbers, and flow charts. The up-dated consumer applications also include unit pricing of food, the annual percentage rate, and managing money.

Outstanding features of *Fundamentals of Mathematics* include the study of ancient and modern numeration systems; contemporary arithmetic; number theory; set theory and notation; logic; statistics; probability; matrices; structure of mathematics; modular arithmetic; a comprehensive modern algebra course with a generous treatment of equations, inequalities, and coordinate geometry; finite number systems; nonmetric geometry; modern interpretation of geometry; enrichment materials such as the Sieve of Eratosthenes, Goldbach's conjecture, Euclid's algorithm, Euler's formula, and Russian peasant and lattice methods of multiplication; and an abundance of everyday practical applications of mathematics and arithmetic practice to develop and maintain skills. The practice material is varied and usually graded by difficulty. Keyed Chapter Reviews and Cumulative Tests are provided throughout the text; and important explanatory material and sample problems are emphasized by color screens.

Fundamentals of Mathematics is a multi-level and multi-purpose textbook, ideal for use with mathematics classes of any grade in the junior high school and with general mathematics classes or specialty classes in the senior high school, including consumer mathematics, commercial arithmetic, and shop mathematics classes. It is excellent for programs calling for acceleration to save a half-year or a year. *Fundamentals of Mathematics* contains so many essential topics that no other supplementary book is necessary. This textbook meets the many and varied requirements of today's programs in mathematics.

The material is presented with clarity and simplicity, requiring no detailed knowledge of contemporary mathematics.

The author acknowledges the assistance of Charlotte Stein Jaffe and Marilyn Stein Lieberman.

EDWIN I. STEIN

Contents

9. UNITS OF MEASURE AND MEASUREMENT

10. SETS

11. ALGEBRA

14. INFORMAL GEOMETRY

Number and Sets

1

1–1 NUMBER AND NUMERAL

Number is an abstract or mental idea. The number symbols we write and see are not numbers but names for numbers which we call *numerals.* Although 19 is commonly called a number, actually it is a numeral or group of number symbols which represent the number named *nineteen.* However, number symbols are generally used to denote both numbers and numerals as "Add the numbers 27 and 36" and "Write the numeral 594."

Word names for any specific number differ because of differences in languages, although the same number symbols are used. In Spanish nineteen is *diez y nueve*; in French, *dix-neuf*; in Italian, *diciannove*; in German, *neunzehn.* A number may have many names; it may be represented by symbols in many ways.

The number named *nineteen* may be represented by any one of the following numerals:

19 XIX $12 + 7$ $32 - 13$ 19×1 $\frac{38}{2}$

Each of the number symbols 0, 1, 2, 3, 4, 5, 6, 7, 8, and 9 is called a *digit* or figure. The numeral 5 contains one digit; the numeral 46 contains two digits; the numeral 979 contains three digits; etc.

EXERCISES

1. Which is larger, numeral 7 or numeral 9?

2. Which is larger, number 7 or number 9?

3. Write a 3-digit numeral containing the digits 5, 4, and 7.
Write a 4-digit numeral containing the digits 3, 8, 9, and 6.

4. Write the numeral 5,273; rewrite the numeral but replace the digit 3 with the digit 8.

5. Write four different number names for each of the following numbers:

a. six **b.** twenty **c.** fourteen **d.** fifty **e.** eleven

1–2 ONE–TO–ONE CORRESPONDENCE

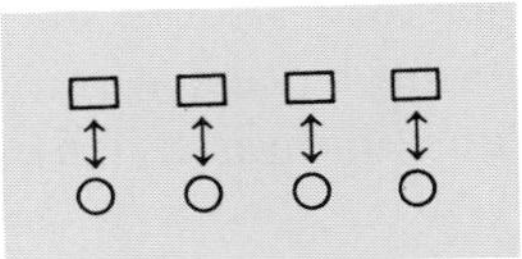

We see at the right that the number of □'s is exactly the same as the number of ○'s. When they are paired off, we find there is one □ for every ○ and one

○ for every □, no more no less. This kind of exact pairing or matching or mapping is called *one-to-one correspondence* and is indicated by a two-headed arrow since the matching is done in both directions. In counting there is one-to-one correspondence between the objects being counted and the set of counting numbers.

EXERCISES

1. Copy, then pair off the objects in the following sets, placing them in one-to-one correspondence by using the two-headed arrow.

A

× × ×

△ △ △

C

□ □ □ □ □

1 2 3 4 5

B

○ ○ ○ ○ ○ ○ ○

1 2 3 4 5 6 7

D

△ △ △ △

□ □ □ □

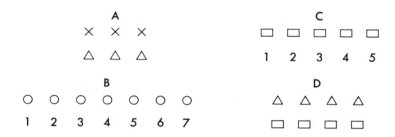

2. If two sets of objects are in one-to-one correspondence, is it true that they are equal in number?

3. In which of the following sets may the objects be paired so that there is one-to-one correspondence?

A

× × × × × × × ×

B

△ △ △ △ △ △

C

□ □ □ □ □ □

D

1 2 3 4 5 6 7 8

1–3 CARDINAL NUMBER AND ORDINAL NUMBER

Number answers the questions of "How many?" and "Which one?". The number that tells how many things are in a collection or its size is called a *cardinal number*. When we count, we match each object to a corresponding number belonging to a standard set of counting numbers and the last number named tells how many objects there are in all. The number that tells the position or order of an object in a collection such as first, second, third, fourth, and so on is called an *ordinal number*. The same number symbols are used to represent both cardinal and ordinal numbers.

EXERCISES

Which of the following are cardinal numbers? Which are ordinal numbers?

1. The class read page 5 of a book containing 360 pages.
2. Scott spent fifty cents for the third day in succession.
3. Stephen had 10 examples to do for homework but had only example number 9 incorrect.
4. The classroom contained 5 rows of 8 seats in each row; Charlotte sat in the seventh seat in the third row.
5. Marilyn ranked number 6 in her class of 130 pupils.

1–4 NATURAL NUMBERS—WHOLE NUMBERS

The numbers 0, 1, 2, 3, 4, 5, 6, 7, 8, 9, 10, 11, 12, 13, 14, etc. are called *whole numbers.* The whole numbers beginning with 1 that are used in counting are called counting or *natural numbers.* Zero is generally not considered to be a natural number.

Instead of the "etc." at the end of the set of numbers three dots may be used to indicate that the set of numbers is unlimited. The set 1, 2, 3, 4, . . . means "one, two, three, four, and so on endlessly." The set 1, 2, 3, . . ., 20 means "one, two, three, and so on up to and including twenty."

EXERCISES

1. Can you name the greatest natural number? Is there a greatest natural number?
2. Is there a last natural number? If so, what is it?
3. Is there a first natural number? If so, what is it?
4. Natural numbers have order; each is followed by another natural number, called the *successor,* which is one (1) greater. The natural number preceding another natural number is called the *predecessor.*
   a. In the set 1, 2, 3, 4, 5, 6, 7, 8, 9, 10, 11, 12, . . ., what is the successor of 6? The predecessor of 9?
   b. Arrange the following numbers in order of size, smallest first:
   14 8 6 11 3 13 26 19 5 31 15 28 20 37 18
5. What whole numbers are between 8 and 11? Between 13 and 21?

1–5 THE NUMBER LINE

A line may be thought of as an endless set of points. A straight line that has its points labeled with numerals so that each point is associated with a number is called a *number line.*

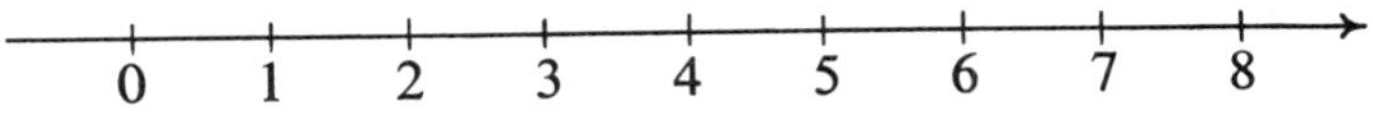

To draw a number line, two points are chosen with the one on the left labeled 0 and the other point labeled 1. With the interval between these points as the unit of measure, points are located to the right of the point marked 1, equally spaced along the line. Each point is assigned a corresponding whole number (*coordinate*) in consecutive order. There is a one-to-one correspondence between these numbers and the points on the line.

We may compare numbers by the number line. The number corresponding to the point on the line farther to the right is the larger number.

EXERCISES

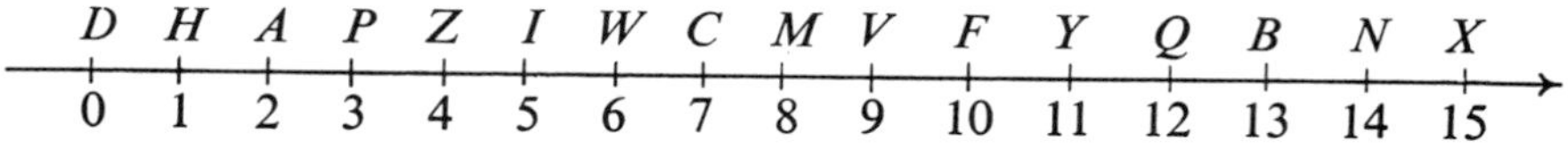

1. What number corresponds to the point marked:

A? *B*? *C*? *D*? *F*? *M*? *N*? *V*? *W*? *X*? *Y*? *Z*?

2. What letter labels the point corresponding to each of the following numbers:

5? 1? 13? 12? 3? 7? 0? 15? 8? 4? 9? 2?

3. Which corresponding point is farther to the right on the number line:

a. 6 or 9? **b.** 15 or 12? **c.** 14 or 8? **d.** 12 or 5? **e.** 0 or 10? **f.** 2 or 11?

1–6 ODD AND EVEN NUMBERS

Whole numbers may be separated into even and odd numbers. An *even number* is a whole number that is divisible (is divided exactly) by two (2). An *odd number* is a whole number that is not divisible by two (2). Zero is considered an even whole number.

EXERCISES

1. Which of the following are odd numbers? Which are even numbers?

18 25 63 70 47 100 709 942 1,836 23,861

2. Is there an odd natural number between every pair of even natural numbers? Is there an even natural number between every pair of odd natural numbers?

3. Does 1 more than any even number make an odd number? Does 1 more than any odd number make an even number?

4. Write the numerals of all one-digit odd numbers. Write the numerals of all the even numbers greater than 11 but less than 23.

5. Select any two odd numbers. Is their sum an odd number or an even number? Is their product an odd number or an even number?

6. Select any two even numbers. Is their sum an odd number or an even number? Is their product an odd number or an even number?
7. Select any odd and any even number. Is their sum an odd number or an even number? Is their product an odd number or an even number?

1–7 PRIME AND COMPOSITE NUMBERS

Whole numbers other than 0 and 1 may be separated into prime and composite numbers.

A *prime number* is a whole number other than 0 and 1 which is divisible (is divided exactly) only by itself and by 1 and by no other whole number.

> 17 is a prime number; it can be divided exactly only by 17 and by 1.

A *composite number* is a whole number other than 0 and 1 which is not a prime number. It can be expressed as a product of two or more smaller whole numbers.

> 10 is a composite number; it can be divided exactly not only by 10 and by 1 but also by 2 and by 5.

A composite number can be expressed as a product of prime numbers. Each composite number has only one set of prime factors but the factors may be arranged in different orders. See page 11.

> 10 may be expressed as 2×5 or 5×2.

A method called the *Sieve of Eratosthenes*, discovered by the Greek mathematician Eratosthenes, may be used to find prime numbers less than a given number.

To find all prime numbers less than 36 by this method, we write the numerals 2 to 35 inclusive (1 is excluded; it is not a prime number) as follows:

2 3 ~~4~~ 5 ~~6~~ 7 ~~8~~ ~~9~~ ~~10~~ 11 ~~12~~ 13 ~~14~~ ~~15~~ ~~16~~ 17 ~~18~~
19 ~~20~~ ~~21~~ ~~22~~ 23 ~~24~~ ~~25~~ ~~26~~ ~~27~~ ~~28~~ 29 ~~30~~ 31 ~~32~~ ~~33~~ ~~34~~ ~~35~~

We cross out numerals representing numbers which are not primes by the following procedure:

(1) 2 is a prime number. We retain numeral 2 but we cross out every second numeral after 2.

(2) 3 is a prime number. We retain numeral 3 but we cross out every third numeral after 3. Some of the numerals are already crossed out but they are included in the count.
(3) Numeral 4 is already crossed out.
(4) 5 is a prime number. We retain numeral 5 but we cross out every fifth numeral after five.
(5) We continue this process until all numerals for numbers other than prime numbers are crossed out.

The prime numbers less than 36 are: 2, 3, 5, 7, 11, 13, 17, 19, 23, 29, 31.

Two prime numbers are called *twin primes* if one number is two more than the other number. 29 and 31 are a pair of twin primes.

Two numbers are said to be *relatively prime* to each other when they have no common factors other than 1. The numbers do not necessarily have to be prime numbers. The numbers 14 and 25 are relatively prime to each other because 1 is the only whole number that will divide into both 14 and 25 exactly.

EXERCISES

1. Which of the following are prime numbers?

45 59 1 101 78 97 87 73 65 51

2. Which of the following are composite numbers?

84 27 79 111 0 53 61 57 91 47

3. Name five composite numbers that are greater than 88 but less than 95.
4. Are there any even prime numbers? If so, what are they?
5. Name all the one-digit prime numbers.
6. Name all the one-digit odd prime numbers.
7. Use the Sieve of Eratosthenes to find all the prime numbers less than 200.
8. What twin primes may be found among numbers less than 100? Between 100 and 200?
9. Goldbach, a German mathematician, conjectured that "any even number greater than 4 can be expressed as the sum of two odd prime numbers." For example:

$6 = 3 + 3$ $\quad 10 = 5 + 5$ or $7 + 3$ $\quad 24 = 5 + 19$ or $7 + 17$ or $11 + 13$

This conjecture has never been proved.
Choose any even number between the following given numbers and find a pair of odd prime numbers that has your chosen even number as their sum.

a. 31 and 41
b. 41 and 61
c. 61 and 81
d. 81 and 101

10. Which of the following are relatively prime?

a. 9 and 21 **b.** 72 and 13 **c.** 24 and 35 **d.** 54 and 45 **e.** 28 and 63

SETS

1–8 SET NOTATION

We are all familiar with collections of objects such as a set of golf clubs, a set of dishes, a set of tools, etc. In mathematics when we use the word *set*, we mean a well-defined collection of objects. Each object of a set is called a *member* or *element* of the set. A pair of braces {} is used to designate a set with the members listed or written inside the braces. The braces mean "the set of" or "the set whose members are."

> The expression {1, 3, 5, 7, 9} is read "the set whose members are one, three, five, seven, nine" and may be described as the set of one-digit odd numbers.

EXERCISES

1. Write each of the following sets listing the elements within braces:

a. 2, 5, 8, 11, and 14. **b.** All months whose names begin with the letter "M."
c. The last ten letters of our alphabet.
d. All two-digit odd numbers between 38 and 48.
e. All the states in the United States that border on the Pacific Ocean.
f. All whole numbers less than 7. **g.** All the subjects you are studying this year.
h. All whole numbers greater than 9 but less than 21.
i. All prime numbers between 20 and 30.
j. The first twelve even natural numbers.

2. Read each of the following:

a. {a, e, i, o, u}
b. {0, 2, 4, 6, 8}
c. {2, 3, 5, 7}
d. {Wyoming, Wisconsin, West Virginia, Washington}
e. {automobile, train, airplane, boat, bus}
f. {New York, Chicago, Los Angeles, Philadelphia}

3. Describe briefly in words each of the sets given above in example 2.

4. Write a set listing: **a.** 3 elements **b.** 7 elements **c.** 5 elements **d.** 6 elements

1–9 SET MEMBERSHIP, EQUAL SETS, EQUIVALENT SETS

Capital letters are used to designate sets:

> $B = \{1, 2, 3, 4, 5, 6, 7, 8, 9\}$ reads "*B* is the set whose members are one, two, three, four, five, six, seven, eight, nine."

The epsilon symbol "ϵ" or the symbol "$\in$" is used to denote that an object is a member of the set.

$5 \in B$ means "five is a member of set B."

The symbol "$\notin$" indicates that an object is *not* a member of the set.

$12 \notin B$ means "twelve is not a member of set B."

Small letters may name members of sets such as: $R = \{a, b, c, d\}$

Sets which contain exactly the same elements are called *equal sets.* {4, 2, 3} and {3, 4, 2} are equal sets. The elements may be listed or written in any order. It is unnecessary to repeat an element when listing them. Sets which contain the same number of elements are called *equivalent sets.* {1, 2, 3, 4} and {1, 3, 5, 7} are equivalent sets.

EXERCISES

1. Read, or write in words, each of the following:

a. $D = \{1, 4, 9, 16, 25\}$
b. $M = \{5, 10, 15, 20, 25, 30, 35\}$
c. $P =$ {Kennedy, Johnson, Nixon, Ford}
d. $S =$ {all natural numbers}

2. Write, using the proper symbols:

a. 9 is an element of set T
b. 11 is not an element of set Z
c. r is not a member of set E
d. n is a member of set A

3. If $B =$ {all even whole numbers}, which of the following are members of set B?

8 9 35 2 0 11 467 596 729 900

4. If $R =$ {all odd numbers greater than 17 but less than 43}, which of the following are *not* members of set R?

26 51 19 36 43 27 33 21 44 13

5. If $T = \{1, 2, 3, 4, 5, 6, 7\}$, which of the following statements are true?

a. $6 \in T$ **b.** $8 \in T$ **c.** $10 \notin T$ **d.** $3 \notin T$ **e.** $5 \in T$ **f.** $0 \notin T$

6. If $C =$ {all prime numbers}, which of the following statements are true?

a. $7 \in C$ **b.** $51 \in C$ **c.** $24 \notin C$ **d.** $97 \notin C$ **e.** $23 \in C$ **f.** $31 \notin C$

7. If $N = \{1, 4, 9, 16, 25, 36, 49, 64, 81\}$, which of the following statements are false?

a. $25 \in N$ **b.** $48 \in N$ **c.** $81 \notin N$ **d.** $35 \notin N$ **e.** $100 \in N$ **f.** $64 \notin N$

8. a. Are {2, 7, 9} and {9, 3, 7} equal sets? Equivalent sets?
b. Are {5, 1, 6, 8, 3} and {8, 6, 1, 3, 5} equal sets? Equivalent sets?
c. Are {4, 8, 12, 16, 20} and {16, 20, 8, 4} equal sets? Equivalent sets?

1–10 NULL SET, FINITE SET, INFINITE SET

Sets may contain no elements, one element, a definite number of elements or an unlimited number of elements. A set containing no elements is called the *null set* or *empty set* and is designated by the symbol "∅" or "{ }". {0} is not an empty set; it contains one element which is zero (0). A set that contains a limited number of elements is called a *finite set*. A set that contains an unlimited number of elements is called an *infinite set*.

When the set contains an unlimited number of elements, it may be described or have a limited number of elements listed followed by three dots.

> The set of all natural numbers may be expressed as:
>
> {all natural numbers} or {1, 2, 3, 4, 5, . . .}

EXERCISES

1. Which of the following sets are finite sets? Which are infinite sets?
   a. {all whole numbers} **b.** {all letters of our alphabet}
   c. {all natural numbers less than 1,000,000,000}
   d. {entire population of the world} **e.** {all odd numbers}
2. Which of the following is a null set?
   a. {all natural numbers between 4 and 6}
   b. {all prime numbers between 7 and 11}
   c. {all whole numbers between 19 and 20}
   d. {all prime numbers less than 3} **e.** {all pupils in your class 4 years old}

FACTORS AND FACTORING

1–11 FACTORS

When we multiply four by five to get the answer twenty (in symbols $4 \times 5 = 20$), the numbers 4 and 5 are called *factors* and the number in the answer is called the *product*.

A *factor* is any one of the numbers used in multiplication to form a product. Thus, one factor × another factor = product.

EXERCISES

Name the factors and the product in each of the following:

a. $7 \times 3 = 21$ **c.** $1 \times 8 = 8$ **e.** $5 \times 0 = 0$ **g.** $12 \times 9 = 108$
b. $5 \times 9 = 45$ **d.** $11 \times 6 = 66$ **f.** $15 \times 15 = 225$ **h.** $25 \times 5 = 125$

We see from $4 \times 5 = 20$, $20 \div 4 = 5$, and $20 \div 5 = 4$, that 20 is divided exactly by factor 4 and by factor 5; from $10 \times 2 = 20$, $20 \div 2 = 10$, and

$20 \div 10 = 2$, that 20 is divided exactly by factor 2 and by factor 10; from $20 \times 1 = 20$, and $20 \div 1 = 20$ and $20 \div 20 = 1$ that 20 is divided exactly by factor 1 and by factor 20.

The factors of 20 are: 1, 2, 4, 5, 10, and 20. This set of factors may be written as:

{1, 2, 4, 5, 10, 20}

A factor of a number divides the number exactly with *no* remainder. Every natural number has as factors 1 and the given number itself. The other factors are whole numbers which divide the given number exactly. Knowledge of the multiplication facts will help determine factors by inspection. Or we may select small numbers and test whether they divide the given number exactly. If any number does, both the divisor and the quotient are factors.

Other pairs of factors of a given number may be found from a known pair of factors, whose product is the given number, by dividing one factor by a number that makes it divisible and then multiplying the other factor by this number. For example, using 8 and 9, a pair of factors of 72, we divide factor 8 by 2 and multiply factor 9 by 2 to get factors 4 and 18; if we divide factor 18 by 3 and multiply factor 4 by 3, we get factors 6 and 12, etc.

EXERCISES

1. Can 24 be divided exactly by 2? by 12? by 3? by 4? by 5? by 6? by 8? by 10? by 18? by 24? Write the set of factors of 24.

2. Is 4 a factor of 36? Is 12 a factor of 60? Is 18 a factor of 54? Is 7 a factor of 54?

3. What is the other factor when:

a 5 is one factor of 30? **c.** 8 is one factor of 56? **e.** 10 is one factor of 1,000?
b. 9 is one factor of 63? **d.** 12 is one factor of 96? **f.** 100 is one factor of 100,000?

4. Find the whole set of factors:

a. Of 96, using the factors 12 and 8 to determine other pairs of factors.
b. Of 144, using the factors 9 and 16 to determine other pairs of factors.

5. Write the whole set of factors of each of the following numbers:

a. 8	**c.** 14	**e.** 54	**g.** 56	**i.** 60	**k.** 120	**m.** 80	**o.** 100
b. 19	**d.** 32	**f.** 90	**h.** 84	**j.** 125	**l.** 64	**n.** 92	**p.** 360

1–12 FACTORING A NATURAL NUMBER

To *factor a natural number* means to replace the number by its whole-number factors expressed as an indicated product. This indicated product may contain two or more whole-number factors. A prime number may be expressed only as the product of the given number and 1. The prime number 29 is expressed as 29×1.

To factor 20 means to replace 20 by any one of the following indicated products of:

(1) Two whole-number factors: 1×20 or 2×10 or 4×5 or 5×4 or 10×2 or 20×1.

(2) Three whole-number factors: $2 \times 2 \times 5$ or $1 \times 2 \times 10$ or $1 \times 4 \times 5$ arranged in different orders.

(3) More than three whole-number factors since the factor 1 may be repeated: $20 \times 1 \times 1 \times 1$ or $4 \times 5 \times 1 \times 1 \times 1 \times 1 \times 1$ or $2 \times 2 \times 5 \times 1 \times 1 \times 1 \times 1 \times 1 \times 1$ etc. However, the factor 1 is usually excluded when there are more than two whole-number factors in the indicated product.

EXERCISES

1. Factor as the product of two whole-number factors in as many ways as possible:

a. 16 **b.** 11 **c.** 42 **d.** 64 **e.** 80 **f.** 100 **g.** 52 **h.** 110

2. Factor as the product of three whole-number factors in as many ways as possible:

a. 30 **b.** 66 **c.** 70 **d.** 18 **e.** 98 **f.** 75 **g.** 63 **h.** 242

3. Factor as the product of four whole-number factors in as many ways as possible:

a. 36 **b.** 90 **c.** 56 **d.** 150 **e.** 132 **f.** 210 **g.** 330 **h.** 441

1–13 COMPLETE FACTORIZATION

A composite number may be expressed as a product of prime numbers. The number is said to be *completely factored* only if the factors are all prime numbers (prime factors).

When 20 is factored as 4×5 or 10×2, it is not completely factored since the 4 and 10 are not prime numbers. However 4×5 may be expressed as $2 \times 2 \times 5$ by factoring the 4; and 10×2 may be expressed as $5 \times 2 \times 2$ or $2 \times 5 \times 2$ by factoring the 10.

Complete factoring of 20 produces $2 \times 2 \times 5$ or $5 \times 2 \times 2$ or $2 \times 5 \times 2$. Observe that the factors are the same 2, 2, and 5 but they appear in different orders. Thus, each composite number has only one set of prime factors but the prime factors may be arranged in different orders.

To factor completely, we first find two factors of the given number and then continue factoring any of the factors which are composite numbers until only prime factors result. For example:

$$20 = 4 \times 5 = 2 \times 2 \times 5 \qquad \text{or} \qquad 20 = 2 \times 10 = 2 \times 2 \times 5$$

Each of these may be arranged as a factor tree:

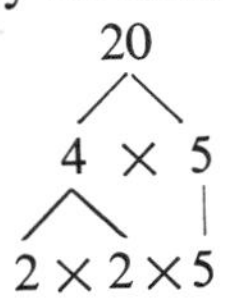

or

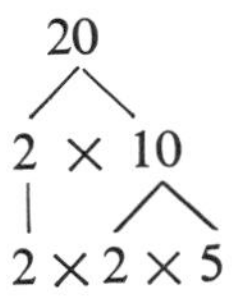

Or we divide the given number and the resulting quotients successively by prime numbers that divide these numbers exactly until a quotient of 1 is obtained. The divisors are the prime factors. Observe in the model example at the right how the quotients are brought down in each row.

2)20
2)10
5) 5
    1

$20 = 2 \times 2 \times 5$

EXERCISES

1. Factor each of the following numbers as a product of prime numbers, using a factor tree:

   a. 42 **b.** 27 **c.** 24 **d.** 64 **e.** 90 **f.** 72 **g.** 100 **h.** 320

2. Factor each of the following numbers as a product of prime numbers:

a. 56	**d.** 60	**g.** 105	**j.** 132	**m.** 300
b. 32	**e.** 36	**h.** 200	**k.** 180	**n.** 480
c. 75	**f.** 144	**i.** 54	**l.** 225	**o.** 720

1–14 COMMON FACTORS

3 is a factor of 6. Factors of 6 are: 1, 2, 3, 6
3 is a factor of 15. Factors of 15 are: 1, 3, 5, 15
Therefore, 3 is a common factor of 6 and 15. The set of common factors of 6 and 15 is {1, 3}.

Any number that is a factor of each of two or more given whole numbers is called a *common factor* of the numbers (sometimes called *common divisor* of the numbers).

Observe that 1 is a common factor of any set of numbers. Some given numbers like 2 and 3 have no common factor except 1; other numbers like 4 and 10 have one common factor (2) other than 1; still other numbers like 18 and 24 have many common factors (2, 3, and 6) other than 1.

EXERCISES

1. What are the factors of 16? What are the factors of 24? What are the common factors of 16 and 24?
2. What are the factors of 8? What are the factors of 12? What are the factors of 20? What are the common factors of 8, 12, and 20?
3. For each of the following sets of numbers, first find the factors of each number, then find their common factors:

   a. 6 and 8 **b.** 9 and 15 **c.** 36 and 60 **d.** 32 and 72 **e.** 48, 96, and 120

4. For each of the following sets of numbers write the set of common factors:

   a. 4 and 10 **b.** 24 and 32 **c.** 35 and 63 **d.** 10, 12, and 16 **e.** 54, 72, and 90

1–15 GREATEST COMMON FACTOR

The factors of 18 are: 1, 2, 3, 6, 9, 18
The factors of 24 are: 1, 2, 3, 4, 6, 8, 12, 24
The common factors of 18 and 24 are 1, 2, 3, and 6. The greatest common factor is 6

The *greatest common factor* of two or more whole numbers is the greatest whole number that will divide all the given numbers exactly.

To determine the greatest common factor (G.C.F.), or greatest common divisor, of two or more given numbers, we factor each given number completely as a product of prime numbers. Then we select all the prime factors that are common to all the given numbers and find their product,

Find the G.C.F. of 24 and 18:

$24 = 2 \cdot 2 \cdot 2 \cdot 3$
$18 = 2 \cdot 3 \cdot 3$

Prime factors 2 and 3 are common.
Therefore the G.C.F. $= 2 \cdot 3$ or 6.

Find the G.C.F. of 36 and 90:

$36 = 2 \cdot 2 \cdot 3 \cdot 3$
$90 = 2 \cdot 3 \cdot 3 \cdot 5$

Prime factors 2, 3, and 3 are common.
Therefore the G.C.F. $= 2 \cdot 3 \cdot 3$ or 18.

Or we may use Euclid's method of finding the G.C.F. of two given numbers as follows:

We divide the larger given number by the smaller number. Then we divide the divisor by the remainder, then we divide the next divisor by the next remainder, continuing in this way until the remainder is zero. The *last non-zero remainder* is the greatest common factor (G.C.F.).

Find the G.C.F. of 24 and 18:

18) 24 (1 R6
6) 18 (3 R0

Answer, G.C.F. = 6

EXERCISES

1. What are the factors of 64? What are the factors of 96? What are the common factors of 64 and 96? What is the greatest common factor of 64 and 96?
2. What are the common factors of 45, 75, and 120? What is the greatest common factor of 45, 75, and 120?
3. Find the greatest common factor of each of the following sets of numbers:

a. 5 and 6
b. 36 and 42
c. 24 and 40
d. 8 and 28
e. 65 and 91
f. 87 and 58
g. 6, 8, and 10
h. 8, 12, and 18
i. 24, 36, and 108

4. Use Euclid's method to find the greatest common factor of:

a. 24 and 60
b. 45 and 100
c. 35 and 84
d. 63 and 90
e. 108 and 132
f. 306 and 414

NUMBER MULTIPLES

1–16 MULTIPLES

A *multiple* of a given whole number is a product of the given number and another whole number factor. Since $0 \times 7 = 0$; $1 \times 7 = 7$; $2 \times 7 = 14$; $3 \times 7 = 21$; $4 \times 7 = 28$; and $5 \times 7 = 35$, the products 0, 7, 14, 21, 28, and 35 are multiples of 7. Each is a product of 7 and another number. A multiple of a given number is divisible by the given number. The above multiples of 7 (0, 7, 14, 21, 28, and 35) are divisible by 7.

The set of all multiples of 7 may be expressed as: {0, 7, 14, 21, 28, 35, 42, . . .}

EXERCISES

1. Name four different multiples of each of the following numbers:

a. 3 **b.** 2 **c.** 9 **d.** 15 **e.** 4 **f.** 20 **g.** 48 **h.** 100

2. Which of the following numbers are multiples of 6?

52 72 28 48 32 36 90 86 100 102

3. Are all whole numbers multiples of 1?

4. Are all even whole numbers multiples of 2?

5. 5 is a factor of 15. Is 15 a multiple of 5? 9 is a factor of 72. Is 72 a multiple of 9? If a number is a factor of a second number, is the second number a multiple of the first number?

6. What are the factors of 32? Is 32 a multiple of each of its factors?

7. Write the set of all multiples of each of the following, listing the first five members:

a. 4	**d.** 12	**g.** 18	**j.** 72	**m.** 200
b. 6	**e.** 11	**h.** 13	**k.** 36	**n.** 125
c. 5	**f.** 30	**i.** 25	**l.** 50	**o.** 320

1–17 COMMON MULTIPLES

Any number which is a multiple of two or more numbers is called the *common multiple* of the numbers. Numbers may have many common multiples.

15 is a multiple of 3; 15 is a multiple of 5; 15 is a common multiple of 3 and 5.
Some of the other common multiples of 3 and 5 include 30, 45, 60, 75, and 90.
The set of all common multiples of 3 and 5 is expressed as: {0, 15, 30, 45, 60, . . .}

EXERCISES

1. Is 24 a multiple of 6? Is 24 a multiple of 8? Is 24 a common multiple of 6 and 8? Is 24 divisible by 6 and by 8? Is the common multiple of two or more given numbers divisible by each of the given numbers?

2. What are the first 15 multiples of 2? of 3? What are the first 5 common multiples of 2 and 3?
3. Write the set of all common multiples of each of the following, listing the first four members:

a. 2 and 5	**d.** 8 and 10	**g.** 18 and 22	**j.** 2, 4, and 8
b. 6 and 4	**e.** 5 and 4	**h.** 15 and 25	**k.** 6, 8, and 12
c. 3 and 7	**f.** 3 and 9	**i.** 2, 3, and 4	**l.** 10, 25, and 100

4. Is the product of 8 and 3 a common multiple of 8 and 3? Is the product of two whole numbers a common multiple of the two numbers?
5. Is the product of 6 and 10 a common multiple of 6 and 10? Is it the smallest common multiple? If not, what is the smallest natural number that can be divided exactly by both 6 and 10?

1–18 LEAST COMMON MULTIPLE

The *least common multiple* (L.C.M.) of two or more numbers is the smallest natural number which is the multiple of all of them. It is the smallest possible natural number that can be divided exactly by all the given numbers.

Zero (0) is excluded when determining the least common multiple, although it is a common multiple of any set of numbers.

The L.C.M. may be found by factoring the given numbers as primes and forming a product of these primes using each the greatest number of times it appears in the factored form of any one number.

The L.C.M. of 8 and 10 is found as follows:

Since $8 = 2 \cdot 2 \cdot 2$
and $10 = 2 \cdot 5$

Therefore, L.C.M. $= \overbrace{2 \cdot 2 \cdot 2}^{8} \cdot 5$, where $\underbrace{2 \cdot 5}_{10}$

Answer, L.C.M. $= 40$

EXERCISES

1. Find the least common multiple for each of the following sets:

a. 12 and 16	**e.** 20 and 24	**i.** 96 and 108	**m.** 6, 10, and 12
b. 3 and 4	**f.** 50 and 100	**j.** 84 and 144	**n.** 2, 3, and 5
c. 10 and 15	**g.** 32 and 56	**k.** 39 and 65	**o.** 4, 12, and 20
d. 14 and 21	**h.** 54 and 72	**l.** 90 and 135	**p.** 18, 27, and 45

2. Find the greatest common factor and the least common multiple of each of the following pairs of numbers. For each pair of numbers compare the product of the G.C.F. and L.C.M. with the product of the two given numbers. What do you find in each case?

a. 10 and 15 **b.** 32 and 20 **c.** 9 and 8 **d.** 72 and 96

CHAPTER REVIEW

1. Write three different numerals naming the number twelve. (1–1)
2. In which of the following sets may the objects be paired so that there is one-to-one correspondence? (1–2)

x △ △ △ △ △ y □ □ □ □ □ □ z ○ ○ ○ ○ ○

3. Does the sentence "Of the 25 players on the softball team, Lisa ranked number 3 in batting" contain a cardinal number? If so, name it. Does it contain an ordinal number? If so, name it. (1–3)
4. Are there any natural numbers between 18 and 23? If so, name them. (1–4)
5. On the number line what point corresponds to 5? to 3?

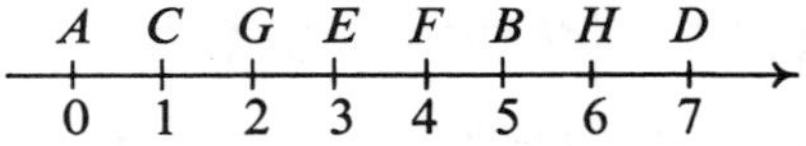

(1–5)

6. Which of the following are odd numbers? Which are even numbers? (1–6)

   221 972 1,050 694 409

7. Use the Sieve of Eratosthenes to find all the prime numbers less than 48. (1–7)
8. Name four composite numbers that are greater than 28 but less than 35. (1–7)
9. Are 15 and 36 relatively prime? (1–7)
10. Write each of the following sets listing the elements within braces: (1–8)

    a. All one-digit even prime numbers.
    b. All whole numbers greater than 6 but less then 13.

11. Write in words "$T = \{1, 2, 3, 4, \ldots, 12\}$." (1–9)
12. If $D = \{\text{all two-digit prime numbers}\}$, which of the following statements are true? (1–9)

    a. $21 \in D$ **b.** $53 \in D$ **c.** $7 \in D$ **d.** $19 \notin D$ **e.** $77 \notin D$

13. Are {5, 13, 20, 31} and {13, 20, 6, 31} equal sets? Equivalent sets? (1–9)
14. Which of the following sets is a null set? (1–10)

    a. {All even numbers}
    b. {All prime numbers between 29 and 31}
    c. {0, 1, 2, 3}

15. Write the whole set of factors of 96. (1–11)
16. What are the common factors of 24 and 36? (1–14)
17. What is the greatest common factor of 75 and 105? (1–15)
18. Write 60 as a product of two whole-number factors in as many ways as possible. (1–12)
19. Factor 48 as the product of prime factors. (1–13)
20. Write the set of multiples of 8, listing the first six members. (1–16)
21. Write the set of common multiples of 6 and 9, listing the first five members. (1–17)
22. What is the least common multiple of 8, 12, and 18? (1–18)

Systems of Numeration

2

A *system of numeration* is a method of naming numbers by writing numerals. Basic number symbols of varying values are arranged according to the principles of grouping used in the specific system.

Although some ancient systems of numeration used different grouping symbols instead of place value, repeating them when necessary, others used the principle of place value where the total value of each number symbol was determined by its position or place in the numeral. In our modern numeration systems we use place value along with the principles of addition and multiplication which were also used in some of the ancient systems.

The *base* of a numeration system is the number it takes in any one place to make *one* (1) in the next higher place; it is the unit of grouping.

ANCIENT SYSTEMS OF NUMERATION

2–1 EGYPTIAN

The ancient Egyptians used the following hieroglyphic numeration:

Symbol 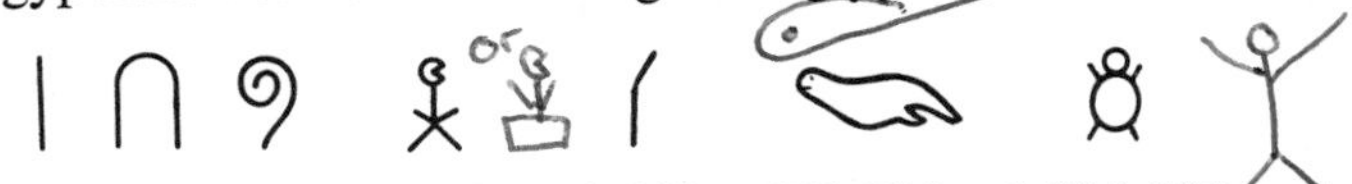

Value 1 10 100 1,000 10,000 100,000 1,000,000

Different grouping symbols were used instead of place value but the base of their system was ten. To write numerals the Egyptians used any number symbol as many as nine times when necessary but used the next higher number symbol if ten of any one symbol were required. The symbols were written next to each other and the sum of their values represented the number.

The Egyptian numeral  **consists of symbols representing 1,000 + 100 + 100 + 10 + 10 + 10 + 10 + 10 + 1 + 1 + 1 or the number 1,253.**

EXERCISES

1. What number does each of the following Egyptian numerals represent?

a. 

f. 

b. 

g. 

c. 

h. 

d. 

i. 

e. 

j. 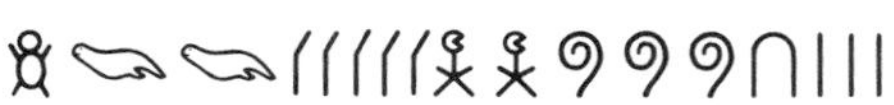

2. Write the Egyptian numeral for each of the following numbers:

a. 14 **b.** 21 **c.** 57 **d.** 136 **e.** 305 **f.** 1,413 **g.** 24,162 **h.** 13,289 **i.** 123,538 **j.** 2,314,675

2–2 BABYLONIAN

Clay tablets have been found containing "cuneiform" number symbols which were written by the Babylonians more than 5,000 years ago. They used a symbol shaped like a wedge  to represent one and the symbol  to represent ten. To write numerals for numbers 1 to 59 inclusive, the Babylonians used their symbol for one as many as nine times when necessary and the symbol for ten as many as five times when necessary. To write numerals for numbers greater than 59 they used the same two symbols but on the base of 60.

The Babylonian numeral  consists of symbols representing $(27 \times 60) + 10 + 1 + 1 + 1 + 1 + 1 + 1$ or the number 1,636.

Observe that the Babylonians used place value, the base of 60, and the principles of repetition, addition, and multiplication.

EXERCISES

1. What number does each of the following Babylonian numerals represent?

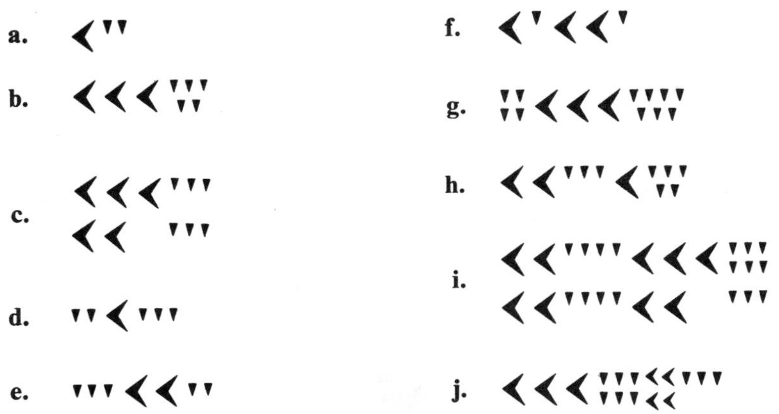

2. Write the Babylonian numeral for each of the following numbers:

a. 15	**c.** 37	**e.** 72	**g.** 185	**i.** 748
b. 21	**d.** 55	**f.** 100	**h.** 276	**j.** 1,529

2–3 CHINESE

The Chinese used the following traditional number symbols:

Symbol	一	二	三	四	五	六
Value	1	2	3	4	5	6
Symbol	七	八	九	十	百	千
Value	7	8	9	10	100	1,000

Chinese numerals are written vertically with the ones on the bottom. To write numerals for numbers from 1 to 10 inclusive, the Chinese used the corresponding symbols shown above. To write numerals for numbers greater than ten, they used symbols to represent powers of ten. They wrote the symbol representing the desired number from one to nine inclusive above the symbol representing the required power of ten to indicate the product of the number by the power.

The Chinese used the base of ten by writing symbols to represent powers of ten. They also used the principles of multiplication and addition.

50 is expressed as 5 times 10.

or 五
十

400 is expressed as 4 times 100.

or 四
百

This numeral	represents
六 千	→ 6 times 1,000 plus
八 百	→ 8 times 100 plus
七 十	→ 7 times 10 plus
三	→ 3

or the number 6,873

EXERCISES

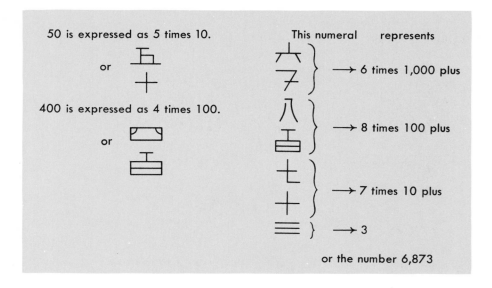

1. What number does each of the following Chinese numerals represent?

a.	**b.**	**c.**	**d.**	**e.**	**f.**	**g.**	**h.**	**i.**	**j.**
七	二	八	三	四	七	六	四	八	九
十	百	百	百	百	千	千	千	千	千
四	四	九	八	二	八	二	七	五	三
	十	十	十	十	百	百	百	百	百
	三	六	一	九	三	八	六	四	七
					十	十	十	十	十
					二	五	七	二	九

2. Write the Chinese numeral for each of the following numbers:

a. 16 **c.** 89 **e.** 746 **g.** 2,582 **i.** 9,628
b. 27 **d.** 253 **f.** 3,491 **h.** 7,354 **j.** 5,879

2–4 MAYAN

The Mayas, an Indian tribe of Central America, combined dots and horizontal bars to represent numbers from 1 to 19. The basic symbols were a dot for one and a bar for five.

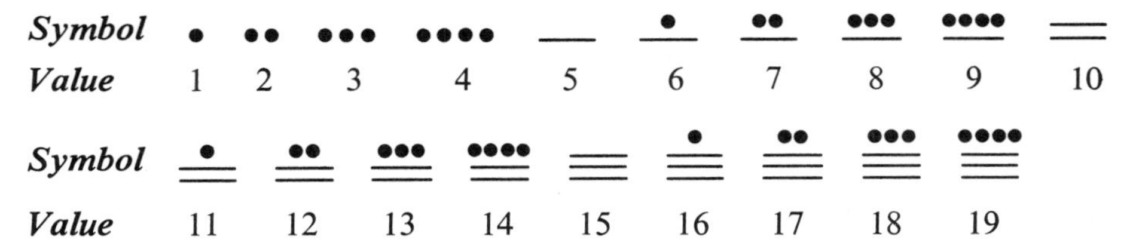

Mayan numerals were written vertically with ones on the bottom. The Mayan system used a base of twenty. The symbol  represented zero and, when written under any of the above symbols for 1 to 17 inclusive, acted as a placeholder.

The Mayan numeral 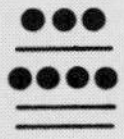 represented (20 × 8) + 14 or the number 174.

The numeral  represented 20 × 8 or 160 since the  was in the 20's place and  represented no ones.

EXERCISES

1. What number does each of the following Mayan numerals represent?

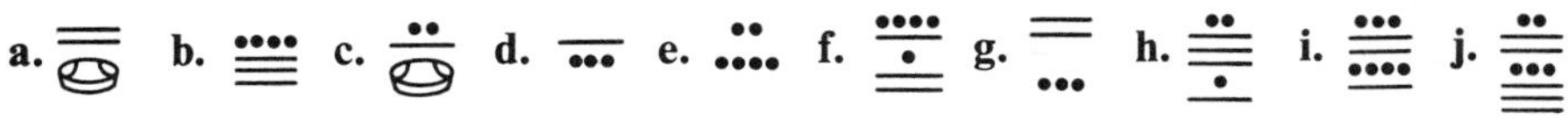

2. Write the Mayan numeral for each of the following numbers:

a. 27	**c.** 60	**e.** 81	**g.** 193	**i.** 346
b. 34	**d.** 102	**f.** 200	**h.** 309	**j.** 257

2–5 GREEK

The Greeks used alphabetic numerals, the first nine letters of their alphabet representing numbers from 1 to 9 inclusive, the next nine letters representing the tens and other letters representing the hundreds.

Symbol	Α	Β	Γ	Δ	Ε	F	Ζ	Η	Θ
Value	1	2	3	4	5	6	7	8	9
Symbol	Ι	Κ	Λ	Μ	Ν	Ξ	Ο	Π	ϙ
Value	10	20	30	40	50	60	70	80	90
Symbol	Ρ	Σ	Τ	Υ	Φ	Χ	Ψ	Ω	ϡ
Value	100	200	300	400	500	600	700	800	900

Letters corresponding to the required values are written next to each other with the ones on the right and increasing to the left. Sometimes an accent mark was written after the letter to indicate that it was a numeral.

> The Greek numeral Φϙ∆ represented 500 + 90 + 4 or the number 594.

EXERCISES

1. What number does each of the following Greek numerals represent?

a. ΛF **c.** ϙΘ **e.** ΡΚΒ **g.** ΨΝΕ **i.** ΩΟΑ

b. ΠΖ **d.** ΜΗ **f.** ΤΞΔ **h.** ϡΙΓ **j.** ΥΛΘ

2. Write the Greek numeral for each of the following numbers:

a. 24 **c.** 69 **e.** 85 **g.** 640 **i.** 833
b. 56 **d.** 17 **f.** 258 **h.** 791 **j.** 499

2–6 HEBREW

The Hebrews, like the Greeks, also used the letters of their alphabet to represent numbers. The Aleph א represented one.

2–7 ROMAN

The Romans used the following number symbols in their system of numeration:

Symbol	I	V	X	L	C	D	M
Value	1	5	10	50	100	500	1,000

They formed Roman numerals by writing from left to right as a sum, first the symbol for the greatest possible value, with symbols I, X, C, and M used as many as three times when necessary, then the symbol for the next smaller value, etc.

When symbols I, X, or C preceded a Roman number symbol of greater value, its value was subtracted from the larger value.

IV = 5 − 1 = 4 XL = 50 − 10 = 40 CD = 500 − 100 = 400
IX = 10 − 1 = 9 XC = 100 − 10 = 90 CM = 1,000 − 100 = 900

MDCXLVII = ?
Since M = 1,000
D = 500
C = 100
XL = 40
V = 5
II = 2
1,647
MDCXLVII = 1,647

178
= 100 + 50 + 20 + 5 + 3
= C + L + XX + V + III
= CLXXVIII

The symbols V, L, and D never preceded a Roman number symbol of greater value and were never used in succession. A bar above the symbol indicated that the value of the symbol was multiplied by 1,000.

EXERCISES

1. What number does each of the following Roman numerals represent?

III	VIII	IV	XI	XV	IX	XVII	XIX
XXV	XXIV	XXVIII	XXX	XXXVI	XXXIX	XXXII	XXXIV
XL	XLVI	LXII	LXXXIX	XC	XCVIII	LIV	XLIX
CXXVI	CCXLI	CDXLIV	DCCLVI	CMVIII	MCXLV	MDCCXI	MCMLIV

2. Write the Roman numeral for each of the following numbers:

2	6	9	13	18	19	14	16
20	31	24	36	29	34	27	35
43	45	56	69	73	84	99	96
104	289	462	596	947	1335	1651	1969

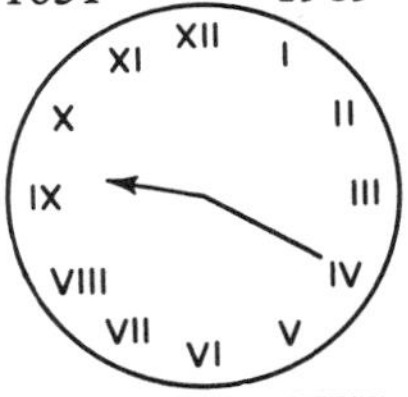

3. What time is indicated on the clock face?

4. A cornerstone is marked MCMXXIV. What date does it represent?

5. What number is represented by the Roman numeral in each of the following:

a. Page XVII **b.** Paragraph XXIX **c.** Chapter LXVI
d. Item XLIV **e.** Page XCIII **f.** Item CCXC

6. In making an outline what Roman numeral should you use following:

a. VIII **b.** XIV **c.** XIX **d.** XXIII **e.** XXXIX **f.** XLIX

REVIEW EXERCISES

1. What number is expressed by each of the following numerals?

a. MDCCLXXXVI

b. 

c. ΦKZ

d. 

e. 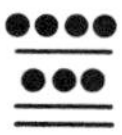

f. 九千八百七十二

2. Express:

a.  by an equivalent Babylonian numeral.

b. CCXLIX by an equivalent Mayan numeral.

c.  by an equivalent Greek numeral.

d. ΩMΘ by an equivalent Roman numeral.

e.  by an equivalent Egyptian numeral.

f. 五千四百六十八 by an equivalent Egyptian numeral.

3. Write  as an equivalent:

a. Roman numeral **b.** Mayan numeral **c.** Chinese numeral

4. Write CCCXIV as an equivalent:

a. Egyptian numeral **b.** Babylonian numeral **c.** Greek numeral

5. Express 135 by an equivalent:

a. Roman numeral
b. Mayan numeral
c. Babylonian numeral
d. Greek numeral
e. Egyptian numeral
f. Chinese numeral

6. Write CXCVIII as an equivalent:

a. Decimal numeral **b.** Chinese numeral **c.** Mayan numeral

7. Express 316 by an equivalent:

a. Babylonian numeral **b.** Mayan numeral **c.** Greek numeral

8. Express 2,749 by an equivalent:

a. Chinese numeral **b.** Roman numeral **c.** Egyptian numeral

MODERN NUMERATION SYSTEMS

DECIMAL SYSTEM OF NUMERATION

2–8 PLACE VALUE

In our system of notation, called the *decimal system*, ten number symbols 0, 1, 2, 3, 4, 5, 6, 7, 8, and 9 are used to represent all numbers. These number symbols are of Hindu-Arabic origin. They were introduced in Europe during the 12th century and are now used generally throughout the world. They were brought to the United States by the colonists.

There is no single number symbol in our numeration system to represent ten or numbers greater than ten. Numerals representing numbers greater than nine are formed by writing two or more number symbols next to each other in different positions or places.

Our system of writing numerals is built on the base ten. It takes ten in any one place to make one in the next higher place. It takes 10 ones to make 1 ten, 10 tens to make 1 hundred, 10 hundreds to make 1 thousand, and so forth.

	Trillions			Billions			Millions			Thousands			Ones		
quadrillions	hundred trillions	ten trillions	trillions	hundred billions	ten billions	billions	hundred millions	ten millions	millions	hundred thousands	ten thousands	thousands	hundreds	tens	ones
3,	3	3	3,	3	3	3,	3	3	3,	3	3	3,	3	3	3

In the number scale the ones (sometimes called units) are located on the right. One place to the left of the ones position is the tens place; one place to the left of the tens position is the hundreds place. These three places form a group or period.

The scale illustrates that the periods in increasing order are: ones, thousands, millions, billions, trillions, and quadrillions. The names of other periods in increasing order are quintillion, sextillion, septillion, octillion, nonillion, and decillion. A decillion is written as:

1,000,000,000,000,000,000,000,000,000,000,000

The decimal system is a positional system. It uses place value to represent each power of ten instead of a special symbol. The value of each place in the decimal numeral is ten times the value of the next place to the right.

Thus, in the numeral the value of each digit depends not only on the symbol but also on its position in the numeral. The digit 8 in the decimal numeral 86 means 8 tens but in the decimal numeral 865 it means 8 hundreds.

The zero symbol indicates "not any" and, when used as a placeholder, it indicates no ones, no tens, etc. A four-digit numeral such as 2,694 indicates 2 thousands, 6 hundreds, 9 tens, 4 ones; or 26 hundreds, 9 tens, 4 ones; or 269 tens, 4 ones; or 2,694 ones.

EXERCISES

1. In the numeral 5,863,295,482,716 what digit is in the:

a. ones place **b.** thousands place **c.** tens place **d.** millions place
e. hundreds place **f.** hundred thousands place **g.** ten thousands place
h. billions place **i.** ten millions place **j.** trillions place **k.** ten billions place

2. What place does the zero hold in each of the following numerals:

a. 98,027 **b.** 406,928 **c.** 3,425,103 **d.** 80,215,896

3. Complete the following:

a. 4,862,357 = ... millions ...... hundred thousands ...... ten thousands
...... thousands ...... hundreds ...... tens ...... ones
b. 7,864 = ... thousands ...... hundreds ...... tens ...... ones
or = ... hundreds ...... tens ...... ones
or = ... tens ...... ones
or = ... ones

4. Write the numeral naming the number:

3 *millions* 4 *hundred thousands* 9 *ten thousands* 0 *thousands* 5 *hundreds* 8 *tens* 2 *ones*

5. What new group does 1 more than 99,999 make? 1 more than 9,999,999?
6. Which number is greater? **a.** 4,862,917 or 4,862,971 **b.** 12,689,754 or 12,698,457
7. Which number is smaller? **a.** 975,062 or 975,026 **b.** 5,748,841 or 5,784,148
8. Name the greatest possible: **a.** 4-place number **b.** 7-place number
9. Name the smallest possible: **a.** 6-place number **b.** 9-place number
10. Write the numeral that names the greatest number possible using the digits:

a. 9, 1, 8 **b.** 4, 7, 2, 5 **c.** 3, 4, 0, 8, 2, 5 **d.** 7, 0, 9, 1, 4, 9, 3

11. Write the numeral that names the smallest number possible using the digits:

a. 7, 6, 2 **b.** 9, 3, 0, 6 **c.** 5, 3, 4, 7, 9, 1 **d.** 8, 0, 5, 0, 3, 1, 6

12. Rearrange the digits in each of the following numerals so that it names the greatest possible number:

a. 204,913 **b.** 46,986 **c.** 4,306,072 **d.** 17,553,421

13. Rearrange the digits in each of the following numerals so that it names the smallest possible number:

    a. 7,146 **b.** 670,392 **c.** 8,680,407 **d.** 51,866,924

14. Arrange the following numbers according to size, writing the numeral for the greatest first:

    437,218 437,128 437,812 437,182

15. Arrange the following numbers according to size, writing the numeral for the smallest first:

    5,986,752 5,986,725 5,698,527 5,869,257

2–9 READING NUMERALS NAMING WHOLE NUMBERS

Before reading a numeral containing four digits or more, it is advisable to separate it by commas into as many groups (or periods) of three digits as possible, starting from the ones place and counting to the left.

To read a numeral, we begin at the left and read each period of digits separately, applying the name of the period indicated by each comma as it is reached. The word "and" is not used between names of periods.

The numeral 78,356,912,405 is read **Seventy-eight billion, three hundred fifty-six million, nine hundred twelve thousand, four hundred five.**

EXERCISES

1. Separate each of the following numerals into proper periods or groups:

   a. 63041 **b.** 58162840 **c.** 3913596804 **d.** 8150276150938 **e.** 2403920005038675

2. Read, or write in words, each of the following:

   a. 908 **c.** 46,867 **e.** 3,729,681 **g.** 78,404,015,263 **i.** 4,394,481,356,672
   b. 5,046 **d.** 800,075 **f.** 520,075,998 **h.** 814,875,000,000 **j.** 19,006,507,093,855

3. Read, or write in words, each of the numerals that appear in the following:

   a. The final presidential election returns showed that one candidate received 46,631,189 votes and the other, 28,422,015 votes.
   b. The area of the earth's surface is approximately 196,940,400 square miles.
   c. There are 2,730,242 farms in the United States covering 1,066,218,650 acres of land.
   d. The public debt of the United States in 1931 was $16,801,281,492 and in 1974 was $475,059,815,732.

2–10 WRITING NUMERALS NAMING WHOLE NUMBERS

We write each period of digits and use commas to represent the names of the periods. Also large whole numbers sometimes are named briefly by a numeral followed by a period name.

> Sixty-nine million, four hundred twenty-five thousand, nine hundred eighty-six written as a numeral is 69,425,986.
>
> 35 billion written as a complete numeral is 35,000,000,000.

EXERCISES

Write the numeral that names each of the following:

1. 54 million **2.** 749 thousand **3.** 9 trillion **4.** 685 billion

5. Twenty-seven thousand, two hundred thirty-two.

6. Three hundred eight thousand, seven hundred eighty-nine.

7. Fifty-four million, six hundred forty-one thousand.

8. Nine million, seven hundred fifty thousand, three hundred twenty-five.

9. Eight hundred thirty-four million, five hundred thousand, seventy-six.

10. Two billion, four hundred eleven million, two hundred eighteen thousand.

11. Forty-five billion, eight hundred thirty thousand, ten.

12. Twenty trillion, fifty-nine billion, four hundred million, one hundred fifty-nine.

2–11 ROUNDING WHOLE NUMBERS

Newspapers often use round numbers in their story leads. An increase in population of 36,927 may be reported as "an increase of almost 37,000" and a bridge costing $2,163,875 may be reported as one "costing more than $2,000,000."

Using the number line to round numbers, we consider a point (which corresponds to the number to be rounded) that is located halfway or

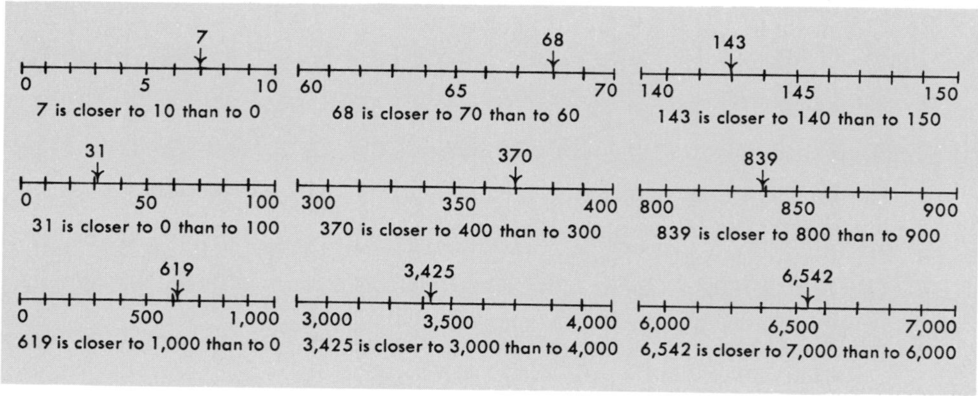

more than halfway between two points of reference to be closer to the point corresponding to the larger number. Numbers may be rounded to the nearest ten, hundred, thousand, etc. depending on what is required.

To round a whole number without using the number line, we find the place in the numeral to which the number is to be rounded (the nearest ten, hundred, thousand, etc.). We then rewrite the given digits to the left of the required place and we write a zero in place of each digit to the right of the required place. If the first digit dropped is 5 or more, we increase the given digit in the required place by 1, otherwise we write the same digit as given. However, as many digits to the left are changed as necessary when the given digit in the required place is 9 and one (1) is added.

4,691,826 rounded to the nearest:	
hundred is	4,691,800
thousand is	4,692,000
million is	5,000,000

EXERCISES

1. Select the numbers that are nearer to:

a. 0 than 10:	3	7	4	6	8
b. 10 than 0:	1	6	2	9	4
c. 70 than 80:	78	74	71	77	72
d. 400 than 300:	317	385	353	348	306
e. 8,000 than 9,000:	8,600	8,125	8,499	8,703	8,514
f. 60,000 than 50,000:	53,000	57,300	54,736	55,002	59,487
g. 300,000 than 400,000:	375,000	304,000	351,003	338,785	349,999
h. 7,000,000 than 6,000,000:	6,225,000	6,630,000	6,409,786	6,745,261	6,300,078

2. Any number from:

a. 45 to 54 rounded to the nearest ten is what number?
b. 650 to 749 rounded to the nearest hundred is what number?
c. 12,500 to 13,499 rounded to the nearest thousand is what number?
d. 9,500,000 to 10,499,999 rounded to the nearest million is what number?
e. 7,500,000,000 to 8,499,999,999 rounded to the nearest billion is what number?

3. Round each of the following numbers to the nearest:

a. Ten:

48	74	26	483	7,965

b. Hundred:

320	592	6,278	1,506	24,814

c. Thousand:

4,750	8,341	15,539	47,852	63,267

d. Ten thousand:

36,000	94,729	71,987	423,275	786,498

e. Hundred thousand:

910,000 252,495 664,019 3,180,531 29,349,653

f. Million:

4,700,000 7,328,000 15,445,625 916,503,064 8,432,198,572

g. Billion:

6,824,000,000 1,050,900,000 8,496,570,000 39,975,201,750 121,263,000,000

4. Round each of the numbers that appear in the following to the nearest:

a. Thousand: The recent census showed there were 98,912,192 males and 104,299,734 females living in the United States.
b. Million: Fire losses for the year amounted to $2,537,200,000.
c. Billion: $79,909,504,000 was spent by the Defense Department.
d. Thousand: The Mediterranean Sea has an area of 966,757 square miles.
e. Million: Motor vehicle registrations for the year reached 125,156,876.

2–12 EXPONENTS

When we use two equal factors in multiplication like 3×3, we may write it in exponential form as 3^2, which is read "three to the second power" or "the second power of three" or "three squared" or the "square of three."

$5 \times 5 \times 5$ may be written as 5^3, which is read "five to the third power" or "five cubed" or "the cube of five."

2^4 represents $2 \times 2 \times 2 \times 2$ or the product 16 and is read "two to the fourth power."

10^5 represents $10 \times 10 \times 10 \times 10 \times 10$ or the product 100,000 and is read "ten to the fifth power."

The small numeral written to the upper right (superscript) of the repeated factor is called an *exponent*. When it names a natural number, it tells how many times the factor is being used in multiplication. The factor that is being repeated is called the *base*. The number 2^4 uses 2 as the base and 4 as the exponent. Numbers such as 2^3, 2^4, 2^5, 2^6, etc. are called *powers* of 2.

EXERCISES

1. Read, or write in words, each of the following:

a. 6^5 **b.** 10^7 **c.** 4^{15} **d.** 2^1 **e.** 14^{10} **f.** 8^3 **g.** 25^8 **h.** 100^2

2. Write as a numeral:

a. Three to the fourth power **c.** Seven squared **e.** Twelve to the sixth power
b. Nine to the eighth power **d.** Four cubed **f.** Twenty to the tenth power

3. What is the exponent in: **a.** 5^9? **b.** 10^{11}? **c.** 3^5? **d.** 8^1? **e.** 15^{18}?
4. What is the base in: **a.** 7^8? **b.** 4^3? **c.** 19^6? **d.** 24^{15}? **e.** 50^9?
5. How many times is the base being used as a factor in:

a. 3^7? **b.** 8^3? **c.** 4^{10}? **d.** 6^2? **e.** 10^{13}?

6. Use the exponential form to write:

a. $5 \times 5 \times 5$
b. $9 \times 9 \times 9 \times 9$
c. 6×6
d. $8 \times 8 \times 8 \times 8 \times 8$
e. $7 \times 7 \times 7 \times 7 \times 7 \times 7 \times 7 \times 7$
f. $3 \times 3 \times 3 \times 3 \times 3 \times 3 \times 3$
g. $4 \times 4 \times 4 \times 4 \times 4 \times 4 \times 4 \times 4 \times 4 \times 4$
h. $2 \times 2 \times 2 \times 2 \times 2 \times 2 \times 2 \times 2 \times 2 \times 2 \times 2$

7. Express each of the following as a product of a repeated factor:

a. $6^3 = 6 \times 6 \times 6$ **b.** 7^6 **c.** 12^4 **d.** 9^{10} **e.** 2^8 **f.** 3^{15} **g.** 5^9 **h.** 8^{12}

8. a. Express 8 as a power of 2.
b. Express 36 as a power of 6.
c. Express 125 as a power of 5.
d. Express 32 as a power of 2.
e. Express 64 as a power of 4.
f. Express 81 as a power of 3.

9. First express each of the following as a product of a repeated factor, then in exponent form:

a. $49 = 7 \times 7 = 7^2$ **b.** 27 **c.** 16 **d.** 121 **e.** 625 **f.** 64 **g.** 144 **h.** 128

10. Find the value of each of the following:

a. 15^2 **b.** 5^3 **c.** 2^7 **d.** 3^5 **e.** 6^3 **f.** 7^4 **g.** 4^6 **h.** 2^{10}

2–13 POWERS OF TEN

Using 10 as a repeated factor, we develop the following table of powers of 10:

$$10 = 10 = 10^1$$
$$100 = 10 \times 10 = 10^2$$
$$1{,}000 = 10 \times 10 \times 10 = 10^3$$
$$10{,}000 = 10 \times 10 \times 10 \times 10 = 10^4$$
$$100{,}000 = 10 \times 10 \times 10 \times 10 \times 10 = 10^5$$
$$1{,}000{,}000 = 10 \times 10 \times 10 \times 10 \times 10 \times 10 = 10^6$$

Observe in the above table:

(1) 10 used one time as a factor may be written with the exponent 1, although the exponent 1 usually is omitted.
(2) The number of zeros found after the digit 1 in the product (first column above) corresponds in each case to the exponent of 10 (third column above). For example, in $1{,}000 = 10^3$ there are 3 zeros in the product (1,000) and the exponent of 10^3 is 3.

Any numeral having all digits zeros except the first digit, such as 20 or 500 or 6,000, may be expressed as a product of a digit and a power of ten.

$$6{,}000 = 6 \times 1{,}000 = 6 \times 10^3$$

EXERCISES

1. Express each of the following as a power of ten:

a. 100,000 **b.** 10,000,000 **c.** 1,000,000,000 **d.** 100,000,000,000 **e.** 1,000,000,000,000

2. Express each of the following as a product of a digit and a power of ten:

a. 30	**e.** 3,000,000	**i.** 2,000,000,000	**m.** 4,000,000,000,000
b. 400	**f.** 700,000	**j.** 70,000,000,000	**n.** 50,000,000,000,000
c. 90,000	**g.** 400,000,000	**k.** 900,000,000,000	**o.** 600,000,000,000,000
d. 5,000	**h.** 60,000,000	**l.** 8,000,000,000	**p.** 3,000,000,000,000,000

2–14 WRITING DECIMAL NUMERALS AS POLYNOMIALS—EXPANDED NOTATION

The value of each place in a decimal numeral may be expressed as a power of ten.

The digit 7 in 79 means 7 tens or 7×10.
The digit 5 in 579 means 5 hundreds or $5 \times 100 = 5 \times 10^2$.
The digit 8 in 8,579 means 8 thousands or $8 \times 1{,}000 = 8 \times 10^3$.

The complete value of each digit in a decimal numeral is equal to the value of the digit itself times its place value.

A decimal numeral may be written as the sum of the products of each digit in the numeral and its place value expressed as a power of ten.

The decimal numeral 8,579

= 8 thousands	+ 5 hundreds	+ 7 tens	+ 9 ones
$= (8 \times 1{,}000)$	$+ (5 \times 100)$	$+ (7 \times 10)$	$+ (9 \times 1)$
$= (8 \times 10^3)$	$+ (5 \times 10^2)$	$+ (7 \times 10^1)$	$+ (9 \times 1)$

Writing a numeral in this expanded form as an indicated sum (*polynomial*) is sometimes called *expanded notation.* Observe that the place values are powers of ten arranged in a decreasing order from left to right.

EXERCISES

1. Express each of the following polynomials as a decimal numeral:

a. $(5 \times 10^4) + (3 \times 10^3) + (7 \times 10^2) + (6 \times 10^1) + (8 \times 1)$
b. $(7 \times 10^5) + (9 \times 10^4) + (2 \times 10^3) + (1 \times 10^2) + (4 \times 10^1) + (3 \times 1)$

2. Write each of the following numerals in expanded form as a polynomial:

a. 96	**d.** 41,965	**g.** 3,248,766	**j.** 8,293,456	**m.** 6,529,104,249
b. 358	**e.** 8,209	**h.** 72,543	**k.** 93,045,327	**n.** 23,865,200,684
c. 2,927	**f.** 617,932	**i.** 459,284	**l.** 205,699,182	**o.** 475,218,836,095

NON-DECIMAL SYSTEMS OF NUMERATION

2–15 GROUPING AND READING

If twenty-three things are grouped:

by tens

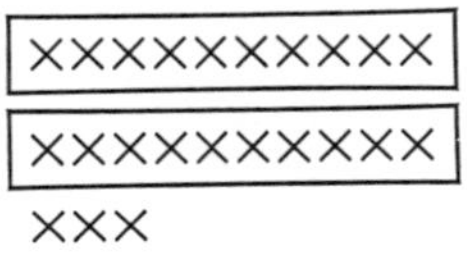

← There are 2 groups of ten and 3 ones, which we record as 23.

by fives

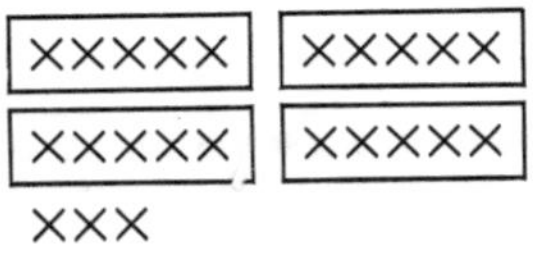

← There are 4 groups of five and 3 ones, which we write as 43_{five} or 43_5. 43_{five} is read "four three, base five" and not "forty-three." The subscript at the lower right of the numeral indicates the *base* of the numeration system and the unit of grouping.

by sevens

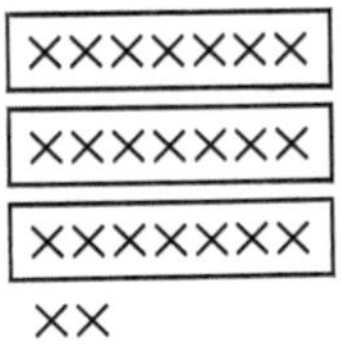

← There are 3 groups of seven and two ones, which we record as 32_{seven} or 32_7. 32_7 is read "three two, base seven."

by eights

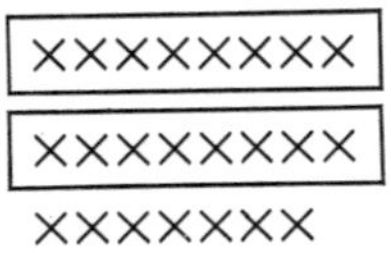

← There are 2 eights and 7 ones, which we record as 27_{eight} or 27_8. 27_8 is read "two seven, base eight."

The numeral 3102_{four} or 3102_4 is read "three one zero two, base four."

If a numeral does not have an identifying subscript or some other identification, it will be considered a base ten numeral.

EXERCISES

1. Read, or write in words, each of the following:

a. 61_{seven} **c.** 101_{two} **e.** 650_{eight} **g.** 98736_{twelve}
b. 20_{six} **d.** 234_{five} **f.** 10011_{two} **h.** 120210_{three}

2. Write each of the following as a numeral:

a. Four two, base nine
b. Three zero, base four
c. One zero zero one, base two
d. Five two three, base seven
e. Two one one two, base three
f. Six five two zero, base eight

3. Copy the x's: ×××××××××××××××××××××××××××

a. Group them by tens. Write a base ten numeral indicating the number of x's.
b. Group them by nines. Write a base nine numeral indicating the number of x's.
c. Group them by sixes. Write a base six numeral indicating the number of x's.

4. Copy the dots: ●●●●●●●●●●●●●●●●●●●●●●●●●●●●●●●●●●●●

a. Group them by eights. Write a base eight numeral indicating the number of dots.
b. Group them by twelves. Write a base twelve numeral indicating the number of dots.
c. Group them by sevens. Write a base seven numeral indicating the number of dots.

5. a. Write as many x's on your paper as are represented by the numeral 24_{five}. Group them to show the meaning of the numeral.
b. Group dots on your paper to show the meaning of each of the following numerals: 31_{four}; 56_{eight}; 11_{two}; 26_{nine}; 32_{five}; 14_{twelve}; 22_{three}; 35_{six}; 40_{seven}

2–16 PLACE VALUE

In each system of notation number symbols from zero up to, but not including, the symbol for the base are arranged by position to represent numbers. *Each system of notation has its own place value.* Any number equal to or greater than the base number is represented by a numeral containing more than one symbol. In the *decimal system of numeration* we found that the value of each place is *ten times* the value of the next place to the right. The base ten system requires ten digits 0, 1, 2, 3, 4, 5, 6, 7, 8, and 9 to write numerals since the grouping is by tens.

In the *quinary* (*base five*) *system of numeration* there are only five digits 0, 1, 2, 3, and 4 required to write all numerals since the grouping is by fives. There is no single digit to represent *five* or numbers greater than five. For these, numerals are formed by writing two or more digits next to each other in different positions or places. The numeral for number 5 is 10_{five} or 10_5 where the digit 1 in the place to the left of the ones place means 1 five. The value of each place, as we move from right to left in a base five numeral, is *five times* the value of the next place to the right. Five of any one group make one of the next larger group. 5 ones make 1 five; 5 fives make 1 twenty-five, and so forth.

2	3	1	4	2_{five}
625's	125's	25's	5's	1's

The numeral 23142_{five} represents 2 groups of 625; 3 groups of 125; 1 group of 25; 4 groups of 5, and 2 ones.

In the *binary* (*or base two*) *system* there are only two digits 0 and 1 since the grouping is by twos. There is no single digit to represent *two* or numbers greater than two. The number 2 is named by the numeral 10_{two} or 10_2 where the digit 1 in the place to the left of the ones place means 1 two. The value of each place is *two times* the value of the next place to the right. Two of any group make one of the next larger group. 2 ones make 1 two; 2 twos make 1 four; 2 fours make 1 eight; and so forth.

1	1	0	1	0	1_{two}
32's	16's	8's	4's	2's	1's

The numeral 110101_{two} represents 1 group of 32; 1 group of 16; 1 group of 4, and 1 one.

In the *duodecimal* (*or base twelve*) *system* twelve digits are required to write numerals. The basic ten digits of the decimal system (0, 1, 2, 3, 4, 5, 6, 7, 8, and 9) and two new symbols, T for ten and E for eleven, are used. The value of each place is *twelve times* the value of the next place to the right. Twelve of any group make one of the next larger group. 12 ones make 1 twelve, 12 twelves make 1 one hundred forty-four, and so forth. There is no single digit to represent *twelve* or numbers greater than twelve. The number 12 is named by the numeral 10_{twelve} or 10_{12} where the digit 1 in the place to the left of the ones place means 1 twelve.

3	T	5	E	9 twelve
20,736's	1,728's	144's	12's	1's

The numeral $3T5E9_{twelve}$ represents 3 groups of 20,736; 10 groups of 1,728; 5 groups of 144; 11 groups of 12, and 9 ones.

Each system of numeration has its own place values. Starting from ones on the right, the values increase going to the left as illustrated by the following table:

NAME OF SYSTEM	BASE	← PLACE VALUES ←				
Binary	2	16	8	4	2	1
Ternary	3	81	27	9	3	1
Quaternary	4	256	64	16	4	1
Quinary	5	625	125	25	5	1
Senary	6	1,296	216	36	6	1
Septenary	7	2,401	343	49	7	1
Octonary	8	4,096	512	64	8	1
Nonary	9	6,561	729	81	9	1
Decimal	10	10,000	1,000	100	10	1
Duodecimal	12	20,736	1,728	144	12	1

EXERCISES

1. Determine the value of the place in which the indicated digit appears in each numeral:

a. Digit 2 in: (1) 120_{three} (2) 5210_{six} (3) 2103_{four} (4) 29467_{ten}
b. Digit 0 in: (1) 4013_{five} (2) 11110_{two} (3) 6405_{seven} (4) $10E52_{twelve}$
c. Digit 3 in: (1) 423_{six} (2) 6352_{eight} (3) 31002_{four} (4) 10234_{five}
d. Digit 5 in: (1) 526_{seven} (2) 6835_{nine} (3) 45183_{twelve} (4) 52443_{six}
e. Digit 1 in: (1) 2143_{five} (2) 31302_{four} (3) 212020_{three} (4) 1000000_{two}
f. Digit 7 in: (1) 572_{nine} (2) $E7T4_{twelve}$ (3) 73524_{ten} (4) 63247_{eight}
g. Digit 8 in: (1) $8T4_{twelve}$ (2) 18526_{ten} (3) 52483_{nine} (4) $83E19_{twelve}$
h. Digit 4 in: (1) 543_{six} (2) 42113_{five} (3) 56634_{seven} (4) 26457_{eight}
i. Digit 9 in: (1) 9017_{ten} (2) $E5T9_{twelve}$ (3) 25986_{twelve} (4) 84293_{ten}

2. Name the different digits that are used to write numerals in:

a. Base six b. Base eight c. Base two d. Base five e. Base twelve

3. What groups do the digits in each of the following numerals represent?

a. 213_{five} b. 110110_{two} c. $ET42_{twelve}$ d. 52136_{eight} e. 73801_{nine} f. 13320_{four} g. 12102_{three} h. 52143_{six} i. 243651_{seven} j. 11011011_{two}

2–17 COUNTING

To count in any base we must remember that in each system of numeration:

(1) Number symbols from zero up to, but not including, the symbol for the base are arranged by position to represent numbers.
(2) Any number less than the base number is represented by its respective symbol.
(3) Any number equal to or greater than the base number is represented by a numeral containing more than one symbol, since no digit whose value is equal to or greater than the value of the digit representing the base number may be used.

In the quinary system the numerals for the first four counting numbers are like those in the decimal system: 1 (one), 2 (two), 3 (three), 4 (four), but since the symbol 5 cannot be used in the quinary system, five is expressed as 10 (one zero). Then six is one more or 11 (one one), seven is one more or 12 (one two), eight is one more or 13 (one three), nine is one more or 14 (one four), but ten cannot be written as 15 since the symbol 5 cannot be used. Therefore ten is written as 20 (two zero) which means 2 fives and no ones. Observe that one is added on to each number to make each succeeding counting number until the largest digit is used, then a new group is formed. One more than 44 in the quinary system is 100; one more than 444 is 1000; etc.

EXERCISES

1. Write the numeral that immediately follows:

a. 1111_{two} b. 324_{five} c. 477_{eight} d. TE_{twelve} e. 2122_{three} f. 6387_{nine} g. 2323_{four} h. 11011_{two} i. 55555_{six} j. 65666_{seven}

2. Write the next five numerals that follow:

a. 101_{two} b. 1221_{three} c. 4434_{five} d. 37667_{eight} e. $4TE9_{twelve}$

3. Write the first fifty base two numerals beginning with 1.
4. Write the first thirty-six base twelve numerals beginning with 1.
5. Make a chart showing the first thirty numerals in base three, base seven, and base eight.

2–18 EXPANDED NOTATION

Non-decimal numerals may be written in expanded form by using the powers of the base expressed in decimal values. For example, when the base is 2, the decimal powers are 2^1, 2^2, 2^3, 2^4, 2^5, etc.; when the base is 5, the decimal powers are 5^1, 5^2, 5^3, 5^4, 5^5 etc.

Or the powers may be written simply as 10^1, 10^2, 10^3, 10^4, 10^5, etc. when expressed by digits of a given base. When binary digits are used, the binary powers are $10^1{}_{two}$, $10^{10}{}_{two}$, $10^{11}{}_{two}$, $10^{100}{}_{two}$, $10^{101}{}_{two}$, etc.; 10_{two} in each case represents the decimal number 2. When quinary digits are used, the quinary powers are $10^1{}_{five}$, $10^2{}_{five}$, $10^3{}_{five}$, $10^4{}_{five}$, $10^{10}{}_{five}$, etc.; 10_{five} in each case represents the decimal number 5.

The complete value of each digit in a numeral is equal to the value of the digit itself times its place value. Thus a numeral may be written as the sum of the products of each digit in the numeral and its place value expressed as a power of the base.

The numeral 2413_{five} is expanded as:

2	4	1	3_{five}
125's	25's	5's	1's

2413_{five} = 2 one hundred twenty-fives + 4 twenty-fives + 1 five + 3 ones

in decimal digits $= (2 \times 5^3) + (4 \times 5^2) + (1 \times 5^1) + (3 \times 1)$

or in digits of base 5 $= (2 \times 10^3{}_{five}) + (4 \times 10^2{}_{five}) + (1 \times 10^1{}_{five}) + (3 \times 1)$

EXERCISES

1. a. Express as a base two numeral: $(1 \times 2^3) + (0 \times 2^2) + (1 \times 2^1) + (1 \times 1)$
   b. Express as a base three numeral:
   $(2 \times 3^4) + (1 \times 3^3) + (1 \times 3^2) + (0 \times 3^1) + (2 \times 1)$
   c. Express as a base eight numeral:
   $(3 \times 8^4) + (4 \times 8^3) + (5 \times 8^2) + (2 \times 8^1) + (4 \times 1)$
   d. Express as a base twelve numeral: $(9 \times 12^2) + (6 \times 12^1) + (4 \times 1)$
   e. Express as a base six numeral:
   $(5 \times 10^4{}_{six}) + (2 \times 10^3{}_{six}) + (3 \times 10^2{}_{six}) + (4 \times 10^1{}_{six}) + (3 \times 1)$

2. Write each of the following numerals in expanded form as a polynomial (using decimal digits):

   a. 234_{six} b. 3705_{nine} c. 101111_{two} d. 2123_{four} e. 63417_{eight} f. 122112_{three} g. 4023_{five} h. $5T9E8_{twelve}$ i. 246981_{ten} j. 63256_{seven}

3. Write each of the following numerals in expanded form as a polynomial (using digits of the base):

a. 231_{four} **b.** 6805_{nine} **c.** 11011_{two} **d.** $E482_{twelve}$ **e.** 20112_{three}

2–19 EXPRESSING NUMERALS AS NUMERALS IN OTHER BASES

To express a numeral given in a base other than ten as a base ten numeral, we multiply the value of each digit by its place value and add these products. Then we write the decimal numeral for the resulting sum.

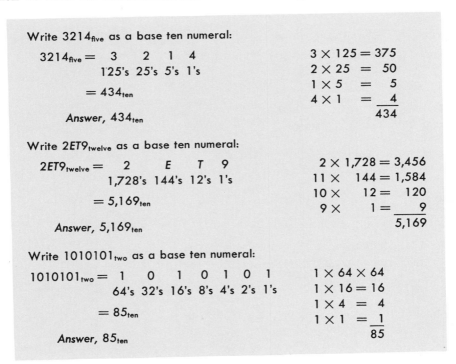

Write 3214_{five} as a base ten numeral:

$3214_{five} =$	3	2	1	4	
	125's	25's	5's	1's	

$= 434_{ten}$

$3 \times 125 = 375$
$2 \times 25 = 50$
$1 \times 5 = 5$
$4 \times 1 = 4$
434

Answer, 434_{ten}

Write $2ET9_{twelve}$ as a base ten numeral:

$2ET9_{twelve} =$	2	*E*	*T*	9
	1,728's	144's	12's	1's

$= 5{,}169_{ten}$

$2 \times 1{,}728 = 3{,}456$
$11 \times 144 = 1{,}584$
$10 \times 12 = 120$
$9 \times 1 = 9$
$5{,}169$

Answer, $5{,}169_{ten}$

Write 1010101_{two} as a base ten numeral:

$1010101_{two} =$	1	0	1	0	1	0	1
	64's	32's	16's	8's	4's	2's	1's

$= 85_{ten}$

$1 \times 64 \times 64$
$1 \times 16 = 16$
$1 \times 4 = 4$
$1 \times 1 = 1$
85

Answer, 85_{ten}

EXERCISES

Express each of the following as an equivalent base ten numeral:

1. **a.** 1011_{two} **b.** 11011_{two} **c.** 1101011_{two} **d.** 11110011_{two} **e.** 11011101111_{two}
2. **a.** 243_{five} **b.** 2034_{five} **c.** 31302_{five} **d.** 21432_{five} **e.** 134024_{five}
3. **a.** $E4_{twelve}$ **b.** $T59_{twelve}$ **c.** 3482_{twelve} **d.** $12E37_{twelve}$ **e.** $24T9E_{twelve}$
4. **a.** 62_{eight} **b.** 375_{eight} **c.** 21264_{eight} **d.** 3427_{eight} **e.** 16354_{eight}
5. **a.** 2102_{three} **b.** 1211_{three} **c.** 11221_{three} **d.** 210122_{three} **e.** 221212_{three}
6. **a.** 425_{six} **b.** 3531_{six} **c.** 5243_{six} **d.** 10455_{six} **e.** 21523_{six}
7. **a.** 231_{four} **b.** 2120_{four} **c.** 3333_{four} **d.** 12321_{four} **e.** 32312_{four}

8. a. 86_{nine}	**b.** 257_{nine}	**c.** 4362_{nine}	**d.** 13278_{nine}	**e.** 24367_{nine}
9. a. 54_{seven}	**b.** 666_{seven}	**c.** 3453_{seven}	**d.** 2645_{seven}	**e.** 15061_{seven}
10. a. 11011101_{two}	**b.** 5245_{six}	**c.** 3342_{five}	**d.** 21873_{nine}	**e.** $T20E5_{\text{twelve}}$

To express a base ten numeral as a numeral in a base other than ten, we may use either the quotients method or the remainders method.

Quotients Method

(1) Divide the largest possible power of the base into the given number. Then divide the remainder by the next lower power of the base. Continue in this manner until the divisor is the base itself.

(2) Take the quotient of each division as the digit for the corresponding position in the required base numeral.

(3) Use the final remainder as the digit for the ones place in the required numeral.

Write 832_{ten} as a base six numeral:

```
216)832(3
    648
 36)184(5
    180
  6)  4(0
      0
      4
```

Answer, 3504_{six}

Remainders Method

(1) Divide the base into the given number. Then divide the base into the quotient, then divide the base into the new quotient, and so forth. Continue until the quotient is zero.

(2) Take the remainders in these divisions as the required digits, using the final remainder as the digit for the greatest place value and the first remainder as the digit for the ones place.

```
6)832(138 Remainder 4  ↑
6)138( 23 Remainder 0  |
6) 23(  3 Remainder 5  |
6)  3(  0 Remainder 3  |
```

Answer, 3504_{six}

EXERCISES

Express each of the following base ten numerals as an equivalent numeral of the indicated base:

1. To base five:	**a.** 56	**b.** 270	**c.** 625	**d.** 1,361	**e.** 4,584
2. To base two:	**a.** 37	**b.** 94	**c.** 125	**d.** 490	**e.** 3,298
3. To base twelve:	**a.** 68	**b.** 197	**c.** 859	**d.** 5,381	**e.** 11,452
4. To base eight:	**a.** 41	**b.** 302	**c.** 1,200	**d.** 6,494	**e.** 10,563
5. To base three:	**a.** 15	**b.** 62	**c.** 225	**d.** 800	**e.** 2,150

6. To base six: **a.** 29 **b.** 75 **c.** 390 **d.** 718 **e.** 3,500
7. To base nine: **a.** 36 **b.** 101 **c.** 645 **d.** 2,389 **e.** 15,692
8. To base four: **a.** 44 **b.** 120 **c.** 900 **d.** 512 **e.** 1,326
9. To base seven: **a.** 98 **b.** 172 **c.** 695 **d.** 1,000 **e.** 5,000

10. a. $610 = (\quad)_{two}$
b. $847 = (\quad)_{eight}$
c. $703 = (\quad)_{five}$
d. $1{,}429 = (\quad)_{four}$
e. $2{,}500 = (\quad)_{nine}$
f. $6{,}144 = (\quad)_{twelve}$
g. $8{,}000 = (\quad)_{seven}$
h. $1{,}312 = (\quad)_{three}$

To express one non-decimal numeral as another non-decimal numeral in a different base, we first change the given non-decimal numeral to a base ten numeral. Then we change the resulting base ten numeral to the numeral of the required base.

Write 1302_{four} as a base five numeral:

$1302_{four} =$ 1 (64's) 3 (16's) 0 (4's) 2 (1's)

$= 114_{ten}$

$= 424_{five}$

$1 \times 64 = 64$
$3 \times 16 = 48$
$2 \times 1 = 2$
Total: 114

25)114(4
100
5) 14(2
10
4

Answer, 1302_{four} expressed as a base ten numeral is 114_{ten}, which is the base five numeral 424_{five}.

EXERCISES

Express each of the following numerals as an equivalent numeral of the indicated base:

1. a. 342_{five} to base two
b. 708_{nine} to base five
c. 647_{eight} to base four
d. $3T4E_{twelve}$ to base seven
e. 1563_{seven} to base six
f. 2385_{nine} to base eight
g. 23130_{four} to base three
h. 45134_{six} to base two
i. $2TT0E_{twelve}$ to base five
j. 12212_{three} to base two

2. a. 11101_{two} to base five
b. 352_{six} to base nine
c. 2201_{three} to base seven
d. 11011100_{two} to base four
e. 764_{eight} to base twelve
f. 4323_{five} to base six
g. 3554_{six} to base eight
h. 3333_{four} to base six
i. 101110111_{two} to base three
j. 14326_{seven} to base nine

3. **a.** 403_{five} to base eight
   b. 2631_{seven} to base two
   c. 1344_{six} to base five
   d. 3765_{nine} to base three
   e. $E48_{twelve}$ to base six
   f. 12212_{three} to base twelve
   g. 3123_{four} to base seven
   h. 25607_{eight} to base nine
   i. 1101101101_{two} to base four
   j. 23014_{five} to base seven

CHAPTER REVIEW

1. What number is expressed by each of the following numerals? (2-1—2-7)

   a. 

   b. MCMXCVII

   c. ΨϘH

   d. 

   e. 

   f. 六千三百八十七

2. Write ΣNΔ as an equivalent: (2-1—2-7)

   a. Babylonian numeral **c.** Mayan numeral **e.** Egyptian numeral
   b. Chinese numeral **d.** Roman numeral

3. Express 191 by an equivalent: (2-1—2-7)

   a. Greek numeral **c.** Mayan numeral **e.** Egyptian numeral
   b. Chinese numeral **d.** Roman numeral **f.** Babylonian numeral

4. Write the numeral that names the smallest number possible using the digits 8, 1, 5, 2, 4, 1. (2-8)
5. Which number is greater: 698,898 or 698,988? (2-8)
6. Write in words 2,875,306,014,009. (2-9)
7. Write the numeral that names: (2-10)
   Forty-three billion, two hundred thirty-six thousand, nineteen.
8. Round: (2-11)

   a. 483,498,627 to the nearest million.
   b. 50,176,038 to the nearest ten thousand.
9. What is the numeral 6 called in 3^6? What is the numeral 3 called? Find the value of 3^6. (2-12)
10. Use exponential notation to write $7 \cdot 7 \cdot 7 \cdot 7 \cdot 7 \cdot 7 \cdot 7 \cdot 7 \cdot 7$. (2-12)
11. Express 600,000 as a product of a digit and a power of ten. (2-13)
12. Express 4,583,219 in expanded form as a polynomial. (2-14)
13. Make as many dots on your paper as are represented by the numeral 23_{five}. Group them to show the meaning of the numeral. (2-15)

14. What is the value of the place where the digit: (2–16)

a. 1 is located in the numeral 1000000_{two}?
b. T is located in the numeral $E5T79_{\text{twelve}}$?
c. 4 is located in the numeral 4322_{five}?
d. 6 is located in the numeral 12065_{seven}?
e. 3 is located in the numeral 32102_{four}?

15. What numeral is the successor of: (2–17)

a. 111111_{two}?
b. 4344_{five}?
c. $T9E_{\text{twelve}}$?
d. 989999_{ten}?
e. 6566_{seven}?
f. 3123_{four}?

16. Write 2143_{six} in expanded form as a polynomial. (2–18)
17. Express 3582_{ten} as an equivalent base five numeral. (2–19)
18. Express $T5E7_{\text{twelve}}$ as an equivalent base ten numeral. (2–19)
19. Express 6205_{eight} as an equivalent base three numeral. (2–19)
20. Express 1101101111_{two} as an equivalent base seven numeral. (2–19)

CUMULATIVE TEST

1. Which of the following are prime numbers: 17, 27, 37, 47, 57? (1–7)
2. If $N = \{0, 2, 4, 6, 8, \ldots\}$, which of the following statements are true? (1–9)

a. $6 \in N$ **b.** $12 = \notin N$ **c.** $21 \notin N$ **d.** $102 \in N$ **e.** $899 \in N$

3. What is the greatest common factor of 54, 90, and 72? (1–13)
4. Factor 108 as a product of prime numbers. (1–15)
5. What is the least common multiple of 24, 36, and 60? (1–18)
6. Write CCXIX as an equivalent: (2–1—2–7)

a. Chinese numeral
b. Egyptian numeral
c. Greek numeral
d. Babylonian numeral
e. Mayan numeral

7. Which number is greater: 1,267,686 or 1,267,668? (2–8)
8. Round 29,999,741 to the nearest thousand. (2–11)
9. Expand each of the following as a polynomial: (2–14)

a. $37{,}526_{\text{ten}}$
b. 12132_{four}

10. Express 13024_{five} as a base two numeral. (2–19)

3 Properties – Operations with Whole Numbers

3–1 BINARY OPERATIONS

When we add 8 and 5 or subtract 19 from 107 or multiply 18 by 62 or divide 72 by 6, we are operating with two numbers to get a third number. We call this a *binary operation.*

In arithmetic we use the binary operations of addition, subtraction, multiplication and division. When we use any one of these operations, the resulting answer is said to be *unique* because it is the "one and only number" that may result from that operation on the given two numbers.

There are certain characteristics or properties that the operations have. A property is not true unless it holds in all cases. Therefore to determine whether a property is not true it is sufficient to show that it does not hold in one case.

3–2 INVERSE OPERATIONS

Operations that undo each other are called *inverse operations.*
If we first add 3 to 5 and then subtract 3 from the answer, we return to the 5.

$$(5 + 3) - 3 = 5$$

Subtraction undoes addition.

If we first subtract 6 from 8 and then add 6 to the answer, we return to the 8.

$$(8 - 6) + 6 = 8$$

Addition undoes subtraction.

Thus addition and subtraction are inverse operations, they undo each other.
If we first multiply 5 by 4 and then divide the answer by 4, we return to the 5.

$$(5 \times 4) \div 4 = 5$$

Division undoes multiplication.

If we first divide 30 by 6 and then multiply the answer by 6, we return to 30.

$$(30 \div 6) \times 6 = 30$$

Multiplication undoes division.

Thus multiplication and division are inverse operations, they undo each other.

EXERCISES

1. Find the missing numbers or symbols of operation as indicated:

a. $(10 \times 3) \div 3 = \square$
b. $(8 + 7) - 7 = ?$
c. $(15 - 9) + 9 = n$
d. $(24 \div 8) \times 8 = \square$
e. $(17 - 5) + ? = 17$
f. $(9 \times 6) \div \square = 9$
g. $(28 \div 7) \times \square = 28$
h. $(14 + 3) - n = 14$
i. $(11 + \square) - 3 = 11$
j. $(20 \div n) \times 4 = 20$
k. $(16 - ?) + 8 = 16$
l. $(6 \times n) \div 5 = 6$
m. $(\square \div 2) \times 2 = 40$
n. $(? - 7) + 7 = 21$
o. $(n + 6) - 6 = 37$
p. $(? \times 10) \div 10 = 60$
q. $(12 \times 4) \, ? \, 4 = 12$
r. $(15 + 12) \, ? \, 12 = 15$
s. $(48 \div 6) \, ? \, 6 = 48$
t. $(50 - 1) \, ? \, 1 = 50$
u. $(32 \, ? \, 8) - 8 = 32$
v. $(13 \, ? \, 5) + 5 = 13$
w. $(4 \, ? \, 3) \div 3 = 4$
x. $(56 \, ? \, 8) \times 8 = 56$

2. a. Is division the inverse operation of subtraction?
b. Is subtraction the inverse operation of addition?
c. Is multiplication the inverse operation of division?
d. Is addition the inverse operation of multiplication?

3–3 COMMUTATIVE PROPERTY

Of Addition

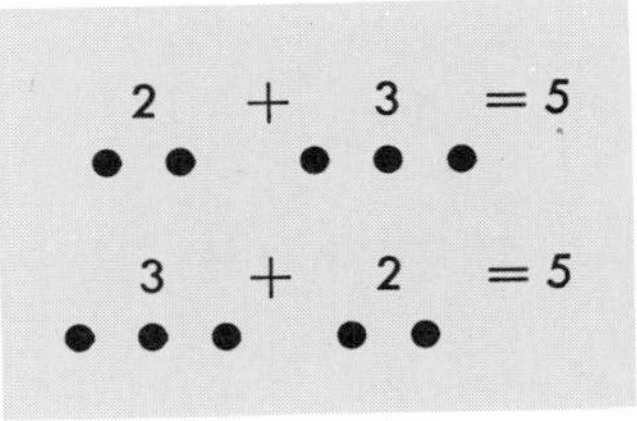

Adding 2 and 3 gives the same sum as adding 3 and 2. That is: $2 + 3 = 3 + 2$. When we add one number to a second number, we get the same sum as when we add the second number to the first number. *The commutative property of addition permits us to change the order of adding two numbers without affecting the sum.*

Of Multiplication

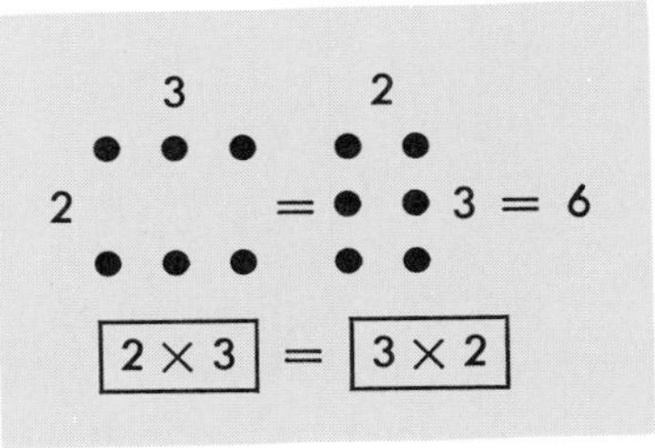

Multiplying 2 and 3 gives the same product as multiplying 3 and 2. That is: $2 \times 3 = 3 \times 2$. See the arrays at the right. They are arrangements of dots in rows and columns. When we multiply one number by a second number, we get the same product as when we multiply the second number by the first number. *The commutative property of multiplication permits us to change the order of multiplying two factors without affecting the product.*

EXERCISES

1. a. Add:

$$\begin{array}{r}6\\ \underline{5}\end{array}\quad \begin{array}{r}5\\ \underline{6}\end{array}\quad \begin{array}{r}9\\ \underline{4}\end{array}\quad \begin{array}{r}4\\ \underline{9}\end{array}$$

c. Multiply:

$$\begin{array}{r}8\\ \underline{4}\end{array}\quad \begin{array}{r}4\\ \underline{8}\end{array}\quad \begin{array}{r}3\\ \underline{6}\end{array}\quad \begin{array}{r}6\\ \underline{3}\end{array}$$

b. Subtract:

$$\begin{array}{r}13\\ \underline{8}\end{array}\quad \begin{array}{r}8\\ \underline{13}\end{array}\quad \begin{array}{r}17\\ \underline{9}\end{array}\quad \begin{array}{r}9\\ \underline{17}\end{array}$$

d. Divide:

$12 \div 4$ $\quad 36 \div 9$
$4 \div 12$ $\quad 9 \div 36$

e. Does interchanging addends (numbers that are added) affect the sum?
f. Does interchanging the minuend and subtrahend affect the answer when we subtract?
g. Does interchanging factors affect the product?
h. Does interchanging the dividend and divisor affect the quotient?
i. Does the commutative property hold for subtraction? For division?

2. Which of the following statements are true?

a. $15 - 19 = 19 - 15$
b. $34 + 10 = 10 + 34$
c. $24 \times 6 = 6 \times 24$
d. $49 \div 7 = 7 \div 49$
e. $158 + 25 = 25 + 158$
f. $35 \times 87 = 87 \times 35$
g. $90 \div 15 = 15 \div 90$
h. $75 - 39 = 39 - 75$
i. $63 + 54 = 54 + 63$

3. Find the numbers that will make the following statements true:

a. $67 + 34 = ? + 67$
b. $59 \times 61 = 61 \times ?$
c. $45 \times ? = 36 \times 45$
d. $? + 60 = 60 + 85$

3–4 ASSOCIATIVE PROPERTY

Of Addition

Since addition is a binary operation, only two numbers may be added at any one time. When there are three addends, we must first select two addends, find their sum, and then add the third addend to this sum.

$2 + 4 + 6$ may be thought of as either:
$(2 + 4) + 6$ which is $6 + 6 = 12$
or $\quad 2 + (4 + 6)$ which is $2 + 10 = 12$.
Thus, $\quad (2 + 4) + 6 = 2 + (4 + 6)$

The associative property of addition permits us to group or associate the first and second numbers and add their sum to the third number or to group or associate the second and third numbers and add their sum to the first number. Either way we get the same final sum.

Of Multiplication

Since multiplication is a binary operation, only two factors may be multiplied at any one time. When there are three factors, we must first select two factors, find their product, and then multiply this product by the third factor.

$2 \times 4 \times 6$ may be thought of as either:
$(2 \times 4) \times 6$ which is $8 \times 6 = 48$
or $2 \times (4 \times 6)$ which is $2 \times 24 = 48$
Thus, $(2 \times 4) \times 6 = 2 \times (4 \times 6)$

The associative property of multiplication permits us to group or associate the first and second numbers and multiply their product by the third number or to group or associate the second and third numbers and multiply their product by the first number. We get the same final product in each case.

EXERCISES

1. Find the value of $(12 + 6) + 4$. Find the value of $12 + (6 + 4)$. Is the statement $(12 + 6) + 4 = 12 + (6 + 4)$ true? Select any three numbers. Check whether the associative property holds for the addition of these numbers.

2. Find the value of $(18 - 5) - 2$. Find the value of $18 - (5 - 2)$. Is the statement $(18 - 5) - 2 = 18 - (5 - 2)$ true? Does the associative property hold for subtraction?

3. Find the value of $(20 \times 5) \times 3$. Find the value of $20 \times (5 \times 3)$. Is the statement $(20 \times 5) \times 3 = 20 \times (5 \times 3)$ true? Select any three numbers. Check whether the associative property holds for multiplication of these numbers.

4. Find the value of $(16 \div 8) \div 2$. Find the value of $16 \div (8 \div 2)$. Is the statement $(16 \div 8) \div 2 = 16 \div (8 \div 2)$ true? Does the associative property hold for division?

5. When we add a column of three addends in a down direction and check by adding in an up direction, what property are we using?

6. Which of the following statements are true?

a. $(9 \times 7) \times 4 = 9 \times (7 \times 4)$
b. $(75 \div 15) \div 5 = 75 \div (15 \div 5)$
c. $(47 - 12) - 10 = 47 - (12 - 10)$
d. $(82 + 35) + 15 = 82 + (35 + 15)$
e. $(24 + 16) + 32 = 24 + (16 + 32)$
f. $(51 \times 8) \times 19 = 51 \times (8 \times 19)$

7. Find the numbers that will make the following statements true:

a. $(6 + 8) + 9 = 6 + (? + 9)$
b. $(12 \times 4) \times ? = 12 \times (4 \times 7)$
c. $(? \times 10) \times 5 = 20 \times (10 \times 5)$
d. $(22 + ?) + 16 = 22 + (8 + 16)$
e. $(36 + 9) + 25 = 36 + (9 + ?)$
f. $(20 \times 50) \times 30 = ? \times (50 \times 30)$

8. Show by using the commutative and associative properties that:

a. $3 + 8 + 7 = (3 + 7) + 8$
b. $5 \times 9 \times 2 = (5 \times 2) \times 9$
c. $9 + 6 + 11 = 6 + (9 + 11)$
d. $4 \times 7 \times 2 = 7 \times (4 \times 2)$

3–5 DISTRIBUTIVE PROPERTY OF MULTIPLICATION OVER ADDITION

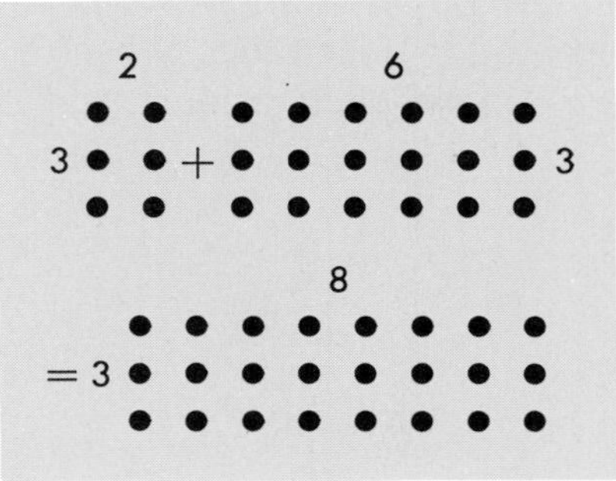

To find the product of $3 \times (2 + 6)$ we may find the sum of 2 and 6 and then multiply this sum (8) by 3.

That is: $3 \times (2 + 6) = 3 \times 8 = 24$

Or we may multiply the 2 by 3 and the 6 by 3 and add the products.

That is: $3 \times (2 + 6) = (3 \times 2) + (3 \times 6) = 6 + 18 = 24$

Either way we get the same final result.

The distributive property of multiplication over addition tells us that when we multiply one number by the sum of a second and a third number we get the same result as when we add the product of the first and second numbers to the product of the first and third numbers. Multiplication is being distributed over addition.

We use the distributive property in computation such as 2×43.

$$2 \times 43 = 2 \times (40 + 3) = (2 \times 40) + (2 \times 3) = 80 + 6 = 86$$

or

$$\begin{array}{r} 43 \\ \underline{\times 2} \end{array} = \begin{array}{r} 40 + 3 \\ \underline{\times 2} \end{array} = \begin{array}{r} 40 \\ \underline{\times 2} \end{array} + \begin{array}{r} 3 \\ \underline{\times 2} \end{array}$$

We also use the distributive property in the following:

$$(7 \times 4) + (7 \times 8) = 7 \times (4 + 8)$$

or

$$(4 \times 7) + (8 \times 7) = (4 + 8) \times 7$$

EXERCISES

1. Find the value of $6 \times (3 + 4)$. Find the value of $(6 \times 3) + (6 \times 4)$. Is the statement $6 \times (3 + 4) = (6 \times 3) + (6 \times 4)$ true? What property is used here?

2. Show how the distributive property is used in computing 3×23.

3. Find the numbers that will make the following statements true:

a. $7 \times (4 + ?) = 7 \times 4 + 7 \times 5$

b. $8 \times (3 + 9) = 8 \times ? + 8 \times 9$

c. $4 \times (12 + 7) = ? \times 12 + ? \times 7$

d. $10 \times 8 + 10 \times 6 = ? \times (8 + 6)$

e. $15 \times 14 + 15 \times ? = 15 \times (14 + 11)$

f. $5 \times (1 + 4) = 5 \times ? + 5 \times ?$

g. $16 \times (3 + 7) = 16 \times ? + ? \times 7$

h. $21 \times (? + ?) = 21 \times 10 + 21 \times 8$

i. $32 \times 9 + 32 \times 6 = ? \times (? + ?)$

j. $50 \times (11 + 17) = ? \times ? + ? \times ?$

4. Find the value of $6 + (3 \times 4)$. Find the value of $(6 + 3)(6 + 4)$. Is the statement $6 + (3 \times 4) = (6 + 3)(6 + 4)$ true? Is addition distributed over multiplication?

5. Is the statement $4 \times (8 - 5) = (4 \times 8) - (4 \times 5)$ true? Is multiplication distributed over subtraction?

3–6 CLOSURE

When we add the two natural numbers such as 4 and 6, is the sum a natural number? Yes, it is 10. Can you find two natural numbers whose sum is not a natural number?

When we multiply the two natural numbers 8 and 5, is the product a natural number? Yes, it is 40. Can you find two natural numbers whose product is not a natural number?

When we subtract the natural number 9 from the natural number 12, is the result a natural number? Can you find two natural numbers whose difference is not a natural number? Try to subtract 8 from 6.

When we divide the natural number 12 by the natural number 4, is the quotient a natural number? Can you find two natural numbers whose quotient is not a natural number? Try to divide 8 by 6.

If we, using all numbers in a given set, add any two numbers (or subtract or multiply or divide) and get as our answer in every case one of the numbers described in the given set, we say the set is closed under that operation. This property is called *closure*.

Not all sets of numbers are closed under all operations. In the above we see that the set of natural numbers is closed under addition and multiplication because the answers in *every* case are natural numbers but is not closed under subtraction and division because the answers in *some* cases are not natural numbers. A property must hold for *all* cases.

EXERCISES

1. Is the set of all prime numbers closed under the operation of addition? Subtraction? Multiplication? Division?

2. Is the set of all even natural numbers closed under the operation of addition? Subtraction? Multiplication? Division?

3. Is the set {5, 10, 15, 20, 25, . . .} closed under the operation of addition? Subtraction? Multiplication? Division?

4. Is the set {0, 1} closed under the operation of addition? Subtraction? Multiplication? Division?

5. Is the set {10, 20, 30, 40, 50, . . .} closed under the operation of addition? Subtraction? Multiplication? Division?

6. Which of the sets at the right are closed under:

a. Addition?

b. Subtraction?

c. Multiplication?

d. Division?

{2, 4, 8}
{0, 1, 2, 3, 4}
{10, 100, 1000, . . .}
{3, 6, 9, 12, . . .}
{1, 3, 5, 7, . . .}
{4, 8, 12, 16, . . .}

3–7 PROPERTIES OF ZERO

When we add zero (0) to any number, the number remains unchanged. This addition property of zero is illustrated by the following:

The sum of 6 and 0 is 6. $6 + 0 = 6$
The sum of 0 and 9 is 9. $0 + 9 = 9$

A number which, when added to a given number, does not change the given number is called the *additive identity* (or identity element for addition). Therefore zero (0) is the additive identity.

The following are other important properties of zero:

(1) Zero subtracted from any number is the number. $8 - 0 = 8$
(2) The difference between any number and itself is zero. $12 - 12 = 0$
(3) When a non-zero number is multiplied by zero, the product is zero. $0 \times 7 = 0$
(4) When 0 is multiplied by a non-zero number, the product is zero. $8 \times 0 = 0$
(5) When zero is multiplied by zero, the product is zero. $0 \times 0 = 0$ Thus, the product of any number and zero is zero.
(6) If the product of two numbers is 0, then one of the factors is zero or both factors are zero.
(7) If zero is divided by any number other than zero, the quotient is zero. $0 \div 5 = 0$
(8) In arithmetic the division by 0 is excluded.

Since multiplication and division are inverse operations, $6 \div 2 = 3$ may be thought of as $3 \times 2 = 6$. Thus $6 \div 0 = ?$ may be thought of as $? \times 0 = 6$. Since the product of any number and zero is zero and not 6, the statement $6 \div 0$ is meaningless; there is no answer. Also $0 \div 0 = ?$ may be thought of as $? \times 0 = 0$. Since the product of *any* number and zero is zero, then the answer could be *any* number and so is indeterminate. Therefore divisions like $6 \div 0$ written also as $0\overline{)6}$ or $\frac{6}{0}$ and $0 \div 0$ written also as $0\overline{)0}$ or $\frac{0}{0}$ are excluded.

EXERCISES

1. Determine the value of each of the following:

a. $7 - 0$
b. $0 + 11$
c. $25 - 25$
d. 50×0
e. $34 + 0$
f. 0×0
g. $0 \div 12$
h. $\frac{16}{16}$
i. $\dfrac{9 - 9}{15 + 14}$
j. 0×18
k. $8 \times 6 \times 4 \times 0 \times 9$
l. $\dfrac{25 - 5}{17 - 17}$

2. Which of the following represent the number zero?

a. $\frac{8}{8}$ **b.** $8 - 8$ **c.** $0 \div 10$ **d.** $\frac{14 - 14}{21}$ **e.** $\frac{27 - 15}{5 - 5}$

3. Which of the following are meaningless or indeterminate?

a. $9 \div 0$ **b.** $0 \div 4$ **c.** $0 \div 0$ **d.** $\frac{23 - 23}{16 - 15}$ **e.** $\frac{6 - 2}{48 - 48}$

3–8 PROPERTIES OF ONE

When we multiply any number by one (1), the number remains unchanged. This multiplicative property of one is illustrated by the following:

1 times 4 is 4. $1 \times 4 = 4$
9 times 1 is 9. $9 \times 1 = 9$

A number which, when multiplied by a given number, does not change the given number is called the *multiplicative identity* (or the identity element for multiplication). Therefore one (1) is the multiplicative identity.

We shall see that the multiplicative identity, one, is used when changing fractions to lower and higher terms, when dividing fractions, and when dividing by a decimal.

The following are other important properties of one:

(1) When any number, except zero, is divided by itself, the quotient is one. Since the fraction bar indicates division, one may be expressed in symbols as follows:

$$\frac{1}{1}, \frac{2}{2}, \frac{3}{3}, \frac{4}{4}, \frac{5}{5}, \frac{6}{6}, \frac{7}{7}, \frac{8}{8}, \frac{9}{9}, \frac{10}{10}, \frac{11}{11}, \text{etc.}$$

(2) One raised to any power is one. For example:

$$1^6 = 1 \times 1 \times 1 \times 1 \times 1 \times 1 = 1$$

(3) When one is added to any whole number, we get the next higher whole number.

(4) When one is added to any even number, we get an odd number. When one is added to any odd number, we get an even number.

EXERCISES

1. Determine the value of each of the following:

a. $\frac{6}{6}$
b. 18×1
c. 1^4
d. 1×53
e. $30 \times 1 \times 1 \times 1$
f. 7×1^{12}
g. $1^{15} \times 1^{24}$
h. $\frac{24 - 15}{4 + 5}$
i. $1 \times 1 \times 1 \times 1 \times 2 \times 1 \times 1 \times 1$

2. Which of the following represent the number 1?

a. $4 - 4$ **b.** $\frac{4}{4}$ **c.** 5^1 **d.** 1^5 **e.** $\frac{28 - 13}{3 \times 5}$

3–9 COMPUTATION—ADDITION

We use addition to put together or combine quantities to find the total number. Only like quantities may be added: ones with ones, tens with tens, hundreds with hundreds, etc. The numbers we add are called the *addends* and the answer is called the *sum*. The symbol used to indicate addition is the *plus* sign (+).

To add whole numbers, we add each column, beginning with the ones column. If the sum of any column is ten or more, we write the last digit of the sum in the answer and carry the other digits to the next column to the left. We check by adding the columns in the opposite direction.

```
356 ↖
 89 ← Addends
293 ↙
---
738 ← Sum
Answer, 738
```

EXERCISES

1. Add and check:

a.	1	7	3	4	6	2	8	3	5	1	0	3	9
	4	2	3	5	1	0	4	6	9	7	8	2	5
b.	3	1	6	0	5	7	4	1	7	2	5	6	4
	4	1	2	4	3	1	6	2	7	2	4	9	1
c.	5	9	3	8	4	2	5	9	4	3	6	0	1
	1	0	7	1	4	5	7	9	2	5	3	0	3
d.	7	9	5	2	6	2	8	2	8	4	6	9	3
	3	6	5	9	5	4	3	1	7	3	0	8	9
e.	4	6	2	4	1	6	5	3	7	1	8	7	5
	9	4	3	7	5	7	0	8	9	6	2	8	6
f.	2	4	3	7	1	3	1	6	9	2	8	9	7
	6	8	1	4	9	0	8	6	3	8	6	2	6
g.	8	0	5	7	6	8	5	9	2	9	8	9	0
	8	7	8	5	8	9	2	1	7	7	5	4	1

2. Find the missing numbers:

a. $5 + 7 = ?$ **c.** $n - 3 = 6$ **e.** $5 + 6 = n$ **g.** $? - 3 = 11$ **i.** $\square + \square = 18$

b. $9 + 8 = \square$ **d.** $? - 9 = 7$ **f.** $\square - 2 = 1$ **h.** $8 + 7 = n$ **j.** $n - 4 = 13$

3. Add each of the following:

a. 2	12	22	52	82	**c.** 9	29	39	69	89
7	7	7	7	7	4	4	4	4	4

b. 53	64	6	1	74	**d.** 27	79	4	5	38
5	2	93	38	4	8	9	67	49	6

4. Add and check:

a.	426	682	29	491	528
	7	43	8	238	325
	983	9	537	475	846
	6	95	86	289	587

b.	3,582	9,734	6,359	2,187	4,139
	165	859	4,869	5,854	8,267
	2,859	26	3,687	3,895	5,918
	406	4,583	5,948	4,669	7,656

c.	9	638	217	82,152	487
	286	46	59	75	2,849
	58	9	4,895	3,946	86,296
	7	28	684	820	908
	526	253	8	1,639	4,356

d.	396	3,688	49,962	62,835	29,157
	9,274	42,916	90,556	91,684	45,209
	833	259	77,385	26,893	63,982
	409	59,837	82,597	80,795	42,867
	84,582	9,479	29,481	39,886	50,786

e.	825	1,369	9,408	38,257	51,863
	4,932	58,473	35,765	989	7,156
	526	257	89,159	8,192	8,399
	39,688	45,164	849	77,266	70,565
	5,897	3,828	98,267	6,097	49,284

f.	92,868	46,556	21,684	74,396	48,395
	57,859	80,908	89,299	58,795	64,857
	65,967	58,385	67,588	26,888	39,848
	42,686	61,579	92,907	41,479	99,059
	78,679	83,438	81,554	78,393	68,586

g.	6,584	4,593	82,584	14,688	89,765
	923	688	93,635	69,526	48,489
	78	46	6,987	83,989	69,473
	92,654	239	5,279	59,675	53,269
	7,369	58,427	38,096	65,114	95,786
	827	948	96,887	82,879	28,698

h. 853,264	4,672	684,597	563,875	486,554
96,587	635,967	483,698	285,973	560,893
9,467	72,584	599,569	864,798	915,921
658	9,593	367,475	975,687	357,865
17,425	298,449	952,388	869,506	831,479
8,269	587,976	876,959	372,452	699,837

5. Add horizontally:

a. 68 + 29 + 64 **c.** 8354 + 918 + 6239 **e.** 615 + 98 + 9 + 968
b. 306 + 85 + 747 **d.** 40 + 56 + 8 + 99 **f.** 9057 + 2892 + 4688 + 5145

6. Arrange in columns, then find the sum of each of the following:

a. 25; 47; 82 **c.** 4,075; 5,385; 1,625 **e.** 8,000; 60; 900; 40
b. 349; 876; 946 **d.** 867; 9; 16; 218 **f.** 563; 89; 7; 6,475; 44

3–10 COMPUTATION—SUBTRACTION

In subtraction a quantity is taken away from another given quantity and the number left over (called the *remainder*) is found. Sometimes subtraction is used to compare two quantities. Here we are finding the difference between two numbers or finding how much more or less one number is than the other. Subtraction is also used in situations where we are required to find how much more is needed. Here we know the sum and one addend and we are required to find the other addend.

The terms that we generally use in subtraction are: *minuend*, the number from which we subtract; *subtrahend*, the number we subtract; and *remainder* or *difference*, the answer in subtraction. Instead of these terms, to simplify the language we can use *sum* as the number from which we subtract, *addend* as the number we subtract and the *missing addend* as the answer. 8 − 5 = ? is thought of as 5 + ? = 8. Subtraction is not considered a principle operation but as an *inverse operation of addition*. The minus sign (−) is the symbol used to indicate subtraction. We can subtract only like quantities: ones from ones, tens from tens, hundreds from hundreds, etc.

To subtract whole numbers, we subtract the digits in the subtrahend from the corresponding digits in the minuend, starting from the ones place. When any digit in the subtrahend is greater than the corresponding digit in the minuend, we increase this digit in the minuend by 10 by taking 1 from the preceding digit in the next higher place and changing it to 10 of the next lower place. This is called *regrouping*. Some people call it *borrowing*.

728	← Minuend
− 453	← Subtrahend
275	← Remainder or Difference

728	← Sum
− 453	← Addend
275	← Addend

Answer, 275

728	=	7 hundreds	2 tens	8 ones	=	6 hundreds 12 tens 8 ones	
453	=	4 hundreds	5 tens	3 ones	=	4 hundreds 5 tens 3 ones	
						2 hundreds 7 tens 5 ones	= 275

We check by adding the remainder to the subtrahend. Their sum should equal the minuend.

EXERCISES

1. Subtract and check:

a.	2	9	8	3	10	11	4	9	13	6	8
	0	5	8	2	4	8	1	8	6	5	6
b.	7	9	12	6	7	10	11	14	5	6	10
	7	6	5	4	6	7	9	7	3	1	9
c.	10	5	15	8	12	16	9	7	16	10	8
	1	2	7	1	8	7	2	3	8	6	2
d.	1	6	7	6	9	17	0	11	12	17	12
	1	3	4	6	4	8	0	5	3	9	7
e.	11	9	4	5	11	9	10	8	13	10	16
	2	1	2	4	4	3	5	3	4	2	9
f.	12	10	13	14	5	12	15	14	11	18	14
	9	8	7	9	1	6	8	5	6	9	6
g.	10	7	4	14	11	9	7	11	4	8	13
	3	1	3	8	7	9	2	3	4	5	8
h.	15	7	6	8	13	12	8	13	5	9	15
	6	5	2	4	5	4	7	9	0	7	9

2. Find the missing numbers:

a. $16 - 9 = \square$ **c.** $? + 6 = 14$ **e.** $n + 7 = 15$ **g.** $8 + n = 13$ **i.** $? + 3 = 4$
b. $13 - 5 = n$ **d.** $4 + \square = 8$ **f.** $12 - 7 = ?$ **h.** $10 - \square = \square$ **j.** $n + 5 = 11$

3. Subtract and check:

a.	59	63	88	95	82	47	86	70
	7	8	23	78	57	27	83	24
b.	347	984	825	617	500	902	316	603
	26	565	179	598	344	496	85	405
c.	5,238	9,386	3,752	8,126	5,174	9,145	6,000	7,594
	4,125	2,931	57	4,077	1,886	9	5,842	3,748
d.	9,285	1,527	8,467	7,418	8,641	4,529	8,506	2,016
	6	49	2,053	3,156	7,936	1,780	4,748	860

e.	62,874	84,592	89,327	46,285	90,526	30,000
	11,562	63,870	639	32,189	64,865	8,503
f.	98,463	35,862	25,174	85,000	91,052	32,067
	72,340	17,927	87	6,958	80,389	15,938
g.	874,506	297,506	619,385	720,094	982,050	
	391,275	8,617	253,419	510,638	875	
h.	4,669,857	6,984,506	4,930,807	8,914,050	4,000,000	
	2,179,427	2,893,715	25,040	5,806,973	3,806,796	
i.	586	4,052	1,537	64,828	891,060	3,101,953
	298	3,987	68	56,179	859,184	9,478
j.	925	3,840	8,545	42,516	650,000	7,430,467
	786	2,761	3,987	27,899	559,077	6,629,408

4. Subtract horizontally:

a.	48 − 29	153 − 78	324 − 169	5,000 − 525	2,800 − 15
b.	85 − 56	247 − 187	900 − 287	8,160 − 5,296	67,000 − 6,700

5. a. From 800 subtract 279.
b. Take 517 from 603.
c. Subtract 3,809 from 7,000.
d. From 4,267 take 1,098.
e. Subtract 43,963 from 51,825.
f. From 70,688 take 9,878.

6. Find the difference between: **a.** 1,680 and 494 **b.** 20,000 and 2,000.

3–11 COMPUTATION—MULTIPLICATION

Multiplication is a short way of adding two or more quantities of the same size. $5 + 5$ becomes 2 fives or 2×5. $8 + 8 + 8$ becomes 3 eights or 3×8. We may think of multiplication as successive addition or repeated addition of equal or identical numbers.

Multiplication is a method of combining groups of equal size or groups containing the same number of things to find how many there are altogether. In order to multiply we must know the number of equal groups and the size or how many things there are in each group.

The terms we generally use in multiplication are: *multiplicand*, the number that we multiply; *multiplier*, the number by which we multiply; *product*, the answer in multiplication; *partial product*, the product obtained by multiplying the multiplicand by any figure in the multiplier containing two or more figures; and *factor*, any number that is multiplied. To simplify the mathematical language, since both the multiplicand and multiplier are factors, we could say:

factor × factor = product
instead of multiplier × multiplicand = product.

The times sign × or the raised dot · are used to indicate multiplication.

One-Digit Multipliers

To multiply by a one-digit multiplier, we multiply each digit in the multiplicand by the one-digit multiplier, starting from the right. If the product or total for any place is ten or more, we write the last digit in the answer and add the other digit to the next product.

$$\begin{array}{rl} 48 & \leftarrow \text{Multiplicand or Factor} \\ \underline{\times\ 2} & \leftarrow \text{Multiplier or Factor} \\ 96 & \leftarrow \text{Product} \end{array}$$

Answer, 96

Multipliers of Two or More Digits

To multiply by a multiplier containing two or more digits, we multiply all the digits in the multiplicand by each digit in the multiplier, starting from the right, to find the partial products. The partial products are written under each other with the numerals placed so that the right-hand digit of each partial product is directly under its corresponding digit in the multiplier. Then we add the partial products. Observe that the distributive principle is used in the model problem. We check by interchanging the multiplier and multiplicand and multiplying again or by division if it has been studied.

$$\begin{array}{r} 12 \\ \underline{\times\ 23} \end{array} = \begin{array}{r} 12 \\ \underline{\times\ 20} \\ 240 \end{array} + \begin{array}{r} 12 \\ \underline{\times\ 3} \\ 36 \end{array}$$

$$\begin{array}{r} 12 \\ \underline{\times\ 23} \\ 36 \\ \underline{240} \\ 276 \end{array} \quad \text{or} \quad \begin{array}{rl} 12 & \\ \underline{23} & \\ 36 & \leftarrow \text{Partial} \\ \underline{24\ \ } & \leftarrow \text{Products} \\ 276 & \end{array}$$

Answer, 276

EXERCISES

1. Multiply and check:

a. 8	3	4	3	9	5	2	1	4	5	6
1	6	0	4	6	7	4	2	3	9	1
b. 6	5	7	2	8	9	4	7	4	3	1
2	3	4	8	4	3	6	7	5	0	5
c. 7	3	1	7	2	8	7	6	2	4	6
3	2	1	6	5	3	0	3	7	2	4
d. 8	1	2	0	7	5	4	0	3	1	7
8	7	3	0	9	1	7	6	8	9	2

e. 3 1	4 8	9 4	7 5	2 6	9 7	8 2	9 8	1 6	6 7	3 5
f. 2 9	8 7	9 1	0 8	5 6	9 5	8 6	6 9	5 8	7 1	8 9
g. 3 7	4 9	2 2	1 4	6 6	5 2	1 8	3 3	0 9	1 3	9 9
h. 8 5	4 4	6 5	5 4	9 2	7 8	0 5	4 1	6 8	5 5	3 9

2. Find the answers to each of the following:

a. 3 times 2, then add 1.
b. 5 times 7, then add 6.
c. 9 times 8, then add 8.
d. 4 times 3, then add 1.
e. 6 times 8, then add 4.
f. 5 times 3, then add 2.
g. 8 times 6, then add 5.
h. 9 times 9, then add 6.
i. 0 times 8, then add 7.
j. 7 times 5, then add 3.

3. Multiply and check:

a. 32 2	72 4	13 6	78 8	197 4	856 7	380 9	703 8
b. 8,016 5	6,080 7	9,354 2	8,005 4	59,687 8	94,758 9	80,103 7	25,914 6

4. Multiply and check:

a. 21 43	69 35	51 87	92 26	56 75	60 97	86 50	70 70	8 69	91 10
b. 231 24	937 85	452 69	750 48	409 36	54 983	805 90	954 99	400 25	568 87
d. 196 217	468 349	259 783	892 495	784 597	969 829	501 467	640 800	793 108	607 906

c. 3,825 79	1,728 81	9,076 90	96 6,080	14,243 12	78,049 67	65,874 56	50,904 38
e. 1,728 456	8,936 659	4,800 307	5,280 178	26,569 827	40,706 905	80,020 204	89,978 698
f. 2,347 3,157	4,339 4,856	8,476 7,288	7,845 9,387	4,018 5,009	6,738 7,654	3,050 3,050	6,159 4,000
g. 95,468 7,986	43,560 2,895	80,070 3,004	90,406 5,634	29,783 49,976	89,612 27,064	93,654 85,917	70,500 90,010

h. 98 7	56 83	497 96	89 320	854 479	325 958	6,080 306	2,586 144	3,928 5,196	26,541 9,085

5. Do as directed in each of the following:

a. 12 times 90	144 times 65	507 times 704
b. 36 × 75	320 × 100	728 × 897
c. 24 × 16 × 8	35 × 42 × 19	46 × 301 × 69
d. Multiply 85 by 72.	Multiply 349 by 60.	Multiply 144 by 700.
e. Find the product of 40 and 70.	Find the product of 624 and 89.	Find the product of 5,280 and 405.

3–12 FACTORIAL

The product of all natural numbers up to and including a given number is called the *factorial* of the given number.

Four factorial, written as 4!, means 1 × 2 × 3 × 4 or the product 24.

One factorial, written as 1!, is equal to 1.

EXERCISES

Find the products represented by the following:

1. 3!	**3.** 2!	**5.** 8!	**7.** 7!	**9.** 9!	**11.** 2!3!	**13.** 2!3!4!
2. 5!	**4.** 6!	**6.** 1!	**8.** 10!	**10.** 12!	**12.** 6!4!	**14.** 3!5!6!

3–13 SQUARING A NUMBER

The *square* of a number is the product obtained when a given number is multiplied by itself.

Thus to square a number we multiply the given number by itself.

Square 9:

9 × 9 = 81

Answer, 81

EXERCISES

Square each of the following numbers:

1. 7	**3.** 28	**5.** 39	**7.** 75	**9.** 100	**11.** 584	**13.** 1,250	**15.** 1,760
2. 12	**4.** 53	**6.** 40	**8.** 89	**10.** 206	**12.** 800	**14.** 3,847	**16.** 9,003

3–14 RUSSIAN PEASANT METHOD OF MULTIPLICATION

In the Russian Peasant method of multiplication one factor is multiplied by 2 (or doubled) and the corresponding other factor is divided by 2 (or halved). The resulting factors are doubled and halved respectively until the halved number 1 is reached. Any fraction associated with the halved number is dropped.

To find the answer, all the doubles corresponding to even halves are first crossed out. Then the remaining doubles are added. The sum, thus found, is the desired answer.

To multiply 125 by 59 using the Russian Peasant method, we first write the given factors 125 and 59 in two columns as shown. We double one factor and halve the other, continuing in this way until 1 is reached in the column of halves. The number 500 is crossed out in the column of doubles since the corresponding halved number is 14, an even number. The sum of the remaining doubles

$$(125 + 250 + 1{,}000 + 2{,}000 + 4{,}000)$$

is found to be 7,375 which is the product of 125 and 59.

Doubles	Halves
125	59
250	29
~~500~~	14
1000	7
2000	3
4000	1
7375	

EXERCISES

Multiply by the Russian Peasant method:

1. 28 × 65	**4.** 51 × 367	**7.** 629 × 564	**10.** 366 × 508
2. 97 × 83	**5.** 434 × 250	**8.** 700 × 803	**11.** 481 × 969
3. 149 × 92	**6.** 893 × 475	**9.** 295 × 389	**12.** 108 × 630

3–15 LATTICE METHOD OF MULTIPLICATION

The Gelosia or lattice method of multiplication was used late in the 15th century. John Napier (1550–1617) used a similar idea to invent a system of rods, called Napier Bones, to multiply numbers.

The lattice is a frame of squares. Each square has a diagonal line drawn through it.

To multiply 9,564 by 287 by the lattice method, one factor (9,564) is written above the frame and the other factor (287) is written at the right, with one digit of each used for each square.

Each digit at the top is used as a factor for each corresponding square in the column below it and each digit at the right is used as a factor for each corresponding square in the row to the left of it. For each square, the corresponding digit at the top and at the right are used as factors and the product is written in the squares (units under the diagonal and tens above the diagonal). If a product consists only of units, then a zero is written in the tens place.

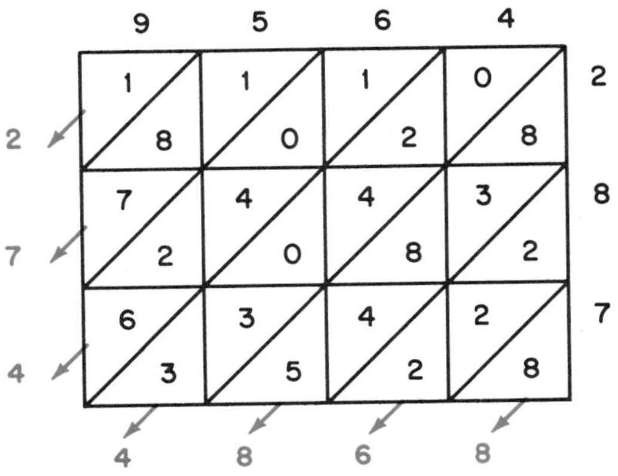

Answer, 2,744,868

Notice in the:

Top row: $9 \times 2 = 18$; $5 \times 2 = 10$; $6 \times 2 = 12$; $4 \times 2 = 8$
Middle row: $9 \times 8 = 72$; $5 \times 8 = 40$; $6 \times 8 = 48$; $4 \times 8 = 32$
Bottom row: $9 \times 7 = 63$; $5 \times 7 = 35$; $6 \times 7 = 42$; $4 \times 7 = 28$

We then add along the diagonals from right to left as we do in ordinary addition, carrying when necessary to the next diagonal.

EXERCISES

Multiply by the lattice method:

1. 67×85
2. 326×49
3. 858×203
4. 926×755
5. 604×907
6. $4{,}318 \times 524$
7. $875 \times 6{,}429$
8. $3{,}508 \times 2{,}706$
9. $7{,}325 \times 4{,}277$
10. $83{,}496 \times 4{,}009$
11. $62{,}051 \times 95{,}902$
12. $59{,}898 \times 83{,}465$

3–16 COMPUTATION—DIVISION

Division is a method of finding how many equal groups can be formed when we know the total number of things and the number of things it takes (sometimes called the size) to make a group. Division is also used to find the number of things in (or the size of) one of the equal groups when we know the number of equal groups and the total number of things. Sometimes division is used to compare one number with another.

In division we generally use the following terms: *dividend*, the number that we divide; *divisor*, the number by which we divide; *quotient*, the answer in division; *remainder*, the number left over when the division is not exact; and *partial dividend,* the first part of the given dividend or digits of the dividend annexed to the remainder. The symbols "÷", ")‾," and the fraction bar "—" are used to indicate division.

Division is not considered a principle operation but as an *inverse operation of multiplication.* Thus division is sometimes thought of as the process of finding the factor which multiplied by a given factor is equal to the given product.

$18 \div 3 = ?$ is considered either as $3 \times ? = 18$ or $? \times 3 = 18$.

Thus,	dividend ÷ divisor = quotient
means	divisor × quotient = dividend
or	factor × factor = product

Dividing a Whole Number by a Whole Number

(1) To divide a whole number by a whole number, we first find the *quotient digit* by dividing the one-digit divisor or the trial divisor, when the divisor contains more than one digit, into the first digit of the dividend. When the divisor contains two or more digits, we use as the trial divisor the first digit of the divisor if the next digit on the right is 0, 1, 2, 3, 4, or 5 and increase the first digit of the divisor by one (1) if the next digit on the right is 6, 7, 8, or 9. The greatest digit that can be used in the quotient at any time is 9.

```
(1) Divide:      (2) Multiply:    (3) Subtract:    (4) Bring down digit:
       2                2                2                2
   32)896           32)896           32)896           32)896
                       64               64               64
                                        25              256
```

(2) We multiply the divisor by the quotient digit and write this product under the corresponding digits in the dividend. This product must be the same or less than the partial dividend. If it is greater, then we use as the trial quotient digit one (1) less than the digit first tried.

(3) We subtract this product from the corresponding numbers in the dividend. The remainder must be less than the whole divisor. If it is not, then we use as the trial quotient digit one (1) more than the digit tried.

```
(5) Divide:
      28
  32)896
     64
     256

(6) Multiply:
      28
  32)896
     64
     256
     256
```

(4) We bring down the next digit of the dividend and annex it to the remainder, if any.

(5) (6) (7) Then using the remainder and annexed numbers as partial dividends, we repeat the above steps for each partial dividend.

We check by multiplying the quotient by the divisor and adding the remainder, if any, to the product. The result should equal the dividend.

(7) Subtract:

```
    28
32)896
   64
   256
   256
   ...
```

Answer, 28

Writing Remainders

When a division is not exact, the remainder may be written in the answer using the letter *R* to indicate it or it may be written as a fraction in lowest terms or it may be indicated (when the divisor is a large number) by a + sign in the answer. In certain social situations remainders are sometimes dropped. For example, the answer when determining how many pieces of ribbon each 5 inches long may be cut from a piece 36 inches long is 7 pieces not $7\frac{1}{5}$ pieces.

```
   3
5)17
  15
   2
```

Answer, 3 *R* 2 or $3\frac{2}{5}$

```
          2
1,760)3,617
      3 520
         97
```

Answer, 2+

```
   7
5)36
  35
   1
```

Answer, 7 pieces

EXERCISES

1. Divide and check:

a. 2)18	8)16	3)15	1)7	5)35	7)21	4)32	8)72	9)54
b. 6)48	3)6	8)40	2)14	1)2	5)40	7)35	3)3	6)24
c. 1)4	4)28	2)6	3)27	4)20	2)2	3)18	7)49	2)12
d. 9)18	1)9	5)45	8)8	7)14	5)10	2)4	7)42	5)15
e. 7)56	4)16	6)18	9)0	5)20	8)64	1)1	2)10	7)28
f. 2)8	5)5	9)45	6)12	1)5	3)12	8)32	7)63	1)3
g. 5)30	6)36	4)8	9)72	8)48	5)25	9)9	7)7	3)24
h. 9)81	4)12	9)36	4)36	6)54	3)21	9)63	3)9	6)30
i. 2)16	8)56	4)4	8)24	1)6	6)6	9)27	6)42	4)24

2. Find the missing numbers:

a. $4 \times ? = 28$ **c.** $7 \times n = 63$ **e.** $? \times 6 = 54$ **g.** $72 \div 9 = \square$ **i.** $? \times 3 = 12$

b. $8 \times \square = 32$ **d.** $\square \times \square = 81$ **f.** $n \times 7 = 42$ **h.** $\square \div 4 = 9$ **j.** $\frac{\square}{5} = 8$

3. Divide and check:

$2\overline{)84}$ $8\overline{)96}$ $4\overline{)872}$ $5\overline{)810}$ $7\overline{)9{,}247}$ $6\overline{)9{,}468}$ $3\overline{)75{,}693}$ $8\overline{)99{,}984}$

4. Divide each of the following. Be careful where you put the first quotient digit:

$7\overline{)427}$ $4\overline{)368}$ $3\overline{)1{,}029}$ $9\overline{)8{,}568}$ $8\overline{)6{,}992}$ $6\overline{)49{,}266}$ $3\overline{)26{,}517}$ $4\overline{)13{,}968}$

5. Watch the zeros in each of the following division problems:

$2\overline{)60}$ $6\overline{)720}$ $5\overline{)4{,}000}$ $3\overline{)6{,}906}$ $5\overline{)510}$ $9\overline{)9{,}045}$ $4\overline{)80{,}120}$ $7\overline{)73{,}542}$

6. Divide each of the following:

a. By 2:	58;	614;	1,706;	73,592;	96,510
b. By 6:	84;	738;	5,442;	42,480;	23,676
c. By 3:	57;	291;	4,215;	26,928;	50,007
d. By 8:	96;	840;	7,648;	56,736;	31,448
e. By 5:	70;	625;	4,300;	81,975;	60,125
f. By 7:	98;	392;	2,695;	56,049;	69,314
g. By 9:	90;	711;	5,682;	27,000;	10,422
h. By 4:	72;	668;	3,160;	97,876;	72,160

7. Find the quotient and the remainder in each of the following:

$8\overline{)59}$ $4\overline{)95}$ $5\overline{)726}$ $6\overline{)8{,}755}$ $8\overline{)6{,}727}$ $3\overline{)44{,}698}$ $8\overline{)56{,}095}$ $6\overline{)91{,}577}$

8. Divide. Write the remainder as a fraction in lowest terms:

$5\overline{)48}$ $6\overline{)46}$ $3\overline{)31}$ $8\overline{)986}$ $6\overline{)8{,}195}$ $9\overline{)31{,}980}$ $7\overline{)84{,}356}$ $4\overline{)33{,}962}$

9. Do as indicated:

a. $3{,}425 \div 5$ **c.** $3{,}628 \div 4$ **e.** $39{,}704 \div 8$ **g.** $81{,}675 \div 9$

b. $8{,}991 \div 9$ **d.** $1{,}770 \div 6$ **f.** $41{,}972 \div 7$ **h.** $32{,}072 \div 8$

10. Divide and check:

a. $10\overline{)40}$ $14\overline{)28}$ $19\overline{)76}$ $63\overline{)189}$ $48\overline{)336}$ $50\overline{)500}$ $96\overline{)864}$ $59\overline{)354}$

b. $18\overline{)612}$ $97\overline{)4{,}462}$ $89\overline{)5{,}162}$ $32\overline{)7{,}648}$ $76\overline{)69{,}768}$

c. $90\overline{)34{,}110}$ $79\overline{)32{,}232}$ $94\overline{)131{,}130}$ $65\overline{)520{,}585}$ $48\overline{)415{,}584}$

d. $31\overline{)722}$ $76\overline{)6{,}565}$ $88\overline{)9{,}039}$ $64\overline{)45{,}467}$ $90\overline{)34{,}733}$

11. Write the remainder as a fraction in the following:

$80\overline{)290}$ $48\overline{)3{,}488}$ $56\overline{)32{,}662}$ $81\overline{)96{,}215}$ $31\overline{)42{,}810}$

12. Divide and check:

a. 298)894   132)3,696   344)60,200   978)644,502   546)221,130

b. 600)491,400   607)549,942   380)189,240   672)565,152   293)830,655

c. 2,240)64,960   6,080)1,270,720   9,006)81,108,036   7,657)6,003,088

13. Find the quotient and the remainder in each of the following:

144)1,368   875)42,375   752)696,935   1,728)39,450   3,600)20,000

14. Do as indicated:

a. 17,856 ÷ 48
b. 522,291 ÷ 969
c. 355,118 ÷ 503
d. 776,640 ÷ 96
e. 144,342 ÷ 297
f. 1,400,000 ÷ 2,240

15. Divide and check:

a. 29)2,107   144)38,629   759)685,436   918)776,899   9,455)595,688

b. 84)5,628   52)49,192   792)331,056   307)303,009   4,298)154,728

3–17 TESTS FOR DIVISIBILITY

A number is said to be *divisible* by another number if it can be divided exactly by the second number with no remainder.

The following are some quick tests to determine whether a number is divisible by 2, 3, 4, 5, 6, 8, 9, 10.

(1) A number is divisible by 2 only if it ends in 0, 2, 4, 6, or 8. All even numbers are divisible by 2.

(2) A number is divisible by 3 only if the sum of its digits is divisible by 3.

> 2,541 is divisible by 3, $2+5+4+1=12$ and 12 is divisible by 3.
> 1,526 is not divisible by 3, $1+5+2+6=14$ and 14 is not divisible by 3.

(3) A number is divisible by 4 only if it is an even number and the number represented by the last two digits (tens and units digits) is divisible by 4. Numbers ending in two zeros are divisible by 4.

> 58,264 is divisible by 4 because *64* is divisible by 4.

(4) A number is divisible by 5 only if it ends in 5 or 0.

> 8,325 ends in 5, thus it is divisible by 5.
> 59,740 ends in 0, thus it is divisible by 5.

(5) A number is divisible by 6 only if it is an even number and the sum of its digits is divisible by 3.

The even number 8,244 is divisible by 6, $8 + 2 + 4 + 4 = 18$ and 18 is divisible by 3.

(6) A number is divisible by 8 only if it is an even number and the number represented by the last three digits (hundreds, tens, and units digits) is divisible by 8. Numbers ending in three zeros are divisible by 8. Use this test only when the number is 1,000 or larger.

The even number 258,496 is divisible by 8 because 496 is divisible by 8.

(7) A number is divisible by 9 only if the sum of its digits is divisible by 9.

86,355 is divisible by 9, $8 + 6 + 3 + 5 + 5 = 27$ and 27 is divisible by 9.

(8) A number is divisible by 10 only if it ends in 0.

6,320 is divisible by 10 because it ends in 0.

EXERCISES

1. **a.** Is 83,592 divisible by 3? by 2? by 6? by 9?
b. Is 578,976 divisible by 4? by 2? by 8? by 6?
c. Is 67,500 divisible by 2? by 5? by 10? by 4?
d. Is 956,000 divisible by 5? by 10? by 6? by 8?
e. Is 468,765 divisible by 9? by 6? by 2? by 3?

2. Determine whether the following numbers are divisible:

a. by 4:	832;	6,794;	8,500;	71,676;	924,635
b. by 3:	558;	7,213;	6,087;	46,881;	483,754
c. by 5:	678;	9,000;	5,265;	28,206;	571,525
d. by 2:	742;	8,504;	7,988;	80,593;	920,000
e. by 6:	285;	9,726;	2,073;	53,262;	727,824
f. by 9:	531;	4,699;	28,386;	47,979;	869,463
g. by 8:	5,952;	8,217;	52,678;	61,000;	644,862
h. by 10:	890;	4,000;	63,505;	80,640;	278,130

3–18 CHECKING BY CASTING OUT NINES

The operations of addition, subtraction, multiplication, and division may be checked by a method called *casting out nines.* This method is not perfect since an incorrect answer sometimes will check.

In this checking process each given number is divided by nine, the nines (quotient) are then cast out (discarded) and only the remainder called the *excess* is used.

Suppose 4,728 is divided by 9. We find there are 525 nines and 3 ones. The remainder 3 indicates there are 3 ones in excess of an exact number of nines (525 nines).

```
    525
 9)4,728
   4 5
     22
     18
      48
      45
       3
```

This excess in a number may be found by an easier method than dividing the given number by 9.

Suppose we add the digits in 4,728; we find the sum is 21. If we add the digits in 21, we find the sum is 3. This sum is the same as the remainder we found when we divided.

$$4 + 7 + 2 + 8 = 21$$
$$2 + 1 = 3$$

Thus we may find the excess ones with respect to the number of groups of nine a number possesses by adding the digits in a given number, then adding the digits in the sum, continuing in this way until the final sum is a one-digit number less than 9. If the sum is 9, it is replaced by a zero since the original number is divisible by 9 and 0 is the remainder. This final sum is the *excess.*

To check addition

We find the excess in each addend and in the sum. The sum of the excesses in the addends should equal the excess in the sum of the addends.[1]

Addition:

```
  238 (4)      (4) + (6) + (3)
  591 (6)      = (13) = (4)
+ 426 (3)
1,255 (4)
```

To check subtraction

We find the excess in the minuend, subtrahend, and answer. The sum of the excesses in the subtrahend and answer should equal the excess in the minuend. When the excess in the minuend is 0, use 9.

Subtraction:

```
  7,523 (8)    (8) − (7) = (1)
− 2,941 (7)
  4,582 (1)    or (7) + (1) = 8
```

[1] If the sum or product of the excesses is a two-digit number, simplify and express as a single digit excess.

To check multiplication

We find the excess in the multiplicand, multiplier, and product. The product of the excesses in the factors should equal the excess in the product.[1]

Multiplication:

46 (1)

× 57 (3) (1) × (3) = (3)

322

2 30

2,622 (3)

To check division

We find the excess in the divisor, dividend, quotient, and remainder if any. The product of the excesses in the divisor and quotient increased by the excess in the remainder should equal the excess in the dividend.[2] When the excess in the divisor is 0, use 9.

Division:

68 (5)

32)2,176 (5))(7)

192

256 (5) × (5) = (25) = (7)

256

EXERCISES

Compute each of the following as directed and check by casting out nines.

1. Add:

654	2,668	78,350	42,837	96,258
8,327	7,157	4,767	95,143	67,442
259	3,826	83,208	88,375	31,597
3,671	5,593	9,315	61,086	87,646
462	8,714	25,785	49,878	32,899

2. Subtract:

6,825	31,057	78,459	387,052	962,584
4,786	2,658	52,847	193,728	638,796

3. Multiply:

325	678	5,687	2,974	83,060
413	745	357	6,758	50,721

4. Divide:

76)61,028	132)3,696	569)49,503	380)189,240	918)776,899

[1,2] If the sum or product of the excesses is a two-digit number, simplify and express as a single digit excess.

3–19 PATTERNS

(1) *Fibonacci Sequence* 1, 1, 2, 3, 5, 8, 13, 21, . . .

This is a sequence of natural numbers in which every number after the second number is equal to the sum of the two preceding numbers.

What are the next five numbers in the sequence after 21?

(2) *Pascal's Triangle*

This is a triangular arrangement of rows of numbers, each row increasing by one number. Each row, except the first, begins and ends in a 1 written diagonally. Beginning with the second row, each number is the sum of the numbers just to the left and right of it in the row above. Note that the numbers are placed midway between the numbers of the row directly above it. Find the numbers that belong in the next four rows.

1
1 1
1 2 1
1 3 3 1
1 4 6 4 1
1 5 10 10 5 1

(3) *Sum of Consecutive Odd Numbers (beginning with 1)*

The sum of 2 odd numbers is 4; of 3 odd numbers is 9; of 4 odd numbers is 16; of 5 odd numbers is 25. What is the relation between the number of addends and the sum? What is the sum of the first 6 consecutive odd numbers without using addition? 9 consecutive odd numbers? 20 consecutive odd numbers? 100 consecutive odd nunbers?

$1+3=4$
$1+3+5=9$
$1+3+5+7=16$
$1+3+5+7+9=25$

(4) Examine the products shown at the right. Does the middle digit in each product correspond to the number of digits in the given factor? Write the product of 11,111 and 11,111 with out multiplying. Of 1,111,111 and 1,111,111. Of 111,111,111 and 111,111,111. Check each multiplication. Will the pattern hold when you multiply 1,111,111,111 by 1,111,111,111?

$1 \times 1 = 1$
$11 \times 11 = 121$
$111 \times 111 = 12{,}321$
$1{,}111 \times 1{,}111 = 1{,}234{,}321$

(5) A *perfect number* is a number that equals the sum of its factors, not including itself. The factors of 6 are 1, 2, 3, and 6. 6 is a perfect number since $6 = 1 + 2 + 3$.

Show that: (a) 28; (b) 496; and (c) 8,128 are perfect numbers.

(6) When you multiply 12,345,679 by 9, what is the product? Without multiplying, what is the product of 12,345,679 and 2×9 (or 18)? Of 12,345,679 by 45 (or 5×9)? Of 12,345,679 by 63?

3–20 FRAME ARITHMETIC

EXERCISES

Find the whole numbers that belong in each frame. Use the same number when the same frame appears more than once.

1. $\square + 6 = 15$
2. $\square - 3 = 8$
3. $4 \times \square = 24$
4. $\square \div 5 = 4$
5. $\square + \square = 28$
6. $\square \times \square = 36$
7. $\square + \square = \square$
8. $\square \times \square = \square$
9. $\square - \square = \square$
10. $\square \div \square = \square$
11. $\square \times 8 = 56$
12. $17 - \square = 10$
13. For which of the above problems can you find more than one correct answer?
14. $\triangle + \square = 10$
15. $\square - \triangle = 2$
16. $\square + \square = \triangle$
17. $\triangle \times \square = 12$
18. $\square \times \square = \triangle$
19. $\square \times \triangle = 18$
20. $\square \div \triangle = 3$
21. $\triangle - \square = \square$
22. $\triangle \times \square = \triangle$
23. $\square \div \triangle = \square$
24. $\square - \square = \triangle$
25. $\square \div \square = \triangle$
26. For which problems 14 to 25 inclusive can you find more than one pair of correct answers?

3–21 PROBLEMS

EXERCISES

1. A recent count showed that in the United States there were 4,620 national banks, 1,147 state banks, 493 mutual savings banks, and 7,919 other banks. Find the total number of banks in the United States.
2. The area of Greenland is 736,518 sq. miles. If the area of Australia is 2,974,581 sq. miles, how much larger is Australia than Greenland?
3. At the school athletic field there are 28 sections each seating 76 persons. What is the total seating capacity?
4. If a pound of grass seed covers 125 sq. feet, how many pounds are needed for a lawn with an area of 2,375 sq. feet?
5. In 1492 there were about 846,000 American Indians in what is now the United States. Recently a census revealed that there are 763,594 American Indians. Find the decrease in population.
6. What is the perimeter of the United States, excluding Alaska and Hawaii, if the northern boundary is 3,987 miles long, eastern boundary 5,565 miles long, southern boundary 5,654 miles long, and western boundary 2,730 miles long?
7. At an average speed of 475 m.p.h. how far can an airplane fly in 15 hours?
8. In Franklin High School there are 5 classes with 39 pupils on roll, 13 classes with 38 pupils, and 12 classes with 37 pupils. Find the total enrollment of the school.
9. Mr. Johnson can buy an automobile for $2,150 cash or $400 down and $80 a month for 24 months. How much can he save by paying cash?
10. Ms. Martinez plans to drive from New York to Chicago, a distance of 840 miles. If she averages 40 m.p.h., how long will it take to make the trip? How many gallons of gasoline will be required if her car averages 14 miles per gallon?

11. During a recent year the mints of the United States manufactured 302,982,760 nickels and 302,876,550 dimes. How many more nickels than dimes were manufactured.

12. Tom's father earns $196 per week. John's mother earns $790 per month. Who earns more per year? How much more?

3–22 COMPUTATION IN NON–DECIMAL BASES

Binary System (Base Two)

In the binary system the *addition facts* are $0 + 0 = 0$; $0 + 1 = 1$; $1 + 0 = 1$; $1 + 1 = 10$ which may be tabulated as shown.

+	0	1
0	0	1
1	1	10

In the binary system the *multiplication facts* are $0 \times 0 = 0$; $0 \times 1 = 0$; $1 \times 0 = 0$; $1 \times 1 = 1$ which may be tabulated as shown.

×	0	1
0	0	0
1	0	1

The *subtraction facts* are $0 - 0 = 0$; $1 - 0 = 1$; $1 - 1 = 0$; $10 - 1 = 1$

The *division facts* are $1\overline{)0}$ quotient 0; $1\overline{)1}$ quotient 1

We add, subtract, multiply and divide binary numbers as we do decimal numbers but we use the number facts of the binary system.

Add the following binary numbers:

$$\begin{array}{r} 1001 \\ 1111 \\ \underline{1101} \\ 100101 \end{array}$$

Ones column—$1 + 1 = 10$; $10 + 1 = 11$. Write 1, carry 1.
Twos column—1 carried $+ 1 = 10$. Write 0, carry 1.
Fours column—1 carried $+ 1 = 10$; $10 + 1 = 11$. Write 1, carry 1.
Eights column—1 carried $+ 1 = 10$; $10 + 1 = 11$; $11 + 1 = 100$. Write 100.

Answer, 100101

The numbers used in the following model examples are binary numbers:

Subtraction:

$$\begin{array}{rcr} 10011 & = & 1^{1}011 \\ \underline{1101} & = & \underline{1\ 101} \\ & & 110 \end{array}$$

Answer, 110

Multiplication:

$$\begin{array}{r} 110 \\ \underline{\times 11} \\ 110 \\ \underline{110} \\ 10010 \end{array}$$

Answer, 10010

Division:

$$\begin{array}{r} 111 \\ 101\overline{)100011} \\ \underline{101} \\ 111 \\ \underline{101} \\ 101 \\ \underline{101} \\ \dots \end{array}$$

Answer, 111

Other Bases

When we add, subtract, multiply, and divide numbers in bases other than ten, we use the same principles as we do with decimal numbers but we use the number facts belonging to each respective base.

EXERCISES

1. Binary System

a. Perform the indicated operations on the following binary numbers:

$1 + 0 = ?$; $1 + 1 = ?$; $0 + 1 = ?$; $0 + 0 = ?$; $0 \times 1 = ?$; $1 \times 1 = ?$;

$1 \times 0 = ?$; $0 \times 0 = ?$; $1 - 0 = ?$; $0 - 0 = ?$; $1 - 1 = ?$; $10 - 1 = ?$;

$1\overline{)1}$; $1\overline{)0}$; $1\overline{)10}$; $11 + 1 = ?$; $11111 + 1 = ?$; $100 - 1 =$; $10000 - 1 = ?$

b. Compute the following binary numbers as directed:

Add:

						101101
				10110	111011	10111
			101	101010	1111	1001110
110	1010	11001	1110	1011	1101101	1111
101	1011	101110	111	111100	11111	100101

Subtract:

100	1101	10000	10100	110101	1100010	11001101
11	1001	1001	10011	1100	1001001	1110110

Multiply:

11	100	110	10101	10111	111101	1001110
11	10	101	1001	11101	110010	110110

Divide:

$10\overline{)1010}$ $11\overline{)110}$ $100\overline{)1100}$ $111\overline{)10101}$

$101\overline{)110010}$ $1111\overline{)1011010}$ $1101\overline{)1110101}$ $1011\overline{)1111001}$

c. Compute the problems in **Exercise 1b** also as follows: First, express each binary numeral as an equivalent base ten numeral. Do the indicated computation, using the base ten numbers. Express answers as binary numerals and compare with the answers you originally obtained in **Exercise 1b.**

2. Quinary System (Base Five)

a. Make a table showing the quinary addition facts. Make a second table showing the quinary multiplication facts.

+	0	1	2	3	4
0					
1					
2					
3					
4					

×	0	1	2	3	4
0					
1					
2					
3					
4					

b. Perform the indicated operations on the following quinary numbers:

$3 + 3 = ?$; $4 + 3 = ?$; $2 + 2 = ?$; $4 + 1 = ?$; $2 + 4 = ?$; $3 + 4 = ?$;

$3 + 2 = ?$; $4 + 4 = ?$; $13 - 4 = ?$; $10 - 3 = ?$; $10 - 1 = ?$; $12 - 3 = ?$;

$11 - 2 = ?$; $4 - 2 = ?$; $12 - 4 = ?$; $11 - 3 = ?$; $2 \times 4 = ?$; $3 \times 1 = ?$;

$4 \times 3 = ?$; $2 \times 3 = ?$; $4 \times 4 = ?$; $2 \times 2 = ?$; $4 \times 2 = ?$; $3 \times 3 = ?$;

$3\overline{)11}$; $4\overline{)31}$; $2\overline{)4}$; $4\overline{)22}$; $3\overline{)14}$; $2\overline{)13}$; $1\overline{)3}$; $4\overline{)13}$

c. Compute the following quinary numbers as directed:

Add:

				1310	4423	21134
		123	203	204	1434	2443
103	324	344	412	2411	2143	32
231	423	424	144	423	3244	14213

Subtract:

433	443	2312	4103	12432	30012	220401
321	204	1222	3214	4323	23134	140032

Multiply:

32	42	324	403	300	434	341
23	34	41	204	432	143	240

Divide:

$12\overline{)211}$ $23\overline{)1313}$ $34\overline{)3311}$ $24\overline{)1133}$ $140\overline{)23320}$ $232\overline{)110002}$

d. Compute the problems in **Exercise 2c** also as follows: First, express each quinary numeral as an equivalent base ten numeral. Do the indicated computation, using the base ten numbers. Express anwers as quinary numerals and compare with the answers you originally obtained in **Exercise 2c.**

3. Duodecimal System (Base Twelve)

a. Make a table showing the duodecimal addition facts. Make a second table showing the duodecimal multiplication facts.

b. Perform the indicated operations on the following duodecimal numbers:

$5 + 6 = ?$; $9 + 8 = ?$; $E + T = ?$; $6 + 9 = ?$; $T + 8 = ?$; $E + 9 = ?$;

$8 + 7 = ?$; $9 + 9 = ?$; $9 - 4 = ?$; $15 - 7 = ?$; $E - 5 = ?$; $10 - 6 = ?$;

$1T - T = ?$; $17 - E = ?$; $13 - 8 = ?$; $8 - 6 = ?$; $3 \times 6 = ?$; $6 \times T = ?$;

$9 \times 4 = ?$; $T \times E = ?$; $8 \times 9 = ?$; $9 \times 7 = ?$; $T \times 3 = ?$; $4 \times 2 = ?$;

$4\overline{)38}$; $2\overline{)10}$; $9\overline{)53}$; $E\overline{)47}$; $4\overline{)48}$; $7\overline{)5T}$; $5\overline{)2E}$; $T\overline{)76}$

c. Compute the following duodecimal numbers as directed:

Add:

				648	5E7T	15E4T
			258	4T36	4832	52068
49	2E	576	6E5	5ET4	1ETE	9583E
57	54	3TE	79T	378E	308E	4TE82

Subtract:

53	TE	406	93T6	9T48	5EOT7	6TE45
45	97	1E9	847	3TE5	6399	3TE69

Multiply:

3E	81	60E	397	9TE	T80	4T7
T2	ET	1T4	803	6T4	7ET	E06

Divide:

$13\overline{)E3}$ $15\overline{)86}$ $22\overline{)TT}$ $3E\overline{)81E}$ $T5\overline{)2647}$ $70\overline{)4TE0}$ $ET\overline{)109T}$

d. Compute the problems in **Exercise 3c** also as follows: First, express each duodecimal numeral as an equivalent base ten numeral. Do the indicated computation, using the base ten numbers. Express answers as duodecimal numerals and compare with the answers you originally obtained in **Exercise 3c.**

4. Other Bases:

Add:

423_{six}	1012_{three}	3517_{eight}	1526_{seven}	32301_{four}	13845_{nine}
310_{six}	2101_{three}	2633_{eight}	4456_{seven}	23232_{four}	21467_{nine}
134_{six}	1222_{three}	4075_{eight}	2565_{seven}	31323_{four}	52273_{nine}
201_{six}	1212_{three}	3246_{eight}	3646_{seven}	12332_{four}	33456_{nine}

Subtract:

3105_{six}	2111_{three}	6403_{eight}	53462_{seven}	23001_{four}	74328_{nine}
1324_{six}	1202_{three}	275_{eight}	36645_{seven}	3210_{four}	65238_{nine}

Multiply:

245_{six}	201_{three}	745_{eight}	463_{seven}	1203_{four}	754_{nine}
324_{six}	221_{three}	326_{eight}	651_{seven}	3312_{four}	863_{nine}

Divide:

$12_{\text{six}}\overline{)200_{\text{six}}}$ $210_{\text{three}}\overline{)110110_{\text{three}}}$ $45_{\text{eight}}\overline{)1635_{\text{eight}}}$

$110_{\text{seven}}\overline{)10230_{\text{seven}}}$ $123_{\text{four}}\overline{)31323_{\text{four}}}$ $58_{\text{nine}}\overline{)6352_{\text{nine}}}$

5. a. Add:

$$\begin{array}{r} 23104_{five} \\ 41223_{five} \\ \underline{24344_{five}} \end{array}$$

b. Subtract:

$$\begin{array}{r} 10110110_{two} \\ \underline{1011011_{two}} \end{array}$$

c. Multiply:

$$\begin{array}{r} 637_{eight} \\ \underline{564_{eight}} \end{array}$$

d. Divide:

$$302_{four}\overline{)102300_{four}}$$

MATHEMATICAL SYSTEMS

3–23 FINITE NUMBER SYSTEMS

A *mathematical system* consists of a set of elements and one or more binary operations. The number system of natural numbers, the number system of whole numbers, and as we shall see, the number system of rational numbers of arithmetic are each a mathematical system. These number systems are infinite number systems since the set of natural numbers, the set of whole numbers, and the set of rational numbers are each an infinite set of numbers.

A *finite number system* is a mathematical system which contains only a limited number of elements. One type of this system may be illustrated by arranging the numbers in the system on a circle as a clockface. The arithmetic on this number circle is called *modular arithmetic.*

Congruent Numbers

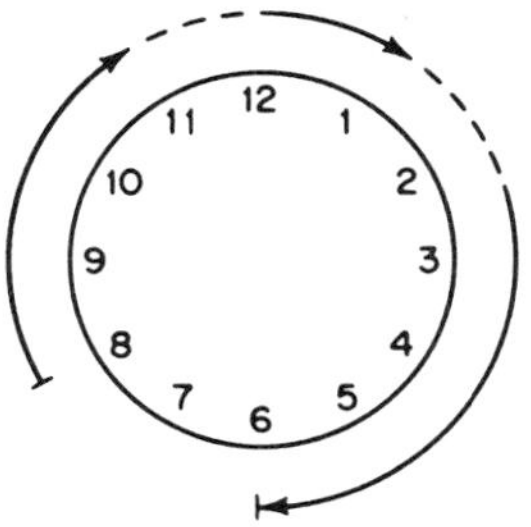

Suppose it is 8 o'clock now. What time will it be 10 hours from now? What time will it be 22 hours from now? Use the 12-hour clock at the right to obtain your answers. We see that it would be 6 o'clock in both cases.

We may write these as:

$$8 + 10 = 6 \quad \text{and} \quad 8 + 22 = 6$$

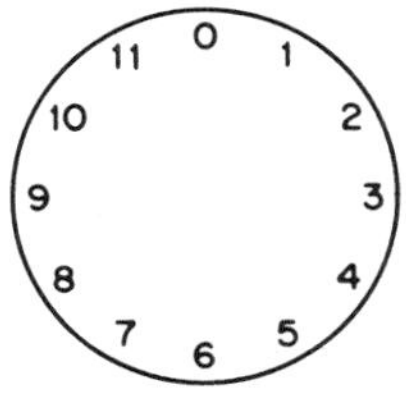

or better

$$8 + 10 \equiv 6 \pmod{12} \quad \text{and} \quad 8 + 22 \equiv 6 \pmod{12}$$

The sentence $8 + 10 \equiv 6 \pmod{12}$ is read "Eight plus ten is congruent to six, modulo twelve." Mod 12 representing "modulo twelve" indicates that 12 numerals are used. The numeral 0 is used instead of the numeral 12 as shown on the lower clock face.

Instead of adding on the clock by counting clockwise, we can determine the answer to $8 + 10 \equiv ? \pmod{12}$ by dividing the sum of 8 and 10 by the modulus number 12 and using only the remainder. The quotient is discarded.

$8 + 10 = 18$; $18 \div 12 = 1$, remainder 6. *Answer*, 6.

Trying this on $8 + 22 \equiv ?$ (mod 12), we find

$8 + 22 = 30$; $30 \div 12 = 2$, remainder 6. *Answer*, 6.

In the above, we see that 18 in mod 12 and 30 in mod 12 both have the remainder 6, expressed as 6 (mod 12). The 18 and 30 in mod 12 are said to be congruent. When two numbers, divided by the same modulus number, result in the same remainder, although the quotients differ, they are called *congruent.* In mod 12, the numbers 6, 18, 30, 42, 54, 66, . . . are congruent. Each, when divided by the modulus number 12, gives the remainder 6.

In modular arithmetic, sometimes called clock arithmetic or circle arithmetic or remainder arithmetic, any number including sums and products, equal to or greater than the modulus number, is usually expressed as a congruent number.

$5 + 8 \equiv ?$ (mod 6)
$5 + 8 = 13$; $13 \div 6 = 2\ R1$
$5 + 8 \equiv 1$ (mod 6)

$10 + 9 \equiv ?$ (mod 11)
$10 + 9 = 19$; $19 \div 11 = 1\ R8$
$10 + 9 \equiv 8$ (mod 11)

$3 + 6 \equiv ?$ (mod 9)
$3 + 6 = 9$; $9 \div 9 = 1\ R0$
$3 + 6 \equiv 0$ (mod 9)

$11 + 17 \equiv ?$ (mod 3)
$11 + 17 = 28$; $28 \div 3 = 9\ R1$
$11 + 17 \equiv 1$ (mod 3)

It should be noted that the remainder of the sum of two numbers may be determined by finding the remainder of the sum of the remainders of the numbers.

$11 + 17 \equiv ?$ (mod 3)
$2 + 2 \equiv ?$ (mod 3)
$2 + 2 \equiv 4$ (mod 3)
4 (mod 3) $\equiv 1$ (mod 3)

To multiply two numbers in modular arithmetic, we find the product of the numbers, then divide this product by the given modulus number and use the remainder as the answer.

$10 \times 11 \equiv ?$ (mod 12)
$10 \times 11 = 110$; $110 \div 12 = 9\ R2$
$10 \times 11 \equiv 2$ (mod 12)

$3 \times 4 \equiv ?$ (mod 5)
$3 \times 4 = 12$; $12 \div 5 = 2\ R2$
$3 \times 4 \equiv 2$ (mod 5)

$4 \times 8 \equiv ? \pmod 9$
$4 \times 8 = 32;\ 32 \div 9 = 3\ R\ 5$
$4 \times 8 \equiv 5 \pmod 9$

$7 \times 9 \equiv ? \pmod 4$
$7 \times 9 = 63;\ 63 \div 4 = 15\ R\ 3$
$7 \times 9 \equiv 3 \pmod 4$

It should be noted that the remainder of the product of two numbers may be determined by finding the remainder of the product of the remainders of the numbers.

$7 \times 9 \equiv ? \pmod 4$
$3 \times 1 \equiv ? \pmod 4$
$3 \times 1 \equiv 3 \pmod 4$

EXERCISES

1. Read, or write in words, each of the following:

a. $23 \equiv 2 \pmod 7$ **b.** $17 \equiv 1 \pmod 4$ **c.** $8 + 6 \equiv 4 \pmod 5$ **d.** $5 \times 7 \equiv 3 \pmod 8$ **e.** $14 \times 2 \equiv 4 \pmod{12}$ **f.** $19 + 32 \equiv 0 \pmod 3$

2. Write the digits and symbols for:

a. Fourteen is congruent to six, modulo eight.
b. Seven plus nine is congruent to four, modulo six.
c. Eleven times eight is congruent to three, modulo five.
d. Fifteen plus twenty-six is congruent to one, modulo two.

3. Find each of the following:

a. $7 \equiv ? \pmod 3$ **b.** $9 \equiv ? \pmod 6$ **c.** $12 \equiv ? \pmod 5$ **d.** $39 \equiv ? \pmod{11}$ **e.** $53 \equiv ? \pmod 9$ **f.** $45 \equiv ? \pmod{12}$ **g.** $33 \equiv ? \pmod 7$ **h.** $10 \equiv ? \pmod 4$ **i.** $15 \equiv ? \pmod 2$ **j.** $71 \equiv ? \pmod 8$ **k.** $19 \equiv ? \pmod 5$ **l.** $26 \equiv ? \pmod 3$

4. Add each of the following on the 4-minute clock:

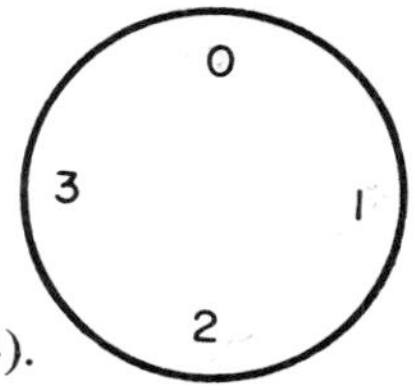

a. $2 + 3 \equiv ?$ **b.** $1 + 2 \equiv ?$ **c.** $3 + 1 \equiv ?$ **d.** $1 + 5 \equiv ?$ **e.** $3 + 4 \equiv ?$ **f.** $2 + 10 \equiv ?$ **g.** $1 + 14 \equiv ?$ **h.** $3 + 19 \equiv ?$ **i.** $2 + 23 \equiv ?$

Add each of the above, using the remainder method (mod 4).

5. a. Are 24 and 36 congruent in mod 3? in mod 8? in mod 12? in mod 7?
b. Are 17 and 25 congruent in mod 2? in mod 3? in mod 7? in mod 4?
c. Are 72 and 60 congruent in mod 5? in mod 4? in mod 8? in mod 6?
d. Are 48 and 56 congruent in mod 8? in mod 9? in mod 6? in mod 3?
e. Are 37 and 29 congruent in mod 4? in mod 7? in mod 5? in mod 9?

6. Add in each of the following, using the remainder method of modular arithmetic:

a. $9 + 8 \equiv ? \pmod 5$ **b.** $5 + 3 \equiv ? \pmod 9$ **c.** $4 + 7 \equiv ? \pmod 2$ **d.** $11 + 5 \equiv ? \pmod 8$ **e.** $13 + 23 \equiv ? \pmod 6$ **f.** $20 + 17 \equiv ? \pmod{11}$ **g.** $9 + 36 \equiv ? \pmod 4$ **h.** $14 + 49 \equiv ? \pmod 3$ **i.** $28 + 59 \equiv ? \pmod 7$ **j.** $42 + 63 \equiv ? \pmod{12}$ **k.** $21 + 54 \equiv ? \pmod 9$ **l.** $37 + 45 \equiv ? \pmod 6$

7. Multiply in each of the following, using the remainder method of modular arithmetic:

a. $7 \times 5 \equiv ?$ (mod 3)
b. $8 \times 9 \equiv ?$ (mod 5)
c. $12 \times 8 \equiv ?$ (mod 4)
d. $11 \times 7 \equiv ?$ (mod 2)
e. $9 \times 3 \equiv ?$ (mod 6)
f. $4 \times 2 \equiv ?$ (mod 9)
g. $6 \times 18 \equiv ?$ (mod 7)
h. $10 \times 9 \equiv ?$ (mod 11)
i. $12 \times 12 \equiv ?$ (mod 8)
j. $15 \times 20 \equiv ?$ (mod 12)
k. $14 \times 16 \equiv ?$ (mod 5)
l. $18 \times 24 \equiv ?$ (mod 9)

8. Add in each of the following, using the remainder of the sum of the remainders of the numbers:

a. $20 + 26 \equiv ?$ (mod 6) **b.** $37 + 25 \equiv ?$ (mod 3) **c.** $19 + 76 \equiv ?$ (mod 8)

9. Multiply in each of the following, using the remainder of the product of the remainders of the numbers:

a. $9 \times 5 \equiv ?$ (mod 4) **b.** $20 \times 30 \equiv ?$ (mod 9) **c.** $35 \times 25 \equiv ?$ (mod 7)

10. Compute as indicated:

a. $7 + 6 \equiv ?$ (mod 5)
b. $3 \times 4 \equiv ?$ (mod 6)
c. $15 \times 12 \equiv ?$ (mod 8)
d. $11 + 29 \equiv ?$ (mod 9)
e. $8 \times 18 \equiv ?$ (mod 12)
f. $2 + 1 \equiv ?$ (mod 3)
g. $5 \times 13 \equiv ?$ (mod 2)
h. $16 + 56 \equiv ?$ (mod 11)
i. $24 \times 10 \equiv ?$ (mod 9)
j. $16 + 34 \equiv ?$ (mod 3)
k. $2 \times 3 \equiv ?$ (mod 8)
l. $21 + 7 \equiv ?$ (mod 4)
m. $12 \times 17 \equiv ?$ (mod 5)
n. $19 \times 50 \equiv ?$ (mod 9)
o. $39 + 23 \equiv ?$ (mod 2)
p. $54 + 88 \equiv ?$ (mod 12)
q. $26 \times 18 \equiv ?$ (mod 6)
r. $99 + 89 \equiv ?$ (mod 7)

Modulo Systems and Their Properties

Suppose we select as a model the modulo 5 system in which only the set of numbers {0, 1, 2, 3, 4} is used. We first determine the addition and multiplication tables for modulo 5, using just this set of numbers. The row at the top and the column at the left in the addition table contain the addends and in the multiplication table this row and column contain the factors.

+	0	1	2	3	4
0	0	1	2	3	4
1	1	2	3	4	0
2	2	3	4	0	1
3	3	4	0	1	2
4	4	0	1	2	3

×	0	1	2	3	4
0	0	0	0	0	0
1	0	1	2	3	4
2	0	2	4	1	3
3	0	3	1	4	2
4	0	4	3	2	1

Observe that the sums and products in these tables are numbers belonging to the set {0, 1, 2, 3, 4}. Thus we can say that the set {0, 1, 2, 3, 4} is closed both under the operation of addition and under the operation of multiplication.

Examination of the addition table for modulo 5 reveals that:

(1) The commutative property holds for addition.

For example: $3 + 4 \equiv 2 \pmod 5$ and $4 + 3 \equiv 2 \pmod 5$.

(2) The associative property holds for addition.

For example: $(1 + 3) + 2 \equiv 1 \pmod 5$ and $1 + (3 + 2) \equiv 1 \pmod 5$.

(3) The additive identity in mod 5 is 0.

For example: $3 + 0 \equiv 3 \pmod 5$.

(4) Each number of the set {0, 1, 2, 3, 4} used in arithmetic mod 5 has an additive inverse (the sum of the number and its additive inverse should equal the additive identity).
The additive inverse of 0 is 0, of 1 is 4, of 2 is 3, of 3 is 2, and of 4 is 1.

(5) As mentioned on page 78, the set of numbers {0, 1, 2, 3, 4} used in arithmetic modulo 5 is closed under the operation of addition since all the sums are numbers belonging to this set.

Examination of the multiplication table for modulo 5 reveals that:

(6) The commutative property holds for multiplication.

For example: $4 \times 2 \equiv 3 \pmod 5$ and $2 \times 4 \equiv 3 \pmod 5$.

(7) The associative property holds for multiplication.

For example: $(2 \times 4) \times 3 \equiv 4 \pmod 5$ and $2 \times (4 \times 3) \equiv 4 \pmod 5$.

(8) The multiplicative identity in mod 5 is 1.

For example: $1 \times 4 \equiv 4 \pmod 5$.

(9) Each non-zero number of the set {0, 1, 2, 3, 4} used in arithmetic modulo 5 has a multiplicative inverse (the product of the number and its inverse should equal the multiplicative identity). This is not true in some modulo systems.
In mod 5 the multiplicative inverse of 1 is 1, of 2 is 3, of 3 is 2, of 4 is 4.

(10) As mentioned on page 78, the set of numbers {0, 1, 2, 3, 4} used in arithmetic modulo 5 is closed under the operation of multiplication since all the products are numbers belonging to this set.

(11) Multiplication is distributed over addition.

For example: $2 \times (3 + 4) = 2 \times (7) \equiv 4 \pmod 5$ and $(2 \times 3) + (2 \times 4) = (6) + (8) \equiv 4 \pmod 5$. Thus in mod 5, $2 \times (3 + 4) = (2 \times 3) + (2 \times 4)$.

EXERCISES

1. Name the numbers used in:

   a. Modulo 3 c. Modulo 7 e. Modulo 12
   b. Modulo 6 d. Modulo 4 f. Modulo 9

2. Develop and write the addition tables and multiplication tables used in the arithmetic:

   a. Modulo 7 c. Modulo 6 e. Modulo 4
   b. Modulo 12 d. Modulo 3 f. Modulo 2

3. Examine the related addition and multiplication tables and determine which of the eleven properties each of the following modulo systems has. For each system illustrate each property and list the multiplicative inverses of that system.

   a. Modulo 12 c. Modulo 3 e. Modulo 2
   b. Modulo 7 d. Modulo 6 f. Modulo 4

3–24 MATRICES

A *matrix* is a rectangular array of numbers. In order to identify a matrix the numerals naming these numbers are written between parentheses or brackets. These numbers are elements or *entries* of the matrix.

The *dimensions* of a matrix are determined and described by the number of rows and number of columns in the matrix.

The matrix $\begin{pmatrix} 6 & 5 \\ 4 & 9 \\ 3 & 8 \end{pmatrix}$ has three rows and two columns. It is therefore a 3×2 (read "three by two") matrix. The first numeral indicates the number of rows and the second numeral indicates the number of columns.

A matrix that has the same number of columns as rows is called a *square matrix*. In this section we shall only deal with the 2×2 matrix and the 3×3 matrix.

Matrix Addition

Only matrices of the same dimensions can be added. In the matrices $\begin{pmatrix} 5 & 3 \\ 11 & 9 \end{pmatrix}$ and $\begin{pmatrix} 4 & 2 \\ 8 & 6 \end{pmatrix}$ observe that numerals 5 and 4 are located in the same relative position in the matrices and are called *corresponding entries.*

The *sum of two given matrices,* both of the same dimensions, is the matrix of these dimensions whose entries are the sums of the corresponding entries of the given matrices.

For example: $\begin{pmatrix} 6 & 1 \\ 3 & 9 \end{pmatrix} + \begin{pmatrix} 2 & 4 \\ 8 & 7 \end{pmatrix} = \begin{pmatrix} 6+2 & 1+4 \\ 3+8 & 9+7 \end{pmatrix} = \begin{pmatrix} 8 & 5 \\ 11 & 16 \end{pmatrix}$

Observe that 6 and 2, 3 and 8, 1 and 4, and 9 and 7 are pairs of corresponding entries. Each entry in the resulting matrix is the sum of a pair of these corresponding entries and is written in the same relative position that they have in the given matrices.

Matrix Multiplication

Two matrices of the same dimensions can be multiplied. However, two matrices of different dimensions can also be multiplied provided that the left matrix has the same number of columns as the right matrix has rows.

To find the product matrix when two 2×2 or two 3×3 matrices are multiplied, we compute the corresponding entries by first multiplying entries of a specific row of the left matrix by entries of a specific column of the right matrix.

For example:

$$\begin{pmatrix} 6 & 1 \\ 3 & 9 \end{pmatrix} \times \begin{pmatrix} 2 & 4 \\ 8 & 7 \end{pmatrix} = \begin{pmatrix} ? & ? \\ ? & ? \end{pmatrix}$$

To determine the entry in the upper left position, which is the first row and first column,

$$\begin{pmatrix} 6 & 1 \\ & \end{pmatrix} \times \begin{pmatrix} 2 & \\ 8 & \end{pmatrix} = \begin{pmatrix} (6 \times 2) + (1 \times 8) & \\ & \end{pmatrix} = \begin{pmatrix} 20 & \\ & \end{pmatrix}$$

we multiply the entries of the first row of the left matrix by the entries of the first column of the right matrix (first entry by first entry, second entry by second entry) and add the products.

Thus, $(6 \times 2) + (1 \times 8) = 12 + 8 = 20$.
20 is the entry in the upper left position.

To determine the entry in the lower left position, which is second row and first column,

$$\begin{pmatrix} & \\ 3 & 9 \end{pmatrix} \times \begin{pmatrix} 2 & \\ 8 & \end{pmatrix} = \begin{pmatrix} & \\ (3 \times 2) + (9 \times 8) & \end{pmatrix} = \begin{pmatrix} & \\ 78 & \end{pmatrix}$$

we multiply the entries of the second row of the left matrix by the entries of the first column of the right matrix (first entry by first entry, second entry by second entry) and add the products.

Thus, $(3 \times 2) + (9 \times 8) = 6 + 72 = 78$.
78 is the entry in the lower left position.

To determine the entry in the upper right position, which is the first row and second column,

$$\begin{pmatrix} 6 & 1 \\ & \end{pmatrix} \times \begin{pmatrix} & 4 \\ & 7 \end{pmatrix} = \begin{pmatrix} & (6 \times 4) + (1 \times 7) \\ & \end{pmatrix} = \begin{pmatrix} & 31 \\ & \end{pmatrix}$$

we multiply the entries of the first row of the left matrix by the entries of the second column of the right matrix (first entry by first entry, second entry by second entry) and add the products.

Thus, $(6 \times 4) + (1 \times 7) = 24 + 7 = 31$.
31 is the entry in the upper right position.

To determine the entry in the lower right position, which is the second row and second column,

$$\begin{pmatrix} & \\ 3 & 9 \end{pmatrix} \times \begin{pmatrix} & 4 \\ & 7 \end{pmatrix} = \begin{pmatrix} & \\ & (3 \times 4) + (9 \times 7) \end{pmatrix} = \begin{pmatrix} & \\ & 75 \end{pmatrix}$$

we multiply the entries of the second row of the left matrix by the entries of the second column of the right matrix (first entry by first entry, second entry by second entry) and add the products.

Thus, $(3 \times 4) + (9 \times 7) = 12 + 63 = 75$.
75 is the entry in the lower right position.

Combining all the above steps, we find that:

$$\begin{pmatrix} 6 & 1 \\ 3 & 9 \end{pmatrix} \times \begin{pmatrix} 2 & 4 \\ 8 & 7 \end{pmatrix} = \begin{pmatrix} 20 & 31 \\ 78 & 75 \end{pmatrix}$$

We usually do the necessary computation and write the complete answer all in a single step. The product matrix when two 3×3 matrices are multiplied is similarly determined.

EXERCISES

1. Find the sum:

a. $\begin{pmatrix} 5 & 2 \\ 4 & 6 \end{pmatrix} + \begin{pmatrix} 1 & 0 \\ 9 & 3 \end{pmatrix}$

b. $\begin{pmatrix} 10 & 7 \\ 3 & 14 \end{pmatrix} + \begin{pmatrix} 5 & 6 \\ 8 & 8 \end{pmatrix}$

c. $\begin{pmatrix} 0 & 4 \\ 1 & 0 \end{pmatrix} + \begin{pmatrix} 1 & 0 \\ 0 & 1 \end{pmatrix}$

d. $\begin{pmatrix} 2 & 5 \\ 7 & 10 \end{pmatrix} + \begin{pmatrix} 2 & 5 \\ 7 & 10 \end{pmatrix}$

e. $\begin{pmatrix} 8 & 4 \\ 16 & 7 \end{pmatrix} + \begin{pmatrix} 11 & 6 \\ 3 & 4 \end{pmatrix}$

f. $\begin{pmatrix} 0 & 9 \\ 0 & 10 \end{pmatrix} + \begin{pmatrix} 6 & 8 \\ 0 & 4 \end{pmatrix}$

g. $\begin{pmatrix} 3 & 15 \\ 11 & 8 \end{pmatrix} + \begin{pmatrix} 0 & 0 \\ 0 & 0 \end{pmatrix}$

h. $\begin{pmatrix} 20 & 17 \\ 13 & 25 \end{pmatrix} + \begin{pmatrix} 24 & 32 \\ 16 & 9 \end{pmatrix}$

i. $\begin{pmatrix} 34 & 5 \\ 19 & 12 \end{pmatrix} + \begin{pmatrix} 8 & 29 \\ 14 & 6 \end{pmatrix}$ j. $\begin{pmatrix} 10 & 18 \\ 0 & 43 \end{pmatrix} + \begin{pmatrix} 6 & 36 \\ 17 & 21 \end{pmatrix}$

2. Find the sum:

a. $\begin{pmatrix} 7 & 4 & 2 \\ 1 & 8 & 5 \\ 9 & 3 & 6 \end{pmatrix} + \begin{pmatrix} 3 & 4 & 8 \\ 5 & 1 & 2 \\ 9 & 2 & 3 \end{pmatrix}$ b. $\begin{pmatrix} 8 & 6 & 12 \\ 10 & 9 & 6 \\ 5 & 3 & 4 \end{pmatrix} + \begin{pmatrix} 11 & 8 & 16 \\ 0 & 17 & 27 \\ 5 & 13 & 9 \end{pmatrix}$

3. a. Does $\begin{pmatrix} 6 & 1 \\ 2 & 9 \end{pmatrix} + \begin{pmatrix} 8 & 5 \\ 7 & 0 \end{pmatrix} = \begin{pmatrix} 8 & 5 \\ 7 & 0 \end{pmatrix} + \begin{pmatrix} 6 & 1 \\ 2 & 9 \end{pmatrix}$?

b. Does $\begin{pmatrix} 9 & 2 & 3 \\ 0 & 1 & 10 \\ 4 & 7 & 6 \end{pmatrix} + \begin{pmatrix} 5 & 4 & 12 \\ 8 & 0 & 7 \\ 3 & 5 & 2 \end{pmatrix} = \begin{pmatrix} 5 & 4 & 12 \\ 8 & 0 & 7 \\ 3 & 5 & 2 \end{pmatrix} + \begin{pmatrix} 9 & 2 & 3 \\ 0 & 1 & 10 \\ 4 & 7 & 6 \end{pmatrix}$?

c. Does it appear that matrix addition is commutative?

4. In each of the following add as indicated, and then compare the sums:

a. $\left[\begin{pmatrix} 2 & 3 \\ 6 & 5 \end{pmatrix} + \begin{pmatrix} 9 & 1 \\ 0 & 4 \end{pmatrix}\right] + \begin{pmatrix} 6 & 0 \\ 1 & 8 \end{pmatrix}$; $\begin{pmatrix} 2 & 3 \\ 6 & 5 \end{pmatrix} + \left[\begin{pmatrix} 9 & 1 \\ 0 & 4 \end{pmatrix} + \begin{pmatrix} 6 & 0 \\ 1 & 8 \end{pmatrix}\right]$

Does $\left[\begin{pmatrix} 2 & 3 \\ 6 & 5 \end{pmatrix} + \begin{pmatrix} 9 & 1 \\ 0 & 4 \end{pmatrix}\right] + \begin{pmatrix} 6 & 0 \\ 1 & 8 \end{pmatrix} = \begin{pmatrix} 2 & 3 \\ 6 & 5 \end{pmatrix} + \left[\begin{pmatrix} 9 & 1 \\ 0 & 4 \end{pmatrix} + \begin{pmatrix} 6 & 0 \\ 1 & 8 \end{pmatrix}\right]$?

b. $\left[\begin{pmatrix} 1 & 2 & 0 \\ 2 & 4 & 1 \\ 3 & 6 & 7 \end{pmatrix} + \begin{pmatrix} 0 & 1 & 3 \\ 5 & 0 & 2 \\ 2 & 1 & 4 \end{pmatrix}\right] + \begin{pmatrix} 2 & 0 & 1 \\ 1 & 3 & 6 \\ 4 & 5 & 2 \end{pmatrix}$;

$\begin{pmatrix} 1 & 2 & 0 \\ 2 & 4 & 1 \\ 3 & 6 & 7 \end{pmatrix} + \left[\begin{pmatrix} 0 & 1 & 3 \\ 5 & 0 & 2 \\ 2 & 1 & 4 \end{pmatrix} + \begin{pmatrix} 2 & 0 & 1 \\ 1 & 3 & 6 \\ 4 & 5 & 2 \end{pmatrix}\right]$

Does $\left[\begin{pmatrix} 1 & 2 & 0 \\ 2 & 4 & 1 \\ 3 & 6 & 7 \end{pmatrix} + \begin{pmatrix} 0 & 1 & 3 \\ 5 & 0 & 2 \\ 2 & 1 & 4 \end{pmatrix}\right] + \begin{pmatrix} 2 & 0 & 1 \\ 1 & 3 & 6 \\ 4 & 5 & 2 \end{pmatrix} =$

$\begin{pmatrix} 1 & 2 & 0 \\ 2 & 4 & 1 \\ 3 & 6 & 7 \end{pmatrix} + \left[\begin{pmatrix} 0 & 1 & 3 \\ 5 & 0 & 2 \\ 2 & 1 & 4 \end{pmatrix} + \begin{pmatrix} 2 & 0 & 1 \\ 1 & 3 & 6 \\ 4 & 5 & 2 \end{pmatrix}\right]$?

c. Does it appear that matrix addition is associative?

5. a. Is the sum of two 2×2 matrices always a 2×2 matrix? If so, tell why and illustrate it. Can we say that the set of 2×2 matrices is closed under the operation of addition?

b. Is the sum of two 3×3 matrices always a 3×3 matrix? If so, tell why and illustrate it. Can we say that the set of 3×3 matrices is closed under the operation of addition?

6. Find the sum of each of the following:

a. $\begin{pmatrix} 5 & 11 \\ 15 & 6 \end{pmatrix} + \begin{pmatrix} 0 & 0 \\ 0 & 0 \end{pmatrix}$

b. $\begin{pmatrix} 0 & 0 \\ 0 & 0 \end{pmatrix} + \begin{pmatrix} 12 & 16 \\ 7 & 0 \end{pmatrix}$

c. $\begin{pmatrix} 8 & 9 & 2 \\ 3 & 5 & 8 \\ 10 & 7 & 13 \end{pmatrix} + \begin{pmatrix} 0 & 0 & 0 \\ 0 & 0 & 0 \\ 0 & 0 & 0 \end{pmatrix}$

d. $\begin{pmatrix} 0 & 0 & 0 \\ 0 & 0 & 0 \\ 0 & 0 & 0 \end{pmatrix} + \begin{pmatrix} 15 & 0 & 30 \\ 7 & 18 & 9 \\ 12 & 21 & 19 \end{pmatrix}$

Can we say that the additive identity for 2×2 matrices is $\begin{pmatrix} 0 & 0 \\ 0 & 0 \end{pmatrix}$?
Is $\begin{pmatrix} 0 & 0 & 0 \\ 0 & 0 & 0 \\ 0 & 0 & 0 \end{pmatrix}$ the additive identity for 3×3 matrices?

***7. a.** Find the sum of each of the following:

(1) $\begin{pmatrix} 4 & 3 \\ 6 & 5 \end{pmatrix} + \begin{pmatrix} -4 & -3 \\ -6 & -5 \end{pmatrix}$

(2) $\begin{pmatrix} 3 & -7 \\ -1 & 10 \end{pmatrix} + \begin{pmatrix} -3 & 7 \\ 1 & -10 \end{pmatrix}$

(3) $\begin{pmatrix} 6 & 5 & 4 \\ 8 & 7 & 1 \\ 4 & 2 & 9 \end{pmatrix} + \begin{pmatrix} -6 & -5 & -4 \\ -8 & -7 & -1 \\ -4 & -2 & -9 \end{pmatrix}$

(4) $\begin{pmatrix} 2 & -1 & 3 \\ -2 & 7 & 8 \\ 11 & -4 & -5 \end{pmatrix} + \begin{pmatrix} -2 & 1 & -3 \\ 2 & -7 & -8 \\ -11 & 4 & 5 \end{pmatrix}$

In above (1) and (2), is the sum of the two given 2×2 matrices the additive identity for the set of 2×2 matrices? Can we call $\begin{pmatrix} 3 & -5 \\ -2 & 4 \end{pmatrix}$ and $\begin{pmatrix} -3 & 5 \\ 2 & -4 \end{pmatrix}$ each the additive inverse of the other? What is true about each pair of corresponding entries?

In above (3) and (4), is the sum of the two given 3×3 matrices the additive identity for the set of 3×3 matrices? Can we call $\begin{pmatrix} 6 & -5 & 7 \\ -9 & 4 & 0 \\ 10 & -8 & -6 \end{pmatrix}$ and $\begin{pmatrix} -6 & 5 & -7 \\ 9 & -4 & 0 \\ -10 & 8 & 6 \end{pmatrix}$ each the additive inverse of the other? What is true about each pair of corresponding entries?

b. Write the additive inverse of each of the following:

(1) $\begin{pmatrix} 6 & 3 \\ 8 & 2 \end{pmatrix}$

(2) $\begin{pmatrix} 5 & -2 \\ -7 & 8 \end{pmatrix}$

(3) $\begin{pmatrix} 0 & -1 \\ 1 & 0 \end{pmatrix}$

(4) $\begin{pmatrix} 0 & 0 \\ 0 & 0 \end{pmatrix}$

(5) $\begin{pmatrix} 3 & -8 & 4 \\ 6 & 11 & -3 \\ -4 & 9 & -7 \end{pmatrix}$

(6) $\begin{pmatrix} 7 & 0 & -4 \\ -2 & -6 & 16 \\ 3 & 0 & -9 \end{pmatrix}$

*Take this problem only if operations with positive and negative numbers have been studied.

8. Determine the product matrix in each of the following:

a. $\begin{pmatrix} 2 & 3 \\ 4 & 5 \end{pmatrix} \times \begin{pmatrix} 6 & 2 \\ 1 & 8 \end{pmatrix}$

b. $\begin{pmatrix} 7 & 4 \\ 3 & 9 \end{pmatrix} \times \begin{pmatrix} 5 & 8 \\ 2 & 10 \end{pmatrix}$

c. $\begin{pmatrix} 8 & 5 \\ 4 & 12 \end{pmatrix} \times \begin{pmatrix} 1 & 0 \\ 0 & 1 \end{pmatrix}$

d. $\begin{pmatrix} 12 & 4 \\ 6 & 7 \end{pmatrix} \times \begin{pmatrix} 0 & 1 \\ 1 & 0 \end{pmatrix}$

e. $\begin{pmatrix} 2 & 0 \\ 0 & 2 \end{pmatrix} \times \begin{pmatrix} \frac{1}{2} & 0 \\ 0 & \frac{1}{2} \end{pmatrix}$

f. $\begin{pmatrix} 10 & 5 \\ 9 & 7 \end{pmatrix} \times \begin{pmatrix} 3 & 8 \\ 6 & 15 \end{pmatrix}$

g. $\begin{pmatrix} 5 & 11 \\ 9 & 4 \end{pmatrix} \times \begin{pmatrix} 0 & 0 \\ 0 & 0 \end{pmatrix}$

h. $\begin{pmatrix} 3 & 0 \\ 0 & 3 \end{pmatrix} \times \begin{pmatrix} 4 & 20 \\ 11 & 6 \end{pmatrix}$

i. $\begin{pmatrix} 7 & 12 \\ 12 & 5 \end{pmatrix} \times \begin{pmatrix} 6 & 8 \\ 10 & 1 \end{pmatrix}$

j. $\begin{pmatrix} 0 & 2 \\ 0 & 8 \end{pmatrix} \times \begin{pmatrix} 0 & 0 \\ 3 & 15 \end{pmatrix}$

k. $\begin{pmatrix} 4 & 0 \\ 0 & 4 \end{pmatrix} \times \begin{pmatrix} 8 & 9 \\ 16 & 30 \end{pmatrix}$

l. $\begin{pmatrix} 1 & 1 \\ 1 & 1 \end{pmatrix} \times \begin{pmatrix} 5 & 6 \\ 6 & 5 \end{pmatrix}$

9. Determine the product matrix in each of the following:

a. $\begin{pmatrix} 3 & 1 & 5 \\ 7 & 8 & 4 \\ 2 & 3 & 6 \end{pmatrix} \times \begin{pmatrix} 2 & 4 & 6 \\ 1 & 0 & 2 \\ 3 & 5 & 1 \end{pmatrix}$

b. $\begin{pmatrix} 12 & 6 & 9 \\ 5 & 1 & 2 \\ 2 & 6 & 10 \end{pmatrix} \times \begin{pmatrix} 12 & 3 & 2 \\ 7 & 8 & 5 \\ 10 & 20 & 4 \end{pmatrix}$

c. $\begin{pmatrix} 8 & 5 & 3 \\ 6 & 10 & 4 \\ 9 & 7 & 2 \end{pmatrix} \times \begin{pmatrix} 1 & 0 & 0 \\ 0 & 1 & 0 \\ 0 & 0 & 1 \end{pmatrix}$

d. $\begin{pmatrix} 0 & 0 & 1 \\ 0 & 1 & 0 \\ 1 & 0 & 0 \end{pmatrix} \times \begin{pmatrix} 9 & 2 & 7 \\ 4 & 5 & 3 \\ 11 & 8 & 6 \end{pmatrix}$

10. a. Does $\begin{pmatrix} 8 & 5 \\ 6 & 7 \end{pmatrix} \times \begin{pmatrix} 2 & 3 \\ 8 & 9 \end{pmatrix} = \begin{pmatrix} 2 & 3 \\ 8 & 9 \end{pmatrix} \times \begin{pmatrix} 8 & 5 \\ 6 & 7 \end{pmatrix}$?

b. Does $\begin{pmatrix} 2 & 8 & 14 \\ 6 & 10 & 3 \\ 4 & 1 & 15 \end{pmatrix} \times \begin{pmatrix} 8 & 9 & 0 \\ 11 & 3 & 5 \\ 6 & 4 & 2 \end{pmatrix} = \begin{pmatrix} 8 & 9 & 0 \\ 11 & 3 & 5 \\ 6 & 4 & 2 \end{pmatrix} \times \begin{pmatrix} 2 & 8 & 14 \\ 6 & 10 & 3 \\ 4 & 1 & 15 \end{pmatrix}$?

c. Does it appear that matrix multiplication is commutative?

11. In each of the following multiply as indicated, then compare the product matrices:

a. $\left[\begin{pmatrix} 6 & 16 \\ 11 & 9 \end{pmatrix} \times \begin{pmatrix} 4 & 8 \\ 3 & 10 \end{pmatrix}\right] \times \begin{pmatrix} 2 & 7 \\ 3 & 8 \end{pmatrix}$; $\begin{pmatrix} 6 & 16 \\ 11 & 9 \end{pmatrix} \times \left[\begin{pmatrix} 4 & 8 \\ 3 & 10 \end{pmatrix} \times \begin{pmatrix} 2 & 7 \\ 3 & 8 \end{pmatrix}\right]$

Does $\left[\begin{pmatrix} 6 & 16 \\ 11 & 9 \end{pmatrix} \times \begin{pmatrix} 4 & 8 \\ 3 & 10 \end{pmatrix}\right] \times \begin{pmatrix} 2 & 7 \\ 3 & 8 \end{pmatrix} = \begin{pmatrix} 6 & 16 \\ 11 & 9 \end{pmatrix} \times \left[\begin{pmatrix} 4 & 8 \\ 3 & 10 \end{pmatrix} \times \begin{pmatrix} 2 & 7 \\ 3 & 8 \end{pmatrix}\right]$?

b. $\left[\begin{pmatrix} 7 & 0 & 5 \\ 2 & 3 & 1 \\ 1 & 8 & 5 \end{pmatrix} \times \begin{pmatrix} 4 & 3 & 10 \\ 2 & 5 & 2 \\ 9 & 7 & 1 \end{pmatrix}\right] \times \begin{pmatrix} 8 & 1 & 2 \\ 0 & 5 & 3 \\ 3 & 6 & 12 \end{pmatrix}$;

$\begin{pmatrix} 7 & 0 & 5 \\ 2 & 3 & 1 \\ 1 & 8 & 5 \end{pmatrix} \times \left[\begin{pmatrix} 4 & 3 & 10 \\ 2 & 5 & 2 \\ 9 & 7 & 1 \end{pmatrix} \times \begin{pmatrix} 8 & 1 & 2 \\ 0 & 5 & 3 \\ 3 & 6 & 12 \end{pmatrix}\right]$

$$\text{Does } \left[\begin{pmatrix}7 & 0 & 5\\2 & 3 & 1\\1 & 8 & 5\end{pmatrix} \times \begin{pmatrix}4 & 3 & 10\\2 & 5 & 2\\9 & 7 & 1\end{pmatrix}\right] \times \begin{pmatrix}8 & 1 & 2\\0 & 5 & 3\\3 & 6 & 12\end{pmatrix} =$$

$$\begin{pmatrix}7 & 0 & 5\\2 & 3 & 1\\1 & 8 & 5\end{pmatrix} \times \left[\begin{pmatrix}4 & 3 & 10\\2 & 5 & 2\\9 & 7 & 1\end{pmatrix} \times \begin{pmatrix}8 & 1 & 2\\0 & 5 & 3\\3 & 6 & 12\end{pmatrix}\right]?$$

c. Does it appear that matrix multiplication is associative?

12. a. Is the product of two 2×2 matrices always a 2×2 matrix? If so, tell why and illustrate it. Can we say that the set of 2×2 matrices is closed under the operation of multiplication?

b. Is the product of two 3×3 matrices always a 3×3 matrix? If so, tell why and illustrate it. Can we say that the set of 3×3 matrices is closed under the operation of multiplication?

13. Find the product matrix of each of the following:

a. $\begin{pmatrix}8 & 4\\2 & 7\end{pmatrix} \times \begin{pmatrix}1 & 0\\0 & 1\end{pmatrix}$

b. $\begin{pmatrix}1 & 0\\0 & 1\end{pmatrix} \times \begin{pmatrix}9 & 6\\12 & 3\end{pmatrix}$

c. $\begin{pmatrix}1 & 0 & 0\\0 & 1 & 0\\0 & 0 & 1\end{pmatrix} \times \begin{pmatrix}7 & 5 & 4\\2 & 8 & 3\\3 & 2 & 9\end{pmatrix}$

d. $\begin{pmatrix}2 & 4 & 3\\1 & 5 & 0\\8 & 6 & 7\end{pmatrix} \times \begin{pmatrix}1 & 0 & 0\\0 & 1 & 0\\0 & 0 & 1\end{pmatrix}$

Can we say that the multiplicative identity for 2×2 matrices is $\begin{pmatrix}1 & 0\\0 & 1\end{pmatrix}$?

Is $\begin{pmatrix}1 & 0 & 0\\0 & 1 & 0\\0 & 0 & 1\end{pmatrix}$ the multiplicative identity for 3×3 matrices?

CHAPTER REVIEW

1. Find the missing numbers: (3–2)

a. $(12 \times 7) \div 7 = \square$ **b.** $(14 - 8) + 8 = \square$

c. What operation is the inverse operation to addition?

2. Which of the following statements are true? (3–3—3–4)

a. $16 + 8 = 8 + 16$

b. $16 - 8 = 8 - 16$

c. $16 \times 8 = 8 \times 16$

d. $16 \div 8 = 8 \div 16$

e. $(16 + 4) + 2 = 16 + (4 + 2)$

f. $(16 - 4) - 2 = 16 - (4 - 2)$

g. $(16 \times 4) \times 2 = 16 \times (4 \times 2)$

h. $(16 \div 4) \div 2 = 16 \div (4 \div 2)$

3. What do we mean by the commutative and associative properties of addition and multiplication? (3–3—3–4)

4. Find the numbers that will make the following statement true: (3–5)

a. $6 \times (3 + 5) = ? \times 3 + ? \times 5$ **b.** $4 \times (8 + 2) = ? \times ? + ? \times ?$

5. Is the set {0, 2, 4, 6, . . .} closed under the operation of addition? Subtraction? Multiplication? Division? (3–6)

6. Determine the value of each of the following: (3–7—3–8)

a. $6 + 0$ **c.** 0×0 **e.** 1^7 **g.** 20×1 **i.** $18 - 0$

b. $17 - 17$ **d.** $\frac{13 - 5}{6 + 2}$ **f.** $\frac{16 + 4}{5 - 5}$ **h.** $\frac{11 - 11}{10}$ **j.** $\frac{17}{17}$

7. Add: **8.** Subtract: **9.** Multiply: (3–9—3–11)

$$\begin{array}{r} 43 \\ 82{,}514 \\ 567{,}827 \\ 9{,}549 \\ \underline{493{,}875} \end{array} \qquad \begin{array}{r} 7{,}040{,}603 \\ \underline{6{,}989{,}594} \end{array} \qquad \begin{array}{r} 90{,}506 \\ \underline{807} \end{array}$$

10. Multiply 562 by 87 using the Russian Peasant method. (3–14)

11. Multiply 3,695 by 418 using the lattice method. (3–15)

12. Divide: **13.** Divide: (3–16)

$904\overline{)289{,}280}$ $3{,}691\overline{)25{,}191{,}075}$

14. Is 71,586 divisible by 2? by 3? by 6? by 9? (3–17)

15. Is 293,000 divisible by 10? by 8? by 5? by 4?

16. Compute each of the following as directed, then check by casting out nines: (3–18)

a. Add: **b.** Subtract: **c.** Multiply: **d.** Divide:

$$\begin{array}{r} 63{,}594 \\ 2{,}855 \\ 19{,}768 \\ 329 \\ \underline{54{,}887} \end{array} \qquad \begin{array}{r} 685{,}104 \\ \underline{638{,}695} \end{array} \qquad \begin{array}{r} 735 \\ \underline{984} \end{array} \qquad 196\overline{)64{,}288}$$

17. Compute as directed: (3–22)

a. Add: **b.** Subtract: **c.** Multiply: **d.** Divide:

$$\begin{array}{r} 11101_{two} \\ 1111_{two} \\ 10111_{two} \\ \underline{1011_{two}} \end{array} \qquad \begin{array}{r} E75_{twelve} \\ \underline{39T_{twelve}} \end{array} \qquad \begin{array}{r} 102_{three} \\ \underline{212_{three}} \end{array} \qquad 324_{five}\overline{)131122_{five}}$$

18. Add or multiply as directed using the remainder method of modular arithmetic: (3–23)

a. Add: $17 + 9 \equiv ? \pmod 6$ **c.** Multiply: $10 \times 7 \equiv ? \pmod 3$

b. Add: $23 + 11 \equiv ? \pmod 5$ **d.** Multiply: $12 \times 15 \equiv ? \pmod 7$

19. Add the following matrices:

$$\begin{pmatrix} 5 & 2 & 0 \\ 3 & 4 & 7 \\ 2 & 1 & 6 \end{pmatrix} + \begin{pmatrix} 6 & 8 & 2 \\ 2 & 1 & 7 \\ 0 & 3 & 4 \end{pmatrix}$$

20. Find the product matrix: (3–24)

$$\begin{pmatrix} 3 & 5 \\ 6 & 8 \end{pmatrix} \times \begin{pmatrix} 4 & 1 \\ 2 & 7 \end{pmatrix}$$

CUMULATIVE TEST

1. If $M = \{\text{all prime numbers greater than 4 but less than 14}\}$, which of the following statements are true? (1–9)

a. $13 \in M$ **b.** $10 \in M$ **c.** $7 \notin M$ **d.** $23 \in M$ **e.** $3 \notin M$

2. What is the greatest common factor of 84 and 63? (1–13)
3. Factor 120 as the product of prime factors. (1–15)
4. What is the least common multiple of 50, 100, and 125? (1–18)
5. Write 499 as a Roman numeral. (2–7)
6. Round 47,309,571,006 to the nearest million. (2–11)
7. Expand 12212_{three} as a polynomial. (2–18)
8. Express 1326_{eight} as an equivalent base two numeral. (2–19)
9. Find the missing numbers: (3–2—3–5)

a. $(8 - 3) + 3 = \square$
b. $11 + 10 = ? + 11$
c. $(6 \times 7) \times 3 = 6 \times (? \times 3)$
d. $15 \times 9 = 9 \times ?$
e. $(8 + 4) + 5 = 8 + (4 + ?)$
f. $2 \times (3 + 4) = ? \times 3 + ? \times 4$

10. Is $\{1\}$ closed under addition? Multiplication? Division? Subtraction? (3–6)

11. Add:

```
 85,876
 76,988
 59,649
 60,795
 98,489
```

12. Subtract:

```
 613,002
 158,075
```

13. Multiply: (3–9—3–11)

```
 798
 869
```

14. Divide: $886\overline{)792,084}$

15. Divide: $1,052\overline{)325,068}$ (3–16)

16. Is 23,954 divisible by 3? by 2? by 4? by 6? (3–17)
17. Multiply 875 by 268. Check by casting out nines. (3–18)

18. Add:

3122_{four}
1332_{four}
2133_{four}

19. Multiply: (3–22)

10111_{two}
1011_{two}

20. Add: $15 + 21 \equiv ? \pmod 8$

21. Multiply: $6 \times 9 \equiv ? \pmod 4$

22. Multiply: $11 \times 10 \equiv ? \pmod 6$ (3–23)

23. Add the following matrices:

$$\begin{pmatrix} 5 & 6 \\ 4 & 3 \end{pmatrix} + \begin{pmatrix} 8 & 2 \\ 0 & 9 \end{pmatrix}$$

24. Find the product matrix (3–24)

$$\begin{pmatrix} 6 & 4 & 1 \\ 2 & 0 & 8 \\ 3 & 7 & 5 \end{pmatrix} \times \begin{pmatrix} 3 & 6 & 5 \\ 4 & 2 & 0 \\ 0 & 1 & 9 \end{pmatrix}$$

Number Sentences – Equalities and Inequalities

4

4–1 NUMBER SENTENCES

Sentences that deal with numbers are called *number sentences* or *mathematical sentences*. The *equality* $4 + 8 = 12$ is a number sentence. It reads "Four plus eight is equal to twelve." The symbol "$=$" is the equality sign. It means "is equal to" and is the verb in the sentence. The expressions on both sides of the equality sign designate the same number.

Number sentences may be true or they may be false. The number sentence $4 + 8 = 12$ is true but the sentence $9 - 5 = 6$ is false. A sentence that is either true or false is called a *statement*. A sentence cannot be both true and false at one time.

When two numbers are compared, one number may be:

(1) Equal to the other.
(2) Greater than the other.
(3) Less than the other.

In any specific case only one of these three possibilities is true.

When one number is greater than or less than a second number, an *inequality* exists. An inequality is a number sentence. It may be true or false. Symbols of inequality include: $\neq$, $>$, $<$, $\not>$, and $\not<$. Each symbol is a verb in a number sentence.

The symbol "$\neq$" means "is not equal to."
$8 - 3 \neq 6$ is read "Eight minus three is not equal to six."

The symbol "$>$" means "is greater than."
$10 > 4$ is read "Ten is greater than four."

The symbol "$<$" means "is less than."
$3 < 7$ is read "Three is less than seven."

The symbol "$\ngtr$" means "is not greater than."
$5 \times 6 \ngtr 40$ is read "Five times six is not greater than forty."

The symbol "$\nless$" means "is not less than."
$2 + 1 \nless 0$ is read "Two plus one is not less than zero."

EXERCISES

1. Read, or write in words, each of the following:

a. $16 > 9$
b. $12 \nless 11$
c. $8 \neq 4 + 2$
d. $4 < 6$
e. $11 \ngtr 15$
f. $6 - 3 < 7$
g. $9 \times 6 \ngtr 54$
h. $20 \div 4 \nless 3$
i. $16 - 9 \neq 8$
j. $30 \times 4 > 100$
k. $8 \times 3 \neq 48 \div 4$
l. $18 + 5 \nless 7 \times 6$
m. $10 \times 9 = 9 \times 10$
n. $35 \div 7 > 11 - 8$
o. $6 \times 0 \nless 15 - 15$
p. $72 \div 8 \ngtr 1 \times 12$
q. $20 - 3 \neq 48 \div 3$
r. $4 - 4 < 4 \div 4$
s. $97 + 3 \nless 10^2$
t. $42 - 7 > 42 \div 7$

2. Write each of the following sentences symbolically:
a. The sum of four and twelve is not equal to fifteen.
b. Sixty-four is less than nine times eight.
c. The product of seven and twelve is not greater than one hundred.
d. Eleven times five is greater than forty plus fourteen.
e. Eighty divided by five is not less than ninety minus seventy-five.

3. Determine which of the following sentences are true and which are false:

a. $9 = 6$
b. $8 > 3$
c. $5 < 1$
d. $4 \neq 8$
e. $1 > 0$
f. $6 + 4 \nless 14 - 4$
g. $12 + 5 < 20 - 5$
h. $7 \times 6 \neq 100 \div 2$
i. $28 \div 7 \ngtr 2^2$
j. $9 \times 4 > 54 - 19$
k. $8 + 4 = 4 + 8$
l. $12 \div 12 \nless 12 - 12$
m. $0 \div 5 > 24 \times 0$
n. $8 \times 12 \ngtr 100 - 5$
o. $3 + 5 + 7 < 7 + 5 + 3$
p. $8 \times (9 - 1) > 9 \times (8 - 1)$
q. $20 \times 18 = 18 \times 20$
r. $8 \times 7 \nless 81 - 28$
s. $2^3 < 3^2$
t. $4 \times (2 + 8) \neq (4 \cdot 2) + (2 \cdot 8)$

4–2 REFLEXIVE, SYMMETRIC, AND TRANSITIVE PROPERTIES

Equality has a *reflexive* property. That is, any number is equal to itself. $a = a$. Inequality has no reflexive property because a number cannot be greater or smaller than itself.

Equality has a *symmetric* property. That is, if one number is equal to a second number, then the second number is equal to the first number. If $a = b$, then $b = a$. Inequality does not have this property. If one number is greater than a second number, the second number cannot be greater than the first number. If one number is smaller than a second number, the second number cannot be smaller than the first number.

Equality and inequality both have the *transitive* property. That is, if one number is equal to a second number and the second number is equal to a third number, then the first number is equal to the third number.

If $a = b$ and $b = c$, then $a = c$.

If one number is greater than a second number and the second number is greater than a third number, then the first number is greater than the third number.

If $a > b$ and $b > c$, then $a > c$.
Similarly if $a < b$ and $b < c$, then $a < c$.

EXERCISES

1. a. Can 8 be greater than itself? Can any number be greater than itself?
b. Can 3 be less than itself? Can any number be less than itself?
c. What property does the sentence $9 = 9$ illustrate? State it.
d. Which one of the following sentences is true: $6 > 6$; $6 = 6$; $6 < 6$?
e. Do inequalities have a reflexive property?

2. Which of the following are symmetric?

a. $8 = 6 + 2$ and $6 + 2 = 8$
b. $12 > 7 + 4$ and $7 + 4 > 12$
c. $10 \div 2 < 6$ and $6 < 10 \div 2$
d. $2 \times 2 = 2 + 2$ and $2 + 2 = 2 \times 2$
e. $8 \times 5 < 6 \times 7$ and $6 \times 7 < 8 \times 5$
f. $21 - 8 > 4 \times 3$ and $4 \times 3 > 21 - 8$
g. What is the symmetric property of equalities?
h. Do inequalities have a symmetric property?

3. Complete by using the transitive property:

a. $8 > 6$ and $6 > 4$, then $8 > ?$.
b. $3 < 51$ and $51 < 70$, then $3 < ?$.
c. $4 \times 5 = 10 + 10$ and $10 + 10 = 20$, then $4 \times 5 = ?$.
d. $2^3 < 4^2$ and $4^2 < 3^4$, then $2^3 < ?$.
e. $12 + 3 > 2 \times 5$ and $2 \times 5 > 8 \div 2$, then $12 + 3 > ?$.
f. $35 - 5 = 27 + 3$ and $27 + 3 = 10 \times 3$ then, $35 - 5 = ?$.
g. State the transitive property. Do equalities have the transitive property? Do inequalities have the transitive property?

5 Fractions

5–1 MEANING OF A FRACTION

When a thing or unit is divided into equal parts, the number expressing the relation of one or more of the equal parts to the total number of equal parts is called a *fraction.*

If a strip of paper 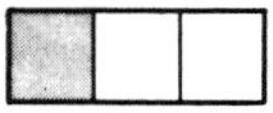 is divided into three equal parts, each part is *one-third* of the whole strip and is represented by the fraction symbol $\frac{1}{3}$.

If the strip of paper is divided into two equal parts 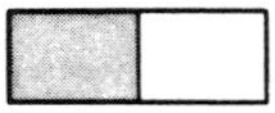, each part is called *one-half,* ($\frac{1}{2}$); four equal parts 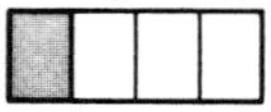, each part is called *one-fourth,* ($\frac{1}{4}$); five equal parts 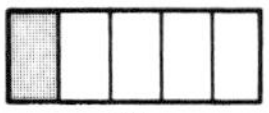, each part is called *one-fifth,* ($\frac{1}{5}$); etc.

Although "fraction" is generally used to mean both the fractional number and the numeral written for the number, some mathematicians use the term "fraction" to mean the symbol or fractional numeral and "rational number" to mean the fractional number. We shall use the term "fraction" to mean both the number and the numeral except in situations where it may be misunderstood.

The symbol for a *common fraction* consists of a pair of numerals, one written above the other, with a horizontal bar between them. The number represented below the fraction bar cannot be zero. In the fraction $\frac{3}{8}$, the numbers 3 and 8 are the *terms* of the fraction. The number above the fraction bar is called the *numerator.* The number below the fraction bar is called the *denominator.* In the fraction $\frac{3}{8}$, 3 is the numerator and 8 is the denominator. The denominator tells us the number of equal parts into which the object is divided. The numerator tells us how many equal parts are being used. The fraction $\frac{3}{8}$ means 3 parts of 8 equal parts.

$$\frac{3}{8}$$

Sometimes a group of things is divided into equal parts. The number expressing the relation of one or more of the equal parts of a group to the total number of equal parts is considered to be a fraction.

A class of 20 pupils is divided into 2 teams of 10 pupils each. Each team is one-half ($\frac{1}{2}$) of the class.

Observe in the following that the *larger* the denominator, the *smaller* the *size* of the *part*.

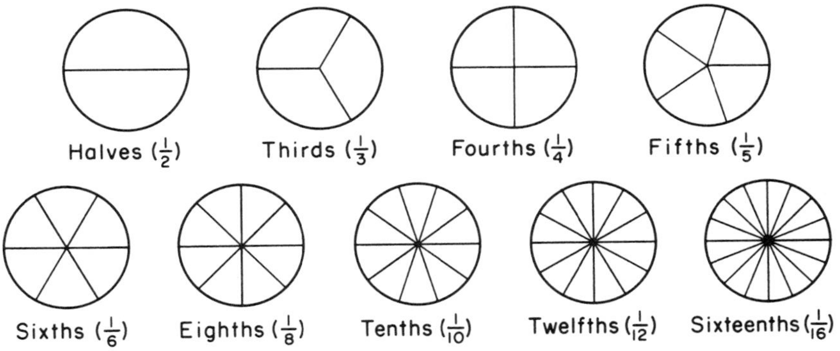

Since the counting numbers and whole numbers are not adequate in making measurements, fractions were invented to meet this need.

Observe in the following that the *smaller the size* of the part (subdivision of the unit), the *more precise* is the measurement.

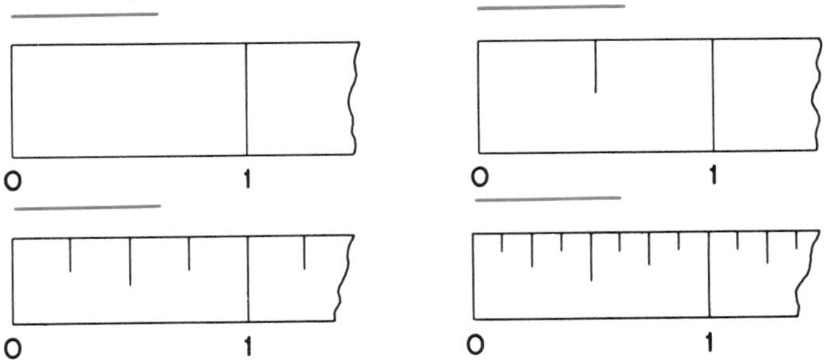

A fraction is used to indicate that one number is divided by another number. Some mathematicians call the number they get when they divide a whole number by any whole number, except zero, a *rational number*.

8 divided by 4 usually written as $8 \div 4$ or $4\overline{)8}$ becomes $\frac{8}{4}$, read "eight over four"; the horizontal bar means "divided by."

7 divided by 4 becomes $\frac{7}{4}$. 3 divided by 4 becomes $\frac{3}{4}$.

The whole number 2 may be considered $\frac{2}{1}$; 3 as $\frac{3}{1}$; 4 as $\frac{4}{1}$, etc.

Thus, if a fraction is an indicated division, the numerator is the number that is divided and the denominator is the number by which you divide.

A fraction may be used to compare two things or two groups of things. For example: a foot is $\frac{1}{3}$ of a yard.

Some mathematicians think of a fraction as an ordered number pair represented by a pair of numerals written within parentheses in a specific order. (7, 8) is an ordered pair of numbers with 7 called the first number and 8 the second number. (8, 7) is a different ordered number pair since 8 is first and 7 is second.

(7, 8) would be associated with the fraction $\frac{7}{8}$ and (8, 7) would be associated with the fraction $\frac{8}{7}$.

The fraction $\frac{7}{8}$ may also mean $7 \times \frac{1}{8}$ or $\frac{1}{8}$ of 7. Generalizing, if a represents the numerator and b the denominator, then the fraction $\frac{a}{b}$ may also mean $a \times \frac{1}{b}$ or $\frac{1}{b} \times a$.

EXERCISES

1. Read, or write in words, each of the following fractions:

a. $\frac{1}{3}$; $\frac{1}{6}$; $\frac{1}{7}$; $\frac{1}{10}$; $\frac{1}{24}$; $\frac{1}{72}$; $\frac{1}{50}$; $\frac{1}{48}$; $\frac{1}{75}$; $\frac{1}{100}$

b. $\frac{5}{8}$; $\frac{4}{5}$; $\frac{3}{4}$; $\frac{11}{16}$; $\frac{5}{12}$; $\frac{9}{10}$; $\frac{8}{6}$; $\frac{3}{100}$; $\frac{17}{20}$; $\frac{15}{8}$

2. Write each of the following as a numeral:

a. one-fifth
b. one-eighth
c. one ninety-third
d. five-sixths
e. fifty-nine hundredths
f. thirty-one fourths

3. Into how many equal parts is the following figure divided?

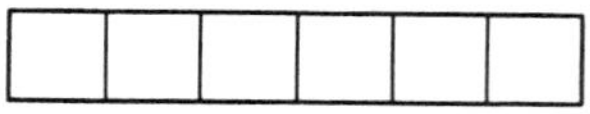

Write as a numeral the size of one of the equal parts.

4. Each of the following figures is divided into equal parts. In each case, write what the size of each part is called, then write it as a numeral.

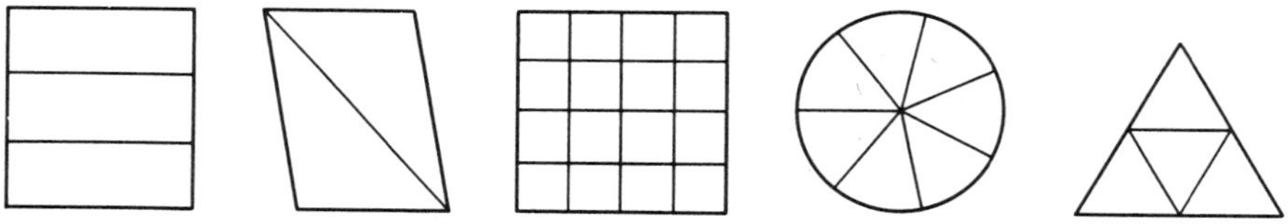

5. Write the numeral that represents the shaded part in each of the following figures:

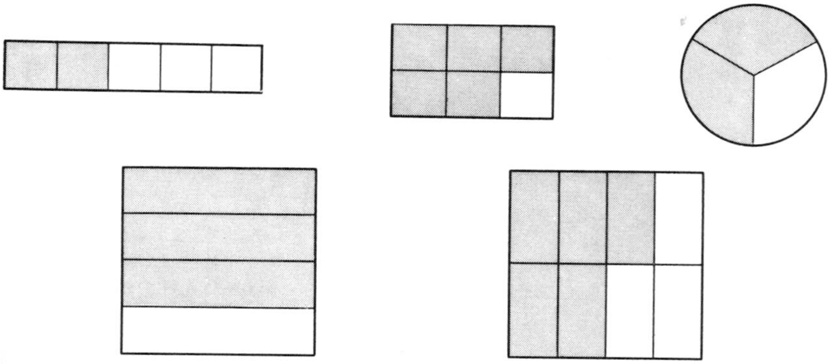

6. Write as a numeral how full each of the following containers is:

a. **b.** **c.** **d.** **e.**

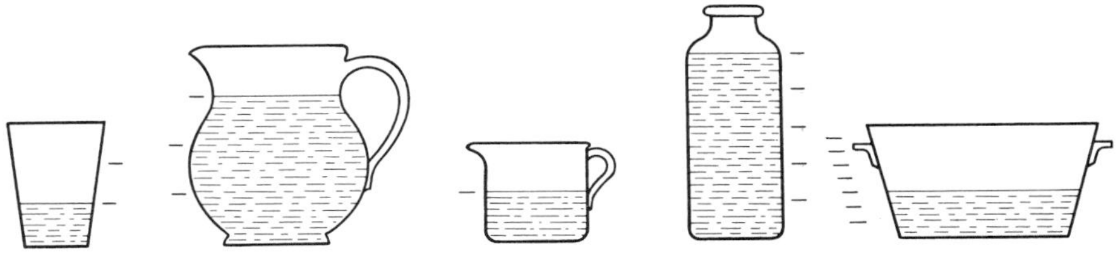

7. Write the numeral naming the fraction indicated in each of the following measuring devices or gauges.

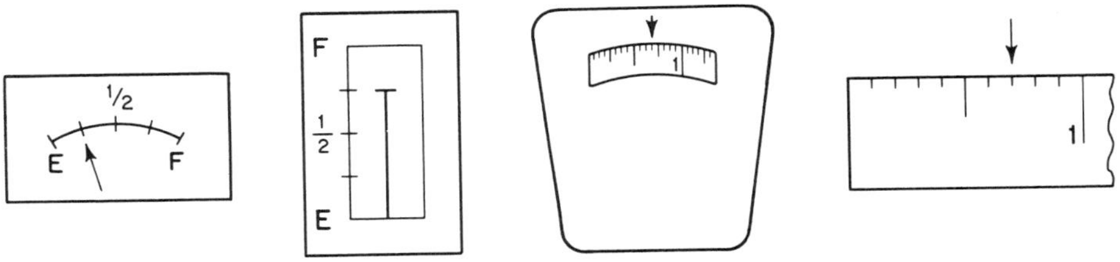

8. Draw figures like each of the following, then divide them into the required number of parts:

a. Divide into eighths **b.** Divide into halves **c.** Divide into twelfths **d.** Divide into thirds **e.** Divide into sixths

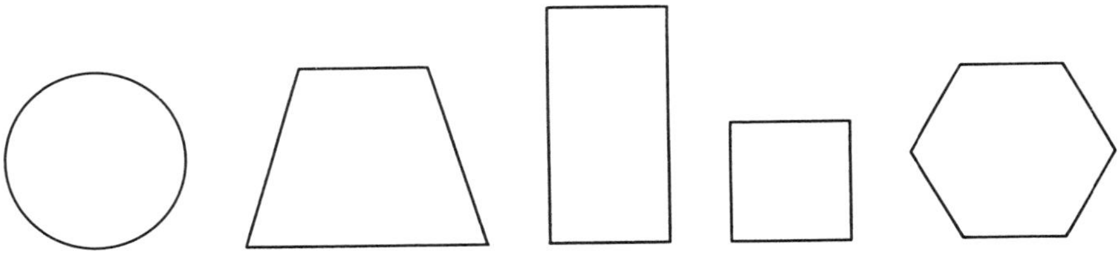

a. **b.** **c.** **d.** **e.**

9. Draw figures like each of the following, then shade the part that represents the given fraction:

a. $\frac{1}{2}$

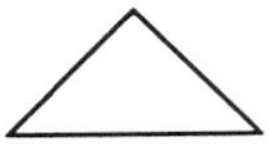

c. $\frac{1}{3}$

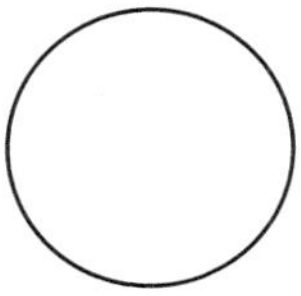

b. $\frac{5}{6}$

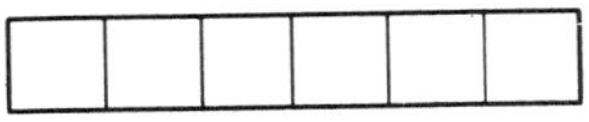

d. $\frac{3}{4}$

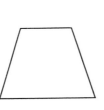

e. $\frac{7}{10}$ 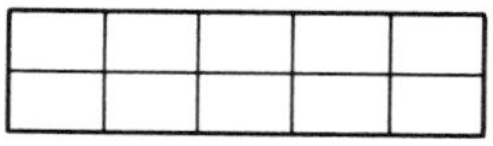

h. $\frac{3}{4}$ 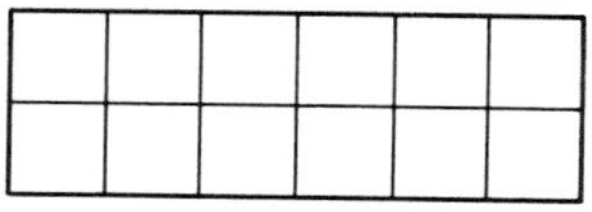

f. $\frac{1}{2}$ 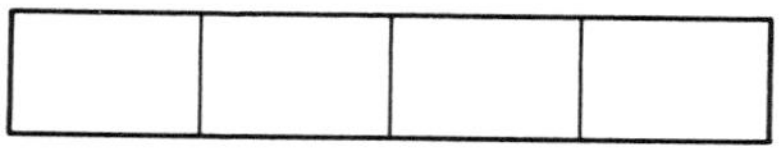

i. $\frac{5}{8}$ 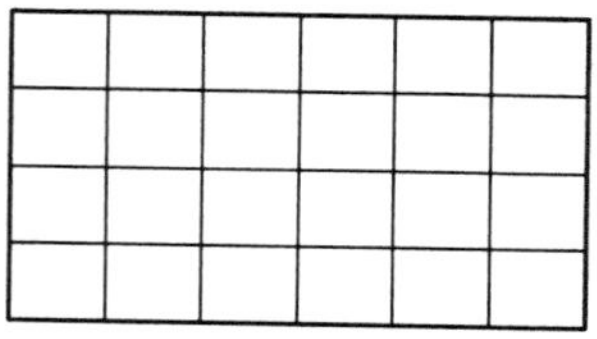

g. $\frac{2}{3}$ 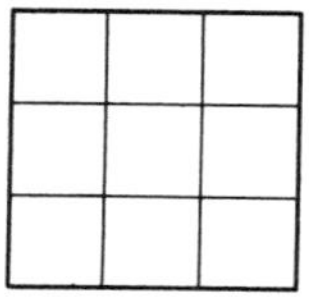

j. $\frac{4}{5}$ 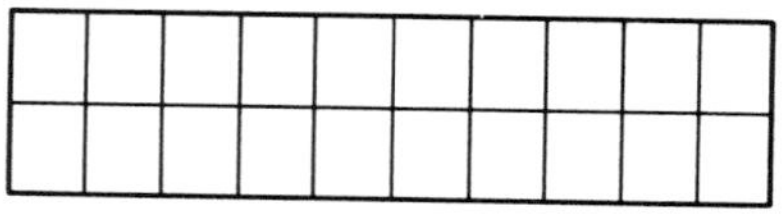

10. a. What part of the pie is left?

b. What fractional part of a dozen eggs was used?

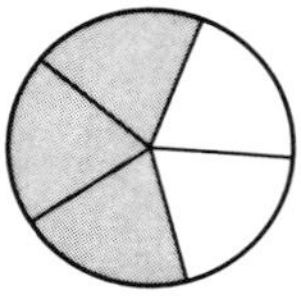

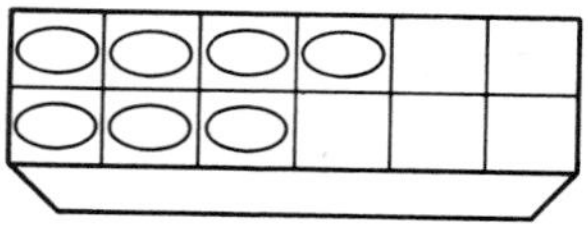

11. What fractional part of the hour past 12 o'clock is indicated on each clock?

12. a. Write the numeral naming the fraction using 4 as the numerator and 5 as the denominator.

b. Write the numeral naming the fraction using 32 as the denominator and 11 as the numerator.

13. Express each of the following as a numeral naming a fraction:

a. 7 divided by 10 **b.** 3 divided by 5 **c.** 2 divided by 3 **d.** 11 divided by 25

e. $3 \div 8$ **f.** $13 \div 16$ **g.** $1 \div 6$ **h.** $25 \div 32$ **i.** $22 \div 7$

j. $2\overline{)1}$ **k.** $5\overline{)4}$ **l.** $10\overline{)3}$ **m.** $24\overline{)19}$ **n.** $32\overline{)27}$

14. a. What does $\frac{5}{6}$ mean? Express $\frac{5}{6}$ in two other ways.

b. What does $\frac{1}{5}$ mean? Express $\frac{1}{5}$ in two other ways.

c. What does $\frac{4}{9}$ mean? Express $\frac{4}{9}$ in two other ways.

15. a. Compare an inch to a foot. An inch is what part of a foot?

b. Compare a quart to a gallon. A quart is what part of a gallon?

c. Compare a nickel to a dime. A nickel is what part of a dime?

d. Compare a day to a week. A day is what part of a week?

e. Compare an ounce to a pound. An ounce is what part of a pound?

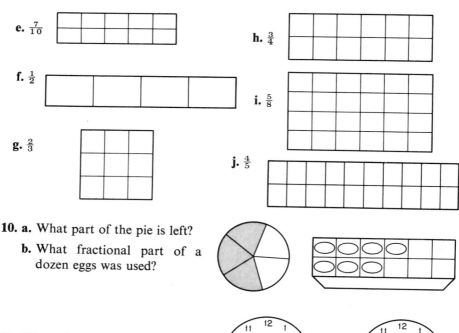

16. Write each of the following ordered pairs of numbers as a numeral naming a fraction:

a. (5, 7) **b.** (9, 4) **c.** (6, 25) **d.** (11, 100) **e.** (17, 24) **f.** (12, 5) **g.** (23, 60) **h.** (37, 100)

17. Write each of the following fractions as an ordered pair:

a. $\frac{6}{11}$ **b.** $\frac{10}{7}$ **c.** $\frac{5}{16}$ **d.** $\frac{11}{40}$ **e.** $\frac{19}{6}$ **f.** $\frac{59}{100}$ **g.** $\frac{25}{16}$ **h.** $\frac{31}{35}$

18. Which of the following:

a. Name $\frac{3}{4}$:	$\frac{1}{4} \times 3$?;	$\frac{1}{3} \times 4$?;	$3 \times \frac{1}{4}$?;	$4 \times \frac{1}{3}$?
b. Name $\frac{12}{7}$:	$\frac{1}{12} \times 7$?;	$12 \times \frac{1}{7}$?;	$\frac{1}{7} \times 12$?;	$7 \times \frac{1}{12}$?
c. Name $\frac{5}{9}$:	$9 \times \frac{1}{5}$?;	$\frac{1}{9} \times 5$?;	$5 \times \frac{1}{9}$?;	$\frac{1}{5} \times 9$?

EQUIVALENT FRACTIONS

Fractions that name the same number are called *equivalent fractions.*

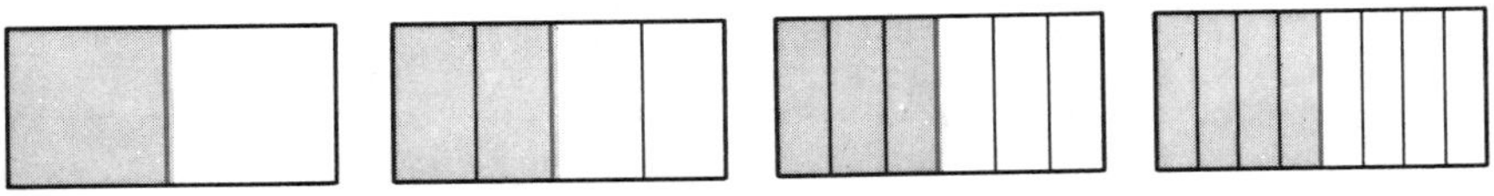

$\frac{1}{2}$ $\frac{2}{4}$ $\frac{3}{6}$ $\frac{4}{8}$

The fractions $\frac{1}{2}$, $\frac{2}{4}$, $\frac{3}{6}$, and $\frac{4}{8}$ are equivalent fractions. They name the same number which in simplest form is $\frac{1}{2}$. This set of equivalent fractions may be written as follows:
$\{\frac{1}{2}, \frac{2}{4}, \frac{3}{6}, \frac{4}{8}, \frac{5}{10}, \frac{6}{12}, \cdots\}$ where the fraction listed first is the name of the fractional number in simplest form.

5–2 EXPRESSING FRACTIONS IN LOWEST TERMS

When the numerator and denominator have no common factor except 1, the fraction is in *simplest form* or in *lowest terms.* When the 4 and 8 of the fraction $\frac{4}{8}$ are each:

(1) Divided by 2, we get the equivalent fraction $\frac{2}{4}$.

(2) Divided by 4, we get the equivalent fraction $\frac{1}{2}$.

$$\frac{4}{8} = \frac{4 \div 2}{8 \div 2} = \frac{2}{4}$$

$$\frac{4}{8} = \frac{4 \div 4}{8 \div 4} = \frac{1}{2}$$

Thus when the numerator and denominator of any fraction are each divided by the same number, except by zero, the result is an equivalent fraction. Therefore to express a fraction in lowest terms:

(a) We factor the numerator and denominator, using as one of the factors the greatest factor common to both. (Use whole numbers other than 1 or 0.)

(b) Then we divide both the numerator and denominator by this common factor.

The multiplicative identity one (1) may be used as shown:

> Express $\frac{30}{48}$ in lowest terms:
>
> $$\frac{30}{48} = \frac{5 \cdot 6}{8 \cdot 6} = \frac{5}{8} \quad \text{or} \quad \frac{30}{48} = \frac{5 \cdot 6}{8 \cdot 6} = \frac{5}{8} \times 1 = \frac{5}{8}$$

EXERCISES

1. Use the section of the ruler to obtain answers to the following:[1]

a. $\frac{2}{4}'' = \frac{\ }{2}''$ $\quad \frac{4}{16}'' = \frac{\ }{4}''$ $\quad \frac{6}{8}'' = \frac{\ }{4}''$ $\quad \frac{8}{16}'' = \frac{\ }{2}''$ $\quad \frac{6}{16}'' = \frac{\ }{8}''$

b. $\frac{4}{8}'' = \frac{\ }{2}''$ $\quad \frac{10}{16}'' = \frac{\ }{8}''$ $\quad \frac{2}{8}'' = \frac{\ }{4}''$ $\quad \frac{14}{16}'' = \frac{\ }{8}''$ $\quad \frac{2}{16}'' = \frac{\ }{8}''$

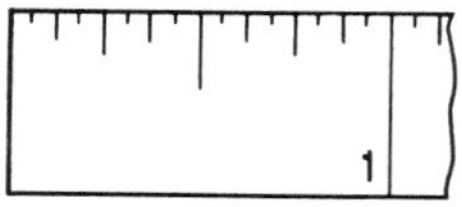

2. Express each of the following in lowest terms:

a. $\frac{5}{10}$ $\quad \frac{3}{24}$ $\quad \frac{2}{32}$ $\quad \frac{3}{15}$ $\quad \frac{5}{60}$ $\quad \frac{11}{77}$ $\quad \frac{2}{12}$ $\quad \frac{7}{56}$ $\quad \frac{6}{42}$ $\quad \frac{9}{90}$

b. $\frac{6}{12}$ $\quad \frac{5}{20}$ $\quad \frac{8}{64}$ $\quad \frac{9}{45}$ $\quad \frac{12}{36}$ $\quad \frac{14}{42}$ $\quad \frac{25}{100}$ $\quad \frac{32}{96}$ $\quad \frac{15}{75}$ $\quad \frac{16}{80}$

c. $\frac{4}{18}$ $\quad \frac{14}{16}$ $\quad \frac{12}{15}$ $\quad \frac{20}{32}$ $\quad \frac{36}{48}$ $\quad \frac{50}{75}$ $\quad \frac{56}{64}$ $\quad \frac{54}{72}$ $\quad \frac{28}{42}$ $\quad \frac{36}{54}$

d. $\frac{10}{50}$ $\quad \frac{30}{90}$ $\quad \frac{60}{100}$ $\quad \frac{210}{700}$ $\quad \frac{400}{1{,}000}$ $\quad \frac{4{,}500}{5{,}000}$ $\quad \frac{6{,}000}{7{,}500}$ $\quad \frac{960}{3{,}600}$ $\quad \frac{5{,}400}{7{,}200}$ $\quad \frac{150}{2{,}400}$

e. $\frac{54}{144}$ $\quad \frac{96}{160}$ $\quad \frac{80}{112}$ $\quad \frac{105}{140}$ $\quad \frac{225}{400}$ $\quad \frac{252}{324}$ $\quad \frac{294}{336}$ $\quad \frac{576}{648}$ $\quad \frac{204}{228}$ $\quad \frac{273}{315}$

3. Which fractions are not expressed in lowest terms?

a. $\frac{4}{7}$ $\quad \frac{5}{8}$ $\quad \frac{4}{9}$ $\quad \frac{6}{9}$ $\qquad$ **c.** $\frac{5}{8}$ $\quad \frac{11}{12}$ $\quad \frac{21}{28}$ $\quad \frac{25}{32}$ $\qquad$ **e.** $\frac{8}{15}$ $\quad \frac{12}{35}$ $\quad \frac{9}{45}$ $\quad \frac{17}{20}$

b. $\frac{9}{10}$ $\quad \frac{7}{10}$ $\quad \frac{5}{10}$ $\quad \frac{3}{10}$ $\qquad$ **d.** $\frac{3}{4}$ $\quad \frac{3}{5}$ $\quad \frac{3}{6}$ $\quad \frac{3}{8}$ $\qquad$ **f.** $\frac{27}{36}$ $\quad \frac{21}{32}$ $\quad \frac{25}{34}$ $\quad \frac{24}{39}$

4. Find which fractions are equivalent in each group:

a. $\frac{6}{8}$ $\quad \frac{12}{15}$ $\quad \frac{18}{24}$ $\quad \frac{9}{12}$ $\qquad$ **c.** $\frac{15}{25}$ $\quad \frac{6}{9}$ $\quad \frac{4}{6}$ $\quad \frac{14}{21}$ $\qquad$ **e.** $\frac{40}{48}$ $\quad \frac{8}{16}$ $\quad \frac{10}{12}$ $\quad \frac{15}{18}$

b. $\frac{16}{32}$ $\quad \frac{4}{8}$ $\quad \frac{32}{64}$ $\quad \frac{8}{16}$ $\qquad$ **d.** $\frac{10}{32}$ $\quad \frac{9}{24}$ $\quad \frac{21}{56}$ $\quad \frac{6}{16}$ $\qquad$ **f.** $\frac{28}{35}$ $\quad \frac{35}{50}$ $\quad \frac{20}{25}$ $\quad \frac{9}{15}$

5–3 EXPRESSING FRACTIONS IN HIGHER TERMS

If we multiply both the 1 and 2 of the fraction $\frac{1}{2}$:

(1) By 2, we get the equivalent fraction $\frac{2}{4}$.

(2) By 4, we get the equivalent fraction $\frac{4}{8}$.

---

[1] The symbol ″ represents "inch."

Thus, when the numerator and denominator of any fraction are each multiplied by the same number, except by zero, the result is an equivalent fraction.

Therefore, to express a fraction in higher terms where the new denominator is specified:

(1) We divide the *new* denominator by the denominator of the given fraction.

(2) Then we multiply *both* the numerator and denominator of the given fraction by the quotient. This is the same as multiplying the fraction by the multiplicative identity one (1).

Change $\frac{2}{3}$ to 12ths:

$$\frac{2}{3}=\frac{2\times 4}{3\times 4}=\frac{8}{12} \quad \text{or} \quad \frac{2}{3}=\frac{2}{3}\times 1=\frac{2\times 4}{3\times 4}=\frac{8}{12}$$

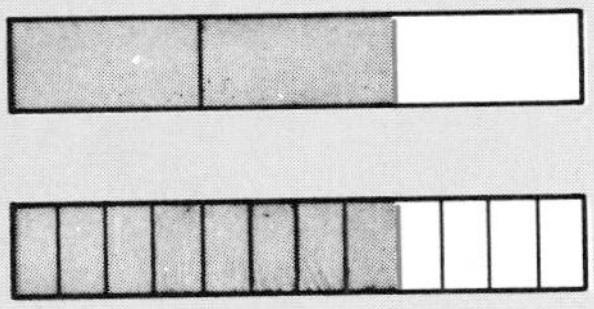

EXERCISES

1. Use the chart to express fractions in higher terms:

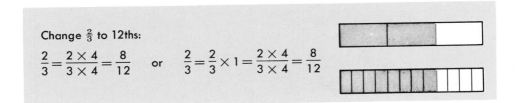

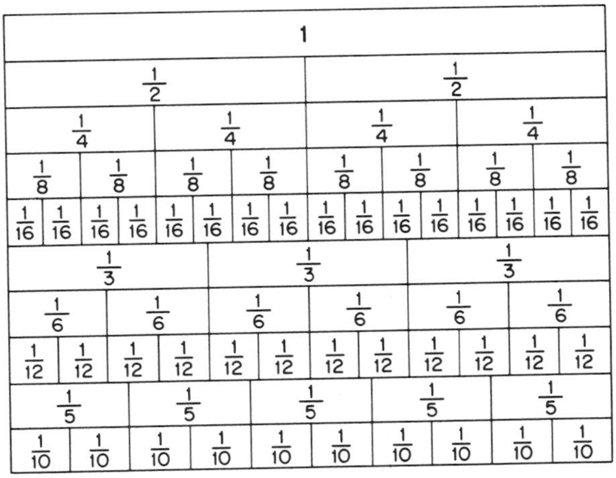

a. $\frac{1}{2}=\frac{\quad}{4}$ $\qquad \frac{1}{4}=\frac{\quad}{16}$ $\qquad \frac{1}{6}=\frac{\quad}{12}$ $\qquad \frac{1}{2}=\frac{\quad}{10}$ $\qquad \frac{1}{3}=\frac{\quad}{6}$

b. $\frac{1}{8}=\frac{\quad}{16}$ $\qquad \frac{1}{5}=\frac{\quad}{10}$ $\qquad \frac{1}{3}=\frac{\quad}{12}$ $\qquad \frac{1}{2}=\frac{\quad}{6}$ $\qquad \frac{1}{4}=\frac{\quad}{12}$

c. $\frac{1}{2}=\frac{\quad}{12}$ $\qquad \frac{1}{2}=\frac{\quad}{8}$ $\qquad \frac{1}{4}=\frac{\quad}{8}$ $\qquad \frac{1}{2}=\frac{\quad}{16}$ $\qquad \frac{3}{4}=\frac{\quad}{8}$

d. $\frac{3}{4}=\frac{\quad}{16}$ $\qquad \frac{2}{3}=\frac{\quad}{6}$ $\qquad \frac{7}{8}=\frac{\quad}{16}$ $\qquad \frac{5}{6}=\frac{\quad}{12}$ $\qquad \frac{5}{8}=\frac{\quad}{16}$

e. $\frac{2}{5}=\frac{\quad}{10}$ $\qquad \frac{3}{4}=\frac{\quad}{12}$ $\qquad \frac{4}{5}=\frac{\quad}{10}$ $\qquad \frac{3}{8}=\frac{\quad}{16}$ $\qquad \frac{2}{3}=\frac{\quad}{12}$

2. Express as fractions in the indicated higher terms:

a. $\frac{1}{2} = \frac{}{24}$ $\frac{1}{8} = \frac{}{32}$ $\frac{1}{10} = \frac{}{50}$ $\frac{1}{6} = \frac{}{48}$ $\frac{1}{5} = \frac{}{20}$ $\frac{1}{16} = \frac{}{32}$ $\frac{1}{3} = \frac{}{15}$ $\frac{1}{12} = \frac{}{72}$

b. $\frac{4}{5} = \frac{}{10}$ $\frac{2}{3} = \frac{}{24}$ $\frac{3}{5} = \frac{}{25}$ $\frac{3}{4} = \frac{}{36}$ $\frac{7}{8} = \frac{}{56}$ $\frac{21}{32} = \frac{}{64}$ $\frac{13}{16} = \frac{}{80}$ $\frac{19}{24} = \frac{}{96}$

c. $\frac{9}{10} = \frac{}{100}$ $\frac{1}{4} = \frac{}{100}$ $\frac{1}{2} = \frac{}{100}$ $\frac{2}{5} = \frac{}{100}$ $\frac{3}{4} = \frac{}{100}$ $\frac{17}{20} = \frac{}{100}$ $\frac{21}{25} = \frac{}{100}$ $\frac{27}{50} = \frac{}{100}$

3. Express:

a. $\frac{1}{6}$ as 24ths c. $\frac{1}{2}$ as 16ths e. $\frac{3}{4}$ as 32nds g. $\frac{4}{5}$ as 20ths i. $\frac{7}{10}$ as 60ths

b. $\frac{2}{3}$ as 36ths d. $\frac{15}{16}$ as 80ths f. $\frac{23}{25}$ as 100ths h. $\frac{11}{12}$ as 72nds j. $\frac{19}{32}$ as 64ths

4. A set of equivalent fractions may be developed from the name of the fractional number in simplest form by expressing it successively in higher terms. Examples are:

$$\{\tfrac{1}{5}, \tfrac{2}{10}, \tfrac{3}{15}, \tfrac{4}{20}, \cdots\} \text{ and } \{\tfrac{2}{3}, \tfrac{4}{6}, \tfrac{6}{9}, \tfrac{8}{12}, \cdots\}.$$

Write the set of fractions equivalent to:

a. $\frac{1}{4}$ c. $\frac{2}{5}$ e. $\frac{1}{6}$ g. $\frac{3}{4}$ i. $\frac{7}{12}$ k. $\frac{1}{3}$ m. $\frac{4}{7}$ o. $\frac{2}{11}$ q. $\frac{1}{20}$ s. $\frac{17}{50}$

b. $\frac{1}{7}$ d. $\frac{3}{8}$ f. $\frac{5}{9}$ h. $\frac{1}{16}$ j. $\frac{5}{6}$ l. $\frac{3}{10}$ n. $\frac{1}{15}$ p. $\frac{9}{13}$ r. $\frac{8}{25}$ t. $\frac{23}{30}$

5. For each of the following write the set of equivalent fractions of which it is a member:

a. $\frac{8}{12}$ b. $\frac{24}{32}$ c. $\frac{49}{56}$ d. $\frac{6}{22}$ e. $\frac{35}{84}$ f. $\frac{45}{72}$ g. $\frac{18}{63}$ h. $\frac{6}{27}$ i. $\frac{99}{144}$ j. $\frac{70}{100}$

5–4 TESTING FOR EQUIVALENT FRACTIONS

We can tell that one fraction is equivalent to another fraction by expressing each given fraction in lowest terms. If the resulting fractions are the same, then the given fractions are equivalent.

Are $\frac{6}{8}$ and $\frac{9}{12}$ equivalent fractions?

Yes, $\frac{6}{8} = \frac{9}{12}$ since $\frac{6}{8} = \frac{3}{4}$ and $\frac{9}{12} = \frac{3}{4}$.

Examination of $\frac{6}{8} = \frac{9}{12}$ will show that the product of the numerator of the first fraction and the denominator of the second fraction (6×12) is equal to the product of the numerator of the second fraction and the denominator of the first fraction (9×8). Since $6 \times 12 = 72$ and $9 \times 8 = 72$, then $6 \times 12 = 9 \times 8$.

Thus, we may say that:
When two fractions are equivalent, the cross products are equal; and when the cross products are equal, the two fractions are equivalent.

Are $\frac{10}{15}$ and $\frac{6}{9}$ equivalent fractions?
$\frac{10}{15} = \frac{6}{9}$ is true if $10 \times 9 = 6 \times 15$.

Since $10 \times 9 = 90$ and $6 \times 15 = 90$, then
$10 \times 9 = 6 \times 15$.

Thus, $\frac{10}{15}$ and $\frac{6}{9}$ are equivalent fractions.

EXERCISES

Test whether each of the following pairs of fractions are equivalent by using the method of:

1. Lowest terms: **a.** $\frac{3}{6}$ and $\frac{8}{12}$ **b.** $\frac{10}{16}$ and $\frac{15}{24}$ **c.** $\frac{6}{10}$ and $\frac{12}{15}$ **d.** $\frac{6}{8}$ and $\frac{12}{18}$ **e.** $\frac{2}{4}$ and $\frac{3}{9}$ **f.** $\frac{15}{25}$ and $\frac{6}{10}$ **g.** $\frac{10}{12}$ and $\frac{15}{18}$ **h.** $\frac{16}{24}$ and $\frac{26}{39}$ **i.** $\frac{21}{27}$ and $\frac{35}{45}$ **j.** $\frac{18}{42}$ and $\frac{14}{35}$
2. Equal cross products: **a.** $\frac{3}{4}$ and $\frac{7}{12}$ **b.** $\frac{10}{16}$ and $\frac{15}{24}$ **c.** $\frac{10}{15}$ and $\frac{12}{18}$ **d.** $\frac{42}{48}$ and $\frac{30}{35}$ **e.** $\frac{8}{20}$ and $\frac{9}{24}$ **f.** $\frac{13}{26}$ and $\frac{4}{8}$ **g.** $\frac{4}{6}$ and $\frac{11}{22}$ **h.** $\frac{14}{21}$ and $\frac{6}{10}$ **i.** $\frac{48}{54}$ and $\frac{14}{16}$ **j.** $\frac{35}{63}$ and $\frac{10}{18}$
3. Either method: **a.** $\frac{4}{5}$ and $\frac{9}{12}$ **b.** $\frac{6}{8}$ and $\frac{10}{15}$ **c.** $\frac{8}{12}$ and $\frac{6}{9}$ **d.** $\frac{8}{10}$ and $\frac{20}{25}$ **e.** $\frac{6}{12}$ and $\frac{8}{14}$ **f.** $\frac{9}{24}$ and $\frac{6}{16}$ **g.** $\frac{7}{16}$ and $\frac{3}{8}$ **h.** $\frac{20}{35}$ and $\frac{28}{49}$ **i.** $\frac{35}{60}$ and $\frac{27}{48}$ **j.** $\frac{15}{40}$ and $\frac{70}{80}$

5–5 PROPER FRACTIONS, IMPROPER FRACTIONS, MIXED NUMBERS

We sometimes use the same number of parts as there are in the whole unit such as: 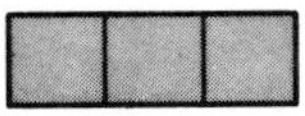 $\frac{3}{3} = 1$ 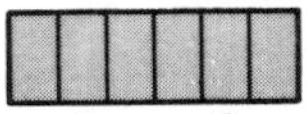 $\frac{6}{6} = 1$
or a number of parts greater than the number of parts in the whole unit such as:

 $\frac{7}{4} = 1\frac{3}{4}$  $\frac{6}{3} = 2$

A fraction whose numerator is smaller than its denominator such as $\frac{3}{4}$ is called a *proper fraction.* The value of a proper fraction is always less than one.

A fraction whose numerator is equal to or greater than its denominator such as $\frac{6}{6}$ or $\frac{7}{4}$ is called an *improper fraction.* The value of an improper fraction is one or more than one.

A *mixed number* consists of a whole number and a fraction such as $3\frac{1}{4}$.

EXERCISES

1. Read or write in words:

   $4\frac{3}{10}$; $1\frac{6}{7}$; $2\frac{5}{8}$; $10\frac{1}{2}$; $3\frac{2}{3}$; $5\frac{29}{32}$; $8\frac{9}{14}$; $12\frac{13}{25}$; $17\frac{11}{12}$; $9\frac{7}{16}$

2. Write as a numeral:

   Fourteen and three-fourths; Three and one-seventh; Forty and five-sixteenths.

3. Which name a whole number?

$\frac{3}{5}$; 4; $5\frac{1}{2}$; 9; $\frac{7}{8}$; $2\frac{3}{4}$; $6\frac{2}{3}$; $\frac{11}{16}$; 1; $\frac{5}{6}$

4. Which name a common fraction?

$\frac{9}{10}$; 6; $8\frac{1}{4}$; $2\frac{3}{8}$; $\frac{1}{3}$; 3; $4\frac{5}{16}$; $\frac{5}{8}$; 10; $\frac{2}{5}$

5. Which name a mixed number?

4; $2\frac{2}{3}$; $\frac{5}{7}$; $4\frac{1}{5}$; 6; $\frac{3}{4}$; 8; $1\frac{9}{16}$; 1; $12\frac{1}{2}$

6. Which name a proper fraction?

$\frac{2}{5}$; $\frac{7}{4}$; $\frac{8}{8}$; $\frac{1}{3}$; $\frac{7}{8}$; $\frac{3}{2}$; $\frac{6}{5}$; $\frac{9}{10}$; $\frac{5}{6}$; $\frac{21}{12}$

7. Which name an improper fraction?

$\frac{1}{2}$; $\frac{10}{10}$; $\frac{18}{6}$; $\frac{5}{8}$; $\frac{8}{5}$; $\frac{2}{3}$; $\frac{22}{7}$; $\frac{9}{4}$; $\frac{15}{20}$; $\frac{32}{32}$

8. Select the numerator of each fraction:

$\frac{2}{3}$; $\frac{11}{8}$; $\frac{1}{2}$; $\frac{13}{16}$; $\frac{5}{6}$; $\frac{7}{12}$; $\frac{19}{20}$; $\frac{37}{100}$; $\frac{5}{10}$; $\frac{14}{5}$

9. Select the denominator of each fraction:

$\frac{3}{4}$; $\frac{7}{10}$; $\frac{1}{8}$; $\frac{22}{7}$; $\frac{11}{16}$; $\frac{8}{3}$; $\frac{25}{32}$; $\frac{5}{6}$; $\frac{49}{100}$; $\frac{43}{64}$

10. a. What measurement is indicated on the ruler by:

A? *B*? *C*? *D*? *E*? *F*? *G*? *H*?

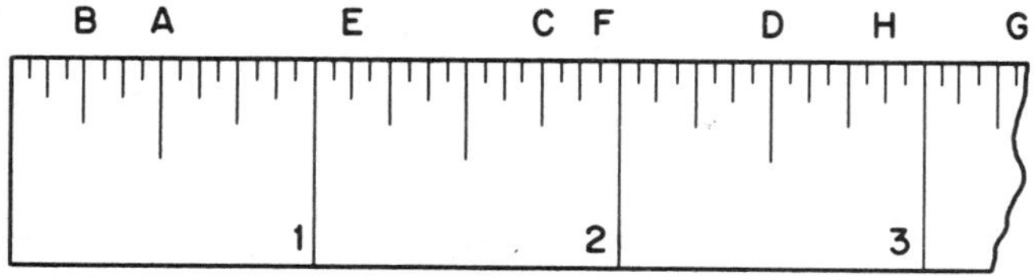

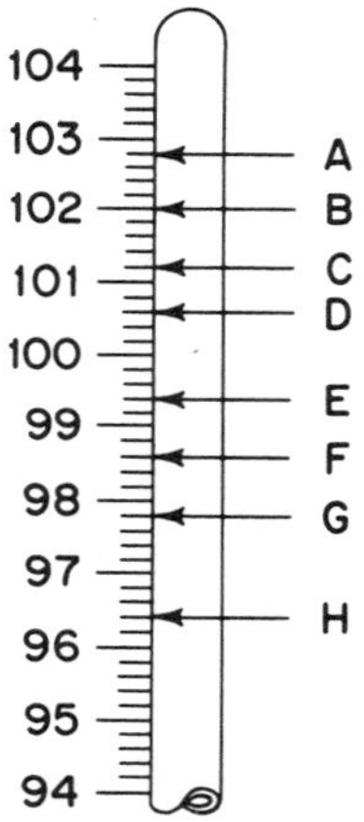

b. What reading on the thermometer is indicated by:

A? *B*? *C*? *D*? *E*? *F*? *G*? *H*?

To express an improper fraction as a whole or mixed number, we divide the numerator by the denominator and write the remainder, if any, as a fraction expressed in lowest terms.

Change $\frac{5}{3}$ to a mixed number:

$\frac{5}{3} = 5 \div 3 = 1\frac{2}{3}$

Answer, $1\frac{2}{3}$

EXERCISES

1. Use a ruler to determine the equivalent whole number or mixed number:

a. $\frac{2}{2}''$ $\frac{4}{4}''$ $\frac{16}{16}''$ $\frac{8}{8}''$ $\frac{4}{2}''$ $\frac{8}{2}''$ $\frac{32}{16}''$ $\frac{24}{8}''$

b. $\frac{5}{4}''$ $\frac{5}{2}''$ $\frac{13}{8}''$ $\frac{25}{16}''$ $\frac{7}{2}''$ $\frac{23}{8}''$ $\frac{13}{4}''$ $\frac{37}{16}''$

c. $\frac{6}{4}''$ $\frac{10}{8}''$ $\frac{28}{16}''$ $\frac{20}{8}''$ $\frac{14}{4}''$ $\frac{42}{16}''$ $\frac{14}{8}''$ $\frac{36}{16}''$

2. Express each of the following as a numeral for a whole number or a mixed number:

a. $\frac{6}{6}$ $\frac{10}{10}$ $\frac{5}{5}$ $\frac{12}{12}$ $\frac{12}{6}$ $\frac{10}{2}$ $\frac{48}{8}$ $\frac{32}{4}$ $\frac{56}{7}$ $\frac{75}{15}$

b. $\frac{4}{3}$ $\frac{16}{9}$ $\frac{11}{6}$ $\frac{23}{16}$ $\frac{9}{2}$ $\frac{18}{5}$ $\frac{35}{8}$ $\frac{27}{10}$ $\frac{22}{7}$ $\frac{35}{18}$

c. $\frac{10}{6}$ $\frac{21}{12}$ $\frac{12}{9}$ $\frac{40}{32}$ $\frac{26}{4}$ $\frac{35}{10}$ $\frac{42}{30}$ $\frac{44}{16}$ $\frac{15}{9}$ $\frac{65}{25}$

To express a whole number as a fraction with a specified denominator, we multiply the whole number by the fraction (equivalent to multiplicative identity one) having the same number in the numerator and denominator as the specified denominator. Also see page 105.

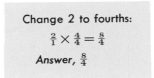

Change 2 to fourths:

$\frac{2}{1} \times \frac{4}{4} = \frac{8}{4}$

Answer, $\frac{8}{4}$

EXERCISES

1. Express 1 as: **a.** twelfths **b.** sixteenths **c.** hundredths **d.** twenty-fourths

2. How many sixths are in 8?

3. How many tenths are in 3?

4. How many: **a.** half inches are in 4 inches? **b.** eighth inches are in 6 inches? **c.** quarter inches are in 12 inches? **d.** sixteenth inches are in 9 inches?

5. Find the missing numbers:

a. $1 = \frac{?}{20}$ **b.** $4 = \frac{?}{8}$ **c.** $2 = \frac{?}{12}$ **d.** $7 = \frac{?}{16}$ **e.** $9 = \frac{?}{5}$

To express a mixed number as an improper fraction, we multiply the whole number by the denominator of the fraction and add the numerator of the fraction to this product. Then we write this sum over the denominator of the fraction.

Express $2\frac{3}{4}$ as an improper fraction:

Briefly: $2\frac{3}{4} = \frac{11}{4}$

or $2\frac{3}{4} = 2 + \frac{3}{4} = \frac{8}{4} + \frac{3}{4} = \frac{11}{4}$

Answer, $\frac{11}{4}$

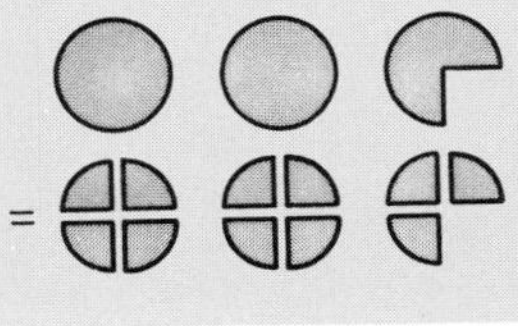

EXERCISES

1. Use a ruler to determine how many:

a. Halves of an inch are in: $2\frac{1}{2}''$; $1\frac{1}{2}''$; $3\frac{1}{2}''$; $5\frac{1}{2}''$
b. Quarters of an inch are in: $\frac{3}{4}''$; $3\frac{1}{4}''$; $2\frac{3}{4}''$; $6\frac{1}{2}''$
c. Eighths of an inch are in: $3\frac{1}{8}''$; $1\frac{7}{8}''$; $2\frac{5}{8}''$; $4\frac{3}{4}''$; $7\frac{1}{2}''$
d. Sixteenths of an inch are in: $\frac{11}{16}''$; $2\frac{13}{16}''$; $3\frac{3}{8}''$; $4\frac{1}{2}''$; $5\frac{3}{4}''$

2. Express each of the following as an improper fraction:

a. $1\frac{1}{5}$	$1\frac{1}{3}$	$1\frac{1}{6}$	$1\frac{1}{7}$	$1\frac{1}{12}$	$1\frac{1}{10}$	$1\frac{1}{8}$	$1\frac{1}{32}$	$1\frac{1}{15}$	$1\frac{1}{18}$
b. $1\frac{5}{8}$	$1\frac{7}{10}$	$1\frac{11}{12}$	$1\frac{5}{6}$	$1\frac{13}{23}$	$1\frac{11}{16}$	$1\frac{7}{9}$	$1\frac{19}{24}$	$1\frac{13}{18}$	$1\frac{23}{50}$
c. $3\frac{1}{7}$	$4\frac{1}{2}$	$3\frac{1}{3}$	$2\frac{1}{10}$	$9\frac{1}{4}$	$6\frac{1}{8}$	$3\frac{1}{5}$	$7\frac{1}{6}$	$10\frac{1}{9}$	$12\frac{1}{3}$
d. $4\frac{5}{6}$	$5\frac{9}{10}$	$7\frac{3}{5}$	$8\frac{15}{16}$	$4\frac{7}{9}$	$12\frac{3}{4}$	$4\frac{5}{8}$	$6\frac{15}{32}$	$15\frac{11}{12}$	$20\frac{5}{7}$

5–6 EXPRESSING FRACTIONS AS EQUIVALENT FRACTIONS WITH COMMON DENOMINATOR

A *common denominator* is a number that can be divided exactly by the denominators of all the given fractions. It is a multiple of the denominators of the given fractions.

The *lowest common denominator* (L.C.D.) is the smallest natural number that can be divided exactly by the denominators of all the given fractions. It is the least common multiple of the given denominators.

The L.C.D. may be found by inspection:

(1) When the L.C.D. is a denominator of one of the given fractions. L.C.D. of $\frac{1}{2}$ and $\frac{1}{8}$ is 8.

(2) When the L.C.D. is the product of the denominators of the given fractions. L.C.D. of $\frac{1}{3}$ and $\frac{1}{5}$ is 15.

When the L.C.D. is greater than any given denominator but smaller than the product of all the denominators, we use the method used in finding the least common multiple (see page 15).

Find the L.C.D of $\frac{7}{12}$, $\frac{5}{6}$ and $\frac{2}{8}$:

$12 = 2 \cdot 2 \cdot 3$ L.C.D. $= 2 \cdot 2 \cdot 3 \cdot 2$
$6 = 2 \cdot 3$ L.C.D. $= 24$
$8 = 2 \cdot 2 \cdot 2$

Answer, 24

To express each given fraction as an equivalent fraction having the lowest common denominator as its denominator, we first find the L.C.D., then we express each fraction as an *equivalent* fraction with the L.C.D. as its denominator.

EXERCISES

First find the lowest common denominator (L.C.D.) of the given fractions in each of the following, then change each fraction to an equivalent fraction with the L.C.D. as its denominator:

1. a. $\frac{1}{4}$ and $\frac{1}{16}$ **c.** $\frac{3}{8}$ and $\frac{3}{4}$ **e.** $\frac{5}{6}$ and $\frac{5}{12}$ **g.** $\frac{13}{16}$ and $\frac{1}{2}$ **i.** $\frac{5}{6}$ and $\frac{13}{24}$
b. $\frac{1}{2}$ and $\frac{5}{6}$ **d.** $\frac{7}{12}$ and $\frac{2}{3}$ **f.** $\frac{7}{8}$ and $\frac{11}{24}$ **h.** $\frac{9}{20}$ and $\frac{3}{5}$ **j.** $\frac{18}{25}$ and $\frac{37}{100}$

2. a. $\frac{1}{3}$ and $\frac{1}{4}$ **c.** $\frac{1}{2}$ and $\frac{4}{5}$ **e.** $\frac{3}{5}$ and $\frac{5}{6}$ **g.** $\frac{7}{9}$ and $\frac{1}{2}$ **i.** $\frac{7}{16}$ and $\frac{2}{3}$
b. $\frac{1}{5}$ and $\frac{1}{8}$ **d.** $\frac{3}{8}$ and $\frac{2}{3}$ **f.** $\frac{1}{3}$ and $\frac{4}{5}$ **h.** $\frac{2}{5}$ and $\frac{11}{12}$ **j.** $\frac{5}{6}$ and $\frac{4}{7}$

3. a. $\frac{1}{8}$ and $\frac{1}{6}$ **c.** $\frac{5}{8}$ and $\frac{9}{10}$ **e.** $\frac{5}{6}$ and $\frac{3}{10}$ **g.** $\frac{15}{16}$ and $\frac{11}{12}$ **i.** $\frac{11}{24}$ and $\frac{13}{18}$
b. $\frac{7}{10}$ and $\frac{3}{4}$ **d.** $\frac{3}{8}$ and $\frac{7}{12}$ **f.** $\frac{5}{12}$ and $\frac{9}{10}$ **h.** $\frac{5}{6}$ and $\frac{7}{16}$ **j.** $\frac{19}{20}$ and $\frac{7}{16}$

4. a. $\frac{1}{2}$, $\frac{1}{4}$ and $\frac{1}{8}$ **e.** $\frac{1}{3}$, $\frac{1}{2}$ and $\frac{1}{5}$ **i.** $\frac{1}{4}$, $\frac{1}{8}$ and $\frac{1}{6}$ **m.** $\frac{1}{10}$, $\frac{1}{8}$ and $\frac{1}{12}$
b. $\frac{1}{3}$, $\frac{1}{6}$ and $\frac{1}{2}$ **f.** $\frac{3}{4}$, $\frac{2}{3}$ and $\frac{2}{5}$ **j.** $\frac{11}{12}$, $\frac{5}{6}$ and $\frac{3}{8}$ **n.** $\frac{2}{3}$, $\frac{5}{8}$ and $\frac{1}{6}$
c. $\frac{3}{4}$, $\frac{5}{6}$ and $\frac{7}{12}$ **g.** $\frac{1}{2}$, $\frac{2}{3}$ and $\frac{1}{4}$ **k.** $\frac{9}{16}$, $\frac{7}{8}$ and $\frac{19}{24}$ **o.** $\frac{7}{10}$, $\frac{15}{16}$ and $\frac{11}{12}$
d. $\frac{11}{16}$, $\frac{5}{8}$ and $\frac{1}{4}$ **h.** $\frac{2}{3}$, $\frac{5}{6}$ and $\frac{3}{5}$ **l.** $\frac{4}{5}$, $\frac{1}{4}$ and $\frac{9}{10}$ **p.** $\frac{3}{8}$, $\frac{3}{10}$ and $\frac{7}{16}$

5–7 NUMBER LINE

There is a point on the number line corresponding to each fraction and mixed number.

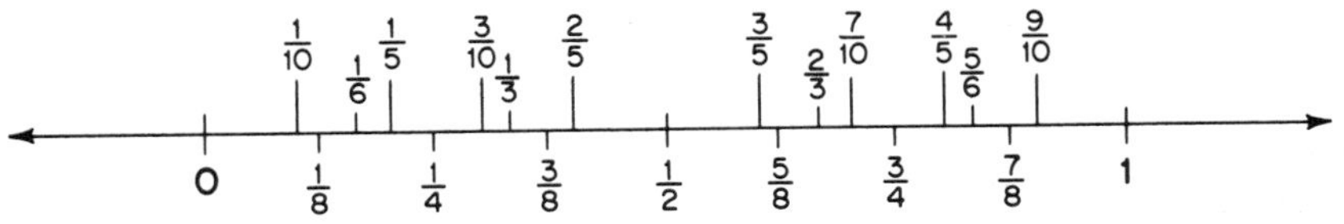

The interval between each whole number on the number line may be divided into halves, thirds, fourths, fifths, etc. so that there is a point on the number line corresponding to each fraction and mixed number. The number of points corresponding to the fractions are *unlimited* or *infinite* because there is always another fraction between any two fractions we select. All we have to do is find the fraction midway between the two given fractions by adding the fractions and dividing the sum by two. We usually say that the *order is dense* if between any two numbers there is an unlimited number of numbers. Thus the order of fractions is said to be dense.

Fractions may be compared by the number line. The fraction corresponding to the point on the line farther to the right is the larger fraction.

EXERCISES

1. a. Copy this line segment. Label each point of division with its corresponding fraction in simplest form.

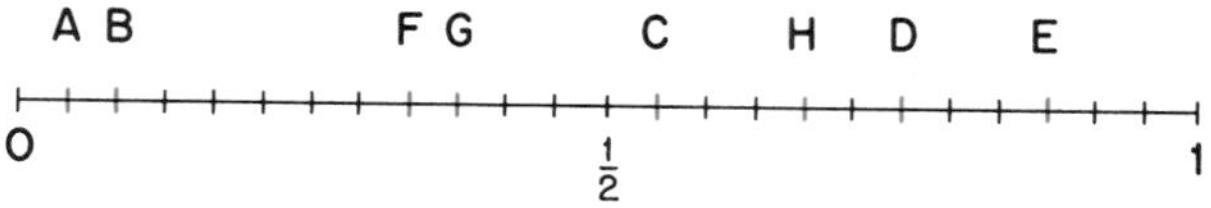

b. What fraction corresponds to point *A*? *B*? *C*? *D*? *E*? *F*? *G*? *H*?

c. Find the midpoint between *A* and *B*. Label it *M*. What fraction is associated with point *M*?

d. Find the midpoint between *A* and *M*. Label it *N*. What fraction is associated with point *N*? What fraction is associated with the midpoint between *N* and *M*?

e. As the parts of the line get smaller and smaller, is there a midpoint for each part? Will this continue endlessly? Do all these points have fractions which correspond? Do you see that there is always another fraction between any two fractions?

f. Under each fraction you used to label the given points on the number line write names of two other equivalent fractions.

5–8 COMPARING FRACTIONS

If two fractions have the same (like) denominator, the fraction with the greater numerator is obviously the greater fraction. $\frac{4}{5}$ is greater than $\frac{2}{5}$.

However, to compare two fractions with different (unlike) denominators, we first express the given fractions as fractions with a common denominator and then take the given fraction that is equivalent to the fraction having the greater numerator and *common denominator* as the greater fraction.

Which is greater $\frac{3}{4}$ or $\frac{2}{3}$?

Since $\frac{3}{4} = \frac{9}{12}$ and $\frac{2}{3} = \frac{8}{12}$, $\frac{3}{4}$ is greater than $\frac{2}{3}$ or $\frac{3}{4} > \frac{2}{3}$.

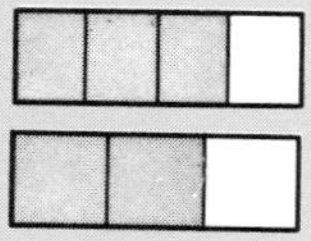

Examination of the sentence $\frac{3}{4} > \frac{2}{3}$ will show that the product of the numerator of the first fraction and the denominator of the second fraction (3×3) is greater than the product of the numerator of the second fraction and the denominator of the first fraction (2×4). If this condition is true when we compare two fractions, we may say that the first fraction is greater than the second fraction.

Examination of the sentence $\frac{2}{3} < \frac{3}{4}$ meaning "$\frac{2}{3}$ is less than $\frac{3}{4}$" will show that the product of the numerator of the first fraction and the denominator of the second fraction (2×4) is less than the product of the numerator of the second fraction and the denominator of the first fraction (3×3). If this condition is true when we compare two fractions, we may say that the first fraction is less than the second fraction.

Whether $\frac{3}{4} > \frac{2}{3}$ is true depends on whether $3 \times 3 > 2 \times 4$ is true. Since $3 \times 3 = 9$ and $2 \times 4 = 8$ and $9 > 8$, then $3 \times 3 > 2 \times 4$ and $\frac{3}{4} > \frac{2}{3}$.	Whether $\frac{2}{3} < \frac{3}{4}$ is true depends on whether $2 \times 4 < 3 \times 3$ is true. Since $2 \times 4 = 8$ and $3 \times 3 = 9$ and $8 < 9$, then $2 \times 4 < 3 \times 3$ and $\frac{2}{3} < \frac{3}{4}$.

Summarizing:

If a represents the numerator of the first fraction; b, the denominator of the first fraction; c, the numerator of the second fraction; and d, the denominator of the second fraction, then:

(1) $\frac{a}{b} = \frac{c}{d}$ if $a \times d = c \times b$ Two fractions are equivalent.

(2) $\frac{a}{b} > \frac{c}{d}$ if $a \times d > c \times b$ First fraction is greater.

(3) $\frac{a}{b} < \frac{c}{d}$ if $a \times d < c \times b$ First fraction is smaller.

EXERCISES

1. Use the common denominator method to determine which of the following statements are true:

a. $\frac{1}{5} > \frac{1}{4}$ **b.** $\frac{1}{8} > \frac{1}{10}$ **c.** $\frac{4}{5} > \frac{9}{10}$ **d.** $\frac{11}{16} > \frac{1}{2}$
e. $\frac{2}{3} > \frac{7}{12}$ **f.** $\frac{2}{5} > \frac{3}{8}$ **g.** $\frac{5}{6} < \frac{7}{8}$ **h.** $\frac{3}{4} < \frac{5}{6}$
i. $\frac{7}{10} < \frac{2}{3}$ **j.** $\frac{11}{12} < \frac{9}{10}$ **k.** $\frac{2}{3} < \frac{5}{8}$ **l.** $\frac{1}{2} < \frac{3}{5}$
m. $\frac{13}{16} > \frac{19}{24}$ **n.** $\frac{5}{9} < \frac{6}{7}$ **o.** $\frac{19}{25} < \frac{73}{100}$ **p.** $\frac{30}{45} > \frac{20}{30}$
q. $\frac{11}{20} < \frac{9}{16}$ **r.** $\frac{17}{18} > \frac{19}{20}$ **s.** $\frac{9}{25} < \frac{3}{8}$ **t.** $\frac{7}{12} > \frac{9}{16}$

2. Use the cross product test to determine which of the following statements are true:

a. $\frac{1}{5} > \frac{1}{4}$ **b.** $\frac{1}{9} > \frac{1}{7}$ **c.** $\frac{5}{8} > \frac{7}{12}$ **d.** $\frac{1}{2} > \frac{5}{8}$
e. $\frac{11}{16} > \frac{2}{3}$ **f.** $\frac{7}{9} > \frac{17}{24}$ **g.** $\frac{3}{4} < \frac{5}{8}$ **h.** $\frac{8}{15} < \frac{9}{16}$
i. $\frac{4}{5} < \frac{7}{8}$ **j.** $\frac{11}{12} < \frac{5}{6}$ **k.** $\frac{3}{4} < \frac{4}{5}$ **l.** $\frac{13}{16} < \frac{21}{25}$
m. $\frac{19}{25} > \frac{3}{4}$ **n.** $\frac{5}{12} > \frac{7}{18}$ **o.** $\frac{11}{15} < \frac{13}{20}$ **p.** $\frac{3}{10} < \frac{7}{16}$
q. $\frac{5}{11} > \frac{2}{5}$ **r.** $\frac{36}{48} < \frac{15}{20}$ **s.** $\frac{9}{25} > \frac{26}{75}$ **t.** $\frac{5}{16} < \frac{19}{64}$

3. Arrange each of the following sets of fractions according to size:
Greatest first:

a. $\frac{5}{12}$, $\frac{11}{12}$, and $\frac{7}{12}$ **b.** $\frac{5}{16}$, $\frac{5}{8}$, and $\frac{5}{12}$ **c.** $\frac{9}{10}$, $\frac{5}{6}$, and $\frac{4}{5}$ **d.** $\frac{3}{5}$, $\frac{5}{8}$, and $\frac{2}{3}$ **e.** $\frac{7}{16}$, $\frac{1}{4}$, and $\frac{3}{8}$

4. Smallest first:

a. $\frac{4}{5}$, $\frac{2}{5}$, and $\frac{3}{5}$ **b.** $\frac{3}{4}$, $\frac{3}{16}$, and $\frac{3}{8}$ **c.** $\frac{5}{12}$, $\frac{3}{10}$, and $\frac{7}{16}$ **d.** $\frac{2}{3}$, $\frac{1}{2}$, and $\frac{3}{4}$ **e.** $\frac{7}{8}$, $\frac{3}{4}$, and $\frac{5}{6}$

5–9 COMPARING PARTS OF UNITS OF DIFFERENT SIZES

(1) Which is larger:
$\frac{1}{2}$ of figure *A* or $\frac{1}{2}$ of figure *B*?

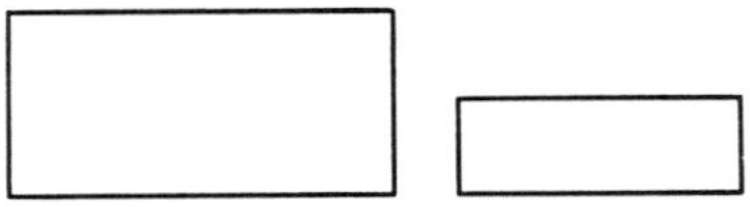

A B

(2) Which is larger:
$\frac{3}{4}$ of figure *C* or $\frac{3}{4}$ of figure *D*?

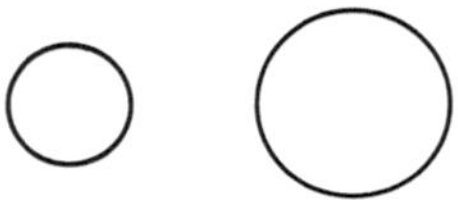

C D

(3) Which is larger:

(a) $\frac{1}{2}$ minute or $\frac{1}{2}$ hour? (c) $\frac{3}{4}$ foot or $\frac{3}{4}$ inch? (e) $\frac{1}{8}$ inch or $\frac{1}{8}$ yard?
(b) $\frac{1}{4}$ ton or $\frac{1}{4}$ pound? (d) $\frac{2}{3}$ gallon or $\frac{2}{3}$ quart? (f) $\frac{5}{8}$ inch or $\frac{5}{8}$ foot?

(4) Is $\frac{1}{6}$ of 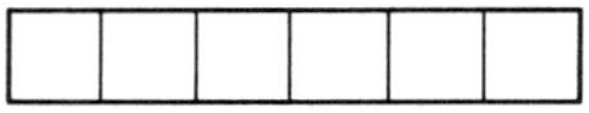 larger, the same as, or smaller than $\frac{1}{3}$ of 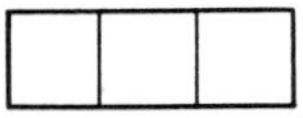 ?

(5) Is $\frac{2}{5}$ of  larger, the same as, or smaller than $\frac{3}{4}$ of 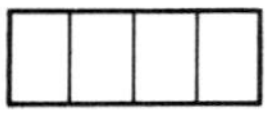 ?

(6) Is $\frac{2}{3}$ of 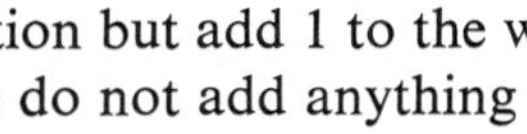 larger, the same as, or smaller than $\frac{1}{2}$ of ?

5–10 ROUNDING MIXED NUMBERS

To round a mixed number to the nearest whole number like $3\frac{5}{8}$ or $7\frac{2}{5}$, we drop the fraction but add 1 to the whole number when the fraction is one-half or more. We do not add anything when the fraction is less than one-half.

> $3\frac{5}{8}$ rounded to nearest whole number is 4.
> $7\frac{2}{5}$ rounded to nearest whole number is 7.

EXERCISES

Round each of the following mixed numbers to the nearest whole number:

1. a. $2\frac{3}{4}$	**b.** $8\frac{2}{3}$	**c.** $7\frac{5}{6}$	**d.** $1\frac{7}{12}$	**e.** $3\frac{9}{16}$	**f.** $4\frac{8}{15}$	**g.** $12\frac{7}{10}$	**h.** $11\frac{1}{2}$
2. a. $5\frac{1}{3}$	**b.** $4\frac{3}{8}$	**c.** $9\frac{2}{5}$	**d.** $6\frac{7}{16}$	**e.** $1\frac{3}{10}$	**f.** $8\frac{3}{7}$	**g.** $14\frac{4}{9}$	**h.** $32\frac{12}{25}$
3. a. $9\frac{1}{4}$	**b.** $7\frac{1}{2}$	**c.** $2\frac{11}{16}$	**d.** $14\frac{3}{5}$	**e.** $25\frac{15}{32}$	**f.** $9\frac{5}{11}$	**g.** $16\frac{13}{24}$	**h.** $10\frac{20}{39}$

5–11 COMPUTATION—ADDITION OF FRACTIONS AND MIXED NUMBERS

(1) To add fractions with *like denominators*, we add the numerators and write the sum over the common denominator. We express the answer in simplest form.

(1) Vertically

$$\begin{array}{r} \frac{5}{16} \\ +\frac{7}{16} \\ \hline \frac{12}{16} = \frac{3}{4} \end{array}$$

Answer, $\frac{3}{4}$

or

Horizontally:

$$\frac{5}{16} + \frac{7}{16} = \frac{5+7}{16} = \frac{12}{16} = \frac{3}{4}$$

Observe that the distributive principle is used here. $\frac{5}{16} = 5 \times \frac{1}{16}$ and $\frac{7}{16} = 7 \times \frac{1}{16}$.
Therefore $\frac{5}{16} + \frac{7}{16} = 5 \times \frac{1}{16} + 7 \times \frac{1}{16} = (5+7)\frac{1}{16} = 12 \times \frac{1}{16} = \frac{12}{16}$.

(2) To add fractions with *unlike denominators*, we express the given fractions as equivalent fractions having a common denominator and add as we do with like denominators.

(2) Add:

$$\begin{array}{r} \frac{1}{2} = \frac{3}{6} \\ \underline{\frac{2}{3}} = \underline{\frac{4}{6}} \\ \frac{7}{6} = 1\frac{1}{6} \end{array}$$

Answer, $1\frac{1}{6}$

(3) To add a fraction and a whole number, we annex the fraction to the whole number.

(3) $4 + \frac{2}{3} = 4\frac{2}{3}$ Answer, $4\frac{2}{3}$

(4) To add a mixed number and a whole number, we add the whole numbers and annex the fraction to this sum.

(4) $2\frac{1}{2} + 3 = 5\frac{1}{2}$ Answer, $5\frac{1}{2}$

(5) To add a mixed number and a fraction, we add the fractions and annex their sum to the whole number.

(5) Add:

$$\begin{array}{r} 2\frac{2}{5} = 2\frac{4}{10} \\ \underline{\frac{1}{2}} = \underline{\frac{5}{10}} \\ 2\frac{9}{10} \end{array}$$

Answer, $2\frac{9}{10}$

(6) To add mixed numbers, we first add the fractions and then add this sum to the sum of the whole numbers. We express answers in simplest form.

> **(6) Add:**
>
> $4\frac{3}{4} = 4\frac{6}{8}$
>
> $2\frac{5}{8} = 2\frac{5}{8}$
>
> $\phantom{2\frac{5}{8} =} 6\frac{11}{8} = 7\frac{3}{8}$
>
> *Answer,* $7\frac{3}{8}$

EXERCISES

1. Use a ruler to determine the sum of each of the following:

a.	$\frac{3}{4}''$	$\frac{9}{16}''$	$\frac{1}{8}''$	$\frac{6}{16}''$	$\frac{1}{4}''$	$\frac{1}{8}''$	$\frac{5}{16}''$	$\frac{3}{4}''$	$\frac{15}{16}''$	$\frac{7}{8}''$	$\frac{7}{16}''$	$\frac{9}{16}''$
	$\frac{1}{4}''$	$\frac{7}{16}''$	$\frac{2}{8}''$	$\frac{7}{16}''$	$\frac{1}{4}''$	$\frac{5}{8}''$	$\frac{7}{16}''$	$\frac{3}{4}''$	$\frac{9}{16}''$	$\frac{7}{8}''$	$\frac{13}{16}''$	$\frac{5}{16}''$
b.	$\frac{1}{2}''$	$\frac{1}{4}''$	$\frac{5}{8}''$	$\frac{1}{2}''$	$\frac{3}{4}''$	$\frac{1}{2}''$	$\frac{7}{8}''$	$\frac{3}{8}''$	$\frac{5}{16}''$	$\frac{1}{2}''$	$\frac{5}{8}''$	$\frac{9}{16}''$
	$\frac{1}{8}''$	$\frac{1}{2}''$	$\frac{1}{4}''$	$\frac{7}{16}''$	$\frac{3}{16}''$	$\frac{3}{4}''$	$\frac{1}{4}''$	$\frac{11}{16}''$	$\frac{3}{4}''$	$\frac{13}{16}''$	$\frac{3}{4}''$	$\frac{3}{8}''$

2. Add on a number line or by use of the fraction chart (see page 91) to determine which of the following statements are true:

a. $\frac{1}{10} + \frac{3}{10} = \frac{2}{5}$ **b.** $\frac{5}{12} + \frac{7}{12} = 1$ **c.** $\frac{1}{3} + \frac{1}{6} = \frac{1}{2}$ **d.** $\frac{3}{10} + \frac{1}{2} = \frac{4}{5}$ **e.** $\frac{1}{2} + \frac{1}{6} = \frac{2}{3}$

Add:

3.	$\frac{1}{5}$	$\frac{4}{7}$	$\frac{2}{3}$	$\frac{15}{32}$	$\frac{9}{20}$	$\frac{8}{16}$	$\frac{1}{6}$	$\frac{3}{10}$	$\frac{5}{12}$
	$\frac{2}{5}$	$\frac{2}{7}$	$\frac{1}{3}$	$\frac{17}{32}$	$\frac{11}{20}$	$\frac{5}{16}$	$\frac{1}{6}$	$\frac{2}{10}$	$\frac{3}{12}$
4.	$\frac{4}{5}$	$\frac{2}{3}$	$\frac{7}{9}$	$\frac{6}{12}$	$\frac{3}{4}$	$\frac{12}{16}$	$\frac{9}{16}$	$\frac{3}{4}$	$\frac{9}{10}$
	$\frac{3}{5}$	$\frac{2}{3}$	$\frac{7}{9}$	$\frac{11}{12}$	$\frac{2}{4}$	$\frac{13}{16}$	$\frac{11}{16}$	$\frac{3}{4}$	$\frac{7}{10}$
5.	$\frac{1}{8}$	$\frac{2}{5}$	$\frac{11}{16}$	$\frac{1}{6}$	$\frac{5}{12}$	$\frac{1}{2}$	$\frac{27}{60}$	$\frac{3}{20}$	$\frac{19}{32}$
	$\frac{1}{2}$	$\frac{3}{10}$	$\frac{1}{4}$	$\frac{2}{3}$	$\frac{5}{24}$	$\frac{1}{6}$	$\frac{7}{15}$	$\frac{41}{100}$	$\frac{5}{8}$
6.	$\frac{1}{2}$	$\frac{3}{4}$	$\frac{1}{4}$	$\frac{3}{5}$	$\frac{3}{8}$	$\frac{2}{3}$	$\frac{4}{5}$	$\frac{2}{3}$	$\frac{7}{8}$
	$\frac{1}{3}$	$\frac{1}{5}$	$\frac{2}{3}$	$\frac{1}{6}$	$\frac{2}{5}$	$\frac{1}{10}$	$\frac{1}{2}$	$\frac{3}{4}$	$\frac{1}{3}$
7.	$\frac{1}{4}$	$\frac{1}{10}$	$\frac{5}{6}$	$\frac{3}{16}$	$\frac{3}{8}$	$\frac{7}{12}$	$\frac{5}{8}$	$\frac{3}{4}$	$\frac{3}{10}$
	$\frac{1}{6}$	$\frac{3}{4}$	$\frac{1}{8}$	$\frac{7}{10}$	$\frac{5}{12}$	$\frac{5}{16}$	$\frac{7}{10}$	$\frac{5}{6}$	$\frac{11}{16}$
8.	$\frac{1}{4}$	$\frac{5}{6}$	$\frac{12}{16}$	$\frac{4}{5}$	$\frac{3}{10}$	$\frac{11}{24}$	$\frac{3}{8}$	$\frac{5}{12}$	$\frac{1}{2}$
	$\frac{1}{4}$	$\frac{1}{6}$	$\frac{9}{16}$	$\frac{2}{5}$	$\frac{9}{10}$	$\frac{12}{24}$	$\frac{1}{2}$	$\frac{2}{3}$	$\frac{3}{4}$
	$\frac{1}{4}$	$\frac{5}{6}$	$\frac{13}{16}$	$\frac{4}{5}$	$\frac{7}{10}$	$\frac{17}{24}$	$\frac{3}{16}$	$\frac{1}{6}$	$\frac{2}{3}$
9.	$\frac{13}{24}$	$\frac{11}{32}$	$\frac{15}{64}$	$\frac{7}{8}$	$\frac{9}{20}$	$\frac{14}{15}$	$\frac{3}{4}$	$\frac{5}{6}$	$\frac{4}{5}$
	$\frac{7}{24}$	$\frac{13}{32}$	$\frac{25}{64}$	$\frac{5}{8}$	$\frac{19}{20}$	$\frac{11}{15}$	$\frac{11}{16}$	$\frac{1}{2}$	$\frac{7}{10}$
10.	$\frac{5}{8}$	$\frac{1}{3}$	$\frac{4}{7}$	$\frac{13}{15}$	$\frac{15}{16}$	$\frac{17}{24}$	$\frac{5}{6}$	$\frac{3}{10}$	$\frac{7}{8}$
	$\frac{4}{5}$	$\frac{9}{10}$	$\frac{8}{9}$	$\frac{9}{10}$	$\frac{11}{12}$	$\frac{9}{16}$	$\frac{7}{8}$	$\frac{2}{5}$	$\frac{7}{12}$
							$\frac{3}{4}$	$\frac{19}{100}$	$\frac{13}{16}$

11. $\frac{1}{8} + \frac{5}{8}$ $\quad \frac{7}{16} + \frac{9}{16}$ $\quad \frac{1}{3} + \frac{1}{5}$ $\quad \frac{3}{8} + \frac{5}{6}$ $\quad \frac{11}{16} + \frac{7}{8}$ $\quad \frac{3}{4} + \frac{2}{3} + \frac{5}{6}$ $\quad \frac{7}{8} + \frac{9}{16} + \frac{1}{2}$

12. 4	$\frac{5}{8}$	$5\frac{3}{4}$	9	$4\frac{2}{5}$	$\frac{2}{3}$	$\frac{5}{12}$	$6\frac{3}{4}$
$\frac{2}{5}$	3	2	$5\frac{13}{16}$	$\frac{1}{5}$	$1\frac{1}{2}$	$5\frac{1}{4}$	$\frac{1}{4}$

13. $5 + \frac{3}{4}$ $\quad 3\frac{2}{3} + 4$ $\quad \frac{7}{8} + 10$ $\quad 9 + 2\frac{13}{16}$ $\quad \frac{7}{12} + 3\frac{5}{12}$ $\quad \frac{11}{16} + 3 + 6\frac{1}{2}$

14. $2\frac{1}{3}$	$6\frac{4}{9}$	$5\frac{3}{8}$	$3\frac{1}{10}$	$7\frac{1}{4}$	$17\frac{5}{12}$	$2\frac{4}{5}$	$7\frac{2}{3}$
$5\frac{1}{3}$	$2\frac{1}{9}$	$1\frac{3}{8}$	$8\frac{3}{10}$	$3\frac{3}{4}$	$10\frac{7}{12}$	$3\frac{2}{5}$	$1\frac{2}{3}$
15. $4\frac{2}{5}$	$3\frac{5}{8}$	$5\frac{1}{4}$	$2\frac{1}{2}$	$1\frac{3}{4}$	$2\frac{7}{12}$	$2\frac{1}{6}$	$12\frac{1}{4}$
$9\frac{3}{10}$	$1\frac{13}{16}$	$8\frac{1}{3}$	$7\frac{2}{5}$	$4\frac{1}{10}$	$8\frac{3}{8}$	$5\frac{1}{3}$	$13\frac{7}{12}$
16. $5\frac{12}{16}$	$4\frac{3}{8}$	$2\frac{7}{10}$	$12\frac{3}{4}$	$7\frac{2}{3}$	$3\frac{5}{6}$	$8\frac{1}{2}$	$12\frac{9}{10}$
$4\frac{9}{16}$	$1\frac{7}{8}$	$5\frac{9}{10}$	$11\frac{3}{4}$	$8\frac{4}{5}$	$5\frac{3}{8}$	$3\frac{5}{6}$	$9\frac{1}{2}$

17. $3\frac{1}{8}$	$4\frac{2}{5}$	$1\frac{9}{10}$	$1\frac{1}{2}$	$2\frac{1}{10}$	$3\frac{2}{3}$
$2\frac{3}{8}$	$3\frac{1}{5}$	$6\frac{3}{10}$	$3\frac{3}{16}$	$7\frac{1}{2}$	$1\frac{3}{4}$
$1\frac{1}{8}$	$7\frac{2}{5}$	$2\frac{7}{10}$	$5\frac{1}{4}$	$4\frac{1}{5}$	$4\frac{5}{6}$
18. $6\frac{7}{8}$	$2\frac{2}{3}$	$8\frac{3}{4}$	$7\frac{13}{16}$	$\frac{5}{8}$	$3\frac{3}{4}$
$3\frac{1}{6}$	$6\frac{1}{2}$	$4\frac{7}{8}$	$2\frac{1}{4}$	$9\frac{2}{3}$	$\frac{1}{2}$
$8\frac{7}{10}$	$7\frac{4}{5}$	$5\frac{5}{6}$	$1\frac{5}{8}$	$\frac{11}{12}$	$5\frac{2}{3}$

19. $3\frac{19}{32} + 1\frac{11}{16} + 5\frac{3}{4}$ $\quad 2\frac{1}{6} + 4\frac{7}{8} + 3\frac{5}{16}$ $\quad 6\frac{3}{4} + 5\frac{5}{8} + 1\frac{1}{6}$ $\quad 7\frac{11}{12} + 2\frac{15}{16} + 3\frac{13}{24}$

20. Which of the following statements are true?

a. $\frac{5}{8} + \frac{2}{8} = \frac{2}{8} + \frac{5}{8}$ **b.** $(\frac{1}{16} + \frac{3}{16}) + \frac{5}{16} = \frac{1}{16} + (\frac{3}{16} + \frac{5}{16})$ **c.** $1\frac{2}{3} + 4\frac{1}{2} = 4\frac{1}{2} + 1\frac{2}{3}$

d. Does the commutative property hold for the addition of fractions?

e. Does the associative property hold for the addition of fractions?

5–12 COMPUTATION—SUBTRACTION OF FRACTIONS AND MIXED NUMBERS

(1) To subtract fractions with *like denominators*, we subtract the numerators and write the difference over the common denominator. We express the answer in lowest terms.

(2) To subtract fractions with *unlike denominators*, we express the fractions as equivalent fractions having a common denominator and subtract as we do with like denominators.

(1) Vertically or Horizontally:

$$\begin{array}{r} \frac{4}{5} \\ -\frac{3}{5} \\ \hline \frac{1}{5} \end{array} \qquad \frac{4}{5} - \frac{3}{5} = \frac{4-3}{5} = \frac{1}{5}$$

Answer, $\frac{1}{5}$

(2) Subtract:

$$\begin{array}{r} \frac{11}{12} = \frac{11}{12} \\ \frac{1}{4} = \frac{3}{12} \\ \hline \frac{8}{12} = \frac{2}{3} \end{array}$$

Answer, $\frac{2}{3}$

(3) To subtract a fraction or mixed number from a whole number, we regroup by taking one (1) from the whole number and express it as a fraction making the numerator and denominator the same. Then we subtract the fractions and we subtract the whole numbers.

> (3) Subtract:
>
> $$\begin{array}{r} 4 = 3\frac{8}{8} \\ \underline{1\frac{5}{8} = 1\frac{5}{8}} \\ 2\frac{3}{8} \end{array}$$
>
> *Answer,* $2\frac{3}{8}$

(4) To subtract a whole number from a mixed number, we find the difference between the whole numbers and annex the fraction to this difference.

> (4) $6\frac{3}{5} - 2 = 4\frac{3}{5}$
>
> *Answer,* $4\frac{3}{5}$

(5) To subtract a fraction or a mixed number from a mixed number, first we subtract the fractions then the whole numbers. If the fraction in the subtrahend is greater than the fraction in the minuend, we regroup by taking one (1) from the whole number in the minuend to increase the fraction of the minuend, then we subtract. We express answers in simplest form.

> (5) Subtract:
>
> $$\begin{array}{r} 6\frac{1}{3} = 6\frac{2}{6} = 5\frac{8}{6} \\ \underline{2\frac{1}{2} = 2\frac{3}{6} = 2\frac{3}{6}} \\ 3\frac{5}{6} \end{array}$$
>
> *Answer,* $3\frac{5}{6}$

EXERCISES

1. Subtract on a ruler:

$\frac{3}{4}''$	$\frac{9}{16}''$	$\frac{5}{8}''$	$\frac{13}{16}''$	$\frac{7}{8}''$	$\frac{11}{16}''$	$\frac{5}{8}''$	$\frac{1}{2}''$	$\frac{13}{16}''$
$\frac{2}{4}''$	$\frac{2}{16}''$	$\frac{1}{8}''$	$\frac{5}{16}''$	$\frac{5}{8}''$	$\frac{5}{16}''$	$\frac{1}{4}''$	$\frac{3}{16}''$	$\frac{3}{4}''$

2. Subtract on a number line or by use of the fraction chart (see page 91) to determine which of the following statements are true:

a. $\frac{5}{6} - \frac{1}{6} = \frac{1}{2}$ **b.** $\frac{11}{12} - \frac{7}{12} = \frac{1}{3}$ **c.** $\frac{9}{10} - \frac{3}{10} = \frac{3}{5}$ **d.** $\frac{1}{2} - \frac{2}{5} = \frac{1}{10}$ **e.** $\frac{11}{12} - \frac{1}{4} = \frac{2}{3}$

Subtract:

3.	$\frac{4}{5}$	$\frac{6}{7}$	$\frac{15}{16}$	$\frac{9}{10}$	$\frac{7}{12}$	$\frac{1}{2}$	$\frac{29}{32}$	$\frac{5}{6}$	$\frac{7}{9}$
	$\frac{2}{5}$	$\frac{4}{7}$	$\frac{5}{16}$	$\frac{3}{10}$	$\frac{5}{12}$	$\frac{1}{2}$	$\frac{17}{32}$	$\frac{5}{6}$	$\frac{4}{9}$
4.	$\frac{1}{3}$	$\frac{11}{12}$	$\frac{3}{5}$	$\frac{1}{2}$	$\frac{9}{10}$	$\frac{5}{6}$	$\frac{3}{4}$	$\frac{7}{8}$	$\frac{2}{3}$
	$\frac{1}{6}$	$\frac{2}{3}$	$\frac{1}{10}$	$\frac{3}{8}$	$\frac{3}{4}$	$\frac{11}{24}$	$\frac{2}{3}$	$\frac{3}{10}$	$\frac{5}{8}$
5.	$6\frac{5}{8}$	$8\frac{1}{2}$	$4\frac{11}{16}$	$9\frac{4}{5}$	$12\frac{9}{20}$	$20\frac{5}{24}$	$5\frac{2}{3}$	$6\frac{7}{8}$	$2\frac{1}{4}$
	3	4	1	7	8	15	5	6	2

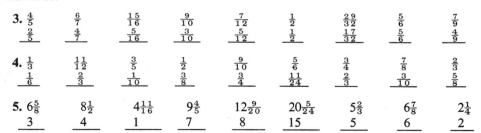

6.	$\frac{9}{10}$	$\frac{13}{15}$	$\frac{17}{20}$	$\frac{15}{16}$	$\frac{5}{6}$	$\frac{11}{12}$	$1\frac{13}{16}$	$9\frac{5}{12}$	$25\frac{1}{6}$
	$\frac{9}{10}$	$\frac{7}{15}$	$\frac{9}{20}$	$\frac{5}{6}$	$\frac{7}{10}$	$\frac{5}{16}$	1	9	25
7.	$3\frac{4}{5}$	$7\frac{5}{6}$	$2\frac{3}{8}$	$1\frac{9}{10}$	$9\frac{11}{12}$	$10\frac{7}{12}$	$6\frac{1}{2}$	$3\frac{4}{5}$	$9\frac{3}{4}$
	$\frac{1}{5}$	$1\frac{1}{6}$	$\frac{3}{8}$	$\frac{3}{10}$	$9\frac{11}{12}$	$10\frac{5}{12}$	$\frac{3}{8}$	$\frac{2}{3}$	$2\frac{1}{6}$
8.	1	1	8	9	4	6	8	12	8
	$\frac{7}{8}$	$\frac{1}{3}$	$\frac{3}{4}$	$3\frac{1}{8}$	$1\frac{5}{6}$	$5\frac{1}{2}$	$6\frac{9}{16}$	$2\frac{3}{5}$	$7\frac{3}{8}$
9.	$6\frac{1}{5}$	$7\frac{3}{8}$	$4\frac{1}{6}$	$5\frac{1}{2}$	$4\frac{2}{5}$	$8\frac{3}{10}$	$1\frac{3}{16}$	$13\frac{7}{8}$	$9\frac{5}{12}$
	$3\frac{3}{5}$	$2\frac{7}{8}$	$1\frac{5}{6}$	$2\frac{5}{8}$	$3\frac{3}{4}$	$2\frac{5}{6}$	$\frac{9}{16}$	$9\frac{11}{12}$	$\frac{2}{3}$
10.	$8\frac{7}{10}$	$15\frac{2}{3}$	$20\frac{9}{10}$	15	10	14	$10\frac{5}{8}$	$4\frac{1}{4}$	$7\frac{3}{8}$
	$4\frac{1}{2}$	$9\frac{1}{4}$	$8\frac{3}{8}$	$8\frac{15}{32}$	$9\frac{1}{3}$	$10\frac{2}{3}$	$1\frac{4}{5}$	$3\frac{1}{3}$	$6\frac{11}{16}$

11. a. $\frac{5}{6} - \frac{3}{8}$ **b.** $7\frac{5}{8} - 7$ **c.** $10\frac{3}{4} - \frac{2}{3}$ **d.** $5 - 4\frac{4}{5}$ **e.** $14\frac{3}{8} - 6\frac{9}{10}$

5–13 COMPUTATION—MULTIPLICATION OF FRACTIONS AND MIXED NUMBERS

To multiply fractions and mixed numbers, we first express each mixed number, if any, as an improper fraction and each whole number, if any, as a fraction by writing the whole number over 1. Then where possible we divide any numerator and denominator by the greatest possible number that is exactly contained in both (greatest common factor). This reduction simplifies computation. Finally we multiply the resulting numerators and multiply the resulting denominators, expressing the answer in simplest form.

$$\frac{1}{5} \times 7 = \frac{1}{5} \times \frac{7}{1} = \frac{7}{5} = 1\frac{2}{5}$$

Answer, $1\frac{2}{5}$

$$10 \times \frac{3}{4} = \frac{\overset{5}{\cancel{10}}}{1} \times \frac{3}{\underset{2}{\cancel{4}}} = \frac{15}{2} = 7\frac{1}{2}$$

Answer, $7\frac{1}{2}$

$$\frac{4}{5} \times \frac{2}{3} = \frac{8}{15}$$

Answer, $\frac{8}{15}$

$$1\frac{9}{16} \times 3\frac{3}{5} = \frac{\overset{5}{\cancel{25}}}{\underset{8}{\cancel{16}}} \times \frac{\overset{9}{\cancel{18}}}{\underset{1}{\cancel{5}}} = \frac{45}{8} = 5\frac{5}{8}$$

Answer, $5\frac{5}{8}$

When we multiply a mixed number and a whole number, the vertical form may also be used.

$$\begin{array}{r} 15 \\ \times\ 2\frac{1}{4} \\ \hline 30 \\ 3\frac{3}{4} \\ \hline 33\frac{3}{4} \end{array}$$

$$\frac{1}{4} \times 15 = \frac{15}{4} = 3\frac{3}{4}$$

Answer, $33\frac{3}{4}$

EXERCISES

Multiply:

1.	$\frac{1}{3} \times 3$	$\frac{5}{8} \times 8$	$\frac{1}{12} \times 60$	$\frac{7}{10} \times 100$	$\frac{1}{16} \times 2$
	$\frac{5}{12} \times 3$	$\frac{1}{10} \times 6$	$\frac{7}{16} \times 12$	$\frac{1}{16} \times 36$	$\frac{7}{8} \times 28$
2.	$6 \times \frac{1}{6}$	$10 \times \frac{7}{10}$	$18 \times \frac{1}{6}$	$16 \times \frac{3}{4}$	$3 \times \frac{1}{12}$
	$2 \times \frac{7}{8}$	$14 \times \frac{1}{16}$	$6 \times \frac{5}{8}$	$35 \times \frac{1}{10}$	$45 \times \frac{5}{6}$
3.	$\frac{1}{8} \times 5$	$\frac{3}{16} \times 1$	$\frac{1}{4} \times 9$	$\frac{2}{3} \times 2$	$\frac{13}{6} \times 5$
	$7 \times \frac{1}{12}$	$3 \times \frac{2}{7}$	$17 \times \frac{1}{6}$	$21 \times \frac{13}{16}$	$12 \times \frac{51}{16}$
4.	$\frac{1}{2} \times \frac{1}{4}$	$\frac{7}{8} \times \frac{5}{6}$	$\frac{1}{3} \times \frac{3}{16}$	$\frac{2}{5} \times \frac{1}{8}$	$\frac{3}{4} \times \frac{5}{24}$
	$\frac{4}{5} \times \frac{15}{16}$	$\frac{21}{32} \times \frac{8}{9}$	$\frac{12}{5} \times \frac{5}{12}$	$\frac{25}{18} \times \frac{9}{10}$	$\frac{22}{7} \times \frac{35}{16}$
5.	$4 \times 2\frac{3}{4}$	$48 \times 3\frac{11}{16}$	$5 \times 3\frac{7}{10}$	$12 \times 2\frac{13}{32}$	$8 \times 5\frac{2}{3}$
	$2\frac{1}{2} \times 2$	$1\frac{1}{3} \times 6$	$6\frac{7}{12} \times 4$	$1\frac{7}{16} \times 20$	$4\frac{3}{8} \times 7$
6.	$\frac{5}{8} \times 3\frac{3}{4}$	$\frac{3}{10} \times 2\frac{1}{6}$	$\frac{9}{16} \times 4\frac{5}{6}$	$\frac{2}{3} \times 1\frac{7}{20}$	$\frac{7}{8} \times 3\frac{1}{7}$
	$2\frac{7}{8} \times \frac{5}{6}$	$1\frac{1}{2} \times \frac{1}{4}$	$2\frac{5}{12} \times \frac{3}{16}$	$5\frac{3}{5} \times \frac{7}{8}$	$6\frac{1}{4} \times \frac{2}{25}$
7.	$2\frac{1}{2} \times 3\frac{1}{2}$	$3\frac{7}{8} \times 1\frac{4}{5}$	$8\frac{1}{3} \times 4\frac{7}{10}$	$2\frac{5}{8} \times 1\frac{5}{6}$	$3\frac{1}{7} \times 6\frac{1}{8}$
	$9\frac{1}{3} \times 1\frac{7}{8}$	$4\frac{4}{5} \times 2\frac{13}{16}$	$1\frac{1}{2} \times 4\frac{2}{3}$	$4\frac{1}{6} \times 1\frac{11}{25}$	$8\frac{1}{10} \times 2\frac{2}{9}$

8.	$\frac{22}{7} \times \frac{9}{16} \times \frac{7}{8}$	$6 \times 2\frac{3}{16} \times \frac{4}{5}$	$1\frac{3}{4} \times 2\frac{1}{2} \times 3\frac{1}{7}$
	$2\frac{5}{6} \times 5\frac{1}{2} \times 3\frac{2}{3}$	$3\frac{3}{8} \times 2\frac{2}{3} \times 4\frac{1}{2}$	$3\frac{1}{7} \times \frac{3}{4} \times 6\frac{1}{8}$

9. Multiply vertically:

$17\frac{2}{3}$	$41\frac{5}{8}$	$5\frac{1}{2}$	$9\frac{5}{6}$	$26\frac{13}{16}$	21	56	8	19	45
9	24	3	14	37	$3\frac{1}{7}$	$73\frac{7}{8}$	$2\frac{1}{3}$	$6\frac{1}{4}$	$89\frac{11}{12}$

10. Which of these statements are true?

a. $\frac{3}{5} \times \frac{7}{8} = \frac{7}{8} \times \frac{3}{5}$ **b.** $(\frac{9}{10} \times 2\frac{1}{2}) \times 3\frac{3}{4} = \frac{9}{10} \times (2\frac{1}{2} \times 3\frac{3}{4})$ **c.** $1\frac{2}{3} \times 6\frac{4}{5} = 6\frac{4}{5} \times 1\frac{2}{3}$

d. Does the commutative property hold for the multiplication of fractions?

e. Does the associative property hold for the multiplication of fractions?

f. Is $\frac{3}{4} \times (8 + 12) = \frac{3}{4} \times 8 + \frac{3}{4} \times 12$ true? Does the distributive property of multiplication over addition hold when fractions are used?

11. Square each of the following:

a. $\frac{2}{3}$ **b.** $\frac{1}{2}$ **c.** $\frac{4}{5}$ **d.** $\frac{5}{8}$ **e.** $\frac{7}{12}$ **f.** $1\frac{1}{4}$ **g.** $3\frac{7}{8}$ **h.** $2\frac{1}{6}$ **i.** $4\frac{3}{5}$ **j.** $5\frac{9}{16}$

5–14 MULTIPLICATIVE INVERSE

If the product of two numbers is one (1), then each factor is called the *multiplicative inverse* or *reciprocal* of the other.

4 and $\frac{1}{4}$ are multiplicative inverses of each other because $4 \times \frac{1}{4} = 1$.
$\frac{3}{2}$ and $\frac{2}{3}$ are multiplicative inverses of each other because $\frac{3}{2} \times \frac{2}{3} = 1$.

Observe that the numerator of one fraction becomes the denominator of its reciprocal and its denominator becomes the numerator of its reciprocal.

Zero has no inverse for multiplication.

EXERCISES

Write the multiplicative inverse or reciprocal of each of the following:

1. **a.** 3 **b.** 5 **c.** 12 **d.** 0 **e.** 6 **f.** 1 **g.** 25 **h.** 100

2. **a.** $\frac{1}{5}$ **b.** $\frac{1}{7}$ **c.** $\frac{1}{10}$ **d.** $\frac{2}{9}$ **e.** $\frac{22}{7}$ **f.** $\frac{25}{32}$ **g.** $\frac{15}{16}$ **h.** $\frac{11}{5}$

3. Find the missing numbers.

a. $6 \times ? = 1$ **c.** $? \times 4 = 1$ **e.** $\frac{1}{3} \times ? = 1$ **g.** $\frac{4}{7} \times ? = 1$ **i.** $8 \times \frac{1}{8} = ?$

b. $9 \times ? = 1$ **d.** $? \times 7 = 1$ **f.** $? \times \frac{1}{2} = 1$ **h.** $? \times \frac{9}{5} = 1$ **j.** $\frac{3}{4} \times \frac{4}{3} = ?$

5–15 COMPUTATION—DIVISION OF FRACTIONS AND MIXED NUMBERS

Division is the inverse operation of multiplication.

When we divide 20 by 5, the quotient is 4. $20 \div 5 = 4$

When we multiply 20 by $\frac{1}{5}$, the product is 4. $20 \times \frac{1}{5} = 4$

Dividing a given number by 5 gives the same answer as multiplying the given number by its multiplicative inverse (or reciprocal) $\frac{1}{5}$.

When we divide 4 by $\frac{1}{3}$, written as $4 \div \frac{1}{3}$, we are finding how many thirds are in four whole units. In each whole unit there are 3 thirds; in 4 units there are 4×3 thirds or 12 thirds. $4 \div \frac{1}{3}$ and 4×3 are both names for the number 12.

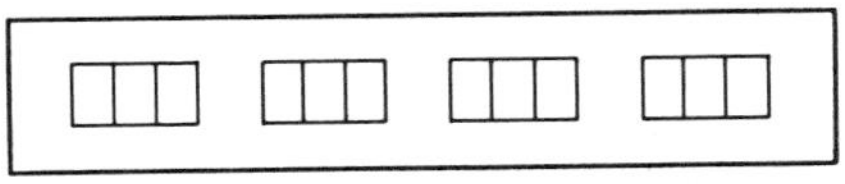

Thus to divide a number by another number, we may instead multiply the first number by the reciprocal of the divisor. Usually this idea is stated briefly as "invert the divisor and multiply."

This principle may be developed by using a complex fraction, the multiplicative inverse and the multiplicative identity one (1) as shown below.

A *complex fraction* is a fraction in which the numerator or denominator or both have a fraction as a term. The complex fraction $\frac{\frac{3}{5}}{\frac{2}{3}}$ may be expressed as $\frac{3}{5} \div \frac{2}{3}$ since they represent the same number. (The fraction bar means division.)

$$\frac{3}{5} \div \frac{2}{3} = \frac{\frac{3}{5}}{\frac{2}{3}} = \frac{\frac{3}{5}}{\frac{2}{3}} \times 1 = \frac{\frac{3}{5} \times \frac{3}{2}}{\frac{2}{3} \times \frac{3}{2}} = \frac{\frac{3}{5} \times \frac{3}{2}}{1} = \frac{3}{5} \times \frac{3}{2} = \frac{9}{10}$$

We multiply both numerator and denominator by $\frac{3}{2}$, the multiplicative inverse of the denominator $\frac{2}{3}$, to get 1 as our denominator.

To divide fractions and mixed numbers by the reciprocal method:

(1) We express each mixed number, if any, as an improper fraction and each whole number, if any, as a fraction by writing the whole number over 1.
(2) We invert the divisor (number after the division sign) to find the reciprocal and multiply as in the multiplication of fractions, using reduction where possible.

$$12 \div \frac{15}{16} = \frac{\overset{4}{\cancel{12}}}{1} \times \frac{16}{\underset{5}{\cancel{15}}} = \frac{64}{5} = 12\tfrac{4}{5}$$

Answer, $12\frac{4}{5}$

$$1\tfrac{1}{3} \div 2\tfrac{1}{4} = \tfrac{4}{3} \div \tfrac{9}{4} = \tfrac{4}{3} \times \tfrac{4}{9} = \tfrac{16}{27}$$

Answer, $\frac{16}{27}$

Sometimes the given fractions are expressed as equivalent fractions having a common denominator and the numerator of the first fraction is divided by the numerator of the second fraction to obtain the answer. For example:

$$\tfrac{3}{5} \div \tfrac{2}{3} = \tfrac{9}{15} \div \tfrac{10}{15} = 9 \div 10 = \tfrac{9}{10}$$

9 fifteenths divided by 10 fifteenths is 9 divided by 10.

Answer, $\frac{9}{10}$

To simplify a complex fraction, we divide the numerator of the complex fraction by its denominator.

$$\frac{\frac{15}{16}}{\frac{5}{6}} = \frac{15}{16} \div \frac{5}{6} = \frac{\overset{3}{\cancel{15}}}{\underset{8}{\cancel{16}}} \times \frac{\overset{3}{\cancel{6}}}{\underset{1}{\cancel{5}}} = \frac{9}{8} = 1\tfrac{1}{8}$$

Answer, $1\frac{1}{8}$

Observe that when a number is divided by a fraction less than 1, the quotient is always greater than the given number. When a number is divided by 1, the quotient is always equal to the given number. When a number is divided by a mixed number or a whole number greater than 1, the quotient is always less than the given number.

EXERCISES

Divide:

1.	$\frac{3}{5} \div 3$	$\frac{15}{16} \div 5$	$\frac{27}{32} \div 3$	$\frac{1}{2} \div 4$	$\frac{1}{3} \div 8$
	$\frac{5}{6} \div 2$	$\frac{3}{4} \div 8$	$\frac{5}{8} \div 10$	$\frac{15}{16} \div 60$	$\frac{21}{32} \div 28$
2.	$2\frac{1}{2} \div 5$	$10\frac{1}{8} \div 9$	$6\frac{2}{3} \div 4$	$13\frac{1}{3} \div 16$	$4\frac{4}{5} \div 20$
	$7\frac{1}{3} \div 4$	$1\frac{7}{8} \div 4$	$2\frac{11}{16} \div 8$	$5\frac{3}{4} \div 2$	$7\frac{7}{8} \div 5$
3.	$\frac{7}{8} \div \frac{1}{8}$	$\frac{1}{3} \div \frac{2}{3}$	$\frac{5}{6} \div \frac{5}{6}$	$\frac{1}{5} \div \frac{1}{2}$	$\frac{3}{4} \div \frac{2}{3}$
	$\frac{5}{8} \div \frac{5}{32}$	$\frac{2}{3} \div \frac{4}{5}$	$\frac{15}{16} \div \frac{9}{10}$	$\frac{25}{32} \div \frac{15}{16}$	$\frac{3}{4} \div \frac{9}{16}$
4.	$8 \div \frac{1}{4}$	$5 \div \frac{5}{6}$	$12 \div \frac{2}{3}$	$2 \div \frac{4}{5}$	$15 \div \frac{9}{10}$
	$7 \div \frac{2}{3}$	$1 \div \frac{1}{6}$	$9 \div \frac{7}{16}$	$1 \div \frac{5}{8}$	$7 \div \frac{21}{32}$
5.	$6\frac{3}{5} \div \frac{1}{5}$	$8\frac{2}{3} \div \frac{2}{3}$	$4\frac{1}{5} \div \frac{7}{8}$	$3\frac{3}{4} \div \frac{1}{2}$	$1\frac{1}{8} \div \frac{3}{4}$
	$2\frac{11}{12} \div \frac{15}{16}$	$3\frac{1}{3} \div \frac{5}{6}$	$2\frac{4}{5} \div \frac{7}{10}$	$2\frac{1}{2} \div \frac{2}{3}$	$5\frac{1}{3} \div \frac{11}{16}$
6.	$\frac{1}{4} \div 1\frac{1}{4}$	$\frac{1}{6} \div 2\frac{5}{6}$	$\frac{7}{12} \div 1\frac{2}{3}$	$\frac{15}{16} \div 3\frac{1}{2}$	$\frac{5}{8} \div 3\frac{3}{4}$
	$\frac{21}{32} \div 4\frac{3}{8}$	$\frac{9}{10} \div 2\frac{2}{5}$	$\frac{2}{3} \div 1\frac{3}{8}$	$\frac{5}{8} \div 2\frac{2}{3}$	$\frac{3}{4} \div 1\frac{3}{5}$
7.	$11 \div 2\frac{3}{4}$	$58 \div 3\frac{5}{8}$	$7 \div 8\frac{3}{4}$	$4 \div 18\frac{2}{3}$	$6 \div 4\frac{1}{2}$
	$18 \div 2\frac{7}{10}$	$4 \div 8\frac{2}{3}$	$9 \div 1\frac{2}{3}$	$12 \div 12\frac{4}{5}$	$1 \div 1\frac{3}{4}$
8.	$3\frac{3}{8} \div 4\frac{2}{5}$	$5\frac{1}{2} \div 2\frac{2}{3}$	$24\frac{1}{2} \div 3\frac{1}{2}$	$18\frac{3}{4} \div 1\frac{9}{16}$	$1\frac{5}{6} \div 2\frac{7}{8}$
	$2\frac{1}{2} \div 3\frac{3}{4}$	$4\frac{1}{4} \div 4\frac{1}{4}$	$5\frac{11}{16} \div 4\frac{1}{2}$	$12\frac{1}{4} \div 1\frac{5}{16}$	$19\frac{1}{3} \div 2\frac{5}{12}$

9. $(\frac{7}{8} \div \frac{3}{16}) \div \frac{2}{3}$ $\quad (8\frac{2}{3} \div 1\frac{5}{8}) \div 1\frac{1}{2}$ $\quad 7\frac{1}{2} \div (2\frac{3}{4} \div \frac{2}{3})$ $\quad 1\frac{7}{8} \div (2\frac{5}{6} \div 2\frac{2}{5})$

10. Simplify:

$$\frac{\frac{9}{16}}{\frac{3}{8}} \qquad \frac{\frac{4}{5}}{\frac{5}{12}} \qquad \frac{\frac{8}{49}}{\frac{22}{7}} \qquad \frac{2\frac{5}{8}}{3\frac{3}{4}} \qquad \frac{3\frac{3}{5}}{\frac{7}{10}}$$

$$\frac{\frac{14}{15}}{7} \qquad \frac{1\frac{1}{2}}{9} \qquad \frac{37\frac{1}{2}}{100} \qquad \frac{8}{\frac{4}{5}} \qquad \frac{9}{2\frac{1}{4}}$$

FRACTIONAL RELATIONSHIPS BETWEEN NUMBERS

Fractions are used to express relationships between numbers. There are 3 types of relationships: (1) finding a fractional part of a number; (2) finding what fractional part one number is of another; and (3) finding a number when a fractional part of it is known.

5–16 FRACTIONAL PART OF A NUMBER

We often find a fractional part of a quantity like $\frac{2}{3}$ of a dozen ($\frac{2}{3}$ of 12). Here we use the process of multiplication, $\frac{2}{3}$ of 12 means $\frac{2}{3} \times 12$. The word "of" used in this way means "times."

$$\frac{2}{\underset{1}{\cancel{3}}} \times \overset{4}{\cancel{12}} = 8$$

Answer, 8

EXERCISES

1. Find: **a.** $\frac{1}{2}$ of 8	**b.** $\frac{1}{3}$ of 17	**c.** $\frac{1}{4}$ of $\frac{4}{5}$	**d.** $\frac{1}{5}$ of 120	**e.** $\frac{1}{6}$ of 3
2. Find: **a.** $\frac{1}{8}$ of 34	**b.** $\frac{1}{10}$ of $8\frac{1}{3}$	**c.** $\frac{1}{12}$ of 96	**d.** $\frac{1}{9}$ of 72	**e.** $\frac{1}{36}$ of 25
3. Find: **a.** $\frac{2}{3}$ of 48	**b.** $\frac{3}{4}$ of 24	**c.** $\frac{7}{8}$ of 6	**d.** $\frac{5}{6}$ of 178	**e.** $\frac{2}{5}$ of 80
4. Find: **a.** $\frac{9}{10}$ of 35	**b.** $\frac{9}{100}$ of 700	**c.** $\frac{5}{12}$ of 60	**d.** $\frac{5}{8}$ of $7\frac{1}{2}$	**e.** $\frac{4}{5}$ of 175
5. Find: **a.** $\frac{3}{8}$ of 50	**b.** $\frac{7}{10}$ of 19	**c.** $\frac{11}{12}$ of 9	**d.** $\frac{3}{5}$ of 2	**e.** $\frac{3}{10}$ of $\frac{7}{8}$
6. Find: **a.** $\frac{7}{12}$ of 360	**b.** $\frac{1}{16}$ of 24	**c.** $\frac{5}{12}$ of $7\frac{1}{5}$	**d.** $\frac{11}{16}$ of 80	**e.** $\frac{5}{6}$ of 4

5–17 WHAT FRACTIONAL PART ONE NUMBER IS OF ANOTHER

A fraction is used to express what part one number is of another. When we find what part of a dozen 8 things are, we are using 8 things out of 12 or $\frac{8}{12}$ of a dozen which, expressed in lowest terms, is $\frac{2}{3}$ of a dozen.

$$\frac{8}{12} = \frac{2 \cdot 4}{3 \cdot 4} = \frac{2}{3}$$

Answer, $\frac{2}{3}$

To find what fractional part one number is of another, we write a numeral naming a fraction with the number of parts used as the numerator and the number of parts in the whole unit or group as the denominator. Where necessary, we express the fraction in lowest terms.

EXERCISES

Find the following:

1. a. What part of 8 is 3?	**b.** What part of 5 is 2?	**c.** What part of 16 is 9?
2. a. 1 is what part of 6?	**b.** 7 is what part of 8?	**c.** 3 is what part of 10?
3. a. What part of 12 is 9?	**b.** What part of 20 is 16?	**c.** What part of 54 is 42?
4. a. 2 is what part of 14?	**b.** 36 is what part of 40?	**c.** 65 is what part of 100?
5. a. What part of 56 is 35?	**b.** 108 is what part of 144?	**c.** What part of 60 is 45?

5–18 FINDING THE NUMBER WHEN A FRACTIONAL PART OF IT IS KNOWN

Suppose we are required to find the number when $\frac{2}{3}$ of it is 8:

Since $\frac{2}{3}$ of the number $= 8$
$\frac{1}{3}$ of the number $= 8 \div 2 = 4$
$\frac{3}{3}$ of the number $= 3 \times 4 = 12$
Therefore the number $= 12$

Note: We divide by 2 (which is the numerator of $\frac{2}{3}$) and we multiply by 3 (which is the denominator of $\frac{2}{3}$). We are doing the same thing when we multiply by the multiplicative inverse $\frac{3}{2}$.

$$\overset{4}{\cancel{8}} \times \frac{3}{\underset{1}{\cancel{2}}} = 12$$

Answer, 12

Therefore to find the number when a fractional part of it is known, we multiply the given number that represents the fractional part of the unknown number by the multiplicative inverse (reciprocal) of the given fraction (or divide the given number by the given fraction as shown below).

$$8 \div \tfrac{2}{3}$$
$$8 \times \tfrac{3}{2} = 12$$

EXERCISES

Find the following:

1. a. $\frac{1}{2}$ of what number is 12? **b.** $\frac{1}{5}$ of what number is 20? **c.** $\frac{1}{8}$ of what number is 7?

2. a. 5 is $\frac{1}{4}$ of what number? **b.** 6 is $\frac{1}{12}$ of what number? **c.** 30 is $\frac{1}{10}$ of what number?

3. a. $\frac{3}{4}$ of what number is 27? **b.** $\frac{2}{3}$ of what number is 48? **c.** $\frac{5}{6}$ of what number is 60?

4. a. 56 is $\frac{7}{8}$ of what number? **b.** 40 is $\frac{2}{5}$ of what number? **c.** 27 is $\frac{9}{10}$ of what number?

5. a. $\frac{5}{12}$ of what number is 100? **b.** $\frac{4}{5}$ of what number is 72? **c.** $\frac{3}{8}$ of what number is 12?

6. a. 85 is $\frac{17}{20}$ of what number? **b.** 60 is $\frac{15}{16}$ of what number? **c.** 54 is $\frac{9}{100}$ of what number?

5–19 PROBLEMS

EXERCISES

1. Joan's mother bought two chickens, one weighing $3\frac{7}{8}$ lb. and the other $4\frac{3}{4}$ lb. What was the total weight of the chickens?

2. How many degrees above normal is a temperature of $101\frac{2}{5}$ degrees if normal body temperature is $98\frac{3}{5}$ degrees?
3. If it takes $1\frac{7}{8}$ yards of goods to make a chair cover, how many yards of goods are needed to make 6 covers?
4. An airplane flew 720 miles in $3\frac{1}{4}$ hours. What was its average ground speed per hour?
5. A merchant cut $4\frac{2}{3}$ yards of cloth from a bolt containing $34\frac{1}{4}$ yards. What length remained in the bolt?
6. A house worth $19,200 was assessed for $\frac{3}{4}$ of its value. What is the assessed value of the house?
7. A stock which closed on Monday at $42\frac{5}{8}$ gained $1\frac{3}{4}$ points on the following day. At what price did it close on Tuesday?
8. How many floor boards $2\frac{1}{4}$ inches wide are needed to cover a floor 15 feet wide?
9. How many miles do $8\frac{3}{16}$ inches represent if the scale 1 inch = 80 miles is used?
10. If you can walk $\frac{3}{4}$ mile in $\frac{2}{5}$ hour, at this rate how far can you walk in one hour?
11. What is the outside diameter of a tube $\frac{3}{8}$ inch thick if the inside diameter is $2\frac{11}{16}$ inches?
12. Ann delivers papers after school each day and on Saturdays, spending $1\frac{1}{4}$ hours for each delivery. How many hours each week does she use to deliver papers?
13. A closed box measuring $9\frac{11}{16}$ inches by $8\frac{7}{8}$ inches by $5\frac{3}{4}$ inches is made of wood $\frac{3}{16}$ inch thick. Find the inside dimensions.
14. Joe built a book case with 3 shelves each $3\frac{1}{2}$ feet long. He also used 2 pieces of wood each $3\frac{3}{4}$ feet long and two side pieces each $4\frac{1}{2}$ feet long. How many feet of wood were required?
15. A family plans to spend $\frac{1}{4}$ of its annual income of $18,600 for shelter, $\frac{1}{3}$ for food and clothing, $\frac{1}{6}$ for general expenses, and $\frac{3}{20}$ for miscellaneous expenses. What fractional part is left for savings? How much is allowed annually for each item?
16. If a car was driven 63 miles in $1\frac{3}{4}$ hours, what was its average speed per hour?
17. How many pieces of wood each $1\frac{1}{3}$ feet long can be cut from a board 16 feet long?
18. Find the perimeter of a triangle if its sides measure $2\frac{5}{8}$ inches, $1\frac{11}{16}$ inches, and $1\frac{3}{4}$ inches.
19. Find the weight of 19 pieces of metal rod each $2\frac{1}{2}$ feet long if the rods weigh $1\frac{7}{8}$ lb. per foot.
20. Which weighs more and how much more, 3 large candy bars each weighing $8\frac{3}{4}$ oz. or 24 small candy bars each weighing $\frac{7}{8}$ oz?

CHAPTER REVIEW

1. Write (8, 11) as a numeral naming a fraction. (5–1)
2. Express $\frac{56}{72}$ in lowest terms. (5–2)
3. Which of the following are not in lowest terms: $\frac{8}{19}$, $\frac{34}{51}$, $\frac{27}{75}$, $\frac{16}{81}$? (5–2)
4. Express $\frac{7}{12}$ as 60ths. (5–3)
5. Are $\frac{54}{90}$ and $\frac{21}{35}$ equivalent fractions? (5–4)

6. Which of the following fractions is not equivalent to the other (5–4)
three fractions: $\frac{14}{21}, \frac{38}{57}, \frac{18}{28}, \frac{24}{36}$? (5–3)
7. Write the set of fractions equivalent to $\frac{5}{8}$. (5–3)
8. Write the set of equivalent fractions of which $\frac{16}{20}$ is a member.
9. Write a numeral naming: **a.** a common fraction (5–5)
b. an improper fraction **c.** a mixed number **d.** a proper fraction. (5–5)
10. Express $\frac{39}{16}$ as a mixed number. (5–5)
11. Express $4\frac{7}{9}$ as an improper fraction.
12. What is the lowest common denominator (L.C.D.) of the fractions (5–6)
$\frac{5}{6}, \frac{5}{8}$, and $\frac{3}{4}$?
(5–8)

13. Which of the following statements are true?

a. $\frac{3}{7} < \frac{4}{9}$ **b.** $\frac{42}{48} = \frac{54}{64}$ **c.** $\frac{7}{12} > \frac{9}{16}$ **d.** $\frac{15}{36} < \frac{11}{24}$ **e.** $\frac{24}{33} > \frac{49}{63}$
(5–8)

14. Arrange $\frac{2}{3}, \frac{7}{11}$, and $\frac{5}{8}$ by size, smallest first. (5–10)
15. Round $3\frac{5}{9}$ to the nearest whole number.
(5–11)

16. Add:

$$\begin{array}{l} 4\frac{3}{5} \\ 2\frac{1}{2} \\ \underline{5\frac{7}{10}} \end{array}$$

17. Add:

$1\frac{1}{4} + \frac{2}{3} + 3\frac{5}{6}$
(5–12)

18. Subtract:

$$\begin{array}{l} 5\frac{1}{3} \\ \underline{2\frac{7}{8}} \end{array}$$

19. Subtract:

$12 - 6\frac{9}{16}$
(5–13)

20. Multiply:

$6 \times 3\frac{3}{4}$

21. Multiply:

$8\frac{1}{3} \times \frac{4}{5} \times 1\frac{1}{2}$

22. Is the product *greater than, equal to,* or *less than* a given number (5–13)
when the given number is multiplied by:

a. 1? **b.** A number greater than 1? **c.** A number less than 1?
(5–14)

23. Write the multiplicative inverse of: **a.** 9 **b.** $\frac{1}{8}$ **c.** $\frac{11}{16}$.
(5–15)

24. Divide:

$15 \div 3\frac{1}{3}$

25. Divide:

$6\frac{1}{8} \div 3\frac{1}{7}$

26. $\dfrac{\frac{5}{6}}{\frac{7}{12}}$

27. Is the quotient *greater than, equal to,* or *less than* a given number (5–15)
when the given number is divided by:

a. 1? **b.** A number greater than 1? **c.** A number less than 1 (zero excluded)?
(5–16)

28. Find $\frac{3}{8}$ of 75. (5–17)
29. What part of 90 is 54? (5–18)
30. $\frac{7}{12}$ of what number is 56?

REVIEW EXERCISES

1. Add:

968	2,597	89,628	82,165	34,079
4,592	8,465	35,977	25,997	88,926
89	3,879	2,849	34,458	81,798
1,649	5,983	93,626	47,429	96,654
7,287	2,654	8,705	93,976	57,973

2. Subtract:

8,591	38,654	48,265	697,000	821,062
6,846	729	9,897	498,915	368,798

3. Multiply:

46	593	608	957	8,529
89	64	700	986	731

4. Divide:

$42\overline{)2{,}814}$ $\quad 85\overline{)81{,}005}$ $\quad 761\overline{)660{,}548}$ $\quad 309\overline{)279{,}954}$ $\quad 2{,}240\overline{)860{,}160}$

5. Add:

$\frac{3}{16}$	$1\frac{1}{4}$	$4\frac{11}{12}$	$3\frac{7}{8}$	$6\frac{5}{6}$
$\frac{3}{4}$	$\frac{2}{3}$	$2\frac{5}{12}$	$5\frac{1}{2}$	$1\frac{3}{8}$

6. Subtract:

$\frac{5}{8}$	$6\frac{13}{16}$	$1\frac{3}{4}$	$7\frac{1}{2}$	$5\frac{7}{10}$
$\frac{1}{3}$	$2\frac{5}{16}$	$\frac{7}{8}$	$1\frac{2}{3}$	$4\frac{1}{6}$

7. Multiply:

$\frac{3}{4} \times \frac{5}{6}$ $\quad 5\frac{1}{2} \times 8$ $\quad \frac{7}{12} \times 4\frac{2}{3}$ $\quad 2\frac{5}{8} \times 3\frac{1}{7}$ $\quad 3\frac{1}{3} \times 1\frac{3}{5}$

8. Divide:

$\frac{7}{8} \div \frac{1}{4}$ $\quad 4\frac{1}{2} \div \frac{5}{16}$ $\quad 9 \div 2\frac{3}{4}$ $\quad 3\frac{3}{16} \div 5\frac{2}{3}$ $\quad 4\frac{5}{6} \div 1\frac{1}{5}$

CUMULATIVE TEST

1. If T = {all fractions whose denominator is 12}, which of the following statements are true? (1–9)

a. $\frac{5}{12} \in T$ **b.** $\frac{12}{7} \in T$ **c.** $\frac{19}{12} \notin T$ **d.** $\frac{12}{19} \in T$ **e.** $\frac{12}{5} \notin T$

2. What is the greatest common factor of 27, 81, and 108? (1–13)

3. Factor 360 as the product of prime factors. (1–15)

4. What is the least common multiple of 40, 64, and 80? (1–18)

5. Write MCMLXIX as a decimal numeral. (2–7)

6. Round 835,482,503,069 to the nearest billion. (2–11)

7. Expand 23514_{six} as a polynomial. (2–18)
8. Express 10110111011_{two} as an equivalent base seven numeral. (2–19)

9. Find the missing numbers: (3–2—3–5)

a. $1\frac{1}{2} + \frac{7}{8} = ? + 1\frac{1}{2}$
b. $(\frac{4}{5} \times 9) \div 9 = \square$
c. $\frac{3}{4} \times 6 = 6 \times ?$
d. $(\frac{2}{3} \times \frac{5}{6}) \times \frac{7}{8} = ? \times (\frac{5}{6} \times \frac{7}{8})$
e. $(2\frac{1}{4} + \frac{1}{8}) + ? = 2\frac{1}{4} + (\frac{1}{8} + \frac{9}{10})$
f. $\frac{1}{2} \times (\frac{3}{4} + \frac{1}{4}) = ? \times \frac{3}{4} + ? \times \frac{1}{4}$

10. Is the set {all odd natural numbers} closed under addition? Multiplication? (3–6)

11. Add: (3–9—3–11)

$$\begin{array}{r} 43{,}938 \\ 95{,}677 \\ 29{,}856 \\ 86{,}969 \\ \underline{91{,}238} \end{array}$$

12. Subtract:

$$\begin{array}{r} 400{,}703 \\ \underline{906} \end{array}$$

13. Multiply:

$$\begin{array}{r} 4{,}908 \\ \underline{5{,}007} \end{array}$$

14. Divide:

$918\overline{)451{,}656}$

15. Divide: (3–16)

$1{,}058\overline{)313{,}168}$

16. Is 635,000 divisible by 4? by 2? by 8? by 10? by 5? (3–17)

17. Add: (3–22)

$$\begin{array}{l} 10111_{two} \\ 11011_{two} \\ 11111_{two} \\ \underline{10110_{two}} \end{array}$$

18. Multiply:

$$\begin{array}{l} 121_{three} \\ \underline{201_{three}} \end{array}$$

(3–23)

19. a. Add:

$17 + 12 \equiv ? \pmod 5$

b. Multiply:

$11 \times 14 \equiv ? \pmod 3$

(3–24)

20. a. Add the following matrices:

$$\begin{pmatrix} 5 & 0 & 3 \\ 2 & 1 & 2 \\ 4 & 6 & 5 \end{pmatrix} + \begin{pmatrix} 3 & 2 & 5 \\ 4 & 0 & 6 \\ 1 & 3 & 0 \end{pmatrix}$$

b. Find the product matrix:

$$\begin{pmatrix} 2 & 0 & 6 \\ 1 & 3 & 1 \\ 0 & 5 & 4 \end{pmatrix} \times \begin{pmatrix} 4 & 1 & 7 \\ 0 & 5 & 4 \\ 3 & 2 & 0 \end{pmatrix}$$

(4–1)

21. Write in words each of the following:

a. $15 > 10$ **b.** $8 \times 5 \ngtr 45$ **c.** $5 - 5 < 5 \div 5$ **d.** $9 \neq 5 + 3$ (4–2)

22. Complete by using the transitive property:

a. $9 > 5$ and $5 > 3$, then $9 > ?$
b. $2^4 < 5^2$ and $5^2 < 3^3$, then $2^4 < ?$

(5–2)
23. Express $\frac{108}{144}$ in lowest terms. (5–3)
24. Express $\frac{3}{5}$ as 40ths. (5–3)
25. Write the set of equivalent fractions of which $\frac{28}{49}$ is a member.

26. Which of the following statements are true? (5–8)

a. $\frac{3}{4} < \frac{5}{8}$ **b.** $\frac{7}{12} > \frac{6}{11}$ **c.** $\frac{26}{39} = \frac{32}{48}$ **d.** $\frac{8}{13} > \frac{9}{15}$ **e.** $\frac{11}{23} < \frac{3}{7}$

27. Arrange $\frac{1}{3}$, $\frac{2}{5}$, and $\frac{3}{8}$ by size, greatest first. (5–8)

28. Add: **29.** Subtract: **30.** Multiply: (5–11—5–13)

$$\begin{array}{r} 4\frac{5}{6} \\ 9\frac{3}{4} \\ \underline{2\frac{1}{2}} \end{array}$$

$11 - 5\frac{3}{8}$

$5\frac{1}{3} \times 4\frac{5}{8}$

31. Divide: **32.** Divide: (5–15)

$3\frac{3}{4} \div \frac{4}{5}$

$6\frac{1}{3} \div 8\frac{1}{3}$

33. Find $\frac{5}{16}$ of 128. (5–16)
34. What part of 84 is 63? (5–17)
35. $\frac{9}{10}$ of what number is 153? (5–18)

Decimal Numeration of Fractions

6

6–1 PLACE VALUE

On the decimal number scale each place has one tenth the value of the next place to the left. By extending the scale to the right of the ones place, we express parts of one. When one whole unit is divided into ten (10) equal parts, the size of each equal part is one tenth ($\frac{1}{10}$).

millions	hundred thousands	ten thousands	thousands	hundreds	tens	ones	tenths	hundredths	thousandths	ten-thousandths	hundred-thousandths	millionths
2,	5	8	9,	6	3	4.	5	1	4	9	8	7

Following the arrangement that each place has one tenth the value of the next place to the left, we express $\frac{1}{10}$ of one whole unit on the number scale by using the next place on the right of the ones place. This place is called "tenths." To separate the whole number from the parts or fraction we use a dot which is called a "decimal point." Thus the numeral for the common fraction $\frac{1}{10}$ and the numeral for the decimal fraction .1 name the same number. Both represent the size of one of the equal parts when a whole unit is divided into 10 equal parts.

$\frac{1}{10}$	$\frac{1}{10}$	$\frac{1}{10}$	$\frac{1}{10}$	$\frac{1}{10}$	$\frac{1}{10}$	$\frac{1}{10}$	$\frac{1}{10}$	$\frac{1}{10}$	$\frac{1}{10}$
.1	.1	.1	.1	.1	.1	.1	.1	.1	.1

When a whole unit is divided into 100 equal parts, the size of each equal part is $\frac{1}{100}$. Since each place on the number scale has one tenth the value of the next place to the left, the second place to the right of the decimal point has

$\frac{1}{10}$ the value of tenths. But $\frac{1}{10}$ of a tenth is one hundredth ($\frac{1}{10} \times \frac{1}{10} = \frac{1}{100}$). Therefore, the value of the second place to the right of the decimal point is "hundredths." Both $\frac{1}{100}$ and .01 name the same number.

The third place to the right of the decimal point expresses thousandths (.001 = $\frac{1}{1,000}$), the fourth place expresses ten-thousandths (.0001 = $\frac{1}{10,000}$), the fifth place expresses hundred-thousandths (.00001 = $\frac{1}{100,000}$), the sixth place expresses millionths (.000001 = $\frac{1}{1,000,000}$), etc.

A *decimal fraction* is a fractional number having as its denominator some power of ten (10; 100; 1,000; etc.) and is named by a numeral in which the denominator is not written as it is in a common fraction but is expressed by place value. Only numerators appear in decimal notation.

A *mixed decimal* is a number containing a whole number and a decimal fraction. The numeral in the above scale names a mixed decimal.

Since multiples of one are expressed in the places to the left of the ones place and parts of one are expressed in the places to the right of the ones place, the ones place is the center of the numeration system and not the decimal point.

The word "decimal" is derived from the Latin word "decem" meaning ten. Our money system is a decimal system. 10 mills = 1 cent ($.01); 10 cents = 1 dime ($.10); 10 dimes = 1 dollar ($1.00 or $1).

EXERCISES

1. Write each of the following as a decimal numeral:

a. $\frac{3}{10}$, $\frac{7}{10}$, $\frac{8}{10}$, $\frac{1}{10}$, $\frac{9}{10}$ **b.** $\frac{25}{100}$, $\frac{9}{100}$, $\frac{87}{100}$, $\frac{60}{100}$, $\frac{43}{100}$

c. $\frac{563}{1,000}$, $\frac{71}{1,000}$, $\frac{4}{1,000}$, $\frac{350}{1,000}$, $\frac{400}{1,000}$

2. Write the decimal numeral that corresponds to the shaded part of each of the following:

a.
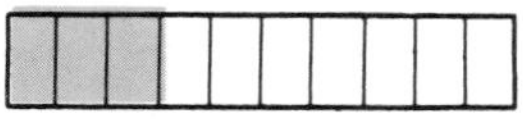

b.
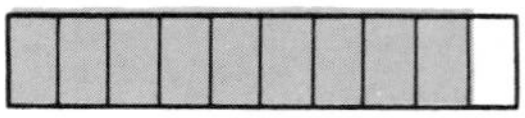

c.
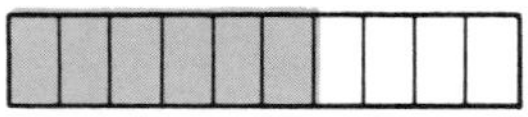

d.
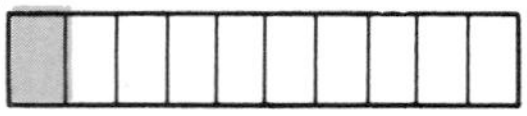

e.
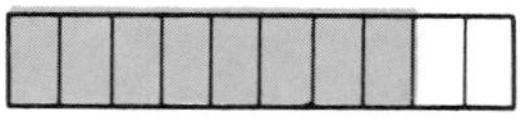

f.
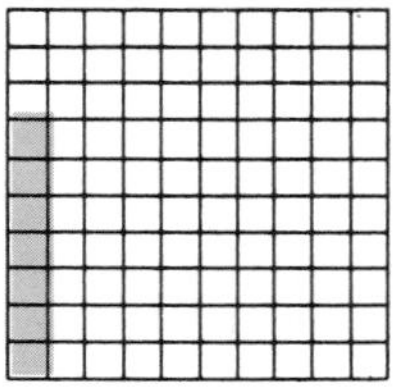

g.
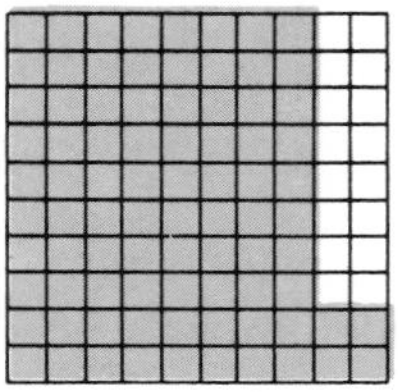

h.
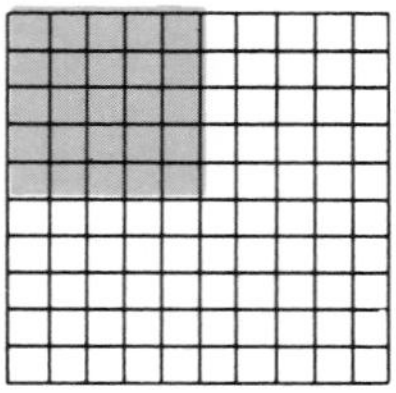

i.

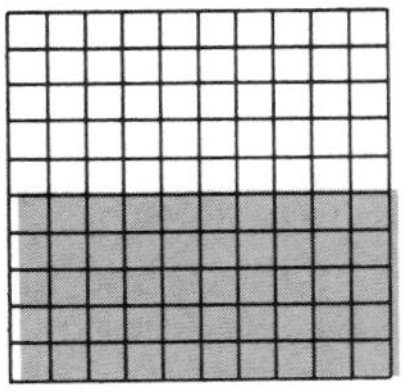

j.

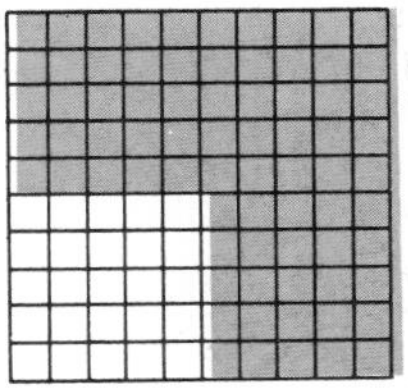

3. The last digit on the right in the following odometer readings registers tenths of a mile. Write the mileage indicated by these readings using decimal numerals.

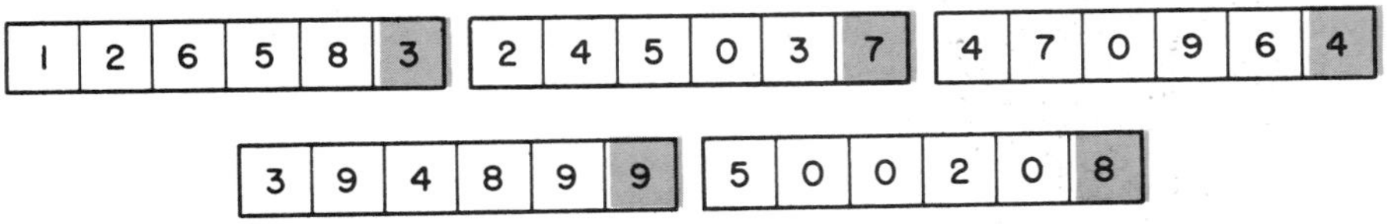

4. Some rulers have each inch subdivided into tenths of an inch. Write as decimal numerals the measurements indicated by *A*, *B*, *C*, *D*, *E*, *F*, *G*, and *H*.

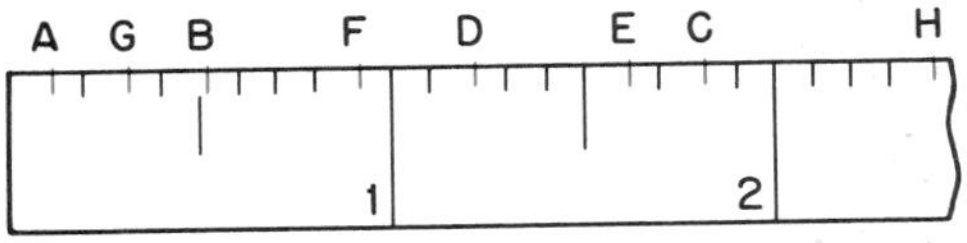

5. In the metric system of measurement, 10 millimeters (mm) = 1 centimeter (cm). Write each of the following measurements as a decimal numeral in terms of centimeters:

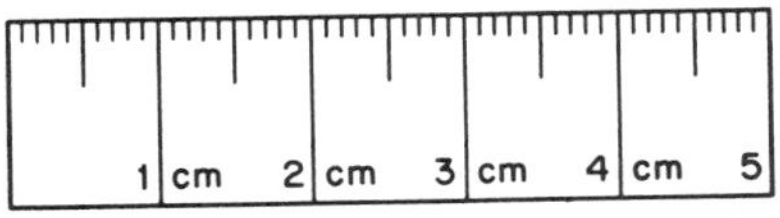

a. 2 mm, 7 mm, 5 mm, 9 mm, 1 mm
b. 14 mm, 21 mm, 39 mm, 43 mm, 17 mm

6. Find the missing values:

a. .25 = __ tenths ____ hundredths or ____ hundredths.
b. .08 = __ tenths ____ hundredths or ____ hundredths.
c. .746 = __ tenths ____ hundredths ______ thousandths or ______ thousandths.
d. .590 = __ tenths ____ hundredths ____ thousandths or ____ thousandths or ____ hundredths.
e. .204 = __ tenths ____ hundredths ______ thousandths or ______ thousandths.

7. Express each of the following as a decimal numeral:

a. 3 dimes **b.** 6 pennies **c.** 8 dimes 4 pennies **d.** 9 dollars 6 dimes 7 pennies **e.** 5 dollars 9 dimes 2 pennies

8. Write each of the following as a decimal numeral:

a. $\frac{17}{10}$, $\frac{36}{10}$, $\frac{278}{100}$, $\frac{906}{100}$, $\frac{1,254}{1,000}$ **b.** $\frac{10}{10}$, $\frac{100}{100}$, $\frac{1,000}{1,000}$, $\frac{20}{10}$, $\frac{300}{100}$

9. Change each of the following to:

Tenths: **a.** .60, .20, .30, .80, .90 **b.** .700, .800, .400, .600, .100 **c.** 5, 9, 7, 3, 8
Hundredths: **a.** .7, .9, .4, .1, .5 **b.** .500, .800, .100, .300, .900 **c.** 1, 6, 3, 9, 4
Thousandths: **a.** .4, .1, .6, .8, .3 **b.** .90, .50, .60, .20, .80 **c.** 2, 8, 1, 7, 6

10. Write the decimal numeral that represents each of the following:

a. 4 tenths 5 hundredths
b. 9 tenths 0 hundredths 4 thousandths
c. 0 tenths 0 hundredths 8 thousandths 7 hundred-thousandths
d. 3 ones 2 tenths
e. 6 hundreds 9 tens 0 ones 5 tenths 1 hundredth

11. What decimal corresponds to point *A*? *B*? *C*? *D*? *E*? *F*? *G*? *H*?

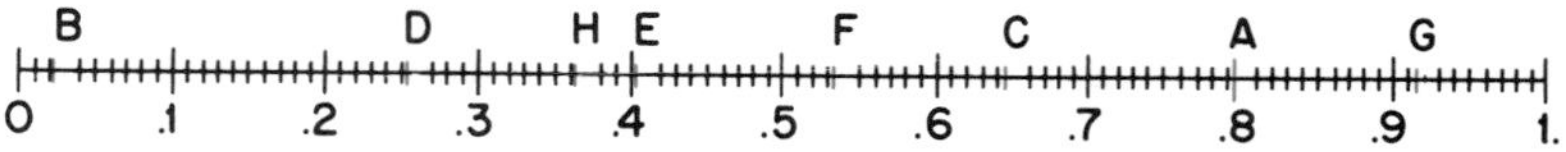

6–2 READING NUMERALS NAMING DECIMAL FRACTIONS

To read a numeral naming a decimal fraction, we read the numeral to the right of the decimal point as we would a numeral for a whole number and use the name that applies to the place value of the last digit on the right. The numeral may have a zero written in the ones place just preceding the decimal point.

> .497 or 0.497 is read: Four hundred ninety-seven thousandths.

To read a numeral naming a mixed decimal, we first read the numeral for the whole number to the left of the decimal point, then the numeral for the decimal fraction, using the word "and" to mark the position of the decimal point.

> 16.52 is read: Sixteen *and* fifty-two hundredths.

EXERCISES

Read, or write in words, each of the following:

1. a. .6; .8; .1; .4; .9 **b.** 0.5; 0.2; 0.7; 0.3; 0.6
2. a. .04; .07; .02; .08; 0.01 **b.** .37; 0.19; .70; 0.46; .85
3. a. 3.5; 8.2; 16.7; 79.3; 210.8 **b.** 1.42; 5.09; 24.75; 98.16; 356.64
4. a. .003; .006; 0.004; .008; .005
b. .017; .082; 0.054; .090; 0.025
c. .375; 0.667; .908; .625; .132
d. 2.016; 9.503; 14.281; 56.875; 400.064

5. a. .0004; .0097; .8603; 0.9375; 9.0841
   b. .00005; 0.04173; .15625; 1.00048; 3.84129
   c. .000007; .050932; 0.998998; 4.000533; 8.014006

6–3 WRITING NUMERALS NAMING DECIMAL FRACTIONS

To write a numeral naming a decimal fraction, we write the digits as we do a numeral naming a whole number and prefix it with a decimal point so that the name of the part corresponds to the place value of the last digit. We insert (prefix) as many zeros as are required between the decimal point and the first digit when it is necessary to make the name of the part and the place value of the last digit correspond.

> Thirty-six thousandths is written as .036 or 0.036

To write a numeral naming a mixed decimal, we first write the numeral for the whole number followed by a decimal point (to indicate the word "and" and to separate the numeral for the whole number from the numeral for the decimal fraction), then we write the numeral naming the decimal fraction.

> One hundred and five tenths is written as 100.5

EXERCISES

Write each of the following as a numeral naming a decimal fraction or mixed decimal:

1. a. Seven tenths; five tenths; one tenth
   b. Two and three tenths; twenty-five and four tenths; two hundred forty-nine and two tenths
2. a. Three hundredths; ninety-six hundredths; forty hundredths
   b. One and seven hundredths; twelve and fourteen hundredths; six hundred and fifty-one hundredths
3. a. Two thousandths; sixty-three thousandths; four hundred sixteen thousandths
   b. Eight and four thousandths; three hundred and eighteen thousandths; one hundred forty and three hundred sixty-two thousandths
4. a. Ninety-seven ten-thousandths; two hundred-thousandths; eighty thousand twenty-four millionths
   b. Four and five ten-thousandths; sixty and thirty-seven millionths; forty-nine and four thousand eight hundred seventy-six hundred-thousandths

6–4 WRITING DECIMALS AS POLYNOMIALS—EXPANDED NOTATION

The value of each place in a decimal numeral may be expressed as a power of ten.

$\frac{1}{10}$ may be written as $\frac{1}{10^1}$; $\frac{1}{100}$ as $\frac{1}{10 \times 10}$ or $\frac{1}{10^2}$; $\frac{1}{1{,}000}$ as $\frac{1}{10 \times 10 \times 10} = \frac{1}{10^3}$; $\frac{1}{10{,}000}$ as $\frac{1}{10 \times 10 \times 10 \times 10}$ or $\frac{1}{10^4}$; etc. It can be shown that $\frac{1}{10} = 10^{-1}$; $\frac{1}{10^2} = 10^{-2}$; $\frac{1}{10^3} = 10^{-3}$; $\frac{1}{10^4} = 10^{-4}$; etc. and that $10^0 = 1$. See page 299.

A decimal numeral may be expressed as the sum of the products of each digit in the numeral and its place value expressed as a power of ten.

$$\begin{aligned} .4358 &= 4 \text{ tenths} + 3 \text{ hundredths} + 5 \text{ thousandths} + 8 \text{ ten-thousandths} \\ &= 4 \times \frac{1}{10} + 3 \times \frac{1}{100} + 5 \times \frac{1}{1{,}000} + 8 \times \frac{1}{10{,}000} \\ &= 4 \times \frac{1}{10^1} + 3 \times \frac{1}{10^2} + 5 \times \frac{1}{10^3} + 8 \times \frac{1}{10^4} \\ &= (4 \times 10^{-1}) + (3 \times 10^{-2}) + (5 \times 10^{-3}) + (8 \times 10^{-4}) \end{aligned}$$

The numeral 532.896 is written in expanded form as:

$$(5 \times 10^2) + (3 \times 10^1) + (2 \times 10^0) + (8 \times 10^{-1}) + (9 \times 10^{-2}) + (6 \times 10^{-3})$$

Observe that the place values are powers of ten arranged in a decreasing order from left to right. Also see page 233.

EXERCISES

1. Express each of the following as a decimal numeral:

a. $(6 \times 10^{-1}) + (4 \times 10^{-2}) + (9 \times 10^{-3})$
b. $(4 \times 10^{-1}) + (5 \times 10^{-2}) + (8 \times 10^{-3}) + (7 \times 10^{-4})$
c. $(3 \times 10^2) + (5 \times 10^1) + (6 \times 10^0) + (2 \times 10^{-1}) + (4 \times 10^{-2})$
d. $(2 \times 10^4) + (8 \times 10^3) + (0 \times 10^2) + (5 \times 10^1) + (9 \times 10^0) + (7 \times 10^{-1}) + (6 \times 10^{-2}) + (1 \times 10^{-3})$

2. Write each of the following numerals in expanded form as a polynomial:

a. .75	**c.** .6041	**e.** 4.7	**g.** 517.26	**i.** 1,256.38
b. .832	**d.** .34952	**f.** 28.39	**h.** 805.934	**j.** 6,492.973

6–5 ROUNDING DECIMALS

Often there is need both in measurement and in money situations for rounding decimals.

To round a decimal, we find the place to which the number is to be rounded, rewriting the given digits to the left of the required place and dropping all the digits to the right of the required place. If the first digit dropped is 5 or more, we increase the given digit in the required place by 1, otherwise we write the same digit as given. However, as many digits to the left are changed as necessary when the given digit in the required place is 9 and one (1) is added.

.32 rounded to the nearest tenth is .3 because .32 is closer to .30 than it is to .40. Note that .3 and .30 are names for the same number.

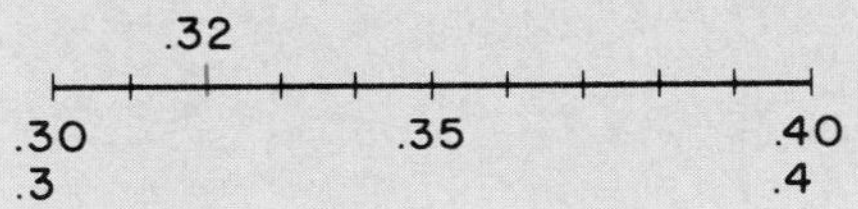

.807 rounded to the nearest hundredth is .81 because .807 is closer to .810 than it is to .800. Note that .81 and .810 are names for the same number.

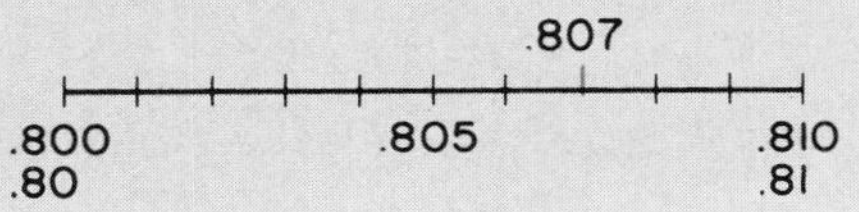

EXERCISES

1. Select the numbers that are nearer to:

a. .5 than .6	.52	.57	.54	.59	.58
b. 0.1 than 0.2	0.16	0.13	0.19	0.11	0.17
c. .87 than .88	.871	.878	.874	.873	.879
d. .90 than .91	.903	.906	.904	.907	.901
e. .246 than .247	.2468	.2460	.2466	.2463	.2467
f. 8 than 9	8.2	8.709	8.49	8.51	8.634

2. Any number from:

a. .35 to .44 rounded to the nearest tenth is what number?
b. .615 to .624 rounded to the nearest hundredth is what number?
c. .8275 to .8284 rounded to the nearest thousandth is what number?
d. 13.5 to 14.4 rounded to the nearest whole number is what number?

3. Round each of the following numbers to the nearest:

a. Tenth:	.56	1.42	6.93	4.07	23.25
	.47	.86	.745	2.33	5.962
b. Hundredth:	.619	6.544	14.238	23.6253	9.4709
	.936	.693	.1925	7.395	16.0694
c. Thousandth:	.1728	4.8273	9.1065	56.83945	18.0507
	.4362	.0546	.38526	9.4384	48.19688
d. Ten-thousandth:	.31625	.49353	7.54196	13.680506	6.58927

e. Hundred-thousandth:	.000047	.000382	.050604	1.350365	4.0906096
f. Millionth:	.0000084	.0093255	.0000049	1.0384563	2.1938427
g. Whole number:	4.6	11.4	29.16	8.09	99.501
	41.3	9.81	64.56	4.475	1.399

4. When required to find an amount correct to the nearest cent or nearest dollar, we follow the same procedure as with decimals. When the amount contains a fractional part of a cent, we drop the fraction but add one cent when the fraction is $\frac{1}{2}$ or more.

$1.574 rounded to nearest cent is $1.57

$.68$\frac{3}{4}$ rounded to nearest cent is $.69

Round each of the following amounts to the nearest:

a. Cent:	$.582	$.098	$1.264	$29.807	$7.5962
	$.48$\frac{1}{2}$	$.56$\frac{2}{5}$	8.18\frac{2}{3}$	5.47\frac{3}{8}$	36.53\frac{5}{6}$
	$.354	$.079	1.45\frac{3}{5}$	82.86\frac{5}{12}$	$34.0649
b. Dollar:	$3.48	$2.37	$8.69	$75.18	$237.75
	$5.25	$9.82	$57.49	$66.98	$540.87

6–6 COMPARING DECIMALS

When necessary, we express the given numbers as numerals containing the same number of decimal places and take the greater number as the greater decimal.

Which is greater .6 or .58?

.6 = .60

.58 = .58

Answer, .6 is greater.

A mixed decimal or whole number is greater than a decimal fraction.

EXERCISES

1. Which is greater:

a. .4 or .39?	**d.** .249 or .25?	**g.** .195 or .1949?	**j.** 2.45 or .253?
b. .72 or .597?	**e.** 3.8 or .38?	**h.** 1.68 or .307?	**k.** 5.009 or 5.0101?
c. .9 or .084?	**f.** .862 or 8.42?	**i.** 7.921 or 7.9209?	**l.** .06 or .009?

2. Which is smaller:

a. .3 or .31?	d. .8 or .799?	g. 4.3 or 4.29?	j. 5.0846 or 5.08461?
b. .60 or .06?	e. .0073 or .008?	h. 7.99 or 8.3?	k. 3.8 or .39?
c. .48 or .5?	f. 6.92 or .692?	i. 3.67 or 3.067?	l. .989 or .9889?

Arrange the following numbers according to size, writing the numeral for:

3. Greatest number first:

a. 5.93, 59.7, .598, .0599
b. .07, .007, .7, .0007
c. 4.78, .0978, .529, .6
d. 1.5, .03, .069, .19
e. 2.02, 1.989, 1.9895, 2.009

4. Smallest number first:

a. 8.5, 85, .85, .085
b. .78, .933, 6.84, .6841
c. 9.06, .91, .9059, 8.95
d. .34, .342, .3401, .3398
e. 1.03, 0.998, 0.9989, 1.016

5. Which of the following are true statements?

a. $.14 < .2$	c. $.9 = 0.9$	e. $.048 > .05$	g. $.299 < .3$	i. $.0004 > .039$
b. $.80 > .8$	d. $3.4 < .34$	f. $.238 < .2381$	h. $16.5 = 1.65$	j. $.999 < 1$

6–7 COMPUTATION—ADDITION OF DECIMALS

Since $.5 = \frac{5}{10}$ and $.3 = \frac{3}{10}$, then $.5 + .3$ and $\frac{5}{10} + \frac{3}{10}$ both name the same number, *eight tenths.*

Decimal Fraction:	Common Fraction:
.5	$\frac{5}{10}$
+.3	$+\frac{3}{10}$
.8	$\frac{8}{10}$

To add decimals:

When it is necessary to write the numerals in columns, we write each addend so that the decimal points are under each other. Zeros may be annexed to the numerals naming decimal fractions so that the addends have the same number of decimal places.

Then we add as in the addition of whole numbers, placing the decimal point in the sum directly under the decimal points in the addends.

When an answer ends in one or more zeros to the right of the decimal point, the zeros may be dropped unless it is necessary to show the exact degree of measurement. We check by adding the columns in the opposite direction.

Add:

6.8 + .47

$$\begin{array}{r} 6.80 \\ .47 \\ \hline 7.27 \end{array}$$

Answer, 7.27

Add:

$$\begin{array}{r} .26 \\ .19 \\ .35 \\ \hline .80 = .8 \end{array}$$

Answer, .8

EXERCISES

1. Use this section of the ruler to find the sum of each of the following:

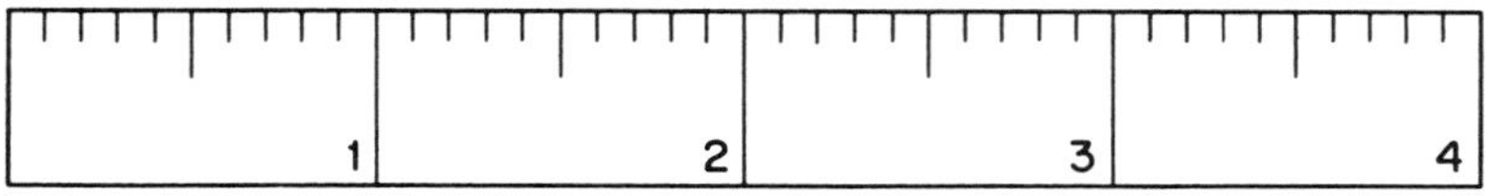

.3″	.1″	.6″	.5″	.9″	1.2″	2.4″	1.6″	1.7″	1.3″
.2″	.8″	.4″	.3″	.1″	2.5″	0.9″	1.8″	0.6″	1.9″

2. In each of the following first add the common fractions or mixed numbers, then the decimals. Compare answers.

a. $\frac{1}{10} + \frac{2}{10}$ $\quad$ $.1 + .2$

b. $\frac{24}{100} + \frac{19}{100}$ $\quad$ $.24 + .19$

c. $2\frac{7}{10} + 3\frac{51}{100}$ $\quad$ $2.7 + 3.51$

3. Add:

a.

.2	.5	.09	.18	.25	3.05	1.2	.48
.4	.7	.03	.23	.82	4.09	.9	7.
.1	.8	.06	.57	.53	1.06	.36	.6

b.

.9	.34	.98	6.5	9.23	1.4	8.403	43.28
.6	.17	.31	7.6	8.42	.82	6.248	36.93
.7	.46	.42	3.6	9.71	3.57	1.352	59.84
.6	.08	.59	2.7	2.83	.6	7.189	20.87

c.

.3	.06	8.3	4.47	.48	6.527	.5684	491.32
.8	.47	9.8	9.06	3.7	2.157	.9748	322.85
.6	.57	3.5	1.59	.5	3.008	.5327	403.26
.5	.28	2.7	3.85	8.	1.282	.7046	915.83
.7	.15	4.8	5.23	1.94	5.309	.8195	226.06

d.

.9	.53	6.4	8.26	.3	85.09	.16928	812.97
.3	.68	9.5	3.05	4.91	19.44	.51107	966.32
.4	.71	2.8	4.50	2.	6.84	.2355	801.49
.7	.27	4.6	9.23	.524	27.75	.83214	602.33
.5	.54	2.5	1.68	3.14	46.81	.5030	579.25
.6	.98	8.2	3.34	.32	.96	.17159	914.08

e.

$.96	$1.75	$2.17	$83.25	$129.75	$ 2.64	$ 54.03	$806.57
.88	.56	8.49	96.71	7.46	14.79	39.29	7.62
.09	3.19	.52	9.48	11.09	.59	615.86	95.26
.76	4.79	.85	.56	347.85	6.08	49.63	136.48
.97	.16	6.74	91.43	2.17	548.67	263.52	345.03

f.

$.38	$8.26	$9.84	$23.37	$445.76	$616.35	$520.88	$268.31
.85	.93	3.68	9.75	386.07	897.42	17.46	985.67
.92	.64	4.25	8.99	91.50	58.09	391.50	29.35
.79	1.39	9.93	54.78	9.86	145.27	69.21	840.58
.08	2.67	5.88	3.26	407.96	9.69	458.68	706.43
.53	5.04	4.26	75.58	85.67	207.34	6.74	149.99

4. Add as indicated:

a. .6 + .2
.07 + 1.3
4.9 + 3.7
18 + .25
.34 + .8

b. 1.84 + 9
.1 + .059
6 + .04
.951 + .24
1.408 + .27

c. .47 + .52 + .85
.35 + 4.2 + 91
2.54 + 83.5 + .927
.005 + .05 + .5
.106 + .9 + .23

d. .35 + .7
.08 + 1.2
9.6 + .58
.4 + .05
.19 + 3

e. 2.84 + .45
1.6 + .397
7.32 + .3
54.9 + .518
40.2 + 8.65

f. 4.23 + .75 + .386
1.69 + .582 + 60.7
85.5 + 9.14 + 208
.756 + .82 + .5
.07 + .017 + .7

g. $.26 + $.57 + $.84
$32.75 + $22.94 + $87.63
$.63 + $5.67 + $23.75
$9.99 + $3.48 + $17.06
$28.85 + $10.39 + $53.97

h. $.86 + $5.28 + $.63
$4.49 + $3.87 + $51.29
$.28 + $15.61 + $7.86
$6.79 + $23.75 + $.57
$50.32 + $84.09 + $65.48

6–8 COMPUTATION—SUBTRACTION OF DECIMALS

Since $.36 = \frac{36}{100}$ and $.15 = \frac{15}{100}$, then $.36 - .15$ and $\frac{36}{100} - \frac{15}{100}$ both name the same number, twenty-one hundredths.

Decimal Fraction:	Common Fraction:
.36	$\frac{36}{100}$
− .15	$-\frac{15}{100}$
.21	$\frac{21}{100}$

To subtract decimals:

When it is necessary to write the numerals, we write the subtrahend under the minuend so that the decimal points are directly under each other. Zeros may be annexed to the decimal fraction or a decimal point and zeros to a whole number in the minuend so that the minuend and subtrahend will have the same number of decimal places.

We subtract as in the subtraction of whole numbers, placing the decimal point in the remainder (or difference) directly under the decimal points of the subtrahend and minuend. When an answer ends in one or more zeros to the right of the decimal point, the zeros may be dropped unless it is necessary to show the exact degree of measurement. We check by adding the remainder to the subtrahend. Their sum should equal the minuend.

Subtract:	25 − 1.64	Subtract:
.56	25.00	25.83
.47	1.64	6.73
.09	23.36	19.10 = 19.1
Answer, .09	*Answer, 23.36*	*Answer, 19.1*

EXERCISES

1. Subtract using the section of the ruler on page 126.

.9″	.8″	.6″	1.0″	1.8″	4.0″	3.6″	1.0″	2.7″	3.5″
− .2″	− .3″	− .5″	− .8″	− 1.4″	− 2.5″	− 1.9″	− 0.6″	− 1.8″	− 0.7″

2. In each of the following first subtract the common fractions or mixed numbers, then the decimals. Compare answers.

a.	$\frac{7}{10}$	.7	**b.**	$8\frac{3}{10}$	8.3	**c.**	$4\frac{9}{10}$	4.9
	$-\frac{4}{10}$	− .4		$-3\frac{6}{10}$	− 3.6		$-1\frac{53}{100}$	− 1.53

3. Subtract:

a.	.8	.37	.84	.09	8.7	9.2	6.1	5.4	.75	.62	8.7	.15
	.5	.14	.29	.03	1.2	3.8	4.7	1.9	.68	.22	3.7	.12

b.	.28	.36	9.4	.432	5.17	.416	.604	53.5	86.54	65.1
	.08	.29	8.4	.257	1.69	.387	.066	8.2	43.95	7.4

c.	4.026	.9685	56.17	8.523	.9608	4.43	.59325	.8624
	1.548	.7949	27.89	6.681	.1599	3.55	.28963	.8619

d.	397.26	61.864	.05693	805.41	9.4257	5.000	8.0000
	178.79	52.879	.05598	319.87	2.4248	3.817	6.0593

e.	9.47	.3	8.	4.1	.05	5.8	7.	1.032
	.6	.154	2.68	.39	.0485	.704	3.002	.28

f.	$6.35	$2.08	$50.00	$58.15	$248.31	$643.06	$300.00	$4,000.00
	4.67	.59	27.36	49.65	209.54	88.97	149.60	2,708.21

4. Subtract as indicated:

a.	**b.**	**c.**
.51 − .27	.64 − .6	5.8 − .32
.048 − .039	.845 − .84	9 − .625
.207 − .087	.407 − .38	2 − 1.964
.4100 − .3608	.36 − .178	3 − .002
8.038 − 4.168	.005 − .0001	8.81 − 1.9009

d. \$.70 – \$.56
\$8.32 – \$4.63
\$19.60 – \$6.85
\$22.86 – \$19.57
\$118.50 – \$23.76

e. \$3.82 – \$.95
\$7.50 – \$.75
\$9.69 – \$2
\$106.23 – \$99
\$6.05 – \$.87

f. \$5 – \$.62
\$10 – \$.74
\$4 – \$1.25
\$9 – \$3.06
\$7 – \$5.91

g. Subtract .67 from .8
Take \$.43 from \$5.
From .071 take .06
From \$6.80 subtract \$5.
Subtract 2.125 from 3
From \$3.17 take \$.98

6–9 COMPUTATION—MULTIPLICATION OF DECIMALS

The product of the common fractions $\frac{3}{10}$ and $\frac{7}{10}$ is $\frac{21}{100}$.
Decimally this is written

$$.3 \times .7 = .21$$

The product of $\frac{2}{10}$ and $\frac{4}{100}$ is $\frac{8}{1,000}$.
Decimally this is written

$$.2 \times .04 = .008$$

Observe that the number of decimal places in the product is equal to the sum of the number of decimal places in the factors.

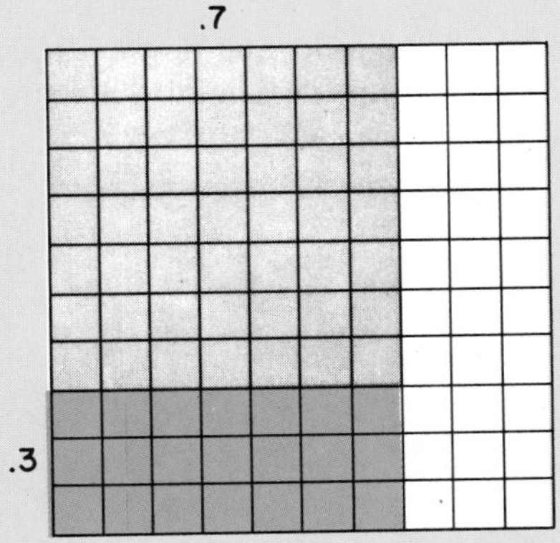

Each square = .01

.7 of the large square is shaded.
.3 of the shaded squares are shaded □.
A count of the squares shaded □ illustrates that .3 × .7 = .21

To multiply decimals:

We write the given numerals and multiply as in the multiplication of whole numbers. The decimal point in one factor does not necessarily have to be directly under the decimal point in the second factor. We point off in the product, counting from right to left, as many decimal places as there are in *both* factors together. When the product contains fewer digits than the required number of decimal places, we prefix as many zeros as are necessary.

When a decimal answer ends in one or more zeros to the right of the decimal point, the zeros may be dropped unless it is necessary to show the exact degree of measurement. We check by interchanging the two factors and multiplying again.

Multiply:	Multiply:	Multiply:	Multiply:
.6	.14	.04	2.3
.3	.5	.02	.005
.18	.070 = .07	.0008	.0115
Answer, .18	*Answer*, .07	*Answer*, .0008	*Answer*, .0115

EXERCISES

1. In each of the following first multiply the common fractions or mixed numbers, then multiply the decimals. Compare answers.

a. $\frac{4}{10} \times \frac{9}{10}$	**b.** $\frac{5}{10} \times \frac{6}{10}$	**c.** $\frac{8}{10} \times \frac{3}{100}$	**d.** $\frac{19}{100} \times \frac{47}{100}$	**e.** $3\frac{1}{10} \times 1\frac{7}{10}$
.4	.5	.8	.19	3.1
× .9	× .6	× .03	× .47	× 1.7

2. Multiply:

a.	.2	3	8	.6	24	.4	9.5	100	.5	144
	3	.3	.9	5	.7	90	56	.6	35	.8
b.	.12	.29	9	.07	3.14	.165	25	.0105	800	6.0075
	6	81	.01	15	6	910	.008	45	.0003	40
c.	.4	.3	.4	.29	.13	.1	.925	.2063	8.92	9.7
	.6	.2	.5	.7	.6	.09	.8	.4	.7	3.8
d.	3.87	4.91	.05	7.28	10.54	2.693	4.107	9.0681	6.823	1.0509
	.5	.6	8.2	5.8	3.9	.5	8.6	2.8	.4	2.7
e.	.46	.32	.14	.01	.8	.068	.004	12.75	2.03	.006
	.53	.33	.07	.01	.50	.75	.02	.06	2.03	.09
f.	3.8	4.7	.5	9.3	15.6	3.084	5.469	6.318	.007	2.9051
	.25	.39	1.75	.87	2.76	.12	.04	5.16	7.83	12.06

g.	.587	.008	.116	.6	14.75	7.004	.0028	1.875
	.625	.014	.129	3.849	2.125	.1055	1.529	1.875
h.	.3937	3.1416	1.0936	.6214	478.3	50.75	6.4516	4.875
	.9	8.5	20.4	.93	.7646	.9144	800.9	2.2046

3. In examples involving money, find products correct to nearest cent:

a. \$.87	\$6.59	\$18.25	\$208.40	\$856	\$9,000	\$328.20
12	60	.03	.29	.625	.045	.007

b. \$20	\$4,000	\$5.26	\$.95	\$280	\$69.25	\$57.90
$.37\frac{1}{2}$	$.05\frac{1}{2}$	$.66\frac{2}{3}$	$.83\frac{1}{3}$	$.04\frac{3}{4}$	$.03\frac{1}{4}$	$.16\frac{2}{3}$

4. Multiply as indicated:

a. $.6 \times .4$	**b.** $\frac{1}{2} \times \$.78$	**c.** $\frac{1}{2} \times \$.59$	**d.** $.75 \times \$4$
$8 \times .005$	$\frac{3}{4} \times \$.92$	$\frac{5}{6} \times \$3.45$	$.6 \times \$.50$
$14 \times .03$	$\frac{7}{8} \times \$1.52$	$\frac{2}{3} \times \$38$	$.02 \times \$8$
$.07 \times 20$	$1\frac{2}{3} \times \$10.47$	$2\frac{3}{4} \times \$10.50$	$.87\frac{1}{2} \times \$24$
$.01 \times .018$	$1\frac{5}{12} \times \$4.08$	$3\frac{5}{8} \times \$7.31$	$.04 \times \$9.50$

5. Square each of the following:

a. .3	**c.** .02	**e.** .48	**g.** 3.14	**i.** 73.8	**k.** 39.37
b. .9	**d.** .001	**f.** 2.37	**h.** .989	**j.** 15.26	**l.** 3.2808

6–10 MULTIPLYING BY POWERS OF TEN

To multiply a decimal by 10; 100; 1,000 or larger powers of ten quickly, we first write the digits of the given numeral, then we move the decimal point as many places to the right of the original position as there are zeros in the given multiplier. Where the product requires them, we use zeros as placeholders.

$10 \times 4.82 = 48.2$
$100 \times .06 = 6$
$1{,}000 \times 7.5 = 7{,}500$

EXERCISES

1. Multiply each of the following numbers by 10:

.7; .16; .945; .08; 6.5; 4.27; .1; .075; .003; 7.50; 41.6; 6.048

2. Multiply each of the following numbers by 100:

.53; .06; .3; .478; .009; 6.48; 1.7; 44.1; 9.225; .015; .4; .29

3. Multiply each of the following numbers by 1,000:

.864; .25; .8; .9018; 6.375; 4.61; .0006; 17.1; 39.37; .0045; .5; .04

4. Multiply by short method:

a. $100{,}000 \times 8.6$	**e.** $100{,}000{,}000 \times 1.8$
b. $9.7 \times 1{,}000{,}000$	**f.** $3.4 \times 1{,}000{,}000{,}000$
c. $10{,}000{,}000 \times 12.1$	**g.** $48.5 \times 10{,}000{,}000{,}000$
d. $36.9 \times 1{,}000{,}000$	**h.** $1{,}000{,}000{,}000 \times 27.3$

6–11 COMPUTATION—DIVISION OF DECIMALS

(1) (2) To divide a decimal fraction or mixed decimal by a whole number, we divide as in the division of whole numbers, placing the decimal point in the quotient directly above the decimal point in the dividend.

(3) To divide a whole number by a whole number where the division is not exact, this division may be carried out to as many decimal places as are required in the quotient by annexing a decimal point and the required zeros to the dividend.

(1)

$$4\overline{)6.4} \quad \text{quotient } 1.6$$

Answer, 1.6

(2)

$$7\overline{).42} \quad \text{quotient } .06$$

Answer, .06

(3)

$$4\overline{)11} = 4\overline{)11.00} \quad \text{quotient } 2.75$$

Answer, 2.75

To divide a decimal fraction or mixed decimal by a decimal fraction or mixed decimal, we make the divisor a whole number by multiplying it by the proper power of ten so that its decimal point is moved to the right of the last digit indicating its new position by a caret ($_\wedge$). We multiply the dividend by the same power of ten so that its decimal point is moved to the right as many places as we moved the decimal point in the divisor and indicate its new position by a caret ($_\wedge$). The reason we may do this is illustrated below (4).

(4)

$$.03\overline{).645} = \frac{.645}{.03} = \frac{.645}{.03} \times 1 = \frac{.645 \times 100}{.03 \times 100} = \frac{64.5}{3} = 3\overline{)64.5}$$

Note that the multiplicative identity 1 is used here.

Thus, $.03\overline{).645}$ becomes $3\overline{)64.5}$

Instead of doing all the work shown above, we may move decimal points directly in the given problem and indicate the new positions by carets.

Then we divide as in the division of whole numbers, placing the decimal point in the quotient directly above the caret ($_\wedge$) in the dividend.

$$.03\overline{).645} \text{ becomes } .03_\wedge\overline{).64_\wedge 5} \quad \text{quotient } 21.5$$

When the dividend contains fewer decimal places than required, we annex as many zeros as are necessary to a decimal fraction dividend, and a decimal point and the required zeros to a dividend containing a whole number.

Divide 119.1 by .003:

$$\begin{array}{r} 39\,700. \\ .003_\wedge\overline{)119.100_\wedge} \end{array}$$

Answer, 39,700

Divide \$5 by \$1.25:

$$\begin{array}{r} 4 \\ \$1.25_\wedge\overline{)\$5.00_\wedge} \end{array}$$

Answer, 4

We check by multiplying the quotient by the divisor and adding the remainder, if any, to the product. The result should be equal to the dividend.

To find the quotient correct to the nearest required decimal place:

We find the quotient to one more than the required number of decimal places, then round it off.

or

We find the quotient to the required number of decimal places, adding 1 to the last digit of the quotient if the remainder is equal to one-half or more of the divisor.

Find quotient to nearest hundredth:

$$\begin{array}{r} 2.883 \\ 3\overline{)8.650} \end{array}$$

2.883 rounded to nearest hundredth is 2.88

Answer, 2.88

Find quotient to nearest tenth:

$$\begin{array}{r} 1.5 \\ 12\overline{)19.1} \\ \underline{12} \\ 7\,1 \\ \underline{6\,0} \\ 1\,1 \end{array}$$

$\frac{1}{2}$ of divisor 12 is 6.

Remainder 11 is greater than 6.

Digit 5 of quotient is increased to 6.

Answer, 1.6

EXERCISES

1. In each of the following annex zeros to the dividend until the division is exact:

$12\overline{)9}$ $\quad$ $80\overline{)52}$ $\quad$ $25\overline{)37}$ $\quad$ $56\overline{)49}$ $\quad$ $125\overline{)1}$

2. Divide:

a.	$3\overline{)6.3}$	$6\overline{).852}$	$8\overline{).0056}$	$38\overline{).266}$	$87\overline{)520.26}$
b.	$8\overline{)40.56}$	$7\overline{)6.013}$	$75\overline{)697.5}$	$48\overline{)36.00}$	$16\overline{)1.0000}$
c.	$8.5\overline{)399.5}$	$.9\overline{)53.01}$	$1.8\overline{).432}$	$6.4\overline{).0512}$	$4.5\overline{).009}$

d.	$.2\overline{)8.18}$	$.5\overline{).004}$	$.9\overline{).0054}$	$7.5\overline{)65.325}$	$3.4\overline{)17.0}$
e.	$.53\overline{).424}$	$.77\overline{)1.9558}$	$6.29\overline{)54.094}$	$.05\overline{).01}$	$.35\overline{)10.5}$
f.	$.03\overline{).00021}$	$.96\overline{)24.00}$	$.87\overline{)2.068}$	$39.37\overline{)338.582}$	$3.14\overline{)20.41}$
g.	$.005\overline{).745}$	$.618\overline{).4944}$	$1.025\overline{)4.1}$	$.134\overline{).67}$	$2.349\overline{)14.094}$
h.	$.006\overline{)19.236}$	$.043\overline{)1.29}$	$.375\overline{)150.000}$	$1.125\overline{)16.875}$	$.018\overline{).01674}$
i.	$.7\overline{)49}$	$9.6\overline{)24}$	$.04\overline{)6}$	$2.75\overline{)22}$	$6.058\overline{)3029}$

3. Find quotient correct to nearest tenth:

$15\overline{)13}$ $\quad 25.4\overline{)54.96}$ $\quad .875\overline{).6}$ $\quad .62\overline{)5.9}$ $\quad 1.1\overline{)20}$

4. Find quotient correct to nearest hundredth:

$9\overline{)4}$ $\quad .7\overline{)3}$ $\quad .625\overline{).52}$ $\quad .87\overline{)63.5}$ $\quad 16.39\overline{)3.94}$

5. Find quotient correct to nearest thousandth:

$12\overline{)85}$ $\quad 2.2\overline{)3{,}000}$ $\quad .039\overline{)4.75}$ $\quad 7.5\overline{).896}$ $\quad 1.47\overline{)5.04}$

6. Find answer correct to nearest cent:

$8\overline{)\$.56}$ $\quad 144\overline{)\$291.62}$ $\quad 96\overline{)\$79.25}$ $\quad 72\overline{)\$8.43}$ $\quad 108\overline{)\$5{,}126.03}$

7. Divide:

$\$.06\overline{)\$30}$ $\quad \$.10\overline{)\$9.70}$ $\quad \$1.75\overline{)\$46}$ $\quad \$.59\overline{)\$18.91}$ $\quad \$2.47\overline{)\$253.36}$

8. Divide as indicated:

$62.5 \div .25$ $\qquad .0141 \div 4.7$ $\qquad \$18 \div \$.24$

6–12 DIVIDING BY POWERS OF TEN

We divide whole numbers and decimals by powers of ten (10; 100; 1,000, etc) quickly by first writing the digits of the given numeral, then we move the decimal point as many places to the left of its original position as there are zeros in the given divisor.

$600 \div 10 = 60$ $\qquad 284.1 \div 100 = 2.841$

$9 \div 100 = .09$ $\qquad 825.4 \div 1{,}000 = .8254$

EXERCISES

1. Divide each of the following numbers by 10:

40; 56; 8; .3; 2.7; 579; .42; .385; 36.9; 800; 6.27; 93.54

2. Divide each of the following numbers by 100:

600; 8,000; 415; 89; 2; .7; .626; 70.4; 831.97; 530; 9.63; 40

3. Divide each of the following numbers by 1,000:
   3,000; 8,375; 486; 57; .281; .5; 868.4; 6; 250,000; 1,230; .01; 4,000.5
4. Divide by the short method:
   a. 3,900,000 ÷ 100,000
   b. 7,800,000 ÷ 10,000
   c. 15,200,000 ÷ 1,000,000
   d. 437,580,000 ÷ 1,000,000
   e. 2,400,000,000 ÷ 1,000,000,000
   f. 225,000,000,000 ÷ 100,000,000

6–13 SHORTENED NAMES FOR LARGE WHOLE NUMBERS

To name a large whole number whose numeral ends in a series of zeros, newspapers and periodicals are now using a numeral followed by the name of the appropriate period of the number scale.

14.9 *million* is the shortened name for 14,900,000.
3.28 *billion* is the shortened name for 3,280,000,000.

To write the complete numeral for the shortened name, we multiply the number named by the given numeral prefix by the power of ten that is equivalent to the value of the period name used. See section 6-10 for short method.

14.9 million = 14.9 × 1,000,000 = 14,900,000
3.28 billion = 3.28 × 1,000,000,000 = 3,280,000,000

To write the shortened name for the numeral ending in zeros, we divide the given number by the power of ten that is equivalent to the value of the period name that we use. See section 6-12 for short method.

14,900,000 = 14.9 million since 14,900,000 ÷ 1,000,000 = 14.9
3,280,000,000 = 3.28 billion since 3,280,000,000 ÷ 1,000,000,000 = 3.28

EXERCISES

1. Write the complete numeral for each of the following:
   a. 57.4 million; 9.25 million; 125.6 million; 6.598 million
   b. 2.3 thousand; 46.59 thousand; 8.827 thousand; 496.4 thousand
   c. 19.9 billion; 105.2 billion; 3.94 billion; 94.703 billion
   d. 5.6 hundred; 45.84 hundred; 957.1 hundred; 33.92 hundred
   e. 8.7 trillion; 17.6 trillion; 6.65 trillion; 254.741 trillion

2. Write the shortened name for each of the following:

a. In billions: 8,500,000,000; 42,900,000,000; 255,840,000,000; 6,967,000,000
b. In millions: 76,400,000; 5,770,000; 349,300,000; 36,828,000
c. In hundreds: 3,910; 2,584; 19,650; 97,437
d. In thousands: 50,700; 9,120; 783,400; 4,236,960
e. In trillions: 9,600,000,000,000; 86,500,000,000,000; 2,730,000,000,000; 514,089,000,000,000; 705,900,000,000,000

6–14 SCIENTIFIC NOTATION

A brief form of writing numerals usually for very large or very small numbers is known as *scientific notation.* A number is expressed in scientific notation when its numeral names a number that is greater than 1 but less than 10 multiplied by some power of ten.

> 6,700,000 is 6.7 million, which may be expressed in scientific notation as 6.7×10^6

To write a numeral by scientific notation, we rewrite the significant digits:

(1) As a numeral for a whole number if there is only one significant digit, as in 6,000,000.
(2) As a numeral for a mixed decimal if there are two or more significant digits, as in 6,700,000 or 6,780,000, using the first digit as a numeral for a whole number and all other digits as the numeral for a decimal fraction.

Then we indicate that this numeral is multiplied by the required power of ten.

The required power of ten may be determined as follows:

By dividing the whole number or mixed decimal into the given number and changing the quotient into a power of ten or by counting the number of places the decimal point is being moved.

When the given number is 10 or greater, a positive integer[1] is used for the exponent.

> Write 73,400,000 by scientific notation:
>
> $73{,}400{,}000 = 7.34 \times ?$
>
> 10,000,000
>
> Divide $7.34_{\wedge})\overline{73{,}400{,}000.00_{\wedge}}$
>
> However $10{,}000{,}000 = 10^7$
>
> $73{,}400{,}000 = 7.34 \times 10^7$
>
> *Answer,* 7.34×10^7

$73{,}400{,}000 = 7.34 \times 10^7$ The decimal point is moved 7 places to the left.

---

[1]See page 230.

When the given number is between 0 and 1, a negative integer[1] is used for the exponent.

> Write .000048 by scientific notation:
>
> $.000048 = 4.8 \times ?$
>
> Divide: $4.8_{\wedge}\overline{).0_{\wedge}00048}$ = .00001
>
> However $.00001 = 10^{-5}$
>
> $.000048 = 4.8 \times 10^{-5}$
>
> *Answer,* 4.8×10^{-5}

$.000048 = 4.8 \times 10^{-5}$

The decimal point is moved 5 places to the right.

EXERCISES

1. Express each of the following numbers by scientific notation:

a. 40 65 900 387 7,000 4,830 6,090

b. 30,000 84,000 460,000 75,000,000 3,274,000

c. 5,000,000,000 6,700,000 198,000,000,000 59,306,000,000,000

d. 860000000000 45200000000000 7000000000000000 38000000000000000

e. 18.5 396.1 27.92 87.017 934.82 6,528.7 3,928.36

2. Express each of the following numbers by scientific notation:

a. .6 .8 .4 .27 .03 .54 .95

b. .076 .004 .591 .0623 .3006 .0058 .9214

c. .00037 .00196 .08285 .000694 .0000075

d. .0000021 .00000583 .000000035 .000000106 .0000000294

e. .000000000064 .0000000000408 .00000000005 .000000000000719

3. Express each of the following numbers by scientific notation:

a. Light travels at a speed of 186,000 miles per second.

b. The earth is 239,000 miles from the moon and 93,000,000 miles from the sun.

c. The weight of the earth is 6,594,000,000,000,000,000,000 tons and its volume is 259,900,000,000 cubic miles.

d. Our Milky Way is about 600,000,000,000,000,000 miles in diameter.

e. A millimicron is .00000003927 inch and an inch is about .000016 mile.

6–15 EXPRESSING NUMERALS NAMING COMMON FRACTIONS AS DECIMAL NUMERALS

Since a common fraction indicates division (for example, $\frac{3}{4}$ means 3 divided by 4), to express a numeral naming a common fraction as a decimal numeral we divide the numerator by the denominator.

---

[1]See page 230.

$$\frac{3}{4} = 3 \div 4 = 4\overline{)3.00}\quad .75$$

Answer, .75

When the common fraction has as its denominator some power of ten (10; 100; 1,000; etc.), we drop the denominator and rewrite the numerator, placing the decimal point so that the name of the part and the place value of the last digit correspond.

$$\frac{489}{1{,}000} = .489$$

To express a numeral naming a mixed number as a decimal numeral, we express the numeral for the common fraction as a numeral for a decimal fraction and annex it to the numeral for the whole number.

$$3\tfrac{1}{2} = 3.5$$
since $\frac{1}{2} = .5$

EXERCISES

1. Express each of the following as a 1-digit numeral naming a decimal fraction:

$\frac{7}{10}$, $\frac{9}{10}$, $\frac{1}{10}$, $\frac{3}{10}$, $\frac{4}{10}$, $\frac{1}{2}$, $\frac{2}{5}$, $\frac{9}{15}$, $\frac{1}{5}$, $\frac{4}{5}$

2. Express each of the following as a 2-digit numeral naming a decimal fraction:

$\frac{21}{100}$, $\frac{7}{100}$, $\frac{59}{100}$, $\frac{1}{4}$, $\frac{3}{4}$, $\frac{4}{25}$, $\frac{9}{50}$, $\frac{17}{20}$, $\frac{27}{36}$, $\frac{14}{56}$

3. Express each of the following as a 3-digit numeral naming a decimal fraction:

$\frac{728}{1{,}000}$, $\frac{3}{8}$, $\frac{1}{8}$, $\frac{5}{8}$, $\frac{7}{8}$, $\frac{45}{72}$, $\frac{7}{56}$, $\frac{18}{48}$, $\frac{84}{96}$, $\frac{61}{1{,}000}$

4. Express each of the following as a 2-digit numeral naming a decimal fraction. Retain the remainder as a common fraction.

$\frac{1}{3}$, $\frac{1}{6}$, $\frac{2}{3}$, $\frac{5}{6}$, $\frac{7}{12}$, $\frac{35}{42}$, $\frac{14}{32}$, $\frac{33}{36}$, $\frac{9}{48}$, $\frac{10}{24}$

5. Express each of the following as a numeral naming a mixed decimal:

$1\frac{1}{2}$, $2\frac{3}{4}$, $5\frac{9}{10}$, $3\frac{5}{8}$, $9\frac{7}{8}$, $\frac{8}{5}$, $\frac{9}{2}$, $\frac{11}{8}$, $\frac{15}{4}$, $\frac{23}{10}$

6–16 REPEATING DECIMALS

To change the fraction $\frac{2}{3}$ to a decimal we divide 2 by 3. However, we observe that the division on the next page is not exact. The remainder at each step is the same (2). Examination of the quotient will show that the digit 6 repeats and will keep repeating endlessly as we extend the division.

When we change $\frac{3}{11}$ to a decimal, we observe that this division also is not exact. The remainder at each second step is the same (3). Examination of the quotient will show that the pair of digits 27 repeats and will keep repeating endlessly as we extend the division.

```
     .666              .2727
  3)2.000          11)3.0000
    1 8                2 2
     20                 80
     18                 77
      20                 30
      18                 22
       2                  80
                          77
                           3
```

We call these decimals which have a digit or group of digits repeating endlessly *repeating decimals*. The first quotient above is written as $.66\overline{6}\ldots$ (or $.\overline{6}\ldots$ or simply $.\overline{6}$) and the second quotient as $.27\overline{27}\ldots$ (or $.\overline{27}\ldots$ or simply $.\overline{27}$), the bar indicating the repeating sequence (period) and the dots indicating that the sequence repeats endlessly.

When we change $\frac{3}{4}$ to a decimal, the exact quotient is .75 which could be written as .750 or .7500 or $.7500\overline{0}\ldots.$ The decimal form .75 is called a *terminating decimal*. However, since it may also be written in the repeating form $.75\overline{0}\ldots$ or $.750\overline{0}\ldots$ or $.7500\overline{0}\ldots$, we see that the terminating decimals are repeating decimals.

```
    .75
  4)3.00
    2 8
     20
     20
      0
```

Summarizing, observe that if a remainder of 0 occurs in the division, the quotient is a terminating decimal. If, however, after a series of divisions, a remainder other than 0 repeats, then the sequence of digits obtained in the quotient between occurrences of this remainder will repeat endlessly.

Since we consider a terminating decimal to be a repeating decimal, then it is possible to express every common fraction as a repeating decimal.

EXERCISES

Express each of the following as a numeral naming a repeating decimal. Indicate the repeating sequence by a horizontal bar.

a. $\frac{1}{3}$	**e.** $\frac{5}{16}$	**i.** $\frac{6}{13}$	**m.** $\frac{19}{21}$	**q.** $\frac{9}{11}$	**u.** $\frac{8}{17}$
b. $\frac{1}{8}$	**f.** $\frac{4}{9}$	**j.** $\frac{15}{33}$	**n.** $\frac{7}{18}$	**r.** $\frac{2}{5}$	**v.** $\frac{7}{13}$
c. $\frac{1}{6}$	**g.** $\frac{7}{12}$	**k.** $\frac{11}{15}$	**o.** $\frac{14}{17}$	**s.** $\frac{13}{15}$	**w.** $\frac{19}{24}$
d. $\frac{4}{7}$	**h.** $\frac{8}{11}$	**l.** $\frac{5}{23}$	**p.** $\frac{8}{9}$	**t.** $\frac{12}{19}$	**x.** $\frac{4}{27}$

6–17 EXPRESSING NUMERALS NAMING DECIMAL FRACTIONS AS NUMERALS NAMING COMMON FRACTIONS

To express a numeral naming a decimal fraction as a numeral naming a common fraction, we write the numeral for a common fraction so that:

The digits of the decimal numeral are the numerator, and a power of ten (10; 100; 1,000; etc.) corresponding to the place value of the last digit of the decimal numeral is the denominator. The fraction is then expressed in lowest terms.

$$.45 = \frac{45}{100} = \frac{9}{20}$$

Answer, $\frac{9}{20}$

To express a numeral naming a mixed decimal as a numeral naming a mixed number, we write the numeral naming the decimal fraction as a numeral naming a common fraction and annex it to the numeral for the whole number.

$$2.5 = 2\frac{5}{10} = 2\frac{1}{2}$$

Answer, $2\frac{1}{2}$

EXERCISES

1. Express each of the following as a numeral naming a common fraction:

a. .3, .4, .6, .2, .8, .57, .25, .85, .75, .68

b. .60, .05, .70, .04, .90, .193, .625, .375, .125, .875

c. $.12\frac{1}{2}$, $.87\frac{1}{2}$, $.06\frac{1}{4}$, $.62\frac{1}{2}$, $.37\frac{1}{2}$, $.16\frac{2}{3}$, $.33\frac{1}{3}$, $.66\frac{2}{3}$, $.83\frac{1}{3}$, $.08\frac{1}{3}$

2. Express each of the following as a numeral naming a mixed number:

1.2, 3.5, 4.25, 9.875, 6.08, $2.87\frac{1}{2}$, $7.66\frac{2}{3}$, $8.12\frac{1}{2}$, $10.33\frac{1}{3}$, $5.06\frac{1}{4}$

6–18 EXPRESSING A REPEATING DECIMAL AS A NUMERAL NAMING A COMMON FRACTION

To change a repeating decimal to a common fraction:

We multiply the given repeating decimal by some power of 10 so that there is a whole number to the left of the decimal point and the repeating sequence begins to the right of the decimal point. If the repeating sequence has 1 digit, we multiply by 10; if 2 digits, we multiply by 100, etc. We subtract the given number from this product to get a whole number for the difference. (1) (2)

(When a repeating decimal begins with digits other than those used in the repeating sequence, another step is necessary as shown in (3).)

Then we solve the resulting equation (see page 295) to find the required common fraction.

Thus we see that every common fraction may be expressed as a repeating decimal and every repeating decimal may be expressed as a common fraction.

(1) $.55\overline{5}\cdots = ?$

Let $n = .555\cdots$
$10n = 5.555\cdots$
$10n - n = 5$
$9n = 5$
$n = \frac{5}{9}$

Answer, $\frac{5}{9}$

(2) $.1818\overline{18}\cdots = ?$

Let $n = .181818\cdots$
$100n = 18.181818\cdots$
$100n - n = 18$
$99n = 18$
$n = \frac{18}{99}$
$n = \frac{2}{11}$

Answer, $\frac{2}{11}$

(3) $.2666\overline{6}\cdots = ?$

Let $n = .2666\cdots$
$100n = 26.666\cdots$
and $10n = 2.666\cdots$
then $90n = 24$
$n = \frac{24}{90}$
$n = \frac{4}{15}$

Answer, $\frac{4}{15}$

EXERCISES

Express each of the following as a numeral naming a common fraction:

a. $.22\overline{2}\ldots$	**d.** $.933\overline{3}\ldots$	**g.** $.\overline{81}$	**j.** $.\overline{153846}$
b. $.\overline{36}\ldots$	**e.** $.08\overline{3}\ldots$	**h.** $.458\overline{3}\ldots$	**k.** $.\overline{285714}$
c. $.\overline{8}$	**f.** $.041\overline{6}$	**i.** $.91\overline{6}\ldots$	**l.** $.\overline{095238}$

DECIMAL RELATIONSHIPS BETWEEN NUMBERS

Decimal fractions like common fractions are used to express relationships between numbers. Similarly there are 3 types of relationships: finding a decimal part of a number, finding what decimal part one number is of another, and finding a number when a decimal part of it is known.

6–19 DECIMAL PART OF A NUMBER

To find a decimal part of a number, we multiply the number by the given decimal.

Find .09 of 25:

$$\begin{array}{r} 25 \\ \times .09 \\ \hline 2.25 \end{array}$$

Answer, 2.25

EXERCISES

1. Find: a. .3 of 36 b. .9 of 400 c. .05 of 257 d. .48 of 350 e. .91 of 1,800 f. .001 of 8 g. .875 of 320 h. .0625 of 6,400

2. Find: a. .2 of .9 b. .1 of .46 c. .7 of .012 d. .05 of .4 e. .38 of .97 f. .82 of .509 g. .625 of .86 h. .004 of .6

3. Find: a. .4 of 7.3 b. .08 of 2.99 c. .7 of 9.25 d. .6 of 80.41 e. .08 of 12.93 f. .25 of 5.37 g. .01 of 183.26 h. .009 of 400.08

6–20 WHAT DECIMAL PART ONE NUMBER IS OF ANOTHER

Since a fraction expresses what part one number is of another and a common fraction may be expressed as an equivalent decimal fraction, to find what decimal part one number is of another, we write the numeral of a common fraction that expresses the fractional part one number is of the other, then we divide the numerator by the denominator.

What decimal part of 25 is 4?

$$\frac{4}{25} = 4 \div 25 = 25\overline{)4.00} = .16$$

$$\begin{array}{r} .16 \\ 25\overline{)4.00} \\ 2\,5 \\ \hline 1\,50 \\ 1\,50 \\ \hline \ldots \end{array}$$

Answer, .16

EXERCISES

Find the following:

1. **a.** What decimal part of 2 is 1? **b.** 2 is what decimal part of 5?
2. **a.** What decimal part of 4 is 3? **b.** 1 is what decimal part of 4?
3. **a.** 5 is what decimal part of 8? **b.** What decimal part of 8 is 7?
4. **a.** What decimal part of 16 is 1? **b.** 9 is what decimal part of 16?
5. **a.** What decimal part of 3 is 1? **b.** 5 is what decimal part of 6?
6. **a.** What decimal part of 15 is 6? **b.** What decimal part of 100 is 80?
7. **a.** 36 is what decimal part of 48? **b.** 49 is what decimal part of 56?
8. **a.** What decimal part of 27 is 18? **b.** 25 is what decimal part of 30?
9. **a.** 84 is what decimal part of 108? **b.** What decimal part of 24 is 10?

6–21 FINDING THE NUMBER WHEN A DECIMAL PART OF IT IS KNOWN

We have found that to determine the number when a fractional part of it is known, we divide the given number that represents the fractional part of the unknown number by the given fraction. Similarly, to find the number when a decimal part of it is known, we divide the given number that represents the decimal part of the unknown number by the given decimal fraction.

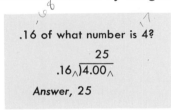

.16 of what number is 4?

$$\begin{array}{r} 25 \\ .16_\wedge \overline{)4.00_\wedge} \end{array}$$

Answer, 25

EXERCISES

Find the following:

1. **a.** .3 of what number is 9? **b.** .4 of what number is 28?
2. **a.** .8 of what number is 6? **b.** .2 of what number is 1?
3. **a.** .09 of what number is 540? **b.** 329 is .47 of what number?
4. **a.** .375 of what number is 900? **b.** .0625 of what number is 45?
5. **a.** .4 of what number is 5? **b.** .06 of what number is 78?
6. **a.** .75 of what number is 29? **b.** .9 of what number is 15?
7. **a.** 60 is .5 of what number? **b.** 2 is .6 of what number?
8. **a.** .25 of what number is 700? **b.** 2,072 is .518 of what number?
9. **a.** .125 of what number is 210? **b.** .875 of what number is 46?

6–22 FRACTION NOTATION IN NON-DECIMAL BASES

The dot that separates the whole number from the fraction in the decimal numeration system is called the *decimal point*. In the binary system it is called the *binary point*, in the quinary system, the *quinary point*, and in the duodecimal system, the *duodecimal point*.

In the decimal system each place is $\frac{1}{10}$ of the next place to the left, but in the binary system it is $\frac{1}{2}$ the value of the next place to the left, in the quinary system, $\frac{1}{5}$ the value, and in the duodecimal system, $\frac{1}{12}$ the value.

Thus the places to the right of the point are:

In the binary system

$$.\ \frac{1}{2}\quad \frac{1}{4}\quad \frac{1}{8}\quad \frac{1}{16}\quad \frac{1}{32}\quad \frac{1}{64} \qquad \text{or} \qquad .\ \frac{1}{2}\quad \frac{1}{2^2}\quad \frac{1}{2^3}\quad \frac{1}{2^4}\quad \frac{1}{2^5}\quad \frac{1}{2^6}$$

In the quinary system

$$.\ \frac{1}{5}\quad \frac{1}{25}\quad \frac{1}{125}\quad \frac{1}{625}\quad \frac{1}{3{,}125} \qquad \text{or} \qquad .\ \frac{1}{5}\quad \frac{1}{5^2}\quad \frac{1}{5^3}\quad \frac{1}{5^4}\quad \frac{1}{5^5}$$

In the duodecimal system

$$.\ \frac{1}{12}\quad \frac{1}{144}\quad \frac{1}{1{,}728}\quad \frac{1}{20{,}736} \qquad \text{or} \qquad .\ \frac{1}{12}\quad \frac{1}{12^2}\quad \frac{1}{12^3}\quad \frac{1}{12^4}$$

EXERCISES

1. Determine the common fraction represented by each of the following:

a. $.01_{\text{two}}$
b. $.1_{\text{twelve}}$
c. $.001_{\text{five}}$
d. $.01_{\text{three}}$
e. $.3_{\text{five}}$
f. $.002_{\text{four}}$
g. $.0E_{\text{twelve}}$
h. $.0005_{\text{eight}}$
i. $.0000001_{\text{two}}$
j. $.00004_{\text{five}}$
k. $.000003_{\text{seven}}$
l. $.0000002_{\text{three}}$

2. Determine the common fraction represented by each of the following:

a. $.1111_{\text{two}}$
b. $.101_{\text{five}}$
c. $.11_{\text{twelve}}$
d. $.201_{\text{three}}$
e. $.134_{\text{five}}$
f. $.615_{\text{seven}}$
g. $.3321_{\text{four}}$
h. $.7203_{\text{eight}}$
i. $.5629_{\text{ten}}$

3. Add:

$.111_{\text{two}}$	$.3214_{\text{five}}$	$.4315_{\text{seven}}$	2.1122_{three}	101.10111_{two}
$.111_{\text{two}}$	$.2031_{\text{five}}$	$.3236_{\text{seven}}$	2.2012_{three}	111.11011_{two}

6–23 PROBLEMS

EXERCISES

1. The length of the United States flag is 1.9 times its width. How long should a flag be if its width is 5 feet?
2. Find the total rainfall for a month during which .12 inch, 1.29 inches, .07 inch, .26 inch, .45 inch, .57 inch, and 1.52 inches of rain fell.
3. Normal atmospheric pressure is 29.92 inches of mercury. How much above normal is a barometric pressure of 30.39 inches?
4. The per capita public debt of the United States in 1860 was $2.06. How many times as great (to the nearest whole number) was the debt in 1974 when it averaged $2,241.81 per person?
5. Steel weighs 7.7 times as much as an equal volume of water. If a cubic foot of water weighs 62.5 lb., how much does a cubic foot of steel weigh?
6. How much do you save on each can if you buy 1 dozen cans of corn for $3.60 instead of $.32 each?
7. Find the new bank balance if the previous monthly balance was $583.06, deposits during the month were $406.58, $93.89, $396.34, $157.95, and $268.43, and checks drawn and cash withdrawals were $309.25, $125.87, $436.81, and $75.64.
8. A television set costs $325 or $32.50 down and 24 monthly payments of $13.75 each. How much do you save by paying cash?
9. An object 50 feet high can be seen at sea for a distance of 8.1 nautical miles. If an object 800 feet high is visible for 32.4 nautical miles, how many times as far can the taller object be seen?
10. A new record of 163.465 m.p.h. was set at the Indianapolis Speedway in 1972. If the previous record was 157.735 m.p.h., how much faster is the new record?
11. An airplane flew 862.5 miles in 1.5 hours. What was its average ground speed?
12. What is the maximum length that is acceptable if a tolerance of .005 inch is permitted on a piece designed to be 1.625 inches long?
13. Workers in manufacturing industries averaged 40.3 hours of work and earnings of $49.25 per week in 1947. In 1974 the work week averaged 40.1 hours and earnings increased to $171.74. Find the increase in average earnings per hour of work.
14. To find the batting average, divide the number of hits by the number of times at bat, finding the quotient to the nearest thousandth. Find the batting average of each of the following players:
    a. Peter, 3 hits out of 10 times at bat.
    b. Joe, 6 hits out of 25 times at bat.
    c. Scott, 15 hits out of 45 times at bat.

d. Carlos, 54 hits out of 186 times at bat.
e. Ed, 49 hits out of 165 times at bat.

15. To find the fielding average, divide the sum of the put-outs and assists by the sum of the errors, put-outs and assists, finding the quotient to the nearest thousandth. Find the fielding average of each of the following players:

a. Maria, 5 errors, 37 put-outs and 49 assists.
b. Debbie, 2 errors, 8 put-outs and 10 assists.
c. Janet, 3 errors, 41 put-outs and 108 assists.
d. Carmela, 6 errors, 198 put-outs and 26 assists.
e. Barbara, 12 errors, 290 put-outs and 425 assists.

CHAPTER REVIEW

1. Write as a numeral: Three hundred and two thousandths. (6–1)
2. Read or write in words: .0584 (6–2) (6–3)
3. Write 5,807.439 as a polynomial. (6–4)
4. Round **a.** 26.4198 to the nearest thousandth. (6–5)
b. $9.8949 to the nearest cent.
5. Arrange 6.6, .066, 66, and .66 by size, greatest number first. (6–6)
6. Which of the following are true statements? (6–6)

a. $.03 < .30$ **b.** $1 > .99$ **c.** $5.0 = 0.5$ **d.** $.806 > .8055$ **e.** $.29 < .3$

7. Add:
3.69 + 58.2 + .197

8. Add: (6–7)

$$\begin{array}{r} \$527.63 \\ 93.59 \\ 4.76 \\ 839.97 \\ \underline{789.58} \end{array}$$

9. Subtract:

.786 − .24

10. Subtract: (6–8)

$63 − $1.28

11. Multiply:

.23 × .004

12. Multiply to the nearest cent: (6–9)

$$\begin{array}{r} \$68.59 \\ \underline{.62\tfrac{1}{2}} \end{array}$$

13. Multiply 58.9276 by 100. (6–10)

14. a. Divide:
$.125\overline{).5}$

b. Divide to the nearest cent: (6–11)
$60\overline{)\$593.27}$

15. Divide 62.59 by 1,000. (6–12)

16. a. Write as a complete numeral: 24.7 billion (6–13)
b. Write the shortened name in trillions: 836,540,000,000,000

17. Write by scientific notation: **a.** 370,000,000,000 **b.** .000046 (6–14)

18. Express $\frac{11}{25}$ as a decimal numeral. (6–15)
19. Express $\frac{7}{9}$ as a repeating decimal. (6–16)
20. Express .74 as a numeral naming a common fraction. (6–17)
21. Express $.\overline{64}\ldots$ as a numeral naming a common fraction. (6–18)
22. Find .06 of 275. (6–19)
23. What decimal part of 72 is 56? (6–20)
24. .15 of what number is 90? (6–21)
25. What common fraction is represented by $.101_{\text{two}}$? (6–22)

REVIEW EXERCISES

1. Add:

66	5,198	28,107	69,428	83,045
527	93,456	9,689	85,274	19,249
4,852	825	57,843	67,855	56,963
9	6,218	18,936	96,589	40,586
275	58,324	2,559	38,417	58,879

2. Subtract:

6,159	84,206	92,345	291,500	580,604
2,949	9,185	87,258	69,809	391,786

3. Multiply:

24	879	307	725	5,280
57	93	409	916	184

4. Divide:

$48\overline{)1,392}$ $\quad 96\overline{)8,928}$ $\quad 144\overline{)121,968}$ $\quad 735\overline{)609,315}$ $\quad 6,080\overline{)358,720}$

5. Add:

$\frac{1}{6}$	$\frac{7}{8}$	$5\frac{1}{3}$	$9\frac{3}{4}$	$7\frac{11}{12}$
$\frac{2}{3}$	$2\frac{13}{16}$	$6\frac{5}{8}$	$4\frac{9}{10}$	$3\frac{5}{6}$

6. Subtract:

$\frac{9}{10}$	$5\frac{3}{8}$	10	$6\frac{3}{4}$	$8\frac{2}{3}$
$\frac{1}{2}$	4	$8\frac{9}{16}$	$4\frac{29}{32}$	$3\frac{4}{5}$

7. Multiply:

$\frac{2}{3} \times \frac{9}{16}$ $\qquad \begin{array}{r} 24 \\ \times\, 4\frac{5}{8} \end{array}$ $\qquad 3\frac{3}{4} \times 7\frac{1}{5}$ $\qquad \frac{7}{8} \times 2\frac{3}{5}$ $\qquad \frac{2}{3} \times 8\frac{1}{4}$

8. Divide:

$\frac{15}{16} \div \frac{5}{8}$ $\qquad 4\frac{1}{2} \div \frac{3}{4}$ $\qquad 12\frac{1}{4} \div 8$ $\qquad 6\frac{2}{3} \div 5\frac{5}{6}$ $\qquad 1\frac{3}{8} \div 2\frac{1}{3}$

9. Add:

.605 + .250 + .105   .8 + .02   .432 + 8   6.49 + 23.1   \$.83 + \$5.94 + \$18.59

10. Subtract:

.825 − .775   34.29 − 3   80 − 29.4   5.854 − 1.63   \$25 − \$18.46

11. Multiply:

75	2.7	9.2	3.1416	\$6,000
.69	.003	4.6	8.5	.045

12. Divide:

$13\overline{).52}$ $.05\overline{)70}$ $.62\overline{).1736}$ $40\overline{)2}$ $2.45\overline{)49}$

CUMULATIVE TEST

1. If R = {all multiples of 4}, which of the following statements are true? (1–9)

a. $24 \in R$ **b.** $42 \in R$ **c.** $98 \notin R$ **d.** $52 \in R$ **e.** $76 \notin R$

2. Factor 1,728 as the product of prime factors. (1–15)
3. Round 43,980,514,306 to the nearest million. (2–11)
4. Expand 23102_{four} as a polynomial. (2–18)
5. Express $2TE_{\text{twelve}}$ as a base five numeral. (2–19)
6. Is {6, 12, 18, 24, . . .} closed under addition? Subtraction? Multiplication? Division? (3–6)

7. Add:

629,849
74,987
258,694
9,038
367,856

8. Subtract:

8,100,052
5,603,994

9. Multiply: (3–9—3–11)

5,086
9,070

10. Divide:

$6,080\overline{)1,854,400}$

11. Divide: (3–16)

$7,096\overline{)745,080}$

12. Add:

1340_{six}
2423_{six}
1435_{six}

13. Multiply: (3–22)

122_{three}
201_{three}

14. a. Add:

$23 + 19 \equiv ? \pmod 9$

b. Multiply: (3–24)

$17 \times 8 \equiv ? \pmod 7$

15. a. Add the following matrices:

$$\begin{pmatrix} 1 & 2 & 0 \\ 0 & 3 & 5 \\ 4 & 8 & 2 \end{pmatrix} + \begin{pmatrix} 5 & 1 & 3 \\ 1 & 7 & 2 \\ 3 & 0 & 4 \end{pmatrix}$$

b. Find the product matrix: (3–24)

$$\begin{pmatrix} 2 & 4 & 3 \\ 1 & 5 & 0 \\ 3 & 0 & 6 \end{pmatrix} \times \begin{pmatrix} 3 & 1 & 5 \\ 0 & 2 & 0 \\ 7 & 4 & 6 \end{pmatrix}$$

16. Write in words each of the following: (4–1)

a. $12 < 25$ **b.** $12 > 10$ **c.** $18 + 6 \nless 7 + 8$ **d.** $20 - 4 \neq 20 \div 4$

17. Express $\frac{49}{196}$ in lowest terms. (5–2)
18. Express $\frac{5}{6}$ as 72nds. (5–3)
19. Write the set of equivalent fractions of which $\frac{6}{16}$ is a member. (5–3)
20. Which of the following statements are true? (5–8)

a. $\frac{3}{5} < \frac{6}{11}$ **b.** $\frac{18}{24} = \frac{63}{84}$ **c.** $\frac{8}{17} > \frac{4}{9}$ **d.** $\frac{11}{13} \ngtr \frac{5}{6}$ **e.** $\frac{13}{15} \nless \frac{6}{7}$

21. Arrange $\frac{15}{24}$, $\frac{27}{48}$, and $\frac{7}{12}$ by size, smallest first. (5–8)

22. Add:

$6\frac{7}{16} + 4\frac{3}{4} + 9\frac{5}{8}$

23. Subtract:

$10\frac{1}{4} - 4\frac{3}{5}$

24. Multiply: (5–11—5–13)

$9\frac{3}{8} \times 2\frac{5}{6}$

25. Divide:

$24 \div 2\frac{2}{3}$

26. Divide: (5–15)

$3\frac{1}{2} \div 21$

27. What fractional part of 52 is 44? (5–17)
28. $\frac{4}{15}$ of what number is 140? (5–18)
29. Write 652.147 as a polynomial. (6–4)
30. Round: (6–5)

a. 6.05024 to the nearest tenth. **b.** $\$18.61\frac{7}{12}$ to the nearest cent.

31. Arrange .41, 4.1, .041, and 41 by size, greatest first. (6–6)
32. Which of the following are true statements? (6–6)

a. $.0800 = .080$ **b.** $.02 > .012$ **c.** $0.989 < 0.998$ **d.** $.56 \ngtr .562$ **e.** $1.2 \nless .122$

33. Add:

\$.85 + \$1.93 + \$24.88

34. Subtract:

$.6 - .509$

35. Multiply: (6–7—6–9)

$.2003 \times .001$

36. Divide:

$\$.04\overline{)\$12}$

37. Divide: (6–11)

$.011\overline{).004048}$

38. a. Write as a complete numeral: 3.59 million (6–13)
b. Write the shortened name in billions: 68,400,000,000
39. Write by scientific notation: (6–14)
a. 8,400,000,000,000 **b.** .00000027
40. Express $\frac{36}{80}$ as a decimal numeral. (6–15)
41. Express $\frac{5}{13}$ as a repeating decimal. (6–16)
42. Express .416 as a numeral naming a common fraction. (6–17)
43. Express $.41\overline{6}\ldots$ as a numeral naming a common fraction. (6–18)
44. Find .625 of 9.36 (6–19)
45. What decimal part of 65 is 39? (6–20)
46. 1.05 of what number is 336? (6–21)

7 Per Cent

7–1 MEANING OF PER CENT

Per cent, represented by the symbol "%," is used extensively in our daily affairs. It is the language of business. We experience the use of per cent in situations like the following:

A bank pays 5% interest on deposits and charges 8% interest on loans. A girl pays 6% tax on telephone service for local and toll calls. A salesman earns 4% on what he sells. A department store advertises a 20% reduction on sales items. A working woman finds that an 18% withholding tax, 5.85% social security tax and 3% state income tax are deducted from her paycheck. A customer pays 10% carrying charges on a new refrigerator. A family receives a 1% discount for paying the real estate taxes on their home in advance. The label in a coat indicates that the material is 40% Dacron and 60% wool. The cost of living has increased 1.3%. A storekeeper makes a profit of 28% on all merchandise sold in the store. A student receives a mark of 86%. The average daily attendance of a class is 95%.

What does per cent mean?

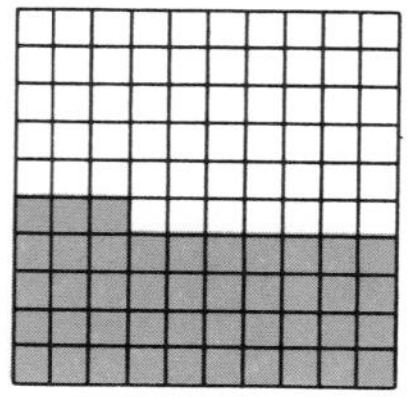

The large square on the right contains 100 small squares. We see that 43 small squares are shaded. But 43 out of a hundred is 43 hundredths. Therefore 43 hundredths of the large square is shaded. 43 hundredths written as a numeral naming a common fraction is $\frac{43}{100}$ and as a numeral naming a decimal fraction is .43. Both represent the part of the large square that is shaded.

There is still another way of indicating that 43 hundredths of the square is shaded. We may say that 43 per cent, written 43%, of the square is shaded.

Thus, per cent means hundredths.

Per cent is a fraction in which the numeral preceding the % symbol is considered the numerator and the % symbol represents the denominator 100. It is equivalent to a two-place decimal fraction or to a common fraction having 100 as its denominator.

57 squares of the hundred squares are not shaded. Therefore 57 hundredths or 57% of the large square is not shaded. If 43% of the square is

shaded and 57% of the large square is not shaded, the whole large square is the sum of the two or 100%.

100% of anything is $\frac{100}{100}$ of it or all of it.

Any per cent less than 100% is equal to a fraction less than 1. 100% is equal to 1. Any per cent more than 100% is equal to a number greater than 1.

A per cent may also be considered a ratio. 85%, which is equivalent to $\frac{85}{100}$, is the ratio of 85 to 100.

EXERCISES

1. In each of the following first find how many small squares out of a hundred are shaded. Then express the answer as a per cent.

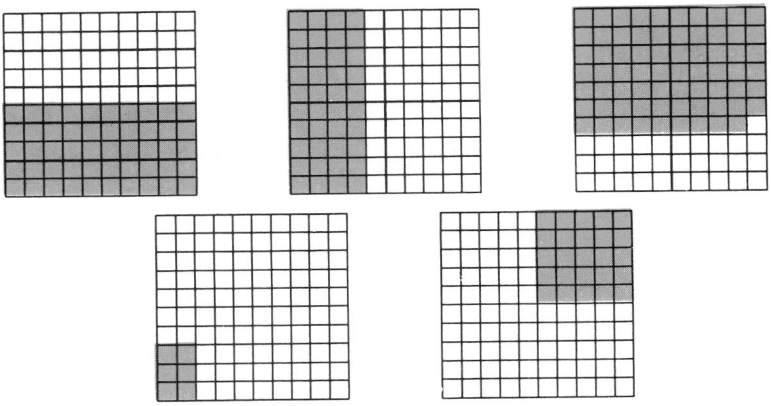

2. Illustrate the following per cents. For each draw a 100-square, then shade the part corresponding to the given per cent.

a. 15% **b.** 83% **c.** 7% **d.** 60% **e.** 100%

3. Express each of the following as a per cent:

a. 21 hundredths	49 hundredths	81 hundredths	70 hundredths
b. 8 hundredths	4 hundredths	9 hundredths	1 hundredth
c. $37\frac{1}{2}$ hundredths	$5\frac{1}{4}$ hundredths	$\frac{1}{2}$ hundredth	$\frac{3}{4}$ hundredth
d. 62.5 hundredths	3.8 hundredths	5.75 hundredths	0.5 hundredth
e. 200 hundredths	150 hundredths	$266\frac{2}{3}$ hundredths	300 hundredths

4. Express each of the following as a per cent:

a. 51 out of 100 **c.** 68 out of 100 **e.** 2 out of 100 **g.** 9 out of 100
b. $12\frac{1}{2}$ out of 100 **d.** $6\frac{3}{4}$ out of 100 **f.** 37.5 out of 100 **h.** 4.25 out of 100

5. Find how many hundredths are in:

a. 6%; 9%; 1%
b. 34%; 95%; 67%
c. $4\frac{3}{4}$%; $87\frac{1}{2}$%; $\frac{1}{4}$%
d. 5.2%; 62.5%; 0.8%

6. Express each of the following as a per cent:

a. The ratio of 11 to 100
b. The ratio of 93 to 100
c. The ratio of 6 to 100
d. The ratio of 59 to 100

PER CENTS, DECIMAL FRACTIONS, AND COMMON FRACTIONS

Since per cent means hundredths, it may be expressed both as a numeral for an equivalent decimal fraction or as a numeral for an equivalent common fraction and vice-versa.

7–2 PER CENTS AS DECIMAL FRACTIONS

The % symbol represents 2 decimal places. When it is removed there should be 2 more decimal places than in the original numeral.

To express a per cent as a numeral for a decimal fraction, we rewrite the digits of the given per cent but drop the per cent symbol. Then we move the decimal point two places to the left. A decimal point is understood after the ones digit in a numeral naming a whole number.

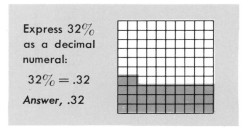

Express 32% as a decimal numeral:

32% = .32

Answer, .32

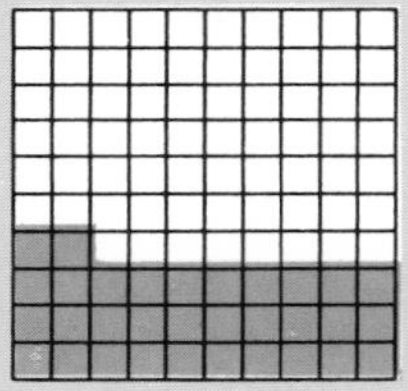

When the per cent is a decimal per cent such as 14.6%, the answer contains more than 2 decimal places.

14.6% = .146

When the per cent is 100% or more, the answer is a numeral naming a whole number or a mixed decimal.

400% = 4; 153% = 1.53

When the per cent is a common fraction less than one per cent such as $\frac{1}{2}\%$, the answer may be written as a two-place decimal numeral $.00\frac{1}{2}$ (or as .005 since $\frac{1}{2}\% = .5\% = .005$).

When the per cent is a mixed number per cent such as $18\frac{3}{4}\%$, the answer may be written as a two-place decimal numeral $.18\frac{3}{4}$ (or as .1875 since $18\frac{3}{4}\% = 18.75\% = .1875$).

EXERCISES

1. Express each of the following per cents as a decimal numeral:

a. 28%, 64%, 75%, 17%, 42%
b. 5%, 3%, 1%, 6%, 8%
c. 30%, 10%, 80%, 50%, 40%
d. 194%, 125%, 103%, 269%, 158%
e. 160%, 100%, 110%, 170%, 220%
f. $62\frac{1}{2}\%$, $3\frac{1}{4}\%$, $5\frac{3}{4}\%$, $33\frac{1}{3}\%$, $\frac{1}{8}\%$
g. 23.8%, 4.9%, 137.51%, 0.2%, 0.875%

2. Express each of the following per cents as a numeral naming a whole number:

500%, 300%, 200%, 800%, 100%

7–3 PER CENTS AS COMMON FRACTIONS

Since the % symbol is equivalent to the denominator 100, the denominator 100 may be used to replace the % symbol.

To express a per cent as a numeral for a common fraction, we write the digits of the given per cent as the numerator over 100 as the denominator. Then, if possible, we express the fraction in lowest terms.

Express 75% as a numeral naming a common fraction:

$75\% = \frac{75}{100} = \frac{3}{4}$

Answer, $\frac{3}{4}$

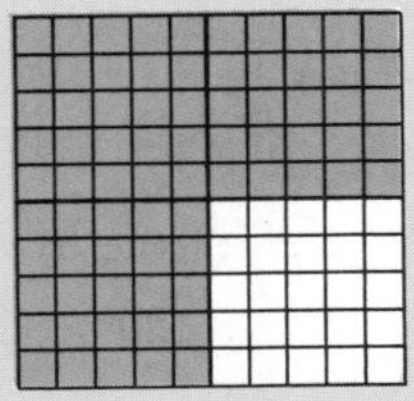

When the per cent is 100% or more, the answer is a numeral naming a whole number or a mixed number.

$150\% = \frac{150}{100} = 1\frac{1}{2}$ or $150\% = 100\% + 50\% = 1 + \frac{1}{2} = 1\frac{1}{2}$

When the per cent is a mixed number per cent, we express the per cent as a decimal fraction, then as a common fraction in its lowest terms.

$12\frac{1}{2}\% = .12\frac{1}{2} = .125 = \frac{125}{1{,}000} = \frac{1}{8}$ or $12\frac{1}{2}\% = \frac{12\frac{1}{2}}{100} = 12\frac{1}{2} \div 100 = \frac{\overset{1}{\cancel{25}}}{2} \times \frac{1}{\underset{4}{\cancel{100}}} = \frac{1}{8}$

EXERCISES

1. What per cent of each of the following squares is shaded? What fractional part of each square is shaded?

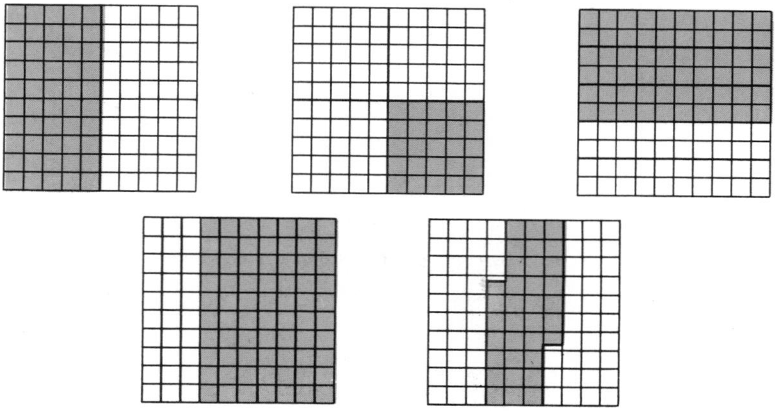

2. Express each of the following per cents as a numeral naming a common fraction in lowest terms:

   a. 50%; 25%; 80%; 30%; 20%; 10%; 40%; 70%; 90%; 60%

   b. 5%; 85%; 4%; 36%; 55%; 96%; 22%; 35%; 74%; 15%

   c. $37\frac{1}{2}\%$; $62\frac{1}{2}\%$; $87\frac{1}{2}\%$; $12\frac{1}{2}\%$; $6\frac{1}{4}\%$; $33\frac{1}{3}\%$; $16\frac{2}{3}\%$; $66\frac{2}{3}\%$; $83\frac{1}{3}\%$; $8\frac{1}{3}\%$

3. Express each of the following per cents as a numeral naming a mixed number:

   125%; 160%; 120%; 210%; 390%; $162\frac{1}{2}\%$; $187\frac{1}{2}\%$; $133\frac{1}{3}\%$; $266\frac{2}{3}\%$; $412\frac{1}{2}\%$

7–4 DECIMAL FRACTIONS AS PER CENTS

Since the % symbol represents two decimal places (hundredths), we can use the symbol to replace two decimal places.

To express a decimal fraction as a per cent, we write the digits of the numeral naming the given decimal. We move the decimal point two places to the right and write the per cent symbol after the numeral.

> Express .57 as a per cent:
>
> .57 = 57%
>
> *Answer,* 57%

In a per cent the decimal point is not written after the last digit. Observe that the per cent has two less decimal places than the equivalent decimal fraction.

> .0004 = .04%; .819 = 81.9%; .7 = .70 = 70%

A mixed decimal is equal to more than 100%.

$$1.94 = 194\%$$

A whole number is equivalent to some multiple of 100%.

$$2 = 2.00 = 200\%, \quad 3 = 3.00 = 300\%, \text{ etc.}$$

EXERCISES

Express each of the following decimals as a per cent:

1. .28, .51, .65, .40, .96
2. .06, .04, .01, .09, .05
3. .4, .8, .5, .2, .3
4. 5.00, 3.00, 9, 7, 4
5. 1.37, 2.05, 1.6, 1.45, 1.9
6. $.62\frac{1}{2}$, $.05\frac{1}{2}$, $.07\frac{3}{4}$, $.70\frac{5}{8}$, $1.16\frac{2}{3}$
7. .875, .0225, 1.206, .004, .0005
8. $.00\frac{1}{2}$, $.00\frac{7}{8}$, $.00\frac{1}{4}$, $.00\frac{3}{4}$, $.00\frac{2}{5}$

7–5 COMMON FRACTIONS AS PER CENTS

Since the % symbol is equivalent to the denominator 100, the % symbol may be used to replace the denominator 100.

$$\tfrac{5}{100} = 5\%$$

Sometimes the fraction has a denominator that can be changed quickly to the denominator 100.

$$\frac{7}{20} = \frac{7 \times 5}{20 \times 5} = \frac{35}{100} = 35\%; \quad \frac{12}{200} = \frac{12 \div 2}{200 \div 2} = \frac{6}{100} = 6\%$$

Generally the fraction has a denominator which cannot be changed easily to the denominator 100. However, a common fraction may be expressed as a decimal fraction and a decimal fraction may be expressed as a per cent.

To express a common fraction as a per cent, we divide the numerator by the denominator, finding the quotient to two decimal places. We rewrite the digits in the quotient, drop the decimal point, and write the per cent symbol % after the numeral.

Express $\frac{7}{8}$ as a per cent:

$$\tfrac{7}{8} = 7 \div 8 = 8\overline{)7.00} \quad .87\tfrac{1}{2} = 87\tfrac{1}{2}\%$$

Answer, $87\frac{1}{2}\%$

If the decimal fraction equivalent is not exact hundredths, we can either write the remainder as a numeral naming a common fraction as shown above or round off the per cent to the degree desired: nearest whole per cent, tenth of a per cent, hundredth of a per cent, etc.

$\frac{2}{3} = .66\frac{2}{3} = 66\frac{2}{3}\%$ or $\frac{2}{3} = .666\frac{2}{3} = 66.7\%$ or $\frac{2}{3} = .6666\frac{2}{3} = 66.67\%$

A mixed number is equal to more than 100%.

$1\frac{3}{4} = 175\%$ since $1 = 100\%$ and $\frac{3}{4} = 75\%$.

EXERCISES

1. Express each of the following fractions as a per cent:

a. $\frac{9}{100}$, $\frac{27}{100}$, $\frac{53}{100}$, $\frac{90}{100}$, $\frac{146}{100}$
b. $\frac{1}{4}$, $\frac{1}{2}$, $\frac{3}{5}$, $\frac{3}{4}$, $\frac{1}{5}$
c. $\frac{7}{10}$, $\frac{2}{5}$, $\frac{9}{10}$, $\frac{4}{5}$, $\frac{1}{10}$
d. $\frac{9}{50}$, $\frac{3}{25}$, $\frac{19}{20}$, $\frac{11}{25}$, $\frac{17}{20}$
e. $\frac{40}{200}$, $\frac{16}{400}$, $\frac{30}{600}$, $\frac{21}{700}$, $\frac{35}{500}$
f. $\frac{1}{8}$, $\frac{1}{3}$, $\frac{1}{6}$, $\frac{5}{8}$, $\frac{2}{9}$
g. $\frac{3}{8}$, $\frac{5}{6}$, $\frac{7}{8}$, $\frac{7}{12}$, $\frac{13}{16}$
h. $\frac{9}{12}$, $\frac{21}{35}$, $\frac{48}{96}$, $\frac{14}{56}$, $\frac{63}{90}$
i. $\frac{19}{57}$, $\frac{48}{56}$, $\frac{70}{112}$, $\frac{60}{72}$, $\frac{84}{126}$
j. $\frac{12}{14}$, $\frac{21}{27}$, $\frac{20}{36}$, $\frac{15}{33}$, $\frac{27}{63}$

2. Express each of the following fractions as a per cent:

a. To nearest whole per cent: $\frac{4}{7}$, $\frac{5}{9}$, $\frac{2}{3}$, $\frac{11}{16}$, $\frac{16}{18}$
b. To nearest tenth of a per cent: $\frac{2}{9}$, $\frac{6}{7}$, $\frac{5}{12}$, $\frac{1}{6}$, $\frac{52}{60}$
c. To nearest hundredth of a per cent: $\frac{5}{7}$, $\frac{11}{12}$, $\frac{7}{9}$, $\frac{19}{24}$, $\frac{14}{49}$

3. Express each of the following mixed numbers as a per cent:

$1\frac{1}{4}$, $1\frac{5}{8}$, $2\frac{1}{2}$, $2\frac{3}{5}$, $1\frac{1}{3}$

4. Express each of the following improper fractions as a per cent:

a. $\frac{34}{34}$, $\frac{41}{41}$, $\frac{112}{112}$, $\frac{16}{8}$, $\frac{27}{3}$
b. $\frac{3}{2}$, $\frac{9}{5}$, $\frac{5}{4}$, $\frac{17}{8}$, $\frac{11}{3}$

Table of Equivalents
Per Cents, Decimals, and Common Fractions

Per Cent	Decimal	Common Fraction
5%	.05	$\frac{1}{20}$
$6\frac{1}{4}\%$	$.06\frac{1}{4}$	$\frac{1}{16}$
$8\frac{1}{3}\%$	$.08\frac{1}{3}$	$\frac{1}{12}$
10%	.10 or .1	$\frac{1}{10}$
$12\frac{1}{2}\%$	$.12\frac{1}{2}$ or .125	$\frac{1}{8}$

Table of Equivalents (*Continued*)
Per Cents, Decimals, and Common Fractions

Per Cent	Decimal	Common Fraction
$16\frac{2}{3}\%$	$.16\frac{2}{3}$	$\frac{1}{6}$
20%	.20 or .2	$\frac{1}{5}$
25%	.25	$\frac{1}{4}$
30%	.30 or .3	$\frac{3}{10}$
$33\frac{1}{3}\%$	$.33\frac{1}{3}$	$\frac{1}{3}$
$37\frac{1}{2}\%$	$.37\frac{1}{2}$ or .375	$\frac{3}{8}$
40%	.40 or .4	$\frac{2}{5}$
50%	.50 or .5	$\frac{1}{2}$
60%	.60 or .6	$\frac{3}{5}$
$62\frac{1}{2}\%$	$.62\frac{1}{2}$ or .625	$\frac{5}{8}$
$66\frac{2}{3}\%$	$.66\frac{2}{3}$	$\frac{2}{3}$
70%	.70 or .7	$\frac{7}{10}$
75%	.75	$\frac{3}{4}$
80%	.80 or .8	$\frac{4}{5}$
$83\frac{1}{3}\%$	$.83\frac{1}{3}$	$\frac{5}{6}$
$87\frac{1}{2}\%$	$.87\frac{1}{2}$ or .875	$\frac{7}{8}$
90%	.90 or .9	$\frac{9}{10}$
100%	1.00 or 1	$\frac{1}{1}, \frac{2}{2}, \frac{3}{3}\ldots$

7–6 COMPUTATION—FINDING A PER CENT OF A NUMBER

In our work with common fractions we studied how to find a fractional part of a number like $\frac{4}{5}$ of 750. Since a per cent is just another way of writing a numeral naming a common fraction, finding 80% of 750 is like finding $\frac{4}{5}$ of 750. When we multiplied decimals, we learned how to find a decimal part of a number like .12 of 67. Since a per cent is another way of writing a numeral naming a decimal fraction, finding 12% of 67 is like finding .12 of 67.

Since a per cent may be written either as a decimal or a common fraction, to find a per cent of a number, we change the per cent to its equivalent decimal fraction or common fraction. Then we multiply the given number by this fraction.

Find 12% of 67:

12% = .12

```
   67
 ×.12
 ----
 1 34
 67
 ----
 8.04
```

Answer, 8.04

Find 80% of 750:

$80\% = \frac{4}{5}$

$\frac{4}{\cancel{5}_1} \times \frac{\cancel{750}^{150}}{1} = 600$

Answer, 600

When finding a mixed number or common fraction per cent of a number, we may use the decimal per cent form.

> $18\frac{1}{2}\%$ of $500 = .18\frac{1}{2} \times 500$ or $.185 \times 500$ since $18\frac{1}{2}\% = 18.5\% = .185$
>
> $\frac{3}{4}\%$ of $120 = .00\frac{3}{4} \times 120$ or $.0075 \times 120$ since $\frac{3}{4}\% = .75\% = .0075$

When finding a per cent of an amount of money, we round the answers to the nearest cent.

> 4% of \$1.28 = .04 × \$1.28 = \$.0512 = \$.05

EXERCISES

Find:

1. a.	48% of 85	26% of 30	11% of 98	57% of 615	83% of 4,000
b.	29% of 140	15% of 98	94% of 632	38% of 497	72% of 3,259
c.	4% of 60	8% of 35	3% of 9	9% of 200	1% of 4,275
d.	5% of 82	6% of 148	2% of 375	7% of 906	4% of 8,647
e.	10% of 20	40% of 7	80% of 468	60% of 250	30% of 5,000
2. a.	6% of 6.8	13% of 7.4	9% of 8.18	36% of 4.25	91% of 8.60
b.	18% of 14.56	4% of 91.05	63% of 42.96	1% or 483.29	88% of 907.43
3. a.	$26\frac{1}{2}\%$ of 54	$94\frac{1}{2}\%$ of 300	$43\frac{1}{4}\%$ of 48	$18\frac{1}{4}\%$ of 964	$65\frac{3}{4}\%$ of 2,000
b.	$5\frac{1}{2}\%$ of 8	$6\frac{1}{4}\%$ of 24	$9\frac{3}{4}\%$ of 672	$1\frac{1}{2}\%$ of 1,287	$4\frac{3}{4}\%$ of 9,175
c.	$34\frac{3}{8}\%$ of 72	$7\frac{5}{8}\%$ of 400	$4\frac{2}{3}\%$ of 969	$25\frac{1}{6}\%$ of 4,200	$9\frac{7}{8}\%$ of 1,000
4. a.	20.6% of 81	17.9% of 650	36.1% of 8	83.2% of 4.75	61.4% of 3,400
b.	8.3% of 40	2.5% of 3	4.9% of 2.91	1.6% of 92.83	7.8% of 5,000
c.	19.28% of 300	80.05% of 2,000	94.62% of 750	57.81% of 6,450	
5. a.	100% of 70	300% of 45	500% of 180	200% of 4.25	800% of 1,000
b.	128% of 9	193% of 87	142% of 256	376% of 8.94	217% of 3,500
c.	106% of 205	101% of 3,000	107% of 75	205% of 12.5	409% of 4,000
d.	150% of 32	120% of 8	360% of 545	110% of 6.8	240% of 7,500
e.	183.6% of 40	100.2% of 1,000	112.84% of 5	246.5% of 18.75	
f.	$117\frac{1}{2}\%$ of 6	$232\frac{1}{4}\%$ of 480	$100\frac{1}{2}\%$ of 6,000	$104\frac{3}{4}\%$ of 47.4	
6. a.	$\frac{1}{2}\%$ of 500	$\frac{1}{4}\%$ of 32	$\frac{3}{4}\%$ of 2,000	$\frac{1}{2}\%$ of 28.46	$\frac{3}{4}\%$ of 105.23
b.	$\frac{3}{8}\%$ of 48	$\frac{7}{8}\%$ of 600	$\frac{2}{3}\%$ of 144	$\frac{5}{8}\%$ of 108.64	$\frac{1}{6}\%$ of 9,000
c.	.08% of 70	0.3% of 2,000	0.5% of 9	0.6% of 26.25	0.1% of 12,000
d.	0.29% of 4	0.73% of 100	0.46% of 85	0.52% of 7.5	0.89% of 4,000

7. Use fractional equivalents for per cents to find:

a. 50% of: 8; 20; 56; 140; 392
b. 25% of: 12; 36; 68; 124; 600
c. 10% of: 50; 90; 140; 480; 1,250
d. 75% of: 4; 24; 76; 136; 860
e. 20% of: 5; 40; 95; 225; 1,000
f. 80% of: 10; 55; 175; 340; 1,425
g. $66\frac{2}{3}\%$ of: 6; 30; 72; 171; 807
h. $87\frac{1}{2}\%$ of: 16; 56; 144; 960; 2,400
i. $12\frac{1}{2}\%$ of: 32; 96; 80; 128; 264
j. $33\frac{1}{3}\%$ of: 3; 48; 120; 729; 1,500
k. $16\frac{2}{3}\%$ of: 24; 60; 96; 282; 2,094
l. $37\frac{1}{2}\%$ of: 32; 104; 800; 3,512; 4,000
m. 40% of: 15; 50; 75; 325; 1,300
n. 70% of: 10; 300; 450; 1,000; 5,600
o. 60% of: 45; 65; 180; 740; 2,225
p. $62\frac{1}{2}\%$ of: 80; 144; 400; 1,296; 7,200
q. 30% of: 20; 70; 130; 500; 1,700
r. $83\frac{1}{3}\%$ of: 18; 42; 132; 864; 2,406
s. 90% of: 60; 100; 250; 960; 8,000
t. $6\frac{1}{4}\%$ of: 64; 112; 192; 1,600; 3,920
u. $8\frac{1}{3}\%$ of: 12; 96; 288; 1,728; 9,000
v. 5% of: 40; 100; 260; 500; 1,380

8. Use fractional equivalents for per cents to find:

a. 10% of 70	40% of 65	50% of 92	80% of 200	30% of 450
b. 25% of 84	75% of 536	90% of 400	20% of 875	60% of 1,250
c. $87\frac{1}{2}\%$ of 72	$12\frac{1}{2}\%$ of 432	$37\frac{1}{2}\%$ of 960	$62\frac{1}{2}\%$ of 584	$6\frac{1}{4}\%$ of 800
d. $33\frac{1}{3}\%$ of 69	$66\frac{2}{3}\%$ of 360	$83\frac{1}{3}\%$ of 144	$16\frac{2}{3}\%$ of 300	$8\frac{1}{3}\%$ of 2,400
e. 50% of 35	10% of 123	$33\frac{1}{3}\%$ of 79	20% of 42	$16\frac{2}{3}\%$ of 21

9. Use mixed numbers for per cents to find:

a. 125% of 28	150% of 354	275% of 260	160% of 800	190% of 1,000
b. $187\frac{1}{2}\%$ of 440	$133\frac{1}{3}\%$ of 960	$162\frac{1}{2}\%$ of 5,600	$183\frac{1}{3}\%$ of 630	$266\frac{2}{3}\%$ of 720

10. Find each of the following to the nearest cent:

a. 26% of $90	33% of $6	54% of $237	86% of $1,250	47% of $4,000
b. 5% of $7	1% of $23	6% of $508	4% of $9,500	3% of $649
c. 69% of $.40	42% of $.50	18% of $9.24	95% of $56.83	70% of $125.75
d. 3% of $.68	9% of $.80	7% of $5.67	6% of $71.09	2% of $482.91
e. $12\frac{1}{2}\%$ of $136	$66\frac{2}{3}\%$ of $.87	$1\frac{1}{2}\%$ of $96.82	$\frac{3}{4}\%$ of $600	$\frac{1}{2}\%$ of $94.85
f. 5.2% of $70	0.7% of $360	16.52% of $40.06	0.1% of $52.87	
g. 120% of $56	136% of $1.20	$100\frac{1}{2}\%$ of $2,000	$287\frac{1}{2}\%$ of $480	

11. Use fractional equivalents for per cents to find each of the following to the nearest cent:

a. 50% of $648	10% of $900	25% of $2,000	$16\frac{2}{3}\%$ of $1,800	$62\frac{1}{2}\%$ of $680
b. 60% of $.50	20% of $8.70	90% of $19.30	75% of $5.24	$33\frac{1}{3}\%$ of $9.54
c. 30% of $1.29	40% of $.63	$66\frac{2}{3}\%$ of $4.75	$83\frac{1}{3}\%$ of $10	70% of $63.89

7–7 COMPUTATION—FINDING WHAT PER CENT ONE NUMBER IS OF ANOTHER

Since per cent is a fraction, the relationship "What fractional part of 8 is 2?" or "What decimal part of 8 is 2?" can also be stated as "What per cent of 8 is 2?"

This relationship as a common fraction is $\frac{1}{4}$ since $\frac{2}{8} = \frac{1}{4}$; as a decimal fraction it is .25 since $\frac{2}{8} = \frac{1}{4} = .25$; and as a per cent it is 25% since $\frac{2}{8} = \frac{1}{4} = .25 = 25\%$.

To find what per cent one number is of another, we find what fractional part one number is of the other expressed in lowest terms. Then we change this fraction to a per cent, using the per cent equivalent if it is known, otherwise we change the fraction first to a 2-place decimal, then to a per cent.

What per cent of 40 is 26?

```
                 .65 = 65%
26/40 = 13/20 = 20)13.00
                   12 0
                    1 00
                    1 00
                    . . .
```

Answer, 65%

Sometimes the fraction has a denominator that can be changed easily to the denominator 100.

$$\frac{8}{25} = \frac{8 \times 4}{25 \times 4} = \frac{32}{100} = 32\% \qquad \frac{24}{300} = \frac{24 \div 3}{300 \div 3} = \frac{8}{100} = 8\%$$

When the per cent is not an exact whole number, we can either show the remainder as a common fraction or round off the per cent to the degree desired: nearest whole per cent, nearest tenth of a per cent, nearest hundredth of a per cent, etc.

9 is what per cent of 32?

```
          .28 1/8                    .2812 1/2
9/32 = 32)9.00            9/32 = 32)9.0000
          6 4                       6 4
          2 60                      2 60
          2 56                      2 56
             4                        40
                                      32
                                      80
                                      64
                                      16
```

Answer, $28\frac{1}{8}\%$, or 28% when rounded to nearest whole per cent, or 28.1% when rounded to nearest tenth of a per cent, or 28.13% when rounded to nearest hundredth of a per cent.

An improper fraction is equal to 100% or more.

What per cent of 5 is 8?

$\frac{8}{5} = 1\frac{3}{5} = 160\%$

Answer, 160%

EXERCISES

Find the answer to each of the following:

1. a. What per cent of 100 is 29?
b. 18 is what per cent of 25?
c. 7 is what per cent of 20?
d. What per cent of 4 is 3?
e. 9 is what per cent of 50?
f. 3 is what per cent of 10?
g. 2 is what per cent of 5?
h. 1 is what per cent of 2?

2. a. What per cent of 3 is 1?
b. What per cent of 6 is 5?
c. What per cent of 8 is 7?
d. 3 is what per cent of 8?

3. a. What per cent of 7 is 5?
b. What per cent of 12 is 11?
c. 19 is what per cent of 24?
d. 4 is what per cent of 9?
e. 9 is what per cent of 16?
f. What per cent of 15 is 13?

4. a. What per cent of 100 is 75?
b. 20 is what per cent of 25?
c. What per cent of 20 is 8?
d. 25 is what per cent of 50?
e. What per cent of 10 is 4?
f. 6 is what per cent of 75?

5. a. What per cent of 200 is 40?
b. 48 is what per cent of 300?
c. What per cent of 800 is 56?
d. 15 is what per cent of 500?
e. What per cent of 400 is 24?
f. 12 is what per cent of 600?

6. a. What per cent of 12 is 6?
b. 18 is what per cent of 45?
c. What per cent of 80 is 32?
d. 63 is what per cent of 72?
e. 9 is what per cent of 36?
f. What per cent of 60 is 42?
g. 45 is what per cent of 75?
h. What per cent of 144 is 108?

7. a. What per cent of 56 is 49?
b. 18 is what per cent of 48?
c. What per cent of 42 is 28?
d. 50 is what per cent of 80?
e. What per cent of 27 is 9?
f. 45 is what per cent of 54?

8. a. What per cent of 43 is 43?
b. What per cent of 100 is 200?
c. 25 is what per cent of 5?
d. 150 is what per cent of 150?
e. 54 is what per cent of 18?
f. What per cent of 300 is 1,200?

9. a. What per cent of 4 is 7?
b. 12 is what per cent of 5?
c. What per cent of 18 is 27?
d. 44 is what per cent of 24?
e. 8 is what per cent of 3?
f. What per cent of 8 is 15?
g. 81 is what per cent of 30?
h. What per cent of 75 is 105?

10. Round to the nearest whole per cent:
a. What per cent of 9 is 5?
b. 28 is what per cent of 32?
c. What per cent of 72 is 15?
d. 3 is what per cent of 16?
e. What per cent of 12 is 11?
f. 20 is what per cent of 14?

11. Round to the nearest tenth of a per cent:
a. 5 is what per cent of 6?
b. What per cent of 36 is 21?
c. 55 is what per cent of 102?
d. What per cent of 18 is 11?
e. 10 is what per cent of 45?
f. What per cent of 48 is 80?

12. Round to the nearest hundredth of a per cent:

a. What per cent of 15 is 4?
b. 3 is what per cent of 9?
c. What per cent of 283 is 100?
d. 17 is what per cent of 32?
e. What per cent of 12 is 2?
f. 50 is what per cent of 38?

13. a. What per cent of 3.6 is 2.7?
b. .8 is what per cent of 2?
c. What per cent of 4.5 is 9?
d. .04 is what per cent of .1?
e. What per cent of 10 is .5?
f. 3.08 is what per cent of 15.4?

14. a. What per cent of 6 is $1\frac{1}{2}$?
b. $\frac{1}{3}$ is what per cent of $\frac{5}{6}$?
c. What per cent of $6\frac{7}{8}$ is $4\frac{1}{8}$?
d. $2\frac{1}{4}$ is what per cent of $3\frac{3}{4}$?
e. What per cent of $\frac{1}{4}$ is $\frac{7}{8}$?
f. 5 is what per cent of $1\frac{2}{3}$?

15. a. What per cent of \$48 is \$12?
b. \$.16 is what per cent of \$.25?
c. What per cent of \$2 is \$.50?
d. \$18 is what per cent of \$22.50?
e. What per cent of \$.60 is \$.96?
f. \$8.70 is what per cent of \$108.75?

7–8 COMPUTATION—FINDING THE NUMBER WHEN A PER CENT OF IT IS KNOWN

When a fractional part or a decimal part of a number is known, the required number is found by dividing the given number by the common fraction or decimal fraction respectively (see pages 119 and 151).

$\frac{3}{4}$ of what number is 18?

$$18 \div \frac{3}{4} = \frac{\overset{6}{\cancel{18}}}{1} \times \frac{4}{\underset{1}{\cancel{3}}} = 24$$

Answer, 24

.75 of what number is 18?

$$.75_\wedge\overline{)18.00_\wedge} = 24.$$

Answer, 24

Since a per cent is a fraction, to find the number when a per cent of it is known, we divide the given number by the common fraction or decimal fraction equivalent of the per cent.

To find the answer to "75% of what number is 18?", we express 75% as $\frac{3}{4}$ or .75 and think of the problem as "$\frac{3}{4}$ of what number is 18?" or ".75 of what number is 18?"

75% of what number is 18?

$75\% = \frac{3}{4}$

$$18 \div \frac{3}{4} = \frac{\overset{6}{\cancel{18}}}{1} \times \frac{4}{\underset{1}{\cancel{3}}} = 24$$

or

$75\% = .75$

$$.75_\wedge\overline{)18.00_\wedge} = 24.$$

Answer, 24

EXERCISES

1. Use decimal equivalents of the per cents to find the following missing numbers:

a. 18% of what number is 9?
b. 6% of what number is 15?
c. 54 is 2% of what number?
d. 40% of what number is 200?
e. 30.6 is 51% of what number?
f. 96 is 32% of what number?
g. 4 is 5% of what number?
h. 75% of what number is 39?
i. 72 is 90% of what number?
j. 3% of what number is 85.29?

2. Use fractional equivalents of the per cents to find the following missing numbers:

a. 50% of what number is 97?
b. 9 is 10% of what number?
c. 75% of what number is 144?
d. $37\frac{1}{2}$% of what number is 225?
e. 408 is $66\frac{2}{3}$% of what number?
f. 41 is 25% of what number?
g. $33\frac{1}{3}$% of what number is 81?
h. 900 is 60% of what number?
i. 1,000 is $62\frac{1}{2}$% of what number?
j. $83\frac{1}{3}$% of what number is 375?

Find the missing numbers:

3. a. 100% of what number is 8?
b. 400% of what number is 36?
c. 52 is 200% of what number?
d. 149 is 100% of what number?
e. 200 is 500% of what number?
f. 300% of what number is 81?

4. a. 125% of what number is 60?
b. 750 is $166\frac{2}{3}$% of what number?
c. 108% of what number is 540?
d. 189 is 150% of what number?
e. $212\frac{1}{2}$% of what number is 34?
f. 94.5 is 135% of what number?

5. a. 4.5% of what number is 36?
b. 54 is 0.6% of what number?
c. 101.2% of what number is 5,060?
d. 283 is 28.3% of what number?
e. 0.1% of what number is 20?
f. 98.75% of what number is 11,850?

6. a. $5\frac{1}{2}$% of what number is 220?
b. $\frac{3}{4}$% of what number is 6?
c. 3,195 is $106\frac{1}{2}$% of what number?
d. 3,740 is $93\frac{1}{2}$% of what number?
e. 680 is $4\frac{1}{4}$% of what number?
f. $\frac{1}{2}$% of what number is 90?

7. a. 4% of what amount is \$100?
b. \$1.60 is $16\frac{2}{3}$% of what amount?
c. 175% of what amount is \$48.30?
d. $\frac{1}{4}$% of what amount is \$250?
e. \$45 is 18% of what amount?
f. 90% of what amount is \$.54?
g. \$27.65 is $3\frac{1}{2}$% of what amount?
h. \$6.85 is 0.5% of what amount?

7–9 PER CENT—SOLVING BY PROPORTIONS

The three basic types of percentage may be treated as one through the use of the proportion. See page 322.

(1) Find 25% of 12.
25% is the ratio of 25 to 100 or $\frac{25}{100}$.
To find 25% of 12 means to determine the number (n) which compared to 12 is the same as 25 compared to 100.

(1) $\frac{n}{12} = \frac{25}{100}$
$100\,n = 300$
$n = 3$
Answer, 3

(2) What per cent of 12 is 3?
To find what per cent of 12 is 3 means to find the number (n) per 100 or the ratio of a number to 100 which has the same ratio as 3 to 12. The proportion $\frac{n}{100} = \frac{3}{12}$ is formed and solved.

(3) 25% of what number is 3?
25% is the ratio of 25 to 100 or $\frac{25}{100}$.
To find the number of which 25% is 3 means to determine the number (n) such that 3 compared to this number is the same as 25 compared to 100. The proportion $\frac{3}{n} = \frac{25}{100}$ is formed and solved.

(2)
$$\frac{n}{100} = \frac{3}{12}$$
$$12n = 300$$
$$n = 25$$
Answer, 25%

(3)
$$\frac{3}{n} = \frac{25}{100}$$
$$25n = 300$$
$$n = 12$$
Answer, 12

EXERCISES

1. Write each of the following as a ratio:

a. 4% **b.** 23% **c.** 60% **d.** 85% **e.** 5% **f.** 130% **g.** 400% **h.** $37\frac{1}{2}$% **i.** $6\frac{3}{4}$% **j.** $\frac{1}{2}$% **k.** 4.8% **l.** $83\frac{1}{3}$% **m.** 0.4% **n.** 175%

2. Find:

a. 6% of 50
b. 80% of 105
c. 75% of 96
d. $62\frac{1}{2}$% of 296
e. 14% of 519
f. 5% of 6.24
g. 2% of \$59
h. 163% of 1,000
i. $3\frac{1}{4}$% of 2,400

3. Find each of the following:

a. 7 is what per cent of 28?
b. What per cent of 50 is 47?
c. What per cent of 120 is 36?
d. 60 is what per cent of 72?
e. 12 is what per cent of 8?
f. What per cent of \$.76 is \$.57?
g. 20 is what per cent of 25?
h. What per cent of \$3 is \$.90?

4. Find each of the following:

a. 35% of what number is 224?
b. 85% of what number is 17?
c. 75% of what number is 18?
d. 52 is 4% of what number?
e. 26 is 26% of what number?
f. 3.2% of what number is 16?
g. 5% of what amount is \$65?
h. \$8.20 is 40% of what amount?

7–10 COMMON FRACTION, DECIMAL FRACTION, AND PER CENT NUMBER RELATIONSHIPS—SOLVING BY EQUATION

The equation[1] may be used to find:

(1) A fractional part or decimal part or per cent of a number.

[1]See pages 309 and 322.

(2) What fractional part or decimal part or per cent one number is of another.
(3) A number when a fractional part or decimal part or per cent of it is known.

To use this method:

We read each problem carefully to find the facts which are related to the missing number. We represent this unknown number by a letter. We form an equation by translating two equal facts, with at least one containing the unknown, into algebraic expressions and writing one expression equal to the other. Where necessary, we change the per cent to a common fraction or to a decimal equivalent.

Then we solve the equation and check.

(1)	(2)	(3)
Find $\frac{1}{4}$ of 12: $\frac{1}{4} \times 12 = n$ $3 = n$ $n = 3$ *Answer,* 3	What fractional part of 12 is 3? $n \times 12 = 3$ $12n = 3$ $\frac{12n}{12} = \frac{3}{12}$ $n = \frac{1}{4}$ *Answer,* $\frac{1}{4}$	$\frac{1}{4}$ of what number is 3? $\frac{1}{4} \times n = 3$ $\frac{1}{4}n = 3$ $4 \cdot \frac{1}{4}n = 4 \cdot 3$ $n = 12$ *Answer,* 12
Find .25 of 12: $.25 \times 12 = n$ $3 = n$ $n = 3$ *Answer,* 3	What decimal part of 12 is 3? $n \times 12 = 3$ $12n = 3$ $n = \frac{1}{4}$ $n = .25$ *Answer,* .25	.25 of what number is 3? $.25 \times n = 3$ $.25n = 3$ $\frac{.25n}{.25} = \frac{3}{.25}$ $n = 12$ *Answer,* 12
Find 25% of 12: $25\% \times 12 = n$ $.25 \times 12 = n$ $3 = n$ $n = 3$ *Answer,* 3	What per cent of 12 is 3? $n\% \times 12 = 3$ $\frac{n}{100} \times 12 = 3$ $\frac{12n}{100} = \frac{3}{1}$ $12n = 300$ $n = 25$ *Answer,* 25%	25% of what number is 3? $25\% \times n = 3$ $.25n = 3$ $\frac{.25n}{.25} = \frac{3}{.25}$ $n = 12$ *Answer,* 12

EXERCISES

1. Find:

a. $\frac{1}{2}$ of 60
b. $\frac{2}{5}$ of 75
c. $\frac{7}{15}$ of 90
d. $\frac{11}{12}$ of 132
e. $\frac{2}{3}$ of 84
f. $\frac{3}{10}$ of 500
g. $\frac{5}{8}$ of 94
h. $\frac{3}{4}$ of 25

2. Find:

a. .2 of 18
b. .75 of 26
c. .04 of 95
d. .33 of 1.49
e. .98 of 6
f. .475 of 840
g. .065 of 8.3
h. .625 of 24

3. Find:

a. 4% of 560
b. 29% of 700
c. 6% of 3,000
d. 132% of 400
e. $66\frac{2}{3}$% of \$.27
f. 75% of \$12.68
g. 200% of 39
h. $4\frac{1}{2}$% of \$5,000

4. Find:

a. $\frac{5}{6}$ of 54
b. 3% of 750
c. .8 of 59
d. $\frac{4}{9}$ of 72
e. $62\frac{1}{2}$% of \$5.76
f. .03 of 4.5
g. $\frac{7}{8}$ of 144
h. 125% of 800

5. Find each of the following:

a. What fractional part of 5 is 3?
b. 6 is what fractional part of 90?
c. 16 is what fractional part of 24?
d. What fractional part of 70 is 42?
e. What fractional part of 91 is 26?
f. 81 is what fractional part of 108?

6. Find each of the following:

a. What decimal part of 12 is 9?
b. 2 is what decimal part of 4?
c. 16 is what decimal part of 25?
d. What decimal part of 40 is 36?
e. 18 is what decimal part of 54?
f. What decimal part of 60 is 50?

7. Find each of the following:

a. 12 is what per cent of 48?
b. 9 is what per cent of 50?
c. What per cent of 600 is 18?
d. What per cent of \$25 is \$17?
e. What per cent of 39 is 39?
f. 20 is what per cent of 16?

8. Find each of the following:

a. What fractional part of 48 is 32?
b. What decimal part of 30 is 24?
c. What per cent of 900 is 63?
d. 84 is what fractional part of 108?
e. \$45 is what per cent of \$54?
f. 120 is what decimal part of 160?

9. Find each of the following:

a. $\frac{1}{6}$ of what number is 37?
b. $\frac{2}{3}$ of what number is 90?
c. $\frac{11}{12}$ of what number is 132?
d. 125 is $\frac{5}{8}$ of what number?
e. 184 is $\frac{8}{9}$ of what number?
f. $\frac{13}{15}$ of what number is 65?

10. Find each of the following:

a. .2 of what number is .12?
b. .04 of what number is 1?
c. .625 of what number is 40?
d. 270 is .75 of what number?
e. 56 is .8 of what number?
f. .375 of what number is 75?

11. Find each of the following:

a. 5 is 10% of what number?
b. 18 is 3% of what number?
c. 24 is 72% of what number?
d. 20% of what number is 9?
e. $87\frac{1}{2}$% of what number is 49?
f. 30 is 125% of what number?

12. Find each of the following:

a. $\frac{4}{5}$ of what number is 40?
b. .05 of what number is 6?
c. 3% of what number is 9?
d. $\frac{3}{7}$ of what number is 84?
e. 17 is .125 of what number?
f. 1.28 is 32% of what number?

13. Find each of the following:

a. 80% of what number is 20?
b. 13 is what per cent of 20?
c. 5% of $207
d. $24 is what per cent of $72?
e. $.97 is 25% of what amount?
f. 115% of 83

14. Find each of the following:

a. $37\frac{1}{2}$% of what number is 18?
b. $4\frac{3}{4}$% of $6,000
c. What per cent of 15 is 21?
d. 36% of $18.53 (to nearest cent)
e. 14 is what per cent of 14?
f. 2% of what amount is $6,280?

7–11 PROBLEMS

EXERCISES

1. How many examples out of 15 may a pupil get wrong and still earn a grade of 80%?
2. If 95% of a class of 40 students were promoted, how many students were left back?
3. Mr. Williams saved $1,152 from his annual income of $12,800. What per cent of his income did he save?
4. During the school year Tom was absent 9 days and was present 171 days. What per cent of the time was he present?
5. What is the cost of a pair of shoes marked $18.50 if a discount of 6% is allowed?
6. Ms. Lee bought an oil burner for $900. She paid 30% in cash. If the balance is to be paid in 18 equal monthly installments, how much must she pay each month?
7. What per cent was a piano reduced if it was marked $600 and sold for $525?
8. A handbag with a 10% tax included sells for $6.49. What is the selling price of the bag without the tax?
9. A salesman receives $75 per week and 3% commission on sales. What are his total earnings if his weekly sales are $3,879?
10. A team won 18 games and lost 6 games. What per cent of the games played did it lose?
11. Mrs. Meade saved $1.25 by buying a blanket at a reduction of 10%. What was the regular price?

12. A house worth \$24,500 is insured for 80% of its value. If the house were destroyed by fire, how much would the owner receive?
13. At what price each must a dealer sell shirts costing \$21.60 a dozen to make a profit of 35% on the cost?
14. Mrs. Jones bought a building for \$18,000 and sold it for \$22,500. What per cent of the cost did she gain?
15. A grocer purchased 24 cans of peas for \$3.84. What per cent of the cost did he gain if he sold them for \$.20 a can?
16. A merchant bought a refrigerator for \$175 and marked it to sell for a profit of 40% of the cost. She sold it for 15% less than the marked price. Find the selling price and the profit.
17. The price of eggs dropped from \$.90 a dozen to \$.81. Find the per cent of decrease in price.
18. Mr. Rossi bought a house for \$33,900. He paid $33\frac{1}{3}\%$ down and gave a mortgage bearing $7\frac{3}{4}\%$ interest for the balance. How much interest does he pay each year?
19. What is the sales tax on a lawn mower costing \$65 if the tax rate is $3\frac{1}{2}\%$?
20. On January 1 a savings account showed a balance of \$280. How much interest will be earned by the end of the year at the annual rate of $5\frac{1}{4}\%$?

CHAPTER REVIEW

1. Express each of the following as a per cent: (7–1)

   a. 92 hundredths **b.** 63 out of 100 **c.** The ratio of 17 to 100

2. Express each of the following per cents as a decimal numeral: (7–2)

   a. 9% **b.** 67% **c.** 180% **d.** $83\frac{1}{3}\%$ **e.** 4.7%

3. Express each of the following per cents as a numeral naming a common fraction: (7–3)

   a. 25% **b.** $16\frac{2}{3}\%$ **c.** 60% **d.** $87\frac{1}{2}\%$ **e.** 48%

4. Express each of the following as a per cent:

   a. .08 .31 1.4 .5793 $.00\frac{1}{2}$ (7–4)
   b. $\frac{1}{2}$ $\frac{51}{68}$ $\frac{36}{64}$ $\frac{13}{25}$ $\frac{88}{100}$ (7–5)

5. Find 39% of 462. (7–6)
6. Find $4\frac{1}{4}\%$ of \$95.87. (7–6)
7. Find 243% of 918.
8. Find 5.9% of 8,000. (7–6)
9. Find $\frac{3}{4}\%$ of \$2,600. (7–6)
10. What per cent of \$57 is \$38? (7–7)
11. 30 is what per cent of 24? (7–7)
12. What per cent of 81 is 18? (7–7)
13. 90% of what number is 72? (7–8)
14. 120 is $62\frac{1}{2}\%$ of what number? (7–8)
15. 5% of what amount is \$1.25? (7–8)

REVIEW EXERCISES

1. Add:

29	3,054	68,947	92,156	75,239
4,638	897	7,052	85,285	18,925
394	52,168	34,369	31,647	38,346
77	36	8,497	52,839	50,182
462	5,785	15,568	49,903	64,697

2. Subtract:

8,500	91,556	46,835	311,524	805,003
698	30,849	9,187	190,615	798,096

3. Multiply:

96	485	708	490	8,249
59	64	891	370	536

4. Divide:

$24\overline{)2,088}$	$36\overline{)23,004}$	$903\overline{)366,618}$	$365\overline{)21,535}$	$1,728\overline{)129,600}$

5. Add:

$4\frac{7}{16}$	$12\frac{2}{3}$	$8\frac{11}{12}$	$2\frac{3}{8}$	$8\frac{1}{2}$
$5\frac{3}{4}$	$6\frac{4}{5}$	$2\frac{5}{8}$	$4\frac{5}{6}$	$5\frac{7}{10}$

6. Subtract:

2	$8\frac{3}{4}$	$10\frac{1}{2}$	$5\frac{2}{3}$	$7\frac{3}{10}$
$1\frac{7}{12}$	$2\frac{13}{16}$	7	$3\frac{1}{4}$	$4\frac{5}{6}$

7. Multiply:

$\frac{3}{10} \times \frac{5}{6}$ $\frac{5}{8} \times 2\frac{3}{4}$ $6 \times 5\frac{7}{8}$ $3\frac{1}{3} \times 2\frac{1}{4}$ $2\frac{11}{12} \times 1\frac{3}{5}$

8. Divide:

$\frac{2}{3} \div 3$ $14 \div 1\frac{5}{16}$ $\frac{4}{5} \div 1\frac{1}{3}$ $5\frac{3}{4} \div \frac{1}{2}$ $3\frac{3}{8} \div 3\frac{3}{5}$

9. Add:

.901 + .899 6.429 + 1.34 20 + 1.8 + .36 .004 + .4 + .04 $6.29 + $.78

10. Subtract:

.403 − .095 6.9 − .42 500 − .235 7.54 − 1.8 $43 − $27.89

11. Multiply:

3.14	4,000	.075	$284.60	$3.84
50	.039	.008	.05	$.62\frac{1}{2}$

12. Divide:

$24\overline{).96}$ $6.1\overline{).0427}$ $.08\overline{)48}$ $.043\overline{)3.01}$ $144\overline{)\$83.52}$

Find each of the following:

13. 2% of $39 53% of 495 $83\frac{1}{3}$% of $5.58 175% of $92 5.9% of 160

14. What per cent of $.12 is $.04? 72 is what per cent of 90? 17 is what per cent of 68? 140 is what per cent of 1,000? 12 is what per cent of 5?

15. 30% of what amount is $18? 35 is 5% of what number? 12% of what number is 16? 120% of what number is 78? 7 is 56% of what number?

CUMULATIVE TEST

1. If B = {all even prime numbers}, which of the following statements are true? (1–9)

a. $4 \in B$ **b.** $28 \in B$ **c.** $7 \notin B$ **d.** $2 \in B$ **e.** $6 \notin B$

2. What is the greatest common factor of 24, 90, and 42? (1–15)
3. Factor 600 as the product of prime factors. (1–13)
4. What is the least common multiple of 6, 8, and 30? (1–18)
5. Write MCMXLVI as a decimal numeral. (2–7)
6. Arrange by size, greatest first: (2–8)

749,876 794,678 749,786 794,687

7. Write as a numeral:

a. Sixty-four million, five hundred six thousand, twenty-nine. (2–10)
b. Seventeen and eleven twelfths. (5–5)
c. Forty and five hundredths. (6–3)
d. Six four three five, base seven. (2–15)

8. Round:

a. 429,604,125 to the nearest million. (2–11)
b. $9\frac{11}{21}$ to the nearest whole number. (5–10)
c. .023509 to the nearest thousandth. (6–5)
d. $87.4068 to the nearest cent. (6–5)

9. Expand each of the following as a polynomial:

a. 6,245,873 **b.** 834.507 (2–14)
c. 12321_{four} **d.** 13412_{five} (2–18)

10. Express 1324_{six} as a base two numeral. (2–19)

11. Find the missing numbers: (3–2—3–5)

a. $15 + 11 = ? + 15$
b. $\frac{3}{4} \times ? = 5 \times \frac{3}{4}$
c. $(9 \times 17) \times 4 = ? \times (17 \times 4)$
d. $(2.9 \times 6) \div 6 = ?$
e. $(.32 + 18) + 4.9 = .32 + (? + 4.9)$
f. $18 \times (6\frac{1}{2} + 7) = ? \times 6\frac{1}{2} + ? \times 7$

12. Is $D = \{9, 18, 27, \ldots\}$ closed under addition? Subtraction? Multiplication? Division? (3–6)

13. Add:

$$\begin{array}{r} 6{,}925 \\ 87{,}694 \\ 529 \\ 98{,}977 \\ \underline{6{,}486} \end{array}$$

14. Subtract:

$$\begin{array}{r} 4{,}050{,}609 \\ \underline{929{,}819} \end{array}$$

15. Multiply: (3–9—3–11)

$$\begin{array}{r} 605 \\ \underline{897} \end{array}$$

16. a. Divide:

$941\overline{)5{,}951{,}825}$

b. Divide: (3–16)

$3{,}007\overline{)24{,}065{,}021}$

17. Is 834,600 divisible by 3? by 2? by 6? by 9? by 4? (3–17)

18. Add:

$$\begin{array}{r} 10111_{two} \\ 11111_{two} \\ 11011_{two} \\ \underline{11111_{two}} \end{array}$$

19. Subtract:

$$\begin{array}{r} 20312_{four} \\ \underline{12221_{four}} \end{array}$$

20. Multiply:

$$\begin{array}{r} 124_{six} \\ \underline{203_{six}} \end{array}$$

21. Divide: (3–22)

$22_{three}\overline{)1012_{three}}$

22. Add:

$33 + 19 \equiv\ ?\ (\text{mod } 8)$

23. Multiply: (3–23)

$13 \times 26 \equiv\ ?\ (\text{mod } 5)$

24. Add the following matrices:

$$\begin{pmatrix} 4 & 3 & 0 \\ 2 & 5 & 6 \\ 0 & 7 & 0 \end{pmatrix} + \begin{pmatrix} 4 & 0 & 10 \\ 1 & 9 & 3 \\ 7 & 2 & 8 \end{pmatrix}$$

25. Find the product matrix: (3–24)

$$\begin{pmatrix} 6 & 0 & 5 \\ 2 & 0 & 4 \\ 1 & 0 & 6 \end{pmatrix} \times \begin{pmatrix} 3 & 7 & 3 \\ 1 & 0 & 2 \\ 0 & 6 & 4 \end{pmatrix}$$

26. Which of the following statements are true?

a. $3 - 3 = \frac{3}{3}$ $\quad 8^2 > 9 \times 7$ $\quad 9 - 0 < 9 + 0$ $\quad 60 + 40 \not> 4 \times 25$ (4–1)
b. $\frac{5}{8} < \frac{5}{9}$ $\quad \frac{25}{60} = \frac{10}{24}$ $\quad \frac{8}{11} > \frac{5}{7}$ $\quad \frac{13}{18} \not< \frac{2}{3}$ (5–8)
c. $.04 > .039$ $\quad .1 < .10$ $\quad .003 = .030$ $\quad .62 \not> .619$ (6–6)

27. Express $\frac{34}{85}$ in lowest terms. (5–2)
28. Express $\frac{19}{24}$ as 96ths. (5–3)
29. Write the set of equivalent fractions of which $\frac{20}{32}$ is a member. (5–3)
30. Arrange by size, greatest first: (5–8)

$\frac{5}{8}, \frac{3}{5}, \frac{1}{2}, \frac{7}{11}, \frac{4}{9}, \frac{2}{3}$

31. Add:

$4\frac{5}{6} + 2\frac{3}{8} + 3\frac{1}{2}$

32. Subtract:

$12\frac{1}{4} - \frac{2}{3}$

33. Multiply: (5–11—5–13)

$3\frac{1}{7} \times 9\frac{4}{5}$

34. a. Divide:

$27 \div 3\frac{3}{4}$

b. Divide: (5–15)

$4\frac{7}{8} \div 5\frac{1}{3}$

35. Find $\frac{5}{12}$ of 168. **36.** What fractional part of 90 is 36? (5-16—5-17)

37. 42 is $\frac{7}{8}$ of what number? **38.** $\frac{3}{7}$ of what number is 81? (5-18)

39. Arrange by size, greatest first: (6-6)

7.07, .077, .770, 7.70

40. Add: **41.** Subtract: **42.** Multiply: (6-7—6-9)

.85 + 9.6 + 43 8.6 − .69 .015 × .008

43. a. Divide: $\$.06\overline{)\$9}$ **b.** Divide: $.023\overline{).000345}$ (6-11)

44. a. Write as a complete numeral: 8.91 trillion (6-13)
b. Write the shortened name in millions: 415,700,000

45. Write in scientific notation: (6-14)

a. 760,000,000,000,000 **b.** .00000000195

46. Express each of the following as a decimal numeral: (6-15)

$\frac{3}{4}$, $\frac{11}{25}$, $\frac{12}{27}$, $\frac{18}{48}$, $\frac{84}{96}$

47. Write $\frac{13}{18}$ as a repeating decimal. (6-16)

48. Express each of the following as a numeral naming a common fraction: (6-17)

.02, .40, .256, .081, $.87\frac{1}{2}$

49. Write $.\overline{72}\ldots$ as a numeral naming a common fraction. (6-18)

50. Find 1.04 of $51. **51.** 28 is what decimal part of 70? (6-19—6-20)

52. 17 is .68 of what number? (6-21)

53. Express each of the following as a decimal numeral: (7-2)

87%, $6\frac{1}{4}$%, 174.2%, 158%, 0.39%

54. Express each of the following as a numeral naming a common fraction: (7-3)

3%, $37\frac{1}{2}$%, 76%, $62\frac{1}{2}$%, $8\frac{1}{3}$%

55. Express each of the following as a per cent: (7-4—7-5)

.07, 1.325, 2.8, $\frac{5}{6}$, $\frac{13}{25}$, $\frac{57}{76}$

56. Find $5\frac{1}{2}$% of 116. **57.** Find 104% of $69.75 (7-6)

58. What per cent of $84 is $63? **59.** 6% of what amount is $3,000? (7-7—7-8)

60. Some time ago Miss Lieberman received a 5% increase on a salary of $140 per week. Recently her salary was reduced 5%. How does her present salary compare with her original salary before the increase? (7-11)

Square Roots – Irrational Numbers

8

The *square root* of a number is that number which when multiplied by itself produces the given number. It is one of the two equal factors of a product.

An *irrational number* is a number that cannot be expressed as a quotient of two whole numbers (with division by zero excluded). A number that is both a non-terminating and non-repeating decimal like the square root of any positive number other than perfect squares (numbers having an exact square root) is an irrational number.

There is a point on the number line that corresponds to each irrational number. To locate the point corresponding to $\sqrt{2}$ on the number line, we construct a square with the side measuring the unit length. By the Rule of Pythagoras we can show that the length of the diagonal of this square is $\sqrt{2}$. Thus, to locate the points corresponding to $\sqrt{2}$ and $^-\sqrt{2}$* on the number line, we describe an arc using the diagonal as the radius. The points where the arc intersects the number line are the required points.

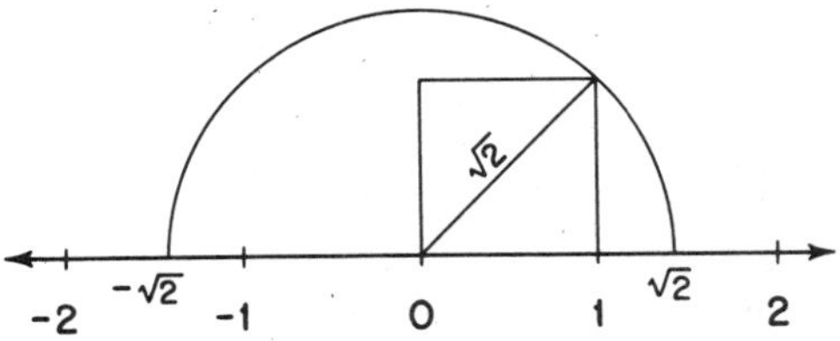

8–1 SQUARE ROOT BY ESTIMATION, DIVISION, AND AVERAGE

We may find the *approximate* square root of a number by:

(1) Estimating the square root of the given number.
(2) Dividing the given number by the estimated square root.
(3) Finding the average of the resulting quotient and the estimated square root.
(4) Dividing the given number by the result of step (3).

---

*See page 235.

(5) Finding the average of the divisor and quotient found in step (4).
(6) Continuing this process to obtain a greater degree of approximation as the divisor and quotient will eventually approximate each other.

Find the square root of 7:

To find the square root of 7 by this method, since the square root of 4 is 2 and the square root of 9 is 3, we estimate that the square root of 7 is between 2 and 3, perhaps 2.5*.

Dividing 7 by 2.5, we get the quotient 2.8. Averaging 2.5 and 2.8, we get 2.65.

Dividing 7 by 2.65, we get the quotient 2.64. Averaging 2.65 and 2.64, we get 2.645.

```
          2.8                 2.64
  2.5∧)7.0∧0       2.65∧)7.00∧00
        5 0               5 30
        2 0 0             1 70 0
        2 0 0             1 59 0
        . . .               11 00
                            10 60
                               40

     2.5                2.65
   + 2.8              + 2.64
  2)5.3(2.65        2)5.29(2.645
```

Answer, 2.645

* 2.6 would be a better estimate and would result in the same answer.

EXERCISES

Find the square root of each of the following by estimation, division, and average:

a. 11	**d.** 54	**g.** 75	**j.** 87	**m.** 138	**p.** 200
b. 28	**e.** 39	**h.** 110	**k.** 19	**n.** 98	**q.** 160
c. 5	**f.** 91	**i.** 44	**l.** 185	**o.** 67	**r.** 320

8–2 SQUARE ROOT—ALTERNATE METHOD

To find the square root of a number, we write its numeral under the square root symbol "$\sqrt{\ }$," and separate the numeral into groups of two digits each to the left and then to the right of the decimal point. If a numeral naming a whole number contains an odd number of digits, there will be one group with one digit on the left. If a numeral naming a decimal fraction contains an odd number of digits, a zero is annexed so that each group contains two digits.

Method:

(1) We find the largest square (49) which can be subtracted from the first group at the left (54) and write it under this group.
(2) We write the square root (7) of this largest square as the first digit in the answer. Each digit in the answer is written directly over its corresponding group.
(3) After we subtract the square, the next group (76) is annexed to the remainder (5).

Find the square root of 5,476:

```
        7  4 .
      √54 76.
       49
  144)  5 76
        5 76
```

Answer, 74

(4) A trial divisor is formed by multiplying the root already found (7) by 2 and annexing a zero which is not written but used mentally.

(5) We divide the dividend (576) by this trial divisor (140) and annex the quotient (4) to the root already found. We also annex it to the trial divisor to form the complete divisor (144).

(6) We multiply the complete divisor by the new digit (4) of the root and subtract this product from the dividend.

(7) This process is continued until all groups are used or the required number of decimal places has been obtained. The decimal point in the answer is placed directly above the decimal point in the given numeral.

Find the square root of 66,049:

2 5 7
$\sqrt{6\ 60\ 49.}$
4
45)2 60
2 25
507) 35 49
35 49

Answer, 257

Find the square root of .09:

.3
$\sqrt{.09}$

Answer, .3

Find the square root of 9.8:

3. 1
$\sqrt{9.80}$
9
61) 80
61
19

Answer, 3.1

EXERCISES

Find the square root of each of the following numbers. If there is a remainder, find answer correct to nearest hundredth:

1. a. 81	**b.** 2,500	**c.** .36	**d.** .0004	**e.** 6,400
2. a. 324	**b.** 961	**c.** 4,624	**d.** 8,649	**e.** 9,409
3. a. 38,416	**b.** 24,964	**c.** 99,856	**d.** 616,225	**e.** 410,881
4. a. 71,824	**b.** 350,464	**c.** 900,601	**d.** 227,529	**e.** 703,921
5. a. 2,399,401	**b.** 6,185,169	**c.** 1,525,225	**d.** 9,771,876	**e.** 8,048,569
6. a. 70,107,129	**b.** 15,952,036	**c.** 35,188,624	**d.** 93,837,969	**e.** 43,454,464
7. a. 11,449	**b.** 824,464	**c.** 424,209	**d.** 902,500	**e.** 253,009
8. a. 49,070,025	**b.** 8,427,409	**c.** 9,072,144	**d.** 64,080,025	**e.** 36,844,900
9. a. .4624	**b.** 3.61	**c.** 8,172.16	**d.** 8	**e.** 1,000
10. a. 49.7	**b.** 6.5	**c.** 193.2	**d.** .234	**e.** 25.963

8–3 SQUARE ROOT BY USE OF TABLE

Using the table of squares and square roots (see page 549), we can find directly the square roots of whole numbers 1 to 99 inclusive and of the perfect squares (squares of whole numbers) given in the table.

To find the square root of any whole number from 1 to 99 inclusive, we first locate the given number in the "No." column and then move to the right to the corresponding "Square Root" column to obtain the required square root.

To find the square root of a perfect square given in the table, we first locate this number in the "Square" column and then move to the left to the corresponding "No." column to obtain the required square root.

EXERCISES

Find the square root of each of the following numbers:

1. **a.** 18 **b.** 53 **c.** 94 **d.** 76 **e.** 39 **f.** 87 **g.** 61 **h.** 45
2. **a.** 26 **b.** 34 **c.** 82 **d.** 63 **e.** 71 **f.** 98 **g.** 41 **h.** 57
3. **a.** 3,844 **b.** 8,281 **c.** 5,929 **d.** 784 **e.** 2,916
4. **a.** 6,561 **b.** 9,025 **c.** 1,849 **d.** 1,024 **e.** 841

8–4 PRINCIPAL SQUARE ROOT*

Since $(^+5) \times (^+5) = 25$ and $(^-5) \times (^-5) = 25$, it follows that 25 has two square roots: $^+5$ and $^-5$. Whole numbers, other than zero, each have two square roots, a positive square root and a negative square root. Zero has only one square root, which is zero (0). The positive square root of a number is called the *principal square root.*

1. Find the two square roots of each of the following numbers:

a. 9 **b.** 100 **c.** 36 **d.** 169 **e** 196 **f.** 256 **g.** 576 **h.** 625

2. What is the principal square root of each of the following numbers?

a. 81 **b.** 4 **c.** 64 **d.** 121 **e** 225 **f.** 900 **g.** 484 **h.** 529

CHAPTER REVIEW

1. Find the square root of each of the following numbers by estimation, division, and average: (8–1)

a. 8 **b.** 17 **c.** 30 **d.** 45 **e.** 78 **f.** 180

2. Find the square root of each of the following numbers by using the table: (8–3)

a. 20 **b.** 43 **c.** 89 **d.** 961 **e.** 9,604 **f.** 3,025

3. Find the square root of each of the following numbers. If there is a remainder, find answer correct to nearest hundredth: (8–2)

a. 4,356 **b.** 56,169 **c.** 6,544,810 **d.** 97 **e.** 725.3

---

* Take this section only if operations with positive and negative numbers have been studied.

Units of Measure and Measurement

9

Many years ago there were no standard units of measure. Early humans used their fingers, feet, and arms to measure length or distance. The width across the open hand at the base of the fingers, called the *palm*, the breadth of a finger, called a *digit*, the greatest stretch of the open hand, called the *span*, and the length of the forearm from the elbow to the end of the middle finger, called the *cubit*, were some of the units used. However, since these measurements varied depending upon the size of the person, they were unsatisfactory.

Today we use metric and customary units of measure which are standard. See pages 551–553 for the tables of measure. Probably in the near future we shall be using in the United States a modernized metric system, the International System of Units, generally known as SI. It should be noted that sometimes the metric unit meter is spelled as metre and liter as litre.

UNITS OF MEASURE

CHANGING TO A SMALLER UNIT OF MEASURE

To change a given number of units of one denomination to units of smaller denomination:

We find the number of units of the smaller denomination that is equivalent to one unit of the larger denomination. This number is sometimes called the *conversion factor*. Then we multiply the given number of units of the larger denomination by this conversion factor.

Since each conversion factor in the metric system is some power of ten, short methods of computation may be used. See section 6-10.

Below are several examples of changing to a smaller unit of measure and references to sections where you will find additional practice exercises.

9–1 Metric—Length	9–2 Metric—Weight	9–3 Metric—Capacity
14 centimeters to millimeters:	9 kilograms to grams:	6 liters to centiliters:
$10 \times 14 = 140$	$1{,}000 \times 9 = 9{,}000$	$100 \times 6 = 600$
Answer, 140 millimeters	*Answer,* 9,000 grams	*Answer,* 600 centiliters

9–4 Metric—Area	9–5 Metric—Volume	9–6 Customary—Length
5 km^2 to m^2:	8 dm^3 to cm^3:	8 feet to inches:
$1{,}000{,}000 \times 5 = 5{,}000{,}000$	$1{,}000 \times 8 = 8{,}000$	$12 \times 8 = 96$
Answer, 5,000,000 m^2	*Answer,* 8,000 cm^3	*Answer,* 96 inches
9–7 Customary—Weight	**9–8 Customary—Liquid**	**9–9 Customary—Dry**
7 pounds to ounces:	18 gallons to quarts:	5 bushels to pecks:
$16 \times 7 = 112$	$4 \times 18 = 72$	$4 \times 5 = 20$
Answer, 112 ounces	*Answer,* 72 quarts	*Answer,* 20 pecks

9–10 Customary—Area	9–11 Customary —Volume
15 sq. yd. to sq. ft.:	6 cu. ft. to cu. in.:
$9 \times 15 = 135$	$1{,}728 \times 6 = 10{,}368$
Answer, 135 sq. ft.	*Answer,* 10,368 cu. in.
9–13 Time	**9–14 Angles and Arcs**
14 hours to minutes:	25 degrees to minutes:
$60 \times 14 = 840$	$60 \times 25 = 1{,}500$
Answer, 840 minutes	*Answer,* 1,500 minutes

CHANGING TO A LARGER UNIT OF MEASURE

To change a given number of units of one denomination to units of a larger denomination:

We find the number of units of the smaller denomination that is equivalent to one of the larger denomination. Then we divide the given number of units of the smaller denomination by this conversion factor.

Since each conversion factor in the metric system is some power of ten, short methods of computation may be used. See section 6-12.

Below are several examples of changing to a larger unit of measure and references to sections where you will find additional practice exercises.

9–1 Metric—Length	9–2 Metric—Weight	9–3 Metric—Capacity
3,000 meters to kilometers:	67 milligrams to centigrams:	945 centiliters to liters:
$3{,}000 \div 1{,}000 = 3$	$67 \div 10 = 6.7$	$945 \div 100 = 9.45$
Answer, 3 kilometers	*Answer,* 6.7 centigrams	*Answer,* 9.45 liters
9–4 Metric—Area	**9–5 Metric—Volume**	**9–6 Customary—Length**
480 mm^2 to cm^2:	250,000 dm^3 to m^3:	15 feet to yards:
$480 \div 100 = 4.8$	$250{,}000 \div 1{,}000 = 250$	$15 \div 3 = 5$
Answer, 4.8 cm^2	*Answer,* 250 m^3	*Answer,* 5 yards

9–7 Customary—Weight	9–8 Customary—Liquid	9–9 Customary—Dry
12,000 lb. to short tons:	14 pints to quarts:	36 pecks to bushels:
12,000 ÷ 2,000 = 6	14 ÷ 2 = 7	36 ÷ 4 = 9
Answer, 6 short tons	*Answer,* 7 quarts	*Answer,* 9 bushels

9–10 Customary—Area	9–11 Customary—Volume
576 sq. in. to sq. ft.:	216 cu. ft. to cu. yd.:
576 ÷ 144 = 4	216 ÷ 27 = 8
Answer, 4 sq. ft.	*Answer,* 8 cu. yd.

9–13 Time	9–14 Angles and Arcs
60 months to years:	240 minutes to degrees:
60 ÷ 12 = 5	240 ÷ 60 = 4
Answer, 5 years	*Answer,* 4 degrees

METRIC SYSTEM

Our monetary system is a decimal system in which the *dollar* is the basic unit. In the metric system, the *meter* (m) is the basic unit of length, the *gram* (g) is the basic unit of weight or mass, and the *liter* (l) is the basic unit of capacity (dry and liquid measures). Other metric units of length, weight, and capacity are named by adding the following prefixes to the basic unit of measure:

Prefix	**Symbol**		**Value**
kilo-	k	meaning	thousand (1,000)
hecto-	h	meaning	hundred (100)
deka-	da	meaning	ten (10)
deci-	d	meaning	one-tenth (.1 or $\frac{1}{10}$)
centi-	c	meaning	one-hundredth (.01 or $\frac{1}{100}$)
milli-	m	meaning	one-thousandth (.001 or $\frac{1}{1,000}$)

The place value chart on the next page shows the relationship among units. Observe the similarity of the metric system of measures to our decimal numeration system and to our monetary system.

It should be noted that just as:

3 dollars 8 dimes 5 cents may be written as a single numeral, 3.85 *dollars* or $3.85; so 3 meters 8 decimeters 5 centimeters may also be written as a single numeral, 3.85 *meters.*

Abbreviations or symbols are written without the period after the last letter. The same symbol is used for both one or more quantities. Thus, *cm* is the symbol for *centimeter* or *centimeters.*

	1,000	100	10	1	.1	.01	.001
Decimal Place Value	thousands	hundreds	tens	ones	tenths	hundredths	thousandths
United States Money	$1,000 bill	$100 bill	$10 bill	dollar	dime	cent	mill
Metric Length	kilometer km	hectometer hm	dekameter dam	meter m	decimeter dm	centimeter cm	millimeter mm
Metric Weight	kilogram kg	hectogram hg	dekagram dag	gram g	decigram dm	centigram cg	milligram mg
Metric Capacity	kiloliter kl	hectoliter hl	dekaliter dal	liter l	deciliter dl	centiliter cl	milliliter ml

The above chart reveals that in the metric system ten (10) of any unit of measure is equivalent to one (1) unit of the next larger size.

The prefix "micro" means one-millionth, and the prefix "mega" means one million. There is a *micrometer* (one-millionth of a meter), a *microgram* (one-millionth of a gram), and a *microliter* (one-millionth of a liter). A *megameter* is one million (1,000,000) meters, a *megagram* is one million grams, and a *megaliter* is one million liters.

9–1 MEASURE OF LENGTH—METRIC

Let us examine the following section of a metric ruler:

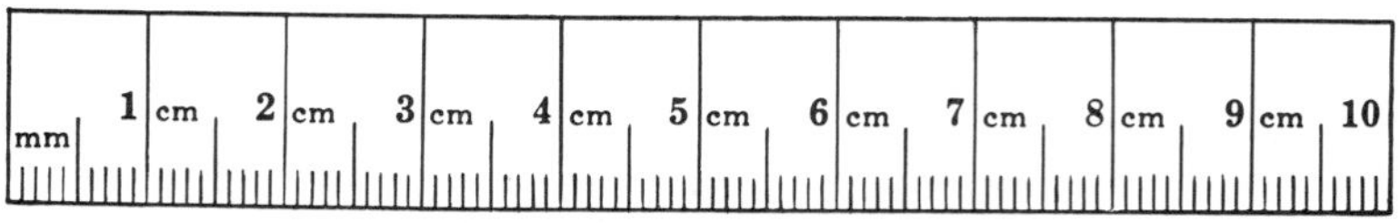

ONE DECIMETER

Each of the smallest subdivisions shown here indicates a measure of 1 *millimeter* (mm). Observe that ten (10) of these millimeter subdivisions form the next larger subdivision, called a *centimeter* (cm), and that ten (10) of the centimeter subdivisions form the next larger subdivision, called a *decimeter* (dm). The meter stick measuring one meter (m), the basic metric unit measuring length, has markings that show 10 decimeter divisions, 100 centimeter divisions, and 1,000 millimeter divisions.

To change a given number of units of one metric denomination to units of another metric denomination, we follow the procedures that are explained on pages 185–187. For short methods of computation, see sections 6-10 and 6-12.

EXERCISES

1. What metric unit of length does each of these symbols represent?

a. dm **b.** dam **c.** km **d.** cm **e.** mm **f.** m **g.** hm

2. Use this section of the metric ruler to answer the following questions:

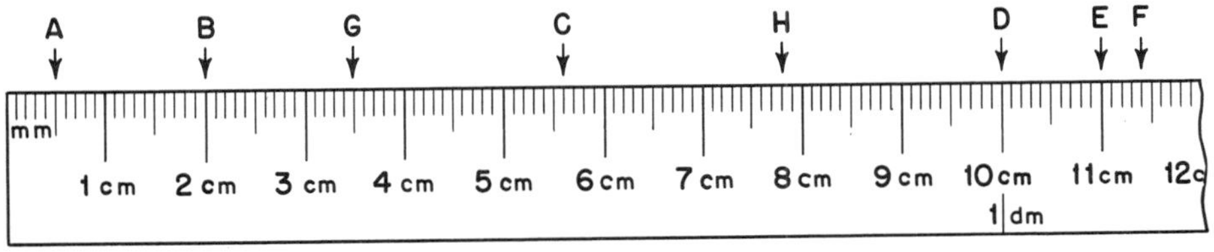

a. What measurement is indicated by the point labeled:

B? *A*? *D*? *G*? *E*? *C*? *F*? *H*?

b. How far from point *D* is point *E*? From point *G* is point *A*? From point *B* is point *H*? From point *F* is point *A*? From point *C* is point *H*?

c. How many centimeters is it from point *B* to point *E*? From point *D* to point *B*? From point *A* to point *G*?

d. How many millimeters is it from point *C* to point *D*? From point *G* to point *F*? From point *A* to point *H*?

e. How many millimeters are in the length from 0 to each of the following markings:

4 cm ? 5 cm 8 mm ? 7 cm 3 mm ? 1 dm 1 cm 4 mm ?

f. In each of the following find the sum of the measurements by locating the mark for the first measurement on the ruler and adding on to this the second measurement. Simplify each sum as indicated:

8 mm + 9 mm = __ mm = __ cm __ mm
42 mm + 23 mm = __ mm = __ cm __ mm
3 cm 4 mm + 2 cm 5 mm = __ cm __ mm
7 cm 8 mm + 1 cm 6 mm = __ cm __ mm = __ cm __ mm
4 cm 6 mm + 2 cm 4 mm = __ cm __ mm = __ cm
1 dm 3 mm + 1 cm 5 mm = __ dm __ cm __ mm
3 cm 9 mm + 6 cm 1 mm = __ cm __ mm = __ cm = __ dm

g. In each of the following locate the marking for the first measurement on the ruler and take away from this the second measurement to find your answer:

16 mm − 9 mm = __ mm
5 cm 6 mm − 2 cm 4 mm = __ cm __ mm
7 cm − 6 mm = __ cm __ mm
2 cm 5 mm − 8 mm = __ cm __ mm
6 cm 3 mm − 5 cm 9 mm = __ cm __ mm
4 cm − 1 cm 7 mm = __ cm __ mm
1 dm − 3 cm 5 mm = __ cm __ mm

h. Find the 106 mm mark on the metric ruler. Subtract from it a measurement of 2.7 cm. What measurement does the mark you reach indicate?

3. Complete each of the following:

a. __ mm = 1 cm	**e.** __ dam = 1 hm	**i.** __ m = 1 dm
b. __ cm = 1 dm	**f.** __ hm = 1 km	**j.** __ dam = 1 m
c. __ dm = 1 m	**g.** __ cm = 1 mm	**k.** __ hm = 1 dam
d. __ m = 1 dam	**h.** __ dm = 1 cm	**l.** __ km = 1 hm

4. Complete each of the following:

a. __ mm = 1 m	**h.** __ m = 1 cm
b. __ cm = 1 m	**i.** __ m = 1 mm
c. __ dm = 1 m	**j.** __ m = 1 dm
d. __ mm = 1 cm	**k.** 1 m = __ dm = __ cm = __ mm
e. __ mm = 1 dm	**l.** 1 km = __ hm = __ dam = __ m
f. __ m = 1 km	**m.** 1 km = __ m = __ dm = __ cm = __ mm
g. __ km = 1 m	**n.** 1 mm = __ m = __ km

5. Express each of the following in meters:

a. 9 km 6 hm 4 dam 8 m = ______ m
b. 6 m 5 dm 1 cm 9 mm = ______ m
c. 5 m 3 cm 2 mm = ______ m
d. 6 km 4 m 2 cm 7 mm = ______ m
e. 2 km 1 hm 7 dam 5 m 2 dm 8 cm 3 mm = ______ m

6. Change each of the following to millimeters:

53 cm; 9.5 m; 2 km; 7.4 cm; 3.7 m

7. Change each of the following to centimeters:

1.8 dm; 46 m; 308 mm; 5 km; 2.9 m

8. Change each of the following to decimeters:

638 m; 590 cm; 4,829 mm; 1.3 km; 18.2 m

9. Change each of the following to meters:

75 dam; 298 mm; 8.6 hm; 40.5 dm; 5,928 cm
.62 km; 19.1 dam; 11,520 mm; 3.475 km; 859 cm

10. Change each of the following to dekameters:

89 m; 43 hm; 6.4 km; 27.8 m; 2,589 cm

11. Change each of the following to hectometers:

40 km; 500 dam; 15.9 m; 3.2 km; 658.45 m

12. Change each of the following to kilometers:

61 hm; 3,582 m; 628,000 cm; 7,150,000 mm; 489.6 dam

13. Find the missing equivalent measurements:

	km	hm	dam	m	dm	cm	mm
a.	?	?	?	500	?	?	?
b.	8	?	?	?	?	?	?
c.	?	?	?	?	?	4,000	?
d.	?	?	?	?	?	?	3,800,000

14. Complete each of the following:

a. 8 cm 3 mm = __ mm
b. 9 km 450 m = __ m
c. 3 m 9 cm = __ cm
d. 7 cm 1 mm = __ mm
e. 6 m 7 mm = __ mm
f. 5 m 18 cm = __ cm

15. Complete each of the following:

a. 2 m 9 dm 4 cm = __ cm
b. 5 m 6 cm 3 mm = __ mm
c. 2 km 5 m 6 cm = __ cm
d. 3 m 7 cm 8 mm = __ mm

16. Complete each of the following:

a. 2 cm 9 mm = __ cm
b. 7 km 8 m = __ km
c. 4 m 7 cm = __ m
d. 6 km 195 m = __ km
e. 8 cm 2 mm = __ cm
f. 3 m 48 mm = __ m

17. Complete each of the following:

a. 6 m 5 cm 2 mm = __ m
b. 8 km 4 m 9 cm = __ km
c. 4 m 9 cm 1 mm = __ m
d. 2 km 18 m 25 cm = __ km

18. Complete each of the following:

a. 2 m 9 cm 4 mm = __ cm
b. 5 km 2 m 1 cm = __ m
c. 75 m 8 cm = __ km
d. 3 cm 9 mm = __ m

19. Sound travels in water at a speed of 1,450 meters per second. How many kilometers does it travel in 10 seconds?

20. If the acceleration of gravity is 9.8 meters per second per second, what is it in terms of centimeters per second per second?

21. Scott has two pieces of wood, one piece measuring 2 m 8 cm and the other three times as long. What is the length of the larger piece of wood? What is the total length of the two pieces of wood? What is the difference between the two pieces of wood? If the larger piece of wood is cut into 8 pieces of equal length, how long is each piece of wood?

22. Which measurement is greater:

a. 6 km or 5,000 m?
b. 8,000 mm or 8 m?
c. 9 m or 2,000 cm?
d. 500 mm or 4 m?
e. 6.1 cm or 74 mm?
f. 2.2 m or 220 mm?

23. Which measurement is smaller:

a. 80 cm or 6 m?
b. 3 km or 40,000 mm?
c. 9 m or 6 km?
d. 58 mm or 4.4 cm?
e. 6.45 m or 645 mm?
f. 8.5 km or 775 m?

24. Arrange the following measurements in order of size (longest first):

2,800 m; 975,000 mm; 5.8 km; 629,600 cm

25. Arrange the following measurements in order of size (shortest first):

345 m; 3.45 km; 3,450,000 cm; 34.5 mm

9–2 MEASURE OF MASS OR WEIGHT—METRIC

Since the weight of one (1) gram is so small, the *kilogram* is generally considered as the practical basic unit of weight. Note also that the weight of 1,000 kilograms is equivalent to one (1) metric ton (t).

To change a given number of units of one metric denomination to units of another denomination, we follow the procedures explained on pages 185–187. For short methods of computation, see sections 6–10 and 6–12.

EXERCISES

1. What metric unit of weight does each of these symbols represent?

a. mg **b.** kg **c.** cg **d.** g **e.** dg **f.** dag **g.** hg

2. Complete each of the following:

a. _ mg = 1 cg	**e.** _ dag = 1 hg	**i.** _ g = 1 dg
b. _ cg = 1 dg	**f.** _ hg = 1 kg	**j.** _ dag = 1 g
c. _ dg = 1 g	**g.** _ cg = 1 mg	**k.** _ hg = 1 dag
d. _ g = 1 dag	**h.** _ dg = 1 cg	**l.** _ kg = 1 hg

3. Complete each of the following:

a. _ mg = 1 g	**g.** 1 kg = _ g = _ mg	**m.** _ mg = 1 kg
b. _ cg = 1 g	**h.** 1 mg = _ g = _ kg	**n.** _ kg = 1 hg
c. _ g = 1 cg	**i.** _ hg = 1 kg	**o.** _ kg = 1 dag
d. _ g = 1 mg	**j.** _ dag = 1 kg	**p.** _ kg = 1 dg
e. _ g = 1 kg	**k.** _ dg = 1 kg	**q.** _ kg = 1 cg
f. _ kg = 1 g	**l.** _ cg = 1 kg	**r.** _ kg = 1 mg

4. Change each of the following to milligrams:

6 cg; 25 g; .375 g; 4.19 kg; 5.6 cg

5. Change each of the following to centigrams:

8.25 dg; 43 g; 23 kg; 5 mg; 9.6 g

6. Change each of the following to decigrams:

17 g; 82 cg; 3,400 mg; 7.9 kg; 4.25 g

7. Change each of the following to grams:

450 mg; 3 kg; 67.5 cg; 920 dg; 80 dag

8. Change each of the following to dekagrams:
2,000 g; 41 kg; 3,500 mg; 7.83 hg; 54.5 cg

9. Change each of the following to hectograms:
30 kg; 5,000 g; 498 dag; 2.7 kg; 325.9 g

10. Change each of the following to kilograms:
75 hg; 2,972 g; 6,100,000 mg; 928,000 cg; 854.7 dag

11. Change each of the following to metric tons:
4,000 kg; 650 kg; 5,100,000 g; 34,000 hg; 92,400 kg

12. Find the missing equivalent weights:

	kg	g	cg	mg
a.	?	8,000	?	?
b.	9.4	?	?	?
c.	?	?	2,500	?
d.	?	?	?	300,000
e.	?	425	?	?

13. Complete each of the following:
a. 8 cg 9 mg = __ mg
b. 5 kg 750 g = __ g
c. 2 g 17 cg = __ cg
d. 6 g 25 mg = __ mg
e. 7 kg 160 g = __ g
f. 4 cg 8 mg = __ mg

14. Complete each of the following:
a. 4 g 5 dg 3 cg = __ cg
b. 9 g 5 cg 3 mg = __ mg
c. 3 kg 8 g 9 cg = __ cg
d. 1 kg 54 g 2 mg = __ mg

15. Complete each of the following:
a. 6 cg 8 mg= __ cg
b. 4 g 3 cg = __ g
c. 9 g 6 mg = __ g
d. 14 kg 250 g = __ kg
e. 6 kg 5 g = __ kg
f. 11 g 92 mg = __ g

16. Complete each of the following:
a. 3 g 8 cg 5 mg = __ g
b. 2 kg 9 g 3 cg = __ kg
c. 5 kg 1 g 7 mg = __ kg
d. 6 g 4 cg 2 mg = __ g

17. Complete each of the following:
a. 4 g 1 cg 8 mg = __ cg
b. 6 kg 5 g 2 cg = __ g
c. 3 cg 5 mg = __ g
d. 250 g 17 cg = __ kg

18. How many 25 mg tablets will be equivalent to a dose of 1 gram?

19. How many kilograms will 325 coins weigh if each coin weighs 8 grams?

20. Which weight is heavier:
a. 5 kg or 5,000 g?
b. 47 mg or 6.3 cg?
c. 25 g or 190 mg?
d. 9,000 cg or 6 kg?
e. 18 mg or 3.1 g?
f. 400 g or 2.5 kg?

21. Which weight is lighter:
a. 40 cg or 1.5 g?
b. 8,400 g or 9 kg?
c. 35 mg or 3.5 cg?
d. 6,100 g or 7.3 kg?
e. 4.8 g or 5,100 mg?
f. 6.53 kg or 7,250 g?

22. Arrange in order of weight (lightest first):
4.7 kg; 53,000 cg; 6,100 g; 750,000 mg

23. Arrange in order of weight (heaviest first):
224 g; 6.25 kg; 4,378,000 mg; 248.5 cg

9–3 MEASURE OF CAPACITY—METRIC

The units that measure capacity also measure volume. See section 9–5 on page 197 where units of cubic measure are studied.

To change a given number of units of one metric denomination to units of another denomination, we follow the procedures explained on pages 185–187. For short methods of computation, see sections 6–10 and 6–12.

EXERCISES

1. What metric unit of capacity does each of these symbols represent?
a. cl **b.** kl **c.** l **d.** ml **e.** dal **f.** hl **g.** dl

2. Complete each of the following:
a. _ ml = 1 cl **d.** _ liter = 1 dal **g.** _ cl = 1 ml **j.** _ dal = 1 liter
b. _ cl = 1 dl **e.** _ dal = 1 hl **h.** _ dl = 1 cl **k.** _ hl = 1 dal
c. _ dl = 1 liter **f.** _ hl = 1 kl **i.** _ liter = 1 dl **l.** _ kl = 1 hl

3. Complete each of the following:
a. _ ml = 1 liter **e.** _ liter = 1 cl
b. _ cl = 1 liter **f.** _ liter = 1 ml
c. _ liters = 1 kl **g.** 1 kl = _ liters = _ ml
d. _ kl = 1 liter **h.** 1 ml = _ liter = _ kl

4. Change each of the following to milliliters:
5 cl; 3.2 liters; 47 dl; .125 liter; 41.7 cl

5. Change each of the following to centiliters:
67 liters; 7 dl; 40 ml; 9.2 liters; 50.6 ml

6. Change each of the following to deciliters:
21 liters; 85 cl; 1.75 liters; 340 ml; 584 cl

7. Change each of the following to liters:
8 dl; 30 hl; 12.75 kl; 600 cl; 9,278 ml

8. Change each of the following to dekaliters:
17 hl; 139 liters; 19.7 liters; 2.5 hl; 468 dl

9. Change each of the following to hectoliters:
9.4 kl; 7.1 dal; 4,305 liters; 51.98 kl; 945.7 liters

10. Change each of the following to kiloliters:
7,853 liters; 26 hl; 685 liters; 49.6 hl; 300 dal

11. Find the missing equivalent capacities:

	liters	dl	cl	ml
a.	6	?	?	?
b.	?	428	?	?
c.	?	?	?	500
d.	9.1	?	?	?
e.	?	?	327.4	?

12. Complete each of the following:

a. 2 cl 9 ml = __ ml
b. 7 liters 4 cl = __ cl
c. 8 liters 5 ml = __ ml
d. 3 kl 400 liters = __ liters
e. 9 liters 56 ml = __ ml
f. 12 liters 35 cl = __ cl

13. Complete each of the following:

a. 1 liter 4 cl 8 ml = __ ml
b. 5 liters 93 cl 3 ml = __ ml

14. Complete each of the following:

a. 4 cl 9 ml = __ cl
b. 6 liters 6 cl = __ liters
c. 2 kl 75 liters = __ kl
d. 2 liters 7 ml = __ liters
e. 8 liters 25 cl = __ liters
f. 9 cl 4 ml = __ cl

15. Complete each of the following:

a. 7 liters 5 cl 8 ml = __ liters
b. 4 liters 9 cl 3 ml = __ liters

16. Complete each of the following:

a. 3 liters 6 cl 7 ml = __ cl
b. 6 liters 8 cl 5 ml = __ cl

17. Marilyn mixed 79 cl of warm water with 68 cl of cold water. How many liters of water did she mix?

18. If a tank holds 1.5 kiloliters of water, how long will it take to fill it if water flows in at the rate of 25 liters per minute?

19. Which capacity is smaller:

a. 30 cl or 56 ml?
b. 4.9 liters or 1,000 cl?
c. 84.3 ml or 1 liter?
d. 4.7 cl or 5 ml?
e. 25 dl or 25 dal?
f. 5.6 liters or 8,300 ml?

20. Arrange the following capacities in order of size (largest capacity first):
54.7 cl; 546 ml; .549 liter; 5.48 dl

21. Arrange the following capacities in order of size (smallest capacity first):
27 liters; 4,325 ml; 38.6 dl; 429 cl

9-4 MEASURE OF AREA—METRIC

We have found in linear measure that ten (10) of any metric unit is equivalent to one (1) of the next higher metric unit. However, in the square measure one hundred (100) of any metric unit is equivalent to one (1) of the next higher unit. As we see in the following table, in the metric measure of area the exponent ^{2} is used to represent the word "square."

100 square millimeters (mm^2)	=	1 square centimeter (cm^2)
100 square centimeters (cm^2)	=	1 square decimeter (dm^2)
100 square decimeters (dm^2)	=	1 square meter (m^2)
100 square meters (m^2)	=	1 square dekameter (dam^2)
100 square dekameters (dam^2)	=	1 square hectometer (hm^2)
100 square hectometers (hm^2)	=	1 square kilometer (km^2)

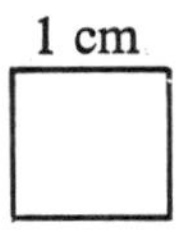

Area = 1 cm^2

Special names are sometimes used. *Centare* may be used instead of square meter, *are* instead of square dekameter, and *hectare* instead of square hectometer. The hectare is used considerably. Thus,

100 centares	=	1 are (a)
100 ares	=	1 hectare (ha)
100 hectares	=	1 square kilometer

To change from one metric unit of square measure to another metric unit of square measure, we follow the procedures on pages 185–187.

EXERCISES

1. What unit of metric square measure does each of these symbols represent?
a. cm^2 **b.** m^2 **c.** km^2 **d.** mm^2 **e.** ha **f.** dm^2 **g.** dam^2

2. **a.** How many square centimeters are in 1 square meter?
b. How many square millimeters are in 1 square meter?
c. How many square meters are in 1 square kilometer?
d. How many square meters are in 1 hectare?
e. How many hectares are in 1 square kilometer?

3. Change each of the following to square millimeters:
7 cm^2; 1.09 m^2; 1.5 dm^2; 23.4 cm^2; .54 m^2

4. Change each of the following to square centimeters:
43 dm^2; 6.5 m^2; 300 mm^2; .89 m^2; 4,500 mm^2

5. Change each of the following to square decimeters:
924 cm^2; 20 m^2; 500,000 mm^2; 6.2 m^2; 430.6 cm^2

6. Change each of the following to square meters (or centares):
25 km^2; 16 dm^2; 50,000 cm^2; 624,000 mm^2; .318 km^2

7. Change each of the following to square dekameters (or ares):
2,000 m^2; 28 hectares; 4.97 m^2; 8.7 hectares; 10,500 m^2

8. Change each of the following to hectares (or square hectometers):
45,000 m^2; 89.6 dam^2; 18 km^2; 275.6 m^2; 3.4 km^2

9. Change each of the following to square kilometers:
300 hectares; 6,000 m^2; 83.3 hectares; 2,960,000 m^2; 527,140 m^2

10. A farm contains 80 hectares of land. How many square meters does it measure? What part of a square kilometer is it?

9-5 MEASURE OF VOLUME—METRIC

The *volume*, also called capacity, is generally the number of units of cubic measure contained in a given space. However, the units of capacity (see section 9-3 on page 194) also are used to measure volume. Refer to the relationship given below between the cubic decimeter and the liter.

In linear measure ten (10) of any metric unit is equivalent to one (1) of the next higher metric unit. In the square measure one hundred (100) of any metric unit is equivalent to one (1) of the next higher unit. However, in the cubic measure one thousand (1,000) of any metric unit is equivalent to one (1) of the next higher unit. As shown below, in the metric measure of volume the exponent ^{3} is used to represent the word "cubic."

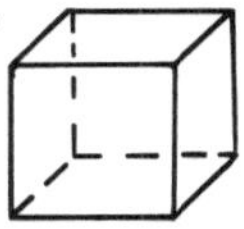

Volume = 1 cm ^{3}

1,000 cubic millimeters (mm^3) = 1 cubic centimeter (cm^3)
1,000 cubic centimeters (cm^3) = 1 cubic decimeter (dm^3)
1,000 cubic decimeters (dm^3) = 1 cubic meter (m^3)

To change from one metric unit of cubic measure to another unit of cubic measure, we follow the procedures as outlined on pages 185–187.

Also note the following relationships:

The volume of one cubic decimeter has the same capacity as one (1) liter. Since 1 cubic decimeter is equivalent to 1,000 cubic centimeters (cm^3) and 1 liter is equivalent to 1,000 milliliters (ml), then the volume of 1 cubic centimeter (cm^3) has the same capacity as 1 milliliter (ml).

Also a gram is the weight of one (1) cubic centimeter (or 1 milliliter) of water at a temperature of 4 degrees Celsius, and a kilogram is the weight of 1,000 cubic centimeters (or 1 liter) of water at a temperature of 4 degrees Celsius.

MISCELLANEOUS EQUIVALENTS

1 liter = 1 cubic decimeter (dm^3) = 1,000 cubic centimeters (cm^3)
1 milliliter (ml) = 1 cubic centimeter (cm^3)
1 liter of water weighs 1 kilogram (kg)
1 milliliter (ml) or cubic centimeter (cm^3) of water weighs 1 gram (g)

EXERCISES

1. What unit of metric cubic measure does each of the following symbols represent?

a. m^3 **b.** cm^3 **c.** mm^3 **d.** km^3 **e.** dam^3 **f.** hm^3 **g.** dm^3

2. a. How many cubic millimeters are in 1 cubic decimeter?
b. How many cubic millimeters are in 1 cubic meter?
c. How many cubic centimeters are in 1 cubic meter?

3. Change each of the following to cubic millimeters:
6 cm^3; 12 m^3; 5.5 cm^3; 23.4 dm^3; 48.5 m^3

4. Change each of the following to cubic centimeters:
24 m^3; 3.59 dm^3; 4,631 mm^3; 9.6 m^3; 850 mm^3

5. Change each of the following to cubic decimeters:
49 m^3; 4,780 cm^3; 14.2 m^3; 52,750 mm^3; 94.8 cm^3

6. Change each of the following to cubic meters:
19,000 dm^3; 8,300,000 mm^3; 26,500 cm^3; 625 dm^3; 500,000 cm^3

7. **a.** 18 milliliters of liquid will occupy a space of how many cubic centimeters?
b. 5 liters of liquid will occupy a space of how many cubic centimeters?
c. A space of 340 cubic centimeters will hold how many milliliters of liquid?
d. A space of 15.7 cubic decimeters will hold how many liters of liquid?
e. A space of 6,800 cubic centimeters will hold how many liters of liquid?
f. 2 liters of water weigh approximately how many kilograms?
g. 85 milliliters of water weigh how many grams?
h. 73 centiliters of water weigh how many grams?
i. 4.5 liters of water weigh how many kilograms?
j. .327 liter of water weighs how many grams?
k. How many liters of water are in a container if the water weighs 9 kilograms?
l. How many milliliters of water weigh 87 grams?
m. How many liters of water are in a container if the water weighs 325 grams?
n. How many cubic decimeters are occupied by 32 kilograms of water?
o. Water weighing 17 grams occupies a space of how many cubic centimeters?
p. Water weighing 4.7 kilograms fills a space of how many cubic centimeters?

8. How much space is occupied by each of the following capacities?

a. In cm^3:	9 ml;	31.4 ml;	25 cl;	6 dl;	.83 liter
b. In dm^3:	4 liters;	97 dl;	367 cl;	19.3 liters;	7.75 liters

9. Find the capacity that will fill each of the following volumes:

a. In ml:	11 cm^3;	2.6 cm^3;	700 mm^3;	8 dm^3;	.05 m^3
b. In cl:	5 cm^3;	8,400 mm^3;	6.3 dm^3;	9.46 cm^3;	2.58 m^3
c. In liters:	6 dm^3;	25.9 dm^3;	3.8 m^3;	40,000 cm^3;	870,000 mm^3

10. Find the weight of each of the following volumes or capacities of water:

a. In kg:	4 dm^3;	87 liters;	5,200 cm^3;	6.92 liters;	156 cm^3
b. In g:	21 cm^3;	6,300 mm^3;	9.7 ml;	3.25 cl;	.367 liter

11. Find the volume or capacity occupied by each of the following weights of water:

a. In cm^3:	7 kg;	40 g;	.3 kg;	850 mg;	4.92 g
b. In liters:	16 kg;	5,100 g;	475 g;	.006 kg;	7,260 mg
c. In ml:	5 g;	157.3 g;	2.6 kg;	9,140 mg;	8,500 cg
d. In cl:	31 g;	8.4 kg;	7,000 mg;	18,300 cg;	425 g

12. A tank holds 25 liters of gasoline. How many cubic decimeters of space are in the tank? How many cubic centimeters?

13. How many liters of water will fill an aquarium if it occupies a space of 4,800 cubic centimeters? What is the weight of the water when the aquarium is full?

14. It takes 200,000 liters of water to fill a swimming pool. How many cubic decimeters does the water occupy when the swimming pool is full? How many kilograms does the water weigh when the swimming pool is three-fourths full?

CUSTOMARY SYSTEM

Since the complete changeover to the metric system is a gradual one, we will find the customary units of measure still being used. Consequently the following sections deal with the customary units of measure for possible use in our everyday affairs.

We find that the conversion factors used in the customary system consist of many different numbers such as 12; 3; 36; 5,280; 1,760; 16; 2,000; etc. When we change from a small customary unit of measure to a larger unit, we sometimes get complicated answers. For example:

49 inches changed to yards is $1\frac{13}{36}$ yards.
39 ounces changed to pounds is $2\frac{7}{16}$ pounds.

However, we have seen that the conversion factors used in the metric system are all some power of ten. When we change from any unit of measure to another unit in the metric system, the computation is quick and easy because we work with the decimal system. For example:

46.2 millimeters = 4.62 centimeters
825 centimeters = 8.25 meters
9,573 milligrams = 9.573 grams

In the following sections 9-6 to 9-11 inclusive, to change a given number of units of one customary denomination to units of another denomination, we follow the procedures that are explained on pages 185–187. See page 552 for tables of measure.

FUNDAMENTAL OPERATIONS WITH DENOMINATE NUMBERS

Any numbers which are expressed in terms of units of measure are called *denominate numbers.*

Addition

To add denominate numbers, we arrange like units in columns, then add each column. Where the sum of any column is greater than the number of units that make the next larger unit, it is simplified as illustrated in the model.

Add:

5 ft. 11 in.
3 ft. 7 in.
8 ft. 18 in. = 9 ft. 6 in.

Answer, 9 ft. 6 in.

Subtraction

To subtract denominate numbers, we arrange like units under each other, then subtract, starting from the right. When the number of units in the subtrahend is greater than the number of corresponding units in the minuend, we take one of the next larger units in the minuend and change it to an equivalent number of smaller units to permit subtraction as shown in the model.

Subtract:

6 hr. 20 min.	=	5 hr. 80 min.
1 hr. 45 min.	=	1 hr. 45 min.
		4 hr. 35 min.

Answer, 4 hr. 35 min.

Multiplication

To multiply denominate numbers, we multiply each unit by the multiplier. Where the product of any column is greater than the number of units that make the next larger unit, it is simplified as shown in the model.

Multiply:

2 yd. 10 in.
4
8 yd. 40 in. = 9 yd. 4 in.

Answer, 9 yd. 4 in.

Division

To divide denominate numbers, we divide each unit by the divisor. If the unit is not exactly divisible, we change the remainder to the next smaller unit and combine with the given number of the smaller unit to form the next partial dividend.

Divide:

```
     2 lb.  7 oz.
6 ) 14 lb. 10 oz.
    12 lb.
     2 lb. 10 oz. = 42 oz.
                    42 oz.
```

Answer, 2 lb. 7 oz.

9–6 MEASURE OF LENGTH—CUSTOMARY

EXERCISES

1. Find the number of inches in:

a.	2 ft.,	19 ft.,	$6\frac{1}{2}$ ft.,	$\frac{3}{4}$ ft.
b.	4 yd.,	27 yd.,	$9\frac{3}{4}$ yd.,	$1\frac{5}{8}$ yd.
c.	1 ft. 6 in.,	5 ft. 11 in.,	1 yd. 9 in.,	6 yd. 23 in.

2. Find the number of feet in:

a.	8 yd.,	21 yd.,	$\frac{2}{3}$ yd.,	$5\frac{1}{2}$ yd.
b.	3 mi.,	15 mi.,	$\frac{5}{8}$ mi.,	$10\frac{1}{4}$ mi.
c.	6 rd.,	34 rd.,	$\frac{1}{2}$ rd.,	$4\frac{3}{4}$ rd.
d.	10 yd. 2 ft.,	45 yd. 1 ft.,	2 mi. 700 ft.,	7 rd. 5 ft.
e.	48 in.,	144 in.,	900 in.,	468 in.

3. Find the number of yards in:

a. 2 mi.,	28 mi.,	$\frac{1}{2}$ mi.,	$3\frac{7}{8}$ mi.
b. 4 rd.,	9 rd.,	$\frac{3}{4}$ rd.,	$8\frac{1}{2}$ rd.
c. 1 mi. 660 yd.,	12 mi. 1,000 yd.,	7 rd. 4 yd.,	10 rd. 2 yd.
d. 108 in.,	504 in.,	720 in.,	648 in.
e. 9 ft.,	42 ft.,	75 ft.,	96 ft.

4. Find the number of rods in:

a. 5 mi.,	12 mi.,	$7\frac{3}{4}$ mi.,	$9\frac{1}{8}$ mi.
b. 33 ft.,	$49\frac{1}{2}$ ft.,	22 yd.,	$16\frac{1}{2}$ yd.

5. Find the number of miles in:

a. 10,560 ft.,	47,520 ft.,	26,400 ft.,	79,200 ft.
b. 7,040 yd.,	17,600 yd.,	21,120 yd.,	44,000 yd.

6. a. What part of a foot is:

6 in., 8 in., 9 in., 5 in., 10 in., 3 in., 1 in.?

b. What part of a yard is:

27 in., 24 in., 30 in., 11 in., 2 ft., 1 ft., $1\frac{1}{2}$ ft.?

c. What part of a mile is:

880 yd., 1,320 ft., 1,540 yd., 80 rd., 110 yd., 3,520 ft.?

7. Add and simplify:

		3 mi. 600 yd.	6 yd. 1 ft. 8 in.
1 ft. 5 in.	2 yd. 18 in.	5 mi. 900 yd.	3 yd. 2 ft. 10 in.
1 ft. 2 in.	3 yd. 26 in.	1 mi. 440 yd.	1 yd. 1 ft. 9 in.

8. Subtract:

8 yd. 9 in.	4 ft.	5 mi. 2,000 ft.	7 yd. 1 ft. 2 in.
5 yd. 3 in.	2 ft. 7 in.	5,000 ft.	3 yd. 2 ft. 6 in.

9. Multiply and simplify:

1 ft. 3 in.	2 mi. 440 yd.	10 yd. 9 in.	4 yd. 2 ft. 8 in.
3	6	4	5

10. Divide:

4)12 mi. 220 yd. 3)7 ft. 9 in. 8)25 yd. 20 in. 5)8 yd. 2 ft. 3 in.

9–7 MEASURE OF WEIGHT—CUSTOMARY

EXERCISES

1. Find the number of ounces in:

a. 7 lb., 18 lb., $4\frac{1}{2}$ lb., $3\frac{3}{4}$ lb. **b.** 1 lb. 9 oz., 2 lb. 14 oz., 5 lb. 3 oz., 9 lb. 10 oz.

2. Find the number of pounds in:

a. 3 s.t.,	25 s.t.,	$8\frac{1}{2}$ s.t.,	$12\frac{2}{5}$ s.t.
b. 5 l.t.,	40 l.t.,	$2\frac{3}{4}$ l.t.,	$6\frac{1}{2}$ l.t.
c. 4 s.t. 500 lb.,	7 l.t. 300 lb.,	10 s.t. 900 lb.,	8 l.t. 1,500 lb.
d. 80 oz.,	128 oz.,	400 oz.,	272 oz.

3. Find the number of short tons in: 8,000 lb., 20,000 lb., 5,000 lb.

4. Find the number of long tons in: 4,480 lb., 20,160 lb., 7,840 lb.

5. **a.** What part of a pound is: 8 oz., 12 oz., 10 oz., 3 oz.?
   b. What part of a short ton is:
   500 lb., 1,000 lb., 200 lb., 1,200 lb.?
   c. What part of a long ton is:
   1,120 lb., 280 lb., 1,680 lb., 1,960 lb.?

6. Add and simplify:

		7 lb. 9 oz.	1 l.t. 800 lb.
6 lb. 4 oz.	4 s.t. 1,500 lb.	8 lb. 11 oz.	5 l.t. 1,600 lb.
3 lb. 8 oz.	2 s.t. 1,000 lb.	3 lb. 12 oz.	2 l.t. 750 lb.

7. Subtract:

5 lb. 11 oz.	8 lb. 4 oz.	3 s.t. 500 lb.	6 lb.
2 lb. 5 oz.	7 lb. 15 oz.	1 s.t. 1,600 lb.	4 lb. 8 oz.

8. Multiply and simplify:

4 lb. 3 oz.	6 s.t. 200 lb.	2 lb. 9 oz.	5 lb. 12 oz.
5	10	6	8

9. Divide:

3)15 lb. 6 oz.   9)27 l.t. 1,800 lb.   7)9 lb. 10 oz.   12)5 lb. 4 oz.

9–8 LIQUID MEASURE—CUSTOMARY

EXERCISES

1. Find the number of ounces in:
   a. 5 pt., 16 pt., $3\frac{1}{2}$ pt., $1\frac{3}{4}$ pt.
   b. 3 qt., 20 qt., $5\frac{1}{4}$ qt., $\frac{1}{2}$ qt.
   c. 2 gal., 5 gal., $\frac{1}{2}$ gal., $1\frac{1}{4}$ gal.
   d. 1 pt. 9 oz., 7 qt. 14 oz., 1 gal. 5 oz.

2. Find the number of pints in:
   a. 6 qt., 40 qt., $8\frac{1}{2}$ qt., $19\frac{1}{4}$ qt.
   b. 2 gal., 9 gal., $5\frac{1}{2}$ gal., $6\frac{3}{4}$ gal.
   c. 3 qt. 1 pt., 8 qt. 1 pt., 10 qt. 1 pt.
   d. 32 oz., 224 oz., 480 oz., 800 oz.

3. Find the number of quarts in:
   a. 2 gal., 15 gal., $9\frac{1}{2}$ gal., $7\frac{3}{4}$ gal.
   b. 5 gal. 3 qt., 1 gal. 2 qt., 8 gal. 1 qt.
   c. 4 pt., 34 pt., 9 pt., 25 pt.
   d. 64 oz., 180 oz., 320 oz., 288 oz.

4. Find the number of gallons in:

a.	12 qt.,	56 qt.,	10 qt.,	19 qt.
b.	16 pt.,	40 pt.,	27 pt.,	82 pt.
c.	128 oz.,	384 oz.,	96 oz.,	192 oz.

5. a. What part of a pint is:

8 oz., 4 oz., 14 oz., 5 oz., 3 oz., 12 oz.?

b. What part of a quart is:

1 pt., 24 oz., 16 oz., $\frac{1}{2}$ pt., $1\frac{1}{2}$ pt., 8 oz.?

c. What part of a gallon is:

3 qt., 2 qt., 1 pt., 64 oz., $2\frac{1}{2}$ qt., 3 pt.?

6. Add and simplify:

		6 gal. 3 qt.	3 pt. 12 oz.
2 gal. 1 qt.	1 qt. 14 oz.	4 gal. 2 qt.	5 pt. 7 oz.
4 gal. 2 qt.	1 qt. 9 oz.	3 gal. 3 qt.	1 pt. 9 oz.

7. Subtract:

3 qt. 1 pt.	1 pt. 12 oz.	7 gal.	3 qt. 1 pt. 5 oz.
2 qt. 1 pt.	15 oz.	4 gal. 3 qt.	1 qt. 1 pt. 12 oz.

8. Multiply and simplify:

5 gal. 1 qt.	3 qt. 7 oz.	1 pt. 9 oz.	4 gal. 3 qt.
3	8	6	9

9. Divide:

3)12 gal. 3 qt.   5)5 pt. 15 oz.   7)9 qt. 6 oz.   10)32 gal. 2 qt.

10. How many ounces of water must be added to:

a. A 6 oz. can of frozen grape concentrate to make $1\frac{1}{2}$ pints of grape juice?
b. A 6 oz. can of frozen lemon concentrate to make 1 quart of lemonade?
c. A 27 oz. can of frozen orange concentrate to make one gallon of juice?

9–9 DRY MEASURE—CUSTOMARY

EXERCISES

1. Find the number of pints in: 6 qt., 25 qt., $9\frac{1}{2}$ qt., 4 qt. 1 pt.

2. Find the number of quarts in:

a.	5 pk.,	18 pk.,	$8\frac{3}{4}$ pt.,	7 pk. 5 qt.
b.	48 pt.,	70 pt.,	1 pt.,	25 pt.

3. Find the number of pecks in:

a.	7 bu.,	24 bu.,	$2\frac{1}{2}$ bu.,	$9\frac{3}{4}$ bu.
b.	3 bu. 3 pk.,	29 bu. 1 pk.,	16 qt.,	28 qt.

4. Find the number of bushels in: 20 pk., 48 pk., 35 pk., 64 qt.

5. a. What part of a bushel is: 2 pk., 3 pk., 1 pk., 4 qt.?

b. What part of a peck is: 4 qt., 6 qt., 3 qt., 1 pt.?

6. Add and simplify:

		5 bu. 3 pk.	3 pk. 4 qt.
3 bu. 2 pk.	2 pk. 3 qt.	1 bu. 2 pk.	7 pk. 2 qt.
6 bu. 1 pk.	5 pk. 4 qt.	3 bu. 3 pk.	2 pk. 3 qt.

7. Subtract:

9 pk. 7 qt.	8 bu.	2 pk. 5 qt.	10 bu. 2 pk.
2 pk. 3 qt.	3 bu. 2 pk.	7 qt.	5 bu. 3 pk.

8. Multiply and simplify:

1 pk. 3 qt.	8 bu. 3 pk.
2	7

9. Divide:

2)14 pk. 6 qt. 5)18 bu. 3 pk.

9–10 MEASURE OF AREA—CUSTOMARY

EXERCISES

Change:

1. To square inches: 17 sq. ft., 4 sq. yd., 20 sq. ft., 5 sq. yd.
2. To square feet: 1,296 sq. in., 15 sq. yd., 42 sq. rd., 4 acres.
3. To square yards: 63 sq. ft., 10,368 sq. in., 8 acres, 5 sq. mi., 117 sq. ft.
4. To square rods: 50 acres, 968 sq. yd., 1,089 sq. yd., 13 acres.
5. To acres: 5,600 sq. rd., 11 sq. mi., 87,120 sq. yd., 2,400 sq. rd.
6. To square miles: 10,880 acres, 18,585,600 sq. yd., 14,720 acres, 3,840 acres.
7. What part of a square foot is: 108 sq. in.? 16 sq. in.? 60 sq. in.?
8. What part of a square yard is: 6 sq. ft.? 5 sq. ft.? 432 sq. in.?
9. What part of a square mile is: 480 acres? 1,936 sq. yd.? 21,780 sq. ft.?

9–11 MEASURE OF VOLUME—CUSTOMARY

EXERCISES

Change:

1. To cubic inches: 15 cu. ft., 7 cu. yd., 40 cu. ft.
2. To cubic inches: 23 cu. yd., $6\frac{1}{2}$ cu. ft., 35 cu. yd.
3. To cubic feet: 30 cu. yd., 16 cu. yd., 6,912 cu. in.
4. To cubic feet: 15,552 cu. in., 34,560 cu. in., 189 cu. yd.
5. To cubic yards: 108 cu. ft., 186,624 cu. in., 513 cu. ft.
6. To cubic yards: 419,904 cu. in., 1,350 cu. ft., 933,120 cu. in.
7. What part of a cubic foot is: 864 cu. in.? 108 cu. in.? 1,296 cu. in.?
8. What part of a cubic yard is: 18 cu. ft.? $13\frac{1}{2}$ cu. ft.? 6 cu. ft.?
9. What part of a cubic foot is: 648 cu. in.? 432 cu. in.? 972 cu. in.?
10. What part of a cubic yard is: 9 cu. ft.? $22\frac{1}{2}$ cu. ft.? 3 cu. ft.?

9–12 VOLUME, CAPACITY, AND WEIGHT RELATIONSHIPS—CUSTOMARY

See page 553 for table of equivalents.

EXERCISES

1. Find the capacity in gallons equal to a volume of 1,848 cu. in.
2. Find the volume in cubic inches equal to a capacity of 15 gallons.
3. Find the capacity in gallons equal to a volume of 60 cu. ft.
4. Find the volume in cubic feet equal to a capacity of 135 gallons.
5. What is the weight in pounds of 42 cu. ft. of sea water?
6. What volume in cubic feet do 5,120 lb. of sea water occupy?
7. What is the weight in pounds of 110 cu. ft. of fresh water?
8. What volume in cubic feet do 4,500 lb. of fresh water occupy?
9. Find the capacity in bushels equal to a volume of 20 cu. ft.
10. Find the volume in cubic feet equal to a capacity of 84 bushels.
11. Find the equivalent:

    a. Capacity in gallons: 3,234 cu. in., 46 cu. ft., 100 cu. ft.
    b. Capacity in bushels: 50 cu. ft., 365 cu. ft., 34 cu. yd.
    c. Volume in cubic feet: 75 gal., 1,200 gal., 236 bu.
    d. Volume in cubic inches: 8 gal., 14 gal., $10\frac{1}{2}$ gal.
    e. Volume in cubic feet: 3,200 lb. of sea water, 7,500 lb. of fresh water.
    f. Weight in pounds: 90 cu. ft. of fresh water, 135 cu. ft. of sea water.

9–13 MEASURE OF TIME

See pages 185–187 for directions to change units of time and pages 199–200 for directions to compute with these units. The table of measure is found on page 553.

EXERCISES

1. Find the number of seconds in:

a.	18 min.,	30 min.,	$7\frac{1}{3}$ min.,	$2\frac{3}{4}$ min.
b.	3 hr.,	12 hr.,	$4\frac{1}{2}$ hr.,	$\frac{2}{3}$ hr.
c.	3 min. 15 sec.,	12 min. 48 sec.,	45 min. 30 sec.,	1 hr. 18 min. 29 sec.

2. Find the number of minutes in:

a.	4 hr.,	17 hr.,	$2\frac{1}{2}$ hr.,	$10\frac{5}{6}$ hr.
b.	7 hr. 45 min.,	9 hr. 23 min.,	14 hr. 56 min.,	21 hr. 4 min.
c.	240 sec.,	3,600 sec.,	90 sec.,	500 sec.

3. Find the number of hours in:

a.	2 da.,	7 da.,	$4\frac{5}{8}$ da.,	$3\frac{2}{3}$ da.
b.	1 da. 9 hr.,	3 da. 14 hr.,	2 da. 23 hr.,	10 da. 5 hr.
c.	180 min.,	1,440 min.,	270 min.,	7,200 sec.

4. Find the number of days in:

a. 5 wk.,	13 wk.,	52 wk.,	$3\frac{4}{7}$ wk.
b.[1] 3 yr.,	2 yr.,	$1\frac{3}{4}$ yr.,	$2\frac{1}{4}$ yr.
c. 1 wk. 3 da.,	7 wk. 5 da.,	1 yr. 100 da.,	2 yr. 263 da.
d. 72 hr.,	216 hr.,	36 hr.,	100 hr.

5. Find the number of months in:

a. 3 yr.,	8 yr.,	$\frac{3}{4}$ yr.,	$4\frac{2}{3}$ yr.
b. 2 yr. 8 mo.,	5 yr. 11 mo.,	7 yr. 7 mo.,	11 yr. 4 mo.

6. Find the number of weeks in:

2 yr., $1\frac{3}{4}$ yr., 1 yr. 40 wk., 2 yr. 29 wk.

7. Find the number of years in:

a. 48 mo.,	120 mo.,	30 mo.,	64 mo.
b. 104 wk.,	156 wk.,	78 wk.,	169 wk.
c. 730 da.,	511 da.,	1,168 da.,	1,461 da.

8. **a.** What part of a minute is: 15 sec., 40 sec., 24 sec., 35 sec.?
b. What part of an hour is: 45 min., 30 min., 5 min., 48 sec.?
c. What part of a day is: 6 hr., 15 hr., 12 hr., 20 hr.?
d. What part of a week is: 3 da., 1 da., 5 da., $3\frac{1}{2}$ da.?
e. What part of a year is: 8 mo., 26 wk., 292 da., 3 mo.?

9. Add and simplify:

		6 yr. 8 mo.	5 hr. 25 min. 32 sec.
48 hr. 28 min.	5 wk. 4 da.	9 yr. 1 mo.	3 hr. 17 min. 51 sec.
8 hr. 39 min.	4 wk. 6 da.	2 yr. 5 mo.	7 hr. 38 min. 47 sec.

10. Subtract:

6 da. 18 hr.	15 yr. 6 mo.	10 min.	8 hr. 19 min. 15 sec.
4 da. 5 hr.	8 mo.	4 min. 36 sec.	2 hr. 30 min. 42 sec.

11. Multiply and simplify:

7 yr. 2 mo.	4 da. 3 hr.	3 hr. 12 min.	1 hr. 38 min. 6 sec.
4	9	10	15

12. Divide:

3)15 yr. 9 mo.   4)18 hr. 32 min.   7)9 da. 15 hr.   15)7 hr. 41 min. 30 sec.

9–14 MEASURE OF ANGLES AND ARCS

See pages 185–187 for directions to change units measuring angles and arcs and pages 199–200 for directions to compute with these units. The table of measure is found on page 553.

[1]Where 4 or more years are involved, use 366 days for every fourth year.

EXERCISES

Change:

1. To minutes: 36°, 18° 25′, 2,580″, 54°
2. To seconds: 51′, 3°, 24′30″, 1° 24′45″
3. To degrees: 840′, 21,660′, 3,600″, 21,660″
4. Add and simplify:

		11° 16′ 30″	82° 29′ 15″
48° 25′	127° 7′ 29″	54° 35′ 25″	36° 50′ 28″
29° 31′	38° 15′ 52″	98° 3′ 18″	60° 40′ 17″

5. Subtract:

160° 25′	90°	81° 34′ 40″	180°
96° 17′	36° 15′	40° 50′ 37″	96° 29′ 46″

6. Multiply and simplify:

26° 14′	7° 38′	47° 12′ 53″	6° 28′ 30″
4	15	3	15

7. Divide:

5)40° 35′	15)10° 30′	6)49° 12′	15)18° 31′ 45″

9–15 RATES OF SPEED

We change the given units to the required units (in the metric model: kilometer to 1,000 meters and hour to 3,600 seconds; in the customary model: mile to 5,280 ft. and hour to 3,600 sec.), then we perform the necessary arithmetical operations. A *knot* is one (1) nautical mile per hour. A nautical mile is approximately equivalent to 6,080 ft. or 1.15 statute miles or 1.85 km. A kilometer is approximately equal to .54 nautical mile. See table of measure on page 552.

Note that in the metric system rates of speed are abbreviated in the following ways: meters per second (m/s); kilometers per hour (km/h); centimeters per second (cm/s); millimeters per minute (mm/min); and so on.

METRIC

Change 72 km/h to m/s:

$$72 \text{ km/h} = \frac{72 \text{ km}}{1 \text{ h}}$$

$$\frac{72 \times 1{,}000}{3{,}600} = 20 \text{ m/s}$$

Answer, 20 meters per second

CUSTOMARY

Change 30 m.p.h. to ft. per sec.:

$$30 \text{ m.p.h.} = \frac{30 \text{ mi.}}{1 \text{ hr.}}$$

$$\frac{30 \times 5{,}280}{3{,}600} = 44 \text{ ft. per sec.}$$

Answer, 44 ft. per sec.

EXERCISES

Change, finding answers to the nearest hundredth wherever necessary:

1. To meters per second:	300 cm/s;	50 km/h;	160 mm/min
2. To meters per minute:	85 mm/s;	240 cm/min;	80 km/h
3. To centimeters per second:	60 km/h;	75 m/min;	630 mm/s
4. To kilometers per hour:	90 m/s;	480 cm/s;	225 knots
5. To knots:	80 km/h;	900 m/s;	350 km/h
6. To feet per second:	300 stat. m.p.h.;	225 stat. m.p.h.;	24 knots
7. To statute miles per hour:	506 ft. per sec.;	140 knots;	858 ft. per min.
8. To nautical miles per hour:	175 knots;	100 stat. m.p.h.;	440 ft. per sec.
9. To knots:	34.8 naut. m.p.h.;	56 stat. m.p.h.;	154 ft. per sec.
10. To feet per minute:	540 ft. per sec.;	165 stat. m.p.h.;	210 knots

9–16 METRIC SYSTEMS AND CUSTOMARY SYSTEMS

Examination of the following section of a customary-metric ruler

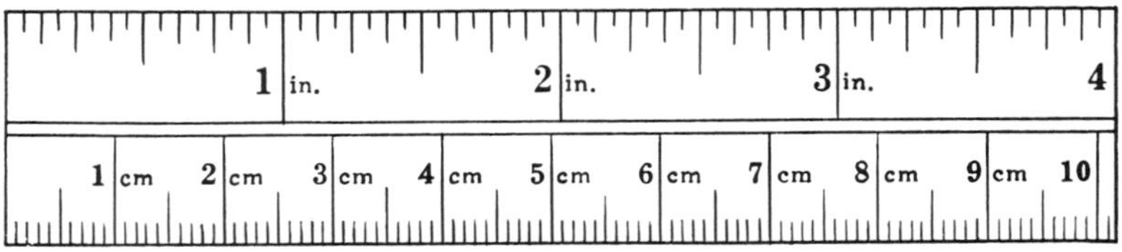

should reveal that:

1 inch is a little longer than $2\frac{1}{2}$ centimeters;
1 centimeter is a little longer than $\frac{3}{8}$ inch; and
1 decimeter is about $3\frac{15}{16}$ inches. Check these observations.

To change a given number of customary units of measure to metric units of measure or to change a given number of metric units of measure to customary units of measure, we first find the number of units of one measure that are equivalent to one unit of the other measure, and then we perform the necessary arithmetical operations. With the exception of the exact equivalents 1 meter = 39.37 inches and 1 inch = 2.54 centimeters all the other customary-metric equivalents used in this exercise are approximate. Generally the symbol "=" is used with exact equivalents, and the symbol "≈" is used with approximate equivalents to represent "is approximately equal to."

Change 148 meters to yards:

$$\begin{array}{r} 148 \\ \times\ 1.09 \\ \hline 13\ 32 \\ 148\ 0 \\ \hline 161.32 \end{array}$$

Answer, 161.32 yards

Change 7 ounces to grams:

$$\begin{array}{r} 28.35 \\ \times\ 7 \\ \hline 198.45 \end{array}$$

Answer, 198.45 grams

EXERCISES

1. Use the ruler on page 208 to find approximately each of the following:

   a. 1 in. ≈ __ mm
   b. 2 in. ≈ __ cm
   c. $2\frac{1}{2}$ in. ≈ __ mm
   d. $3\frac{1}{4}$ in. ≈ __ cm
   e. $1\frac{3}{8}$ in. ≈ __ mm
   f. 4 cm ≈ __ in.
   g. 31 mm ≈ __ in.
   h. 2.6 cm ≈ __ in.
   i. 87 mm ≈ __ in.
   j. 6.4 cm ≈ __ in.
   k. $3\frac{9}{16}$ in. ≈ __ mm
   l. 7 cm ≈ __ in.
   m. $2\frac{7}{8}$ in. ≈ __ cm
   n. 49 mm ≈ __ in.
   o. $1\frac{13}{16}$ in. ≈ __ mm

2. If 1 meter is exactly equal to 39.37 inches in length, determine each of the following (round the approximate equivalent, indicated by the symbol ≈, correct to the nearest hundredth):

   a. 1 cm = __ in. ≈ __ in.
   b. 1 mm = __ in. ≈ __ in.
   c. 1 m ≈ __ ft.
   d. 1 m ≈ __ yd.
   e. 1 dm ≈ __ ft.
   f. 1 km ≈ __ ft.
   g. 1 km ≈ __ yd.
   h. 1 km ≈ __ mi.

3. If 1 inch is exactly equal to 2.54 centimeters in length, determine each of the following (round the approximate equivalent, indicated by the symbol ≈, correct to the nearest hundredth):

   a. 1 in. = __ mm
   b. 1 in. = __ dm ≈ __ dm
   c. 1 ft. = __ cm
   d. 1 ft. = __ mm
   e. 1 ft. = __ dm ≈ __ dm
   f. 1 ft. = __ m ≈ __ m
   g. 1 yd. = __ cm
   h. 1 yd. = __ mm
   i. 1 yd. = __ dm ≈ __ dm
   j. 1 yd. = __ m ≈ __ m
   k. 1 mi. = __ m ≈ __ m
   l. 1 mi. = __ km ≈ __ km

4. In the following problems use the exact equivalent if it contains two or less decimal places, otherwise use the approximate equivalent rounded to the nearest hundredth. Change:

   a. To inches: 16 m; 80 mm; 42 cm; 137 mm; 9.75 m
   b. To feet: 7 m; 316 m; 27.9 km; 680 cm; 4,600 mm
   c. To yards: 49 m; 600 m; 530 cm; 9.4 km; .388 km
   d. To miles: 70 km; 147 km; 300 m; 25.1 km; 1,700 m
   e. To meters: 160 in.; 660 yd.; 72 ft.; 2,000 yd.; 18 in.
   f. To centimeters: 8 in.; 7 ft.; 51 in.; 50 yd.; $4\frac{1}{2}$ ft.
   g. To millimeters: 9 in.; 4 ft.; $6\frac{1}{2}$ in.; $1\frac{3}{4}$ ft.; 5 yd.
   h. To kilometers: 48 mi.; 190 mi.; 5,000 yd.; 12,000 ft.; 7.4 mi.

5. In each of the following the distance by automobile between the two cities is given in kilometers. Determine what this distance is, correct to the nearest mile.

   a. Madrid to Zurich, 1,711 km
   b. Paris to Berlin, 1,092 km
   c. Rome to Amsterdam, 1,735 km
   d. Athens to Lisbon, 4,581 km
   e. Stockholm to Oslo, 348 km
   f. London to Istanbul, 3,416 km

6. The speed limit on some European highways is 100 kilometers per hour and in some European cities 40 kilometers per hour. Express these speed limits in miles per hour.

7. How many inches wide is film 8 mm wide? 16 mm wide? 35 mm wide?
8. Which is longer, the 200-meter dash or the 220-yard dash? How much longer?
9. Using the following approximate equivalents:

1 liquid quart ≈ .95 liter	1 liter ≈ 1.06 liquid quarts
1 dry quart ≈ 1.1 liters	1 liter ≈ .91 dry quart

Determine each of the following:

a. 1 liquid pint ≈ __ liter
b. 1 gallon ≈ __ liters
c. 1 liquid quart ≈ __ ml
d. 1 liquid pint ≈ __ ml
e. 8 fluid ounces ≈ __ ml
f. 12 fluid ounces ≈ __ ml
g. 6 fluid ounces ≈ __ ml
h. 1 fluid ounce ≈ __ ml
i. 1 peck ≈ __ liters
j. 1 bushel ≈ __ liters
k. 1 liter ≈ __ liquid pints
l. 1 liter ≈ __ gallon
m. 1 liter ≈ __ peck
n. 1 liter ≈ __ bushel

10. Change (round answer to nearest hundredth):

a. To liquid quarts: 14 liters; 60 liters; 5,000 ml; 10.5 liters; 183 cl
b. To liters: 6 liquid qt.; 31 dry qt.; 47 liquid pt.; 15 gallons; 23 bushels

11. Some labels on canned and bottled food state the capacity of the contents both in metric and customary measures. Check the accuracy of each of the following:

a. A can of apple juice, 1 liquid quart or 946 ml
b. A bottle of vinegar, 16 fluid ounces or 473 cm^3
c. A jar of pickles, 1 pint 10 fluid ounces or .77 liter
d. A can of fruit juice, 1 quart 14 fluid ounces or 1.36 liters
e. A can of tomato juice, 12 fluid ounces or .356 liter

12. A tank holds 600 liters of water. How many gallons does it hold?
13. Using the equivalent of weight 1 ounce ≈ 28.35 grams, determine each of the following (round answer to nearest hundredth):

a. 1 pound ≈ __ grams ≈ __ kg
b. 1 gram ≈ __ ounce
c. 1 kilogram ≈ __ pounds

14. Change (round answer to nearest hundredth):

a. To pounds: 7 kg; 4.65 kg; 9,230 g; .672 kg; 588 g
b. To kilograms: 12 lb.; 100 lb.; 40 oz.; 9.2 lb.; 152 oz.
c. To ounces: 24 g; 1.4 kg; 3,800 mg; 6.5 g; .74 kg
d. To grams: 4 oz.; 2 lb.; 31 oz.; 1 lb. 3 oz.; 2 lb. 14 oz.

15. Some labels on canned and bottled food state the weight of the contents both in metric and customary measures. Check the accuracy in each of the following:

a. A can of corn, weighing 1 lb. 1 oz. or 482 g
b. A can of peaches, weighing 1 lb. 13 oz. or 822 g
c. A can of pears, weighing 15 oz. or 425 g
d. A bottle of ketchup, weighing 1 lb. 4 oz. or 567 g
e. A can of salmon, weighing $7\frac{3}{4}$ oz. or 220 g

16. How many pounds does a 10-kg bag of apples weigh?

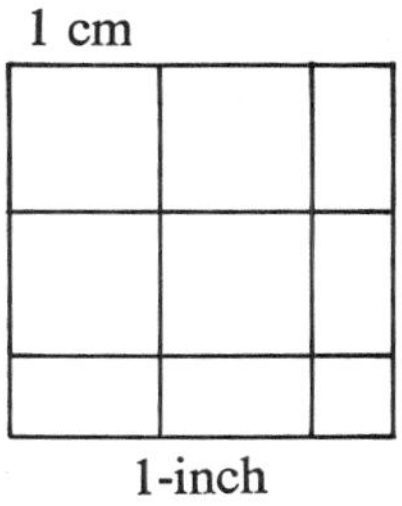

17. If we draw a square one inch on the side and then divide it into squares each measuring one centimeter on a side, we see that it takes approximately $6\frac{1}{4}$ to $6\frac{1}{2}$ square centimeters to make 1 square inch.

Use the approximate equivalent, 1 square inch ≈ 6.45 square centimeters, to determine each of the following:

a. 1 sq. ft. ≈ __ cm^2
b. 1 sq. ft. ≈ __ m^2
c. 1 sq. yd. ≈ __ m^2
d. 1 sq. mi. ≈ __ m^2
e. 1 sq. mi. ≈ __ km^2
f. 1 cm^2 ≈ __ sq. in.
g. 1 mm^2 ≈ __ sq. in.
h. 1 m^2 ≈ __ sq. ft.
i. 1 m^2 ≈ __ sq. yd.
j. 1 km^2 ≈ __ sq. mi.

18. Change (round answer to nearest hundredth):

a. 42 cm^2 to sq. in.
b. 700 mm^2 to sq. in.
c. 125 m^2 to sq. ft.
d. 16.3 m^2 to sq. yd.
e. 200 km^2 to sq. mi.
f. 39 sq. in. to cm^2
g. 8 sq. in. to mm^2
h. 560 sq. ft. to m^2
i. 75 sq. yd. to m^2
j. 100 sq. mi. to km^2

19. A space of one cubic inch measures approximately 16.39 cubic centimeters. Use this approximate equivalent to find each of the following:

a. 1 cu. ft. ≈ __ cm^3
b. 1 cu. ft. ≈ __ m^3
c. 1 cu. yd. ≈ __ m^3
d. 1 cu. ft. ≈ __ dm^3
e. 1 cu. ft. ≈ __ liters
f. 1 cm^3 ≈ __ cu. in.
g. 1 m^3 ≈ __ cu. ft.
h. 1 m^3 ≈ __ cu. yd.
i. 1 dm^3 ≈ __ cu. ft.
j. 1 liter ≈ __ cu. ft.

20. Change (round answer to nearest hundredth):

a. 64 cm^3 to cu. in.
b. 143.9 m^3 to cu. ft.
c. 8 m^3 to cu. yd.
d. 93 dm^3 to cu. ft.
e. 2.5 liters to cu. ft.
f. 180 cu. in. to cm^3
g. 59 cu. ft. to m^3
h. 300 cu. yd. to m^3
i. 6.7 cu. ft. to dm^3
j. 520 cu. ft. to liters

9–17 TEMPERATURE

The three thermometer scales that are used to measure temperature are the *Celsius*, the *Fahrenheit*, and the *Kelvin*. The Celsius scale (formerly called centigrade) is based on 100 divisions, each called a *degree*. On the Celsius scale the freezing point of water is indicated as 0° and the boiling point as 100°. On the Fahrenheit scale the freezing point and the boiling point of water are indicated as 32° and 212° respectively for a 180° interval. It now appears that the Celsius scale will soon replace the Fahrenheit scale in the United States.

The Kelvin temperature scale, used in SI measurement, is related to the Celsius scale. One degree Celsius is exactly equal to one *kelvin* (the name used to mean degree Kelvin). The reading of a specific temperature on the Kelvin scale is approximately 273 kelvins more than its reading on the Celsius scale.

To change a Celsius temperature reading to a kelvin reading, we add 273° to the Celsius reading or use the formula: $K = C + 273$.

To change a kelvin temperature reading to a Celsius reading, we subtract 273° from the kelvin reading or use the formula: $C = K - 273$.

To change a Celsius temperature reading to a Fahrenheit reading, we add 32° to nine-fifths of the Celsius temperature reading or use the formula: $F = \frac{9}{5}C + 32$.

To change a Fahrenheit temperature reading to a Celsius reading, we subtract 32° from the Fahrenheit temperature reading and take five-ninths of this answer or use the formula: $C = \frac{5}{9}(F - 32)$.

EXERCISES

1. Change each of the following Celsius temperature readings to a corresponding kelvin reading:

   a. 40° C **b.** 98° C **c.** 0° C **d.** 9° C **e.** 200° C

2. Change each of the following kelvin temperature readings to a corresponding Celsius reading:

   a. 275 K **b.** 358 K **c.** 400 K **d.** 323 K **e.** 381 K

3. Change each of the following Fahrenheit readings to a corresponding Celsius temperature reading:

   a. 77° F **b.** 212° F **c.** 185° F **d.** 32° F **e.** 100° F

4. Change each of the following Celsius temperature readings to a corresponding Fahrenheit temperature reading:

   a. 40° C **b.** 100° C **c.** 7° C **d.** 65° C **e.** 0° C

5. The classroom thermostat is set at 68° F. What would it be on the Celsius temperature scale? What is it in kelvins?
6. Copper has a melting point of 1,083° C and a boiling point of 2,336° C, silver has a melting point of 960.8° C and a boiling point of 1,950° C and gold has a melting point of 1,063° C and a boiling point of 2,600° C. Express each of these as a Fahrenheit temperature reading. Also express in kelvins.
7. The average temperatures in some of the larger cities of Europe during the month of April are as follows: Paris, 10° C; London, 9° C; Rome, 14° C;

Madrid, 12° C; Lisbon, 15° C; Helsinki, 3° C; and Athens, 16° C. Express each of these Celsius temperature readings as a Fahrenheit temperature reading to the nearest degree. Also express in kelvins.

8. If the temperature of standard air at sea level is 59°F, what is the corresponding temperature reading on the Celsius scale? What is it in kelvins?

9–18 TWENTY–FOUR HOUR CLOCK

To tell time on a 24-hour clock, a 4-digit numeral is used, the first two digits indicating the hour and the last two the minutes. 1 P.M. is thought of as the 13th hour, 2 P.M. the 14th hour, etc. Thus 1200 is added to the given digits when expressing time from 1 P.M. to 12 midnight. The numeral 12 representing the hour is replaced by two zeros when expressing the time from midnight to 1 A.M.

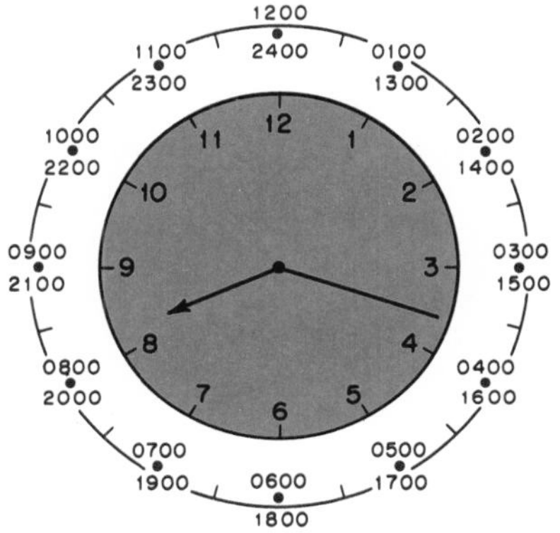

8:43 A.M. = 0843	5:09 P.M. = 1709	12:37 A.M. = 0037	12:28 P.M. = 1228

EXERCISES

1. Using the 24-hour system express each of the following:

a. 11:42 A.M.
b. 2:59 A.M.
c. 6:00 A.M.
d. 9:00 P.M.
e. 3:24 P.M.
f. 10:05 P.M.
g. 12:14 A.M.
h. 12:40 P.M.
i. 10:30 A.M.
j. 4:15 P.M.
k. 12:23 P.M.
l. 5:50 A.M.

2. Express each of the following in A.M. or P.M. time:

a. 0125
b. 1106
c. 0400
d. 1800
e. 2130
f. 0028
g. 2336
h. 1217
i. 0754
j. 1000
k. 0002
l. 1745

3. Find the difference in times between:

a. 0043 and 0800 **b.** 0935 and 1710 **c.** 1122 and 2309

4. An aircraft carrier left port at 0517 and arrived at its destination at 2005. How long did it take the aircraft carrier to reach its destination?

5. An airplane took off from Miami at 0945 and arrived in Boston at 1320. A second airplane traveled the same route, taking off at 2155 and landing at 0043 the next day. Which airplane made the faster time and how much faster?

9–19 TIME ZONES

There are four standard *time belts* in the United States, excluding Alaska and Hawaii: Eastern (EST), Central (CST), Mountain (MST), and Pacific (PST). Central time is one hour earlier than Eastern time, Mountain time is one hour earlier than Central time, and Pacific time is one hour earlier than Mountain time. The meridians at 75°, 90°, 105°, and 120° west longitude are used to determine the time in these zones.

Alaska has four time zones determined by the meridians at 120°, 135°, 150°, and 165° west longitude. Hawaii uses the time zone of the meridian at 150° west longitude. Parts of Canada east of Maine are in another time zone, the Atlantic Standard Time Zone, which is one hour later than Eastern Standard Time.

EXERCISES

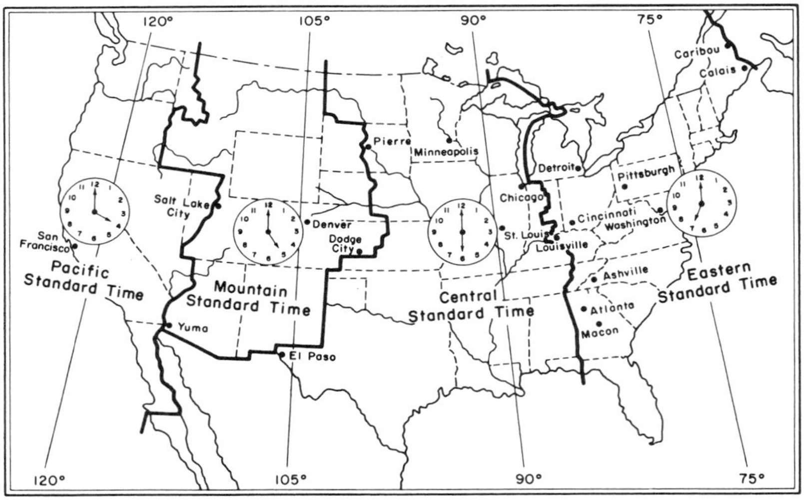

1. In what time zone is your city located?
2. If it is 1 P.M. in the Mountain Time Zone, what time is it in:
   a. Central Zone? b. Pacific Zone? c. Eastern Zone?
3. If it is 8 A.M. in Chicago, what time is it in:
   a. San Francisco? d. Washington? g. Atlanta?
   b. Minneapolis? e. Denver? h. Los Angeles?
   c. Cincinnati? f. Boston? i. Your city?
4. If a nationwide program is telecast at 5 P.M. from New York City, what time is it seen in:
   a. Pittsburgh? c. St. Louis? e. Detroit?
   b. Los Angeles? d. Salt Lake City? f. Your city?

5. How long did it take an airplane to fly from San Francisco to Philadelphia if it left San Francisco at 10:15 A.M. (PST) and arrived in Philadelphia at 6:27 P.M. (EST)?
6. If, exactly at midnight, you place a telephone call in Pittsburgh for Denver, what time is it in Denver?

MEASUREMENT

9–20 PRECISION AND ACCURACY IN MEASUREMENT

Measurement is never exact, it is approximate. The same length measured on each of the following scales shows a measurement of:

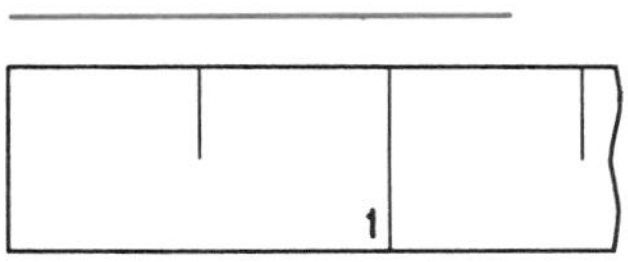

$1\frac{1}{2}''$ to nearest $\frac{1}{2}''$

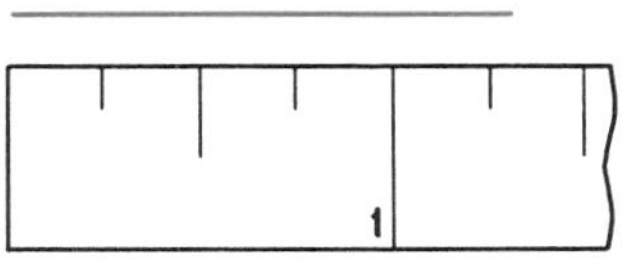

$1\frac{1}{4}''$ to nearest $\frac{1}{4}''$

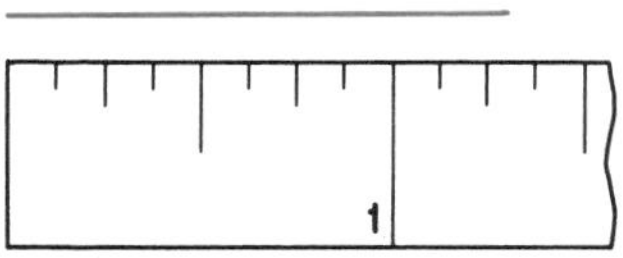

$1\frac{3}{8}''$ to nearest $\frac{1}{8}''$

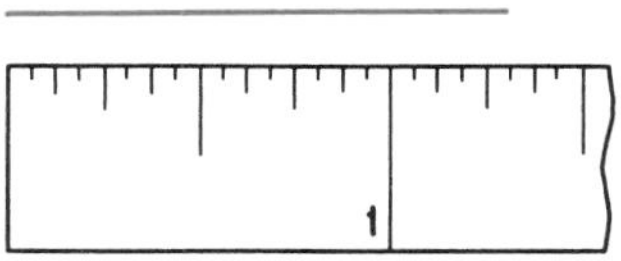

$1\frac{5}{16}''$ to nearest $\frac{1}{16}''$

Observe that $1\frac{1}{2}''$ is precise to the nearest $\frac{1}{2}''$, $1\frac{1}{4}''$ is precise to the nearest $\frac{1}{4}''$; $1\frac{3}{8}''$ is precise to the nearest $\frac{1}{8}''$; $1\frac{5}{16}''$ is precise to the nearest $\frac{1}{16}''$. The smaller the unit, the more precise is the measurement. The measurement $1\frac{5}{16}''$ is the most precise of the above measurements.

Precision is the closeness to the true measurement. It is determined by the smallest unit of measure used to make the measurement.

The precision of a measurement named by a numeral for a whole number ending in zeros is indicated by an underlined zero as shown below. (Also see page 217.)

$25{,}00\underline{0}$ kilometers is precise to the nearest kilometer.

$25{,}0\underline{0}0$ kilometers is precise to the nearest 10 kilometers.

$25{,}\underline{0}00$ miles is precise to the nearest 100 miles.

25,000 miles is precise to the nearest 1,000 miles.

25,000.0 miles is precise to the nearest tenth of a mile.

In measuring the above length $1\frac{5}{16}''$, the unit used was $\frac{1}{16}$ of an inch. The real length is between $1\frac{9}{32}''$ and $1\frac{11}{32}''$. The greatest possible error between the measurement $1\frac{5}{16}''$ and the true measurement can only be $\frac{1}{32}''$ since the *greatest possible error* is one-half of the unit used to measure.

The *relative error* is the ratio of the greatest possible error to the measurement. The *accuracy* of the measurement is determined by the relative error. The smaller the relative error, the more accurate is the measurement.

The measurement $1\frac{5}{16}''$ has the greatest possible error of $\frac{1}{32}''$.	$\frac{1}{32}$ to $1\frac{5}{16}$
The ratio of $\frac{1}{32}''$ to $1\frac{5}{16}''$ is $\frac{1}{42}$. The relative error is $\frac{1}{42}$.	$\frac{1}{32} \div 1\frac{5}{16}$
	$\frac{1}{32} \times \frac{16}{21} = \frac{1}{42}$

Sometimes to make quick comparisons the relative error is expressed as a per cent. To do this we multiply the relative error by 100 or see page 155.

The allowance for error in measurement is called *tolerance*. A required measurement of 8.2 inches with a tolerance of .05 inch allowed in both directions would be indicated as $8.2'' \pm .05''$ and would represent any measurement from 8.15 inches to 8.25 inches inclusive.

EXERCISES

1. To what unit of measure are these measurements precise?

a. $1\frac{7}{8}$ in.
b. $5\frac{22}{32}$ in.
c. $8\frac{3}{4}$ lb.
d. $2\frac{1}{2}$ pt.
e. $6\frac{2}{3}$ hr.
f. $1\frac{5}{6}$ yd.
g. .05 mm
h. .324 in.
i. 12.8 mi.
j. .070 cm
k. 8.6 lb.
l. 4.89 ft.
m. 6 kg 3 g
n. 9 hr. 17 min.
o. 2 ft. $4\frac{1}{2}$ in.

2. Find the unit of measure or precision of each of the following measurements:

a. 237 mi.
b. 15,0$\underline{0}$0 ft.
c. 18 g
d. 40,$\underline{0}$00 tons
e. 930 sec.
f. 27$\underline{0}$,000 persons
g. 2,583 cl
h. 5,60$\underline{0}$,000,000 mi.

3. Which measurement in each of the following is more precise?

a. .065 m or .0126 m?
b. $9\frac{3}{4}$ in. or $7\frac{1}{8}$ in.?
c. 8.0 in. or 4.05 in.?
d. $4\frac{3}{16}$ mi. or $8\frac{1}{2}$ in.?
e. $6\frac{5}{8}$ lb. or 3 lb. 5 oz.?
f. $2\frac{1}{4}$ ft. or 2 ft. 3 in.?

4. Find the greatest possible error in each of these measurements:

a. $3\frac{1}{2}$ in.
b. $4\frac{3}{4}$ in.
c. $1\frac{7}{8}$ in.
d. $2\frac{9}{32}$ in.
e. $6\frac{11}{16}$ lb.
f. $1\frac{3}{5}$ hr.
g. 2.5 liters
h. 1.06 cm
i. .098 in.
j. 34 lb.
k. 942 m
l. 680 yd.
m. 4,$\underline{0}$00 ft.
n. 12$\underline{0}$,000 km
o. 3,$\underline{0}$00,000 mi.
p. 1 ft. 8 in.
q. $4\frac{10}{16}$ in.
r. 1 hr. 4$\underline{0}$ min.
s. 2.0014 in.
t. $5\frac{28}{64}$ in.

5. Find the relative error in each of the measurements given in Exercise **4**.

6. Find the relative error as a per cent for each of the following measurements:

a. 0.8 lb.
b. 40 km
c. $6\frac{1}{4}$ in.
d. $2\frac{2}{3}$ hr.
e. 25 mi.

7. Which measurement in each of the following is more accurate?

a. $2\frac{7}{8}$ in. or $3\frac{5}{16}$ in.? **c.** 5.4 mm or .54 mm? **e.** $30{,}\underline{0}00$ mi. or $1\frac{1}{4}$ in.?
b. 18.5 in. or 1.92 in.? **d.** 69 ft. or 6.9 in.? **f.** .02 ft. or 230 mi.?

8. Which is more precise: 0.9 mm or 0.03 mm? Which is more accurate?

9. Arrange the following measurements in order of precision, least precise first, and also in order of accuracy, least accurate first:

a. .04 mm, .040 mm, 4.0 mm, 40 mm, 4 mm, .004 mm
b. 8.45 mi., 6.9 in., 5.207 ft., 290 yd.

10. What range of measurements does each of the following represent?

a. 4.6″ ± .05″ **c.** $3\frac{7}{8}$″ ± $\frac{3}{16}$″ **e.** 17.5″ ± .07″ **g.** $9\frac{3}{4}$″ ± $\frac{19}{64}$″
b. $5\frac{1}{2}$″ ± $\frac{1}{4}$″ **d.** 6.09 mm ± .025 mm **f.** 2.934″ ± .0001″ **h.** 1″ ± .001″

9–21 SIGNIFICANT DIGITS

Digits are *significant* in an approximate number when they indicate the precision which is determined by the value of the place of the last significant digit on the right.

In 46.25, the significant digit 5 indicates precision to the nearest hundredth.
In $94{,}0\underline{0}0$, the underlined 0 indicates precision to the nearest ten.

The digits described below in items (1) to (4) inclusive are significant, those described in items (5) and (6) are not significant except as noted in (5).

(1) All non-zero digits are significant.	(1) 825 has 3 significant digits.
(2) Zeros located between non-zero digits are significant.	(2) 2,007 has 4 significant digits.
(3) When a decimal or mixed decimal ends in zeros, these zeros are significant.	(3) 43.270 has 5 significant digits.
(4) When numbers are expressed in scientific notation, all the digits in the first factor are significant.	(4) In 8.59×10^6, digits 8, 5, and 9 are significant.

(5) When a number ends in zeros, these zeros are not significant unless they are specified as being significant or indicated as significant by a line drawn under them.

(5) 35,000 has 2 significant digits (3,5)
$35{,}\underline{0}00$ has 3 significant digits (3,5,0)
$35{,}00\underline{0}$ has 5 significant digits (3,5,0,0,0)
2,070 has 3 significant digits (2,0,7)

(6) In a decimal fraction (when the number is between 0 and 1) the zeros immediately following the decimal point are not significant.

(6) .00572 has 3 significant digits (5,7,2) but
5.0072 has 5 significant digits (5,0,0,7,2).
.0050 has 2 significant digits (5,0).

EXERCISES

Determine the number of significant digits in each of the following:

1. 47
2. 9.386
3. 590
4. 6,080,000
5. 940,500,000
6. 5
7. .1468
8. .002
9. .0163
10. .30
11. .00060
12. .010020
13. 84.09
14. 6.00325
15. $96{,}\underline{0}00$
16. $420{,}0\underline{0}0$
17. 9.8×10^8
18. 3.05×10^{12}
19. 5,006
20. 5,600
21. .0465
22. 4.0650
23. $250{,}00\underline{0}$
24. 6.7×10^4
25. 65,273
26. 82.040
27. .0009
28. 3,910
29. $50{,}\underline{0}00$
30. .00000013
31. 4.000208
32. 1.93×10^7
33. 24.0
34. 3.1416
35. 5.987×10^9
36. $42\underline{0}{,}000$
37. 37.004
38. .000025
39. 9,248,017
40. 32.680

9–22 APPROXIMATE NUMBERS

Numbers may be exact or approximate. Numbers used to count are exact; numbers used to measure are approximate. Estimated numbers are approximate numbers. An approximate number is a number that is almost equal to the true number.

Since measurement is approximate, we may be required to compute with approximate numbers arising through measurement. A measurement of 6.4 inches could mean any measurement from 6.35 inches to 6.45 inches. When written as $6.4'' \pm .05''$, it shows the measurement plus and minus the greatest possible error. The result of a computation with approximate numbers cannot be more accurate than the least accurate approximate number involved in the computation.

There are several methods of computing with approximate numbers. Observe the alternative methods at the bottom of this page.

Addition

To add approximate numbers we add as we usually do, then round the sum using the unit of the least precise addend.

Add:

```
  1.82
  4.363
+ 3.1
------
  9.283 = 9.3
```

Subtraction

To subtract approximate numbers we subtract as we usually do, then round the difference using the unit of the less precise of the given numbers.

Subtract:

```
  9.621
- 4.21
------
  5.411 = 5.41
```

Multiplication

To multiply approximate numbers we multiply as we usually do, then round the product so that it contains the same number of significant digits as the factor having the smaller number of significant digits.

Multiply:

```
  2.53
   1.7
------
 1 771
 2 53
------
 4.301 = 4.3
```

Division

To divide approximate numbers we divide as we usually do, then round the quotient so that it contains the same number of significant digits as there are in either the dividend or divisor, whichever is less.

Divide:

```
         1.77 = 1.8
2.5^)4.4^25
     25
     ---
     192
     175
     ---
      175
      175
      ---
      . . .
```

Alternate Methods

To add or subtract we first round the given approximate numbers to the least precise number, then perform the required operation.

To multiply when one of two factors contains more significant digits than the other, we round the factor which has more significant digits so that it contains only one more significant digit than the other. Then we multiply and round the product so that it contains the same number of significant digits as the factor having the smaller number of significant digits.

EXERCISES

Compute the following approximate numbers as indicated:

1. Add:

a. 4.6 + 2.52
b. 9.385 + 3.4 + 8.19
c. .0038 + .005 + .01
d. 16 + .93 + 6.4
e. 185.2 + 4.0094 + 17.83

2. Subtract:

a. 68.953 − 7.82
b. 19.526 − 8.6
c. 75 − 2.3
d. 84.6 − .0087
e. 6.2509 − 1.39256

3. Multiply:

a. 65 × .807
b. 4.3 × 9.6
c. 12.84 × 8.3015
d. .0092 × 1.04
e. 4.751 × .0283

4. Divide:

a. $24\overline{)9.35}$
b. $1.9\overline{)842.03}$
c. $7\overline{).015}$
d. $.023\overline{)10.2471}$
e. $3.86\overline{)958.3059}$

9–23 ESTIMATING ANSWERS

EXERCISES

For each of the following select your nearest estimate:

1. 42 times 71 is approximately:	280;	3,000;	28,000;	2,500
2. $4\frac{7}{8}$ times $3\frac{1}{2}$ is approximately:	7;	12;	17;	25
3. 5% of 3,956 is approximately:	18,000;	200;	1,900;	15
4. 890 divided by 15 is approximately:	450;	1,000;	30;	60
5. .061 divided by .2 is approximately:	35;	.3;	305;	.035
6. 8 divided by $\frac{3}{4}$ is approximately:	6;	11;	24;	35
7. 20% of 324 is approximately:	628;	65;	6;	.06
8. 4,483 divided by 50 is approximately:	890;	2,000;	30;	90
9. 798 times 699 is approximately:	42,000;	1,500;	560,000;	9,000
10. .03 times 9.8 is approximately:	.3;	270;	.01;	29
11. 90% of 437 is approximately:	500;	400;	43;	5
12. 78,926 divided by 198 is approximately:	700;	150;	400;	1,200
13. 47 divided by .6 is approximately:	8;	800;	.8;	80
14. 2.9 times .11 is approximately:	30;	290;	.3;	2.99
15. $19\frac{1}{2}$ divided by $2\frac{1}{2}$ is approximately:	38;	8;	40;	17
16. $66\frac{2}{3}$% of 4,152 is approximately:	4,000;	50;	2,800;	$\frac{2}{3}$
17. $10\frac{1}{4}$ times $5\frac{1}{8}$ is approximately:	600;	$\frac{1}{32}$;	50;	2

18. 117 times 76 is approximately:	190;	9,000;	10,000;	2,000
19. $\frac{7}{8}$ divided by 3 is approximately:	3;	$\frac{1}{3}$;	$\frac{3}{4}$;	1
20. $12\frac{1}{2}\%$ of 325 is approximately:	360;	40;	20;	1,000
21. $3\frac{1}{7}$ times 16 is approximately:	5;	13;	28;	50
22. 66 divided by $1\frac{7}{8}$ is approximately:	30;	60;	120;	15
23. .7 divided by .04 is approximately:	.28;	18;	2;	1.7
24. 3% of $14.85 is approximately:	$45;	$4.48;	$.45;	$14.85
25. $6\frac{1}{3}$ times $9\frac{5}{6}$ is approximately:	60;	54;	27;	18
26. 156,918 divided by 406 is approximately:	700;	60;	2,500;	400
27. .26 times 728 is approximately:	500;	200;	14;	1.6
28. $\frac{3}{4}$ of 587 is approximately:	500;	120;	450;	60
29. 60% of 2,150 is approximately:	120;	1,300;	1,500;	13
30. $4\frac{5}{8}$ times $2\frac{7}{8}$ is approximately:	8;	11;	13;	16
31. 825,000 divided by 195 is approximately:	5,000;	450;	4,000;	30
32. 2% of $8.91 is approximately:	$.18;	$1.60;	$17.82;	$.02
33. 102 divided by .4 is approximately:	2,500;	25;	2.5;	250
34. 913 times 69 is approximately:	5,400;	63,000;	500;	100,000
35. $\frac{4}{5}$ of 247 is approximately:	50;	100;	200;	500

CHAPTER REVIEW

1. Change: 9.6 m to mm; 530 m to km (9–1)

2. Change: 380 g to kg; 4.5 cg to mg (9–2)

3. How many: cl are in 12 liters? dl are in 1,500 ml? (9–3)

4. How many: hectares are in 25 km^2? cm^2 are in 600 m^2? (9–4)

5. 1.25 liters of water occupies a space of ___ cm^3 and weighs ___ kg. (9–5)

6. Change: 5.9 m^3 to cm^3; 4,750 cm^3 to dm^3 (9–5)

7. How many feet are in $5\frac{1}{4}$ miles? (9–6)

8. How many ounces are in 8 pounds? (9–7)

9. How many quarts are in 7 gallons? (9–8)

10. What part of a peck is 6 quarts? (9–9)

11. Change 1,152 sq. in. to sq. ft. (9–10)

12. Change 14 cu. ft. to cu. in. (9–11)

13. From 8 hr. 30 min. subtract 3 hr. 17 min. 35 sec. (9–13)

14. Multiply $40°15'$ by 7. Simplify your answer. (9–14)

15. Change 144 km/h to m/s. (9–15)

16. Change: 8.2 m to in.; 14.5 liters to gallons (9–16)

17. Change: 6.75 kg to lb.; 70 km to mi. (9–16)

18. Is a temperature of 84° F warmer than a temperature of 30° C? (9–17)

19. Express as A.M. or P.M. time: **a.** 1830 **b.** 0015 (9–18)

20. If it is 10 P.M. in Los Angeles, what time is it in New York City? (9–19)

21. Find the greatest possible error and the relative error in each of the following measurements: (9–20)

a. $2\frac{1}{2}$ in. **b.** $9\frac{3}{8}$ lb. **c.** 6.25 g **d.** 150,000 km **e.** 500 m

22. Which is more precise, 0.02 in. or 0.020 in.? Which is more accurate? (9–20)

23. Arrange the following measurements in order of precision, most precise first; and also in order of accuracy, most accurate first: (9–20)

50 yd., 2.93 ft., 8.5 in., 760,000 mi.

24. How many significant digits are in each of the following? (9–21)

a. .00010 **b.** 9,040 **c.** 2.6×10^8 **d.** 8.0297 **e.** 400,000

25. Compute the following approximate numbers as indicated: (9–22)

a. $1.674 + 2.53 + 4.6$
b. $78.562 - 9.42$
c. $6.91 \times .452$
d. $8.9\overline{).9534}$

Sets

10

10–1 SUBSETS

If every member of one set is also a member of a second set, then the first set is said to be a *subset* of the second set. Suppose set A includes only the ninth-year pupils at Martin Luther King High School and set B includes all the pupils at the same school. Since every ninth-year pupil at Martin Luther King High School is a member of the school, we can say A is a subset of B.

Any subset of a set that is not the whole set is called a *proper subset.* A proper subset contains some, but not all, members of the related set. However, any set is assumed to be a subset of itself, and the null set or empty set is assumed to be a subset of every set. Therefore, the subsets of any set include the null set, all proper subsets, and the whole set itself.

The subsets of $\{1, 2, 3\}$ are: the proper subsets $\{1, 2\}$, $\{1, 3\}$, $\{2, 3\}$, $\{1\}$, $\{2\}$, and $\{3\}$, the whole set $\{1, 2, 3\}$, and the null set $\varnothing$ or $\{\ \}$.

When a particular set in a discussion has one or more subsets, this overall set is generally called the *universal set* or *universe.* Not all universal sets are the same since their membership depends on the problem discussed. The symbol U is used to designate the universal set.

We may use to represent a subset either the same capital letter that is used for the entire set but with a subscript or any other capital letter.

The subsets of $S = \{a, b, c, d\}$ are:

$S_1 = \{a, b, c, d\}$ $S_2 = \{a, b, c\}$ $S_3 = \{a, b, d\}$ $S_4 = \{a, c, d\}$ $S_5 = \{b, c, d\}$

$S_6 = \{a, b\}$ $S_7 = \{a, c\}$ $S_8 = \{a, d\}$ $S_9 = \{b, c\}$ $S_{10} = \{b, d\}$ $S_{11} = \{c, d\}$

$S_{12} = \{a\}$ $S_{13} = \{b\}$ $S_{14} = \{c\}$ $S_{15} = \{d\}$ $S_{16} = \varnothing$

To indicate that "A is a proper subset of B" we write
$A \subset B$ or $B \supset A$ which is read: "A is properly contained in B."

$A \subset B$ or $B \supset A$ may also be thought of as indicating that B is a superset of A. The symbols $\subset$ and $\supset$ are used with proper subsets only, while the symbols $\subseteq$ and $\supseteq$ are used with all sets.

$A \subseteq B$ and $B \supseteq A$ is read "A is contained in B."

We can determine how many subsets a set has by the number of members it contains.

A set with 1 member like $\{a\}$ has 2 (or 2^1) subsets: $\{a\}$, $\varnothing$.

A set with 2 members like $\{a, b\}$ has 4 (or 2^2) subsets: $\{a, b\}$, $\{a\}$, $\{b\}$, $\varnothing$.

A set with 3 members (see page 220) has 8 or (2^3) subsets and a set with 4 members (see page 220) has 16 (or 2^4) subsets. Thus, a set with n members has 2^n subsets.

EXERCISES

1. Read, or write in words, each of the following:

a. $M \subset R$ **b.** $E \supset D$ **c.** $B \subseteq T$ **d.** $\{1, 9, 25\} \subset \{1, 4, 9, 16, 25, 36\}$

2. a. Is $\{2, 7, 12\}$ a subset of $\{1, 2, 5, 6, 9, 11, 12\}$?
b. Is $\varnothing$ a subset of $\{0\}$?
c. Is $\{1, 2, 3, \ldots, 19\}$ a subset of $\{1, 2, 3, \ldots, 25\}$?
d. Is $\{3, 8, 11\}$ a subset of $\{11, 3, 8\}$?
e. Is {all prime numbers} a subset of {all odd numbers}?

3. If $F = \{2, 3, 4, 6, 7, 9, 10, 11, 12\}$, $G = \{3, 6, 9, 12\}$, and $H = \{2, 4, 6, 8, 10, 12\}$,

a. Is H a subset of F? **b.** Is G a subset of F? **c.** Is G a subset of H?

4. If $A = \{0, 1, 2, 3, \ldots, 25\}$, what is the subset of A that lists all the prime numbers in set A?

5. If $B = \{1, 2, 3, \ldots, 18\}$, what is the subset of B that lists all the numbers in set B that are divisible by 3?

6. Which of the following are subsets of L if L = {all the one-digit prime numbers}?

a. $\{2, 3, 4\}$
b. $\{2, 3, 7\}$
c. $\{5\}$
d. $\{3, 7, 9\}$
e. $\{2\}$
f. $\{1, 3, 5, 7\}$

7. Which of the following are true?

a. $\{c, e\} \subset \{a, b, c, d, e\}$
b. $\{1, 3, 4, 6, 9\} \supset \{1, 2, 3, 4, 6, 8, 9, 10\}$
c. $\{6, 12, 16, 24\} \subset$ {all multiples of 6}

8. If $A = \{1, 2, 3, \ldots, 15\}$, $B = \{4, 8, 12, 15\}$, $D = \{4, 8, 9, 12, 15\}$, and $E = \{1, 2, 4, 7, 8, 9, 12, 15\}$, which of the following are true?

a. $B \subset E$ **b.** $D \subset A$ **c.** $A \subset E$ **d.** $D \subset B$ **e.** $B \subset A$ **f.** $D \subset E$

9. Write all the possible subsets of:

a. $\{1\}$ **b.** $\{0, 1\}$ **c.** $\{x, y, z\}$ **d.** $\{1, 2, 3, 4, 5\}$

10. If M = {all whole numbers}, N = {10, 20, 30, 40, 50}, S = {all whole numbers divisible by 5}, and T = {0, 2, 4, 6, . . .}, which of the following are false?

a. $N \subset M$ **b.** $S \subset N$ **c.** $M \subset S$ **d.** $N \subset T$ **e.** $T \subset M$ **f.** $S \subset T$

10–2 OPERATIONS WITH SETS

Operations with sets include intersection, union, and complement.

The *intersection* of two sets is the set composed of the common elements that belong to both sets. The symbol used to indicate the operation of intersection is $\cap$.

$A \cap B$ is read "A intersection B" or "the intersection of A and B" or "A cap B".

If A = {1, 2, 3, 4} and B = {2, 4, 6, 8}, then $A \cap B$ = {2, 4}.

Only elements 2 and 4 appear in both sets; therefore the intersection of A and B represented by $A \cap B$ is the set {2, 4}.

The *union* of two sets is the set composed of those elements which are in either of the two given sets or in both sets. The symbol used to indicate the operation of union is $\cup$.

$A \cup B$ is read "A union B" or "the union of A and B" or "A cup B".

If A = {1, 2, 3, 4} and B = {2, 4, 6, 8}, then $A \cup B$ = {1, 2, 3, 4, 6, 8}.

Elements 1, 2, 3, 4 appear in A and elements 2, 4, 6, 8 appear in B. The union of A and B is the set containing all the elements of both A and B. Therefore the union of A and B represented by $A \cup B$ is the set {1, 2, 3, 4, 6, 8}. The common elements are written only once in the union set.

The *complement* of a set with respect to a given universe is the set of elements in the universe which are not in the given set.

The symbol for the complement is a bar over the letter, or the symbol $'$ on the upper right of the letter, or the symbol $\sim$ preceding the letter. The complement of A is $\bar{A}$ or A' or $\sim A$. $\bar{A}$ is read "A bar."

If U represents all the boys at the Central High School and A represents the boys on the Central High School football team, then $\bar{A}$ represents all the boys at Central High School who are not on the football team.

Two sets are said to be *disjoint* if they have no elements in common. Their intersection is the null set.

> $A = \{1, 2, 3, 4\}$ and $B = \{5, 6, 7, 8\}$ are disjoint sets because there is no common element. Thus, $A \cap B = \varnothing$

EXERCISES

1. If $A = \{1, 3, 5, 7, 9, 11, 13\}$, $B = \{2, 4, 6, 8, 10, 12, 14\}$, $C = \{4, 8, 12\}$, $D = \{3, 6, 9, 12\}$, $E = \{9\}$, and $F = \varnothing$, write the resulting set listing the elements for each of the following:

a. $C \cap D$ **c.** $D \cap B$ **e.** $C \cap E$ **g.** $F \cap E$ **i.** $D \cap A$
b. $A \cap B$ **d.** $E \cap A$ **f.** $B \cap C$ **h.** $A \cap C$ **j.** $B \cap F$

2. If $M = \{a, b, c, d\}$, $N = \{e, f, g, h\}$, $P = \{b, d, f, h\}$, $R = \{c, e\}$, $S = \{g\}$, and $T = \varnothing$, write the resulting set listing the elements for each of the following:

a. $M \cup N$ **c.** $R \cup N$ **e.** $S \cup P$ **g.** $N \cup S$ **i.** $T \cup M$
b. $N \cup P$ **d.** $P \cup T$ **f.** $P \cup M$ **h.** $M \cup R$ **j.** $P \cup R$

3. If $U = \{1, 2, 3, \ldots, 16\}$, $A = \{2, 5, 9, 11, 13, 15\}$, $D = \{1, 4, 7, 8, 10, 12, 14, 15, 16\}$, $E =$ {all natural numbers less than 7}, $F =$ {all even whole numbers greater than 6 and less than 14}, and $G =$ {all one-digit odd prime numbers}, find:

a. $\overline{A}$ **b.** $\overline{D}$ **c.** $\overline{E}$ **d.** $\overline{F}$ **e.** $\overline{G}$

4. Which of the following are disjoint sets?

a. $\{a, b, e, g\}$ and $\{c, d, f, h\}$
b. $\{1, 2, 3, 4, 5, 6\}$ and $\{4, 5, 6, 7, 8, 9\}$
c. $\{5, 11, 14, 28, 35\}$ and {all whole numbers divisible by 4}
d. $M =$ {all natural numbers} and $N =$ {all even prime numbers}
e. $R =$ {all whole numbers greater than 5} and $S =$ {all whole numbers less than 6}

5. Write the resulting set for each of the following operations:

a. $\{a, b, c, d, e, f, g\} \cup \{a, e, i, o, u\}$
b. $\{10, 20, 30, 40, 50\} \cap \{5, 10, 15, 20, 25, 30\}$
c. $\{5, 7, 9, 11, 13, 15\} \cup \{3, 9, 15, 21, 27\}$
d. $\{r, s, t, u\} \cap \{m, n, o, p\}$

6. If $U = \{0, 1, 2, \ldots, 12\}$, $N = \{1, 5, 6, 8, 9, 11\}$, and $S = \{0, 1, 3, 7, 8, 10, 12\}$, write the resulting set listing the elements for each of the following:

a. $N \cup S$ **b.** $\overline{N}$ **c.** $S \cap N$ **d.** $\overline{S}$ **e.** $S \cup N$ **f.** $N \cap S$

7. Use $B = \{1, 2, 3, 4, 8\}$, $G = \{3, 4, 5, 7\}$, and $H = \{2, 4, 8\}$ to show that:

a. $B \cap G = G \cap B$ **c.** $B \cap H = H \cap B$
b. $G \cup H = H \cup G$ **d.** $G \cup B = B \cup G$

8. If $U = \{1, 2, 3, \ldots, 19\}$, $A = \{3, 9, 11, 13\}$, $D = \{1, 5, 13, 15\}$, $N = \{2, 6, 10, 14\}$, $R = \{3, 7, 9, 11\}$, $S = \{5, 11, 15, 19\}$, and $T = \varnothing$, find each of the following:

a. $A \cup N$ **b.** $D \cap S$ **c.** $N \cap R$
d. $D \cup A$ **e.** $\overline{A}$ **f.** $T \cap D$
g. $S \cap N$ **h.** $R \cup A$ **i.** $\overline{S}$
j. $R \cup S$ **k.** $\overline{D}$ **l.** $A \cap S$
m. $\overline{N}$ **n.** $S \cup T$ **o.** $R \cap D$

9. Use $R = \{1, 5, 6, 8, 9, 12\}$, $S = \{2, 4, 6, 8, 10, 12\}$, and $T = \{1, 4, 6, 8, 9\}$ to show that:

a. $(R \cap S) \cap T = R \cap (S \cap T)$
b. $(R \cup S) \cup T = R \cup (S \cup T)$
c. $(S \cap T) \cap R = S \cap (T \cap R)$
d. $(S \cup T) \cup R = S \cup (T \cup R)$

10. Use $A = \{0, 3, 4, 7, 8, 9\}$, $E = \{1, 3, 5, 7, 9\}$, and $R = \{0, 2, 4, 7, 8\}$ to show that:

a. $A \cup (E \cap R) = (A \cup E) \cap (A \cup R)$
b. $A \cap (E \cup R) = (A \cap E) \cup (A \cap R)$
c. $E \cup (R \cap A) = (E \cup R) \cap (E \cup A)$
d. $E \cap (R \cup A) = (E \cap R) \cup (E \cap A)$

10–3 VENN DIAGRAMS

We may illustrate sets, subsets, and operation with sets pictorially by Venn diagrams. A Venn diagram is one that represents members of a set as points of a plane placed inside a closed curve.

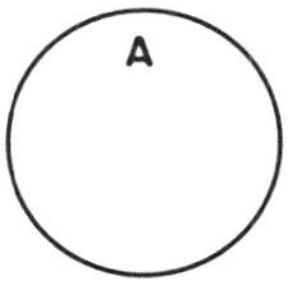

Set A

Sets are represented as sections of a plane. When one set is a subset of another set, the section of the plane representing the subset is placed inside the section representing the other set.

$A \subset B$

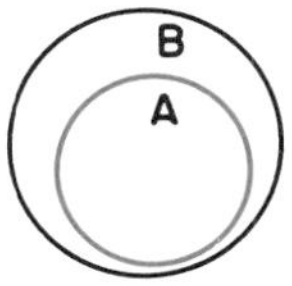

In a Venn diagram the universal set U is represented by all points within and on a rectangle. Each subset of the universal set may be represented by a circle placed inside the rectangle. The members of the subset are represented by points within and on this circle.

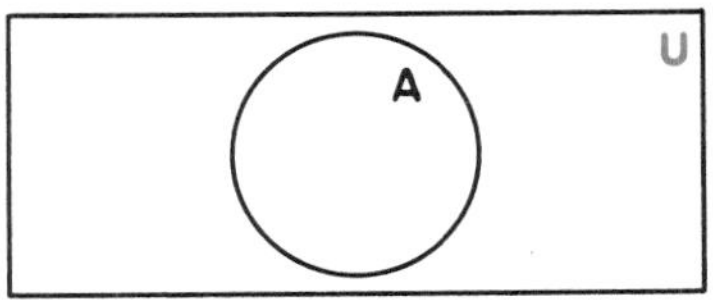

Set A

The intersection of two sets is pictured by each of the following Venn diagrams. The common region is shaded.

$A \cap B$

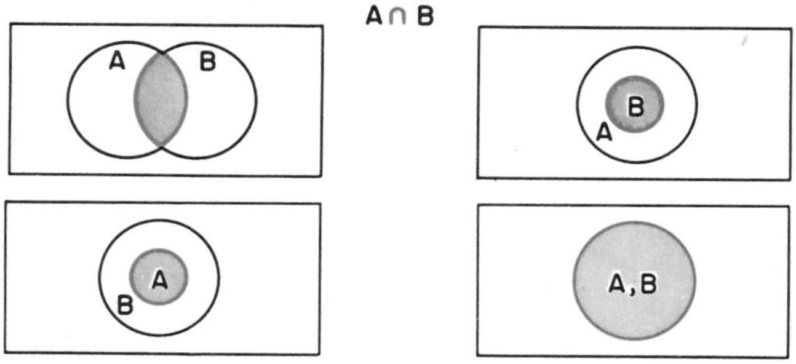

The union of two sets is pictured by each of the following Venn diagrams. All circles are shaded.

$A \cup B$

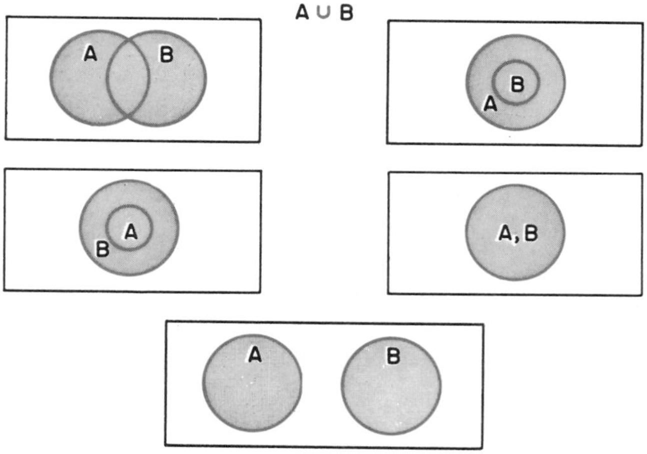

Two disjoint sets are pictured with no shading since there is no common region.

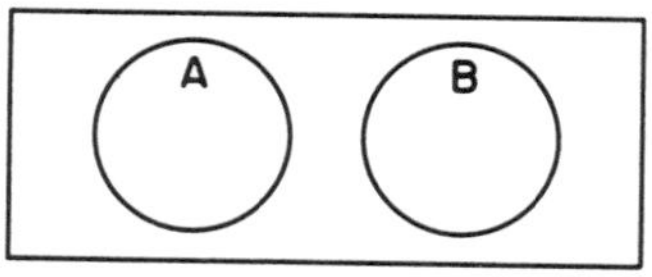

The complement of a set is pictured by drawing a rectangle to represent the universal set and a circle within the rectangle to represent the given set. The section of the rectangle that does not include the circle and its interior represents the complement and is shaded to show this.

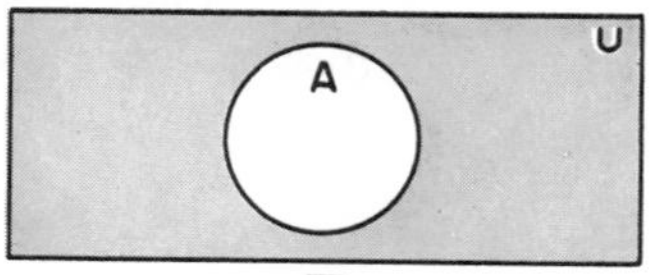

$\overline{A}$

EXERCISES

1. 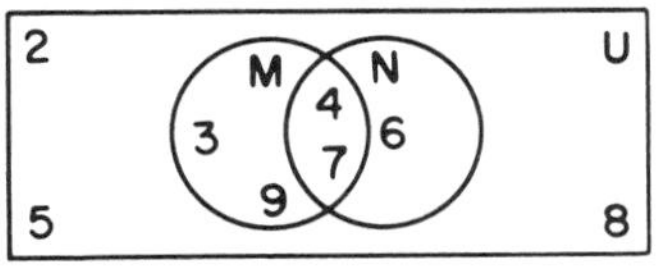

What elements belong to set U? to set M? to set N? Write the resulting set, listing the elements for:

a. $M \cup N$ **b.** $\overline{N}$ **c.** $N \cap M$ **d.** $\overline{M}$

2. If $U = \{0, 1, 2, 3, \ldots, 9\}$, $A = \{2, 4, 6, 7\}$, $B = \{1, 3, 7\}$, and $E = \{3, 4, 7, 9\}$, use a Venn diagram to illustrate of the following:

a. $A \cap E$ **c.** $\overline{B}$ **e.** $E \cup A$ **g.** $B \cap E$ **i.** $\overline{A}$

b. $E \cup B$ **d.** $B \cap A$ **f.** $\overline{E}$ **h.** $A \cup B$ **j.** $E \cap A$

3. If $U = \{1, 2, 3, \ldots, 10\}$, $M = \{1, 6, 9, 10\}$, $R = \{2, 4, 6, 9\}$, and $S = \{3, 4, 6, 10\}$, illustrate each of the following by a Venn diagram:

a. $(M \cup R) \cup S$ **b.** $M \cup (R \cup S)$ **c.** $(M \cap R) \cap S$ **d.** $M \cap (R \cap S)$

4. The set of common factors of two numbers is often described as the intersection of the set of factors of one number and the set of factors of the second number. For each of the following draw a Venn diagram illustrating the set of common factors resulting from the intersection of the sets of factors of:

a. 8 and 6 **b.** 12 and 20 **c.** 30 and 24 **d.** 56 and 42

5. Of 98 students in the senior class, 18 are members of the school gymnastic team, 12 are members of the school volleyball team, and of these students 7 are members of both the gymnastic and volleyball teams. Draw a Venn diagram to illustrate this. Also determine how many students in the senior class are not members of either the gymnastic team or the volleyball team or of both teams.

CHAPTER REVIEW

1. **a.** Is (5, 9, 16, 21} a subset of {2, 5, 6, 9, 14, 17, 21, 22}? (10–1)
b. Is {9, 18, 27, 36, 45} a subset of {0, 3, 6, . . .}?

2. If $D = \{$all even prime numbers$\}$, $E = \{0, 4, 8, \ldots, 60\}$, $R = \{0, 2, 4, \ldots, 60\}$, and $S = \{12, 14, 16, \ldots, 48\}$, which of the following are true? (10–1)

a. $S \subset E$ **b.** $D \subset S$ **c.** $D \subset E$ **d.** $R \subset E$ **e.** $S \subset R$ **f.** $D \subset R$

3. If $U = \{0, 1, 2, \ldots, 20\}$, $G = \{2, 5, 8, 11, 13, 14, 17, 18\}$, $H = \{2, 3, 5, 7, 11, 13, 17, 19\}$, and $M = \{8, 9, 10, \ldots, 15\}$, find each of the following: (10–2)

a. $G \cup H$ **b.** $H \cap M$ **c.** $\overline{H}$ **d.** $H \cap G$ **e.** $\overline{M}$ **f.** $M \cup G$

4. If $U = \{1, 2, 3, \ldots, 15\}$, $A = \{5, 8, 9, 11, 14, 15\}$, $B = \{1, 3, 5, 6, 7, 9, 12, 14\}$, and $N = \{2, 4, 6, 9, 11, 12, 14\}$, illustrate each of the following by a Venn diagram: (10–2)

a. $A \cap N$ **b.** $B \cup A$ **c.** $\overline{B}$ **d.** $N \cup B$ **e.** $B \cap A$ **f.** $\overline{N}$

5. 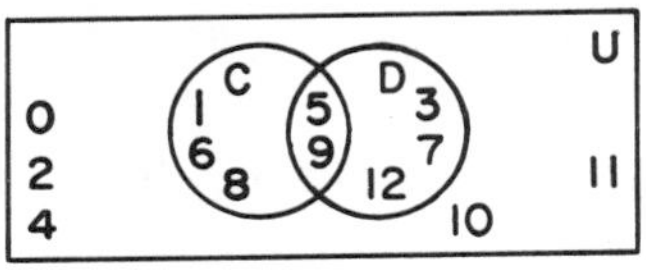

Which elements belong to set U? to set C? to set D? Write the resulting set, listing the elements for: (10–3)

a. $C \cap D$ **b.** $D \cup C$ **c.** $\overline{C}$ **d.** $\overline{D}$

11 Algebra

EXTENSION OF THE NUMBER SYSTEM

When we add any two whole numbers, we find that the sum is always a whole number. Thus, the set of whole numbers is said to be *closed* under the operation of addition.

However, when we subtract one whole number from another whole number, the difference is not always a whole number.

For example:

In each of the following subtractions the difference is a whole number.

$$\begin{array}{r} 5 \\ -1 \\ \hline 4 \end{array} \qquad \begin{array}{r} 5 \\ -2 \\ \hline 3 \end{array} \qquad \begin{array}{r} 5 \\ -3 \\ \hline 2 \end{array} \qquad \begin{array}{r} 5 \\ -4 \\ \hline 1 \end{array} \qquad \begin{array}{r} 5 \\ -5 \\ \hline 0 \end{array}$$

But in each of the following subtractions the difference cannot be a whole number.

$$\begin{array}{r} 5 \\ -6 \\ \hline ? \end{array} \qquad \begin{array}{r} 5 \\ -7 \\ \hline ? \end{array} \qquad \begin{array}{r} 5 \\ -8 \\ \hline ? \end{array} \qquad \begin{array}{r} 5 \\ -9 \\ \hline ? \end{array} \qquad \begin{array}{r} 5 \\ -10 \\ \hline ? \end{array}$$

Thus we see that the set of whole numbers is not closed under the operation of subtraction. In order to be able to subtract in problems like those above, let us now enlarge our number system.

11–1 INTEGERS

In section 1–5 we found that the numerals naming whole numbers may be arranged in a definite order on the number line so that they correspond one-to-one with points on the number line to the right of the point labeled 0.

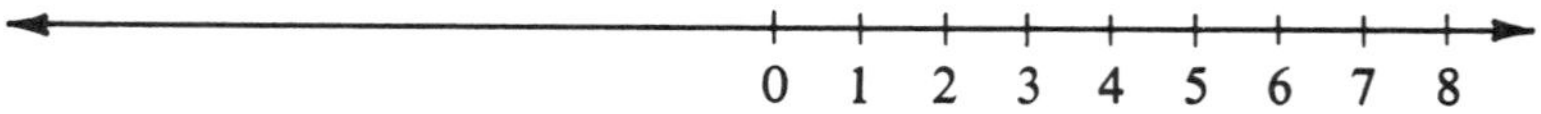

Now let us extend the number line to the left of the point marked 0. Using the interval between 0 and 1 as the unit of measure, we locate equally spaced points to the left of 0. The first new point is labeled $^-1$, read "negative one";

the second point $^{-}2$, read "negative two"; the third point $^{-}3$, etc.

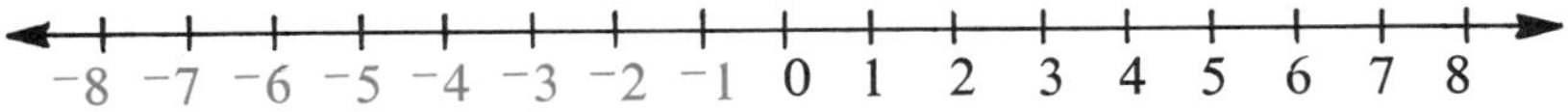

The numbers corresponding to the points to the left of 0 are *negative numbers* and their numerals are identified by the raised dash ($^{-}$) symbols that precede the digits. The numbers corresponding to the points to the right of 0 are *positive numbers* and their numerals may either contain no sign or for emphasis are identified by the raised ($^{+}$) symbols that precede the digits, in which case the number line would look like this:

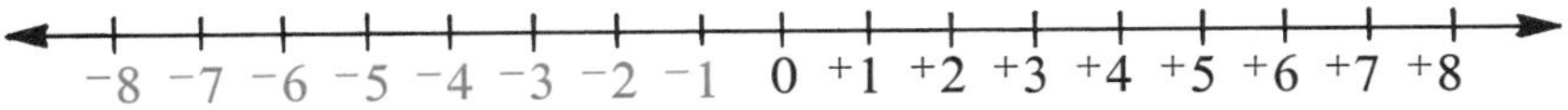

Positive and negative numbers are sometimes called *signed numbers* or *directed numbers.* The numeral $^{-}8$ names the number *negative eight.* The numeral $^{+}7$ or simply 7 names the number *positive seven.* 0 is neither positive nor negative. All numbers greater than 0 are positive numbers.

Examination of the number line below will show that the points that correspond to $^{+}6$ and $^{-}6$ fall on opposite sides of the point marked 0 but are the same distance from 0.

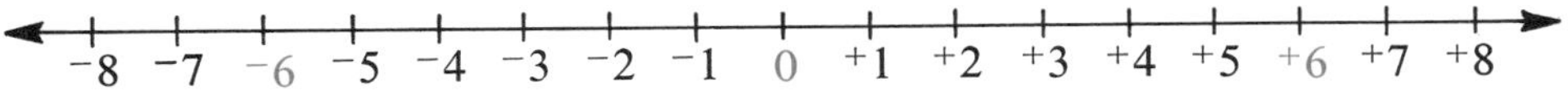

A pair of numbers, one positive and the other negative, such as $^{+}6$ and $^{-}6$, which have the same absolute value (see section 12–3) are called *opposites.* $^{-}6$ is the opposite of $^{+}6$ and $^{+}6$ is the opposite of $^{-}6$. Each is the opposite of the other. The opposite of zero is zero.

The set consisting of all the whole numbers and their opposites is called the set of *integers.* Both $\{\ldots, {}^{-}2, {}^{-}1, 0, 1, 2, \ldots\}$ and $\{\ldots, {}^{-}2, {}^{-}1, 0, {}^{+}1, {}^{+}2, \ldots\}$ describe the set of integers. Observe that three dots are used at both ends since the integers continue in both directions without ending. Thus, $\{1, 2, 3, \ldots\}$ describes the set of *positive integers* (or the set of natural numbers); $\{0, 1, 2, 3, \ldots\}$ describes the set of *non-negative integers* (or the set of whole numbers); $\{\ldots, {}^{-}3, {}^{-}2, {}^{-}1\}$ describes the set of *negative integers*; and $\{\ldots, {}^{-}3, {}^{-}2, {}^{-}1, 0\}$ describes the set of *non-positive integers.*

The set of *even integers* consists of all the integers that are divisible by two (2). The set of even integers is described by $\{\ldots, {}^{-}4, {}^{-}2, 0, 2, 4, \ldots\}$. The set of *odd integers* consists of all the integers that are not divisible by two (2). The set of odd integers is described by $\{\ldots, {}^{-}3, {}^{-}1, 1, 3, \ldots\}$.

EXERCISES

1. Read or write in words, each of the following:

a. $^{-}10$ **b.** $^{+}8$ **c.** $^{-}4$ **d.** $^{+}11$ **e.** $^{-}27$ **f.** $^{+}51$ **g.** $^{+}72$ **h.** $^{-}125$

2. Write symbolically:

   a. Positive four c. Negative twelve e. Positive fifteen g. Negative nineteen
   b. Negative two d. Positive one f. Negative twenty h. Positive sixty-six

3. What number is the opposite of each of the following?

   a. $^{+}3$ b. $^{-}9$ c. $^{+}12$ d. 0 e. $^{-}18$ f. $^{+}37$ g. $^{-}64$ h. $^{-}100$

4. a. Does $^{-}7$ name a whole number? An integer?
   b. Does 16 name a whole number? An integer?

5. List all the elements described by each of the following sets:

   a. $\{^{-}10, {}^{-}9, {}^{-}8, \ldots, 4\}$ b. $\{^{-}8, {}^{-}6, {}^{-}4, \ldots, 6\}$ c. $\{^{-}11, {}^{-}9, {}^{-}7, \ldots, 11\}$

6. a. Is $\{^{-}9, {}^{-}5, 0\}$ a subset of $\{\ldots, {}^{-}3, {}^{-}2, {}^{-}1\}$?
   b. Is $\{^{-}3, {}^{-}1, 0, 1, 3\}$ a subset of $\{^{-}8, {}^{-}7, {}^{-}6, \ldots, 6\}$?
   c. Is $\{\ldots, {}^{-}3, {}^{-}2, {}^{-}1\}$ a subset of $\{\ldots, {}^{-}3, {}^{-}2, {}^{-}1, 0\}$?
   d. Is $\{^{-}12, {}^{-}10, {}^{-}8, \ldots, 2\}$ a subset of $\{\ldots, {}^{-}4, {}^{-}2, 0, 2, 4\}$?
   e. Is $\{^{-}17, {}^{-}15, {}^{-}13, \ldots, 1\}$ a subset of $\{\ldots, {}^{-}3, {}^{-}1, 1, 3, 5\}$?

7. If $U = \{^{-}11, {}^{-}10, {}^{-}9, \ldots, 11\}$, $A = \{^{-}9, {}^{-}8, {}^{-}7, \ldots, 0\}$, $B = \{^{-}6, {}^{-}5, {}^{-}4, \ldots, 8\}$, and $C = \{^{-}10, {}^{-}6, {}^{-}1, 0, 3, 7, 11\}$, write the resulting set listing the elements for each of the following:

   a. $A \cap B$ c. $\overline{C}$ e. $\overline{B}$ g. $C \cup B$
   b. $B \cup A$ d. $A \cup C$ f. $C \cap A$ h. $B \cap C$

8. If $R = \{\ldots, {}^{-}3, {}^{-}2, {}^{-}1, 0\}$ and $S = \{0, 1, 2, 3, \ldots\}$, write the resulting set listing the elements for $R \cup S$. For $R \cap S$.
9. If $G = \{\ldots, {}^{-}4, {}^{-}2, 0, 2, 4\}$ and $H = \{\ldots, {}^{-}3, {}^{-}1, 1, 3, \ldots\}$, write the resulting set listing the elements for $G \cap H$. For $G \cup H$.
10. If $U = \{^{-}6, {}^{-}5, {}^{-}4, \ldots, 6\}$, $M = \{^{-}5, {}^{-}4, {}^{-}3, \ldots, 1\}$, and $N = \{^{-}2, {}^{-}1, 0, 3, 5, 6\}$, draw a Venn diagram illustrating each of the following:

    a. $M \cap N$ b. $M \cup N$ c. $\overline{M}$ d. $\overline{N}$

11–2 RATIONAL NUMBERS; REAL NUMBERS

In section 5–7 we found that numerals naming the fractional numbers of arithmetic may be arranged in a definite order on the number line to the right of the point labeled 0. There is a point on the number line corresponding to each fractional number.

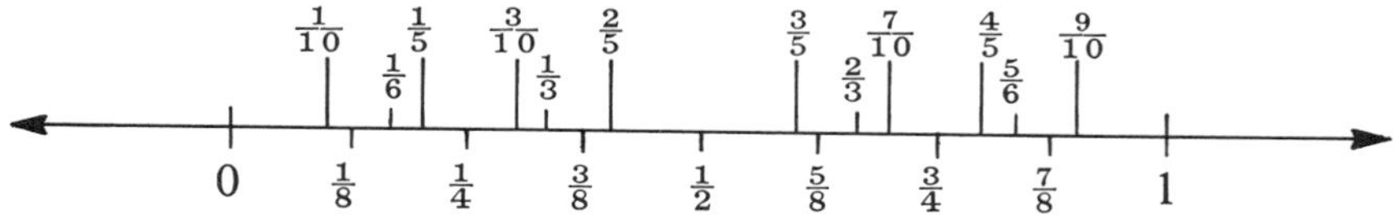

Examination of the number line below will show that there are fractional subdivisions to the left of zero matching those to the right of zero with each point of division corresponding to a fractional number. Since there are an infinite number of positive and negative fractional numbers, only a few are indicated. However, there is a point on the number line corresponding to each positive and negative fractional number.

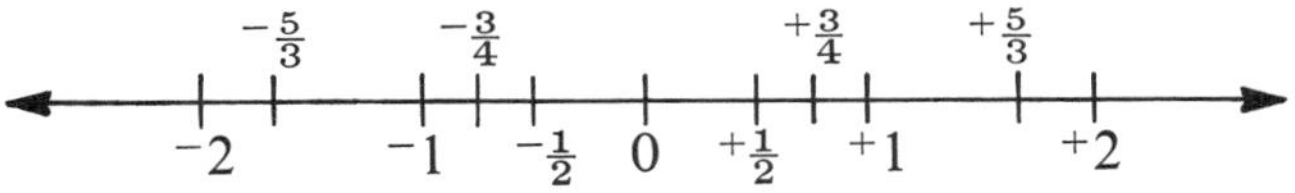

The numeral $^+\frac{3}{4}$ or $^+(\frac{3}{4})$ or just $\frac{3}{4}$ names the number *positive three fourths.*

The numeral $^-\frac{1}{2}$ or $^-(\frac{1}{2})$ names the number *negative one half.*

Observe in numerals like $^-\frac{1}{2}$ or $^+\frac{3}{4}$ where parentheses are not used the $^-$ or $^+$ symbol belongs to the entire fraction although it appears that the symbol belongs to the numerator alone.

Each positive fractional positive number has an opposite negative fractional number and each negative fractional number has an opposite positive fractional number.

$^-\frac{4}{5}$ is the opposite of $^+\frac{4}{5}$. $\quad$ $^+1\frac{5}{8}$ is the opposite of $^-1\frac{5}{8}$.

The set consisting of all the fractional numbers and their opposites is called the set of *rational numbers.* This set includes the integers since each integer may be named in fraction form. For example $^-5 = \frac{^-5}{1}$, $^+3 = \frac{^+3}{1}$, and $0 = \frac{0}{1} = \frac{0}{2} = \frac{0}{3}$, etc. A rational number may be described as a number named by a numeral that expresses a quotient of two integers with division by zero excluded.

An *irrational number* is a number that is both a non-terminating and non-repeating decimal. See pages 147 and 181. An irrational number cannot be named by a quotient of two integers with division by zero excluded. The set consisting of all the positive and negative irrational numbers is called the set of *irrational numbers.*

The set consisting of all the rational numbers and all the irrational numbers is called the set of *real numbers.* It includes all the integers, all the positive and negative fractional numbers, and all the positive and negative irrational numbers. There are an infinite number of real numbers in our number system.

EXERCISES

1. Read, or write in words, each of the following:

a. $^-\frac{7}{8}$ **b.** $^+\frac{6}{5}$ **c.** $^-2\frac{3}{4}$ **d.** $^+.85$ **e.** $^-7.4$ **f.** $^+\frac{1}{16}$ **g.** $^-\frac{17}{4}$ **h.** $^-.001$

2. Write symbolically:

a. Negative one fourth

b. Positive three tenths

c. Positive seven hundredths

d. Negative two and five tenths

e. Negative seven and fifty-two hundredths

f. Positive one hundred forty-six thousandths

3. What number is the opposite of each of the following?

a. $^{+}\frac{3}{16}$ b. $^{-}\frac{7}{2}$ c. $^{+}.03$ d. $^{-}1.98$ e. $^{-}4\frac{5}{9}$ f. $^{+}.006$ g. $^{+}3\frac{8}{11}$ h. $^{-}47.2$

4. a. Is $^{-}\frac{9}{10}$ a whole number? An integer? A rational number? A real number?
b. Is $^{+}6\frac{1}{8}$ a whole number? An integer? A rational number? A real number?
c. Is $^{+}.25$ a whole number? An integer? A rational number? A real number?
d. Is $^{-}7.6$ a whole number? An integer? A rational number? A real number?
e. Is $^{-}5$ a whole number? An integer? A rational number? A real number?
f. Is $^{-}\sqrt{23}$ a whole number? An integer? A rational number? A real number?

5. Which of the following numerals name integers? Which name rational numbers? Which name irrational numbers? Which name real numbers?

a. $^{-}14$; $^{+}.07$; $^{-}\frac{3}{10}$; $^{-}\sqrt{41}$; $\frac{12}{3}$; $^{-}8\frac{1}{7}$
b. $^{+}\frac{8}{15}$; $^{-}200$; $^{-}1\frac{4}{5}$; $^{-}2.01$; $^{+}\sqrt{17}$; $^{-}\frac{27}{9}$
c. $^{-}.125$; $^{-}\frac{1}{16}$; $^{+}\sqrt{25}$; $^{-}\frac{56}{8}$; $^{-}84$; $^{+}9\frac{3}{4}$
d. $^{+}29$; $^{-}\sqrt{3}$; $^{-}\frac{11}{32}$; $^{-}.67$; $^{-}1\frac{5}{12}$; $^{+}\frac{19}{6}$

6. a. Is $\{^{-}\frac{1}{2}, \frac{1}{2}\}$ a subset of $\{^{-}\frac{3}{4}, ^{-}\frac{2}{4}, ^{-}\frac{1}{4}, 0, \frac{1}{4}, \frac{2}{4}, \frac{3}{4}\}$?
b. Is $\{^{-}\frac{4}{3}, ^{-}\frac{2}{3}, 0, \frac{2}{3}, \frac{4}{3}\}$ a subset of {all rational numbers whose denominator is 3}?
c. Is {all integers} a subset of {all rational numbers}?
d. Is {all real numbers} a subset of {all integers}?

11–3 ABSOLUTE VALUE

The *absolute value* of any number is the value of the corresponding arithmetic number which has no sign. The absolute value of $^{-}3$ is 3 and the absolute value of $^{+}3$ is 3. The absolute value of 0 is 0.

A pair of vertical bars $|\quad|$ is the symbol used to designate absolute value. $|^{-}6|$ is read "the absolute value of negative 6."

EXERCISES

1. Read, or write in words, each of the following:

a. $|17|$ b. $|^{-}1|$ c. $|^{+}26|$ d. $|^{-}\frac{6}{11}|$ e. $|^{-}.08|$ f. $|^{+}4\frac{3}{8}|$

2. Write symbolically, using the absolute value symbol:

a. The absolute value of negative eleven
b. The absolute value of positive six sevenths
c. The absolute value of negative nine and three tenths
d. The absolute value of positive seven thousandths

3. Find the absolute value of each of the following:

a. $|^{-}4|$ c. $|^{-}9|$ e. $|^{+}.6|$ g. $|^{-}1\frac{3}{4}|$ i. $|^{-}12|$ k. $|^{+}\frac{12}{5}|$
b. $|^{+}8|$ d. $|0|$ f. $|^{-}\frac{5}{8}|$ h. $|^{+}16|$ j. $|^{-}.53|$ l. $|^{-}100|$

In each of the following first find the absolute value of each number as required, then apply the necessary operation to obtain the answer:

4. a. $\lvert {}^{-}6 \rvert + \lvert {}^{-}5 \rvert$	**b.** $\lvert {}^{+}\frac{1}{2} \rvert + \lvert {}^{-}\frac{1}{2} \rvert$	**c.** $\lvert {}^{-}1\frac{3}{4} \rvert + \lvert {}^{+}2\frac{1}{3} \rvert$	**d.** $\lvert {}^{-}.82 \rvert + \lvert {}^{-}1.6 \rvert$
5. a. $\lvert {}^{-}15 \rvert - \lvert {}^{+}6 \rvert$	**b.** $\lvert {}^{+}\frac{7}{8} \rvert - \lvert {}^{-}\frac{2}{3} \rvert$	**c.** $\lvert {}^{-}5\frac{1}{5} \rvert - \lvert {}^{-}4\frac{3}{8} \rvert$	**d.** $\lvert {}^{+}1.9 \rvert - \lvert {}^{-}.05 \rvert$
6. a. $\lvert {}^{+}7 \rvert \times \lvert {}^{-}8 \rvert$	**b.** $\lvert {}^{-}\frac{3}{4} \rvert \times \lvert {}^{-}\frac{9}{10} \rvert$	**c.** $\lvert {}^{+}2\frac{5}{6} \rvert \times \lvert {}^{-}12 \rvert$	**d.** $\lvert {}^{-}4.16 \rvert \times \lvert {}^{+}.04 \rvert$
7. a. $\lvert {}^{-}56 \rvert \div \lvert {}^{+}14 \rvert$	**b.** $\lvert {}^{+}\frac{5}{6} \rvert \div \lvert {}^{-}\frac{3}{5} \rvert$	**c.** $\lvert {}^{-}1\frac{2}{3} \rvert \div \lvert {}^{-}2\frac{11}{12} \rvert$	**d.** $\lvert {}^{+}7.5 \rvert \div \lvert {}^{-}.03 \rvert$

11–4 THE REAL NUMBER LINE; GRAPH OF A SET OF NUMBERS

The *real number line* is the complete set of points which corresponds to the set of all real numbers. The real number line is endless in both directions and only a part of it is shown at any one time. There are an infinite number of points on the real number line. However, there is one and only one point that corresponds to each real number and one and only one real number that corresponds to each point on the real number line.

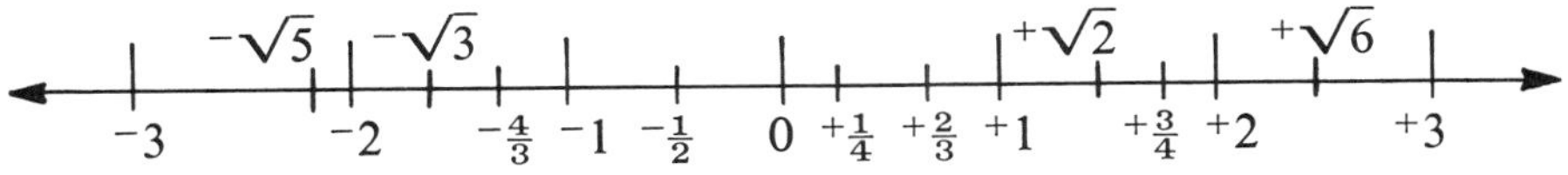

Usually the real number line is labeled only with the numerals naming the integers.

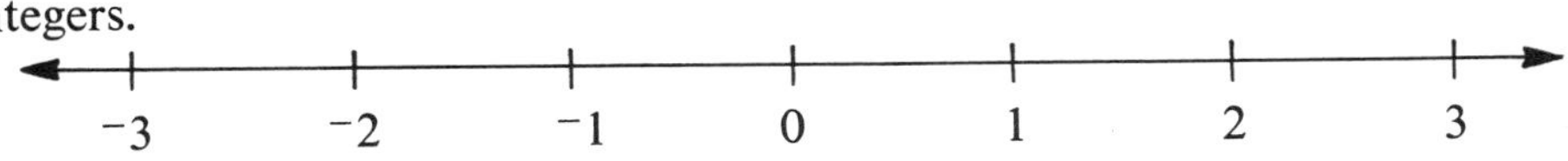

Each point on the number line is called the *graph* of the number to which it corresponds and each number is called the *coordinate* of the related point on the line. Capital letters are generally used to identify particular points.

Point C is the graph of ⁻3. ⁻3 is the coordinate of point C.

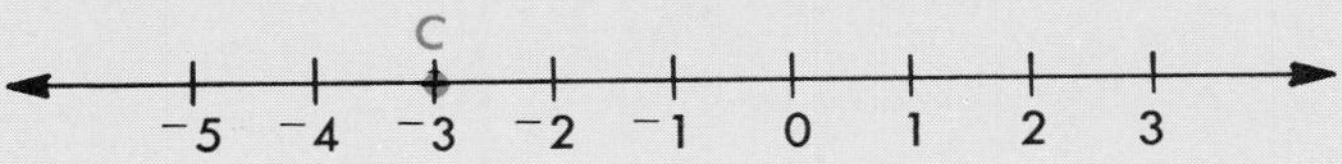

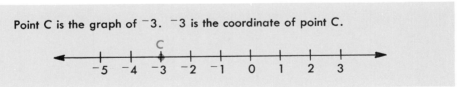

The graph of a number is a point on the number line whose coordinate is the number.

The *graph of a set of numbers* is the set of points on the number line whose coordinates are the numbers.

The set of points {B, D, E, G} is the graph of the set of numbers {⁻3, ⁻1, 0, 2}.

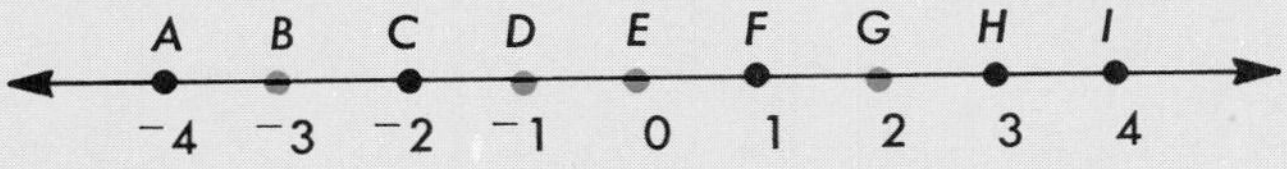

To draw the graph of a set of numbers, we first draw an appropriate number line and locate the point or points whose coordinate or coordinates are listed in the given set of numbers. We use heavy solid or colored dots to indicate these points.

The graph of $\{^-1, 2, 5\}$ is:

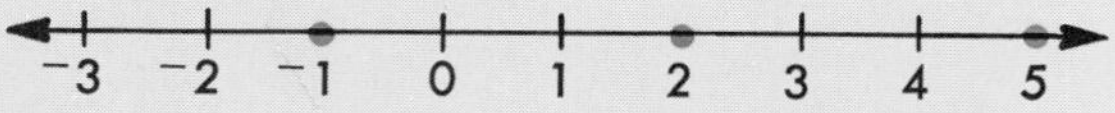

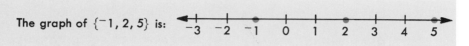

Sometimes a restricted number line is required. To construct this kind of number line, we draw a straight line and, using a convenient unit of measure, we locate the required points of division. These points of division are then properly labeled.

The number line that shows the points of division associated with the set $\{^-4, ^-3, ^-2, ^-1, 0, 1, 2\}$ is:

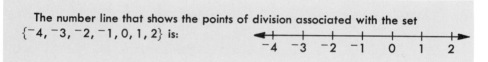

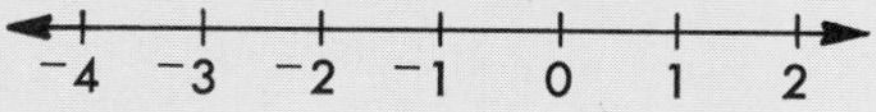

EXERCISES

1.

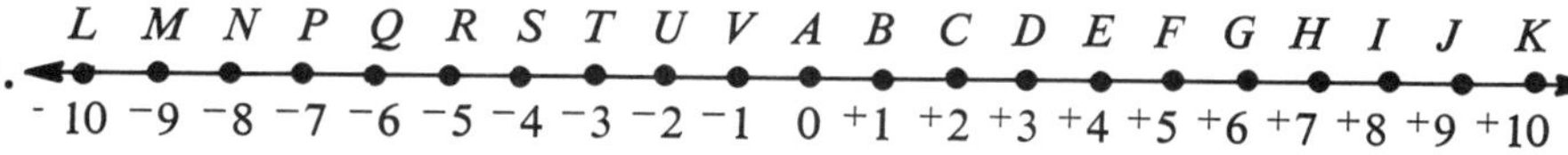

a. What number corresponds to point *K*? *E*? *P*? *D*? *V*? *L*? *J*? *T*? *S*? *A*?

b. What letter labels the point corresponding to each of the following numbers?

$^-5$ $\quad ^+6$ $\quad ^-2$ $\quad ^+2$ $\quad ^+8$ $\quad 0$ $\quad ^-9$ $\quad ^+7$ $\quad ^+1$ $\quad ^-6$

2. Draw the graph of each of the following sets of numbers on a number line:

a. $\{^-5)$

b. $\{^-7, ^+3\}$

c. $\{^-8, ^-4, 0, ^+2\}$

d. $\{^-3, ^-1, 3, 5, 7\}$

e. $\{^-5, ^-4, ^-1, 0, 1, 4\}$

f. $\{^-2, ^+6, ^-3\}$

g. $\{^+3, ^-5, ^+2, ^-8, ^+9, ^+4, ^-1\}$

h. $\{^-9, 0, 1, ^-6, ^-2, 8, ^-4, 7\}$

i. $\{^-6, ^-5, ^-4, \ldots, 3\}$

j. $\{^-10, ^-8, ^-6, \ldots, 4\}$

3. Draw the graph of each of the following on a number line:

a. {all integers less than $^+1$ and greater than $^-1$}

b. {all integers greater than $^-3$ and less than 0}

c. {all integers less than $^+4$ and greater than $^-4$}

4. Write the set of coordinates of which each of the following is the graph:

a.

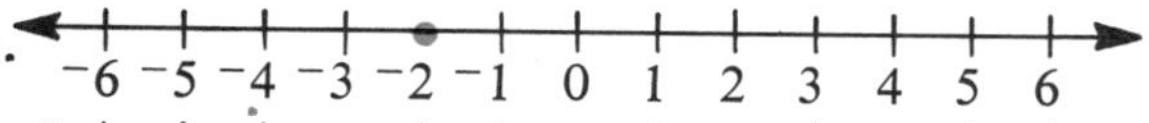

b.

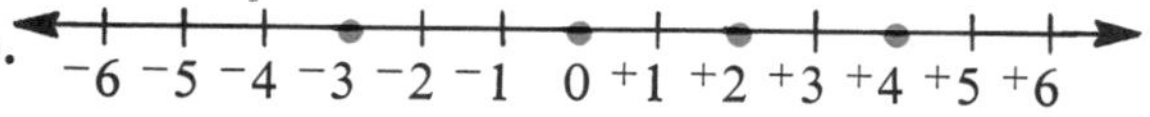

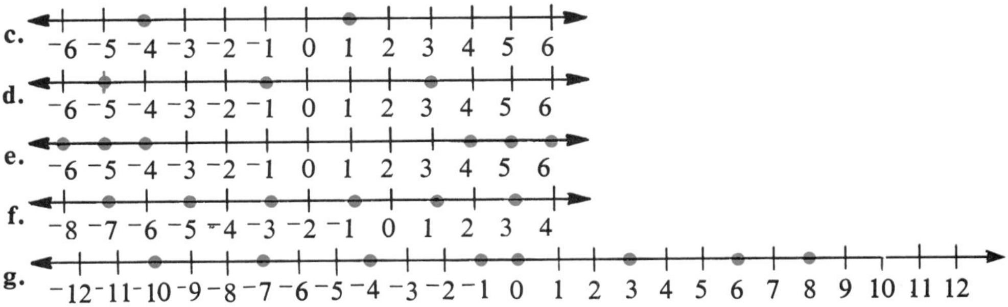

11–5 COMPARING INTEGERS

The number line may be drawn horizontally or vertically.

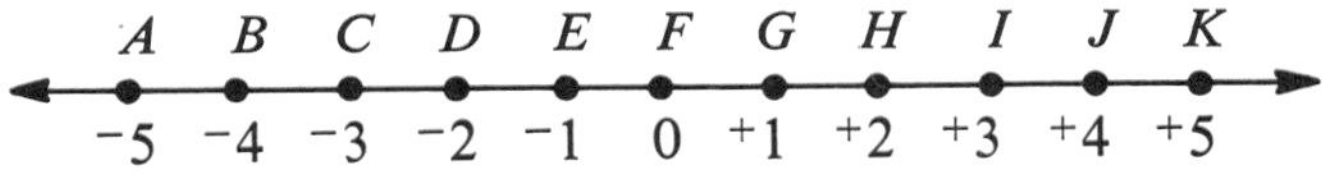

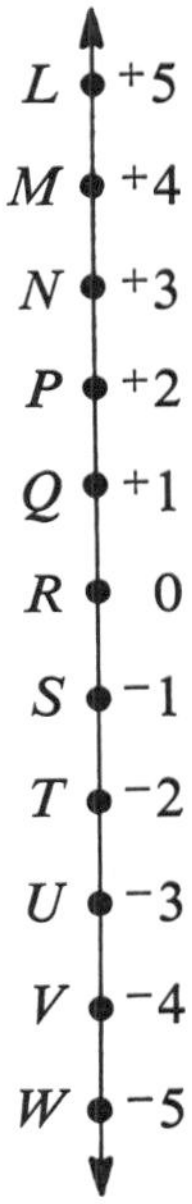

On the horizontal number line, of two numbers the number corresponding to the point on the line *farther to the right* is the greater number.

On the vertical number line, the numbers *above* zero are positive and those *below* zero are negative. Any number corresponding to a point is greater than any number corresponding to a point located below it.

EXERCISES

1. On the horizontal number line, which point corresponds to the greater number:

a. Point *J* or point *B*?
b. Point *C* or point *H*?
c. Point *E* or point *D*?
d. Point *F* or point *A*?

2. On the vertical number line, which point corresponds to the greater number:

a. Point *P* or point *T*?
b. Point *W* or point *N*?
c. Point *R* or point *M*?
d. Point *V* or point *S*?

3. On the horizontal number line is the point corresponding to $^{-}1$ to the left or to the right of the point corresponding to $^{-}5$? Which number is greater, $^{-}1$ or $^{-}5$?

4. On the vertical number line is the point corresponding to $^{-}4$ above or below the point corresponding to $^{-}2$? Which number is greater, $^{-}2$ or $^{-}4$?

5. Which of the following sentences are true?

a. $^{+}6 > {}^{+}9$
b. $^{-}6 > {}^{-}9$
c. $^{+}2 < 0$
d. $^{+}3 < {}^{-}8$
e. $0 > {}^{-}7$
f. $^{-}5 > {}^{-}5$
g. $^{-}1 < {}^{+}1$
h. $^{-}2 \not> {}^{-}4$
i. $^{-}7 \not< {}^{-}7$
j. $^{+}8 < {}^{-}10$
k. $^{-}4 \not> {}^{-}1$
l. $^{-}3 \not< 0$
m. $7 > {}^{-}50$
n. $^{-}12 \not< {}^{-}6$
o. $^{+}20 \not> {}^{-}45$

6. Rewrite each of the following and insert between the two numerals the symbol ($=$, $<$, or $>$) which will make the sentence true:

a. $^{-}6$? $^{-}2$
b. $^{+}9$? $^{+}7$
c. $^{-}5$? $^{+}3$
d. $^{-}8$? $^{-}8$
e. $^{+}17$? $^{-}20$
f. $^{-}10$? 0
g. $^{-}4$? $^{-}6$
h. 0 ? $^{-}9$
i. $^{-}16$? $^{+}1$

7. Name the following numbers in order of size (smallest first):

a. $^{-}2$, $^{+}9$, $^{-}10$, $^{+}14$, $^{-}6$, $^{+}5$, 0, $^{+}10$, $^{-}3$, $^{-}1$
b. $^{+}7$, $^{-}4$, $^{-}8$, 0, $^{+}5$, $^{-}12$, $^{+}1$, $^{-}2$, $^{-}7$, $^{+}3$

8. Name the following numbers in order of size (greatest first):

a. $^{-}8$, $^{+}5$, $^{-}3$, $^{+}9$, 0, $^{-}4$, $^{+}4$, $^{-}11$, $^{+}15$, $^{-}6$
b. $^{+}6$, $^{-}7$, $^{+}2$, $^{-}1$, $^{+}12$, $^{-}5$, $^{+}3$, 0, $^{-}2$, $^{-}13$

9. Which is greater:

a. The absolute value of $^{-}7$ or the absolute value of $^{+}2$?
b. The absolute value of $^{+}10$ or the absolute value of $^{-}10$?

10. Which has the greater opposite number?

a. $^{+}9$ or $^{+}5$
b. $^{-}4$ or $^{-}7$
c. $^{-}6$ or $^{+}8$
d. $^{+}3$ or $^{-}5$

11-6 OPPOSITE MEANINGS

Positive and negative numbers are used in science, statistics, weather reports, stock reports, sports, and many other fields to express opposite meanings or directions. For example:

If $^{+}3\%$ indicates an *increase* of 3% in the cost of living, $^{-}4\%$ would indicate a 4% *decrease* in the cost of living.

EXERCISES

1. a. If $^{+}1\frac{1}{2}$ points represents a $1\frac{1}{2}$-point gain in a stock, what does $^{-}2$ points represent?
b. If $^{-}15°$ indicates a temperature of 15° below zero, how can a temperature of 23° above zero be indicated?
c. If 50° west longitude is represented by $^{-}50°$, what does $^{+}65°$ represent?
d. If $^{+}58{,}000$ people represents an increase of 58,000 in population, how can a decrease of 17,000 in population be represented?

e. If $^{-}10$ amperes means a discharge of 10 amperes of electricity, what does $^{+}8$ amperes mean?

f. If $^{+}25$ m.p.h. indicates a tail wind of 25 m.p.h., how can a head wind of 30 m.p.h. be indicated?

g. If $^{+}1{,}500$ feet represents 1,500 feet above sea level, what does $^{-}700$ feet represent?

h. If an excess of 1.75 inches of rainfall is indicated by $^{+}1.75$ inches, how can a deficiency of 2.38 inches be indicated?

i. If 9 degrees below normal is represented by $^{-}9°$, how can 17 degrees above normal be represented?

j. If 36° north latitude is indicated by $^{+}36°$, how can 51° south latitude be indicated?

11–7 OPPOSITE DIRECTIONS; VECTORS

Positive and negative numbers, used as directed numbers, indicate movements in opposite directions. A movement to the right of a particular point is generally considered as moving in a positive direction from the point and a movement to the left of a point as moving in a negative direction from the point.

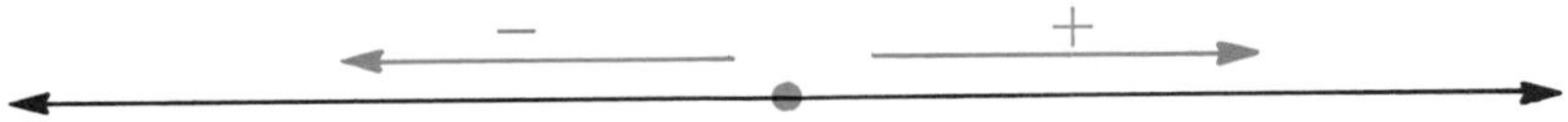

The sign in the numeral naming a directed number indicates the direction, and the absolute value of the directed number represents the magnitude (distance in units) of the movement.

For example,

$^{+}7$ means moving 7 units to the right.
$^{-}4$ means moving 4 units to the left.

An arrow that represents a directed line segment is called a *vector*. It is used to picture the size and the direction of the movement that is indicated by a signed number. The absolute value of the signed number indicates the length of the arrow and the sign indicates the direction.

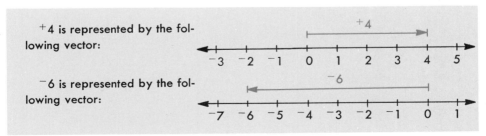

$^{+}4$ is represented by the following vector:

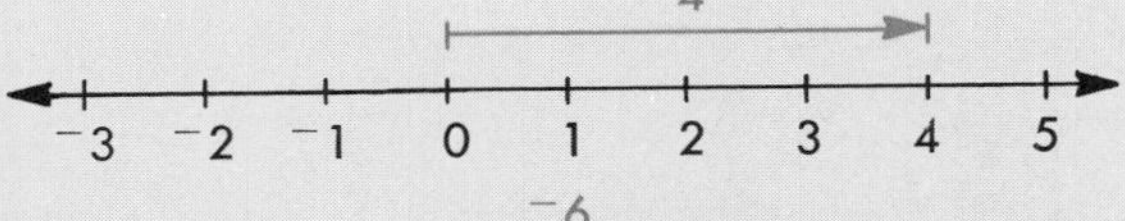

$^{-}6$ is represented by the following vector:

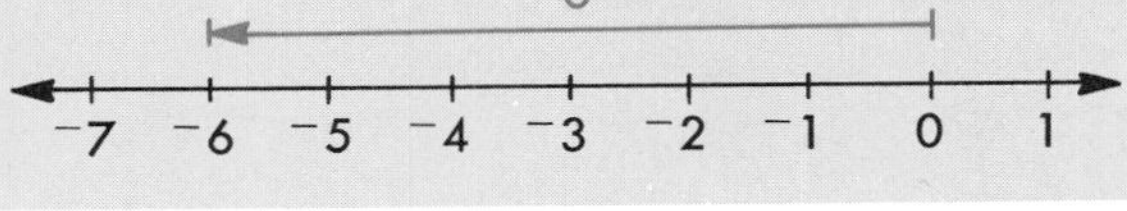

To determine the numeral that describes the movement from one point to another point on the number line, we draw the vector between the points with the arrowhead pointing in the direction of the movement. We use the number of units of length in the vector as the absolute value of the numeral. We select the sign according to the direction of the vector. If it points to the right, the sign is positive ($^+$), if it points to the left, the sign is negative ($^-$).

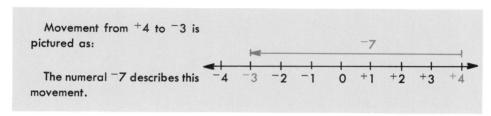

Movement from $^+4$ to $^-3$ is pictured as:

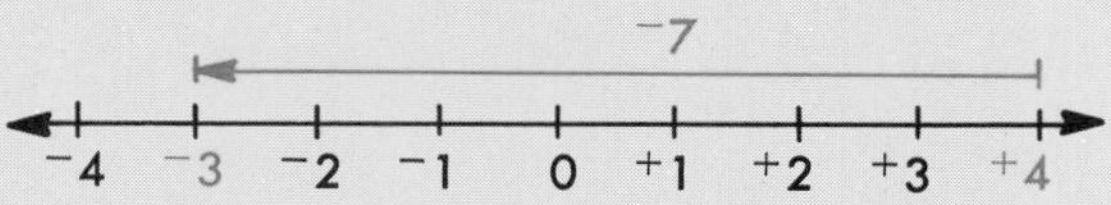

The numeral $^-7$ describes this movement.

EXERCISES

1. A movement of how many units of distance and in what direction is represented by each of the following numbers?

a. $^-8$ **b.** $^+10$ **c.** $^-1$ **d.** $^+3$ **e.** 18 **f.** $^-2\frac{1}{4}$ **g.** $^-3.75$ **h.** $^+\frac{5}{6}$

2. Represent each of the following movements by a numeral naming a signed number:

a. Moving 5 units to the left
b. Moving 9 units to the right
c. Moving $10\frac{3}{4}$ units to the right
d. Moving 8.5 units to the left

3. Write the numeral that is represented by each of the following vectors?

a.
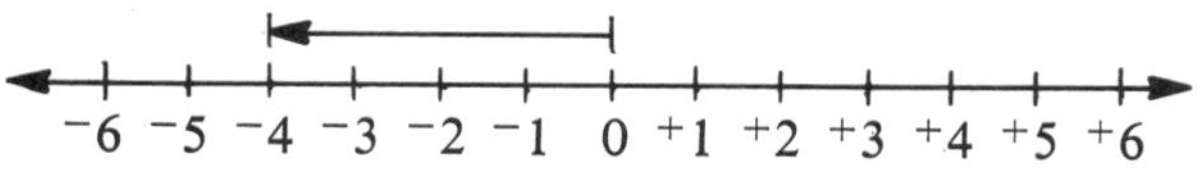

b.
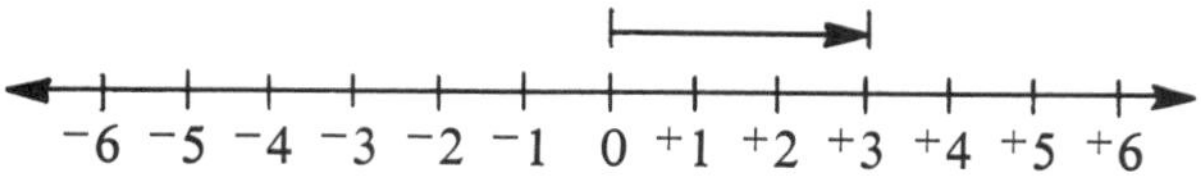

c.
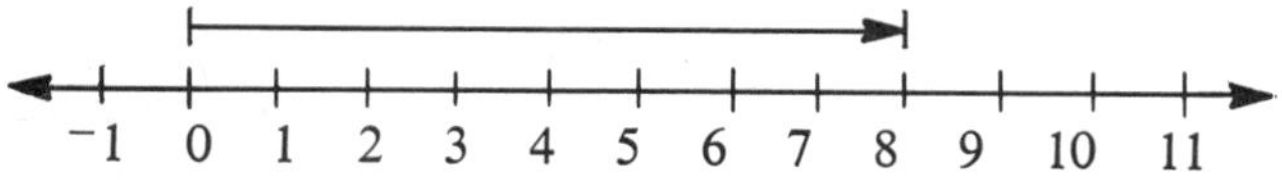

d.
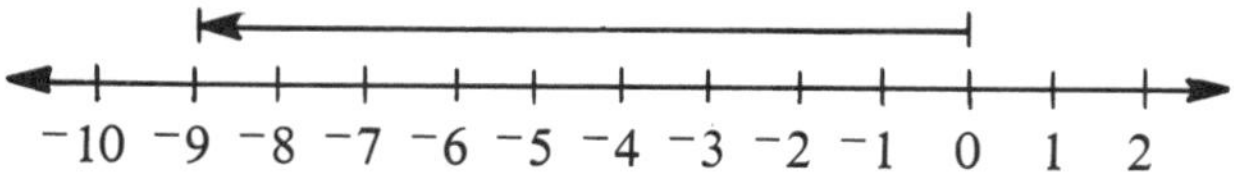

4. Using the number line as a scale, draw a vector representing:

a. $^-3$ **b.** $^+6$ **c.** $^-7$ **d.** 5 **e.** $^+1\frac{1}{2}$ **f.** $^-4.6$ **g.** $^-10$ **h.** $^+3\frac{3}{4}$

5. Write the numeral that is represented by each of the following vectors:

a.
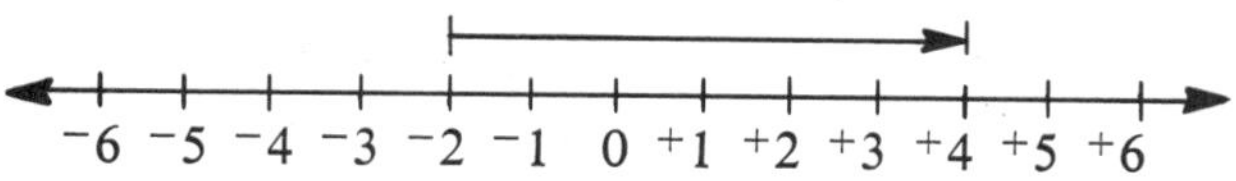

b. 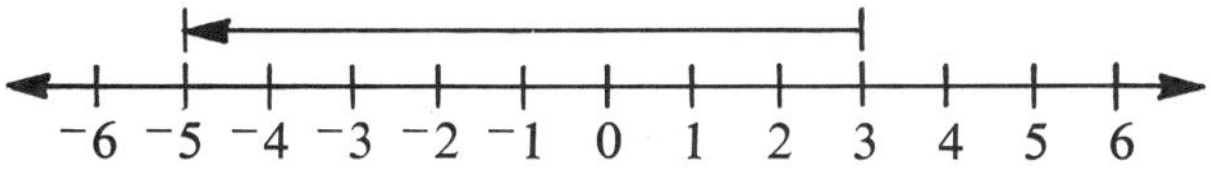

c. 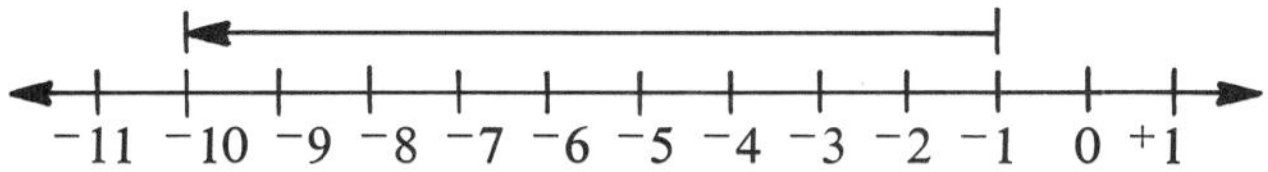

d. 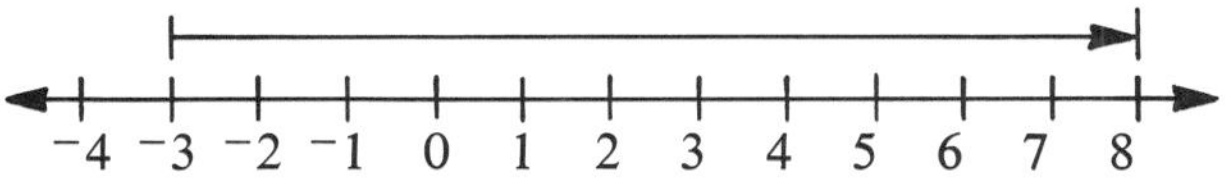

6. Use the number line as a scale to draw a vector that illustrates each of the following movements. Write the numeral that is represented by each of these vectors.

a. From 0 to $^+8$
b. From 0 to $^-6$
c. From $^-5$ to 0
d. From $^+4$ to 0
e. From $^+5$ to $^+9$
f. From $^+11$ to $^+1$
g. From $^-2$ to $^-7$
h. From $^-5$ to $^-1$
i. From $^+3$ to $^-3$
j. From $^-8$ to $^+11$
k. From $^+10$ to $^-9$
l. From $^+5$ to $^+15$
m. From $^-4$ to $^-8$
n. From $^-8$ to $^-4$
o. From $^+1$ to $^-3$
p. From $^-12$ to $^+12$
q. From $^+8$ to $^-2$
r. From $^+9$ to $^+2$

11–8 NEGATIVE OF A NUMBER; ADDITIVE INVERSE

Sometimes the *opposite of a number* is called either the *negative of the number* or the *additive inverse of the number.* When the sum of two numbers is zero, each addend is said to be the *additive inverse* of the other addend. We shall see in sections 11–9 and 11–10 that the sum of a number and its opposite is zero.

Therefore we may call $^-5$ the opposite of or the negative of or the additive inverse of $^+5$ (or 5) and $^+5$ (or 5) the opposite of or the negative of or the additive inverse of $^-5$. Observe that the negative of a negative number is a positive number. Thus not all negatives of numbers are negative numbers.

A centered dash symbol is used in algebra to indicate the opposite of a number. The sentence $-(^-5) = {}^+5$ is read "the opposite of negative five is positive five" or sometimes "the negative of negative five is positive five." The sentence $-5 = {}^-5$ is read "the opposite of five is negative five."

EXERCISES

1. Read, or write in words, each of the following:

a. -10
b. $-(^+7)$
c. $-(^-\frac{3}{8})$
d. $-(-6)$
e. $-(^-3) = {}^+3$
f. $-(^+12) = {}^-12$
g. $-20 = {}^-20$
h. $-(-8.2) = 8.2$

2. Write symbolically using the "opposite symbol":

a. The opposite of negative one
b. The negative of negative sixteen
c. The opposite of forty-seven

d. The opposite of positive nine
e. The negative of positive sixteen
f. The opposite of twelve is negative twelve
g. The negative of negative two is positive two
h. The opposite of the opposite of fifty-three is fifty-three
i. The negative of the negative of eighty-four is eighty-four
j. The opposite of negative forty is positive forty

3. Write the opposite of:

a. $^{+}11$ b. $^{-}7$ c. $^{-}\frac{5}{6}$ d. $^{-}.27$ e. $^{+}\frac{13}{5}$ f. 0 g. $^{-}2\frac{3}{4}$ h. $^{+}25$

4. Write the negative of:

a. $^{-}8$ b. $^{+}100$ c. $^{-}63$ d. $^{+}.9$ e. $^{+}\frac{1}{8}$ f. $^{-}\frac{7}{5}$ g. 0 h. $^{+}3\frac{1}{3}$

5. Write the additive inverse of:

a. $^{-}4$ b. 0 c. $^{+}58$ d. $^{-}\frac{7}{16}$ e. $^{+}1.04$ f. $^{+}\frac{2}{3}$ g. $^{-}6\frac{5}{8}$ h. $^{-}\frac{9}{4}$

6. Find the:

a. Opposite of the opposite of: $^{+}1$; $^{-}9$; $^{+}\frac{5}{12}$; $^{-}.7$; $^{-}5\frac{3}{8}$
b. Negative of the negative of: $^{-}15$; $^{+}27$; $^{-}\frac{4}{5}$; $^{+}.003$; $^{+}6\frac{1}{2}$
c. What is always true about the opposite of the opposite of a number?

7. Are all negatives of numbers negative numbers? If not, illustrate.

COMPUTATION WITH POSITIVE AND NEGATIVE NUMBERS

11–9 ADDITION ON THE NUMBER LINE

If we move 5 units to the right from the point marked 0, we reach the point whose coordinate is $^{+}5$. If we then move 3 units to the right of this point, we reach the point whose coordinate is $^{+}8$. The sum of these two movements is the same as a single movement of 8 units to the right.

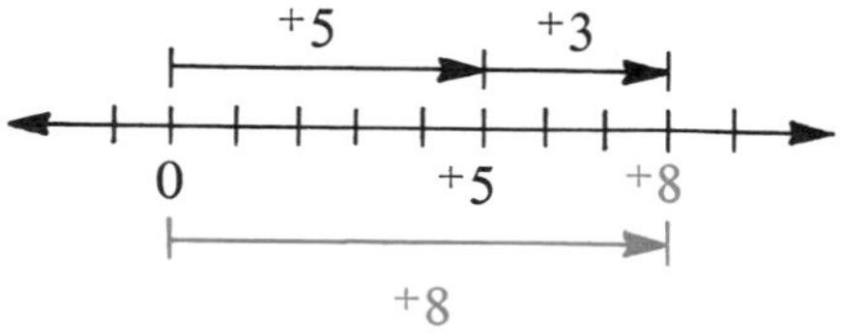

$^{+}5 + {}^{+}3 = {}^{+}8$

Observe in the illustration that a vector is drawn to represent the sum and each addend. To avoid confusion, we draw the vectors representing the addends above the number line and the vector representing the sum below the number line. The vector representing the first addend begins at the point 0 and the second vector representing the second addend is drawn from the point reached by the first vector. The resultant vector that represents the sum is drawn from the point 0 to the point reached by the vector representing the second addend.

If we move 5 units to the left from the point marked 0, we reach the point whose coordinate is $^{-}5$. If we then move 3 units to the left of this point, we reach the point whose coordinate is $^{-}8$. The sum of these two movements is the same as a single movement of 8 units to the left. Observe how vectors are used to illustrate this.

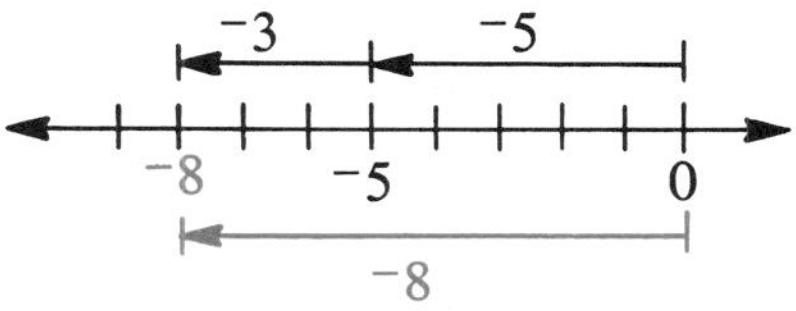

$$^{-}5 + ^{-}3 = ^{-}8$$

If we first move 5 units to the right from the point marked 0 and then from the point reached we move 3 units to the left, we reach the point whose coordinate is $^{+}2$. The sum of these two movements is the same as a single movement of 2 units to the right.

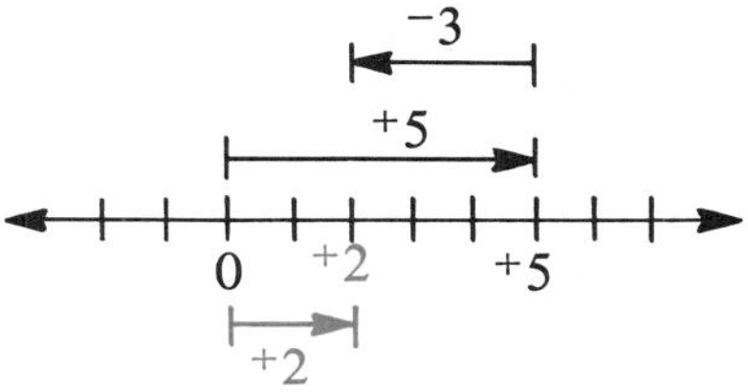

$$^{+}5 + ^{-}3 = ^{+}2$$

If we first move 5 units to the right from the point marked 0 and then from the point reached we move 8 units to the left, we reach the point whose coordinate is $^{-}3$. The sum of these two movements is the same as a single movement of 3 units to the left.

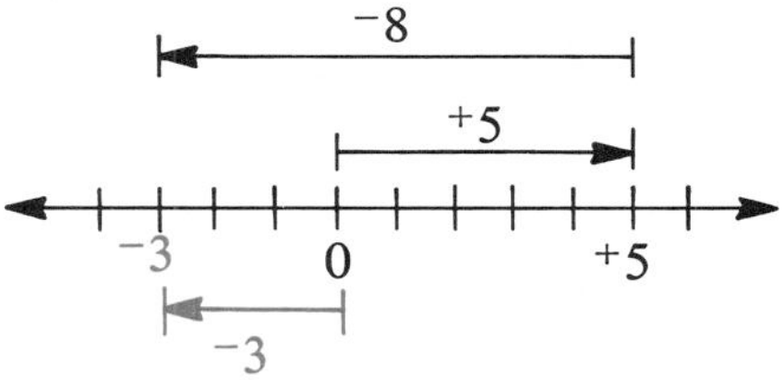

$$^{+}5 + ^{-}8 = ^{-}3$$

If we first move 5 units to the left from the point marked 0 and then from the point reached we move 3 units to the right, we reach the point whose coordinate is $^{-}2$. The sum of these two movements is the same as a single movement of 2 units to the left.

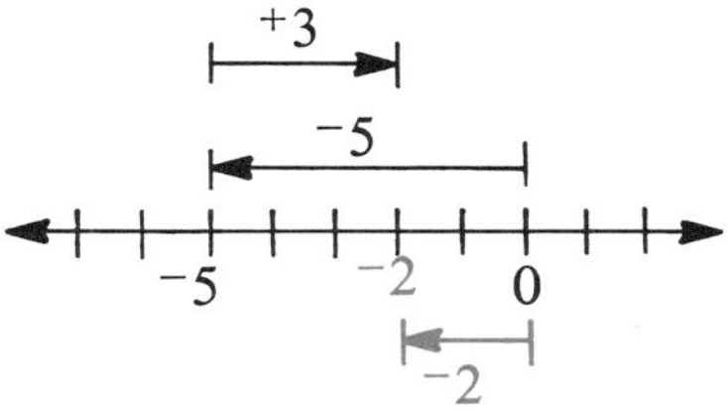

$$^{-}5 + ^{+}3 = ^{-}2$$

If we first move 5 units to the left from the point marked 0 and then from the point reached we move 8 units to the right, we reach the point whose coordinate is $^{+}3$. The sum of these two movements is the same as a single movement of 3 units to the right.

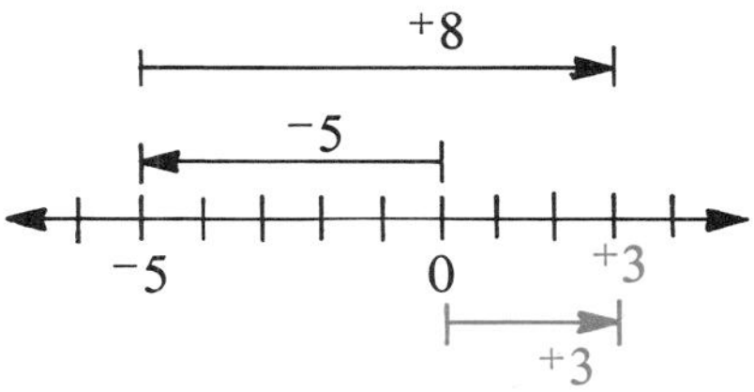

$$^{-}5 + ^{+}8 = ^{+}3$$

If we first move 5 units to the right from the point marked 0 and then from the point reached we move 5 units to the left, we reach the point whose coordinate is 0. The first movement is offset by an equal opposite movement. The sum of these two movements is a movement of zero units.

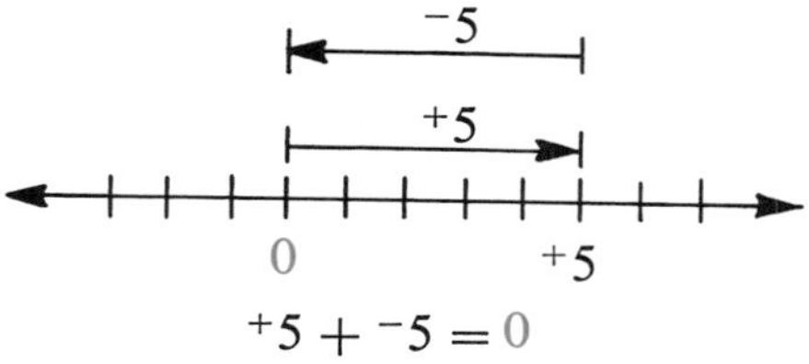

We may add rational numbers on the number line. For example:

If we first move $2\frac{1}{2}$ units to the right from the point marked 0 and then from the point reached we move $6\frac{1}{2}$ units to the left, we reach a point whose coordinate is $^{-}4$. The sum of these two movements is the same as a single movement of 4 units to the left.

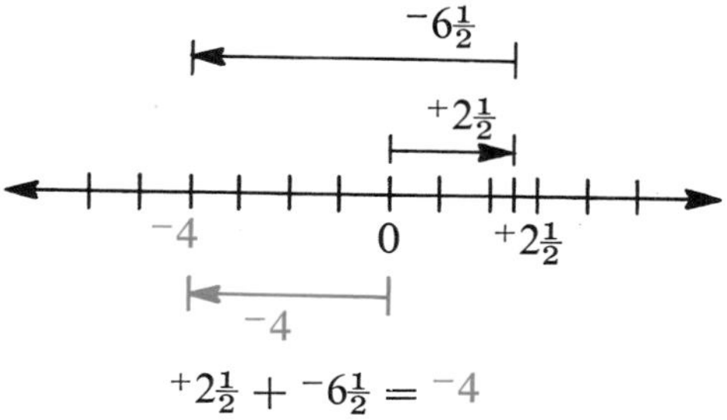

Observe that when vectors are used, the sum of two positive numbers or the sum of two negative numbers is represented by a vector whose length is equal to the sum of the lengths of the vectors representing the two addends. This resultant vector has a positive direction when the two addends are both positive numbers and a negative direction when the two addends are both negative numbers.

However, the sum of a positive number and a negative number or the sum of a negative number and a positive number is represented by a vector whose length is equal to the difference in the lengths of the vectors representing the two addends and its direction is determined by the direction of the longer of the two vectors representing the two addends.

EXERCISES

In each of the following add as indicated, using the number line to find the sum:

1. a. $^{+}2 + {}^{+}3$ **c.** $^{+}4 + {}^{+}4$ **e.** $^{+}9 + {}^{+}2$ **g.** $^{+}3 + {}^{+}6$
b. $^{+}6 + {}^{+}1$ **d.** $^{+}7 + {}^{+}3$ **f.** $^{+}5 + {}^{+}5$ **h.** $^{+}8 + {}^{+}4$

In each of the above problems do you see that the absolute value of the sum is the sum of the absolute values of the two addends and that the sum of two positive integers is a positive integer?

2. a. $^{-}5 + {}^{-}2$ **c.** $^{-}1 + {}^{-}9$ **e.** $^{-}6 + {}^{-}5$ **g.** $^{-}2 + {}^{-}2$
b. $^{-}3 + {}^{-}4$ **d.** $^{-}2 + {}^{-}8$ **f.** $^{-}4 + {}^{-}1$ **h.** $^{-}7 + {}^{-}6$

In each of the above problems do you see that the absolute value of the sum is the sum of the absolute values of the two addends and that the sum of two negative integers is a negative integer?

3. a. $^+6 + ^-1$ b. $^+8 + ^-3$ c. $^+4 + ^-7$ d. $^+2 + ^-9$ e. $^+5 + ^-2$ f. $^+1 + ^-6$ g. $^+10 + ^-7$ h. $^+3 + ^-11$

In each of the above problems do you see that the absolute value of the sum is the difference in the absolute values of the two addends and that the sign in the sum, when we add a positive number and a negative number, corresponds to the sign of the addend having the greater absolute value?

4. a. $^-7 + ^+2$ b. $^-5 + ^+9$ c. $^-1 + ^+3$ d. $^-4 + ^+8$ e. $^-8 + ^+11$ f. $^-2 + ^+1$ g. $^-10 + ^+7$ h. $^-6 + ^+12$

In each of the above problems do you see that the absolute value of the sum is the difference in the absolute values of the two addends and that the sign in the sum, when we add a negative number and a positive number, corresponds to the sign of the addend having the greater absolute value?

5. a. $^+3 + ^-3$ b. $^+6 + ^-6$ c. $^-2 + ^+2$ d. $^-7 + ^+7$ e. $^+10 + ^-10$ f. $^-8 + ^+8$ g. $^-11 + ^+11$ h. $^+9 + ^-9$

In each of the above problems do you see that the sum of an integer and its opposite is zero?

6. a. $0 + ^-4$ b. $^+3 + 0$ c. $0 + ^+6$ d. $^-2 + 0$ e. $^+8 + 0$ f. $0 + ^+9$ g. $^-10 + 0$ h. $0 + ^-7$

In each of the above problems do you see that the sum of any non-zero integer and zero is the non-zero integer?

7. a. $^+\frac{1}{2} + ^+1\frac{1}{2}$ b. $^-2\frac{3}{4} + ^-2\frac{3}{4}$ c. $^+5\frac{1}{2} + ^-6$ d. $^-3\frac{2}{3} + ^+4\frac{1}{3}$ e. $^-2\frac{7}{8} + 0$ f. $^+4\frac{1}{4} + ^-4\frac{1}{4}$ g. $^-3\frac{1}{3} + ^-\frac{5}{6}$ h. $^-5 + ^+7\frac{1}{2}$

In the above problems do you see that the same procedures are used to add rational numbers that are used to add integers?

8. a. $^+3 + ^+4$ b. $^+7 + ^-2$ c. $^-4 + ^+4$ d. $^-1 + ^-6$ e. $^+8 + ^-10$ f. $0 + ^-7$ g. $^-6 + ^-5$ h. $^-4 + ^+2$ i. $^-12 + ^+9$ j. $^+1 + ^+8$ k. $^+1\frac{3}{8} + ^-\frac{3}{4}$ l. $^-2\frac{1}{4} + ^-3\frac{1}{2}$ m. $^+1 + ^-1$ n. $^-6 + ^-6$ o. $^-2 + ^+10$ p. $^+5 + ^-13$

11–10 ADDITION OF INTEGERS; PROPERTIES

Our observations in adding on the number line lead us to conclude that:

To add two positive integers, we find the sum of their absolute values and prefix the numeral with a positive sign. Observe that addition may be indicated horizontally or vertically.

In $^+5 + ^+3 = ^+8$, the absolute value of the sum (8) is the sum of the absolute values of the addends $(5 + 3)$ and the sign is the common sign, + in this case.

Horizontally:

$^+5 + ^+3 = ^+(5 + 3) = ^+8$

Answer, $^+8$

Vertically:

$$\begin{array}{r} ^+5 \\ ^+3 \\ \hline ^+8 \end{array}$$

To add two negative integers, we find the sum of their absolute values and prefix the numeral with a negative sign.

In ${}^-5 + {}^-3 = {}^-8$, the absolute value of the sum (8) is the sum of the absolute values of the addends $(5 + 3)$ and the sign is the common sign, ${}^-$ in this case.

Horizontally: ${}^-5 + {}^-3 = {}^-(5 + 3) = {}^-8$

Vertically:
$$\begin{array}{r} {}^-5 \\ \underline{{}^-3} \\ {}^-8 \end{array}$$

Answer, ${}^-8$

To add a positive integer and a negative integer or *a negative integer and a positive integer,* we subtract the smaller absolute value from the greater absolute value of the two addends and prefix the difference with the sign of the number having the greater absolute value.

In ${}^+5 + {}^-3 = {}^+2$, the absolute value of the sum (2) is equal to the difference in the absolute values of the addends $(5 - 3)$ and the sign of the sum $({}^+)$ is the same as the sign of the addend with the greater absolute value (the sign of ${}^+5$).

Horizontally: ${}^+5 + {}^-3 = {}^+(5 - 3) = {}^+2$

Vertically:
$$\begin{array}{r} {}^+5 \\ \underline{{}^-3} \\ {}^+2 \end{array}$$

Answer, ${}^+2$

In ${}^+5 + {}^-8 = {}^-3$, the absolute value of the sum (3) is equal to the difference in the absolute values of the addends $(8 - 5)$ and the sign of the sum $({}^-)$ is the same as the sign of the addend with the greater absolute value (the sign of ${}^-8$).

Horizontally: ${}^+5 + {}^-8 = {}^-(8 - 5) = {}^-3$

Vertically:
$$\begin{array}{r} {}^+5 \\ \underline{{}^-8} \\ {}^-3 \end{array}$$

Answer, ${}^-3$

In ${}^-5 + {}^+3 = {}^-2$, the absolute value of the sum (2) is equal to the difference in the absolute values of the addends $(5 - 3)$ and the sign of the sum $({}^-)$ is the same as the sign of the addend with the greater absolute value (the sign of ${}^-5$).

Horizontally: ${}^-5 + {}^+3 = {}^-(5 - 3) = {}^-2$

Vertically:
$$\begin{array}{r} {}^-5 \\ \underline{{}^+3} \\ {}^-2 \end{array}$$

Answer, ${}^-2$

In ${}^-5 + {}^+8 = {}^+3$, the absolute value of the sum (3) is equal to the difference in the absolute values of the addends $(8 - 5)$ and the sign of the sum $({}^+)$ is the same as the sign of the addend with the greater absolute value (the sign of ${}^+8$).

Horizontally: ${}^-5 + {}^+8 = {}^+(8 - 5) = {}^+3$

Vertically:
$$\begin{array}{r} {}^-5 \\ \underline{{}^+8} \\ {}^+3 \end{array}$$

Answer, ${}^+3$

The sum of any integer and its opposite is zero.

$^{+}5 + {}^{-}5 = 0$ *Answer,* 0
$^{-}8 + {}^{+}8 = 0$ *Answer,* 0

The sum of any non-zero integer and zero is the non-zero integer.

$^{-}7 + 0 = {}^{-}7$ *Answer,* $^{-}7$
$0 + {}^{+}1 = {}^{+}1$ *Answer,* $^{+}1$

The numeral naming a positive integer does not require the positive ($^{+}$) sign.

$6 + {}^{-}11 = {}^{-}5$ *Answer,* $^{-}5$
$^{-}4 + 7 = 3$ *Answer,* 3

To add three or more integers, we either add successively as shown in the model or we add first the positive numbers, then the negative numbers, and finally the two answers.

$^{+}6 + {}^{-}9 + {}^{+}2$
$= {}^{-}3 + {}^{+}2$
$= {}^{-}1$ *Answer,* $^{-}1$

EXERCISES

Add as indicated:

1. a. $^{+}5 + {}^{+}7$ **b.** $^{+}36 + {}^{+}8$ **c.** $^{+}19 + {}^{+}43$ **d.** $^{+}127 + {}^{+}99$
2. a. $^{-}6 + {}^{-}11$ **b.** $^{-}18 + {}^{-}49$ **c.** $^{-}56 + {}^{-}25$ **d.** $^{-}34 + {}^{-}67$
3. a. $^{+}8 + {}^{-}2$ **b.** $^{+}21 + {}^{-}14$ **c.** $^{+}33 + {}^{-}27$ **d.** $^{+}61 + {}^{-}42$
4. a. $^{+}3 + {}^{-}9$ **b.** $^{+}6 + {}^{-}20$ **c.** $^{+}46 + {}^{-}62$ **d.** $^{+}59 + {}^{-}85$
5. a. $^{-}7 + {}^{+}1$ **b.** $^{-}45 + {}^{+}12$ **c.** $^{-}51 + {}^{+}35$ **d.** $^{-}101 + {}^{+}74$
6. a. $^{-}2 + {}^{+}10$ **b.** $^{-}17 + {}^{+}40$ **c.** $^{-}29 + {}^{+}74$ **d.** $^{-}87 + {}^{+}120$
7. a. $^{+}4 + 0$ **b.** $0 + {}^{-}13$ **c.** $0 + {}^{+}21$ **d.** $^{-}52 + 0$
8. a. $^{+}15 + {}^{-}15$ **b.** $^{-}47 + {}^{+}47$ **c.** $^{+}92 + {}^{-}92$ **d.** $^{-}66 + {}^{+}66$
9. a. $8 + {}^{-}14$ **b.** $^{-}31 + 16$ **c.** $17 + {}^{-}25$ **d.** $^{-}42 + 80$

10. a. $^{+}3 + {}^{+}2 + {}^{+}6$ **d.** $^{+}20 + {}^{-}18 + {}^{-}5$ **g.** $^{-}3 + {}^{+}4 + {}^{-}7 + {}^{+}2$
b. $^{-}8 + {}^{+}9 + {}^{-}1$ **e.** $^{+}19 + {}^{+}8 + {}^{-}24$ **h.** $^{-}11 + {}^{+}16 + {}^{-}3 + {}^{-}5$
c. $^{-}11 + {}^{-}6 + {}^{-}28$ **f.** $^{-}25 + {}^{-}47 + {}^{+}56$ **i.** $^{+}17 + {}^{-}23 + {}^{-}35 + {}^{+}46$

11. a. $^{-}16 + {}^{-}5$ **e.** $^{-}64 + {}^{+}73$ **i.** $^{-}26 + {}^{+}75 + {}^{-}48$
b. $^{-}8 + {}^{+}13$ **f.** $^{+}32 + {}^{-}32$ **j.** $^{+}15 + {}^{-}6 + {}^{-}37$
c. $^{+}21 + {}^{-}9$ **g.** $^{-}99 + {}^{-}89$ **k.** $^{+}83 + {}^{-}19 + {}^{-}67 + {}^{+}8$
d. $^{+}56 + {}^{+}75$ **h.** $0 + {}^{-}24$ **l.** $^{-}55 + {}^{+}28 + {}^{-}36 + {}^{+}57$

12. a. $({}^{+}6 + {}^{+}5) + {}^{-}3$ **c.** $({}^{-}11 + {}^{+}16) + {}^{-}6$ **e.** $^{-}20 + ({}^{+}37 + {}^{-}15)$
b. $^{+}8 + ({}^{-}9 + {}^{-}7)$ **d.** $^{-}12 + ({}^{-}7 + {}^{-}25)$ **f.** $({}^{+}26 + {}^{-}40) + {}^{+}34$

13. Add:

a.	$^{+}3$	$^{-}7$	$^{-}16$	$^{+}34$	$^{-}42$	$^{+}31$	$^{-}20$	$^{+}11$
	$^{+}9$	$^{-}17$	$^{+}8$	$^{-}19$	$^{+}42$	0	$^{-}20$	$^{-}52$

b.	$^{+}28$	$^{-}81$	$^{-}67$	$^{+}95$	0	$^{+}44$	$^{-}39$	$^{-}19$
	$\underline{^{-}67}$	$\underline{^{+}25}$	$\underline{^{-}67}$	$\underline{^{+}107}$	$\underline{^{-}75}$	$\underline{^{-}5}$	$\underline{^{+}46}$	$\underline{^{-}1}$
c.	$^{-}35$	$^{-}28$	$^{+}43$	$^{-}55$	$^{+}29$	$^{+}82$	0	$^{-}101$
	$\underline{^{-}49}$	$\underline{^{+}8}$	$\underline{^{-}71}$	$\underline{^{+}63}$	$\underline{^{-}29}$	$\underline{^{+}19}$	$\underline{^{-}56}$	$\underline{^{+}63}$
d.	$^{+}3$	$^{-}12$	$^{-}27$	$^{+}6$	$^{-}27$	$^{-}6$	$^{+}103$	$^{-}38$
	$^{-}9$	$^{+}8$	$^{+}53$	$^{-}8$	$^{-}9$	$^{+}34$	$^{-}96$	$^{-}53$
	$\underline{^{+}5}$	$\underline{^{-}2}$	$\underline{^{-}19}$	$^{+}1$	$^{-}15$	$^{-}57$	$^{-}42$	$^{-}67$
				$\underline{^{-}2}$	$\underline{^{+}60}$	$\underline{^{+}31}$	$\underline{^{+}29}$	$\underline{^{-}95}$

14. Find the missing numbers:

a. $^{+}3 + {}^{-}9 = ?$
b. $^{-}13 + {}^{+}8 = \square$
c. $^{+}21 + {}^{-}31 = n$
d. $^{-}5 + {}^{-}16 = \square$
e. $^{-}17 + {}^{+}17 = n$
f. $^{+}34 + {}^{+}43 = ?$
g. $^{-}26 + {}^{+}19 = n$
h. $^{-}38 + {}^{-}56 = n$
i. $^{+}45 + {}^{-}27 = n$

15. Find the missing numbers:

a. $? + {}^{+}6 = {}^{+}8$
b. $\square + {}^{-}4 = {}^{-}11$
c. $n + {}^{-}2 = {}^{-}2$
d. $\square + {}^{-}19 = {}^{-}3$
e. $? + {}^{+}21 = {}^{+}6$
f. $n + {}^{-}32 = {}^{+}8$
g. $n + {}^{+}5 = {}^{-}18$
h. $n + {}^{-}43 = 0$
i. $n + {}^{-}59 = {}^{-}22$

16. Find the missing numbers:

a. $^{+}7 + ? = {}^{+}12$
b. $^{-}2 + \square = {}^{-}9$
c. $^{+}6 + n = {}^{-}1$
d. $^{-}13 + \square = {}^{-}4$
e. $^{+}25 + n = 0$
f. $^{-}18 + ? = {}^{-}2$
g. $^{+}36 + n = {}^{+}7$
h. $^{-}53 + n = {}^{-}53$
i. $^{-}41 + n = {}^{+}14$

Properties

In the following exercises we determine the properties that hold for the set of integers under the operation of addition.

1. Commutative Property

a. What do we mean when we say that the commutative property holds for the addition of a certain set of numbers?

b. In each of the following find the two sums by adding on the number line using vectors. Compare the two sums by comparing their vectors.

(1) $^{+}2 + {}^{+}5$; $^{+}5 + {}^{+}2$. Does $^{+}2 + {}^{+}5 = {}^{+}5 + {}^{+}2$?
(2) $^{-}6 + {}^{+}4$; $^{+}4 + {}^{-}6$. Does $^{-}6 + {}^{+}4 = {}^{+}4 + {}^{-}6$?

c. In each of the following add as indicated, then compare the sums:

(1) $^{+}7 + {}^{+}8$; $^{+}8 + {}^{+}7$. Does $^{+}7 + {}^{+}8 = {}^{+}8 + {}^{+}7$?
(2) $^{-}3 + {}^{-}1$; $^{-}1 + {}^{-}3$. Does $^{-}3 + {}^{-}1 = {}^{-}1 + {}^{-}3$?
(3) $^{+}9 + {}^{-}4$; $^{-}4 + {}^{+}9$. Does $^{+}9 + {}^{-}4 = {}^{-}4 + {}^{+}9$?
(4) $^{-}6 + {}^{+}5$; $^{+}5 + {}^{-}6$. Does $^{-}6 + {}^{+}5 = {}^{+}5 + {}^{-}6$?

d. Can you find two integers for which the commutative property of addition does not hold?

e. Can we say that the commutative property holds for the addition of integers?

2. Associative Property

a. What do we mean when we say that the associative property holds for the addition of a certain set of numbers?

b. In each of the following find the two sums by adding on the number line using vectors. Compare the two sums by comparing their vectors.

(1) $(^-6 + {^+7}) + {^+3}$; $\quad {^-6} + ({^+7} + {^+3})$.
Does $(^-6 + {^+7}) + {^+3} = {^-6} + ({^+7} + {^+3})$?

(2) $(^+8 + {^-4}) + {^-5}$; $\quad {^+8} + ({^-4} + {^-5})$.
Does $(^+8 + {^-4}) + {^-5} = {^+8} + ({^-4} + {^-5})$?

c. In each of the following add as indicated, then compare the sums:

(1) $(^-3 + {^-8}) + {^+14}$; $\quad {^-3} + ({^-8} + {^+14})$.
Does $(^-3 + {^-8}) + {^+14} = {^-3} + ({^-8} + {^+14})$?

(2) $(^+2 + {^-10}) + {^-1}$; $\quad {^+2} + ({^-10} + {^-1})$.
Does $(^+2 + {^-10}) + {^-1} = {^+2} + ({^-10} + {^-1})$?

d. Can you find three integers for which the associative property of addition does not hold?

e. Can we say that the associative property holds for the addition of integers?

3. Closure

a. What do we mean when we say that a certain set of numbers is closed under the operation of addition?

b. Is the sum of each of the following an integer?

(1) $^+4 + {^-7}$	(4) $^-5 + {^+3}$	(7) $^+9 + 0$
(2) $^-2 + {^-2}$	(5) $0 + {^-10}$	(8) $^-15 + {^+7}$
(3) $^+8 + {^+12}$	(6) $^-11 + {^+11}$	(9) $^+20 + {^-20}$

c. Can you find two integers for which the sum is not an integer?

d. Is the sum of two integers always an integer?

e. Can we say that the set of integers is closed under the operation of addition?

4. Additive Identity

a. What do we mean when we say that a certain set of numbers has an additive identity?

b. What is the sum of each of the following?

(1) $^-7 + 0$	(3) $^+8 + 0$	(5) $0 + {^+16}$	(7) $^+11 + 0$
(2) $0 + {^+3}$	(4) $0 + {^-1}$	(6) $^-14 + 0$	(8) $0 + {^-2}$

c. In each case in problem **4b** is the sum of the integer and zero the integer itself?

d. Can you find any integer which when added to zero does not have the given integer as the sum?

e. Does the set of integers have an additive identity? If so, what is it?

5. Additive Inverse

a. What do we mean by an additive inverse of an integer?

b. What is the sum of each of the following?

(1) $^-2 + ^+2$	(3) $^+13 + ^-13$	(5) $0 + 0$	(7) $^+82 + ^-82$
(2) $^+7 + ^-7$	(4) $^-5 + ^+5$	(6) $^+27 + ^-27$	(8) $^-100 + ^+100$

c. Are we adding in each case in problem **5b** an integer and its additive inverse? What is the sum in each case?

d. Can you find any integer which when added to its additive inverse does not have zero as the sum?

e. Does each integer in the set of integers have an additive inverse?

11–11 ADDITION OF RATIONAL NUMBERS; PROPERTIES

Examination of addition on the number line reveals that to add positive and negative fractional numbers, we follow the same procedures as we do when we add integers. See sections 11–9 and 11–10. Since in the addition of rational numbers we sometimes add the absolute values and at other times subtract the absolute values, we should review at this time both the addition and the subtraction of the fractional numbers of arithmetic. See sections 5–11 and 5–12. Although the model problems below only illustrate common fractions, rational numbers may also be named by decimal fraction numerals. Thus sections 6–7 and 6–8 should also be reviewed at this time.

To add two positive rational numbers, we find the sum of their absolute values and prefix the numeral with a positive sign.

To add two negative rational numbers, we find the sum of their absolute values and prefix the numeral with a negative sign.

Add: $\frac{^-4}{9} + \frac{^-2}{9} = ?$

$$\frac{^-4}{9} + \frac{^-2}{9} = {}^-\left(\left|\frac{^-4}{9}\right| + \left|\frac{^-2}{9}\right|\right) = {}^-\left(\frac{4}{9} + \frac{2}{9}\right) = {}^-\left(\frac{4+2}{9}\right) = \frac{^-6}{9} = \frac{^-2}{3}$$

or simply: Vertically:

$$\begin{array}{r} \frac{^-4}{9} \\ \frac{^-2}{9} \\ \hline \frac{^-6}{9} = \frac{^-2}{3} \end{array}$$

Horizontally:

$$\frac{^-4}{9} + \frac{^-2}{9} = \frac{^-4 + ^-2}{9} = \frac{^-6}{9} = \frac{^-2}{3}$$

Answer, $\frac{^-2}{3}$

To add a positive rational number and a negative rational number or *a negative rational number and a positive rational number*, subtract the smaller absolute value from the greater absolute value of the two addends and prefix the difference with the sign of the number having the greater absolute value.

Add: $\frac{^+3}{4} + \frac{^-2}{3} = ?$

$$\frac{^+3}{4} + \frac{^-2}{3} = {}^+\left(\left|\frac{^+3}{4}\right| - \left|\frac{^-2}{3}\right|\right) = {}^+\left(\frac{3}{4} - \frac{2}{3}\right) = {}^+\left(\frac{3}{4}\times\frac{3}{3} - \frac{2}{3}\times\frac{4}{4}\right)$$

$$= {}^+\left(\frac{9}{12} - \frac{8}{12}\right) = \frac{^+1}{12}$$

or simply:

Vertically:

Add:

$$\frac{^+3}{4} = \frac{^+3}{4} \times \frac{3}{3} = \frac{^+9}{12}$$

$$\frac{^-2}{3} = \frac{^-2}{3} \times \frac{4}{4} = \frac{^-8}{12}$$

$$\frac{^+1}{12}$$

Horizontally:

$$\frac{^+3}{4} + \frac{^-2}{3} = \frac{^+3 \times 3}{4 \times 3} + \frac{^-2 \times 4}{3 \times 4} = \frac{^+9}{12} + \frac{^-8}{12} = \frac{^+9 + {}^-8}{12} = \frac{^+1}{12}$$

Answer, $\frac{^+1}{12}$

Observe in the model problem below that sometimes it is necessary to interchange the absolute values to make this subtraction possible.

Add: $\frac{^+1}{2} + \frac{^-7}{8} = ?$

$$\frac{^+1}{2} + \frac{^-7}{8} = {}^-\left(\left|\frac{^-7}{8}\right| - \left|\frac{^+1}{2}\right|\right)^* = {}^-\left(\frac{7}{8} - \frac{1}{2}\right) = {}^-\left(\frac{7}{8} - \frac{1}{2}\times\frac{4}{4}\right) = {}^-\left(\frac{7}{8} - \frac{4}{8}\right) = \frac{^-3}{8}$$

*(Since the smaller absolute value happens to be in the first addend in this problem, we interchange absolute values in order to subtract.)

or simply:

Vertically:

Add:

$$\frac{^+1}{2} = \frac{^+4}{8}$$

$$\frac{^-7}{8} = \frac{^-7}{8}$$

$$\frac{^-3}{8}$$

Horizontally:

$$\frac{^+1}{2} + \frac{^-7}{8} = \frac{^+4}{8} + \frac{^-7}{8} = \frac{^+4 + {}^-7}{8} = \frac{^-3}{8}$$

Answer, $\frac{^-3}{8}$

The sum of any rational number and its opposite is zero.

The sum of any non-zero rational number and zero is the non-zero rational number.

The numeral naming a positive rational number does not require the positive ($^{+}$) sign.

To add three or more rational numbers, we either add successively or first add the positive numbers, then the negative numbers, and finally the two answers.

$$
\begin{aligned}
& {}^{+}1\tfrac{1}{2} + {}^{-}3\tfrac{1}{4} + {}^{+}2\tfrac{5}{6} \\
&= {}^{+}1\tfrac{6}{12} + {}^{-}3\tfrac{3}{12} + {}^{+}2\tfrac{10}{12} \\
&= {}^{+}1\tfrac{6}{12} + {}^{+}2\tfrac{10}{12} + {}^{-}3\tfrac{3}{12} \\
&= {}^{+}3\tfrac{16}{12} + {}^{-}3\tfrac{3}{12} \\
&= {}^{+}\tfrac{13}{12} = {}^{+}1\tfrac{1}{12}
\end{aligned}
$$

Answer, ${}^{+}1\frac{1}{12}$

EXERCISES

Add as indicated:

1. a. ${}^{+}\frac{1}{8} + {}^{+}\frac{3}{8}$ **b.** ${}^{+}\frac{2}{3} + {}^{+}\frac{7}{12}$ **c.** ${}^{+}1\frac{1}{4} + {}^{+}3\frac{5}{6}$ **d.** ${}^{+}7\frac{9}{16} + {}^{+}4\frac{3}{10}$

2. a. ${}^{+}.7 + {}^{+}.9$ **b.** ${}^{+}3.6 + {}^{+}.15$ **c.** ${}^{+}.019 + {}^{+}.03$ **d.** ${}^{+}20.69 + {}^{+}5.386$

3. a. ${}^{-}\frac{5}{6} + {}^{-}\frac{1}{6}$ **b.** ${}^{-}\frac{4}{5} + {}^{-}\frac{1}{2}$ **c.** ${}^{-}2\frac{7}{8} + {}^{-}1\frac{5}{8}$ **d.** ${}^{-}6\frac{11}{12} + {}^{-}8\frac{5}{16}$

4. a. ${}^{-}.84 + {}^{-}.5$ **b.** ${}^{-}.07 + {}^{-}.17$ **c.** ${}^{-}16 + {}^{-}2.8$ **d.** ${}^{-}9.62 + {}^{-}84.9$

5. a. ${}^{+}\frac{3}{4} + {}^{-}\frac{1}{8}$ **b.** ${}^{+}2\frac{1}{3} + {}^{-}3\frac{1}{4}$ **c.** ${}^{+}6\frac{3}{8} + {}^{-}\frac{11}{16}$ **d.** ${}^{+}7\frac{1}{2} + {}^{-}9\frac{7}{12}$

6. a. ${}^{+}1.4 + {}^{-}2.5$ **b.** ${}^{+}.9 + {}^{-}1.1$ **c.** ${}^{+}.359 + {}^{-}.29$ **d.** ${}^{+}27.6 + {}^{-}2.76$

7. a. ${}^{-}\frac{2}{5} + {}^{+}\frac{7}{100}$ **b.** ${}^{-}8\frac{1}{6} + {}^{+}1\frac{2}{3}$ **c.** ${}^{-}3\frac{1}{2} + {}^{+}7\frac{1}{16}$ **d.** ${}^{-}1\frac{3}{10} + {}^{+}1\frac{7}{16}$

8. a. ${}^{-}9.6 + {}^{+}6.9$ **b.** ${}^{-}.71 + {}^{+}.8$ **c.** ${}^{-}.004 + {}^{+}.0037$ **d.** ${}^{-}1.16 + {}^{+}11.3$

9. a. ${}^{+}1\frac{1}{2} + {}^{-}1\frac{1}{2}$ **b.** ${}^{-}\frac{11}{12} + {}^{+}\frac{11}{12}$ **c.** ${}^{+}7\frac{2}{3} + {}^{-}7\frac{2}{3}$ **d.** ${}^{-}5\frac{13}{16} + {}^{+}5\frac{13}{16}$

10. a. ${}^{-}3.9 + {}^{+}3.9$ **b.** ${}^{+}.7 + {}^{-}.7$ **c.** ${}^{-}.046 + {}^{+}.046$ **d.** ${}^{+}9.183 + {}^{-}9.183$

11. a. ${}^{-}5\frac{7}{9} + 0$ **b.** $0 + {}^{+}8\frac{2}{3}$ **c.** ${}^{-}1.6 + 0$ **d.** $0 + {}^{+}.055$

12. a. ${}^{-}9\frac{1}{4} + 5\frac{3}{16}$ **b.** $7\frac{1}{2} + {}^{-}1\frac{3}{8}$ **c.** ${}^{-}.63 + 1.4$ **d.** $5.81 + {}^{-}3.142$

13. a. ${}^{+}\frac{3}{5} + {}^{-}\frac{1}{2} + {}^{-}\frac{7}{10}$ **c.** ${}^{-}5\frac{1}{4} + {}^{-}4\frac{2}{3} + {}^{-}2\frac{5}{8}$ **e.** ${}^{-}6.29 + {}^{-}.518 + {}^{+}73.4$

b. ${}^{+}2\frac{2}{3} + {}^{-}9\frac{5}{6} + {}^{+}6\frac{1}{2}$ **d.** ${}^{-}1\frac{1}{6} + {}^{+}2\frac{7}{8} + {}^{-}\frac{3}{4}$ **f.** ${}^{+}.193 + {}^{-}.3621 + {}^{+}.27$

14. a. $({}^{-}\frac{2}{3} + {}^{-}\frac{11}{12}) + {}^{-}\frac{5}{6}$ **d.** $({}^{-}\frac{4}{5} + {}^{-}1\frac{5}{6}) + {}^{+}3$ **g.** $({}^{-}.7 + {}^{+}.82) + {}^{-}.1$

b. $({}^{+}3\frac{1}{2} + {}^{+}7\frac{3}{4}) + {}^{+}4\frac{1}{3}$ **e.** ${}^{+}6\frac{1}{4} + ({}^{-}4\frac{7}{12} + {}^{+}2\frac{3}{16})$ **h.** ${}^{-}2.5 + ({}^{-}.96 + {}^{-}1.4)$

c. ${}^{-}1\frac{5}{8} + ({}^{+}2\frac{1}{4} + {}^{+}\frac{9}{16})$ **f.** ${}^{-}4\frac{1}{6} + ({}^{+}5\frac{1}{2} + {}^{-}10\frac{7}{8})$ **i.** $({}^{+}8.17 + {}^{-}91.3) + {}^{+}.796$

15. Add:

${}^{-}\frac{11}{16}$	${}^{+}\frac{5}{6}$	${}^{-}\frac{4}{5}$	${}^{+}1\frac{1}{2}$	${}^{+}7\frac{5}{12}$	${}^{-}6\frac{9}{10}$	${}^{-}4\frac{9}{16}$
${}^{-}\frac{9}{16}$	${}^{-}\frac{17}{18}$	${}^{+}\frac{3}{4}$	${}^{-}\frac{5}{8}$	${}^{+}3\frac{1}{4}$	${}^{+}8\frac{5}{6}$	${}^{-}9\frac{7}{12}$

16. Add:

${}^{-}\frac{7}{8}$	${}^{+}\frac{2}{3}$	${}^{-}\frac{1}{4}$	${}^{+}4\frac{3}{4}$	${}^{-}3\frac{2}{3}$	${}^{+}9\frac{3}{8}$	${}^{-}5\frac{13}{16}$
${}^{-}\frac{5}{16}$	${}^{-}\frac{1}{2}$	${}^{-}\frac{11}{12}$	${}^{+}2\frac{1}{2}$	${}^{-}1\frac{3}{5}$	${}^{-}3\frac{7}{12}$	${}^{+}4\frac{11}{12}$
${}^{+}\frac{3}{4}$	${}^{+}\frac{3}{8}$	${}^{-}\frac{9}{16}$	${}^{-}6\frac{7}{8}$	${}^{+}5\frac{1}{2}$	${}^{-}7\frac{5}{6}$	${}^{-}2\frac{5}{8}$

17. Add:

${}^{-}8.7$	${}^{+}.126$	${}^{-}5.98$	${}^{-}17.5$	${}^{-}7.48$	${}^{-}.045$	${}^{-}6.5$
${}^{+}4.9$	${}^{-}.318$	${}^{-}3.02$	${}^{+}28.2$	${}^{+}3.91$	${}^{-}.039$	${}^{+}7.12$
			${}^{-}10.4$	${}^{+}4.86$	${}^{+}.024$	${}^{-}1.839$

18. Find the missing numbers:

a. ${}^{+}\frac{7}{8} + {}^{-}\frac{1}{8} = ?$ **c.** ${}^{-}2\frac{3}{4} + {}^{-}1\frac{5}{16} = \square$ **e.** ${}^{-}1\frac{7}{10} + {}^{+}2\frac{5}{8} = n$
b. ${}^{-}\frac{4}{5} + {}^{+}\frac{2}{3} = n$ **d.** ${}^{+}5\frac{5}{6} + {}^{+}3\frac{7}{12} = n$ **f.** ${}^{+}7\frac{1}{3} + {}^{-}5\frac{3}{4} = n$

19. Find the missing numbers:

a. ${}^{-}.9 + {}^{-}.3 = \square$ **c.** ${}^{-}3.4 + {}^{+}2 = ?$ **e.** ${}^{+}5.4 + {}^{-}.54 = n$
b. ${}^{+}.14 + {}^{-}2.5 = n$ **d.** ${}^{+}6.7 + {}^{-}6.7 = n$ **f.** ${}^{-}.825 + {}^{+}.91 = n$

20. Find the missing numbers:

a. $? + {}^{+}\frac{2}{3} = {}^{+}1\frac{1}{6}$ **c.** $\square + {}^{-}2\frac{5}{6} = {}^{-}8\frac{1}{3}$ **e.** $n + {}^{+}1\frac{3}{4} = {}^{+}\frac{11}{12}$
b. $n + {}^{-}\frac{7}{8} = {}^{-}\frac{11}{16}$ **d.** $n + {}^{-}7\frac{4}{9} = 0$ **f.** $n + {}^{-}9\frac{4}{5} = {}^{-}4\frac{2}{3}$

21. Find the missing numbers:

a. $? + {}^{+}.6 = {}^{+}1.3$ **c.** $n + {}^{+}9.6 = {}^{+}1.82$ **e.** $n + {}^{-}.469 = {}^{-}.469$
b. $n + {}^{-}2.7 = {}^{-}5.2$ **d.** $\square + {}^{-}10 = {}^{-}16.7$ **f.** $n + {}^{+}7.25 = {}^{-}2.986$

22. Find the missing numbers:

a. ${}^{-}\frac{1}{2} + \square = {}^{-}\frac{1}{4}$ **c.** ${}^{+}2\frac{2}{3} + ? = {}^{+}7\frac{5}{6}$ **e.** ${}^{-}3\frac{3}{8} + n = {}^{-}9\frac{3}{4}$
b. ${}^{+}1\frac{1}{8} + n = {}^{-}1\frac{1}{2}$ **d.** ${}^{-}4\frac{11}{16} + ? = {}^{+}4\frac{11}{16}$ **f.** ${}^{-}7\frac{2}{5} + n = {}^{+}5$

23. Find the missing numbers:

a. ${}^{+}.8 + ? = {}^{+}1.4$ **c.** ${}^{-}3.6 + \square = 0$ **e.** ${}^{-}4.8 + n = {}^{+}3.1$
b. ${}^{-}1.7 + n = {}^{-}2$ **d.** ${}^{+}.5 + n = {}^{+}.27$ **f.** ${}^{+}.699 + n = {}^{-}.0001$

Properties

In the following exercises we determine the properties that hold for the set of rational numbers under the operation of addition.

1. Commutative Property

a. What do we mean when we say that the commutative property holds for the addition of a certain set of rational numbers?

b. In each of the following find the two sums by adding on the number line, using vectors. Compare the two sums by comparing their vectors.

(1) ${}^{+}\frac{1}{2} + {}^{-}\frac{3}{4}$; ${}^{-}\frac{3}{4} + {}^{+}\frac{1}{2}$. Does ${}^{+}\frac{1}{2} + {}^{-}\frac{3}{4} = {}^{-}\frac{3}{4} + {}^{+}\frac{1}{2}$?
(2) ${}^{-}2\frac{1}{2} + {}^{-}1\frac{1}{4}$; ${}^{-}1\frac{1}{4} + {}^{-}2\frac{1}{2}$. Does ${}^{-}2\frac{1}{2} + {}^{-}1\frac{1}{4} = {}^{-}1\frac{1}{4} + {}^{-}2\frac{1}{2}$?

c. In each of the following add as indicated, then compare the sums:

(1) ${}^{-}\frac{2}{3} + {}^{+}\frac{5}{6}$; ${}^{+}\frac{5}{6} + {}^{-}\frac{2}{3}$. Does ${}^{-}\frac{2}{3} + {}^{+}\frac{5}{6} = {}^{+}\frac{5}{6} + {}^{-}\frac{2}{3}$?
(2) ${}^{+}3\frac{7}{8} + {}^{+}2\frac{3}{8}$; ${}^{+}2\frac{3}{8} + {}^{+}3\frac{7}{8}$. Does ${}^{+}3\frac{7}{8} + {}^{+}2\frac{3}{8} = {}^{+}2\frac{3}{8} + {}^{+}3\frac{7}{8}$?

(3) $^{+}4\frac{3}{4} + {}^{-}5\frac{11}{16}$; ${}^{-}5\frac{11}{16} + {}^{+}4\frac{3}{4}$. Does $^{+}4\frac{3}{4} + {}^{-}5\frac{11}{16} = {}^{-}5\frac{11}{16} + {}^{+}4\frac{3}{4}$?
(4) ${}^{-}1.48 + {}^{-}.357$; ${}^{-}.357 + {}^{-}1.48$. Does ${}^{-}1.48 + {}^{-}.357 = {}^{-}.357 + {}^{-}1.48$?

d. Can you find two rational numbers for which the commutative property of addition does not hold?

e. Can we say that the commutative property holds for the addition of rational numbers?

2. Associative Property

a. What do we mean when we say that the associative property holds for the addition of a certain set of numbers?

b. In each of the following find the two sums by adding on the number line, using vectors. Compare the two sums by comparing their vectors.

(1) $({}^{-}\frac{3}{4} + {}^{-}\frac{3}{4}) + {}^{-}\frac{1}{2}$; ${}^{-}\frac{3}{4} + ({}^{-}\frac{3}{4} + {}^{-}\frac{1}{2})$.
Does $({}^{-}\frac{3}{4} + {}^{-}\frac{3}{4}) + {}^{-}\frac{1}{2} = {}^{-}\frac{3}{4} + ({}^{-}\frac{3}{4} + {}^{-}\frac{1}{2})$?
(2) $({}^{+}1\frac{2}{3} + {}^{-}3\frac{1}{3}) + {}^{+}\frac{2}{3}$; ${}^{+}1\frac{2}{3} + ({}^{-}3\frac{1}{3} + {}^{+}\frac{2}{3})$.
Does $({}^{+}1\frac{2}{3} + {}^{-}3\frac{1}{3}) + {}^{+}\frac{2}{3} = {}^{+}1\frac{2}{3} + ({}^{-}3\frac{1}{3} + {}^{+}\frac{2}{3})$?

c. In each of the following add as indicated, then compare the sums:

(1) $({}^{-}2\frac{5}{8} + {}^{+}1\frac{3}{4}) + {}^{-}5\frac{1}{2}$; ${}^{-}2\frac{5}{8} + ({}^{+}1\frac{3}{4} + {}^{-}5\frac{1}{2})$.
Does $({}^{-}2\frac{5}{8} + {}^{+}1\frac{3}{4}) + {}^{-}5\frac{1}{2} = {}^{-}2\frac{5}{8} + ({}^{+}1\frac{3}{4} + {}^{-}5\frac{1}{2})$?
(2) $({}^{-}5.2 + {}^{+}4.25) + {}^{+}.84$; ${}^{-}5.2 + ({}^{+}4.25 + {}^{+}.84)$.
Does $({}^{-}5.2 + {}^{+}4.25) + {}^{+}.84 = {}^{-}5.2 + ({}^{+}4.25 + {}^{+}.84)$?

d. Can you find three rational numbers for which the associative property of addition does not hold?

e. Can we say that the associative property holds for the addition of rational numbers?

3. Closure

a. What do we mean when we say that a certain set of numbers is closed under the operation of addition?

b. Is the sum of each of the following a rational number?

(1) ${}^{+}\frac{3}{5} + {}^{+}\frac{4}{5}$
(2) ${}^{-}\frac{7}{8} + {}^{+}\frac{3}{4}$
(3) ${}^{-}\frac{2}{3} + {}^{-}\frac{7}{16}$
(4) ${}^{-}5\frac{1}{6} + 0$
(5) ${}^{-}4\frac{3}{8} + {}^{+}1\frac{11}{12}$
(6) ${}^{+}2\frac{2}{5} + {}^{-}7\frac{1}{4}$
(7) ${}^{-}.276 + {}^{-}3.84$
(8) ${}^{+}.04 + {}^{-}.004$
(9) ${}^{-}6.08 + {}^{+}2.5$

c. Can you find two rational numbers for which the sum is not a rational number?

d. Is the sum of two rational numbers always a rational number?

e. Can we say that the set of rational numbers is closed under the operation of addition?

4. Additive Identity

a. What do we mean when we say that a certain set of numbers has an additive identity?

b. What is the sum of each of the following?

(1) $^+\frac{7}{8} + 0$	(3) $0 + {}^+3.9$	(5) $^-\frac{11}{4} + 0$	(7) $0 + {}^-.6$
(2) $0 + {}^-1\frac{1}{3}$	(4) $^-.01 + 0$	(6) $0 + {}^+3.72$	(8) $^-9\frac{11}{16} + 0$

c. In each case in problem **4b**, is the sum of the rational number and zero the rational number itself?

d. Can you find any rational number which when added to zero does not have the given rational number as the sum?

e. Does the set of rational numbers have an additive identity? If so, what is it?

5. Additive Inverse

a. What do we mean by an additive inverse of a rational number?

b. What is the sum of each of the following?

(1) $^-\frac{7}{12} + {}^+\frac{7}{12}$	(3) $^+.9 + {}^-.9$	(5) $^-\frac{15}{8} + {}^+\frac{15}{8}$	(7) $^+10\frac{3}{4} + {}^-10\frac{3}{4}$
(2) $^+2\frac{1}{2} + {}^-2\frac{1}{2}$	(4) $^-6.33 + {}^+6.33$	(6) $^+.002 + {}^-.002$	(8) $^-9.6 + {}^+9.6$

c. Are we adding in each case in problem **5b** a rational number and its additive inverse? What is the sum in each case?

d. Can you find a rational number which when added to its additive inverse does not have zero as the sum?

e. Does each rational number in the set of rational numbers have an additive inverse?

11–12 SUBTRACTION ON THE NUMBER LINE

Since subtraction is the inverse operation of addition, to subtract we find the addend which when added to the given addend (subtrahend) will equal the given sum (minuend).

$$^+5 - {}^-3 = ? \quad \text{is thought of as} \quad {}^-3 + ? = {}^+5$$

Using this idea of inverse operation, to subtract on the number line we locate the points reached by the movements represented by the given sum and by the given addend, moving in both cases from the point marked 0.

We draw (1) a vector below the number line to represent the movement indicated by the given sum and (2) a vector above the number line to

$^+5 - {}^-3 = ?$

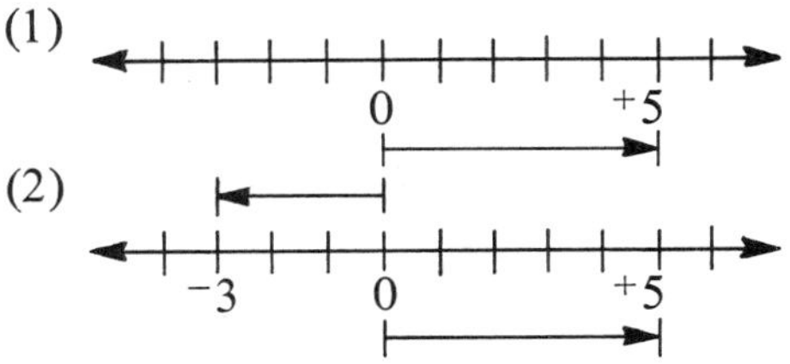

represent the movement indicated by the given addend.

To find the missing addend, we draw (3) a vector *from* the point reached by the vector representing the given addend *to* the point reached by the vector representing the sum. The length of this vector is the absolute value and the direction of the vector indicates the sign of the missing addend.

(3)

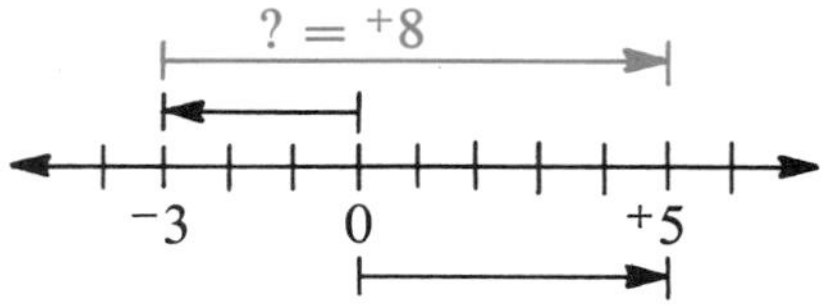

The length of the vector from $^-3$ to $^+5$ is 8 units. The direction of this vector is to the right or the positive direction.
Thus, $^+5 - {^-3} = {^+8}$

EXERCISES

In each of the following subtract as indicated on the number line, using vectors to find the difference:

1. a. $^+6 - {^+3}$	**b.** $^+4 - {^+2}$	**c.** $^+9 - {^+5}$	**d.** $^+8 - {^+6}$
2. a. $^+4 - {^+7}$	**b.** $^+2 - {^+8}$	**c.** $^+3 - {^+6}$	**d.** $^+1 - {^+9}$
3. a. $^-8 - {^-5}$	**b.** $^-6 - {^-1}$	**c.** $^-7 - {^-3}$	**d.** $^-10 - {^-2}$
4. a. $^-1 - {^-7}$	**b.** $^-5 - {^-9}$	**c.** $^-2 - {^-11}$	**d.** $^-4 - {^-6}$
5. a. $^+5 - {^-4}$	**b.** $^+7 - {^-8}$	**c.** $^+9 - {^-6}$	**d.** $^+8 - {^-10}$
6. a. $^-7 - {^+2}$	**b.** $^-1 - {^+12}$	**c.** $^-4 - {^+4}$	**d.** $^-9 - {^+12}$
7. a. $^-6 - {^-6}$	**b.** $^+3 - {^-3}$	**c.** $^-8 - {^-8}$	**d.** $^+4 - {^+4}$
8. a. $0 - {^-4}$	**b.** $0 - {^+7}$	**c.** $^-3 - 0$	**d.** $^+8 - 0$
9. a. $^-2\frac{1}{2} - {^+3\frac{1}{2}}$	**b.** $^-4 - {^-5\frac{3}{4}}$	**c.** $^+3\frac{1}{3} - {^-2\frac{2}{3}}$	**d.** $^-6\frac{1}{4} - {^-6\frac{1}{4}}$

11–13 SUBTRACTION OF INTEGERS; PROPERTIES

In the addition of integers we learned that $^+5 + {^+3} = {^+8}$. In subtracting integers on the number line we found that $^+5 - {^-3} = {^+8}$.

A comparison of $\quad ^+5 - {^-3} = {^+8}$
and $\quad ^+5 + {^+3} = {^+8}$

shows that subtracting $^-3$ from $^+5$ gives the same answer ($^+8$) as adding $^+3$ to $^+5$. Since $^+3$ is the opposite of $^-3$, we may say in this case that *subtracting a number* gives the same answer as *adding its opposite*. Let us determine whether this is true in other cases.

a. $^+2 - {^+6} = ?$ means "What number added to $^+6$ equals $^+2$?" The answer is $^-4$. Thus, $^+2 - {^+6} = {^-4}$.

b. $^-7 - {^+3} = ?$ means "What number added to $^+3$ equals $^-7$?" The answer is $^-10$. Thus, $^-7 - {^+3} = {^-10}$.

c. $^-4 - {^-9} = ?$ means "What number added to $^-9$ equals $^-4$?" The answer is $^+5$. Thus, $^-4 - {^-9} = {^+5}$.

d. $0 - {}^-6 = ?$ means "What number added to $^-6$ equals 0?" The answer is $^+6$. Thus, $0 - {}^-6 = {}^+6$.

When we compare:

a.	b.	c.	d.
$^+2 - {}^+6 = {}^-4$	$^-7 - {}^+3 = {}^-10$	$^-4 - {}^-9 = {}^+5$	$0 - {}^-6 = {}^+6$
$^+2 + {}^-6 = {}^-4$	$^-7 + {}^-3 = {}^-10$	$^-4 + {}^+9 = {}^+5$	$0 + {}^+6 = {}^+6$

we conclude that both subtracting a number and adding the opposite of the number which is to be subtracted give the same answer.

Thus, *to subtract any integer from another integer*, we add to the minuend the opposite (or additive inverse) of the integer which is to be subtracted. The problem first could be rewritten, arranged horizontally or vertically. For example:

Subtract: $^-6 - {}^+4 = ?$

Horizontally:

$^-6 - {}^+4 = {}^-6 + {}^-4 = {}^-10$

Vertically:

Subtract:		Add:	
$^-6$		$^-6$	
$^+4$	→	$^-4$	
		$^-10$	Answer, $^-10$

Subtract: $^-1 - {}^-3 = ?$

Horizontally:

$^-1 - {}^-3 = {}^-1 + {}^+3 = {}^+2$

Vertically:

Subtract:		Add:	
$^-1$		$^-1$	
$^-3$	→	$^+3$	
		$^+2$	Answer, $^+2$

Subtract: $0 - {}^+2 = ?$

Horizontally:

$0 - {}^+2 = 0 + {}^-2 = {}^-2$

Vertically:

Subtract:		Add:	
0		0	
$^+2$	→	$^-2$	
		$^-2$	Answer, $^-2$

The numeral naming a positive number does not require the positive ($^+$) sign.

$8 - 11 = 8 + {}^-11 = {}^-3$
Answer, $^-3$

EXERCISES

Subtract as indicated:

1. a. $^+7 - {}^+2$	**b.** $^+11 - {}^+5$	**c.** $^+33 - {}^+19$	**d.** $^+57 - {}^+27$
2. a. $^+1 - {}^+8$	**b.** $^+9 - {}^+32$	**c.** $^+16 - {}^+54$	**d.** $^+49 - {}^+90$

3. a. $^{-}10 - {}^{-}5$ **b.** $^{-}61 - {}^{-}28$ **c.** $^{-}73 - {}^{-}36$ **d.** $^{-}110 - {}^{-}42$

4. a. $^{-}2 - {}^{-}9$ **b.** $^{-}12 - {}^{-}40$ **c.** $^{-}27 - {}^{-}65$ **d.** $^{-}76 - {}^{-}103$

5. a. $^{+}14 - {}^{-}6$ **b.** $^{+}27 - {}^{-}39$ **c.** $^{+}60 - {}^{-}43$ **d.** $^{+}51 - {}^{-}51$

6. a. $^{-}24 - {}^{+}18$ **b.** $^{-}48 - {}^{+}65$ **c.** $^{-}94 - {}^{+}38$ **d.** $^{-}169 - {}^{+}250$

7. a. $^{+}9 - 0$ **b.** $^{-}17 - 0$ **c.** $^{-}56 - 0$ **d.** $^{+}31 - 0$

8. a. $0 - {}^{-}6$ **b.** $0 - {}^{+}24$ **c.** $0 - {}^{-}61$ **d.** $0 - {}^{+}108$

9. a. $^{-}12 - {}^{-}12$ **b.** $^{+}92 - {}^{+}92$ **c.** $^{-}75 - {}^{-}75$ **d.** $^{-}29 - {}^{-}29$

10. a. $6 - 11$ **b.** $2 - 28$ **c.** $19 - 54$ **d.** $67 - 100$

11. a. $({}^{-}9 - {}^{+}6) - {}^{+}10$ **b.** $({}^{+}8 - {}^{+}14) - {}^{-}7$ **c.** $^{-}24 - ({}^{+}11 - {}^{+}16)$ **d.** $^{+}18 - ({}^{-}5 - {}^{-}21)$ **e.** $({}^{+}10 - {}^{-}5) - {}^{+}15$ **f.** $^{-}1 - ({}^{-}45 - {}^{+}45)$

12. Subtract in each of the following:

a.	$^{+}17$	$^{-}97$	$^{-}40$	$^{+}9$	$^{-}18$	0	$^{-}85$	29
	$^{+}28$	$^{+}39$	$^{-}15$	$^{-}56$	$^{-}72$	$^{-}34$	$^{+}114$	52
b.	$^{+}65$	$^{-}91$	$^{-}54$	$^{+}18$	$^{-}87$	8	$^{-}99$	$^{-}45$
	$^{-}35$	$^{-}76$	$^{+}6$	$^{+}30$	$^{+}87$	43	0	$^{-}78$
c.	$^{-}42$	$^{-}26$	$^{+}62$	0	93	$^{+}7$	$^{-}59$	$^{-}32$
	$^{-}81$	$^{+}19$	$^{+}75$	$^{+}10$	200	$^{-}38$	$^{-}59$	$^{+}32$

13. Find the missing numbers:

a. $^{+}6 - {}^{+}11 = ?$ **b.** $^{-}19 - {}^{-}12 = \square$ **c.** $^{-}31 - {}^{+}45 = n$ **d.** $0 - {}^{-}63 = \square$ **e.** $^{+}18 - {}^{-}18 = ?$ **f.** $^{-}50 - {}^{-}50 = n$ **g.** $^{-}24 - {}^{-}57 = n$ **h.** $^{+}100 - {}^{-}78 = n$ **i.** $^{+}45 - {}^{+}82 = n$

14. Find the missing numbers:

a. $? - {}^{+}3 = {}^{+}9$ **b.** $\square - {}^{-}14 = {}^{-}20$ **c.** $n - {}^{-}1 = {}^{+}8$ **d.** $\square - {}^{+}6 = {}^{-}13$ **e.** $? - {}^{-}15 = 0$ **f.** $n - 9 = {}^{-}5$ **g.** $n - {}^{-}12 = 10$ **h.** $n - 0 = {}^{-}36$ **i.** $n - {}^{+}23 = {}^{-}18$

15. Find the missing numbers:

a. $^{-}7 - \square = {}^{-}15$ **b.** $^{+}11 - ? = {}^{-}2$ **c.** $^{-}4 - n = {}^{+}12$ **d.** $^{-}24 - ? = 0$ **e.** $0 - n = {}^{-}19$ **f.** $^{-}35 - \square = {}^{+}25$ **g.** $^{+}100 - n = {}^{+}275$ **h.** $^{-}49 - n = {}^{-}50$ **i.** $^{+}63 - n = {}^{-}36$

Properties

In the following exercises we determine the properties that hold for the set of integers under the operation of subtraction.

1. Commutative Property

a. If the commutative property holds for the subtraction of integers, what would have to be true?

b. In each of the following subtract as indicated, then compare answers:

(1) $^{+}9 - {}^{+}5$; $^{+}5 - {}^{+}9$. Does $^{+}9 - {}^{+}5 = {}^{+}5 - {}^{+}9$?

(2) $^{-}16 - {}^{-}21$; $^{-}21 - {}^{-}16$. Does $^{-}16 - {}^{-}21 = {}^{-}21 - {}^{-}16$?

(3) $^{+}7 - {}^{-}9$; $^{-}9 - {}^{+}7$. Does $^{+}7 - {}^{-}9 = {}^{-}9 - {}^{+}7$?

(4) $^{-}6 - {}^{+}3$; $^{+}3 - {}^{-}6$. Does $^{-}6 - {}^{+}3 = {}^{+}3 - {}^{-}6$?

c. Does the commutative property hold for the subtraction of integers?

2. Associative Property

a. If the associative property holds for the subtraction of integers, what would have to be true?

b. In each of the following subtract as indicated, then compare answers:

(1) $(^{-}10 - {}^{+}8) - {}^{-}6$; $\quad {}^{-}10 - ({}^{+}8 - {}^{-}6)$.
Does $(^{-}10 - {}^{+}8) - {}^{-}6 = {}^{-}10 - ({}^{+}8 - {}^{-}6)$?

(2) $(^{-}3 - {}^{-}9) - {}^{+}2$; $\quad {}^{-}3 - ({}^{-}9 - {}^{+}2)$.
Does $(^{-}3 - {}^{-}9) - {}^{+}2 = {}^{-}3 - ({}^{-}9 - {}^{+}2)$?

c. Does the associative property hold for the subtraction of integers?

3. Closure

a. What do we mean when we say that a certain set of numbers is closed under the operation of subtraction?

b. When you subtract in each of the following, is the difference in each case an integer?

(1) $^{-}8 - {}^{-}1$ (4) $^{+}12 - {}^{-}12$ (7) $4 - 9$

(2) $^{+}7 - {}^{+}10$ (5) $0 - {}^{+}5$ (8) $^{+}3 - {}^{-}11$

(3) $^{-}1 - {}^{+}6$ (6) $^{-}2 - 0$ (9) $^{-}6 - {}^{-}6$

c. Can you find two integers for which the difference is not an integer?

d. Is the difference of two integers always an integer?

e. Can we say that the set of integers is closed under the operation of subtraction?

f. Is the set $\{^{-}1, 0, 1\}$ closed under the operation of subtraction?

11–14 SUBTRACTION OF RATIONAL NUMBERS; PROPERTIES

Examination of subtraction on the number line in section 11–12 and the development in section 11–13, that subtracting a number and adding the opposite of the number which is to be subtracted both give the same answer, lead us to conclude that:

To subtract any rational number from another rational number, we add to the minuend the opposite (or additive inverse) of the rational number which is to be subtracted. The problem first could be rewritten, arranged horizontally or vertically. For example:

Subtract: ${}^{+}\frac{1}{5} - {}^{+}\frac{2}{5} = ?$

Horizontally:

$${}^{+}\frac{1}{5} - {}^{+}\frac{2}{5} = {}^{+}\frac{1}{5} + {}^{-}\frac{2}{5} = \frac{{}^{+}1 + {}^{-}2}{5} = {}^{-}\frac{1}{5}$$

Vertically:

$$\begin{array}{ccc} \text{Subtract:} & & \text{Add:} \\ {}^{+}\frac{1}{5} & & {}^{+}\frac{1}{5} \\ \underline{{}^{+}\frac{2}{5}} & \longrightarrow & \underline{{}^{-}\frac{2}{5}} \\ & & {}^{-}\frac{1}{5} \end{array}$$

Answer, ${}^{-}\frac{1}{5}$

Subtract: ${}^{-}3\frac{1}{2} - {}^{+}2\frac{3}{8} = ?$

Horizontally:

$${}^{-}3\tfrac{1}{2} - {}^{+}2\tfrac{3}{8} = {}^{-}3\tfrac{1}{2} + {}^{-}2\tfrac{3}{8} = {}^{-}3\tfrac{4}{8} + {}^{-}2\tfrac{3}{8} = {}^{-}5\tfrac{7}{8}$$

Vertically:

$$\begin{array}{lcl} \text{Subtract:} & & \text{Add:} \\ {}^{-}3\frac{1}{2} & & {}^{-}3\frac{1}{2} = {}^{-}3\frac{4}{8} \\ \underline{{}^{+}2\frac{3}{8}} & \longrightarrow & \underline{{}^{-}2\frac{3}{8}} = \underline{{}^{-}2\frac{3}{8}} \\ & & \qquad\quad {}^{-}5\frac{7}{8} \end{array}$$

Answer, ${}^{-}5\frac{7}{8}$

Subtract: $0 - {}^{-}6.8 = ?$

Horizontally:

$$0 - {}^{-}6.8 = 0 + {}^{+}6.8 = {}^{+}6.8$$

Vertically:

$$\begin{array}{rcr} \text{Subtract:} & & \text{Add:} \\ 0 & & 0 \\ \underline{{}^{-}6.8} & \longrightarrow & \underline{{}^{+}6.8} \\ & & {}^{+}6.8 \end{array}$$

Answer, ${}^{+}6.8$

It may be necessary to review at this time both the addition and the subtraction of the rational numbers of arithmetic (common fractions, mixed numbers, and decimal fractions). See sections 5–11, 5–12, 6–7, and 6–8.

EXERCISES

1. a. ${}^{+}\frac{2}{3} - {}^{+}\frac{1}{3}$ **b.** ${}^{+}\frac{5}{6} - {}^{+}\frac{7}{12}$ **c.** ${}^{+}4\frac{7}{8} - {}^{+}1\frac{3}{8}$ **d.** ${}^{+}6\frac{1}{3} - {}^{+}3\frac{3}{4}$

2. a. ${}^{+}\frac{1}{9} - {}^{+}\frac{5}{9}$ **b.** ${}^{+}\frac{9}{16} - {}^{+}\frac{3}{4}$ **c.** ${}^{+}1\frac{1}{2} - {}^{+}3\frac{4}{5}$ **d.** ${}^{+}\frac{11}{2} - {}^{+}1\frac{2}{3}$

3. a. ${}^{+}.63 - {}^{+}.59$ **b.** ${}^{+}.8 - {}^{+}.08$ **c.** ${}^{+}1.6 - {}^{+}.19$ **d.** ${}^{+}.832 - {}^{+}.83$

4. a. ${}^{+}.85 - {}^{+}1.2$ **b.** ${}^{+}.62 - {}^{+}.7$ **c.** ${}^{+}.43 - {}^{+}4.3$ **d.** ${}^{+}.1689 - {}^{+}.169$

5. a. ${}^{-}\frac{5}{7} - {}^{-}\frac{3}{7}$ **b.** ${}^{-}\frac{5}{8} - {}^{-}\frac{1}{6}$ **c.** ${}^{-}2\frac{2}{3} - {}^{-}2\frac{1}{4}$ **d.** ${}^{-}7\frac{1}{2} - {}^{-}5\frac{11}{12}$

6. a. ${}^{-}\frac{1}{10} - {}^{-}\frac{7}{10}$ **b.** ${}^{-}\frac{3}{4} - {}^{-}\frac{4}{5}$ **c.** ${}^{-}1\frac{1}{6} - {}^{-}5\frac{3}{8}$ **d.** ${}^{-}4\frac{2}{3} - {}^{-}6\frac{11}{16}$

7. a. ${}^{-}8.6 - {}^{-}1.4$ **b.** ${}^{-}9 - {}^{-}.83$ **c.** ${}^{-}.106 - {}^{-}.0081$ **d.** ${}^{-}25.6 - {}^{-}.907$

8. a. ${}^{-}.05 - {}^{-}.2$ **b.** ${}^{-}4.8 - {}^{-}11.4$ **c.** ${}^{-}6.96 - {}^{-}27.1$ **d.** ${}^{-}.0004 - {}^{-}.1$

9. a. ${}^{+}\frac{13}{16} - {}^{-}\frac{3}{16}$ **b.** ${}^{+}\frac{5}{6} - {}^{-}\frac{1}{8}$ **c.** ${}^{+}3\frac{1}{2} - {}^{-}1\frac{2}{3}$ **d.** ${}^{+}1\frac{11}{12} - {}^{-}4\frac{3}{4}$

10. a. ${}^{+}.25 - {}^{-}.2$ **b.** ${}^{+}8.9 - {}^{-}.105$ **c.** ${}^{+}4 - {}^{-}.04$ **d.** ${}^{+}.517 - {}^{-}.25$

11. a. ${}^{-}\frac{4}{5} - {}^{+}\frac{2}{3}$ **b.** ${}^{-}\frac{9}{16} - {}^{+}\frac{7}{12}$ **c.** ${}^{-}8\frac{1}{6} - {}^{+}7\frac{5}{8}$ **d.** ${}^{-}10\frac{1}{2} - {}^{+}6\frac{29}{32}$

12. **a.** ${}^{-}.79 - {}^{+}.157$ **b.** ${}^{-}9.584 - {}^{+}1.269$ **c.** ${}^{-}9.47 - {}^{+}.9$ **d.** ${}^{-}2.7 - {}^{+}5$

13. **a.** ${}^{-}\frac{11}{12} - 0$ **b.** ${}^{+}4\frac{2}{3} - 0$ **c.** ${}^{+}2.8 - 0$ **d.** ${}^{-}.039 - 0$

14. **a.** $0 - {}^{-}\frac{19}{25}$ **b.** $0 - {}^{+}8\frac{15}{16}$ **c.** $0 - {}^{+}.735$ **d.** $0 - {}^{-}5.26$

15. **a.** ${}^{-}\frac{9}{11} - {}^{-}\frac{9}{11}$ **b.** ${}^{+}5\frac{3}{4} - {}^{+}5\frac{3}{4}$ **c.** ${}^{-}.6 - {}^{-}.6$ **d.** ${}^{+}7.834 - {}^{+}7.834$

16. **a.** $\frac{3}{8} - \frac{7}{8}$ **b.** $2\frac{13}{16} - 9$ **c.** $.56 - 1.3$ **d.** $.092 - .25$

17. **a.** $({}^{-}\frac{1}{4} - {}^{-}\frac{2}{3}) - {}^{-}\frac{1}{6}$ **c.** ${}^{-}1\frac{1}{3} - ({}^{-}2\frac{2}{5} - {}^{+}4\frac{3}{4})$ **e.** $({}^{+}.07 - {}^{-}.4) - {}^{-}.203$
b. ${}^{+}6\frac{1}{2} - ({}^{-}3\frac{5}{8} - {}^{-}4\frac{11}{16})$ **d.** $({}^{-}3.5 - {}^{+}8.2) - {}^{+}5.4$ **f.** ${}^{+}.632 - ({}^{+}.92 - {}^{+}.2)$

18. Subtract in each of the following:

a.	${}^{+}\frac{5}{8}$	${}^{-}\frac{1}{2}$	${}^{+}\frac{3}{8}$	${}^{-}4\frac{3}{4}$	${}^{-}5$	${}^{+}6\frac{5}{6}$	$2\frac{3}{10}$
	${}^{+}\frac{1}{8}$	${}^{-}\frac{5}{6}$	${}^{-}\frac{1}{5}$	${}^{+}\frac{15}{16}$	${}^{-}1\frac{11}{12}$	${}^{-}3\frac{1}{4}$	$8\frac{2}{3}$
b.	$6\frac{1}{3}$	${}^{+}5\frac{1}{3}$	${}^{-}1\frac{9}{16}$	${}^{-}6\frac{1}{2}$	$3\frac{7}{8}$	${}^{+}6\frac{7}{12}$	${}^{-}10\frac{7}{32}$
	$9\frac{2}{3}$	${}^{-}3\frac{2}{5}$	${}^{-}2\frac{5}{8}$	${}^{+}4\frac{5}{6}$	$7\frac{11}{12}$	${}^{-}5\frac{13}{16}$	${}^{-}2\frac{3}{4}$
c.	${}^{-}.8$	${}^{+}1.6$	$.035$	${}^{-}7.495$	${}^{-}10.27$	2.87	${}^{-}.67$
	${}^{-}.9$	${}^{-}.7$	$.04$	${}^{+}1.53$	${}^{-}.8$	5	${}^{-}.8924$

19. Find the missing numbers:

a. ${}^{+}\frac{4}{5} - {}^{+}\frac{1}{5} = ?$ **c.** ${}^{+}1\frac{1}{2} - {}^{+}\frac{5}{6} = \square$ **e.** ${}^{-}6\frac{1}{3} - {}^{-}7\frac{5}{12} = n$
b. ${}^{-}\frac{11}{16} - {}^{-}\frac{13}{20} = n$ **d.** ${}^{+}2\frac{7}{8} - {}^{-}1 = n$ **f.** $4\frac{5}{8} - 9\frac{3}{16} = n$

20. Find the missing numbers:

a. ${}^{+}.6 - {}^{-}.1 = \square$ **c.** ${}^{-}8 - {}^{+}.07 = ?$ **e.** ${}^{+}5.07 - {}^{-}.829 = n$
b. ${}^{-}8.7 - {}^{-}4.9 = n$ **d.** $.35 - .91 = n$ **f.** ${}^{-}.8 - {}^{-}3.58 = n$

21. Find the missing numbers:

a. $? - {}^{-}\frac{5}{16} = {}^{+}\frac{13}{16}$ **c.** $\square - {}^{-}2\frac{1}{4} = {}^{-}6\frac{1}{2}$ **e.** $n - {}^{-}7\frac{5}{8} = 10$
b. $n - {}^{+}\frac{5}{6} = {}^{-}\frac{1}{3}$ **d.** $n - 5\frac{7}{12} = 3\frac{5}{6}$ **f.** $n - 1\frac{3}{4} = \frac{11}{16}$

22. Find the missing numbers:

a. $? - {}^{-}.4 = {}^{+}.7$ **c.** $n - {}^{-}.59 = {}^{-}1.2$ **e.** $n - {}^{-}.225 = {}^{-}.001$
b. $n - {}^{+}1.6 = {}^{-}8$ **d.** $\square - {}^{+}6.3 = {}^{-}3.27$ **f.** $n - 30.8 = 6.54$

23. Find the missing numbers:

a. ${}^{+}\frac{1}{12} - \square = {}^{+}\frac{7}{12}$ **c.** ${}^{-}8\frac{1}{2} - ? = {}^{-}4\frac{3}{8}$ **e.** ${}^{-}3\frac{11}{16} - n = {}^{-}4\frac{2}{3}$
b. ${}^{-}\frac{9}{10} - n = {}^{+}\frac{2}{5}$ **d.** $11 - n = {}^{-}\frac{5}{6}$ **f.** ${}^{-}9\frac{5}{12} - n = 5\frac{3}{4}$

24. Find the missing numbers:

a. ${}^{-}.7 - ? = 0$ **c.** $.925 - \square = 7.6$ **e.** ${}^{-}16.2 - n = 6.8$
b. ${}^{+}4.2 - n = {}^{-}3.1$ **d.** ${}^{-}.489 - n = {}^{-}5$ **f.** $.375 - n = .2635$

Properties

In the following exercises we determine the properties that hold for the set of rational numbers under the operation of subtraction.

1. Commutative Property

a. If the commutative property holds for the subtraction of rational numbers, what would have to be true?

b. In each of the following subtract as indicated, then compare answers:

(1) $^{-}\frac{2}{3} - {}^{+}\frac{1}{6}$; $^{+}\frac{1}{6} - {}^{-}\frac{2}{3}$. Does $^{-}\frac{2}{3} - {}^{+}\frac{1}{6} = {}^{+}\frac{1}{6} - {}^{-}\frac{2}{3}$?

(2) $^{-}1\frac{3}{8} - {}^{-}2\frac{1}{4}$; $^{-}2\frac{1}{4} - {}^{-}1\frac{3}{8}$. Does $^{-}1\frac{3}{8} - {}^{-}2\frac{1}{4} = {}^{-}2\frac{1}{4} - {}^{-}1\frac{3}{8}$?

(3) $^{+}6 - {}^{+}\frac{13}{16}$; $^{+}\frac{13}{16} - {}^{+}6$. Does $^{+}6 - {}^{+}\frac{13}{16} = {}^{+}\frac{13}{16} - {}^{+}6$?

(4) $^{+}8.7 - {}^{-}9.3$; $^{-}9.3 - {}^{+}8.7$. Does $^{+}8.7 - {}^{-}9.3 = {}^{-}9.3 - {}^{+}8.7$?

c. Does the commutative property hold for the subtraction of rational numbers?

2. Associative Property

a. If the associative property holds for the subtraction of rational numbers, what would have to be true?

b. In each of the following subtract as indicated, then compare answers:

(1) $({}^{+}\frac{4}{5} - {}^{-}\frac{7}{10}) - {}^{-}\frac{1}{2}$; $^{+}\frac{4}{5} - ({}^{-}\frac{7}{10} - {}^{-}\frac{1}{2})$.
Does $({}^{+}\frac{4}{5} - {}^{-}\frac{7}{10}) - {}^{-}\frac{1}{2} = {}^{+}\frac{4}{5} - ({}^{-}\frac{7}{10} - {}^{-}\frac{1}{2})$?

(2) $({}^{-}2\frac{3}{4} - {}^{-}1\frac{5}{8}) - {}^{+}\frac{11}{12}$; $^{-}2\frac{3}{4} - ({}^{-}1\frac{5}{8} - {}^{+}\frac{11}{12})$.
Does $({}^{-}2\frac{3}{4} - {}^{-}1\frac{5}{8}) - {}^{+}\frac{11}{12} = {}^{-}2\frac{3}{4} - ({}^{-}1\frac{5}{8} - {}^{+}\frac{11}{12})$?

c. Does the associative property hold for the subtraction of rational numbers?

3. Closure

a. What do we mean when we say that a certain set of numbers is closed under the operation of subtraction?

b. When you subtract in each of the following, is the difference in each case a rational number?

(1) $^{+}\frac{3}{4} - {}^{-}\frac{5}{6}$
(2) $^{-}\frac{7}{12} - {}^{+}1\frac{1}{4}$
(3) $^{-}3\frac{1}{2} - {}^{-}5\frac{1}{3}$
(4) $^{-}5\frac{3}{8} - 0$
(5) $^{-}.07 - {}^{-}.003$
(6) $^{+}8\frac{7}{16} - {}^{-}8\frac{7}{16}$
(7) $^{+}1\frac{2}{5} - {}^{+}3\frac{2}{3}$
(8) $0 - {}^{-}4\frac{3}{7}$
(9) $^{-}3.8 - {}^{+}.17$

c. Can you find two rational numbers for which the difference is not a rational number?

d. Can we say that the set of rational numbers is closed under the operation of subtraction?

11–15 MULTIPLICATION OF INTEGERS; PROPERTIES

To develop the procedure of determining the product of two positive integers (like $^{+}4 \times {}^{+}5$), or of two negative integers (like $^{-}4 \times {}^{-}5$), or of a positive integer and a negative integer (like $^{+}4 \times {}^{-}5$), or of a negative integer and a positive integer (like $^{-}4 \times {}^{+}5$), let us analyze the following:

Representing a bank deposit of \$5 by $^{+}5$, a withdrawal of \$5 by $^{-}5$, 4 weeks from now by $^{+}4$, and 4 weeks ago by $^{-}4$, let us find the answers to the following situations using signed numbers.

a. If we deposit in our bank account \$5 each week for 4 weeks, how will our bank account 4 weeks from now compare with our present bank account?

There will be \$20 more in the account 4 weeks from now. $^{+}4 \times {}^{+}5 = {}^{+}20$

b. If we deposited in our bank account \$5 each week for the past 4 weeks, how did our bank account 4 weeks ago compare with our present bank account?

There was \$20 less in the account 4 weeks ago. $^{-}4 \times {}^{+}5 = {}^{-}20$

c. If we withdraw from our bank account \$5 each week for the next 4 weeks, how will our bank account 4 weeks from now compare with our present bank account?

There will be \$20 less in the account 4 weeks from now. $^{+}4 \times {}^{-}5 = {}^{-}20$

d. If we withdrew from our bank account \$5 each week for the past 4 weeks, how did our bank account 4 weeks ago compare with our present bank account?

There was \$20 more in the account 4 weeks ago. $^{-}4 \times {}^{-}5 = {}^{+}20$

From the above it appears that the product of two positive integers is a positive integer, the product of two negative integers is a positive integer, and the product of a positive integer and a negative integer or a negative integer and a positive integer is a negative integer. Let us check this further.

Multiplication may be thought of as repeated addition. *Four times three*, expressed as 4×3, is the sum of four 3's: $(3 + 3 + 3 + 3)$ or 12. The expression four times three may be written as 4×3, $4 \cdot 3$, $(4) \times (3)$, $4(3)$, $(4)3$, or $(4)(3)$. Observe that the $\times$ symbol may be replaced by a raised dot or the factors may be written next to each other within parentheses without any multiplication symbol.

Since a numeral without a sign represents a positive number, 4×3 may be expressed as $^{+}4 \times {}^{+}3$ or $(^{+}4)(^{+}3)$.

Thus, $\quad 4 \times 3$ or $^{+}4 \times {}^{+}3 = {}^{+}3 + {}^{+}3 + {}^{+}3 + {}^{+}3 = {}^{+}12.$

That is: $^{+}4 \times {}^{+}3 = {}^{+}12$

This is illustrated on the number line as:

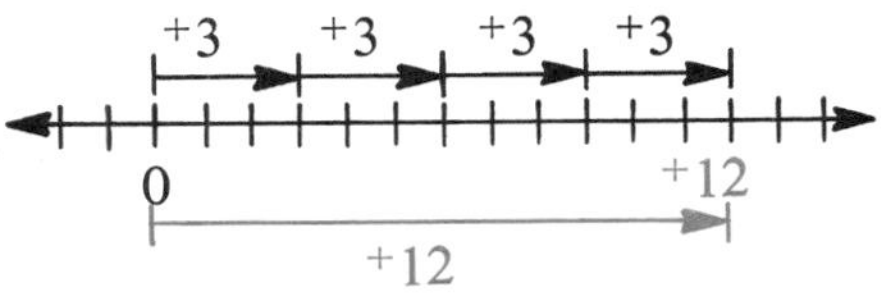

Thus, the product of two positive integers is a positive integer whose absolute value is the product of the absolute values of the two given factors.

The product of zero and any integer is zero.

$^{-}3 \times 0 = 0$	$^{+}8 \times 0 = 0$
$0 \times {}^{+}7 = 0$	$0 \times {}^{-}2 = 0$

The indicated product $^{+}4 \times {}^{-}3$ means four $^{-}3$'s, or the sum of $^{-}3 + {}^{-}3 + {}^{-}3 + {}^{-}3$, which is $^{-}12$.

That is: $^{+}4 \times {}^{-}3 = {}^{-}12$

This is illustrated on the number line as:

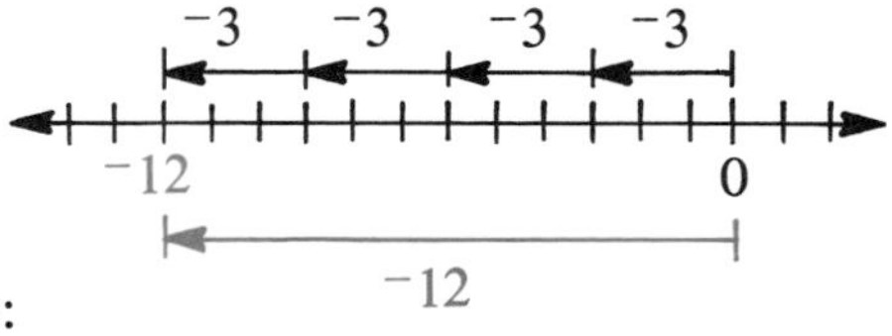

Let us examine the following sequence:

$$
\begin{aligned}
{}^{+}4 \times {}^{+}3 &= {}^{+}12\\
{}^{+}4 \times {}^{+}2 &= {}^{+}8\\
{}^{+}4 \times {}^{+}1 &= {}^{+}4\\
{}^{+}4 \times 0 &= 0\\
{}^{+}4 \times {}^{-}1 &= {}^{-}4\\
{}^{+}4 \times {}^{-}2 &= {}^{-}8\\
{}^{+}4 \times {}^{-}3 &= {}^{-}12
\end{aligned}
$$

Observe that each time the second factor decreases 1, the product decreases 4.

Thus, the product $^{+}4 \times {}^{-}1$ should be 4 less than 0, the product of $^{+}4 \times 0$. This is $^{-}4$.

The product of $^{+}4 \times {}^{-}2$ should be 4 less than $^{-}4$, the product of $^{+}4 \times {}^{-}1$. This is $^{-}8$.

The product of $^{+}4 \times {}^{-}3$ should be 4 less than $^{-}8$, the product of $^{+}4 \times {}^{-}2$. This is $^{-}12$.

Thus, $^{+}4 \times {}^{-}3 = {}^{-}12$.

This also illustrates that $^{+}4 \times {}^{-}3 = {}^{-}12$.

Thus, the product of a positive integer and a negative integer is a negative integer whose absolute value is the product of the absolute values of the two given factors.

Let us examine the following sequence:

$$
\begin{aligned}
{}^{+}4 \times {}^{+}3 &= {}^{+}12\\
{}^{+}3 \times {}^{+}3 &= {}^{+}9\\
{}^{+}2 \times {}^{+}3 &= {}^{+}6\\
{}^{+}1 \times {}^{+}3 &= {}^{+}3\\
0 \times {}^{+}3 &= 0\\
{}^{-}1 \times {}^{+}3 &= {}^{-}3\\
{}^{-}2 \times {}^{+}3 &= {}^{-}6\\
{}^{-}3 \times {}^{+}3 &= {}^{-}9\\
{}^{-}4 \times {}^{+}3 &= {}^{-}12
\end{aligned}
$$

Observe that each time the first factor decreases 1, the product decreases 3.

Thus, the product of $^{-}1 \times {}^{+}3$ should be 3 less than 0, the product of $0 \times {}^{+}3$. This is $^{-}3$.

The product of $^{-}2 \times {}^{+}3$ should be 3 less than $^{-}3$, the product of $^{-}1 \times {}^{+}3$. This is $^{-}6$.

The product of $^{-}3 \times {}^{+}3$ should be 3 less than $^{-}6$, the product of $^{-}2 \times {}^{+}3$. This is $^{-}9$.

The product of $^{-}4 \times {}^{+}3$ should be 3 less than $^{-}9$, the product of $^{-}3 \times {}^{+}3$. This is $^{-}12$.

Thus, $^{-}4 \times {}^{+}3 = {}^{-}12$.

This illustrates that $^{-}4 \times {}^{+}3 = {}^{-}12$.

Thus, the product of a negative integer and a positive integer is a negative integer whose absolute value is the product of the absolute values of the two given factors.

Let us examine the following sequence:

$$
\begin{aligned}
{}^{-}4 \times {}^{+}3 &= {}^{-}12\\
{}^{-}4 \times {}^{+}2 &= {}^{-}8
\end{aligned}
$$

Observe that each time the second factor decreases 1, the product increases 4.

$^-4 \times {}^+1 = {}^-4$	Thus, the product of $^-4 \times {}^-1$ should be 4 more than 0, the product $^-4 \times 0$. This is $^+4$.
$^-4 \times 0 = 0$	The product of $^-4 \times {}^-2$ should be 4 more than $^+4$, the product of $^-4 \times {}^-1$. This is $^+8$.
$^-4 \times {}^-1 = {}^+4$	The product $^-4 \times {}^-3$ should be 4 more than $^+8$, the product of $^-4 \times {}^-2$. This is $^+12$.
$^-4 \times {}^-2 = {}^+8$	Thus, $^-4 \times {}^-3 = {}^+12$.
$^-4 \times {}^-3 = {}^+12$	

This illustrates that $^-4 \times {}^-3 = {}^+12$.

Thus, the product of two negative integers is a positive integer whose absolute value is the product of the absolute values of the two given factors.

Summarizing, *to multiply two positive integers* or *two negative integers*, we find the product of their absolute values and prefix the numeral for this product with a positive sign.

Multiply: $^+7 \times {}^+6 = ?$

$^+7 \times {}^+6 = {}^+(7 \times 6) = {}^+42$ or $\begin{array}{r} ^+7 \\ ^+6 \\ \hline ^+42 \end{array}$

Answer, $^+42$

Multiply: $^-4 \times {}^-9 = ?$

$^-4 \times {}^-9 = {}^+(4 \times 9) = {}^+36$ or $\begin{array}{r} ^-4 \\ ^-9 \\ \hline ^+36 \end{array}$

Answer, $^+36$

To multiply a positive integer by a negative integer or *a negative integer by a positive integer*, we find the product of their absolute values and prefix the numeral for this product with a negative sign.

Multiply: $^+8 \times {}^-3 = ?$

$^+8 \times {}^-3 = {}^-(8 \times 3) = {}^-24$ or $\begin{array}{r} ^+8 \\ ^-3 \\ \hline ^-24 \end{array}$

Answer, $^-24$

Multiply: $^-5 \times {}^+7 = ?$

$^-5 \times {}^+7 = {}^-(5 \times 7) = {}^-35$ or $\begin{array}{r} ^-5 \\ ^+7 \\ \hline ^-35 \end{array}$

Answer, $^-35$

The numeral naming a positive integer does not require the positive ($^+$) sign.

$^-8 \times 7 = {}^-56$ Answer, $^-56$

The product of zero and any integer is zero.

$0 \times {}^-4 = 0$ Answer, 0

$^+6 \times 0 = 0$ Answer, 0

To multiply three or more factors, we multiply successively. Observe that an odd number of negative factors produces a negative number as the product, while an even number of negative factors produces a positive number as the product, provided zero is not a factor. In each case the absolute value of the product is equal to the product of the absolute values of the given factors.

> Multiply:
> $(^{-}2)(^{+}4)(^{-}1)(^{-}6) = {^{-}48}$
> *Answer,* $^{-}48$
> $(^{-}1)(^{-}3)(^{-}2)(^{-}4) = {^{+}24}$
> *Answer,* $^{+}24$

EXERCISES

Multiply as indicated:

1. **a.** $^{+}9 \times {^{+}7}$ **b.** $^{+}5 \times {^{+}11}$ **c.** $^{+}3 \times {^{+}25}$ **d.** $^{+}42 \times {^{+}58}$

2. **a.** $^{-}8 \times {^{-}6}$ **b.** $^{-}2 \times {^{-}16}$ **c.** $^{-}30 \times {^{-}14}$ **d.** $^{-}1 \times {^{-}125}$

3. **a.** $^{+}4 \times {^{-}9}$ **b.** $^{+}15 \times {^{-}6}$ **c.** $^{+}27 \times {^{-}100}$ **d.** $^{+}64 \times {^{-}75}$

4. **a.** $^{-}7 \times {^{+}8}$ **b.** $^{-}4 \times {^{+}17}$ **c.** $^{-}10 \times {^{+}36}$ **d.** $^{-}53 \times {^{+}29}$

5. **a.** $^{-}6 \times 6$ **b.** $^{-}12 \times 5$ **c.** $8 \times {^{-}47}$ **d.** $34 \times {^{-}18}$

6. **a.** $0 \times {^{-}19}$ **b.** $^{+}28 \times 0$ **c.** $^{-}16 \times 0$ **d.** $0 \times {^{+}41}$

7. **a.** $(^{-}1)(^{-}2)$ **b.** $(^{-}9)(^{+}5)$ **c.** $(^{+}11)(^{-}4)$ **d.** $(^{-}3)(^{-}10)$ **e.** $(^{+}8)(^{+}8)$
f. $(^{-}4)(^{-}5)$ **g.** $(0)(^{-}1)$ **h.** $(^{-}14)(3)$ **i.** $(^{-}6)(^{-}12)$ **j.** $(8)(^{-}7)$
k. $(^{-}13)(^{-}13)$ **l.** $(^{-}45)(20)$ **m.** $(106)(^{-}51)$ **n.** $(^{-}59)(^{-}200)$ **o.** $(^{-}325)(^{-}198)$
p. $^{-}7(^{-}4)$ **q.** $^{-}9(10)$ **r.** $8(^{-}5)$ **s.** $12(^{-}6)$ **t.** $^{-}16(^{-}12)$

8. Multiply in each of the following:

$^{-}4$	$^{+}8$	$^{-}9$	$^{+}6$	$^{-}14$	$^{+}11$	$^{-}1$	$^{-}25$
$\underline{^{-}3}$	$\underline{^{-}5}$	$\underline{^{+}9}$	$\underline{^{+}7}$	$\underline{^{-}6}$	$\underline{^{-}8}$	$\underline{^{-}4}$	$\underline{^{+}2}$
$^{+}7$	$^{-}19$	0	$^{-}45$	18	$^{-}67$	$^{-}29$	$^{-}100$
$\underline{^{-}12}$	$\underline{^{-}9}$	$\underline{^{-}23}$	$\underline{^{+}1}$	$\underline{^{-}6}$	$\underline{34}$	$\underline{^{-}48}$	$\underline{^{-}100}$

9. **a.** $(^{-}4 \times {^{-}6}) \times {^{+}2}$ **b.** $(^{+}8 \times {^{-}3}) \times {^{-}1}$ **c.** $^{+}7 \times (^{-}2 \times {^{+}5})$ **d.** $^{-}3 \times (^{-}9 \times {^{-}2})$ **e.** $(^{-}4 \times {^{-}4}) \times {^{-}4}$ **f.** $^{+}8 \times (^{-}5 \times {^{-}3})$

10. **a.** $^{+}5 \times {^{-}8} \times {^{+}3}$ **b.** $^{+}7 \times {^{-}4} \times {^{-}5}$ **c.** $^{-}10 \times {^{-}10} \times {^{-}10}$ **d.** $^{-}15 \times {^{+}4} \times {^{-}1}$ **e.** $^{-}3 \times {^{+}6} \times {^{-}1} \times {^{-}2}$ **f.** $^{-}2 \times {^{-}2} \times {^{-}2} \times {^{-}2}$

11. **a.** $(^{-}2)(3)(^{-}1)$ **b.** $(^{-}1)(^{-}1)(^{-}1)$ **c.** $(5)(2)(^{-}3)(4)$ **d.** $(^{-}3)(6)(^{-}4)(^{-}2)$ **e.** $(4)(^{-}3)(2)(^{-}1)(5)$ **f.** $(^{-}2)(^{-}3)(^{-}1)(^{-}3)(^{-}4)$

12. Find the missing numbers:

a. $^{-}9 \times {^{-}6} = ?$ **b.** $7 \times {^{-}3} = \square$ **c.** $^{-}8 \times 4 = n$
d. $6 \times {^{-}10} = \square$ **e.** $^{-}15 \times {^{-}7} = ?$ **f.** $^{-}12 \times 9 = n$
g. $20 \times {^{-}1} = n$ **h.** $^{-}36 \times 8 = n$ **i.** $^{-}3 \times {^{-}32} = n$

13. Find the missing numbers:

a. $3 \times ? = {^{-}12}$ **b.** $5 \times n = {^{-}30}$ **c.** $^{-}12 \times \square = {^{-}72}$

d. $^-1 \times \square = 16$
e. $9 \times ? = {}^-63$
f. $^-8 \times n = 0$
g. $^-20 \times n = {}^-100$
h. $7 \times n = 42$
i. $^-2 \times n = 34$

14. Find the missing numbers:

a. $? \times 4 = {}^-20$
b. $\square \times {}^-3 = 30$
c. $n \times {}^-6 = {}^-48$
d. $\square \times {}^-8 = 64$
e. $n \times {}^-5 = {}^-45$
f. $? \times 12 = 72$
g. $n \times {}^-7 = 56$
h. $n \times 2 = {}^-66$
i. $n \times {}^-25 = {}^-200$

Properties

In the following exercises we determine the properties that hold for the set of integers under the operation of multiplication.

1. Commutative Property

a. What do we mean when we say that the commutative property holds for the multiplication of a certain set of numbers?

b. In each of the following multiply as indicated, then compare the products:

(1) $^+7 \times {}^+5$; $^+5 \times {}^+7$. Does $^+7 \times {}^+5 = {}^+5 \times {}^+7$?
(2) $^-6 \times {}^-8$; $^-8 \times {}^-6$. Does $^-6 \times {}^-8 = {}^-8 \times {}^-6$?
(3) $^+12 \times {}^-3$; $^-3 \times {}^+12$. Does $^+12 \times {}^-3 = {}^-3 \times {}^+12$?
(4) $^-9 \times {}^+9$; $^+9 \times {}^-9$. Does $^-9 \times {}^+9 = {}^+9 \times {}^-9$?

c. Can you find two integers for which the commutative property of multiplication does not hold?

d. Can we say that the commutative property holds for the multiplication of integers?

2. Associative Property

a. What do we mean when we say that the associative property holds for the multiplication of a certain set of numbers?

b. In each of the following multiply as indicated, then compare the products:

(1) $({}^+8 \times {}^-2) \times {}^-3$; $\quad {}^+8 \times ({}^-2 \times {}^-3)$.
Does $({}^+8 \times {}^-2) \times {}^-3 = {}^+8 \times {}^-2 \times {}^-3$?
(2) $({}^-5 \times {}^-9) \times {}^+7$; $\quad {}^-5 \times ({}^-9 \times {}^+7)$.
Does $({}^-5 \times {}^-9) \times {}^+7 = {}^-5 \times ({}^-9 \times {}^+7)$?

c. Can you find three integers for which the associative property of multiplication does not hold?

d. Can we say that the associative property holds for the multiplication of integers?

3. Distributive Property

a. What do we mean when we say that the distributive property of multiplication over addition holds for the set of integers?

b. In each of the following perform the indicated operations, then compare answers:

(1) $^+6 \times (^-7 + {}^+5)$; $(^+6 \times {}^-7) + (^+6 \times {}^+5)$.
Does $^+6 \times (^-7 + {}^+5) = (^+6 \times {}^-7) + (^+6 \times {}^+5)$?

(2) $^-3 \times (^+8 + {}^-9)$; $(^-3 \times {}^+8) + (^-3 \times {}^-9)$.
Does $^-3 \times (^+8 + {}^-9) = (^-3 \times {}^+8) + (^-3 \times {}^-9)$?

c. Can you find any integers for which multiplication is not distributed over addition?

d. Can we say that the distributive property of multiplication over addition holds for the set of integers?

4. Closure

a. What do we mean when we say that a certain set of numbers is closed under the operation of multiplication?

b. Is the product of each of the following an integer?

(1) $^-4 \times {}^+2$	(4) $^+6 \times {}^+12$	(7) $0 \times {}^+3$
(2) $^+10 \times {}^-9$	(5) $^-7 \times 0$	(8) $^-11 \times {}^-11$
(3) $^-8 \times {}^-8$	(6) $^-1 \times {}^-15$	(9) $^-24 \times {}^+1$

c. Can you find two integers for which the product is not an integer?

d. Is the product of two integers always an integer?

e. Can we say that the set of integers is closed under the operation of multiplication?

5. Multiplicative Identity

a. What do we mean when we say that a certain set of numbers has a multiplicative identity?

b. What is the product of each of the following?

(1) $^-9 \times 1$	(3) $^+13 \times 1$	(5) $^-1 \times 1$	(7) $1 \times {}^+35$
(2) $1 \times {}^+7$	(4) $1 \times {}^-20$	(6) 1×0	(8) $^-100 \times 1$

c. In each case in problem **5b** is the product of the integer and one (1) the integer itself?

d. Can you find any integer which when multiplied by one (1) does not have the given integer as the product?

e. Does the set of integers have a multiplicative identity? If so, what is it?

11–16 MULTIPLICATION OF RATIONAL NUMBERS; PROPERTIES

We follow the same procedures to multiply rational numbers as we do to multiply integers. See section 11–15. It may be necessary to review at this time the multiplication of the rational numbers of arithmetic (common fractions, mixed numbers, and decimal fractions). See sections 5–13 and 6–9.

To multiply two positive rational numbers or two negative rational numbers, we find the product of their absolute values and prefix the numeral for this product with a positive sign.

Multiply: $^{+}\frac{1}{3} \times {}^{+}\frac{5}{8} = ?$

$^{+}\frac{1}{3} \times {}^{+}\frac{5}{8} = {}^{+}(\frac{1}{3} \times \frac{5}{8}) = {}^{+}\frac{5}{24}$ *Answer,* $^{+}\frac{5}{24}$

Multiply: $^{-}\frac{3}{4} \times {}^{-}1\frac{1}{2} = ?$

$^{-}\frac{3}{4} \times {}^{-}1\frac{1}{2} = {}^{+}(\frac{3}{4} \times 1\frac{1}{2}) = {}^{+}(\frac{3}{4} \times \frac{3}{2}) = {}^{+}\frac{9}{8} = {}^{+}1\frac{1}{8}$

Answer, $^{+}1\frac{1}{8}$

Multiply: $^{+}2\frac{2}{3} \times {}^{-}1\frac{1}{5} = ?$

$^{+}2\frac{2}{3} \times {}^{-}1\frac{1}{5} = {}^{-}(2\frac{2}{3} \times 1\frac{1}{5}) = {}^{-}(\frac{8}{3} \times \frac{6}{5}) = {}^{-}\frac{48}{15} = {}^{-}3\frac{1}{5}$

Answer, $^{-}3\frac{1}{5}$

Multiply: $^{-}1.2 \times {}^{+}.7 = ?$

$^{-}1.2 \times {}^{+}.7 = {}^{-}(1.2 \times .7) = {}^{-}.84$ *Answer,* $^{-}.84$

To multiply a positive rational number by a negative rational number or a negative rational number by a positive rational number, we find the product of their absolute values and prefix the numeral for this product with a negative sign. The numeral naming a positive rational number does not require the positive ($^{+}$) sign.

The product of zero and any rational number is zero.

$0 \times {}^{-}2\frac{7}{8} = 0$ *Answer,* 0

$^{+}3.095 \times 0 = 0$ *Answer,* 0

To multiply three or more factors, we multiply successively. The absolute value of the product is equal to the product of the absolute values of the given factors. An odd number of negative factors produces a negative number as the product, while an even number of negative factors produces a positive number as the product, provided zero is not a factor.

Multiply:

$(^{-}.6)(^{-}9)(^{-}.02) = {}^{-}.108$ *Answer,* $^{-}.108$

$(^{-}\frac{2}{3})(^{+}\frac{1}{2})(^{-}\frac{3}{4}) = {}^{+}\frac{1}{4}$ *Answer,* $^{+}\frac{1}{4}$

EXERCISES

Multiply as indicated:

1. a. $^{+}\frac{1}{2} \times {}^{+}\frac{3}{5}$	**b.** $^{+}\frac{3}{4} \times {}^{+}16$	**c.** $^{+}\frac{5}{6} \times {}^{+}2\frac{1}{10}$	**d.** $^{+}4\frac{1}{2} \times {}^{+}1\frac{5}{12}$
2. a. $^{+}.9 \times {}^{+}.4$	**b.** $^{+}67 \times {}^{+}.03$	**c.** $^{+}1.4 \times {}^{+}2.5$	**d.** $^{+}.375 \times {}^{+}.48$
3. a. $^{-}\frac{5}{8} \times {}^{-}\frac{16}{25}$	**b.** $^{-}5 \times {}^{-}\frac{7}{10}$	**c.** $^{-}\frac{9}{16} \times {}^{-}1\frac{1}{6}$	**d.** $^{-}3\frac{3}{4} \times {}^{-}2\frac{2}{3}$
4. a. $^{-}.02 \times {}^{-}.3$	**b.** $^{-}.4 \times {}^{-}50$	**c.** $^{-}.68 \times {}^{-}8.1$	**d.** $^{-}.001 \times {}^{-}.001$
5. a. $^{+}\frac{2}{3} \times {}^{-}\frac{1}{2}$	**b.** $^{+}24 \times {}^{-}3\frac{1}{8}$	**c.** $^{+}6\frac{2}{3} \times {}^{-}\frac{11}{12}$	**d.** $^{+}3\frac{1}{7} \times {}^{-}5\frac{5}{6}$
6. a. $^{+}.125 \times {}^{-}.8$	**b.** $^{+}100 \times {}^{-}.002$	**c.** $^{+}.01 \times {}^{-}.005$	**d.** $^{+}6.2 \times {}^{-}83.6$
7. a. $^{-}\frac{9}{10} \times {}^{+}\frac{15}{16}$	**b.** $^{-}5\frac{1}{4} \times {}^{+}2$	**c.** $^{-}2\frac{3}{16} \times {}^{+}\frac{4}{7}$	**d.** $^{-}5\frac{1}{3} \times {}^{+}2\frac{1}{4}$
8. a. $^{-}.009 \times {}^{+}.003$	**b.** $^{-}24 \times {}^{+}8.5$	**c.** $^{-}3.27 \times {}^{+}.42$	**d.** $^{-}1.05 \times {}^{+}.0015$
9. a. $8 \times {}^{-}\frac{11}{16}$	**b.** $1\frac{2}{3} \times {}^{-}30$	**c.** $^{-}.7 \times 5.6$	**d.** $.04 \times {}^{-}.928$

10. **a.** $0 \times {}^{-}\frac{7}{8}$ **b.** ${}^{+}6\frac{3}{5} \times 0$ **c.** ${}^{-}.63 \times 0$ **d.** $0 \times {}^{-}8.25$

11. **a.** $({}^{-}\frac{1}{2})({}^{-}\frac{1}{2})$
b. $({}^{-}\frac{2}{3})(\frac{3}{4})$
c. $(\frac{1}{4})({}^{-}\frac{7}{8})$
d. $(24)({}^{-}\frac{15}{16})$
e. $({}^{-}\frac{11}{12})({}^{-}60)$
f. $({}^{-}1\frac{1}{2})({}^{-}2\frac{1}{3})$
g. $(36)({}^{-}3\frac{5}{6})$
h. $(\frac{7}{10})({}^{-}6\frac{1}{4})$
i. $({}^{-}1\frac{9}{16})({}^{-}4\frac{2}{3})$
j. $({}^{-}3\frac{1}{8})(1\frac{3}{5})$
k. $({}^{-}.3)({}^{-}.3)$
l. $(.8)({}^{-}.14)$
m. $({}^{-}4.2)(.57)$
n. $({}^{-}.009)({}^{-}.06)$
o. $(2.5)({}^{-}.309)$
p. $\frac{1}{2}({}^{-}4\frac{1}{4})$
q. ${}^{-}\frac{3}{8}(70)$
r. ${}^{-}\frac{2}{3}({}^{-}7\frac{2}{5})$
s. ${}^{-}.9(8.35)$
t. $3.4({}^{-}.001)$

12. Multiply in each of the following:

$^{-}.02$	$^{-}57$	$.3$	$^{-}.006$	$^{-}200$	$^{-}1.1$	$^{-}.3197$
$^{-}.2$	$.09$	$^{-}.47$	$^{-}.001$	$.005$	$^{-}.11$	2.5

13. **a.** $({}^{+}\frac{5}{6} \times {}^{-}\frac{1}{4}) \times {}^{+}\frac{2}{5}$
b. ${}^{+}\frac{3}{10} \times ({}^{-}\frac{5}{8} \times {}^{+}\frac{2}{3})$
c. $({}^{-}1\frac{1}{8} \times {}^{-}3\frac{1}{3}) \times {}^{-}2\frac{1}{2}$
d. ${}^{-}2\frac{4}{7} \times ({}^{+}7\frac{5}{16} \times {}^{-}2\frac{2}{3})$
e. $({}^{-}.8 \times {}^{+}.12) \times {}^{-}.05$
f. ${}^{-}3.6 \times ({}^{-}.25 \times {}^{-}7.4)$

14. **a.** ${}^{-}\frac{1}{3} \times \frac{1}{2} \times {}^{-}\frac{1}{4}$
b. ${}^{-}\frac{5}{6} \times {}^{-}\frac{3}{8} \times {}^{-}\frac{4}{5}$
c. ${}^{-}4\frac{1}{5} \times {}^{-}\frac{5}{8} \times 3\frac{1}{7}$
d. ${}^{-}6\frac{2}{3} \times {}^{-}2\frac{1}{2} \times {}^{-}1\frac{4}{5}$
e. $.08 \times {}^{-}4.3 \times .5$
f. ${}^{-}1.25 \times {}^{-}.004 \times {}^{-}.039$

15. **a.** $(\frac{3}{10})({}^{-}\frac{1}{2})(\frac{5}{6})$
b. $({}^{-}\frac{3}{8})({}^{-}\frac{3}{4})({}^{-}\frac{2}{3})$
c. $({}^{-}1\frac{3}{5})(10)({}^{-}\frac{15}{16})$
d. $({}^{-}2\frac{1}{4})({}^{-}\frac{5}{6})(3\frac{1}{3})({}^{-}36)$
e. $({}^{-}.5)({}^{-}.1)({}^{-}.2)$
f. $({}^{-}1.4)({}^{-}.32)({}^{-}.25)({}^{-}.008)$

16. Find the missing numbers:

a. ${}^{-}\frac{3}{4} \times {}^{-}\frac{3}{4} = ?$
b. $\frac{5}{8} \times {}^{-}\frac{2}{3} = n$
c. ${}^{-}\frac{7}{12} \times {}^{-}90 = \square$
d. ${}^{-}3\frac{1}{8} \times 1\frac{3}{5} = n$
e. ${}^{-}4\frac{7}{8} \times {}^{-}2\frac{1}{2} = n$
f. $20 \times {}^{-}8\frac{1}{6} = n$

17. Find the missing numbers:

a. $.1 \times {}^{-}.02 = \square$
b. ${}^{-}.006 \times {}^{-}.006 = n$
c. ${}^{-}.15 \times 40 = ?$
d. $3.75 \times {}^{-}.132 = n$
e. ${}^{-}.0062 \times {}^{-}.05 = n$
f. ${}^{-}.0015 \times 2.8 = n$

Properties

In the following exercises we determine the properties that hold for the set of rational numbers under the operation of multiplication.

1. Commutative Property

a. What do we mean when we say that the commutative property holds for the multiplication of a certain set of numbers?

b. In each of the following multiply as indicated, then compare the products:

(1) ${}^{-}\frac{3}{8} \times {}^{-}\frac{4}{5}$; ${}^{-}\frac{4}{5} \times {}^{-}\frac{3}{8}$. Does ${}^{-}\frac{3}{8} \times {}^{-}\frac{4}{5} = {}^{-}\frac{4}{5} \times {}^{-}\frac{3}{8}$?
(2) ${}^{-}2\frac{1}{2} \times {}^{+}16$; ${}^{+}16 \times {}^{-}2\frac{1}{2}$. Does ${}^{-}2\frac{1}{2} \times {}^{+}16 = {}^{+}16 \times {}^{-}2\frac{1}{2}$?
(3) ${}^{+}3\frac{5}{8} \times {}^{+}1\frac{1}{3}$; ${}^{+}1\frac{1}{3} \times {}^{+}3\frac{5}{8}$. Does ${}^{+}3\frac{5}{8} \times {}^{+}1\frac{1}{3} = {}^{+}1\frac{1}{3} \times {}^{+}3\frac{5}{8}$?
(4) ${}^{+}8.7 \times {}^{-}.04$; ${}^{-}.04 \times {}^{+}8.7$. Does ${}^{+}8.7 \times {}^{-}.04 = {}^{-}.04 \times {}^{+}8.7$?

c. Can you find two rational numbers for which the commutative property of multiplication does not hold?

d. Can we say that the commutative property holds for the multiplication of rational numbers?

2. Associative Property

a. What do we mean when we say that the associative property holds for the multiplication of a certain set of numbers?

b. In each of the following multiply as indicated, then compare the products:

(1) $(^{-}\frac{3}{5} \times {}^{-}\frac{2}{3}) \times {}^{+}\frac{5}{6}$; $\quad {}^{-}\frac{3}{5} \times (^{-}\frac{2}{3} \times {}^{+}\frac{5}{6})$.
Does $(^{-}\frac{3}{5} \times {}^{-}\frac{2}{3}) \times {}^{+}\frac{5}{6} = {}^{-}\frac{3}{5} \times (^{-}\frac{2}{3} \times {}^{+}\frac{5}{6})$?

(2) $(^{+}2\frac{1}{4} \times {}^{+}1\frac{3}{5}) \times {}^{-}4\frac{1}{2}$; $\quad {}^{+}2\frac{1}{4} \times (^{+}1\frac{3}{5} \times {}^{-}4\frac{1}{2})$.
Does $(^{+}2\frac{1}{4} \times {}^{+}1\frac{3}{5}) \times {}^{-}4\frac{1}{2} = {}^{+}2\frac{1}{4} \times (^{+}1\frac{3}{5} \times {}^{-}4\frac{1}{2})$?

c. Can you find three rational numbers for which the associative property of multiplication does not hold?

d. Can we say that the associative property holds for the multiplication of rational numbers?

3. Distributive Property

a. What do we mean when we say that the distributive property of multiplication over addition holds for the set of rational numbers?

b. In each of the following perform the indicated operations, then compare answers:

(1) ${}^{-}\frac{1}{2} \times (^{+}\frac{7}{8} + {}^{-}\frac{1}{4})$; $(^{-}\frac{1}{2} \times {}^{+}\frac{7}{8}) + (^{-}\frac{1}{2} \times {}^{-}\frac{1}{4})$.
Does ${}^{-}\frac{1}{2} \times (^{+}\frac{7}{8} + {}^{-}\frac{1}{4}) = (^{-}\frac{1}{2} \times {}^{+}\frac{7}{8}) + (^{-}\frac{1}{2} \times {}^{-}\frac{1}{4})$?

(2) ${}^{+}3\frac{1}{3} \times (^{-}1\frac{1}{2} + {}^{+}1\frac{1}{5})$; $\quad (^{+}3\frac{1}{3} \times {}^{-}1\frac{1}{2}) + (^{+}3\frac{1}{3} \times {}^{+}1\frac{1}{5})$.
Does ${}^{+}3\frac{1}{3} \times (^{-}1\frac{1}{2} + {}^{+}1\frac{1}{5}) = (^{+}3\frac{1}{3} \times {}^{-}1\frac{1}{2}) + (^{+}3\frac{1}{3} \times {}^{+}1\frac{1}{5})$?

c. Can you find any rational numbers for which multiplication is not distributed over addition?

d. Can we say that the distributive property of multiplication over addition holds for the set of rational numbers?

4. Closure

a. What do we mean when we say that a certain set of numbers is closed under the operation of multiplication?

b. Is the product of each of the following a rational number?

(1) ${}^{+}\frac{7}{8} \times {}^{+}\frac{3}{4}$	(4) ${}^{-}\frac{5}{12} \times {}^{-}48$	(7) ${}^{+}0.6 \times {}^{-}5.8$
(2) ${}^{+}1\frac{1}{2} \times {}^{-}\frac{5}{6}$	(5) ${}^{-}1 \times {}^{+}7\frac{1}{3}$	(8) ${}^{-}1.75 \times {}^{-}3.04$
(3) ${}^{-}4\frac{2}{3} \times {}^{+}3\frac{1}{7}$	(6) $0 \times {}^{-}5\frac{11}{16}$	(9) ${}^{-}7.3 \times {}^{+}.001$

c. Can you find two rational numbers for which the product is not a rational number?

d. Is the product of two rational numbers always a rational number?

e. Can we say that the set of rational numbers is closed under the operation of multiplication?

5. Multiplicative Identity

a. What do we mean when we say that a certain set of numbers has a multiplicative identity?

b. What is the product of each of the following?

(1) $1 \times {}^{+}\frac{5}{6}$	(3) $1 \times {}^{-}2\frac{1}{8}$	(5) ${}^{-}7\frac{3}{4} \times 1$	(7) $1 \times {}^{-}3.69$
(2) ${}^{-}\frac{11}{12} \times 1$	(4) ${}^{+}6\frac{2}{5} \times 1$	(6) $1 \times {}^{+}5\frac{2}{3}$	(8) ${}^{+}10.5 \times 1$

c. In each case in problem **5b,** is the product of the rational number and one (1) the rational number itself?

d. Can you find any rational number which when multiplied by one (1) does not have the given rational number as the product?

e. Does the set of rational numbers have a multiplicative identity? If so, what is it?

6. Multiplicative Inverse

a. What do we mean by a multiplicative inverse of a rational number?

b. What is the multiplicative inverse of:

${}^{-}9$? ${}^{+}\frac{1}{10}$? ${}^{-}\frac{5}{8}$? ${}^{+}\frac{11}{3}$? ${}^{-}1\frac{3}{4}$?

c. Is each of the following given indicated products a product of a rational number and its multiplicative inverse? Multiply each of the following:

(1) ${}^{+}\frac{2}{3} \times {}^{+}\frac{3}{2}$	(3) ${}^{-}\frac{5}{16} \times {}^{-}\frac{16}{5}$	(5) ${}^{+}1\frac{1}{6} \times {}^{+}\frac{6}{7}$	(7) ${}^{+}12 \times {}^{+}\frac{1}{12}$
(2) ${}^{-}\frac{1}{6} \times {}^{-}6$	(4) ${}^{+}\frac{5}{4} \times {}^{+}\frac{4}{5}$	(6) ${}^{-}2\frac{1}{4} \times {}^{-}\frac{4}{9}$	(8) ${}^{-}\frac{3}{8} \times {}^{-}2\frac{2}{3}$

d. In each case is the product of the two non-zero rational numbers one (1)?

e. Can you find any non-zero rational number that does not have a multiplicative inverse?

f. Does each non-zero rational number in the set of rational numbers have a multiplicative inverse?

11–17 DIVISION OF INTEGERS; PROPERTIES

Since division is the inverse operation of multiplication, to divide we find the factor (quotient) which multiplied by the given factor (divisor) will equal the given product (dividend).

To divide ${}^{+}20$ by ${}^{+}5$ means to find the number which multiplied by ${}^{+}5$ will equal ${}^{+}20$.

That is: ${}^{+}20 \div {}^{+}5 = ?$ becomes ${}^{+}5 \times ? = {}^{+}20$.

Since ${}^{+}5 \times {}^{+}4 = {}^{+}20$,

then ${}^{+}20 \div {}^{+}5 = {}^{+}4$.

${}^{+}20 \div {}^{+}5 = {}^{+}4$ is also written as $\frac{{}^{+}20}{{}^{+}5} = {}^{+}4$

Thus, the quotient of a positive integer divided by a positive integer is a positive number whose absolute value is the quotient of the absolute values of the two given integers.

To divide $^{-}20$ by $^{-}5$ means to find the number which multiplied by $^{-}5$ will equal $^{-}20$.

That is: $^{-}20 \div {}^{-}5 = ?$ becomes $^{-}5 \times ? = {}^{-}20$.

Since $^{-}5 \times {}^{+}4 = {}^{-}20$,

then $^{-}20 \div {}^{-}5 = {}^{+}4$.

$^{-}20 \div {}^{-}5 = {}^{+}4$ is also written as $\frac{^{-}20}{^{-}5} = {}^{+}4$

Thus, the quotient of a negative integer divided by a negative integer is a positive number whose absolute value is the quotient of the absolute values of the two given integers.

To divide $^{-}20$ by $^{+}5$ means to find the number which multiplied by $^{+}5$ will equal $^{-}20$.

That is: $^{-}20 \div {}^{+}5 = ?$ becomes $^{+}5 \times ? = {}^{-}20$.

Since $^{+}5 \times {}^{-}4 = {}^{-}20$,

then $^{-}20 \div {}^{+}5 = {}^{-}4$.

$^{-}20 \div {}^{+}5 = {}^{-}4$ is also written as $\frac{^{-}20}{^{+}5} = {}^{-}4$

Thus, the quotient of a negative integer divided by a positive integer is a negative number whose absolute value is the quotient of the absolute values of the two given integers.

To divide $^{+}20$ by $^{-}5$ means to find the number which multiplied by $^{-}5$ will equal $^{+}20$.

That is: $^{+}20 \div {}^{-}5 = ?$ becomes $^{-}5 \times ? = {}^{+}20$.

Since $^{-}5 \times {}^{-}4 = {}^{+}20$,

then $^{+}20 \div {}^{-}5 = {}^{-}4$.

$^{+}20 \div {}^{-}5 = {}^{-}4$ is also written as $\frac{^{+}20}{^{-}5} = {}^{-}4$

Thus, the quotient of a positive integer divided by a negative integer is a negative number whose absolute value is the quotient of the absolute values of the two given integers.

Summarizing, *to divide a positive integer by a positive integer* or *a negative integer by a negative integer*, we divide their absolute values and prefix the numeral for this quotient with a positive sign.

Divide: $^{+}36 \div {}^{+}3 = ?$

$^{+}36 \div {}^{+}3 = {}^{+}(36 \div 3) = {}^{+}12$ or $\frac{^{+}36}{^{+}3} = {}^{+}12$

Answer, $^{+}12$

Divide: $^{-}45 \div {}^{-}5 = ?$

$^{-}45 \div {}^{-}5 = {}^{+}(45 \div 5) = {}^{+}9$ or $\frac{^{-}45}{^{-}5} = {}^{+}9$

Answer, $^{+}9$

To divide a positive integer by a negative integer or *a negative integer by a positive integer*, we divide their absolute values and prefix the numeral for this quotient with a negative sign.

Divide: $^{-}30 \div {}^{+}6 = ?$

$^{-}30 \div {}^{+}6 = {}^{-}(30 \div 6) = {}^{-}5$ or $\frac{^{-}30}{^{+}6} = {}^{-}5$

Answer, $^{-}5$

Divide: $^{+}24 \div {}^{-}3 = ?$

$^{+}24 \div {}^{-}3 = {}^{-}(24 \div 3) = {}^{-}8$ or $\frac{^{+}24}{^{-}3} = {}^{-}8$

Answer, $^{-}8$

Any non-zero integer divided by itself is one.

Divide: $^{-}2 \div {}^{-}2 = ?$

$^{-}2 \div {}^{-}2 = {}^{+}(2 \div 2) = {}^{+}1$ or $\frac{^{-}2}{^{-}2} = {}^{+}1$

Answer, $^{+}1$

Zero divided either by a positive integer or by a negative integer is zero. We cannot divide a positive integer or a negative integer by zero.

$0 \div {}^{+}8 = 0$ Answer, 0

$\frac{0}{^{-}3} = 0$ Answer, 0

The numeral naming a positive number does not require the positive ($^{+}$) sign.

$^{-}16 \div 8 = {}^{-}2$ or $\frac{^{-}16}{8} = {}^{-}2$

Answer, $^{-}2$

Sometimes when an integer is divided by a non-zero integer, the quotient is not an integer but a rational number.

Divide: $9 \div {}^{-}2 = ?$

$9 \div {}^{-}2 = {}^{-}(9 \div 2) = {}^{-}\frac{9}{2} = {}^{-}4\frac{1}{2}$

Answer, $^{-}4\frac{1}{2}$

Divide: $^{-}8 \div {}^{-}12 = ?$

$^{-}8 \div {}^{-}12 = {}^{+}(8 \div 12) = {}^{+}\frac{8}{12} = {}^{+}\frac{2}{3}$

Answer, $^{+}\frac{2}{3}$

EXERCISES

Divide as indicated:

1. a. $^{+}16 \div {}^{+}2$ **b.** $^{+}55 \div {}^{+}5$ **c.** $^{+}36 \div {}^{+}4$ **d.** $^{+}100 \div {}^{+}20$
2. a. $^{-}32 \div {}^{-}8$ **b.** $^{-}90 \div {}^{-}10$ **c.** $^{-}27 \div {}^{-}9$ **d.** $^{-}84 \div {}^{-}6$
3. a. $^{-}42 \div {}^{+}7$ **b.** $^{-}56 \div {}^{+}4$ **c.** $^{-}144 \div {}^{+}12$ **d.** $^{-}72 \div {}^{+}18$
4. a. $^{+}35 \div {}^{-}5$ **b.** $^{+}64 \div {}^{-}8$ **c.** $^{+}19 \div {}^{-}1$ **d.** $^{+}150 \div {}^{-}25$
5. a. $0 \div {}^{-}3$ **b.** $0 \div {}^{+}11$ **c.** $0 \div {}^{-}24$ **d.** $0 \div {}^{+}60$
6. a. $^{-}80 \div 10$ **b.** $28 \div {}^{-}7$ **c.** $^{-}65 \div 13$ **d.** $132 \div {}^{-}12$
7. a. $^{-}9 \div {}^{-}9$ **b.** $^{-}15 \div 15$ **c.** $21 \div {}^{-}21$ **d.** $^{-}100 \div {}^{-}100$
8. a. $7 \div {}^{-}3$ **b.** $^{-}14 \div {}^{-}5$ **c.** $^{-}21 \div 9$ **d.** $^{-}40 \div {}^{-}6$
9. a. $^{-}2 \div {}^{-}5$ **b.** $^{-}4 \div 24$ **c.** $6 \div {}^{-}15$ **d.** $^{-}56 \div 63$

10. Divide in each of the following:

a.	$\frac{^{-}15}{^{-}3}$	$\frac{^{+}21}{^{-}7}$	$\frac{^{-}56}{^{+}8}$	$\frac{^{-}60}{^{-}12}$	$\frac{^{+}75}{^{-}5}$	$\frac{^{+}110}{^{+}10}$	$\frac{^{-}96}{^{+}24}$	$\frac{^{-}350}{^{-}50}$
b.	$\frac{^{-}16}{^{+}2}$	$\frac{^{+}81}{^{-}9}$	$\frac{^{-}47}{^{-}1}$	$\frac{^{-}78}{^{+}6}$	$\frac{^{+}92}{^{+}4}$	$\frac{^{-}80}{^{-}16}$	$\frac{^{+}108}{^{-}12}$	$\frac{^{-}68}{^{+}68}$
c.	$\frac{40}{^{-}5}$	$\frac{^{-}104}{8}$	$\frac{^{-}48}{^{-}3}$	$\frac{^{-}370}{10}$	$\frac{126}{^{-}18}$	$\frac{17}{^{-}17}$	$\frac{^{-}99}{^{-}11}$	$\frac{775}{^{-}25}$
d.	$\frac{^{-}48}{4}$	$\frac{^{-}33}{1}$	$\frac{98}{^{-}7}$	$\frac{^{-}180}{9}$	$\frac{56}{^{-}2}$	$\frac{^{-}91}{^{-}13}$	$\frac{165}{^{-}15}$	$\frac{^{-}520}{^{-}40}$

11. a. $(^{-}12 \div {}^{+}3) \div {}^{-}2$ **c.** $(^{-}36 \div {}^{-}12) \div {}^{-}3$ **e.** $^{-}100 \div (^{-}10 \div {}^{-}5)$
b. $^{+}60 \div (^{+}6 \div {}^{-}2)$ **d.** $^{-}80 \div (^{-}20 \div {}^{+}4)$ **f.** $(^{-}64 \div {}^{+}8) \div {}^{+}4$

12. Find the missing numbers:

a. $^{-}18 \div 6 = ?$ **d.** $^{-}40 \div {}^{-}5 = \square$ **g.** $96 \div {}^{-}8 = n$
b. $36 \div {}^{-}4 = \square$ **e.** $^{-}72 \div 9 = ?$ **h.** $^{-}85 \div {}^{-}17 = n$
c. $\frac{7}{^{-}1} = n$ **f.** $\frac{^{-}56}{^{-}7} = n$ **i.** $\frac{^{-}180}{12} = n$

13. Find the missing numbers:

a. $? \div 2 = {}^{-}8$ **d.** $\square \div {}^{-}6 = {}^{-}11$ **g.** $n \div {}^{-}13 = 0$
b. $\square \div {}^{-}1 = 19$ **e.** $? \div 3 = {}^{-}1$ **h.** $n \div {}^{-}9 = {}^{-}108$
c. $\frac{n}{^{-}7} = 4$ **f.** $\frac{n}{8} = {}^{-}9$ **i.** $\frac{n}{^{-}2} = 29$

14. Find the missing numbers:

a. $20 \div ? = {}^{-}10$ **d.** $^{-}76 \div n = {}^{-}19$ **g.** $^{-}5 \div n = 5$
b. $^{-}16 \div \square = 4$ **e.** $^{-}100 \div ? = 10$ **h.** $42 \div n = {}^{-}6$
c. $\frac{^{-}15}{n} = 1$ **f.** $\frac{35}{n} = {}^{-}7$ **i.** $\frac{^{-}50}{n} = {}^{-}2$

Properties

In the following exercises we determine the properties that hold for the set of integers under the operation of division.

1. Commutative Property

a. If the commutative property holds for the division of integers, what would have to be true?

b. In each of the following divide as indicated, then compare quotients:

(1) $^+18 \div {}^+3$; $^+3 \div {}^+18$. Does $^+18 \div {}^+3 = {}^+3 \div {}^+18$?
(2) $^-21 \div {}^-7$; $^-7 \div {}^-21$. Does $^-21 \div {}^-7 = {}^-7 \div {}^-21$?
(3) $^+56 \div {}^-8$; $^-8 \div {}^+56$. Does $^+56 \div {}^-8 = {}^-8 \div {}^+56$?
(4) $^-63 \div {}^+9$; $^+9 \div {}^-63$. Does $^-63 \div {}^+9 = {}^+9 \div {}^-63$?

c. Does the commutative property hold for the division of integers?

2. Associative Property

a. If the associative property holds for the division of integers, what would have to be true?

b. In each of the following divide as indicated, then compare quotients:

(1) $({}^+24 \div {}^-8) \div {}^-2$; $\quad {}^+24 \div ({}^-8 \div {}^-2)$.
Does $({}^+24 \div {}^-8) \div {}^-2 = {}^+24 \div ({}^-8 \div {}^-2)$?
(2) $({}^-90 \div {}^-6) \div {}^+3)$; $\quad {}^-90 \div ({}^-6 \div {}^+3)$.
Does $({}^-90 \div {}^-6) \div {}^+3 = {}^-90 \div ({}^-6 \div {}^+3)$?

c. Does the associative property hold for the division of integers?

3. Distributive Property

a. If the distributive property of division over addition holds for the set of integers, what would have to be true?

b. In each of the following perform the indicated operations, then compare answers:

(1) $^+48 \div ({}^+12 + {}^-4)$; $\quad ({}^+48 \div {}^+12) + ({}^+48 \div {}^-4)$.
Does $^+48 \div ({}^+12 + {}^-4) = ({}^+48 \div {}^+12) + ({}^+48 \div {}^-4)$?
(2) $({}^-30 + {}^-25) \div {}^-5$; $({}^-30 \div {}^-5) + ({}^-25 \div {}^-5)$.
Does $({}^-30 + {}^-25) \div {}^-5 = ({}^-30 \div {}^-5) + ({}^-25 \div {}^-5)$?

c. Is division distributed over addition in problem **3b** (1)? In problem **3b** (2)? When is division distributed over addition in the set of integers?

4. Closure

a. What do we mean when we say that a certain set of numbers is closed under the operation of division?

b. When you divide in each of the following, is the quotient in each case an integer?

(1) $^-20 \div {}^-4$
(2) $^+6 \div {}^+9$
(3) $0 \div {}^-5$
(4) $^-1 \div {}^+6$
(5) $^+17 \div {}^-25$
(6) $^-72 \div {}^+18$
(7) $^+9 \div {}^-10$
(8) $^-36 \div {}^-1$
(9) $^-21 \div {}^+14$

c. Can you find two integers for which the quotient is not an integer provided that division by zero is excluded?

d. Can we say that the set of integers is closed under the operation of division provided that division by zero is excluded?

11–18 DIVISION OF RATIONAL NUMBERS; PROPERTIES

We follow the same procedures to divide rational numbers as we do when we divide integers. See section 11–17. However, it may be necessary to review at this time the division of the rational numbers of arithmetic (common fractions, mixed numbers, and decimal fractions). See sections 5–15 and 6–11.

To divide a positive rational number by a positive rational number or a negative rational number by a negative rational number, we divide their absolute values and prefix the numeral for this quotient with a positive sign.

Divide: $^-\frac{2}{3} \div {}^-\frac{4}{5} = ?$

$^-\frac{2}{3} \div {}^-\frac{4}{5} = {}^+(\frac{2}{3} \div \frac{4}{5}) = {}^+(\frac{2}{3} \times \frac{5}{4}) = {}^+\frac{5}{6}$

Answer, $^+\frac{5}{6}$

To divide a positive rational number by a negative rational number or a negative rational number by a positive rational number, we divide their absolute values and prefix the numeral for this quotient with a negative sign.

Divide: $^-1\frac{7}{8} \div {}^+1\frac{1}{2} = ?$

$^-1\frac{7}{8} \div {}^+1\frac{1}{2} = {}^-(1\frac{7}{8} \div 1\frac{1}{2})$

$= {}^-(\frac{15}{8} \times \frac{2}{3}) = {}^-\frac{5}{4} = {}^-1\frac{1}{4}$

Answer, $^-1\frac{1}{4}$

Any non-zero rational number divided by itself is one (1).

Divide: $^-.7 \div {}^-.7 = ?$

$^-.7 \div {}^-.7 = {}^+(.7 \div .7) = {}^+1$ or $\frac{^-.7}{^-.7} = {}^+1$

Answer, $^+1$

Zero divided either by a positive rational number or by a negative rational number is zero. We cannot divide a positive rational number or a negative rational number by zero.

$0 \div {}^-2\frac{1}{2} = 0$ *Answer,* 0

$0 \div {}^-8.5 = 0$ *Answer,* 0

The numeral naming a positive rational number does not require the positive (⁺) sign.

EXERCISES

Divide as indicated:

1. a. $^+\frac{13}{16} \div {}^+\frac{7}{8}$ **b.** $^+12 \div {}^+\frac{3}{5}$ **c.** $^+4\frac{1}{2} \div {}^+27$ **d.** $^+9\frac{1}{3} \div {}^+1\frac{1}{6}$

2. a. ${}^{+}.8 \div {}^{+}.4$ b. ${}^{+}15 \div {}^{+}.5$ c. ${}^{+}5.6 \div {}^{+}.14$ d. ${}^{+}.005 \div {}^{+}.0002$
3. a. ${}^{-}\frac{3}{4} \div {}^{-}\frac{3}{8}$ b. ${}^{-}\frac{5}{6} \div {}^{-}2$ c. ${}^{-}1\frac{1}{8} \div {}^{-}\frac{3}{32}$ d. ${}^{-}7\frac{1}{2} \div {}^{-}1\frac{1}{4}$
4. a. ${}^{-}7.6 \div {}^{-}.4$ b. ${}^{-}6.9 \div {}^{-}3$ c. ${}^{-}.004 \div {}^{-}.02$ d. ${}^{-}27 \div {}^{-}.09$
5. a. ${}^{-}\frac{9}{10} \div {}^{+}\frac{1}{3}$ b. ${}^{-}7 \div {}^{+}\frac{1}{6}$ c. ${}^{-}\frac{9}{16} \div {}^{+}1\frac{3}{5}$ d. ${}^{-}2\frac{1}{16} \div {}^{+}1\frac{3}{8}$
6. a. ${}^{-}.72 \div {}^{+}.08$ b. ${}^{-}.0028 \div {}^{+}.7$ c. ${}^{-}14 \div {}^{+}5.6$ d. ${}^{-}.168 \div {}^{+}24$
7. a. ${}^{+}\frac{5}{8} \div {}^{-}\frac{5}{6}$ b. ${}^{+}\frac{11}{12} \div {}^{-}2$ c. ${}^{+}10\frac{5}{8} \div {}^{-}5$ d. ${}^{+}1\frac{11}{16} \div {}^{-}2\frac{1}{4}$
8. a. ${}^{+}8.4 \div {}^{-}1.2$ b. ${}^{+}2 \div {}^{-}.8$ c. ${}^{+}7.686 \div {}^{-}.06$ d. ${}^{+}.0016 \div {}^{-}.4$
9. a. $0 \div {}^{-}\frac{11}{16}$ b. $0 \div {}^{-}3\frac{2}{3}$ c. $0 \div {}^{-}.7$ d. $0 \div {}^{-}4.9$
10. a. ${}^{-}\frac{2}{3} \div \frac{5}{6}$ b. $3\frac{3}{4} \div {}^{-}4\frac{2}{5}$ c. ${}^{-}2.4 \div .03$ d. $90 \div {}^{-}1.5$
11. a. ${}^{-}\frac{3}{4} \div {}^{-}\frac{3}{4}$ b. ${}^{-}7\frac{5}{12} \div 7\frac{5}{12}$ c. $.68 \div {}^{-}.68$ d. ${}^{-}3.517 \div {}^{-}3.517$

12. Divide in each of the following:

a. $\dfrac{{}^{-}\frac{3}{4}}{\frac{7}{8}}$ $\dfrac{{}^{-}10}{{}^{-}\frac{15}{16}}$ $\dfrac{\frac{2}{3}}{{}^{-}12}$ $\dfrac{{}^{-}\frac{15}{16}}{{}^{-}1\frac{1}{8}}$ $\dfrac{3\frac{3}{8}}{{}^{-}4\frac{1}{2}}$ $\dfrac{{}^{-}\frac{7}{12}}{2\frac{1}{3}}$ $\dfrac{{}^{-}5\frac{1}{4}}{{}^{-}3\frac{2}{3}}$

b. $\dfrac{{}^{-}62.5}{.5}$ $\dfrac{1.08}{{}^{-}9}$ $\dfrac{{}^{-}6}{{}^{-}1.2}$ $\dfrac{{}^{-}.1}{.4}$ $\dfrac{{}^{-}.02}{{}^{-}.004}$ $\dfrac{9.07}{{}^{-}9.07}$ $\dfrac{{}^{-}.003}{{}^{-}.15}$

13. a. $({}^{-}\frac{7}{8} \div {}^{-}\frac{9}{16}) \div {}^{+}\frac{2}{3}$ c. $({}^{-}2\frac{1}{3} \div {}^{+}5\frac{1}{2}) \div {}^{+}1\frac{3}{4}$ e. $({}^{+}1.2 \div {}^{-}.2) \div {}^{-}.3$
b. ${}^{-}\frac{3}{4} \div ({}^{-}\frac{4}{5} \div {}^{-}\frac{9}{10})$ d. ${}^{-}4\frac{1}{6} \div ({}^{+}20 \div {}^{-}3\frac{3}{4})$ f. ${}^{-}14.4 \div ({}^{-}7.2 \div {}^{-}.04)$

14. Find the missing numbers:

a. ${}^{-}\frac{11}{16} \div {}^{-}\frac{5}{8} = ?$ d. ${}^{-}\frac{7}{8} \div {}^{-}\frac{7}{8} = \Box$ g. $2\frac{5}{12} \div {}^{-}2 = n$
b. $\dfrac{7}{12} \div \dfrac{{}^{-}9}{10} = \Box$ e. ${}^{-}1\frac{9}{16} \div \frac{5}{6} = ?$ h. ${}^{-}3\frac{3}{8} \div {}^{-}4\frac{1}{2} = n$
c. $\dfrac{{}^{-}\frac{1}{2}}{\frac{3}{4}} = n$ f. $\dfrac{5}{{}^{-}3\frac{1}{8}} = n$ i. $\dfrac{{}^{-}1\frac{1}{6}}{{}^{-}9\frac{1}{3}} = n$

15. Find the missing numbers:

a. ${}^{-}.8 \div {}^{-}.2 = \Box$ d. $.3 \div {}^{-}.12 = ?$ g. $64.8 \div {}^{-}.27 = n$
b. ${}^{-}3.24 \div 6 = ?$ e. ${}^{-}.504 \div {}^{-}6.3 = \Box$ h. ${}^{-}.006 \div .006 = n$
c. $\dfrac{7.6}{{}^{-}.4} = n$ f. $\dfrac{{}^{-}10}{.25} = n$ i. $\dfrac{{}^{-}.3}{{}^{-}.01} = n$

Properties

In the following exercises we determine the properties that hold for the set of rational numbers under the operation of division.

1. Commutative Property

a. If the commutative property holds for the division of rational numbers, what would have to be true?

b. In each of the following divide as indicated, then compare quotients:

(1) ${}^{+}\frac{3}{4} \div {}^{-}\frac{1}{4}$; ${}^{-}\frac{1}{4} \div {}^{+}\frac{3}{4}$. Does ${}^{+}\frac{3}{4} \div {}^{-}\frac{1}{4} = {}^{-}\frac{1}{4} \div {}^{+}\frac{3}{4}$?

(2) ${}^{-}1\frac{1}{2} \div {}^{-}\frac{3}{8}$; ${}^{-}\frac{3}{8} \div {}^{-}1\frac{1}{2}$. Does ${}^{-}1\frac{1}{2} \div {}^{-}\frac{3}{8} = {}^{-}\frac{3}{8} \div {}^{-}1\frac{1}{2}$?

(3) $^{-}6\frac{1}{4} \div {}^{+}3\frac{1}{3}$; $^{+}3\frac{1}{3} \div {}^{-}6\frac{1}{4}$. Does $^{-}6\frac{1}{4} \div {}^{+}3\frac{1}{3} = {}^{+}3\frac{1}{3} \div {}^{-}6\frac{1}{4}$?
(4) $^{+}.8 \div {}^{+}.02$; $^{+}.02 \div {}^{+}.8$. Does $^{+}.8 \div {}^{+}.02 = {}^{+}.02 \div {}^{+}.8$?

c. Does the commutative property hold for the division of rational numbers?

2. Associative Property

a. If the associative property holds for the division of rational numbers, what would have to be true?

b. In each of the following divide as indicated, then compare quotients:

(1) $({}^{-}\frac{5}{6} \div {}^{+}\frac{1}{2}) \div {}^{-}\frac{2}{3}$; $^{-}\frac{5}{6} \div ({}^{+}\frac{1}{2} \div {}^{-}\frac{2}{3})$.
Does $({}^{-}\frac{5}{6} \div {}^{+}\frac{1}{2}) \div {}^{-}\frac{2}{3} = {}^{-}\frac{5}{6} \div ({}^{+}\frac{1}{2} \div {}^{-}\frac{2}{3})$?
(2) $({}^{+}3\frac{1}{8} \div {}^{-}1\frac{1}{4}) \div {}^{-}2\frac{1}{2}$; $^{+}3\frac{1}{8} \div ({}^{-}1\frac{1}{4} \div {}^{-}2\frac{1}{2})$.
Does $({}^{+}3\frac{1}{8} \div {}^{-}1\frac{1}{4}) \div {}^{-}2\frac{1}{2} = {}^{+}3\frac{1}{8} \div ({}^{-}1\frac{1}{4} \div {}^{-}2\frac{1}{2})$?

c. Does the associative property hold for the division of rational numbers?

3. Distributive Property

a. If the distributive property of division over addition holds for the set of rational numbers, what would have to be true?

b. In each of the following perform the indicated operations, then compare answers:

(1) $^{-}\frac{3}{4} \div ({}^{+}\frac{9}{16} + {}^{-}\frac{3}{8})$; $({}^{-}\frac{3}{4} \div {}^{+}\frac{9}{16}) + ({}^{-}\frac{3}{4} \div {}^{-}\frac{3}{8})$.
Does $^{-}\frac{3}{4} \div ({}^{+}\frac{9}{16} + {}^{-}\frac{3}{8}) = ({}^{-}\frac{3}{4} \div {}^{+}\frac{9}{16}) + ({}^{-}\frac{3}{4} \div {}^{-}\frac{3}{8})$?
(2) $({}^{+}4\frac{1}{2} + {}^{-}6\frac{2}{3}) \div {}^{-}1\frac{1}{4}$; $({}^{+}4\frac{1}{2} \div {}^{-}1\frac{1}{4}) + ({}^{-}6\frac{2}{3} \div {}^{-}1\frac{1}{4})$.
Does $({}^{+}4\frac{1}{2} + {}^{-}6\frac{2}{3}) \div {}^{-}1\frac{1}{4} = ({}^{+}4\frac{1}{2} \div {}^{-}1\frac{1}{4}) + ({}^{-}6\frac{2}{3} \div {}^{-}1\frac{1}{4})$?

c. Is division distributed over addition in problem **3b** (1)? In problem **3b** (2)? When is division distributed over addition in the set of rational numbers?

4. Closure

a. What do we mean when we say that a certain set of numbers is closed under the operation of division?

b. When you divide in each of the following, is the quotient in each case a rational number?

(1) $^{-}\frac{5}{8} \div {}^{-}\frac{1}{8}$	(4) $^{-}2\frac{3}{5} \div {}^{+}4\frac{1}{2}$	(7) $^{+}1.28 \div {}^{-}.04$
(2) $^{+}\frac{2}{3} \div {}^{-}\frac{4}{5}$	(5) $^{+}\frac{13}{16} \div {}^{-}1\frac{7}{8}$	(8) $^{+}.06 \div {}^{+}.15$
(3) $0 \div {}^{-}\frac{9}{16}$	(6) $^{-}3\frac{5}{12} \div {}^{-}4$	(9) $^{-}10 \div {}^{-}.1$

c. Can you find two rational numbers for which the quotient is not a rational number provided that division by zero is excluded?

d. Can we say that the set of rational numbers is closed under the operation of division provided that division by zero is excluded?

11-19 COMPUTATION—USING NUMERALS WITH CENTERED SIGNS

The number that is the *opposite of six*, named by the numeral -6, and the number *negative six*, named by the numeral $^{-}6$ are one and the same number. Since there is no need to have two symbols that differ so slightly to name the same number, to simplify matters we shall discard the use of the raised sign and henceforth use only numerals with centered signs to name numbers. The numeral -6 represents both the *opposite of six* and the number *negative six*. The numeral $+6$ or 6 names the number *positive six*.

While both sign locations are correct in naming signed numbers, the numerals having the signs centered are more generally used.

To compute with signed numbers named by numerals with centered signs, we follow the same procedures as with numbers named by numerals having raised signs.

Also, to simplify an algebraic expression such as $8 - 10 + 7 - 2$, we may take it to mean $(+8) + (-10) + (+7) + (-2)$ where the given signs tell you which numbers are positive and which are negative but the operation is considered to be addition.

$$8 - 10 + 7 - 2 = (+8) + (-10) + (+7) + (-2) = +3$$

EXERCISES

1. Add:

a.	$+9$	-7	$+6$	$+1$	-8	-10	-4	$+2$	-15	-5
	$+4$	-8	-3	-5	$+2$	$+16$	-7	-12	$+14$	$+21$
b.	$+7$	-9	-8	0	-3	$+\frac{2}{5}$	$+\frac{7}{8}$	$-\frac{3}{4}$	$-.5$	-3.94
	-7	$+9$	$+8$	$+11$	0	$+\frac{1}{5}$	$-\frac{3}{8}$	$-\frac{2}{3}$	$+1.8$	-1.06

2. Add as indicated:

a. $(+3) + (-9)$

b. $(-2) + (+8) + (-5)$

c. $[(-6) + (-10)] + (-4)$

d. $(+1) + [(+7) + (-9)]$

e. $(-11) + (+4) + (+9) + (-1)$

f. $[(15) + (-8)] + [(+2) + (-9)]$

3. Simplify:

a. $8 + 4$
b. $6 - 7$
c. $-5 + 6$
d. $-4 - 10$
e. $9 - 15$
f. $-2 - 4$
g. $7 - 2 - 3$
h. $12 + 7 - 9$
i. $16 - 8 + 2$
j. $5 - 8 + 1$
k. $6 - 9 - 3$
l. $-2 + 5 - 2$
m. $11 - 4 + 8 - 10$
n. $-3 + 5 + 2 - 5$
o. $1 - 8 + 2 - 4 + 3$
p. $5 + 2 - 9 + 6 - 4$

4. Subtract:

a.	$+21$	$+8$	11	2	-5	-1	$+9$	-3	-8	$+12$
	$+18$	$+9$	3	10	-2	-11	-4	$+15$	$+1$	-12

b.										
	$+7$	-2	0	7	0	$+\frac{3}{4}$	$-.18$	$-1\frac{1}{2}$	$+1.05$	$-4\frac{1}{2}$
	$+7$	-2	$+6$	0	-1	$-\frac{1}{4}$	$-.25$	$+\frac{2}{3}$	$+2.75$	$+2\frac{1}{2}$

5. Subtract as indicated:

a. $(-9)-(+7)$ **b.** $(-8)-(-3)$ **c.** $(0)-(8)$

d. $(+6)-(-5)$ **e.** $(6-5)-(2-9)$ **f.** $(8-11)-(4-5)$

g. $[(-9)+(-3)]-(-15)$ **h.** $(+6)-[(-8)-(+4)]$ **i.** $[(-11)-(+6)]-(-2)$

6. a. From -8 subtract 4. **b.** Subtract 7 from 0. **c.** Take 11 from 9. **d.** From -1 take -1. **e.** Subtract 12 from 5. **f.** From 8 take -6.

7. Multiply:

a.	6	$+9$	-7	-6	-8	-4	$+7$	$+16$	-1	$+12$
	7	$+8$	-7	-9	6	$+4$	-8	-3	-1	-12
b.	0	-8	0	$+5$	0	$+.3$	$+.04$	$-1\frac{1}{2}$	$-3\frac{3}{4}$	-1.5
	9	0	-4	0	$+6$	$-.2$	$+.78$	-6	$+4\frac{1}{3}$	$-.07$

8. Multiply as indicated:

a. $(-9)\times(-8)$ **b.** $(+6)\cdot(-7)$ **c.** $(-2)(+2)$ **d.** $-5(4-10)$

e. $(+6)(-.8)$ **f.** $(-\frac{2}{3})(-9)$ **g.** $(-11)(0)(-2)$ **h.** $[(+16)(-1)]\times(-4)$

i. $(-9)\times[(-2)(-5)]$ **j.** $-6(2-9)-(2-3)$ **k.** $(-3)(-4)(+1)(-4)$ **l.** $(-5)(-2)(-1)(-3)$

9. Find the value of each of the following: **a.** $(-2)^2$ **b.** $(+5)^2$ **c.** $(-6)^3$ **d.** $(+3)^4$ **e.** $(-1)^5$ **f.** $(-5)^4$ **g.** $(-1)^6$

10. Divide as indicated:

a. $\frac{+18}{+3}$ $\frac{-30}{-5}$ $\frac{+21}{-7}$ $\frac{-56}{+8}$ $\frac{-72}{-18}$ $\frac{+54}{-9}$ $\frac{-80}{+20}$ $\frac{+35}{+7}$ $\frac{-90}{+18}$ $\frac{-60}{-4}$

b. $\frac{-6}{-6}$ $\frac{+18}{-18}$ $\frac{-29}{+29}$ $\frac{-8}{-1}$ $\frac{-25}{+1}$ $\frac{+7}{-1}$ $\frac{0}{+5}$ $\frac{0}{-6}$ $\frac{0}{+8}$ $\frac{0}{-3}$

11. Divide as indicated:

a. $(-24)\div(+6)$ **b.** $(+40)\div(-8)$

c. $(-12)\div(-\frac{2}{3})$ **d.** $(+.36)\div(-.9)$

e. $(-25)\div(-25)$ **f.** $(+4\frac{1}{6})\div(-4\frac{1}{6})$

12. Simplify by performing the indicated operations:

a. $\frac{6-16}{2}$ **b.** $\frac{10-2(3)}{-4}$

c. $\frac{9(-2)+3(-6)}{6(-1-1)}$ **d.** $\frac{8(1-4)-6(9-1)}{-5(14-2)}$

e. $\frac{(-5)^2-(-3)^2}{(-2)^2}$ **f.** $\frac{-3(-1-3)}{5(-4)-2(2-9)}$

13. a. Does $(-5)+(-4)=|-5|+|-4|$?

b. Does $(+6)+(-9)=|+6|+|-9|$?

c. Does $|-2|+|-7|=|(-2)+(-7)|$?

d. Does $|+8|+|-3|=|(+8)+(-3)|$?

e. Does $(-6)(-2)=|-6|\times|-2|$?

f. Does $(-3)(+7)=|-3|\times|+7|$?

LANGUAGE OF ALGEBRA

11-20 SYMBOLS—REVIEW

In algebra we continue to use the operational symbols of arithmetic.

The signs of operation may assume any one of several meanings, as follows:

The *addition* symbol or plus symbol "+" means sum, add, more than, increased by, exceeded by.

The *subtraction* symbol or minus symbol "−" means difference, subtract, take away, less, less than, decreased by, diminished by.

The *multiplication* symbol "×" or the raised dot "·" means product, multiply, times.

The *division* symbol "÷" or "$\overline{)\quad}$" means quotient, divide. The fraction bar, as in $\frac{a}{b}$, is generally used in algebra to indicate division. The word "over" is sometimes used to express the relationship of the numerator to the denominator.

Other symbols used in both algebra and arithmetic include exponents, the square root symbol, parentheses, the verbs of mathematical sentences such as $=$, $\neq$, $>$, $\not>$, $<$, $\not<$, $\geq$, $\not\geq$, $\leq$, $\not\leq$, and the symbols of sets such as $\{\}$, $\in$, $\notin$, $\cup$, $\cap$, $\subset$, $\supset$.

An *exponent* is the small numeral written to the upper right (superscript) of the base. When it names a natural number, it tells how many times the factor is being used in multiplication. 8^2 represents 8×8 and is read "the square of eight" and 8^3 represents $8 \times 8 \times 8$ and is read "the cube of eight." The factor that is being repeated is called the *base*. The number 8^4 uses 8 as the base and 4 as the exponent. Numbers such as 8^2, 8^3, 8^4, 8^5, etc. are called *powers* of 8.

The *square root* symbol "$\sqrt{\quad}$" written over the numeral indicates the square root of the corresponding number. The *square root* of a number is that number which when multiplied by itself produces the given number.

Parentheses "( )" are generally used to group together two or more numerals so that they are treated as a single quantity. Sometimes parentheses are used to set off a numeral from another so that the meaning will not be misunderstood. $(9 + 5)$ may be read as "the quantity nine plus five."

A numeral is sometimes called a *constant* because it names a definite number.

A *variable* is a letter (small letter, capital letter, letter with a subscript such as "b_1," read "b sub one," or letter with a prime mark such as S', read "S prime") or a frame (such as □, ▭, ○, and △) or a blank which holds a place open for a number. Sometimes the first letter of a key word is used as the variable. A variable may represent any number, but under certain conditions it represents a specific number or numbers.

$-x$ means the opposite of the variable x and not negative x or minus x. $-x$ is not necessarily a negative number. When x is a positive number, $-x$ is a negative number; but when x is a negative number, $-x$ is a positive number.

EXERCISES

1. Read, or write in words, each of the following:

a. $19 - 8$	**d.** $51 + 47$	**g.** $(11) \times (72)$	**j.** $6 \cdot 45$	**m.** $(72 + 54)$
b. 10×36	**e.** 15^2	**h.** $(32 - 29)$	**k.** 44^2	**n.** $(9 \times 8) \times 12$
c. $\frac{21}{7}$	**f.** $\sqrt{62}$	**i.** 56^3	**l.** $\sqrt{102}$	**o.** $4 + (3 + 11)$

2. What symbol represents the variable in each of the following?

a. $\square - 9 = 31$	**c.** $x + 7 = 31$	**e.** $T' - 32 = 18$	**g.** $15 + 29 = \triangle$
b. $28 + ? = 56$	**d.** $9\,y = 54$	**f.** $\frac{n}{3} = 8$	**h.** $48 = b_2 - 11$

3. In each of the following, what is the base? What is the exponent? How many times is the base being used as a factor?

a. 6^{11} **b.** 10^7 **c.** 3^9 **d.** 15^{10} **e.** 21^{16} **f.** 1^{100}

11–21 MATHEMATICAL PHRASES OR MATHEMATICAL EXPRESSIONS

A *numerical expression* or *numerical phrase* consists of a single numeral with or without operational symbols like 15 or 2^5 or two or more numerals with operational symbols like $8 + 9$; $63 - 27$; 8×36; $54 \div 9$; $6 \times (5 + 7)$; etc.

An *algebraic expression* or *algebraic phrase* may be a numerical expression as described above or an expression containing one or more variables joined by operational symbols like a; $c - 3\,d$; $5\,x^2 - 3\,x + 7$; etc.

Both numerical expressions and algebraic expressions are *mathematical expressions*.

In an algebraic expression no multiplication symbol is necessary when the factors are two letters (variables) or a numeral and a letter. In the latter case the numeral always precedes the letter or the variable. This numeral is usually called the *numerical coefficient* of the variable. a times b may be expressed as $a \times b$, $a \cdot b$, ab (preferred), $(a)(b)$, $a(b)$, or $(a)b$. In $4\,a^3$, the 4 is the numerical coefficient of a^3.

To write algebraic expressions, we write numerals, variables, and operational symbols as required in the proper order.

The sum of y and six. *Answer,* $y + 6$

The product of n and three. *Answer,* $3n$

The square of the edge (e). *Answer,* e^2

The difference between the length (l) and the width (w). *Answer,* $l - w$

The area (A) divided by pi (π). *Answer,* $\frac{A}{\pi}$

The square root of the distance (d). *Answer,* $\sqrt{d}$

Seven times the sum of eleven and x. *Answer,* $7(11 + x)$

EXERCISES

1. Write each of the following as an algebraic expression:

a. Eighteen added to ten.
b. From fifteen subtract nine.
c. Twenty times seventeen.
d. Sixty divided by twelve.
e. The sum of seven and five.
f. The difference between sixty and twenty-one.
g. The product of six and eight.
h. The quotient of fourteen divided by seven.
i. The square of ten.
j. The cube of twenty-one.
k. The square root of six.
l. Nine times the sum of five and two.

2. Write each of the following as an algebraic expression:

a. c added to b.
b. From m subtract r.
c. t times x.
d. a divided by d.
e. The sum of p and i.
f. The difference between l and m.
g. The product of b and h.
h The quotient of d divided by r.
i. The square of V.
j. The cube of s.
k. The square root of A.
l. Two times the sum of l and w.
m. The product of s and six.
n. Twenty less than x.

3. Write each of the following as an algebraic expression:

a. The sum of angles E, F, G, and H.
b. 180° decreased by angle B.
c. The product of the length (l), width (w), and height (h).
d. The square root of the difference between the squares of the hypotenuse (h) and the base (b).
e. Six times the square of the side (s).
f. 273° more than the Celsius temperature reading (C).
g. Twice the product of pi and the radius (r) times the sum of the radius (r) and the height (h).

h. The product of force (F) and distance (d).
i. The sum of a, b, and c divided by 3.
j. The number of teeth (t) of the driving gear multiplied by the number of revolutions it makes per minute (R).
k. The gross weight (W) divided by the wing area (A).
l. The number of articles (n) multiplied by the price of one article (p).
m. The true course ($T.C.$) increased by west variation ($W.V.$).
n. The electromotive force (E) divided by the resistance (R).
o. The product of the weight of the body (W) and the square of the velocity (v) divided by the product of the acceleration of gravity (g) and the radius of the circle (r).
p. The profit (p) divided by the selling price (s).
q. The sum of the wing drag (D_w) and the parasite drag (D_p).
r. The quotient of the atomic weight ($A.W.$) divided by the equivalent weight ($E.W.$).
s. The span of the wing (s) multiplied by the chord (c).
t. Pi times radius (r) times sum of radius and slant height (l).

4. Read, or write in words, each of the following:

a. $x + 10$	**d.** b^2	**g.** $90 - A$	**j.** xyz	**m.** $\frac{1}{6}\pi d^3$
b. $8n$	**e.** $z - 5$	**h.** ab	**k.** $2\pi r^2$	**n.** $h(b + b')$
c. $\dfrac{a}{15}$	**f.** $\sqrt{t}$	**i.** $R_1 + R_2$	**l.** $\dfrac{d}{t}$	**o.** $16(y - 1)$

11–22 MATHEMATICAL SENTENCES—OPEN SENTENCES; EQUATIONS; INEQUALITIES

Sentences such as:

"Some number n increased by eight is equal to twelve"
expressed in symbols as $n + 8 = 12$

or "Nine times each number b is greater than fifty"
expressed in symbols as $9\,b > 50$

or "Each number x decreased by six is less than fourteen"
expressed in symbols as $x - 6 < 14$

are *mathematical sentences.* A mathematical sentence that contains a variable is called an *open sentence.* $n + 8 = 12$, $9\,b > 50$, and $x - 6 < 14$ are open sentences. A verbal (word) sentence may be expressed as an equivalent algebraic sentence and vice-versa.

An open sentence that has the equality sign "$=$" as its verb is called an *equation.* If the sentence uses $\neq$, $>$, $<$, $\not>$, $\not<$, $\geq$, $\leq$, $\not\geq$, or $\not\leq$ as its verb, it is called an *inequality.*

The sentence $n + 8 = 12$ is an equation; the sentences $9\,b > 50$ and $x - 6 < 14$ are inequalities.

Symbolic verbs used in mathematical sentences include:

$=$ meaning "is equal to."
$\neq$ meaning "is not equal to."
$<$ meaning "is less than."
$\nless$ meaning "is not less than."
$>$ meaning "is greater than."
$\ngtr$ meaning "is not greater than."
$\leq$ or $\leqq$ meaning "is less than or equal to."
$\nleq$ or $\nleqq$ meaning "is not less than or not equal to."
$\geq$ or $\geqq$ meaning "is greater than or equal to."
$\ngeq$ or $\ngeqq$ meaning "is not greater than or not equal to."

Observe that the symbol $\nless$ is equivalent to symbol $\geq$; symbol $\ngtr$ is equivalent to symbol $\leq$; symbol $\nleq$ is equivalent to symbol $>$; and symbol $\ngeq$ is equivalent to symbol $<$.

An open sentence is not a statement. We cannot tell whether it is true or false. It is only after we substitute a number for the variable that the open sentence becomes a statement. Then we can determine whether the sentence is true or false.

The sentence "$9\,x \leq 63$" is the shortened form of "$9\,x < 63$ or $9\,x = 63$" and is read "Nine times each number x is less than or equal to sixty-three."

The sentence "$4 < y < 7$" is the shortened form of "$4 < y$ and $y < 7$" and is read "Four is less than each number y which is less than seven" or "Each number y is greater than four and less than seven."

The sentence "$10 \geq a \geq 2$" is read "Ten is greater than or equal to each number a which is greater than or equal to two" or "Each number a is less than or equal to ten and greater than or equal to two."

EXERCISES

1. Read, or write in words, each of the following:

a. $n < 15$
b. $5\,c > 30$
c. $b - 8 > 11$
d. $x + 9 < 20$
e. $4\,t \nless 36$
f. $3\,x + 5 \neq x - 10$
g. $8\,n \ngtr 21 - 6\,n$
h. $\frac{d}{9} = 14$
i. $y - 7 < 4\,y + 6$
j. $11\,b - b \neq 25 - 2\,b$
k. $16\,a + 5 \ngtr 8\,a - 3$
l. $m \geq 39$
m. $12\,x \leq 72$
n. $b + 1 \geq 6$
o. $3\,n \nleq 54$
p. $7\,c - 1 \ngeq 3\,c + 4$
q. $3 < x < 8$
r. $7 > n > 1$
s. $5 < a < 32$
t. $17 > 4\,c > 0$
u. $2 \leq y \leq 56$
v. $14 \geq d \geq 10$
w. $3 < r \leq 8$
x. $1 \leq b < 15$

2. Write each of the following as an open sentence symbolically:

a. Some number r decreased by five is equal to twenty-one.
b. Each number n increased by ten is greater than sixteen.
c. Eleven times each number t plus fifteen is not equal to ninety.
d. Four times each number x is less than forty-eight.
e. Nine times each number b is not greater than eighty-one.
f. Each number y increased by one is not less than fourteen.
g. Each number a divided by six is greater than or equal to twenty.
h. Eight times each number d decreased by seven is less than or equal to nine.
i. Ten times each number s is greater than thirty and less than seventy.
j. Each number c is less than eleven and greater than six.

11–23 FORMULAS

A special kind of equation, called a *formula*, is a mathematical rule expressing the relationship of two or more quantities by means of numerals, variables, and operating symbols. The formula contains algebraic expressions.

To express mathematical and scientific principles as formulas, we write numerals, operating symbols, and letters (variables) representing the given quantities in the required order to show the relationship between quantities. A quantity may be represented by the first letter of a key word.

Write as a formula:

The selling price (s) is equal to the cost (c) increased by the margin (m).

Answer, $s = c + m$

To translate a formula to a word statement, we write a word rule stating the relationship expressed by the formula.

Translate:

$c = \pi d$ where c = circumference of circle, π = pi or 3.14, and d = diameter of circle.

Answer, The circumference of a circle equals pi times the diameter of the circle.

EXERCISES

1. Express each of the following as a formula:

a. The area of a rectangle (A) is equal to the product of the length (l) and width (w).

b. The sum of angles A, B, and C of triangle ABC is 180°.
c. The circumference of a circle (c) is equal to twice the product of pi and the radius (r).
d. The central angle of a regular polygon (a) equals 360° divided by the number of sides (n).
e. The perimeter of a rectangle (p) is twice the sum of the length (l) and width (w).
f. The area of a circle (A) is equal to one-fourth the product of pi and the square of the diameter (d).
g. The hypotenuse (h) of a right triangle is equal to the square root of the sum of the squares of the altitude (a) and the base (b).
h. The volume of a sphere (V) is four-thirds the product of pi and the cube of the radius (r).
i. The distance (d) a freely falling body drops is one-half the product of the acceleration due to gravity (g) and the square of the time of falling (t).
j. The temperature reading on the Fahrenheit scale (F) is equal to nine-fifths of the reading on the Celsius scale (C) increased by 32°.
k. The displacement of the piston (D) equals the area of the piston (A) times the stroke (s).
l. The pitch of a roof (p) is equal to the rise (r) divided by the span (s).
m. The cutting speed (s) of a bandsaw in feet per minute is equal to pi times the diameter (d) in feet times the number of revolutions per minute (R).
n. The capital (C) of a business is the difference between the assets (A) and the liabilities (L).
o. The selling price (s) is equal to the sum of the cost (c) and the profit (p).
p. The rate of discount (r) is equal to the discount (d) divided by the list price (l).
q. The tip speed of a propeller ($T.S.$) is equal to pi times the diameter (d) times the number of revolutions per second (N).
r. The horsepower ($H.P.$) required for the wing of an airplane equals the product of the drag of the wing (D) and the velocity (V) divided by 550.
s. The lift (L) of an airplane wing is the product of the lift coefficient (C), half of the air density (d), the wing area (A), and the square of the air speed or velocity (V).
t. The aspect ratio (r) of a wing of an airplane is equal to the span of the wing (s) divided by the chord (c).

2. Express each of the following formulas as a word statement:

a. $d = rt$ where d = distance traveled, r = average rate of speed, and t = time of travel.
b. $A = p + i$ where A = amount, p = principal, and i = interest.
c. $C = K - 273°$ where C = Celsius temperature reading and K = Kelvin temperature reading.
d. $d = \frac{m}{v}$ where d = density, m = mass, and v = volume.
e. $p = 2l + 2w$ where p = perimeter of rectangle, l = length, and w = width.
f. $A = \pi r^2$ where A = area of circle, π = pi or 3.14, and r = radius.

EVALUATION

11-24 ALGEBRAIC EXPRESSIONS

To evaluate an algebraic expression means to find the value of the algebraic expression which depends upon the values of the variables. If these values change, the value of the expression usually changes.

(1) To evaluate an algebraic expression, we copy the expression and substitute the given value for each variable. We then perform the necessary operations as indicated in the expression.
Observe that the value of each term (part of expression connected to other parts by either a plus or minus sign) is found and these values are combined.

(1) Find the value of $2x + 3y$ when $x = 8$ and $y = 5$.

$$\begin{aligned} &2x + 3y \\ &= 2\cdot 8 + 3\cdot 5 \\ &= 16 + 15 \\ &= 31 \end{aligned}$$

Answer, 31

(2) If the expression is a fraction, we simplify both the numerator and denominator separately, and then express the fraction in simplest terms.

(2) Find the value of $\frac{a+b}{a-b}$ when $a = 6$ and $b = 3$.

$$\frac{a+b}{a-b} = \frac{6+3}{6-3} = \frac{9}{3} = 3$$

Answer, 3

(3) If there are parentheses in the expression, we find the value of the expressions in each set of parentheses, and then perform the necessary operations to get the answer.

(3) Find the value of $2(l + w)$ when $l = 10$ and $w = 7$.

$$\begin{aligned} &2(l + w) \\ &= 2(10 + 7) \\ &= 2(17) \\ &= 34 \end{aligned}$$

Answer, 34

(4) If there are exponents in the expression, we raise each quantity to the proper power usually before performing the other operations.

(4) Find the value of $a^2 + b^2$ when $a = 5$ and $b = 4$.

$$\begin{aligned} &a^2 + b^2 \\ &= 5^2 + 4^2 \\ &= 25 + 16 \\ &= 41 \end{aligned}$$

Answer, 41

EXERCISES

Find the value of each of the following algebraic expressions:

When $a = 6$ and $c = 3$:

1. $a + c$ **2.** $a - c$ **3.** ac **4.** $\frac{a}{c}$ **5.** $a + 9c$

When $x = 12$, $y = 4$, and $z = 2$:

6. $3x + 5y$ **7.** $4xyz - 3xy$ **8.** $4xz - 8z - 11$ **9.** $\frac{4x - 5y}{3y + x}$

When $c = 10$, $n = -8$, and $y = -4$:

10. $4n + 5y$ **11.** $2c - 6n$ **12.** $\frac{2n}{3y}$ **13.** $cn + 2ny - 4y$

When $b = 8$, $c = 5$, and $x = 2$:

14. $b(c + x)$ **15.** $4b + (c - x)$ **16.** $5b + c(b + x)$ **17.** $(b + c)(b - x)$

When $m = 8$, $n = 4$, and $y = 1$:

18. $m(n - y)$ **19.** $3mn(m + y)$ **20.** $my + m(3m - 5n)$ **21.** $\frac{6(mn - 2y)}{11 + (m - ny)}$

When $a = -6$, $x = 4$, and $y = -4$:

22. $a - (x - y)$ **23.** $(a - x)(a + y)$ **24.** $\frac{3(y - 2a)}{x - 2(y + 2)}$ **25.** $a + y(x - a)$

When $a = 5$, $b = 4$, and $c = 2$:

26. $3a^2$ **27.** $b^3 - c^2$ **28.** $4a^2 + 3b^2$ **29.** $a^2 - 3a + 7$ **30.** $\frac{(a + b)^2}{a^2 + b^2}$

When $d = 2$, $n = 3$, and $x = 6$:

31. $n^2 - d^2$ **32.** $5x^2 - 2x + 9$ **33.** $8d^2 - 5dn$ **34.** $\frac{2n^2 + 5n + 6}{3n^2 - 7n + 2}$

When $b = -10$, $c = 5$, and $d = -2$:

35. $6d^2$ **36.** $c^2 - 2d^2$ **37.** $9b^3 - 4b^2 - 3b + 8$ **38.** $2b^2 - 3bd + 5d^2$

When $a = 10$, $b = 4$, $c = 3$, $d = 2\frac{1}{2}$, and $n = 1.5$:

39. $5a + 3c + 4d$ **40.** $a + b(b - n)$ **41.** $a^2 - 3ab + 2b^2$ **42.** $n^2 - \frac{n}{a}$

11–25 FORMULAS

To find the required value, we copy the given formula, substitute the given values for the letters, and perform the necessary operations.

Find the value of F when $C = 20$, using the formula $F = 1.8\,C + 32$:

$F = 1.8\,C + 32$
$F = 1.8 \times 20 + 32$
$F = 36 + 32$
$F = 68$ *Answer*, $F = 68$

EXERCISES

Find the value of:

1. p when $s = 8$, using the formula $p = 3\,s$.
2. A when $l = 12$ and $w = 7$, using the formula $A = lw$.
3. E when $I = 9$ and $R = 15$, using the formula $E = IR$.
4. W when $F = 90$ and $d = 8$, using the formula $W = Fd$.
5. A when $p = 125$ and $i = 19$, using the formula $A = p + i$.
6. K when $C = 38$, using the formula $K = C + 273$.
7. F when $C = 40$, using the formula $F = 1.8\,C + 32$.
8. v when $V = 15$, $g = 32$, and $t = 3$, using the formula $v = V + gt$.
9. B when $A = 53$, using the formula $B = 90 - A$.
10. C when $A = 9{,}800$ and $L = 1{,}250$, using the formula $C = A - L$.
11. b when $p = 23$ and $e = 8$, using the formula $b = p - 2\,e$.
12. C when $A = 42$ and $B = 76$, using the formula $C = 180 - A - B$.
13. a when $n = 8$, using the formula $a = \frac{360}{n}$.
14. r when $d = 171$ and $t = 9$, using the formula $r = \frac{d}{t}$.
15. I when $E = 110$ and $R = 22$, using the formula $I = \frac{E}{R}$.
16. W when $w = 75$, $l = 12$, and $L = 6$, using the formula $W = \frac{wl}{L}$.
17. A when $d = 50$, using the formula $A = .7854\,d^2$.
18. s when $t = 7$, using the formula $s = 16\,t^2$.
19. W when $I = 6$ and $R = 15$, using the formula $W = I^2R$.
20. V when $\pi = \frac{22}{7}$, $r = 14$, and $h = 35$, using the formula $V = \pi r^2 h$.
21. A when $p = 140$, $r = .06$, and $t = 5$, using the formula $A = p(1 + rt)$.
22. B when $A = 83$ and $C = 28$, using the formula $B = 180 - (A + C)$.
23. l when $a = 6$, $n = 9$, and $d = 3$, using the formula $l = a + (n - 1)d$.
24. C when $F = 50$, using the formula $C = \frac{5}{9}(F - 32)$.
25. E when $I = 7$, $r = 16$, and $R = 23$, using the formula $E = Ir + IR$.
26. k when $m = 14$ and $v = 5$, using the formula $k = \frac{1}{2}\,mv^2$.
27. A when $\pi = \frac{22}{7}$, $r = 7$, and $h = 18$, using the formula $A = 2\,\pi r(r + h)$.
28. K when $C = -15$, using the formula $K = C + 273$.
29. C when $F = -4$, using the formula $C = \frac{5}{9}(F - 32)$.

OPERATIONS

A *monomial* is a number or a variable or a product of variables or a product of a number and a variable or variables. 6, b, x^2y^3z, $9\,c$, and $-5\,a^4x$ are monomials.

A *polynomial* is a monomial or a sum of monomials. $7 + n$, $4\,x - 5$,

$2b^2 - 5ab + 8b^2$, and $4a^3 - 2a^2 + a - 9$ are polynomials.

Like terms are terms which have a common factor of identical variables raised to identical powers. $9b^2d^3$ and $-7b^2d^3$ are like terms. x and x^2 are not like terms.

Observe that an unexpressed numerical coefficient of 1 is understood in both x meaning $1x$ or $+1x$ and $-x$ meaning $-1x$. Of course $0x = 0$.

11–26 ADDITION

Monomials

To add like terms, we use the distributive property of multiplication over addition or briefly:

We find the algebraic sum of the numerical coefficients and prefix it to their common factor of variables.

To add unlike terms like $5b$ to $3a$, we indicate their sum as $3a + 5b$ since it cannot be expressed as a single term.

The sum of 2 x and 5 x:

$$2x + 5x = (2+5)x = 7x$$

or

$$\begin{array}{r} 2x \\ \underline{5x} \\ 7x \end{array}$$

Answer, $7x$

$$\begin{array}{r} -3a^3 \\ \underline{-3a^3} \\ -6a^3 \end{array} \qquad \begin{array}{r} +7xy \\ \underline{-8xy} \\ -xy \end{array}$$

Answer, $-6a^3$ Answer, $-xy$

EXERCISES

1. Add:

a.

$+3x$	$-7a$	$+6b$	$-4cd$	$+8n$	$-x$	$+6c$
$+2x$	$-5a$	$-8b$	$-cd$	$-13n$	$-x$	$-6c$

b.

$-1.4d$	$-4\frac{2}{3}xy$	$-6y^2$	$-8c^3d$	$7b^4$	$2a^3bc^2$	$-2x^2$
$+2.3d$	$+2\frac{1}{3}xy$	$+5y^2$	$-3c^3d$	$-b^4$	$4a^3bc^2$	$-2x^2$

c.

$-5x$	$-2bc$	$-8mn$	$2x^2y$	$-a$	$+6c$	$+7a^2x^3$
$+6x$	$-3bc$	$-3mn$	$-4x^2y$	$+2a$	$-7c$	$-4a^2x^3$
$-8x$	$+5bc$	$-mn$	$6x^2y$	$-a$	$-9c$	$+3a^2x^3$
				$+3a$	$+10c$	$-8a^2x^3$

2. Add as indicated:

a. $(6c) + (-9c)$

b. $(-2d) + (-d)$

c. $(-x) + (5x)$

d. $(-8y^3) + (-4y^3)$

e. $(-a) + (+a) + (-a)$

f. $(-4b) + (-b) + (5b)$

g. $(2xy) + (-5xy) + (-8xy)$

h. $(-7d^4) + (4d^4) + (-8d^4) + (-d^4)$

i. $(-m^2x) + (m^2x) + (3m^2x) + (-9m^2x)$

j. $(3a^2b^2) + (-5a^2b^2) + (-6a^2b^2) + (4a^2b^2)$

3. Simplify:

a. $3n + 7n + 4n$
b. $2a - 8a + a - 3a$
c. $-d - 2d - 3d - d$
d. $5b + 6b - 8b - 9b + b$
e. $4x - x - 2x + 3x - 5x$
f. $6ax - 3ax + ax - 2ax$
g. $-9bc - 3bc + 2bc + 4bc$
h. $-3c^2 + 4c^2 - c^2 - 10c^2$
i. $8a^2b^2 - a^2b^2 - 5a^2b^2 - 4a^2b^2 + 2a^2b^2$
j. $10y^3 - 3y^3 - 8y^3 + y^3 - 8y^3$

4. Add:

a. $4a$ and $5x$ **b.** $7c$ and $-5d$ **c.** $9x^2$ and $2x$ **d.** $-4bx$ and $-2cx$

Polynomials

To add two or more polynomials, we write the polynomials so that like terms are under each other and we add each column.

$$\begin{array}{rrr} 4a^2 & -\ ab & +\ 3b^2 \\ 3a^2 & -\ 7ab & -\ 4b^2 \\ \hline 7a^2 & -\ 8ab & -\ b^2 \end{array}$$

Answer, $7a^2 - 8ab - b^2$

Polynomials should be expressed in descending or ascending order of the powers of the variable.

$x^2 + 5 - 3x$ is written as $x^2 - 3x + 5$ or $5 - 3x + x^2$

EXERCISES

1. Add:

a.
$$\begin{array}{r} 3x+2 \\ \underline{7x+6} \end{array} \quad \begin{array}{r} 5a+8b \\ \underline{9a+b} \end{array} \quad \begin{array}{r} 7c-d \\ \underline{c-4d} \end{array} \quad \begin{array}{r} 4x^2-3y^2 \\ \underline{-2x^2+3y^2} \end{array} \quad \begin{array}{r} -7ax-2y \\ \underline{-4ax+y} \end{array} \quad \begin{array}{r} -3ac^3+9ac^2 \\ \underline{+3ac^3-9ac^2} \end{array}$$

b.
$$\begin{array}{r} 9a-4b+5c \\ \underline{3a+4b-5c} \end{array} \quad \begin{array}{r} 7a^2-3ab-4b^2 \\ \underline{2a^2-5ab+b^2} \end{array} \quad \begin{array}{r} 6m^2-mn \\ \underline{3mn+2n^2} \end{array}$$

$$\begin{array}{r} 2a-3b \\ -4a+7b \\ \underline{2a-5b} \end{array} \quad \begin{array}{r} 3b^2-5b-8 \\ 3b^2-7b-9 \\ \underline{-5b^2+2b-4} \end{array} \quad \begin{array}{r} x^2-4xy-y^2 \\ -5x^2-9xy+2y^2 \\ \underline{-4x^2-xy-y^2} \end{array}$$

2. Add as indicated:

a. $(x + 5) + (x + 2)$
b. $(c - 3d) + (4c - 5d)$
c. $(x^2 - 2y^2) + (y^2 - x^2)$
d. $(m + 3n) + (3m - n) + (2m - 5n)$
e. $(b^2 - 4bc) + (2bc - c^2) + (c^2 - b^2)$
f. $(a^2 - 5a + 9) + (2a^2 - a - 7) + (4 - 2a - 3a^2)$

3. Simplify:

a. $6x + 3y + 2x + 7y$
b. $3a - 1 + 4 - 5a$
c. $2c^2 - 3c + c^2$
d. $7x^2 - 4x + 8 + 9x - 4 - 2x^2$
e. $4c^2 - 5cd - d^2 + 8c^2 - 4d^2 - 9cd - c^2 + d^2$
f. $8b^3 - 2b^2 + b - 3 - 2b^2 - 10b^3 - b + 7 - b^2 - 1$

11–27 SUBTRACTION

Monomials

To subtract like terms, we use the distributive property of multiplication with respect to subtraction or briefly:

We add the additive inverse (or opposite) of the subtrahend to the minuend.

From 4 *ab* take 7 *ab*:

$(4ab) - (7ab) = (4 - 7)ab = -3ab$

$$\begin{array}{lrcr} & \text{Subtract:} & & \text{Add:} \\ \text{or} & 4ab & & +4ab \\ & \underline{7ab} & \longrightarrow & \underline{-7ab} \\ & & & -3ab \end{array}$$

Answer, $-3ab$

To subtract unlike terms, like subtracting $3x$ from $-5b$, we indicate their difference as $-5b - 3x$ since it cannot be expressed as a single term.

EXERCISES

1. Subtract:

a.	$7x$	$2a$	$+5x^2y$	$-6y$	$-2b^3$	$+9b$	$+ax$
	$6x$	$8a$	$+7x^2y$	$-2y$	$-11b^3$	$-3b$	$-4ax$
b.	$-5abc$	$-2r^2s$	$-8c$	0	$9b^2x$	$+6mn$	$-4abc$
	$+3abc$	$+10r^2s$	$-7c$	$+3cd$	0	$+6mn$	$-4abc$

2. Subtract as indicated:

a. $(6m) - (-5m)$
b. $(-3c^2x) - (-c^2x)$
c. $(-8dy) - (-8dy)$
d. $(-2cd^2) - (+3cd^2)$
e. $(10abx) - (11abx)$
f. $(a^2y^4) - (-a^2y^4)$

3. Do as indicated:

a. From $2n$ subtract $7n$.
b. Take $5y$ from $-9y$.
c. Subtract $-3x^2y^5$ from $-x^2y^5$.
d. From 0 take $-4cx$.

4. Do as indicated:

a. From $8x$ take $3y$.
b. Subtract $-2b$ from a.
c. From $-m^2$ subtract $-mn$.
d. Take $4x^2$ from 25.

Polynomials

To subtract one polynomial from another, we write the polynomials so that like terms are under each other and we subtract each column.

$$\begin{array}{r} 7x^2 - 4x + 6 \\ 2x^2 - 5x - 4 \\ \hline 5x^2 + x + 10 \end{array}$$

Answer, $5x^2 + x + 10$

EXERCISES

1. Subtract:

$$\begin{array}{r} 3c + 7 \\ c + 6 \\ \hline \end{array} \quad \begin{array}{r} 2x^2 - 3y^2 \\ 8x^2 + 7y^2 \\ \hline \end{array} \quad \begin{array}{r} 7a^2 - 9a - 5 \\ 2a^2 + 9a - 5 \\ \hline \end{array} \quad \begin{array}{r} 3cd + 6 \\ 5c^2 - 5cd \\ \hline \end{array} \quad \begin{array}{r} a^2 - 7ab - 2b^2 \\ -6a^2 - 8ab + b^2 \\ \hline \end{array}$$

2. Subtract as indicated:

a. $(6a - y) - (8a - 6y)$
b. $(x - y) - (-a - x)$
c. $(x^2 - 4x - 8) - (3x^2 - 5x + 7)$
d. $(4m^2 - 5mn - 2n^2) - (4n^2 - 8mn - m^2)$

3. a. From $4a - 7b$ take $2a + 9b$.
b. Subtract $9x^2 - 5x - 6$ from $3x^2 - 6x + 11$.
c. Take $6c^2$ from $2c^2 - 5$.
d. From $3b^3 - 4b^2 + 6$ subtract $9b - 8$.
e. What must be added to $12y^2 - 8y + 1$ to get $8y^2 - 6y - 2$?
f. From zero subtract $2x^2 - 3y^2$.
g. From the sum of $8m^2 - 7m + 2$ and $3m^2 - 9m - 7$ subtract $12 - 11m^2$.
h. Subtract the sum of $9a^2 + 5ab - 10b^2$ and $6b^2 - a^2$ from $4a^2 - 2ab$.
i. From the sum of $4n^2 - 8n - 5$ and $8n + 5$ subtract the sum of $3n - 4$ and $9 - 6n - n^2$.

11–28 MULTIPLICATION

Law of Exponents for Multiplication

Since $a^5 = a \cdot a \cdot a \cdot a \cdot a$ and $a^3 = a \cdot a \cdot a$
then $a^5 \cdot a^3 = \underline{a \cdot a \cdot a \cdot a \cdot a} \cdot \underline{a \cdot a \cdot a} = a^8$
Thus $a^5 \cdot a^3 = a^{5+3} = a^8$

In general, $a^m \cdot a^n = a^{m+n}$ where a represents any number and m and n are positive integers. Observe that the exponent of the variable in the product is the sum of the exponents of that variable in the given factors. This is not true for the product of two different variables. For example, the product of b^5 and c^3 is b^5c^3. Here the variables are written alphabetically.

Monomials

To multiply monomials, we use the commutative and associative properties and the law of exponents for multiplication, or, briefly:

We multiply the numerical coefficients as we do signed numbers and find the product of the variables using the law of exponents.

We prefix the numerical product to the product of the variables arranged in alphabetical order.

Note: $x = 1\,x^1$

Multiply $8\,c^3d^2$ by $3\,c^4d^6$:

$$(8\,c^3d^2)(3\,c^4d^6)$$
$$= (8 \cdot 3)(c^3 \cdot c^4)(d^2 \cdot d^6)$$
$$= \quad 24 \qquad c^7 \qquad d^8$$

Answer, $24\,c^7d^8$

Multiply:

$$\begin{array}{r} -\,7\,m^4x^2 \\ +\,8\,m^2y^3 \\ \hline -\,56\,m^6x^2y^3 \end{array}$$

Answer, $-\,56\,m^6x^2y^3$

EXERCISES

1. Write in shortened form using exponents:

a. $c \cdot c \cdot c \cdot x \cdot x$
b. $a \cdot a \cdot a \cdot y \cdot y \cdot y \cdot y$
c. $b \cdot b \cdot b \cdot b \cdot b \cdot d \cdot d \cdot d \cdot d$
d. $x \cdot x \cdot x \cdot x \cdot x \cdot x \cdot x \cdot x \cdot y$

2. Multiply:

a. $n^5 \cdot n^7$
b. $x^8 \cdot x^{11}$
c. $t^5 \cdot t$
d. $10^5 \times 10^8$
e. $10^{12} \times 10^9$
f. $6^2 \times 6^7$
g. $b^9 \cdot a^{10}$
h. $c^2 \cdot y$
i. $n^4 \cdot d^5$
j. $a^5x^2 \cdot ax^8$
k. $c^7d^2 \cdot c^2d^5$
l. $x^9y \cdot y^7$

3. Multiply:

a.

a^4	m^{15}	$-\,c$	$6\,a^6$	$-\,5\,x$	$+\,8\,m^3$	$-\,2\,b^4$
a^7	m^{10}	$-\,c$	$4\,a^4$	$-\,6\,x$	$-\,8\,m^3$	$+\,5\,b^9$

b.

$6\,m^2n$	$-\,4\,b^7c$	$8\,a^3b^2$	$-\,7\,x$	$-\,y$	$-\,1$	$-\,cy$
$3\,mn^5$	$9\,b^4c^5$	$6\,a^2b^8$	$-\,xy$	$-\,8$	$4\,d^3$	$-\,5\,bx^9$

c.

$-\,s$	$6\,c$	$-\,7\,x^3y^5$	$-\,3\,bcd$	$-\,9\,ax^3$	$-\,10\,dt^5$	$8\,m^7n^6r^8$
$5\,t$	$-\,cx$	$-\,4\,x^2y^9$	8	$-\,2\,a^4y^2$	$-\,10\,dt^5$	$-\,m^3n^4r^8$

4. Multiply as indicated:

a. $7\,y \cdot y$
b. $6\,ax^2 \cdot 2\,a^5x^3$
c. $(-\,x)(-\,4\,x^6)$
d. $(7\,c^4d^5)(-\,8\,c^3d^7)$
e. $(-\,9\,b^2x)(-\,6\,x^5y^6)$
f. $6(8\,a)$
g. $-\,5\,x^3(x^4)$
h. $-\,c\,(-\,d)$
i. $-\,8\,a^4x^3(4\,a^5x^9)$
j. $10\,b^7y^5(-\,3\,b^6y^8)$
k. $4\,y^3 \cdot 3\,y^2 \cdot 2\,y^7$
l. $5\,a \cdot 2\,b \cdot 9\,c$
m. $(-\,b)(-\,b)(-\,b)$
n. $(-\,3\,c^2)(-\,3\,c^2)(-\,3\,c^2)(-\,3\,c^2)$
o. $(2\,x^2y)(-\,5\,x^3)(-\,3\,xy^2)(-\,y^7)$

5. Multiply as indicated:

a. $(c^3y)^2$
b. $(-\,5\,b^4x^3)^2$
c. $(-\,3\,x)^3$
d. $(-\,4\,m^3x^5)^4$
e. $(7\,a^5c^6d)^4$
f. $(-\,2\,b^2x^4y^8)^5$

Polynomials

To multiply a polynomial by a monomial, we use the distributive property by multiplying each term of the polynomial by the monomial.

$$2a^3x(3a^2 + 5ax + 7x^2) = (2a^3x \cdot 3a^2) + (2a^3x \cdot 5ax) + (2a^3x \cdot 7x^2)$$
$$= 6a^5x + 10a^4x^2 + 14a^3x^3$$

Answer, $6a^5x + 10a^4x^2 + 14a^3x^3$

EXERCISES

1. Multiply:

a. $-5(x + 7)$
b. $6(3a - 9b)$
c. $y(y + 1)$
d. $-a^3(a^2b + b^5)$
e. $-6x^5y^2(-4x^2 - 3y^4)$
f. $4c^2(6c^2 - 4c + 3)$
g. $-9b^3d^2(2b^2 - 5bd - 6d^2)$
h. $-3b^4(7b^2 - 6b + 9)$
i. $-10m^7n^3(8m^4 + 3m^2n^2 - 5n^4)$
j. $7ab^3x^5(4a^5b^2 - 6a^3x^6 + bx^7)$

2. Multiply:

$$\begin{array}{r} 3x - 7 \\ 4 \\ \hline \end{array} \qquad \begin{array}{r} 12b^2 - 5bc \\ -b \\ \hline \end{array} \qquad \begin{array}{r} 6a^2 + ax - 4x^2 \\ -2ax \\ \hline \end{array} \qquad \begin{array}{r} x^3 - 5xy + 6xy^2 - 8y^3 \\ 9x^2y^3 \\ \hline \end{array}$$

11–29 DIVISION

Law of Exponents for Division

Since $a^8 = a \cdot a \cdot a \cdot a \cdot a \cdot a \cdot a \cdot a$ and $a^3 = a \cdot a \cdot a$

then $\dfrac{a^8}{a^3} = \dfrac{\overset{1}{\cancel{a}} \cdot \overset{1}{\cancel{a}} \cdot \overset{1}{\cancel{a}} \cdot a \cdot a \cdot a \cdot a \cdot a \cdot}{\underset{1}{\cancel{a}} \cdot \underset{1}{\cancel{a}} \cdot \underset{1}{\cancel{a}}} = a^5$

Thus $\dfrac{a^8}{a^3} = a^{8-3} = a^5$

In general $a^m \div a^n = \dfrac{a^m}{a^n} = a^{m-n}$ where a represents any non-zero number and m and n are positive integers such that $m > n$. Observe that for any given variable its exponent in the quotient is equal to its exponent in the dividend (numerator) minus its exponent in the divisor (denominator).

Monomials

To divide monomials, we divide the numerical coefficients as we do signed numbers and find the quotient of the variables using the law of exponents for division. We prefix the numerical quotient to the product of the quotients of the variables arranged in alphabetical order.

Divide $-48\,b^9x^8$ by $6\,b^3x^4$:

$$\frac{-48\,b^9x^8}{6\,b^3x^4} = \frac{-48}{6} \cdot \frac{b^9}{b^3} \cdot \frac{x^8}{x^4}$$
$$= -8 \quad b^6 \quad x^4$$

Answer, $-8\,b^6x^4$

EXERCISES

1. Divide:

a. x^7 by x^4
b. d^9 by d^3
c. y^{12} by y^2
d. 10^8 by 10^2
e. 10^{11} by 10^5
f. 8^6 by 8^3
g. m^9 by m^9
h. b^6c^3 by c^2
i. x^7y^{15} by xy^8
j. a^5b^9 by a^4b^6
k. $m^{12}x^8$ by m^5
l. $r^{10}s^8t^4$ by rs^7t^2

2. Divide:

a. $\dfrac{6\,a^{12}}{3}$ $\quad \dfrac{-35\,x^8}{-7}$ $\quad \dfrac{9\,b^5}{-1}$ $\quad \dfrac{-8\,y^6}{-8}$ $\quad \dfrac{54\,c^3d^9}{9\,c^2d^3}$ $\quad \dfrac{-72\,m^6x^8}{-8\,m^2x^4}$ $\quad \dfrac{+18\,b^3c}{-6\,b^2}$

b. $\dfrac{+32\,gt^2}{-16\,g}$ $\quad \dfrac{-36\,a^7x^8y}{4\,a^5x^2y}$ $\quad \dfrac{-25\,b^6c^5}{-25\,b^2c}$ $\quad \dfrac{80\,m^7x^3}{-5\,m^7x^3}$ $\quad \dfrac{-x}{-1}$ $\quad \dfrac{24\,c^6x^5y^3}{4\,c^4y^3}$ $\quad \dfrac{-2\,m^2x^4}{2\,m^2x^4}$

3. Divide as indicated:

a. $x^6\overline{)-x^{12}}$
b. $-3\overline{)-12\,a^4b^5}$
c. $-5\,c^2d\overline{)20\,c^7d^8}$
d. $-6\,a^6\overline{)-30\,a^{18}y^4}$
e. $-2\,x^3y^4\overline{)-8\,x^{10}y^6}$
f. $(a^6b^9) \div (a^3b)$
g. $(32\,b^5x^7) \div (-4\,x^2)$
h. $(-72\,s^{10}) \div (-9\,s^2)$
i. $(30\,b^8t^7) \div (-5\,b^5t^7)$

Polynomials

To divide a polynomial by a monomial, we use the distributive property by dividing each term of the polynomial by the monomial.

Divide $4\,c^9x^3 - 10c^6x^5 + 12\,c^4x^7$ by $-2\,c^4x^2$:

$$\frac{4\,c^9x^3 - 10c^6x^5 + 12\,c^4x^7}{-2\,c^4x^2} = \frac{4\,c^9x^3}{-2\,c^4x^2} + \frac{-10\,c^6x^5}{-2\,c^4x^2} + \frac{12\,c^4x^7}{-2\,c^4x^2}$$
$$= -2\,c^5x + 5\,c^2x^3 - 6\,x^5$$

Answer, $-2\,c^5x + 5\,c^2x^3 - 6\,x^5$

EXERCISES

1. Divide:

a. $\frac{24a + 16}{8}$ $\quad \frac{6a - 12x}{-3}$ $\quad \frac{14c - 7}{-7}$ $\quad \frac{30b - 5}{5}$ $\quad \frac{2m^2 + 7mn - n^2}{-1}$

b. $\frac{x^4 - x^3}{-x}$ $\quad \frac{8d^2 - d}{-d}$ $\quad \frac{a^4b^3 - a^3b^4 + a^2b^5}{a^2b^2}$ $\quad \frac{16x^5y^3 - 20x^4y^4 + 36x^3y^5}{-4x^3y^2}$

2. Divide as indicated:

a. $-9\overline{)27x^4 - 18}$

b. $5b^2\overline{)20b^4 - 15b^3 + 35b^2}$

c. $-2x^2y^4\overline{)8x^{10}y^6 - 4x^6y^8 - 10x^4y^{12}}$

d. $(8a - 12b) \div (-4)$

e. $(12x^7 - 9x^5 + 6x^3) \div (-3x^2)$

f. $(36a^9b^7 + 30a^8b^6 - 42a^5b^4) \div (6a^5b^3)$

11–30 ZERO EXPONENTS; NEGATIVE EXPONENTS

The law of exponents for division (see section 11–29) indicates that $a^m \div a^n = \frac{a^m}{a^n} = a^{m-n}$ where a represents any non-zero number and m and n are positive integers such that $m > n$. Here we use exponents which name natural numbers indicating how many times the factor is being used in multiplication as a repeated factor. We found that for any given variable its exponent in the quotient is equal to its exponent in the dividend (numerator) minus its exponent in the divisor (denominator).

For example: $\frac{7^9}{7^5} = 7^{9-5} = 7^4$ Answer, 7^4

and when $b \neq 0$, $\frac{b^{12}}{b^4} = b^{12-4} = b^8$ Answer, b^8

In order to give meaning to zero exponents and negative exponents, let us extend the law of exponents for division to include $\frac{a^m}{a^n}$ both for $m = n$ and $m < n$ when $a \neq 0$.

For example: $\frac{a^6}{a^6} = \frac{\overset{1}{\cancel{a}} \cdot \overset{1}{\cancel{a}} \cdot \overset{1}{\cancel{a}} \cdot \overset{1}{\cancel{a}} \cdot \overset{1}{\cancel{a}} \cdot \overset{1}{\cancel{a}}}{\underset{1}{\cancel{a}} \cdot \underset{1}{\cancel{a}} \cdot \underset{1}{\cancel{a}} \cdot \underset{1}{\cancel{a}} \cdot \underset{1}{\cancel{a}} \cdot \underset{1}{\cancel{a}}} = \frac{1}{1} = 1$

By law of exponents, $\frac{a^6}{a^6} = a^{6-6} = a^0$

Since in general $\frac{a^m}{a^m} = a^{m-m} = a^0$ and $\frac{a^m}{a^m} = 1$, then we may define a^0 to equal 1 when a represents any non-zero number. That is: $a^0 = 1$.

The expression 0^0 is meaningless.

We have found that $\frac{a^8}{a^3} = \frac{\overset{1}{\cancel{a}} \cdot \overset{1}{\cancel{a}} \cdot \overset{1}{\cancel{a}} \cdot a \cdot a \cdot a \cdot a \cdot a}{\underset{1}{\cancel{a}} \cdot \underset{1}{\cancel{a}} \cdot \underset{1}{\cancel{a}}} = a^5$

and by law of exponents that $\frac{a^8}{a^3} = a^{8-3} = a^5$ when $a \neq 0$.

Here we see that the exponent five (5) in the quotient a^5 indicates that the numerator has five (5) more of the repeated factor than the denominator.

In a similar manner, we find that $\frac{a^3}{a^8} = \frac{\overset{1}{\cancel{a}} \cdot \overset{1}{\cancel{a}} \cdot \overset{1}{\cancel{a}}}{\underset{1}{\cancel{a}} \cdot \underset{1}{\cancel{a}} \cdot \underset{1}{\cancel{a}} \cdot a \cdot a \cdot a \cdot a \cdot a} = \frac{1}{a^5}$

and by law of exponents that $\frac{a^3}{a^8} = a^{3-8} = a^{-5}$.

That is: $a^{-5} = \frac{1}{a^5}$ when $a \neq 0$.

Here we see that the exponent negative five (−5) in the quotient a^{-5} indicates that the denominator has five (5) more of the repeated factor than the numerator.

Thus in general,

$$a^{-1} = \frac{1}{a};\ a^{-2} = \frac{1}{a^2};\ a^{-3} = \frac{1}{a^3};\ a^{-4} = \frac{1}{a^4};\ \text{etc. when } a \neq 0.$$

EXERCISES

1. Find the numerical value of each of the following:

a. 10^0 **b.** 3^0 **c.** 100^0 **d.** 19^0 **e.** 35^0 **f.** 147^0 **g.** 500^0

2. Find the numerical value of each of the following:

a. 3×10^0 **b.** 8×10^0 **c.** 5×10^0 **d.** 4×6^0 **e.** $5^0 \times 7^0$ **f.** $12^0 \times 3$ **g.** $9^0 \times 4^0$

3. Find the numerical value of each of the following when the variables are not zero:

a. m^0 **b.** d^0 **c.** r^0 **d.** x^0y^0 **e.** $(4c)^0$ **f.** $10n^0$ **g.** $7z^0$

4. Divide, then determine the numerical value when the variables are not zero:

a. $\frac{b^4}{b^4}$ b. $\frac{m^8}{m^8}$ c. $\frac{t^9}{t^9}$ d. $\frac{r^3s^2}{r^3s^2}$ e. $\frac{ab^3c^5}{ab^3c^5}$ f. $\frac{x^2y^4z^6}{x^2y^4z^6}$ g. $\frac{b^9c^6d^7}{b^9c^6d^7}$

5. Name each of the following, using negative exponents:

a. $\frac{1}{10^3}$ c. $\frac{1}{10^1}$ e. $\frac{1}{10^8}$ g. $\frac{1}{3^4}$ i. $\frac{1}{8^6}$ k. $\frac{1}{7^5}$
b. $\frac{1}{10^2}$ d. $\frac{1}{10^5}$ f. $\frac{1}{10^{10}}$ h. $\frac{1}{5^1}$ j. $\frac{1}{12^9}$ l. $\frac{1}{6^{14}}$

6. Name each of the following as a fraction, using a positive exponent:

a. 10^{-1} c. 10^{-4} e. 10^{-12} g. 8^{-5} i. 12^{-2} k. 100^{-15}
b. 10^{-6} d. 10^{-9} f. 10^{-7} h. 6^{-3} j. 15^{-7} l. 70^{-10}

7. Find the numerical value of each of the following:

a. 10^{-2} c. 10^{-3} e. 10^{-10} g. 2^{-2} i. 3^{-4} k. 5^{-6}
b. 10^{-4} d. 10^{-7} f. 10^{-9} h. 1^{-7} j. 4^{-3} l. 7^{-2}

8. Multiply, expressing each product as a power of the given base:

a. $x^4 \cdot x^6$ c. $b^3 \cdot b^{-3}$ e. $n^{-9} \cdot n^{-1}$ g. $y^{-2} \cdot y^2$ i. $w^{-3} \cdot w$
b. $x^{-4} \cdot x^{-6}$ d. $c^{-8} \cdot c^5$ f. $d^0 \cdot d^{-6}$ h. $a^{-1} \cdot a^{-1}$ j. $x^{-5} \cdot x^6$

9. Multiply, expressing first each product as a power of the given base, then find its numerical value:

a. $10^3 \times 10^6$ d. $10^6 \times 10^{-2}$ g. $10^0 \times 10$ j. $3^2 \times 3^{-3}$ m. $1^{12} \times 1^{-8}$
b. $10^{-1} \times 10^{-4}$ e. $10^{-11} \times 10^8$ h. $10^{-2} \times 10^{-4}$ k. $5^{-7} \times 5^4$ n. $3^{-5} \times 3^9$
c. $10^{-5} \times 10^3$ f. $10^{-5} \times 10^7$ i. $10^{-6} \times 10^6$ l. $6^{-2} \times 6^{-2}$ o. $2^{-8} \times 2^{-1}$

10. Divide, expressing each quotient as a power of the given base:

a. $\frac{n^2}{n^5}$ d. $m^9 \div m^{17}$ g. $\frac{s^2}{s^3}$ j. $b^4 \div b^{11}$ m. $\frac{x^5}{x^{15}}$
b. $\frac{b^9}{b^{14}}$ e. $d^5 \div d^5$ h. $\frac{a^8}{a^{15}}$ k. $z \div z^{10}$ n. $\frac{y^3}{y^{11}}$
c. $\frac{r}{r^6}$ f. $t^{12} \div t^{11}$ i. $\frac{c}{c^{10}}$ l. $w^4 \div w^0$ o. $\frac{n^0}{n^2}$

11. Divide, expressing first each quotient as a power of the given base, then find its numerical value:

a. $\frac{10^2}{10^6}$ d. $\frac{10^{-8}}{10^{12}}$ g. $\frac{10^{-5}}{10^{-6}}$ j. $\frac{10^{-16}}{10^{-7}}$ m. $\frac{3^{-7}}{3^{-5}}$
b. $\frac{10^9}{10^{16}}$ e. $\frac{10^7}{10^{-5}}$ h. $\frac{10^{-3}}{10^{-1}}$ k. $\frac{10^{11}}{10^0}$ n. $\frac{2^8}{2^{-1}}$
c. $\frac{10^{-2}}{10^2}$ f. $\frac{10^6}{10^{-9}}$ i. $\frac{10^0}{10^{-7}}$ l. $\frac{10^{-6}}{10^{-6}}$ o. $\frac{5^0}{5^{-3}}$

12. Compute, expressing each answer as a decimal numeral:

a. 3×10^0	**d.** 5×10^{-5}	**g.** 4×10^{-12}	**j.** 6×5^0	**m.** 5×9^{-1}
b. 8×10^{-2}	**e.** 7×10^{-3}	**h.** 9×10^{-15}	**k.** 7×2^{-2}	**n.** 4×6^{-2}
c. 9×10^{-6}	**f.** 6×10^{-4}	**i.** 2×10^{-20}	**l.** 8×3^{-3}	**o.** 2×7^{-4}

13. Show that:

a. $\frac{1}{2^{-3}} = 2^3$ **b.** $\frac{1}{10^{-7}} = 10^7$ **c.** $\frac{a^{-3}}{b^{-4}} = \frac{b^4}{a^3}$ **d.** $\frac{c^2x^{-5}}{d^3y^{-2}} = \frac{c^2y^2}{d^3x^5}$

14. In the following, simplify each product by scientific notation:

Multiply:

a. 2×10^5 by 3×10^8	**c.** 3×10^{12} by 5×10^{16}	**e.** 8×10^{19} by 6×10^{13}
b. 2×10^{11} by 2×10^{15}	**d.** 9×10^{21} by 4×10^{14}	**f.** 7×10^{15} by 9×10^9

15. Divide:

a. $\frac{6 \times 10^8}{3 \times 10^4}$ **b.** $\frac{8 \times 10^{15}}{2 \times 10^7}$ **c.** $\frac{3.4 \times 10^9}{1.7 \times 10^2}$ **d.** $\frac{9.6 \times 10^{11}}{1.2 \times 10^7}$ **e.** $\frac{5.4 \times 10^{25}}{1.8 \times 10^{14}}$

OPEN SENTENCES IN ONE VARIABLE

11–31 INTRODUCTION TO THE SOLUTION OF EQUATIONS IN ONE VARIABLE

To solve an equation in one variable means to find the number represented by the variable which, when substituted for the variable, will make the sentence true.

Any number that makes the sentence true is said to *satisfy* the equation and to be the *root* or the *solution* of the open sentence.

The equation $n + 2 = 8$ indicates that some number n plus two equals eight. When 6 is substituted for the variable n, the resulting sentence $6 + 2 = 8$ is a true sentence. Thus 6 is said to be the root of the equation $n + 2 = 8$.

The set of all numbers which satisfy the sentence is called the *solution set* or the *truth set* of the sentence. Since there is only one number, 6, which satisfies the equation $n + 2 = 8$, then we say that the solution set of the equation $n + 2 = 8$ contains only one element, 6, and is written as $\{6\}$.

Checking an equation is testing whether some number belongs to the solution set of this equation by substituting the number for the variable in the equation.

If the resulting sentence is true, the number belongs to the solution set; if not, the number does not belong to the solution set.

> 6 belongs to the solution set of $n + 2 = 8$ because $6 + 2 = 8$. 5 does not belong because $5 + 2 \neq 8$.

The expressions at the left and at the right of the equality sign in an equation are called *members* or *sides* of the equation.

> In the equation $n + 2 = 8$, the expression $n + 2$ is the left member and the expression 8 is the right member.

Equations which have exactly the same solution sets are called *equivalent equations.* The equation $n + 2 = 8$ and $n = 6$ are equivalent equations because they both have the same solution set $\{6\}$. An equation is considered to be in its *simplest form* when one member contains only the variable itself and the other member is a constant term. The equation $n = 6$ is an equation in the simplest form.

EXERCISES

1. Which of the following numbers will make the sentence $x + 3 = 9$ true?

a. $x = 7$ **b.** $x = 3$ **c.** $x = 5$ **d.** $x = 6$ **e.** $x = 12$

2. Which of the following numbers will make the sentence $n - 4 = 12$ true?

a. $n = 8$ **b.** $n = 3$ **c.** n = 4 **d.** $n = 18$ **e.** $n = 16$

3. Which of the following numbers will make the sentence $6b = 30$ true?

a. $b = 24$ **b.** $b = 6$ **c.** $b = \frac{1}{5}$ **d.** $b = 5$ **e.** $b = 36$

4. Which of the following numbers will make the sentence $\frac{c}{8} = 4$ true?

a. $c = 2$ **b.** $c = 4$ **c.** $c = 32$ **d.** $c = \frac{1}{2}$ **e.** $c = 1$

5. Which of the following numbers is the root of the equation $r + 5 = 5$?

a. $r = 10$ **b.** $r = 1$ **c.** $r = 0$ **d.** $r = -1$ **e.** $r = -10$

6. Which of the following numbers is the root of the equation $8n = 48$?

a. $n = 40$ **b.** $n = 56$ **c.** $n = 6$ **d.** $n = 14$ **e.** $n = 26$

7. Which of the following numbers is the root of the equation $y - 5 = 1$?

a. $y = 5$ **b.** $y = -1$ **c.** $y = 4$ **d.** $y = 6$ **e.** $y = 3$

8. Which of the following numbers is the root of the equation $\frac{d}{2} = 6$?

a. $d = 3$ b. $d = 4$ c. $d = 8$ d. d = 12 e. $d = \frac{1}{3}$

9. Which of the following sets is the solution set of the equation $s + 11 = 33$?

a. {44} b. {3} c. {4} d. {22} e. {11}

10. Which of the following sets is the solution set of the equation $\frac{m}{15} = 12$?

a. {160} b. {3} c. {180} d. {27} e. {200}

11. Which of the following sets is the solution set of the equation $9v = 54$?

a. {45} b. {63} c. {7} d. {6} e. {− 9}

12. Which of the following sets is the solution set of the equation $w - 8 = 8$?

a. {0} b. {88} c. {− 16} d. {1} e. {16}

13. a. Write the right member of the equation $12x - 9 = 31$.
b. Write the left member of the equation $4y + 17 = 5$.

14. Which of the following equations have {7} as the solution set? Which are equivalent equations?

a. $n + 13 = 6$ b. $n - 3 = 4$ c. $12n = 84$ d. $n = 7$ e. $\frac{n}{7} = 1$

15. Which of the following equations have {8} as the solution set? Which are equivalent equations? Which equation of the equivalent equations is in the simplest form?

a. $x + 7 = 15$ b. $\frac{x}{4} = 2$ c. $x - 3 = 5$ d. $x = 8$ e. $10x = 80$

11–32 INVERSE OPERATIONS; PROPERTIES OF EQUALITY

In section 1–8 we studied that *inverse operations undo each other*. That is: subtraction undoes addition, addition undoes subtraction, division undoes multiplication, and multiplication undoes division.

We found that:

(1) When we first add a number to the given number n and then subtract this number from the sum, we return to the given number n.

That is: $(n + 1) - 1 = n$; $(n + 2) - 2 = n$; $(n + 3) - 3 = n$; etc.

(2) When we first subtract a number from the given number n and then add this number to the answer, we return to the given number n.

That is: $(n - 1) + 1 = n$; $(n - 2) + 2 = n$; $(n - 3) + 3 = n$; etc.

(3) When we first multiply the given number n by a number and then we divide the product by this number, we return to the given number n.

That is: $(n \times 2) \div 2 = n$ or $\frac{2n}{2} = n$; $(n \times 3) \div 3 = n$ or $\frac{3n}{3} = n$;

$(n \times 4) \div 4 = n$ or $\frac{4n}{4} = n$; etc.

(4) When we first divide the given number n by a number and then we multiply the quotient by this number, we return to the given number n.

That is: $(n \div 2) \times 2 = n$ or $\frac{n}{2} \times 2 = n$; $(n \div 3) \times 3 = n$ or $\frac{n}{3} \times 3 = n$;

$(n \div 4) \times 4 = n$ or $\frac{n}{4} \times 4 = n$; etc.

The *properties of equality* which state that the results are equals when equals are increased or decreased or multiplied or divided by equals, with division by zero excluded, are called *axioms.*

These axioms allow us in each of the following cases to obtain an equation in the simplest form.

a. When we subtract the same number from both sides of a given equation.
For example: Given: $n + 5 = 30$
Subtracting 5 from each side: $(n + 5) - 5 = 30 - 5$
we get: $n = 25$

b. When we add the same number to both sides of a given equation.
For example: Given: $n - 5 = 30$
Adding 5 to each side: $(n - 5) + 5 = 30 + 5$
we get: $n = 35$

c. When we divide both sides of given equation by the same non-zero number.
For example: Given: $5n = 30$
Dividing each side by 5: $\frac{5n}{5} = \frac{30}{5}$
we get: $n = 6$

d. When we multiply both sides of a given equation by the same non-zero number.
For example: Given: $\frac{n}{5} = 30$
Multiplying each side by 5: $5 \cdot \frac{n}{5} = 5 \cdot 30$
we get: $n = 150$

EXERCISES

In each of the following, find the missing number and, where required, the missing operation:

1. a. $(15 + 6) - \square = 15$ b. $(18 + 43)\ ?\ \square = 18$ c. $(n + 5)\ ?\ \square = n$
2. a. $(24 - 2) +\ ? = 24$ b. $(52 - 17)\ ?\ \square = 52$ c. $(x - 9)\ ?\ \square = x$
3. a. $(7 \times 3) \div \square = 7$ b. $(26 \times 18)\ ?\ \square = 26$ c. $(a \times 8)\ ?\ \square = a$
4. a. $(11 \times 8) \div \square = 8$ b. $(41 \times 72)\ ?\ \square = 72$ c. $(10y)\ ?\ \square = y$
5. a. $(12 \div 2) \times \square = 12$ b. $(15 \div 4)\ ?\ \square = 15$ c. $(d \div 5)\ ?\ \square = d$
6. a. $\frac{5}{6} \times\ ? = 5$ b. $\frac{11}{4} \times\ ? = 11$ c. $\frac{n}{8} \times\ ? = n$
7. a. $(r - 10)\ ?\ \square = r$ c. $(x + 45)\ ?\ \square = x$ e. $(17m)\ ?\ \square = m$
   b. $(16h)\ ?\ \square = h$ d. $(y - 32)\ ?\ \square = y$ f. $\frac{a}{50}\ ?\ \square = a$
8. In each of the following, what number do you subtract from both sides of the given equation to get an equivalent equation in simplest form? Also write this equivalent equation.

   a. $n + 9 = 19$ b. $c + 3 = 11$ c. $r + 36 = 53$ d. $t + 90 = 117$
9. In each of the following, what number do you add to both sides of the given equation to get an equivalent equation in simplest form? Also write this equivalent equation.

   a. $w - 4 = 9$ b. $x - 13 = 7$ c. $m - 6 = 0$ d. $p - 28 = 28$
10. In each of the following, by what number do you divide both sides of the given equation to get an equivalent equation in simplest form? Also write this equivalent equation.

    a. $11c = 55$ b. $42a = 21$ c. $7c = 3$ d. $16y = 38$
11. In each of the following, by what number do you multiply both sides of the given equation to get an equivalent equation in simplest form? Also write this equivalent equation.

    a. $\frac{x}{6} = 12$ b. $\frac{n}{8} = 2$ c. $\frac{b}{16} = 13$ d. $\frac{w}{9} = 9$
12. What operation with what number do you use on both sides of each of the following equations to get an equivalent equation in simplest form? Also write this equivalent equation.

    a. $8c = 24$
    b. $n + 7 = 12$
    c. $y - 4 = 4$
    d. $9d = 45$
    e. $\frac{w}{7} = 6$
    f. $13r = 60$
    g. $m - 21 = 8$
    h. $z + 37 = 37$
    i. $\frac{n}{5} = 1$
    j. $18t = 12$
    k. $h - 110 = 39$
    l. $3y = 2$
    m. $r + 87 = 125$
    n. $50 = c - 19$
    o. $24 = \frac{b}{8}$
    p. $r - 83 = 0$
    q. $20t = 75$
    r. $\frac{v}{12} = 9$
    s. $91 = p + 62$
    t. $6 = 18t$

11–33 SOLVING EQUATIONS IN ONE VARIABLE

To solve an equation in one variable, we transform the given equation to the simplest equivalent equation which has only the variable itself as one member and a constant naming a number as the other member. To make this transformation we use the properties of equality (axioms) with inverse operations or with the additive inverse or multiplicative inverse as required.

An equation may be compared to balanced scales. To keep the equation in balance, any change on one side of the equality sign must be balanced by an equal change on the other side of the equality sign.

Observe in the models on this page and on pages 305, 306, and 307 that the following four basic types of equations:

Type I $n + 2 = 8$ **Type III** $2n = 8$

Type II $n - 2 = 8$ **Type IV** $\frac{n}{2} = 8$

are solved both by inverse operations and by the additive inverse or the multiplicative inverse. When the inverse operation method is used, the operation indicated in the given equation by the variable and its connected constant is undone by performing the inverse operation on both sides of the equation.

To check whether the number found is a member of the solution set, we substitute this number for the variable in the given equation. If the resulting sentence is true, then the number belongs to the solution set. A check mark is sometimes used to indicate that the resulting sentence is true. In this section the solution sets will contain only one member.

Basic Type I

Solve and check: $n + 2 = 8$

Solution by inverse operation:
The indicated operation of $n + 2$ is addition. To find the root we subtract 2 from each member.

$$n + 2 = 8$$
$$n + 2 - 2 = 8 - 2$$
$$n = 6$$

The root is 6.
Answer, Solution set $\{6\}$

Solution by additive inverse:
The additive inverse of $+2$ is -2. We add the additive inverse to both members, using the addition axiom.

$$n + 2 = 8$$
$$n + 2 + (-2) = 8 + (-2)$$
$$n = 6$$

Answer, Solution set $\{6\}$

Check:
$n + 2 = 8$
$6 + 2 = 8$
$8 = 8\checkmark$

Observe that the equation $n + 2 = 8$ may also appear as $2 + n = 8$ or $8 = n + 2$ or $8 = 2 + n$. These are forms of the same equation. In each case the constant 2 must be eliminated in order to get an equivalent equation in the simplest form. This is $n = 6$ when solving $n + 2 = 8$ or $2 + n = 8$, and $6 = n$ when solving $8 = n + 2$ or $8 = 2 + n$. The equation $6 = n$ may be rewritten as $n = 6$.

EXERCISES

Solve and check:

1. $x + 3 = 13$	**10.** $65 = y + 19$	**19.** $93 = y + 37$	**28.** $-8 = x + 3$
2. $n + 29 = 54$	**11.** $38 = x + 38$	**20.** $68 = 9 + b$	**29.** $\$15 = a + \$2.75$
3. $b + 36 = 95$	**12.** $80 = n + 42$	**21.** $b + \frac{3}{4} = 9$	**30.** $y + 14 = 43$
4. $a + 75 = 120$	**13.** $8 = 5 + x$	**22.** $2\frac{7}{8} + x = 4\frac{1}{4}$	**31.** $36 + x = 144$
5. $11 + y = 34$	**14.** $32 = 17 + c$	**23.** $y + .9 = 6$	**32.** $78 = 78 + r$
6. $8 + c = 27$	**15.** $55 = 26 + n$	**24.** $c + \$.05 = \$.83$	**33.** $12 = n + 5\frac{3}{8}$
7. $29 + x = 46$	**16.** $94 = 56 + T$	**25.** $a + 6 = 4$	**34.** $c + 5 = 2.3$
8. $51 + s = 69$	**17.** $m + 45 = 45$	**26.** $r + 5 = -2$	**35.** $6 = s + 3.4$
9. $21 = a + 7$	**18.** $39 + d = 59$	**27.** $7 + s = 5$	**36.** $m + 12 = 0$

Basic Type II

Solve and check: $n - 2 = 8$

Solution by addition axiom:
The indicated operation of $n - 2$ is subtraction. To find the root we add 2 to both members.

$$n - 2 = 8$$
$$n - 2 + 2 = 8 + 2$$
$$n = 10$$

The root is 10.

Answer, Solution set $\{10\}$

Solution by additive inverse:
The additive inverse of -2 is $+2$. We add the additive inverse to both members using the addition axiom.

$$n - 2 = 8$$
$$n + (-2) = 8$$
$$n + (-2) + (+2) = 8 + (+2)$$
$$n = 10$$

Answer, Solution set $\{10\}$

Check:

$$n - 2 = 8$$
$$10 - 2 = 8$$
$$8 = 8 \checkmark$$

Observe that the equation $n - 2 = 8$ may also appear as $8 = n - 2$ but not as $2 - n = 8$ or $8 = 2 - n$. When solving $n - 2 = 8$ or $8 = n - 2$, the constant 2 must be eliminated in each case in order to get an equivalent equation in the simplest form. This is $n = 10$ when solving $n - 2 = 8$, or $10 = n$ when solving $8 = n - 2$. The equation $10 = n$ may be rewritten as $n = 10$.

EXERCISES

Solve and check:

1. $a - 3 = 8$	**5.** $15 = c - 4$	**9.** $y - 7 = 7$	**13.** $x - \frac{1}{2} = \frac{3}{4}$
2. $x - 12 = 9$	**6.** $6 = s - 26$	**10.** $39 = t - 39$	**14.** $a - 7\frac{1}{3} = 6$
3. $b - 25 = 52$	**7.** $64 = x - 33$	**11.** $n - 24 = 0$	**15.** $8 = y - 2\frac{3}{8}$
4. $n - 60 = 19$	**8.** $72 = y - 85$	**12.** $0 = m - 93$	**16.** $5\frac{1}{4} = n - 5\frac{1}{4}$

17. $n - .9 = 2.6$	**21.** $x - 6 = -8$	**25.** $s - \$1.45 = \$.75$	**29.** $31 = c - 20$
18. $54 = b - 1.5$	**22.** $x - 21 = -35$	**26.** $a - 49 = 85$	**30.** $x - 12 = 0$
19. $8.3 = x - 6$	**23.** $y - 12 = -9$	**27.** $b - .3 = 4.7$	**31.** $6\frac{1}{2} = y - \frac{7}{8}$
20. $a - \$.08 = \$.64$	**24.** $a - 4 = -4$	**28.** $n - 1\frac{1}{2} = -3\frac{1}{2}$	**32.** $a - \$.27 = \$.15$

Basic Type III

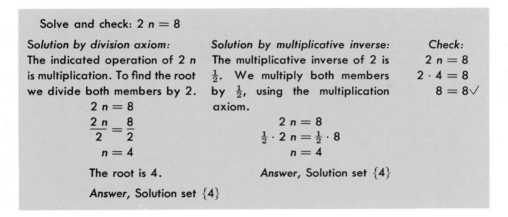

Solve and check: $2n = 8$

Solution by division axiom:
The indicated operation of $2n$ is multiplication. To find the root we divide both members by 2.

$$2n = 8$$
$$\frac{2n}{2} = \frac{8}{2}$$
$$n = 4$$

The root is 4.

Answer, Solution set $\{4\}$

Solution by multiplicative inverse:
The multiplicative inverse of 2 is $\frac{1}{2}$. We multiply both members by $\frac{1}{2}$, using the multiplication axiom.

$$2n = 8$$
$$\frac{1}{2} \cdot 2n = \frac{1}{2} \cdot 8$$
$$n = 4$$

Answer, Solution set $\{4\}$

Check:

$$2n = 8$$
$$2 \cdot 4 = 8$$
$$8 = 8\checkmark$$

Observe that the equation $2n = 8$ may also appear as $8 = 2n$. When solving $2n = 8$ or $8 = 2n$, the numerical coefficient 2 must be eliminated in each case in order to get an equivalent equation in simplest form. This is $n = 4$ when solving $2n = 8$, or $4 = n$ when solving $8 = 2n$. The equation $4 = n$ may be rewritten as $n = 4$.

EXERCISES

Solve and check:

1. $8n = 56$	**13.** $4n = 9$	**25.** $\frac{1}{3}a = 27$	**37.** $16T = 400$
2. $10x = 70$	**14.** $12x = 20$	**26.** $\frac{5}{8}x = 35$	**38.** $180 = 20x$
3. $9y = 27$	**15.** $17 = 5c$	**27.** $40 = \frac{4}{5}b$	**39.** $19r = 19$
4. $12a = 84$	**16.** $26 = 8N$	**28.** $2\frac{1}{2}c = 15$	**40.** $16b = 30$
5. $12 = 3a$	**17.** $6y = 1$	**29.** $9x = -27$	**41.** $25x = 15$
6. $54 = 6d$	**18.** $4 = 7T$	**30.** $-6b = 54$	**42.** $1.05c = 420$
7. $63 = 7n$	**19.** $8b = 6$	**31.** $-42 = -7y$	**43.** $1\frac{1}{4}n = 8\frac{3}{4}$
8. $60 = 15y$	**20.** $10 = 18y$	**32.** $8d = -48$	**44.** $15T = 8$
9. $5x = 5$	**21.** $8x = .24$	**33.** $-x = 8$	**45.** $9b = -54$
10. $14 = 14b$	**22.** $.2a = 16$	**34.** $\frac{2}{3}n = -18$	**46.** $-12x = 80$
11. $6y = 0$	**23.** $9c = \$.45$	**35.** $14b = -4$	**47.** $21 = 24c$
12. $0 = 9m$	**24.** $\$.06n = \$1.32$	**36.** $-24y = -12$	**48.** $.04p = 96$

Basic Type IV

Solve and check: $\frac{n}{2} = 8$

Solution by inverse operation:

The indicated operation of $\frac{n}{2}$ is division. To find the root we multiply both members by 2.

$$\frac{n}{2} = 8$$
$$2 \cdot \frac{n}{2} = 2 \cdot 8$$
$$n = 16$$

The root is 16.
Answer, Solution set $\{16\}$

Solution by multiplicative inverse:

Since $\frac{n}{2} = \frac{1}{2}n$, we use the multiplicative inverse of $\frac{1}{2}$, which is 2. We multiply both members by 2, using the multiplication axiom.

$$\frac{n}{2} = 8$$
$$2 \cdot \frac{1}{2}n = 2 \cdot 8$$
$$n = 16$$

Answer, Solution set $\{16\}$

Check:

$$\frac{n}{2} = 8$$
$$\frac{16}{2} = 8$$
$$8 = 8\checkmark$$

Observe that the equation $\frac{n}{2} = 8$ may also appear as $8 = \frac{n}{2}$. When solving $\frac{n}{2} = 8$ or $8 = \frac{n}{2}$, the denominator 2 must be eliminated in order to get an equivalent equation in simplest form. This is $n = 16$ when solving $\frac{n}{2} = 8$, or $16 = n$ when solving $8 = \frac{n}{2}$. The equation $16 = n$ may be rewritten as $n = 16$.

EXERCISES

Solve and check:

1. $\frac{x}{2} = 7$	**7.** $16 = \frac{d}{20}$	**13.** $\frac{t}{5} = -9$	**19.** $\$1.25 = \frac{1}{8}m$
2. $\frac{c}{5} = 3$	**8.** $\frac{s}{10} = 10$	**14.** $\frac{b}{-4} = 8$	**20.** $\frac{1}{4}r = 9$
3. $\frac{y}{6} = 0$	**9.** $\frac{c}{9} = 1.8$	**15.** $\frac{a}{18} = -6$	**21.** $\frac{b}{9} = -3$
4. $9 = \frac{a}{4}$	**10.** $\frac{n}{1.04} = 60$	**16.** $\frac{d}{-3} = -15$	**22.** $\frac{c}{-5} = 12$
5. $\frac{x}{3} = 15$	**11.** $\frac{1}{2}n = 27$	**17.** $\frac{n}{7} = 1$	**23.** $30 = \frac{n}{6}$
6. $\frac{b}{12} = 4$	**12.** $\frac{1}{3}a = \$.54$	**18.** $\frac{r}{2} = 24$	**24.** $\frac{y}{5} = 5$

The solution of more difficult equations may involve the use of more than one axiom or the combining of terms as shown below.

Solve and check: $7n + 6 = 41$

$$7n + 6 = 41$$
$$7n + 6 - 6 = 41 - 6$$
$$7n = 35$$
$$\frac{7n}{7} = \frac{35}{7}$$
$$n = 5$$

Check:

$$7n + 6 = 41$$
$$(7 \cdot 5) + 6 = 41$$
$$35 + 6 = 41$$
$$41 = 41 \checkmark$$

Answer, Solution set $\{5\}$

Solve and check: $3x + 2x = 20$

$$3x + 2x = 20$$
$$5x = 20$$
$$\frac{5x}{5} = \frac{20}{5}$$
$$x = 4$$

Check:

$$3x + 2x = 20$$
$$(3 \cdot 4) + (2 \cdot 4) = 20$$
$$12 + 8 = 20$$
$$20 = 20 \checkmark$$

Answer, Solution set $\{4\}$

Solve and check: $6y + 5 = 29 - 2y$

$$6y + 5 = 29 - 2y$$
$$6y + (2y) + 5 = 29 + (-2y) + (2y)$$
$$8y + 5 = 29$$
$$8y + 5 + (-5) = 29 + (-5)$$
$$8y = 24$$
$$y = 3$$

Check:

$$6y + 5 = 29 - 2y$$
$$(6 \cdot 3) + 5 = 29 - (2 \cdot 3)$$
$$18 + 5 = 29 - 6$$
$$23 = 23 \checkmark$$

Answer, Solution set $\{3\}$

EXERCISES

Solve and check:

1. $3n + 7 = 31$
2. $9 + 8x = 57$
3. $71 = 5d + 6$
4. $80 = 8 + 9c$
5. $6y + 13 = 35$
6. $59 = 12m + 35$
7. $26 + 11b = 70$
8. $51 = 16d + 3$
9. $8x - 5 = 27$
10. $15x - 44 = 16$
11. $9z - 36 = 0$
12. $29 = 6n - 25$
13. $10b - 17 = 13$
14. $45 = 18a - 45$
15. $4x - 23 = 19$
16. $0 = 7x - 10$
17. $21 = 15T - 39$
18. $1 - 9a = 10$
19. $n + n = 38$
20. $5x + 9x = 84$
21. $7y - 3y = 28$
22. $8b = 19 + 29$
23. $x + 2x + 3x = 96$
24. $11c - c = 95$
25. $5n = 52 - 27$
26. $68 = 6a + 11a$
27. $9d - 4d + 2d = 63$
28. $p + .12p = 560$
29. $a - .04a = 384$
30. $1.8c + 32 = 68$
31. $30 + .6T = 45$
32. $l - .25l = \$.81$
33. $\frac{2}{5}n = 8$
34. $\frac{7}{8}b = 21$
35. $\frac{3}{4}y = \$8.46$
36. $m + \frac{1}{2}m = 4\frac{1}{2}$
37. $\frac{5}{8}n - 8 = 22$
38. $6x + 53 = 11$
39. $15a - 34 = 41$
40. $c + .18c = 236$
41. $\frac{2}{3}a = 74$
42. $9 + 8n = 13$
43. $15 = 12y - 57$
44. $3x - x = -10$
45. $6n + 15 = 9$
46. $6x - 11x = 35$
47. $9y - 15y = -24$
48. $5x + 6 = 2x + 27$
49. $11y - 16 = 40 - 3y$
50. $6n + 2 = 4n - 14$
51. $8x + 5x - 9 = 30$
52. $9y - 7y + 8 = 6y - 28$
53. $10x + 5(x + 4) = 230$
54. $9a - 43 = 2(a + 3)$

11–34 SOLUTION SET NOTATION AND SET-BUILDER

The solution set of $7n + 6 = 41$ consists of all the numbers which make the equation $7n + 6 = 41$ true. We have found that the solution set consists of only one number (5) which may be expressed as $\{5\}$.

The expression $\{n \mid 7n + 6 = 41\}$ is read as:

"The set of all *n* such that seven *n* plus six is equal to forty-one."

The notation $\{n \mid 7n + 6 = 41\}$ represents a set of numbers which have been selected by the condition that $7n + 6 = 41$ and which make this sentence true if they replace the variable. Thus $\{n \mid 7n + 6 = 41\}$ is a method of describing the solution set of $7n + 6 = 41$ which in listed form is $\{5\}$.

Briefly: $\{n \mid 7n + 6 = 41\} = \{5\}$

The expression $\{n \mid \quad\}$ is called the *set-builder*.

EXERCISES

1. Read, or write in words, each of the following:

a. $\{n \mid 2n = 14\}$
b. $\{x \mid x + 6 = 21\}$
c. $\{a \mid a - 3 = 42\}$
d. $\left\{t \,\middle|\, \frac{t}{4} = 9\right\}$
e. $\{y \mid 3y + 9 = 24\}$
f. $\{b \mid 4b - 2b = -12\}$

2. Write each of the following symbolically using set-builder notation:

a. The set of all *x* such that *x* plus seven is equal to twenty.
b. The set of all *z* such that *z* divided by six equals eight.
c. The set of all *s* such that seven times *s* minus five equals nine.

11–35 SOLVING EQUATIONS WHEN THE REPLACEMENT SET IS RESTRICTED

A *replacement set* is a defined set of numbers which may replace the variable. The description of the replacement set is sometimes called the *domain* of the variable. This may be a listing of possible replacements such as the very limited set $\{0, 1, 2, 3, 4, 5\}$ or a description like "the set of all prime numbers" or "the set of all integers," etc.

The solution set of an equation depends not only on the equation but also on how restricted the replacement set is. The same equation with different replacement sets may have different solution sets.

The solution set of $3x = 12$ when the replacement set is the set of all natural numbers is $\{4\}$; when the replacement set is the set of all even numbers, it is $\{4\}$; when the replacement set is the set of all prime numbers, it is the empty set $\varnothing$ or $\{\ \}$ since there is no solution.

The equation $4y = 3$ whose solution set in the system of real numbers is $\{\frac{3}{4}\}$ would have no solution or a solution set with no members (the empty or null set) when the replacement set is the set of whole numbers. There are no fractions in the replacement set of whole numbers.

When the replacement set for a variable is not given in this book, use the set of all real numbers as the replacement set.

EXERCISES

1. Find the solution set of each of the following equations:

a. When the replacement set is the set of all natural numbers:

$n + 12 = 9$ $\quad n - 6 = 6$ $\quad 5d = 20$ $\quad \frac{b}{4} = 7$

b. When the replacement set is $\{0, 1, 2, 3, 4, 5, 6, 7, 8, 9\}$:

$6y = 42$ $\quad n - 4 = 1$ $\quad 2x + 5 = 11$ $\quad \frac{n}{3} = 6$

c. When the replacement set is the set of all integers:

$-5y = -20$ $\quad c + 8 = 8$ $\quad 6b - 4 = 15$ $\quad \frac{c}{2} = -4$

d. When the replacement set is the set of all prime numbers:

$b + 6 = 8$ $\quad n - 4 = 13$ $\quad \frac{t}{3} = 2$ $\quad 20s = 60$

e. When the replacement set is the set of all non-positive integers:

$-7x = -21$ $\quad \frac{c}{8} = -5$ $\quad n + 9 = 9$ $\quad 8a - 11 = -43$

f. When the replacement set is $\{-3, -2, -1, 0, 1, 2, 3\}$:

$r - 2 = 2$ $\quad -18b = 6$ $\quad \frac{m}{2} = -1$ $\quad 5c + 18 = 3$

g. When the replacement set is the set of all one-digit even whole numbers:

$d + 13 = 17$ $\quad 6a - 4a = 24$ $\quad -8y = -56$ $\quad \frac{r}{7} = 4$

h. When the replacement set is $\{0, 1, 2, 3, 4, \ldots\}$:

$h - 9 = 16$ $\quad -11t = -79$ $\quad \frac{x}{9} = 1$ $\quad 3m + 28 = 13$

i. When the replacement set is the set of all odd whole numbers:

$x + 8 = 13$ $\quad \frac{n}{5} = 9$ $\quad 15y = 90$ $\quad 10b - 7 = 13$

j. When the replacement set is the set of all two-digit prime numbers less than 25:

$t - 17 = 6$ $\quad 2x = 58$ $\quad \frac{d}{12} = 8$ $\quad 9c + 4c = 169$

2. Do not solve the following equations. Substitute each number of the given set to determine which is the root of the open sentence.

Open Sentence	*Replacement Set*	*Open Sentence*	*Replacement Set*
a. $n + 5 = 6$	$\{2, -1, 1\}$	d. $6n + 9 = -3$	$\{5, -5, 2, -2\}$
b. $t - 1 = 8$	$\{-7, 0, 9\}$	e. $17y - y = 32$	$\{-2, -1, 0, 1, 2\}$
c. $9x = 72$	$\{4, 6, 8, 10\}$	f. $\frac{b}{2} = 3$	$\{2, 4, 6, 8\}$

11–36 EQUATIONS INVOLVING ABSOLUTE VALUES

Since $|+2| = 2$ and $|-2| = 2$, replacement of the variable in the equation $|x| = 2$ by either $+2$ or -2 will make the sentence true. Therefore, the roots of $|x| = 2$ are $+2$ and -2 or its solution set is $\{+2, -2\}$.

We use the equality axioms to solve equations of the basic types:

$$|x| + 7 = 12$$
$$|x| + 7 - 7 = 12 - 7$$
$$|x| = 5$$
$$x = +5 \text{ or } -5$$

Answer, Solution set $\{+5, -5\}$

EXERCISES

Find the solution set of each of the following equations when the replacement set is the set of all real numbers:

1. $|x| = 8$
2. $|y| = 12$
3. $|s| + 3 = 4$
4. $|t| - 7 = 15$
5. $8 \times |d| = 72$
6. $|n| + 9 = 27$
7. $4 \times |n| = 40$
8. $\frac{|a|}{6} = 5$
9. $16 - |x| = 5$
10. $\frac{|c|}{9} = 3$

11–37 EVALUATING FORMULAS

To determine the value of any variable in a formula when the values of the other variables are known, we copy the formula, substitute the given values for the variables, and perform the necessary operations.

Find the value of t when $d = 60$ and $r = 15$, using the formula $d = rt$:

$$d = rt$$
$$60 = 15t$$
$$4 = t$$
$$t = 4$$

Answer, $t = 4$

We then solve the resulting equation for the value of the required variable.

EXERCISES

Find the value of:

1. p when $C = 63$ and $n = 9$, using the formula $C = np$.
2. h when $V = 168$, $l = 7$, and $w = 6$, using the formula $V = lwh$.
3. t when $i = 24$, $p = 80$, and $r = .06$, using the formula $i = prt$.
4. I when $W = 260$ and $E = 65$, using the formula $W = IE$.
5. a when $v = 105$ and $t = 15$, using the formula $v = at$.
6. p when $i = 80$, $r = .02$ and $t = 8$, using the formula $i = prt$.
7. A when $F = 720$, $H = 5$, and $D = 24$, using the formula $F = AHD$.
8. r when $i = 36$, $p = 90$, and $t = 10$, using the formula $i = prt$.
9. A when $B = 34$, using the formula $A + B = 90$.
10. D_w when $D_t = 93$ and $D_p = 28$, using the formula $D_t = D_w + D_p$.
11. C when $A = 7{,}500$ and $L = 1{,}800$, using the formula $A = L + C$.
12. N when $M = 109$, using the formula $M + N = 180$.
13. A when $p = 85$ and $i = 6$, using the formula $p = A - i$.
14. K when $C = 47$, using the formula $C = K - 273$.
15. V when $v = 123$, $g = 32$, and $t = 3$, using the formula $v = V + gt$.
16. a when $l = 57$, $n = 13$, and $d = 4$, using the formula $l = a + (n - 1)\,d$.
17. l when $p = 92$ and $w = 21$, using the formula $p = 2\,l + 2\,w$.
18. C when $F = 176$, using the formula $F = 1.8\,C + 32$.
19. r when $A = 496$, $p = 400$, and $t = 4$, using the formula $A = p + prt$.
20. h when $A = 2{,}816$, $\pi = \frac{22}{7}$, and $r = 14$, using the formula $A = 2\,\pi r^2 + 2\,\pi rh$.
21. W when $H.P. = 3.5$, using the formula $H.P. = \dfrac{W}{746}$.
22. E when $I = 4$ and $R = 55$, using the formula $I = \dfrac{E}{R}$.
23. N when $S = 1{,}884$, $\pi = 3.14$, and $D = 36$, using the formula $S = \dfrac{\pi DN}{12}$.
24. A when $L = 1{,}250$, $C = .5$, $d = .002$, and $V = 100$, using the formula $L = \dfrac{CdAV^2}{2}$.
25. F when $P = 160$, $d = 8$, and $t = 6$, using the formula $P = \dfrac{Fd}{t}$.
26. p when $A = 310$, $r = .04$, and $t = 6$, using the formula $A = p + prt$.
27. I when $E = 96$, $r = 7$, and $R = 9$, using the formula $E = Ir + IR$.
28. V when $P = 52$, $P' = 28$, and $V' = 13$, using the formula $PV = P'V'$.
29. C when $F = -13$, using the formula $F = 1.8\,C + 32$.
30. T_2 when $P_1 = 18$, $T_1 = 40$, and $P_2 = 9$, using the formula $P_1T_2 = T_1P_2$.

11–38 ALGEBRAIC REPRESENTATION

We write an algebraic expression using the given variables, numerals, and proper symbols to show the relationship involved.

Observe that when there are two related quantities involved, one quantity may be a given number times the second quantity or a given number more or less than the second quantity, or the sum of the two quantities may be given.

(1) There are n degrees in one angle. If a second angle is 6 times as large, how many degrees does it measure? *Answer,* $6n$

(2) Pierre is x years old. How old will he be in 4 years? *Answer,* $x + 4$

(3) Joe and Helen together have 20 books. If Joe has B books, how many books does Helen have? *Answer,* $20 - B$

EXERCISES

Write an algebraic expression for each of the following:

1. Jane is n years old. How old was she 8 years ago?
2. How many hours did it take an automobile to travel x miles if it averaged 50 m.p.h.?
3. Tom has b dollars in the bank. If he deposits c dollars, how many dollars does he then have in the bank?
4. Rose has s cents. She spends 35 cents for lunch. How many cents does she have left?
5. Richard has 3 dollars more than John. If John has d dollars, how many dollars does Richard have?
6. Joe is x years old. Mary is 8 times as old as Joe. How old is Mary?
7. If Mrs. Kono saves D dollars each week, how many weeks will it take her to save $500?
8. How many pencils can be bought for n cents if one pencil costs c cents?
9. A streamliner travels R miles per hour. How many miles can it go in h hours?
10. The length of a rectangle is 4 times the width. If the width is m inches, what is, the length in inches?
11. The sum of two numbers is 25. If one number is n, what is the other number?
12. The base of a triangle is t inches long. The altitude is 4 inches less than the base. How many inches is the altitude?
13. If the radius of a circle is y inches, how long is the diameter?
14. A grocer has 60 lb. of $.99 and $1.19 grapes. If she has b lb. of $1.19 grapes how many lb. of $.99 grapes has she?
15. Harry has x $5 bills but 6 times as many $1 bills. How many $5 and $1 bills does he have in all?
16. Ingrid has n dollars. Arnold has $7 less than 3 times as many dollars as Ingrid. How many dollars does Arnold have?
17. One of two angles has d degrees and their sum is 75°. How many degrees are in the second angle?
18. How many miles can an automobile travel on 15 gallons of gasoline if it averages b miles per gallon?
19. There are n pupils in a class. If there are x boys, how many girls are in the class?

20. A board 16 feet long is divided into 2 pieces. If one piece is f feet long, how long is the other piece?
21. At the rate of r m.p.h., how many hours will it take an airplane to fly s miles?
22. Angle A is 5 times as large as angle B. Angle B has h degrees. How many degrees are in angle A?
23. John is 18 years old. How old will he be in y years?
24. Nancy and Maureen together sold 240 magazines. If Nancy sold n magazines, how many did Maureen sell?
25. Mr. Wilson has r 13¢ stamps. He has 9 more 1¢ stamps than 13¢ stamps and 4 times as many 10¢ as 13¢ stamps. How many stamps does he have in all?

11–39 PROBLEMS

To solve problems by the use of algebraic techniques, we first read the problem carefully to determine what is required and any facts which are related to this unknown value.

When there is only one required value, we represent it by a variable, usually a letter. When there are two required values, we represent the smaller unknown value by some letter and express the other unknown value in terms of that letter.

How long should each piece be if a ribbon 42 inches long is to be cut into two pieces, one twice as long as the other?

Solution:

Let x = length of smaller piece in inches
and $2x$ = length of larger piece in inches

$$x + 2x = 42$$
$$3x = 42$$
$$\frac{3x}{3} = \frac{42}{3}$$

Check:

14″
28″
42″

$$14''\overline{)28''}\ \ 2$$

x = 14 inches, length of smaller piece
$2x$ = 28 inches, length of larger piece

Answer, 14 inches and 28 inches

We then find two facts, at least one involving the unknown value, which are equal to each other. We translate them into algebraic expressions and form an equation by writing one algebraic expression equal to the other.

We solve the equation to find the required value. We check the answer directly with the given problem.

EXERCISES

Solve the following problems:

1. Tom asked Kurt to guess his age. He said, "If you add 16 to six times my age, you get 100." How old is Tom?
2. $\frac{7}{8}$ of what number is 42?
3. 4% of what number is 59?
4. How much money must be invested at 6% to earn $1,500 per year?

5. There are 11 more girls than boys in the graduating class of 167 students. How many girls are in the class?
6. What mark must Renee get in a test to bring her average up to 85 if her other test marks are 79 and 87?
7. The net price of a radio is \$54 when a 10% discount is allowed. What is its list price?
8. How much must a salesclerk sell each week on a 15% commission basis to earn \$72 weekly?
9. Two girls together sold 60 tickets for the school show. If one girl sold 4 times as many as the other, how many tickets did each girl sell?
10. A dealer sold a rug for \$108, making a profit of 35% on the cost. How much did it cost?
11. A board 12 feet long is to be cut into two pieces, one 6 feet longer than the other. Find the length of each piece.
12. A man is 3 times as old as his son. Together the sum of their ages is 52. What are their ages?
13. What is the selling price of a television set if it cost \$175 and the profit is 30% of the selling price?
14. Find the regular price of a suit that sold for \$41.25 at a 25% reduction sale.
15. The perimeter of a rectangle is 126 ft. Its length is 6 times its width. Find the dimensions.
16. Find the measures of the angles of a triangle if the first angle measures 3 times the second angle and the third angle measures twice the second angle.
17. Joe has 8 more dimes than nickels. If he has 26 coins in all, how many of each does he have?
18. Find two consecutive integers whose sum is 71.

11–40 GRAPHING AN EQUATION IN ONE VARIABLE ON THE NUMBER LINE

To draw the graph of an equation in one variable on the number line, we locate the point or points whose coordinate or coordinates are numbers belonging to the solution set of the equation. The graph of the equation is the graph of its solution set. We indicate these points by heavy or colored dots.

The solution sets of the equations studied thus far consisted mostly of one element although some contained no elements and others two elements.

(1) The solution set of the equation $n + 2 = 8$ is $\{6\}$. The graph of $n + 2 = 8$ is the point whose coordinate is 6.

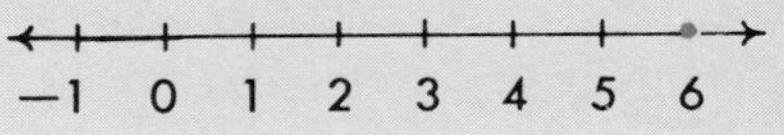

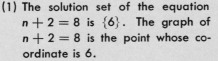

(2) The solution set of the equation $7n + 6 = 41$ is $\{5\}$. The graph of $7n + 6 = 41$ is the point whose coordinate is 5.

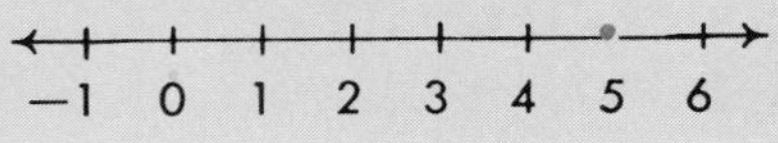

(3) The solution set of the equation $x = x + 3$ is $\varnothing$.

We have no points to draw the graph.

(4) The solution set of $|x| = 2$ is $\{-2, +2\}$. The graph of $|x| = 2$ is the set of points whose coordinates are -2 and $+2$.

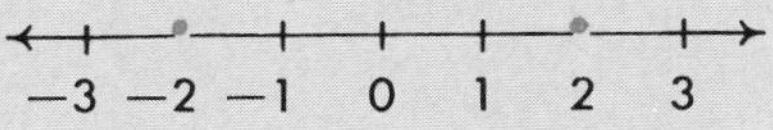

(5) The solution set of $3x + 12 = 3(x + 4)$ is an infinite set since the equation (identity) is satisfied by every real number. The graph of $3x + 12 = 3(x + 4)$ is the entire number line indicated by a heavy or colored line with an arrowhead in each direction to show that it is endless.

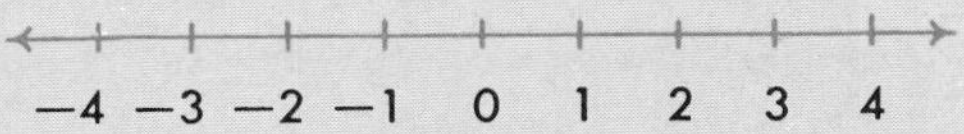

(6) The graph

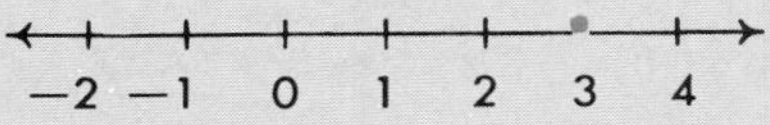

is a picture of the equation $x = 3$ or any equivalent equation such as $4x = 12$, $x + 2 = 5$, $x - 1 = 2$, etc.

EXERCISES

1. On the number line below what is the coordinate of point D? J? P? B? I? L? A? K? N? E?

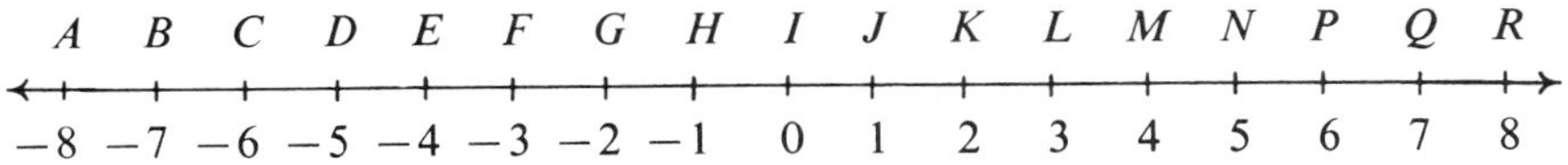

2. On the number line above what point is the graph of -3? 8? -6? 0? 4? -2? 7? -1? 1? -4?

3. For each of the following equations draw an appropriate number line, then graph its solution set. The replacement set is the set of all real numbers.

a. $y = 3$ **d.** $7a = 42$ **g.** $8x - x = 63$ **j.** $|x| = 6$

b. $n = -2$ **e.** $b - 4 = 1$ **h.** $4x + 3 = 3$ **k.** $y = y + 5$

c. $x + 8 = 15$ **f.** $2c + 7 = 5$ **i.** $\frac{x}{5} = 1$ **l.** $9x + 27 = 9(x + 3)$

4. On a number line (see section 11–46) draw the graph of:

a. $6x = 12$ when the replacement set is the set of all prime numbers.

b. $\frac{x}{2} = 4$ when the replacement set is the set of all natural numbers.

c. $b + 7 = 2$ when the replacement set is the set of all integers.

d. $8y - 4 = 44$ when the replacement set is the set of all multiples of 3.

e. $4x - x = -12$ when the replacement set is $\{-6, -4, -2, 0, 2, 4, 6\}$.

5. Write a corresponding equation which is pictured by each of the following graphs:

a. 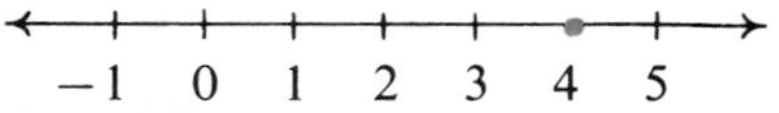

c. 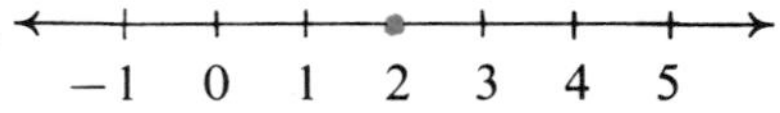

b. 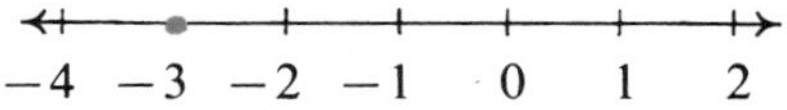

d. 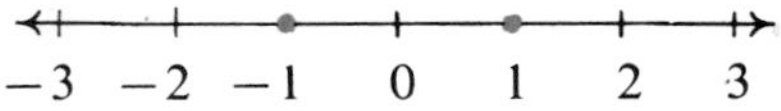

11–41 RATIO

We may compare two numbers by subtraction or by division. When we compare 20 with 5, we find by subtraction that 20 is fifteen more than 5 and by division that 20 is four times as large as 5. When we compare 5 with 20, we find by subtraction that 5 is fifteen less than 20 and by division that 5 is one-fourth as large as 20.

The answer we obtain when we compare two quantities by division is called the *ratio* of the two quantities.

When 20 is compared to 5, 20 is divided by 5 as $\frac{20}{5}$ which equals $\frac{4}{1}$. The ratio is $\frac{4}{1}$ or 4 to 1.

When 5 is compared to 20, 5 is divided by 20 as $\frac{5}{20}$ which equals $\frac{1}{4}$. The ratio is $\frac{1}{4}$ or 1 to 4.

A ratio has two terms, the number that is being compared (found in the numerator) and the number to which the first number is being compared (found in the denominator). The ratio in fraction form is usually expressed in lowest terms.

The ratio of 5 miles to 6 miles is $\frac{5}{6}$. The fraction $\frac{5}{6}$ is an indicated division meaning $5 \div 6$. Usually the ratio is expressed as a common fraction but it may also be expressed as a decimal fraction or a per cent.

The ratio $\frac{5}{6}$ is sometimes written as 5 : 6. The colon may be used instead of the fraction bar. This ratio in either form is read "5 to 6" "(five to six)."

If the quantities compared are denominate numbers, they must first be expressed in the same units. The ratio is an abstract number; it contains no unit of measurement.

A ratio may be used to express a rate. The average rate of speed of an automobile when it travels 150 miles in 3 hours may be expressed by the ratio which is $\frac{150}{3}$ or 50 miles per hour. If pears sell at the rate of 3 for 19¢, the ratio expressing this rate is $\frac{3}{19}$ and is read "3 for 19."

When a ratio is used to express a rate, the two quantities have different names like miles and hours, pears and cents, etc.

Ratios such as $\frac{2}{6}$, $\frac{3}{9}$, $\frac{4}{12}$, $\frac{5}{15}$, although written with different number pairs, express the same comparison or rate. These ratios are called *equivalent ratios.* In simplest form they are the same, the ratio $\frac{1}{3}$.

EXERCISES

1. Find the ratio of:

   a. 4 to 12
   b. 8 to 10
   c. 3 to 5
   d. 15 to 60
   e. 21 to 28
   f. 12 to 6
   g. 7 to 4
   h. 18 to 8
   i. 25 to 5
   j. 40 to 24
   k. 15 to 3
   l. 6 to 24
   m. 16 to 18
   n. 30 to 12
   o. 36 to 16

2. Find the ratio of:

   a. 6 in. to 9 in.
   b. 8 cm to 1 m
   c. 30 in. to 1 yd.
   d. 3 mm to 2 cm
   e. 1 ft. 6 in. to 4 yd.
   f. 10 lb. to 35 lb.
   g. 3 kg to 8 g
   h. 2 cl to 1 liter
   i. 3 gal. to 3 qt.
   j. 1 gal. to 1 pt.
   k. 20 min. to 45 min.
   l. 30 sec. to 2 min.
   m. 2 hr. to 40 min.
   n. 1 hr. 15 min. to 25 min.
   o. 18 mo. to 2 yr.

3. What is the ratio of:

   a. A dime to a dollar?
   b. A quarter to a nickel?
   c. A half dollar to a dime?
   d. 4 cents to 2 nickels?
   e. 6 dollars to 3 dimes?
   f. 3 things to a dozen?
   g. 10 things to a dozen?
   h. 1 dozen to 8 things?
   i. 3 dozen to 9 dozen?
   j. $1\frac{1}{2}$ dozen to 4 things?

4. Express each of the following rates as a ratio:

   a. 5 oranges for 48 cents
   b. $7 for 3 shirts
   c. 850 miles per hour
   d. 81¢ for 27 kw. hr.
   e. 91 miles on 7 gal.
   f. 2,400 liters in 30 minutes
   g. 325 kilometers in 5 hours
   h. 7,200 revolutions in 6 minutes
   i. 128 meters in 4 seconds
   j. 7 books for $10

5. There are 16 girls and 12 boys in a mathematics class. What is the ratio of: girls to boys? boys to girls? girls to entire class? boys to entire class?
6. The distribution of final marks in Science showed 4 A's, 6 B's, 12 C's, 8 D's, and 2 E's. What is the ratio of the number of pupils receiving A to those receiving E? B's to C's? D's to A's? C's to E's? A's to B's?
7. In a school where there are 988 students and 38 teachers, what is the ratio of the number of pupils to the number of teachers?
8. The pitch of a roof is the ratio of the rise to the span. What is the pitch of a roof with a rise of 6 feet and a span of 18 feet?
9. Find the ratio of a 24-tooth gear to a 36-tooth gear.
10. The aspect ratio is the ratio of the length of an airplane wing to its width. What is the aspect ratio of a wing 84 ft. long and 12 ft. wide?
11. The lift-drag ratio is the ratio of the lift of an airplane to its drag. Find the lift-drag ratio if the lift is 1,980 lb. and the drag is 165 lb.
12. The compression ratio is the ratio of the total volume of a cylinder to its clearance volume. Find the compression ratio of a cylinder of an engine if its total volume is 80 cu. in. and its clearance volume is 16 cu. in.

11–42 PROPORTIONS

A *proportion* is a mathematical sentence which states that two ratios are equivalent.

Using the equivalent ratios $\frac{4}{8}$ and $\frac{12}{24}$ we may write the proportion $\frac{4}{8} = \frac{12}{24}$. This proportion may also be expressed as $4 : 8 = 12 : 24$. The proportion in both forms is read "4 is to 8 as 12 is to 24."

There are four terms in a proportion as shown:

$$\begin{array}{r} \text{first} \rightarrow 4 \\ \text{second} \rightarrow 8 \end{array} \;\; \frac{}{} = \;\; \begin{array}{l} 12 \leftarrow \text{third} \\ 24 \leftarrow \text{fourth} \end{array}$$

$$\begin{array}{cccccc} \text{1st} & & \text{2nd} & & \text{3rd} & \text{4th} \\ \downarrow & & \downarrow & & \downarrow & \downarrow \\ 4 & : & 8 & = & 12 : & 24 \end{array}$$

In the proportion $4 : 8 = 12 : 24$, the first term (4) and the fourth term (24) are called the *extremes*; the second term (8) and the third term (12) are called the *means*. Observe that the product of the extremes (4×24) is equal to the product of the means (8×12). In the form $\frac{4}{8} = \frac{12}{24}$ observe that the cross products are equal ($4 \times 24 = 8 \times 12$). These products are equal *only* when the ratios are equivalent. We can check whether two ratios are equivalent by either expressing each ratio in lowest terms or by determining whether the cross products are equal.

If any three of the four terms of the proportion are known quantities, the fourth may be determined. We use the idea that the product of the extremes is equal to the product of the means to transform the given proportion to a simpler equation.

Thus we find the cross products by multiplying the numerator of the first fraction by the denominator of the second fraction and the numerator of the second fraction by the denominator of the first fraction. We write one product equal to the other, then we solve the resulting equation.

Solve and check: $\frac{n}{54} = \frac{5}{6}$

Solution:

$$\frac{n}{54} = \frac{5}{6}$$
$$6 \times n = 5 \times 54$$
$$6n = 270$$
$$\frac{6n}{6} = \frac{270}{6}$$
$$n = 45$$

Answer, Solution set $\{45\}$

Check:

$$\frac{45}{54} = \frac{5}{6}$$
$$\frac{5}{6} = \frac{5}{6} \checkmark$$

EXERCISES

Solve and check:

1. $\frac{x}{12} = \frac{3}{4}$
2. $\frac{n}{36} = \frac{5}{9}$
3. $\frac{c}{27} = \frac{2}{3}$
4. $\frac{b}{6} = \frac{11}{15}$
5. $\frac{a}{5} = \frac{7}{8}$
6. $\frac{2}{5} = \frac{b}{20}$
7. $\frac{3}{8} = \frac{x}{56}$
8. $\frac{5}{6} = \frac{y}{42}$
9. $\frac{7}{12} = \frac{c}{16}$
10. $\frac{14}{18} = \frac{w}{27}$

11. $\frac{9}{c} = \frac{3}{53}$ 13. $\frac{10}{d} = \frac{5}{16}$ 15. $\frac{6}{b} = \frac{4}{8}$ 17. $\frac{7}{9} = \frac{28}{n}$ 19. $\frac{36}{54} = \frac{12}{d}$

12. $\frac{2}{y} = \frac{1}{12}$ 14. $\frac{18}{x} = \frac{96}{64}$ 16. $\frac{16}{7} = \frac{96}{a}$ 18. $\frac{72}{180} = \frac{8}{x}$ 20. $\frac{4}{13} = \frac{15}{m}$

21. Eric saved \$25 in 9 weeks. At that rate how long will it take him to save \$175?
22. A motorist travels 480 kilometers in 6 hours. How long will it take her at that rate to travel 640 kilometers?
23. A tree casts a shadow of 26 feet, while a 5-foot post nearby casts a shadow of 2 feet. Find the height of the tree.
24. Miss Wilson pays \$450 taxes on a house assessed at \$7,500. Using the same tax rate, find the taxes on a house assessed at \$9,200.
25. A picture $2\frac{1}{2}$ inches wide and $3\frac{1}{2}$ inches high is to be enlarged so that the height will be 7 inches. How wide will it be?
26. Find the value of V if $V' = 9$, $P' = 4$, and $P = 12$, using the formula, $\frac{V}{V'} = \frac{P'}{P}$
27. Find the value of r if $c = 132$, $c' = 88$, and $r' = 14$, using the formula, $\frac{c}{c'} = \frac{r}{r'}$
28. Find the value of R_2 if $R_1 = 450$, $R_3 = 600$, and $R_4 = 720$, using the formula, $\frac{R_1}{R_2} = \frac{R_3}{R_4}$
29. Find the value of L if $W = 120$, $w = 80$, and $l = 6$, using the formula, $\frac{W}{w} = \frac{l}{L}$

11–43 SOLVING INEQUALITIES IN ONE VARIABLE

The expression $\{n \mid n < 8\}$ represents the set of numbers that satisfies the inequality $n < 8$. The $n < 8$ is the condition on which the solution set is based.

> $\{n \mid n < 8\}$ is read:
>
> "The set of all *n* such that each *n* is less than eight."

EXERCISES

1. Read, or write in words, each of the following:

a. $\{n \mid n \neq 1\}$
b. $\{x \mid x > 4\}$
c. $\{y \mid 5y < 20\}$
d. $\{a \mid a \geq 9\}$
e. $\{b \mid b \leq -3\}$
f. $\{s \mid s + 6 < 11\}$
g. $\{m \mid m - 4 \geq 12\}$
h. $\{x \mid 9x - 1 \neq 0\}$
i. $\{y \mid 12y + 3 < -2\}$
j. $\left\{c \,\middle|\, \frac{c}{2} \neq 5\right\}$
k. $\{n \mid 4n - 8 < 15\}$

2. Write each of the following symbolically, using set-builder notation:

a. The set of all *x* such that each *x* is not equal to twenty.
b. The set of all *n* such that each *n* is greater than negative four.
c. The set of all *c* such that each *c* is less than nine.

d. The set of all b such that each b is greater than or equal to zero.
e. The set of all y such that each y is less than or equal to negative one.

To solve an inequality means to find the set of all numbers which make the sentence true. This solution set may consist of an unlimited number of elements, a finite number of one or more elements, or no elements.

In order to be able to complete the solutions of inequalities, observe the interpretations of the following basic types in their simplest forms:

(1) The solution set of $x < 5$ in the system of natural numbers consists of every natural number less than five.

Therefore $x = 1, 2, 3,$ or 4. Written in set form: $\{x \mid x < 5\} = \{1, 2, 3, 4\}$.

(2) The solution set of $a > 10$ in the system of whole numbers consists of every whole number greater than 10.

Therefore $a = 11, 12, 13, 14, \ldots$

Written in set form: $\{a \mid a > 10\} = \{11, 12, 13, 14, \ldots\}$.

(3) The solution set of $n \neq 8$ in the system of real numbers consists of every real number except 8.

Therefore $n =$ every real number except 8.

Written in set form: $\{n \mid n \neq 8\}$ or {every real number except 8}.

(4) The solution set of $d \geq -2$ in the system of integers consists of -2 and all integers greater than -2.

Therefore $d = -2, -1, 0, 1, 2, 3, \ldots$

Written in set form: $\{d \mid d \geq -2\} = \{-2, -1, 0, 1, 2, 3, \ldots\}$.

(5) The solution set of $x \leq 4$ when the replacement set is $\{5, 6, 7, 8, 9, 10\}$ contains no elements.

Therefore x has no solution. Written in set form: $\{x \mid x \leq 4\} = \varnothing$ or $\{\}$.

(6) The solution set of $n \ngtr 1$ in the system of integers is the same as the solution set of $n \leq 1$ which consists of 1 and all integers less than 1.

Therefore $n = \ldots, -2, -1, 0, 1$.

Written in set form: $\{n \mid n \ngtr 1\} = \{\ldots, -2, -1, 0, 1\}$.

(7) The solution set of $x \nless 2$ in the system of real numbers is the same as the solution set of $x \geq 2$ which consists of 2 and all real numbers greater than 2.

Therefore $x = 2$ or any real number greater than 2.

Written in set form: $\{x \mid x \nless 2\} = \{2$, all real numbers greater than $2\}$.

Observe that we generally list elements of the solution set where convenient, especially when the replacement set is restricted.

We solve inequalities as we do equations using transformations based on the axioms and on the additive and multiplicative inverses to get an equivalent inequality in each case which has the same solution set as the given inequality. However, the equivalent inequality is expressed with only the variable itself as one member of the inequality and a numeral for the other member.

The axioms used to solve inequalities are separated into the following three sections:

I. When the same number is added to or subtracted from both members of an inequality, another inequality of the *same order* results.

Solve $n + 3 > 5$ in the system of natural numbers:

$$n + 3 > 5$$
$$n + 3 - 3 > 5 - 3$$
$$n > 2$$
$$n = 3, 4, 5, \ldots$$

or $\{n \mid n + 3 > 5\} = \{3, 4, 5, \ldots\}$

Answer. Solution set $\{3, 4, 5, \ldots\}$

Solve $b - 6 < 4$ in the system of whole numbers:

$$b - 6 < 4$$
$$b - 6 + 6 < 4 + 6$$
$$b < 10$$
$$b = 0, 1, 2, \ldots, 9$$

or $\{b \mid b - 6 < 4\} = \{0, 1, 2, \ldots, 9\}$

Answer, Solution set $\{0, 1, 2, \ldots, 9\}$

EXERCISES

Find the solution set of each of the following inequalities when the replacement set is the set of all the real numbers:

1. **a.** $x + 6 > 9$ **b.** $n + 7 > 7$ **c.** $t + .4 > 1.3$ **d.** $y + 2 > 1$ **e.** $21 + b > 54$

2. **a.** $n - 4 > 7$ **b.** $r - 10 > 2$ **c.** $a - \frac{2}{3} > 1\frac{1}{3}$ **d.** $b - 4 > -6$ **e.** $x - 18 > 35$

3. **a.** $c + 2 < 11$ **b.** $m + 5 < 2$ **c.** $y + 1\frac{1}{2} < 4$ **d.** $n + 9 < 19$ **e.** $x + 30 < 45$

4. **a.** $d - 6 < 15$ **b.** $g - 7 < 3$ **c.** $n - .8 < .25$ **d.** $y - 1 < -4$ **e.** $x - 20 < 12$

5. **a.** $y + 7 \geq 10$ **b.** $r + 1 \geq 9$ **c.** $a + 2\frac{1}{4} \geq 4\frac{3}{4}$ **d.** $x + 24 \nless 0$ **e.** $s + 8 \geq 8$

6. **a.** $z - 4 \geq 6$ **b.** $t - 13 \geq 5$ **c.** $y - 8.2 \geq 6.4$ **d.** $m - 17 \nless 21$ **e.** $p - 5 \geq -5$

7. **a.** $c + 1 \leq 11$ **b.** $n + 15 \leq 14$ **c.** $w + .6 \leq -.3$ **d.** $x + 9 \ngtr 9$ **e.** $r + 32 \leq 50$

8. **a.** $x - 5 \leq 7$ **b.** $n - 2 \leq -8$ **c.** $t - 3\frac{1}{3} \leq -2\frac{5}{6}$ **d.** $y - 6 \ngtr 4$ **e.** $a - 10 \leq 6$

9. **a.** $g + 3 \neq 15$ **b.** $x + 8 \neq 22$ **c.** $d + 1.5 \neq 1.1$ **d.** $y + 18 \neq -7$ **e.** $n + 9 \neq 16$

10. **a.** $b - 2 \neq 2$ **b.** $h - 1 \neq 10$ **c.** $n - \frac{1}{2} \neq \frac{3}{4}$ **d.** $x - 14 \neq 6$ **e.** $y - 7 \neq 21$

11. a. $s + 4 < 11$
b. $d - 5 > 8$
c. $x - 3 \neq 2$
d. $b + 9 \leq 5$
e. $r - 13 \geq 1$
f. $y + 4 \neq 4$
g. $b - 12 < 0$
h. $w + 20 > 15$
i. $x - 8 \nless -17$

Find the solution set of each of the following inequalities:

12. When the replacement set is the set of all one-digit prime numbers:
a. $x + 2 < 5$ b. $n - 4 > 2$ c. $y + 8 \leq 15$ d. $c - 7 \nless 1$

13. When the replacement set is $\{-5, -3, -1, 1, 3, 5\}$:
a. $y + 7 > 3$ b. $t - 5 < 1$ c. $m + 2 \neq 7$ d. $x - 1 \geq 3$

14. When the replacement set is the set of all natural numbers:
a. $n - 4 \neq 4$ b. $w + 8 > 8$ c. $y - 8 \ngtr 2$ d. $t + 6 < 7$

15. When the replacement set is the set of all integers:
a. $t - 9 < 1$ b. $x + 17 > 20$ c. $a - 1 \leq -1$ d. $n + 5 \neq 5$

16. When the replacement set is the set of all non-positive integers:
a. $n + 4 > 3$ b. $x + 6 \nless 2$ c. $a - 7 \leq 2$ d. $b - 4 < -4$

17. When the replacement set is the set of all multiples of 3 greater than 5 and less than 25:
a. $y - 4 \neq 8$ b. $c + 5 > 19$ c. $d - 9 < 12$ d. $n + 7 \ngtr 11$

18. When the replacement set is $\{-4, -2, 0, 2, 4\}$:
a. $x - 1 < 3$ b. $m + 5 \neq 1$ c. $r + 3 \geq 5$ d. $y - 2 > 2$

19. When the replacement set is the set of all negative integers greater than -6:
a. $n - 3 > -6$ b. $y + 8 < 3$ c. $t - 10 \neq -14$ d. $x + 9 \leq 0$

II. When both members of an inequality are either multiplied or divided by the same *positive* number, another inequality of the *same order* results.

Solve $4b > 20$ in the system of real numbers:

$4b > 20$

$\frac{4b}{4} > \frac{20}{4}$

$b > 5$

$b =$ all real numbers greater than 5

or $\{b \mid 4b > 20\} = \{b \mid b > 5\}$

$=$ {all real numbers greater than 5}

Answer, Solution set {all real numbers greater than 5}

Solve $6a - 5 \leq 13$ when the replacement set is $\{1, 3, 5, 7, 9\}$:

$6a - 5 \leq 13$

$6a - 5 + 5 \leq 13 + 5$

$6a \leq 18$

$a \leq 3$

$a = 1$ or 3

or $\{a \mid 6a - 5 \leq 13\}$

$= \{a \mid a \leq 3\} = \{1,3\}$

Answer, Solution set $\{1, 3\}$

EXERCISES

Find the solution set of each of the following inequalities when the replacement set is the set of all the real numbers:

1. **a.** $3a > 21$ **b.** $8y > 56$ **c.** $6a > 15$ **d.** $.4n > 5$ **e.** $10t > 8$ **f.** $4y > -14$ **g.** $\frac{1}{2}v > 7$ **h.** $9z > -54$ **i.** $12b > 60$ **j.** $7x > 42$

2. **a.** $\frac{x}{6} > 2$ **b.** $\frac{n}{8} > 5$ **c.** $\frac{y}{9} > 12$ **d.** $\frac{w}{3} > -3$ **e.** $\frac{b}{10} > 6$

3. **a.** $9n < 72$ **b.** $5r < 40$ **c.** $3w < 11$ **d.** $18x < 15$ **e.** $.05s < 1$ **f.** $6b < -30$ **g.** $\frac{2}{3}y < 12$ **h.** $4m < -28$ **i.** $20t < 100$ **j.** $32g < 96$

4. **a.** $\frac{m}{3} < 15$ **b.** $\frac{x}{12} < 1$ **c.** $\frac{t}{20} < 0$ **d.** $\frac{a}{8} < -2$ **e.** $\frac{n}{9} < 4$

5. **a.** $8m \geq 48$ **b.** $6r \geq 114$ **c.** $21x \geq 14$ **d.** $7b \nless 91$ **e.** $1.05y \geq 10.5$ **f.** $\frac{3}{4}c \geq 36$ **g.** $11a \geq 99$ **h.** $5y \geq 42$ **i.** $18t \nless 90$ **j.** $25x \geq 300$ **k.** $4y \nless -24$ **l.** $3n \geq -81$

6. **a.** $\frac{n}{4} \geq 1$ **b.** $\frac{c}{12} \geq 2$ **c.** $\frac{r}{2} \geq -4$ **d.** $\frac{v}{5} \nless 7$ **e.** $\frac{x}{10} \geq 5$ **f.** $\frac{t}{4} \nless -3$

7. **a.** $4b \leq 64$ **b.** $12y \leq 56$ **c.** $2d \leq -10$ **d.** $5m \ngtr 100$ **e.** $9x \leq -63$ **f.** $54n \leq 18$ **g.** $.01w \leq 1$ **h.** $14c \leq 84$ **i.** $24y \ngtr 72$ **j.** $30z \leq 120$ **k.** $.7b \ngtr 1.4$ **l.** $8a \leq -56$

8. **a.** $\frac{x}{11} \leq 2$ **b.** $\frac{d}{4} \leq 8$ **c.** $\frac{g}{2} \leq 19$ **d.** $\frac{n}{5} \ngtr -6$ **e.** $\frac{x}{16} \leq 3$ **f.** $\frac{r}{6} \leq -9$

9. **a.** $3x \neq 57$ **b.** $12c \neq -12$ **c.** $7y \neq 49$ **d.** $8m \neq 144$ **e.** $48b \neq 36$ **f.** $11z \neq -132$ **g.** $.2a \neq 5$ **h.** $\frac{5}{8}t \neq 30$ **i.** $9g \neq 51$ **j.** $4s \neq 104$

10. **a.** $\frac{c}{5} \neq 2$ **b.** $\frac{m}{3} \neq -4$ **c.** $\frac{r}{10} \neq 0$ **d.** $\frac{x}{6} \neq 8$ **e.** $\frac{w}{2} \neq -10$

11. **a.** $8a > 96$ **b.** $10g < 400$ **c.** $13x \neq 39$ **d.** $\frac{h}{6} > 9$ **e.** $16n \geq 12$ **f.** $5b \leq -40$ **g.** $\frac{r}{8} < 8$ **h.** $\frac{7}{10}a > 14$ **i.** $\frac{c}{9} \leq -2$ **j.** $17m \nless -51$ **k.** $\frac{s}{8} \neq 0$ **l.** $\frac{x}{14} \geq 2$ **m.** $3d \ngtr 26$ **n.** $17y < 0$

Find the solution set of each of the following inequalities:

12. When the replacement set is the set of all natural numbers:

a. $2x > 14$ **b.** $4y < -4$ **c.** $10n \leq 80$ **d.** $\frac{x}{6} > 7$

13. When the replacement set is the set of all integers:

a. $9d < -45$ **b.** $16n \not> 80$ **c.** $\frac{5}{6}y \neq 20$ **d.** $\frac{x}{4} \geq -1$

14. When the replacement set is $\{-8, -6, -4, -2, 0, 2, 4, 6, 8\}$:

a. $6b > -12$ **b.** $8x \not< 48$ **c.** $\frac{n}{2} < -3$ **d.** $3t \leq 0$

15. When the replacement set is the set of all odd prime numbers less than 15:

a. $25n < 25$ **b.** $\frac{s}{5} \not> 3$ **c.** $7x \leq 63$ **d.** $24y \neq 120$

III. When both members of an inequality are either multiplied or divided by the same *negative* number, an inequality of the *reverse order* results.

Solve $-3c > 21$ in the system of integers:

$$-3c > 21$$
$$\frac{-3c}{-3} < \frac{21}{-3}$$
$$c < -7$$
$$c = \ldots, -10, -9, -8$$

or $\{c \mid -3c > 21\} = \{c \mid c < -7\}$
$= \{\ldots, -10, -9, -8\}$

Observe that -6 which is greater than -7 makes $-3c > 21$ false $(18 > 21)$ but -8 makes it true $(24 > 21)$.

Answer, Solution set $\{\ldots, -10, -9, -8\}$

Solve $-5x < -20$ in the system of integers:

$$-5x < -20$$
$$\frac{-5x}{-5} > \frac{-20}{-5}$$
$$x > 4$$
$$x = 5, 6, 7, \ldots$$

or $\{x \mid -5x < -20\}$
$= \{x \mid x > 4\} = \{5, 6, 7, \ldots\}$

Observe that 3 which is less than 4 makes $-5x < -20$ false $(-15 < -20)$ but 5 makes it true $(-25 < -20)$.

Answer, Solution set $\{5, 6, 7, \ldots\}$

EXERCISES

Find the solution set of each of the following inequalities when the replacement set is the set of all real numbers:

1. **a.** $-3x > 15$
b. $-8y > 80$
c. $-.7d > 21$
d. $-12a > 108$
e. $-25n > 100$

2. **a.** $-2y > -14$
b. $-5d > -23$
c. $-\frac{1}{3}t > -60$
d. $-14n > -84$
e. $-9r > -72$

3. **a.** $-4m < 48$
b. $-6c < 0$
c. $-\frac{3}{5}x < 12$
d. $-10z < 4$
e. $-16s < 64$

4. **a.** $-5b < -10$
b. $-9m < -63$
c. $-1.2y < -6$
d. $-6n < -15$
e. $-7x < -56$

5. **a.** $-4n \geq 52$
b. $-x \geq 1$
c. $-.9b \geq 5.4$
d. $-3a \not< 19$
e. $-18y \geq 90$

6. **a.** $-9z \geq -36$
b. $-15d \geq -105$
c. $-\frac{5}{7}x \geq -20$
d. $-n \not< -1$
e. $-8a \geq -128$

7. **a.** $-6s \le 120$
b. $-24c \le 16$
c. $-.05n \le 2.5$
d. $-17m \not> 51$
e. $-5x \le 0$

8. **a.** $-9b \le -117$
b. $-10y \le -35$
c. $-\frac{3}{4}a \le -24$
d. $-7n \not> -147$
e. $-4x \le -56$

9. **a.** $-8x > -16$ **b.** $-3y < 81$ **c.** $-4z \ge -72$
d. $-9t \le -90$ **e.** $-7m < 23$ **f.** $-6p > 0$
g. $-10n \not< 30$ **h.** $-5a \not> 18$ **i.** $-g < -1$
j. $-12h \le 48$ **k.** $-2w > -1$ **l.** $-24c \ge -144$

10. **a.** $\frac{b}{-2} < 6$ **b.** $\frac{n}{-5} > 4$ **c.** $\frac{c}{-3} \le 7$
d. $\frac{a}{-4} > -2$ **e.** $\frac{x}{-6} < -8$ **f.** $\frac{d}{-8} \ge 9$
g. $\frac{n}{-10} < -2$ **h.** $\frac{y}{-7} \not< -10$ **i.** $\frac{a}{-9} > 5$
j. $\frac{x}{-12} < 7$ **k.** $\frac{w}{-3} \not> -1$ **l.** $\frac{y}{-15} > 12$

Find the solution set of each of the following inequalities:

11. When the replacement set is the set of all integers:

a. $-7x > 42$ **b.** $-12c < -24$ **c.** $-4t \not< 20$ **d.** $\frac{n}{-9} \not> 2$

12. When the replacement set is the set of all natural numbers:

a. $-15b < 45$ **b.** $-y > -1$ **c.** $-3a \not> 0$ **d.** $\frac{w}{-3} \not< -5$

13. When the replacement set is $\{-2, -1, 0, 1, 2\}$:

a. $-4c > -4$ **b.** $-6t \not> 12$ **c.** $-10n \ge 30$ **d.** $\frac{x}{-1} < 1$

14. When the replacement set is the set of all non-negative integers:

a. $-2y < 8$ **b.** $-6x > -12$ **c.** $-x \le 0$ **d.** $\frac{v}{-9} \not< 1$

15. When the replacement set is the set of integers greater than -5 and less than 2:

a. $-5x \ge -5$ **b.** $-3a < 0$ **c.** $-4y \not> -8$ **d.** $\frac{b}{-4} > -1$

MISCELLANEOUS EXERCISES

1. Find the solution set of the following inequalities when the replacement set is the set of all real numbers:

a. $x + 6 < 17$
b. $w - 11 > 13$
c. $9y \ne 90$
d. $-6s > 84$
e. $\frac{n}{4} < -3$
f. $3x + 9 \le 18$
g. $-8y > 56$
h. $11a - 4a < 14$
i. $4m - 5 \ge -21$
j. $\frac{d}{-5} \not> 2$
k. $-5c < -10$
l. $2x - x > -1$
m. $-7z \not< 0$
n. $10y + 3 \ne 8$
o. $\frac{b}{9} > 6$
p. $d + 7d \ge -64$
q. $24n - 3 < 93$
r. $8h + 6 \ne 6$
s. $4x + 7 > -11$
t. $\frac{x}{-2} \not< -9$

2. Find the solution set of the following inequalities when the replacement set is the set of all real numbers:

a. $12b + 7 < 43$
b. $5c - 3 > 22$
c. $8m - 4 \leq 60$
d. $18y - 8 \neq 16$
e. $3n + 17 \geq 5$
f. $7c + 9c > 32$
g. $6n + 12n < -9$
h. $2b - 8b \not> -54$
i. $10x - x \geq 63$
j. $14c + 11c \leq 10$
k. $4m - 7 < 35$
l. $x + x > -8$
m. $16y + 5 \not> 37$
n. $6t - 11t \neq 72$
o. $3c - 12 \not< 0$

3. Find the solution set of each of the following inequalities when the replacement set is the set of all natural numbers:

a. $6x + 3 > 3$ **b.** $9n - 5 < 13$ **c.** $12a + 2a > 42$ **d.** $11b - b < 50$

4. Find the solution set of each of the following inequalities when the replacement set is the set of all integers:

a. $7n - n < -48$ **b.** $4y + 10 > -2$ **c.** $d + d \leq 0$ **d.** $3x - 1 \geq 2$

5. Find the solution set of each of the following inequalities when the replacement set is the set of all non-positive integers:

a. $2x - 7x > 10$ **b.** $8n - 5 < -19$ **c.** $3b + 6 > 15$ **d.** $13w - 4w < 0$

11–44 BASIC GRAPHS OF INEQUALITIES IN SIMPLEST FORM

The graph of the numbers in a solution set of a sentence in one variable is the graph of the set of all numbers that satisfy or make the sentence true. Therefore the graph of an inequality in one variable is the set of all points on the number line whose coordinates are the numbers belonging to the solution set of the inequality.

The set of real numbers is used in the following graphs:

The graph of a set of all real numbers greater than a given number is a half-line extending to the right along the number line.

Graph of $x > 3$

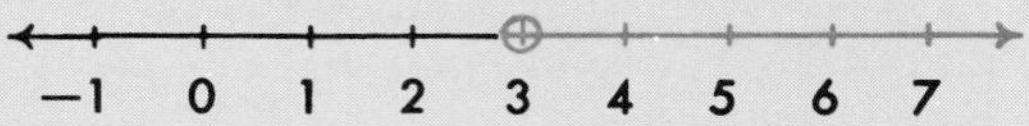

(A line is a set of points. A point on the line separates the line into two half-lines. The half-line extends indefinitely in one direction only and does not include the endpoint separating it from the other half-line. An open dot indicates the exclusion of this endpoint and the arrowhead indicates the half-line is endless.)

The graph of a set of all real numbers less than a given number is a half-line extending to the left along the number line.

Graph of $x < 3$

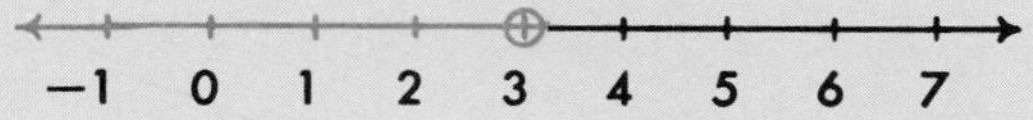

The graph of a set of all real numbers greater than or equal to a given number is a ray extending to the right along the number line. A *ray* is a half-line which includes one endpoint. A solid dot indicates the inclusion of this endpoint.

Graph of $x \geq 3$

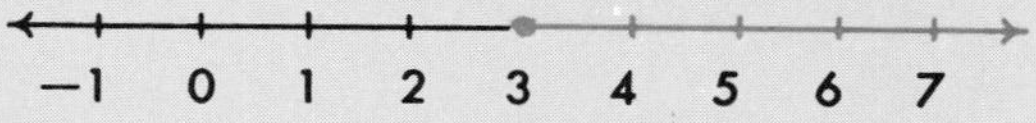

The graph of a set of all real numbers less than or equal to a given number is a ray extending to the left along the number line.

Graph of $x \leq 3$

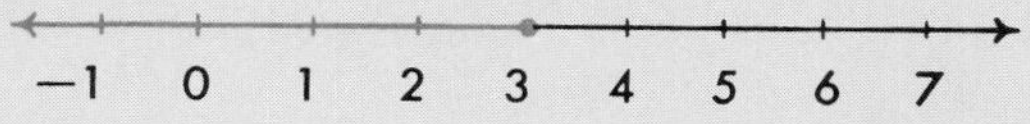

The graph of a set of all real numbers with the exception of one number is the entire real number line excluding the point corresponding to the excluded number—or two half-lines.

Graph of $x \neq 2$

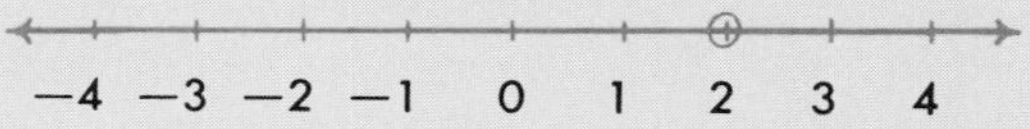

The graph of a set of numbers between two given numbers and including the two given numbers is a line segment which is a definite part of a line including the two endpoints.

Graph of $-2 \leq x \leq 3$

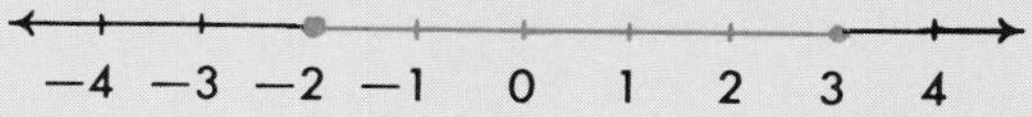

The graph of a set of numbers between the two given numbers but not including the two given numbers is an interval which is a definite part of a line excluding the two endpoints.

Graph of $-2 < x < 3$

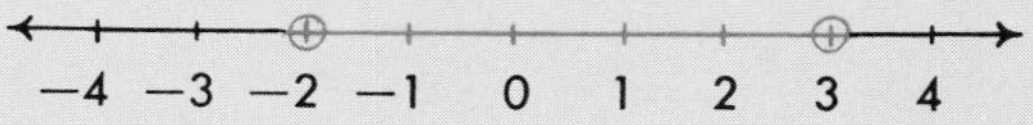

EXERCISES

For each of the following inequalities draw an appropriate number line, then graph its solution set. The replacement set is the set of all real numbers.

1. $x > 1$
2. $n < -3$
3. $x \neq 0$
4. $y \geq 2$
5. $-1 < x < 1$
6. $b \not> 2$
7. $-4 \leq n \leq 3$
8. $a \not< -4$
9. $0 \leq x < 5$
10. $z \leq 0$
11. $x > -4$
12. $-3 < n \leq 3$
13. $y \neq -2$
14. $|x| < 1$
15. $n \geq -1$
16. $-2 < y < 4$
17. $|y| > 2$
18. $-5 \leq x \leq -1$
19. $\{x \mid x < -2\}$
20. $\{y \mid y > 0\}$
21. $\{n \mid n \neq -1\}$
22. $\{x \mid x \leq 2\}$
23. $\{z \mid z \geq -4\}$
24. $\{a \mid 1 < a < 4\}$

11–45 GRAPH OF AN INEQUALITY

Since the graph of an inequality in one variable is the set of all points on the number line whose coordinates are the numbers belonging to its solution set, to draw the graph of an inequality not in its simplest form, we first find its solution set and then draw the graph of the solution set on the number line.

The graph of $5x - 6 > 4$, when the replacement set is the set of all real numbers, is the graph of its solution, $x > 2$, or the graph of its solution set:

Graph of $5x - 6 > 4$
is graph of: $x > 2$

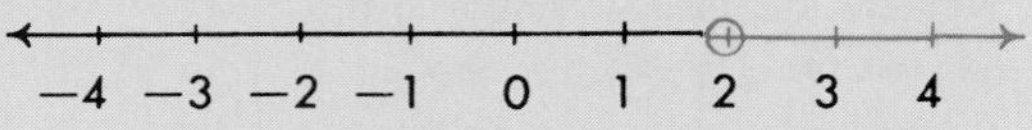

$\{x \mid 5x - 6 > 4\} = \{x \mid x > 2\} = \{\text{all real numbers greater than } 2\}$.

EXERCISES

1. For each of the following inequalities draw an appropriate number line, then graph its solution set. The replacement set is the set of all real numbers.

a. $x + 3 < 6$ **f.** $-3y > 12$ **k.** $c - 5 \leq 7$ **p.** $3y - 8y < 10$
b. $n - 1 > 4$ **g.** $6x - x \not< 15$ **l.** $-2x > -4$ **q.** $t - 2 \not> 4$
c. $5b \leq 20$ **h.** $y + y \not> 6$ **m.** $11x + 9 < -13$ **r.** $9n + 6 > 6$
d. $a + 1 \geq 5$ **i.** $4x + 5 \neq 9$ **n.** $7z - 14 \neq 0$ **s.** $4w - 10 \not< -2$
e. $\frac{x}{2} < 3$ **j.** $8n - 12 < 4$ **o.** $\frac{n}{5} \geq -1$ **t.** $12y - 7 < 5$

2. Write the corresponding inequality which is pictured by each of the following graphs:

a.
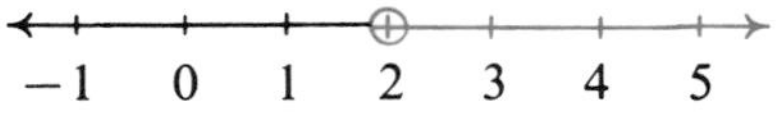

d.
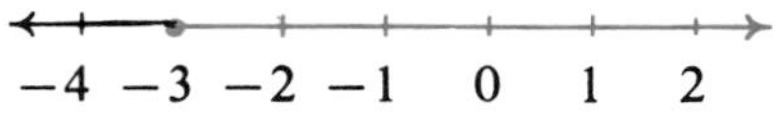

b.
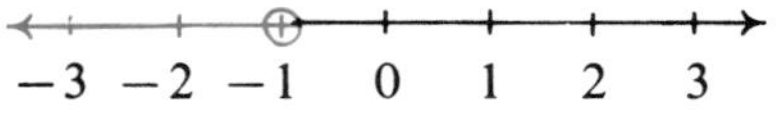

e.
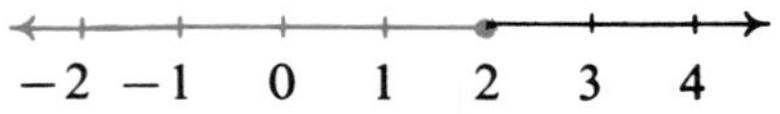

c.
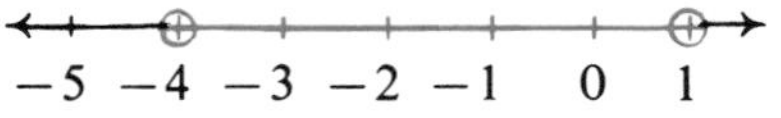

f.
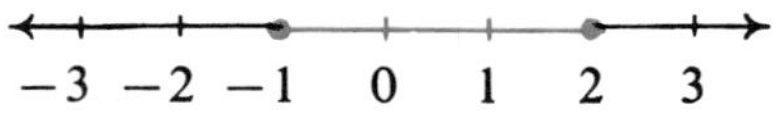

3. Draw the graph of each of the following:

a. $n + 3 > n$ **b.** $x - 2 < x$ **c.** $y + 2 < y$ **d.** $n - 5 > n$

11–46 RESTRICTED REPLACEMENT SETS

When a replacement set is restricted, we mark on the number line those points whose coordinates are numbers belonging to the replacement set and then indicate by heavy or colored dots the points belonging to the graph.

The graph of $x > 3$ when the replacement set is $\{0, 1, 2, 3, 4, 5, 6, 7, 8\}$ is shown at the right.

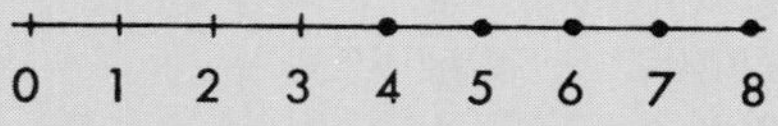

EXERCISES

For each of the following inequalities draw an appropriate number line, then graph its solution set:

1. When the replacement set is $\{-5, -4, -3, -2, -1, 0, 1, 2, 3, 4, 5\}$:

a. $x > -2$ **b.** $y + 3 < 4$ **c.** $a - 1 \geq -4$ **d.** $2n + 5 \neq 7$

2. When the replacement set is $\{-4, -2, 0, 2, 4\}$:

a. $-x < 0$ **b.** $n - 2 > -1$ **c.** $4d + d \leq 5$ **d.** $7n - 1 \nless -15$

3. When the replacement set is $\{-1, 0, 1\}$:

a. $-9w > 9$ **b.** $x - 5 < -5$ **c.** $3y - y \ngtr -2$ **d.** $6x + 9 \geq 15$

4. When the replacement set is the set of one-digit prime numbers:

a. $y + y < 10$ **b.** $9x - 2 > 16$ **c.** $8c \neq 56$ **d.** $12n + 5 \leq 41$

5. When the replacement set is the set of integers greater than -3 but less than 4:

a. $4x - 6 > -10$ **b.** $n - 5n < -8$ **c.** $9s - 3s \leq -12$ **d.** $7y + 11 \geq 11$

COORDINATE GEOMETRY SYSTEMS OF EQUATIONS

11–47 ORDERED PAIRS

The equation $x + y = 4$ is a sentence with two variables x and y. It has an infinite number of solutions in the system of real numbers but, as the replacement set is restricted, the number of solutions becomes limited. In the system of real numbers there is an unlimited number of pairs of positive and negative whole numbers and fractions which may be used to satisfy the equation $x + y = 4$. Examples are $x = 1$ and $y = 3$; $x = 1\frac{1}{2}$ and $y = 2\frac{1}{2}$; $x = 0.6$ and $y = 3.4$; $x = -1$ and $y = 5$. These and other pairs of numbers that satisfy the equation may be found by selecting values for one variable (x) from the infinite set of real numbers which may be substituted for it and computing the corresponding values of the other variable (y). In the system of *natural numbers* the only solutions of $x + y = 4$ are: $x = 1$ and $y = 3$, $x = 2$ and $y = 2$; $x = 3$ and $y = 1$ since they are the only pairs of natural numbers that satisfy the equation.

An *ordered pair of numbers* is a pair of numbers expressed in a definite order so that one number is first (first component) and the other number is second (second component). The numerals for an ordered pair are usually written in parentheses with numeral for the first number written first and for the second number written second separated by a comma. $x = 1$ and $y = 3$ is written as (1, 3), the x representing the first component and the y the second component. The solution set of $x + y = 4$ in the system of natural numbers consists of the ordered number pairs (1, 3), (2, 2), (3, 1) because they are the only pairs that satisfy the equation. Observe that (1, 3) and (3, 1) are *different* ordered pairs of numbers.

The *solution set of a sentence in two variables* is the set of all ordered pairs of numbers which make the sentence true.

In set notation the solution set of $x + y = 4$ is expressed as:

$$\{(x, y) \mid x + y = 4 \text{ and } x \text{ and } y \text{ are natural numbers}\} = \{(1, 3), (2, 2), (3, 1)\}.$$

> The expression $\{(x, y) \mid x + y = 4\}$ is read:
>
> "The set of all ordered pairs (x, y) such that each x plus each y is equal to four."

A sentence in two variables expresses a *relation*. We may think of a relation between two variables x and y such as $y = 2x$ or $y < 2x$ or $y > 2x$ or $x + y = 4$ etc. as a set of ordered pairs of numbers since in each case we may substitute for x any of the numbers given in the replacement set and obtain a corresponding number for y.

Sometimes the replacement set for x is called the *domain* of the relation and the corresponding set of numbers for y is called the *range* of the relation.

EXERCISES

1. Write the ordered pair that has:

a. 2 as the first component and 5 as the second component.
b. − 3 as the first component and 0 as the second component.

2. Write all the ordered pairs which have:

a. 1 or 2 as the first component and 1, 2, or 3 as the second component.
b. − 2, 0, 2 as the first component and − 3, − 1, 1, or 3 as the second component.
c. 1, 2, 3, 4, or 5 as the first component and 1, 2, 3, or 4 as the second component.

3. Write the set listing all the ordered pairs which have:

a. 0, 1, or 2 as the first component and 0, 1, or 2 as the second component.
b. 1, 2, 3, or 4 as the first component and 1, 2, 3, or 4 as the second component.
c. − 2, − 1, or 0 as the first component and 1 or 2 as the second component.

4. Read, or write in words, each of the following:

a. $\{(x, y) \mid x + y = 7\}$ **c.** $\{(x, y) \mid x - 2y > 1\}$ **e.** $\{(x, y) \mid 2x \geq 3y\}$
b. $\{(x, y) \mid x - y = 3\}$ **d.** $\{(x, y) \mid 3x + 5y < 6\}$ **f.** $\{(x, y) \mid 4x + y \leq 0\}$

5. Write each of the following symbolically using set notation:

a. The set of all ordered pairs (x, y) such that each x minus each y is equal to nine.

b. The set of all ordered pairs (x, y) such that eight times each x is less than five times each y.

c. The set of all ordered pairs (x, y) such that two times each x is less than or equal to seven times each y.

d. The set of all ordered pairs (x, y) such that ten times each x plus three times each y is greater than or equal to twelve.

11–48 GRAPHING AN EQUATION IN TWO VARIABLES

The graph of an equation in two variables pictures the relationship of the two variables.

The graph of an equation in two variables is drawn on a real number plane or, when the replacement set is limited, on a restricted number plane. When we located a point in a number line, we required only one number. However, to locate a point in a number plane we use an ordered pair of numbers because the point is located with respect to a *pair of coordinate axes.* These axes, one the horizontal axis or X-axis and the other the vertical axis or Y-axis, are two perpendicular number lines. They intersect at the point (0, 0) which is called the *origin.*

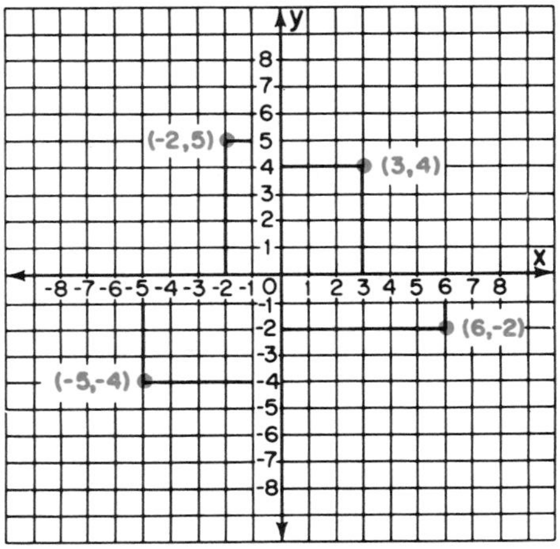

To locate a point in a plane, two perpendicular lines are drawn from the point, one to the X-axis and the other to the Y-axis. The number associated with the point where the perpendicular line intersects the X-axis is called the *x-coordinate* or *abscissa* of the point. The number associated with the point where the perpendicular line intersects the Y-axis is the *y-coordinate* or *ordinate* of the point. These coordinates are expressed as an ordered pair of numbers with the x-coordinate written first and the y-coordinate second. There is a one-to-one correspondence between all the points in the real number plane and the set of all ordered pairs of real numbers.

The graph of a sentence in two variables is the set of all points of the number plane whose coordinates are the ordered pairs of numbers belonging to the solution set of the sentence. In short, *the graph of a sentence is the graph of its solution set.* The graph of $x + y = 4$ is the graph of $\{(x, y) \mid x + y = 4\}$.

The graph of an equation in x and y is the set of all points whose coordinates (x, y) satisfy the equation. Therefore to draw the graph of the equation $x + y = 4$, we first determine at least three ordered pairs of numbers, each pair consisting of an x-coordinate and the corresponding y-coordinate which satisfy the given equation. These ordered pairs are found by selecting three different values for the x-coordinate, substituting each value for x in the equation, and computing the corresponding value of the y, each of which becomes the corresponding y-coordinate. We then plot the points by locating their coordinates with respect to the X-axis and Y-axis. If the real number plane is being used, we draw a line through the plotted points. The graph of $x + y = 4$ is a straight line (see below). Equations such as $x + y = 4$ are called *linear equations.*

$x + y = 4$

$y = 4 - x$

x	0	2	4
y	4	2	0

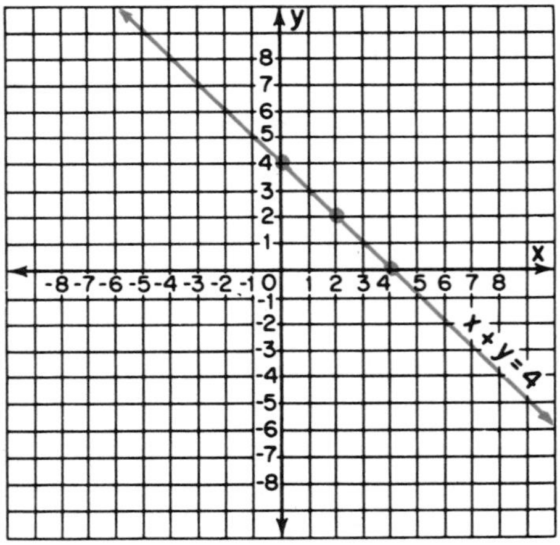

The set of points of the plane from which the set of points of the graph may be selected is sometimes restricted. If the set of points forming the number plane corresponds only to coordinates which are ordered pairs of natural numbers, then the points forming the graph will also have only coordinates which are ordered pairs of natural numbers.

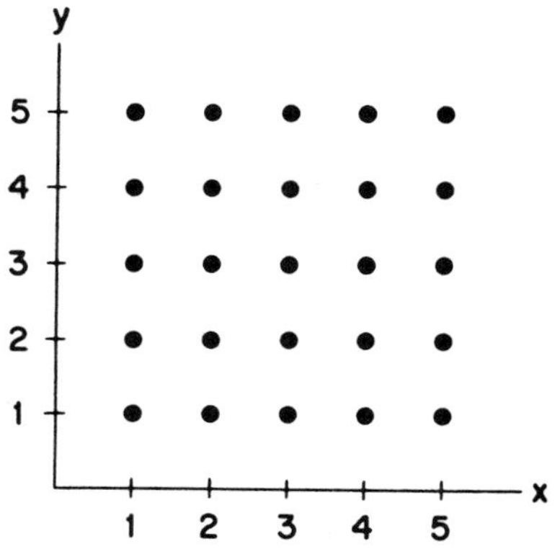

Suppose the number plane is limited by a finite set of numbers such as $\{1, 2, 3, 4, 5\}$. Each axis would be scaled 1, 2, 3, 4, 5 and the only possible points on the number plane would be points whose coordinates contained 1, 2, 3, 4, and 5 as the x-coordinate and 1, 2, 3, 4, and 5 as the y-coordinate. This restricts the graph to a selection from a set of twenty-five points whose coordinates form the set of the following ordered pairs of numbers: $\{(1, 1), (1, 2), (1, 3), (1, 4), (1, 5), (2, 1), (2, 2), (2, 3), (2, 4), (2, 5),$
$(3, 1), (3, 2), (3, 3), (3, 4), (3, 5), (4, 1), (4, 2), (4, 3), (4, 4), (4, 5), (5, 1),$
$(5, 2), (5, 3), (5, 4), (5, 5)\}$.

This is the set of all ordered pairs of numbers which can be formed by using the elements of U when $U = \{1, 2, 3, 4, 5\}$. The set of all ordered pairs whose elements belong to U is described as $U \times U$ (read "U cross U") and is called the *Cartesian product* of U.

The graph of the Cartesian product $U \times U$ shows the set of all points in the plane from which the points belonging to the graph of a given sentence may be selected. The graph of $U \times U$ is the real number plane when U is the set of all real numbers but is a restricted plane consisting of a lattice of points represented by a square or rectangular array of dots when U is a limited set. Each lattice point of the graph of $U \times U$ corresponds to an ordered pair of numbers belonging to $U \times U$.

When $U = \{1, 2, 3, 4, 5\}$, then the graph of $U \times U$ is a square array (see above).

When $C = \{1, 2, 3, 4\}$ and $D = \{1, 2, 3\}$, then the Cartesian product $C \times D$ is the set of all ordered pairs that can be formed by using each element of C as the first component and each element of D as the second component. The graph of $C \times D$ is represented by a rectangular array of dots as illustrated on page 338 (left illustration).

To graph a sentence when U is limited, heavy or colored dots can be used to indicate points whose coordinates are ordered pairs of numbers which belong to the solution set of the sentence. The graph of $x + y = 4$ when

$$U = \{1, 2, 3, 4, 5\}$$

is the graph of $\{(1, 3), (2, 2), (3, 1)\}$, as illustrated on page 338 (right illustration).

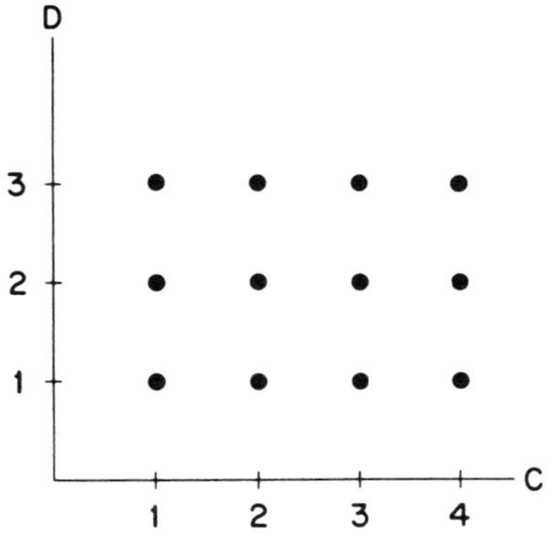

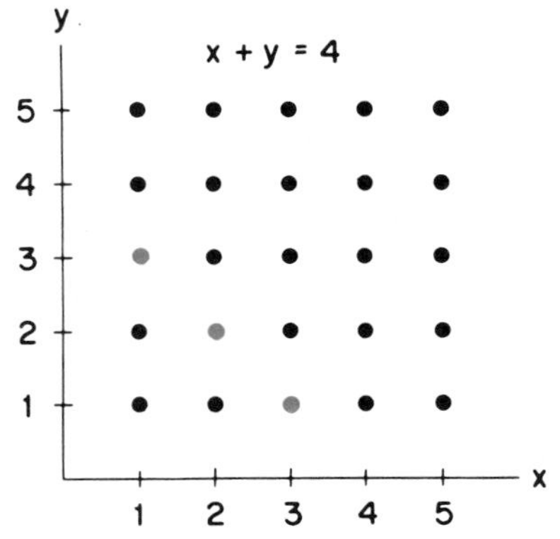

EXERCISES

1. Draw a set of coordinate axes, then plot points whose coordinates are:

a. (2, 1)
b. (1, 5)
c. (4, 3)
d. (3, − 1)
e. (6, − 4)
f. (1, − 8)
g. (− 1, 2)
h. (− 5, 4)
i. (− 6, 8)
j. (− 2, − 2)
k. (− 5, − 3)
l. (− 4, − 7)
m. (0, − 2)
n. (7, 0)
o. (0, 0)
p. (3, 6)
q. (− 7, − 2)
r. (− 3, 4)
s. (9, − 6)
t. (0, − 5)

2. Using the set of all real numbers, draw the graph of each of the following equations:

a. $y = x + 2$
b. $y = x - 1$
c. $y = x$
d. $y = 2x$
e. $y = -x$
f. $y = -4x$
g. $y = 5 + x$
h. $y = 7 - x$
i. $y = -5$
j. $x + y = 3$
k. $x - y = 4$
l. $x - y = -6$
m. $y = 8 - 3x$
n. $2y = 7 - 5x$
o. $4x + y = 4$
p. $3x + 4y = 0$
q. $2x + 3y = 10$
r. $5x - 2y = 3$
s. $x = 7$
t. $3x + 2y = 6$

3. Using the set of all real numbers, draw the graph of each of the following solution sets:

a. $\{(x, y) \mid y = x + 3\}$
b. $\{(x, y) \mid y = -2x\}$
c. $\{(x, y) \mid x + y = 6\}$

4. In each of the following first write the set listing all the ordered pairs of numbers contained in $U \times U$, then draw the graph of $U \times U$:

a. $U = \{1, 2\}$
b. $U = \{1, 2, 3, 4\}$
c. $U = \{0, 1, 2, 3, 4, 5, 6\}$
d. $U = \{-4, -3, -2, -1, 0\}$
e. $U = \{-2, -1, 0, 1, 2\}$
f. $U = \{-1, 0, 1, 2, 3\}$

5. In each of the following first write the set listing all the ordered pairs of numbers contained in $C \times D$, then draw the graph of $C \times D$:

a. $C = \{1, 2\}$ and $D = \{1\}$
b. $C = \{1, 2, 3, 4\}$ and $D = \{1, 2, 3, 4, 5\}$
c. $C = \{-2, -1, 0\}$ and $D = \{-1, 0\}$
d. $C = \{-3, -2, -1, 0\}$ and $D = \{-2, -1, 0\}$

6. In each of the following substitute each given element of U for x in the given equation to determine the corresponding value for y. Select, then write the set listing all ordered pairs of numbers belonging to the solution set of:

a. $y = 2x$ when $U = \{1, 2, 3\}$
b. $y = x + 3$ when $U = \{1, 2, 3, \ldots, 6\}$
c. $y = 5 - x$ when $U = \{1, 2, 3, 4, 5\}$
d. $x + y = 2$ when $U = \{0, 1, 2, 3\}$
e. $x - y = 4$ when $U = \{1, 2, 3, \ldots, 9\}$
f. $2x + y = 7$ when $U = \{0, 1, 2, 3, 4, 5\}$
g. $x - 2y = 0$ when $U = \{0, 1, 2, \ldots, 8\}$
h. $3x + 2y = 6$ when $U = \{-3, -2, -1, 0, 1, 2, 3\}$
i. $4x - 3y = 12$ when $U = \{-4, -3, -2, \ldots, 4\}$
j. $2x - y = 3$ when $U = \{0, 1, 2, \ldots, 9\}$

7. First draw the graph of $U \times U$, then use heavy or colored dots to show the graph of the solution set of each of the following:

a. $y = x$ when $U = \{1, 2, 3, 4\}$
b. $y = -3x$ when $U = \{-3, -2, -1, \ldots, 3\}$
c. $y = 6 - x$ when $U = \{1, 2, 3, 4, 5\}$
d. $y = 2x + 1$ when $U = \{1, 2, 3, \ldots, 9\}$
e. $x + y = 7$ when $U = \{0, 1, 2, \ldots, 10\}$
f. $x - y = 3$ when $U = \{-4, -3, -2, \ldots, 4\}$
g. $2x - 3y = 0$ when $U = \{1, 2, 3\}$
h. $4x - y = 7$ when $U = \{0, 1, 2, \ldots, 7\}$
i. $x + y = -2$ when $U = \{-2, -1, 0, 1, 2\}$
j. $3x - 2y = 2$ when $U = \{0, 1, 2, \ldots, 6\}$

11–49 GRAPHS OF FORMULAS

Reading from a Graph

To determine a value directly from a graph, we locate the point on the scale corresponding to the given value and at this point we erect a perpendicular line which intersects the graph. Then at this point of intersection we erect a line perpendicular to the first perpendicular line. The value of the point on the second scale intersected by the second perpendicular line is the required value.

EXERCISES

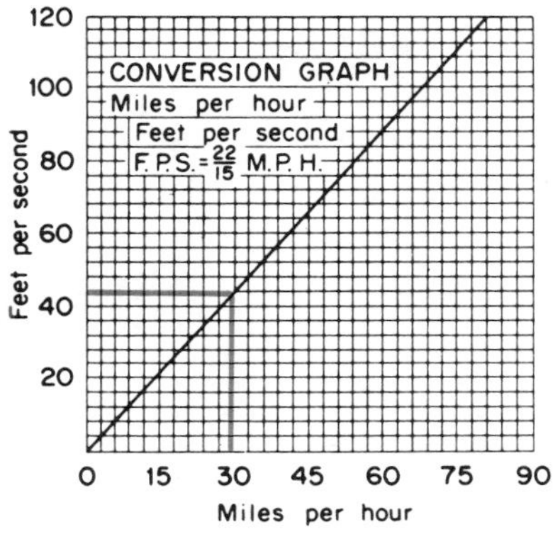

Answer directly from the graph:

1. Express in feet per second:

a. 30 m.p.h.
b. 45 m.p.h.
c. 20 m.p.h.
d. 50 m.p.h.

2. Express in miles per hour:

a. 88 ft. per sec.
b. 60 ft. per sec.
c. 100 ft. per sec.
d. 42 ft. per sec.

Drawing a Graph

To draw a graph of a formula, we first construct a table of values showing the relationship between the variables in the formula just as we do in preparing to draw the graph of an equation in two variables. We select scales for each variable, using the bottom of the paper for the horizontal scale and the left side for the vertical scale. We label the scales. Using the values in the table, we plot points corresponding to the ordered pairs of numbers. We draw a line through these points and print the formula and title. This line is the *graph* of the formula.

EXERCISES

Construct graphs of each of the following formulas:

1. Perimeter and length of a side of a square. Formula: $p = 4s$
2. Annual simple interest at 6% and principals ranging from \$0 to \$1,000. Formula: $i = .06p$
3. Fahrenheit-Celsius temperature readings. Formula: $F = 1.8C + 32$

11–50 GRAPHING AN INEQUALITY IN TWO VARIABLES

To draw the graph of an inequality in two variables using the set of real numbers, we first draw the graph of the equation which can be formed from the inequality by changing only the inequality symbol to an equality symbol. Thus, to draw the graphs of $x + y > 4$ or $x + y < 4$ or $x + y \ngtr 4$ or $x + y \nless 4$, we would change each to $x + y = 4$ and draw the graph of $x + y = 4$. However, the graph of this equation is only the boundary line which divides the plane into two half-planes one of which is the graph or part of the graph of the above inequalities. The region of the plane corresponding to the graph is generally shaded. Sometimes the boundary line is part of the graph. If it is, a solid or colored line is used to indicate this, if it is not, then a dashed line is used.

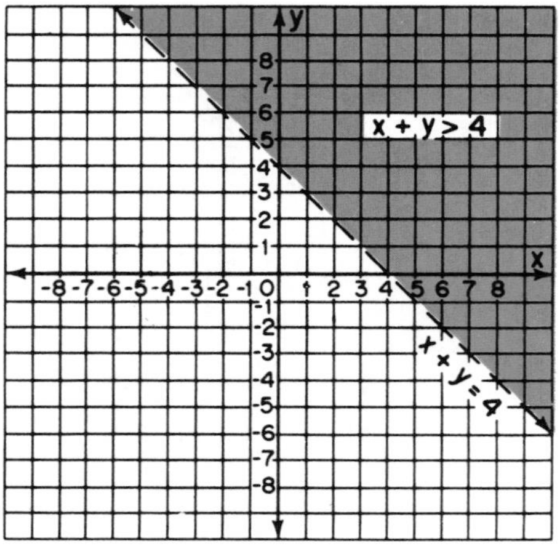

The graph of $x + y > 4$, shown partially, is the half-plane above and to the right of the boundary line. It does not include the boundary line. This graph is the graph of the solution set $\{(x, y) \mid (x + y > 4)\}$ which is the set of all ordered pairs of numbers which are coordinates of the points located in this upper half-plane.

The graph of $x + y \ngtr 4$ is like the graph of $x + y > 4$ but it includes

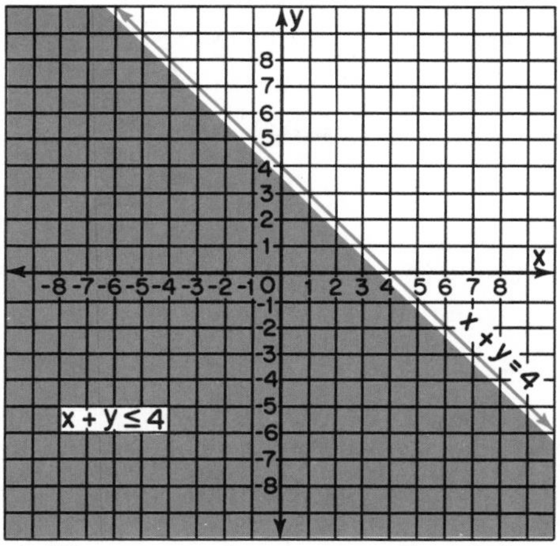

the boundary line, indicated by a solid or colored line instead of a dashed line.

The graph of $x + y < 4$ is the lower half-plane to the left and below the boundary line. Since this graph does not include the boundary line, a dashed line would be used.

The graph of $x + y \leq 4$, shown on the right, is like the graph of $x + y < 4$ but it includes the boundary line. A solid or colored line is used instead of a dashed line.

When the plane is restricted (Figures *A–D*), we use heavy or colored dots to indicate the points whose coordinates are ordered pairs of numbers which belong to the solution set of the inequality. To graph an inequality we graph its solution set.

A

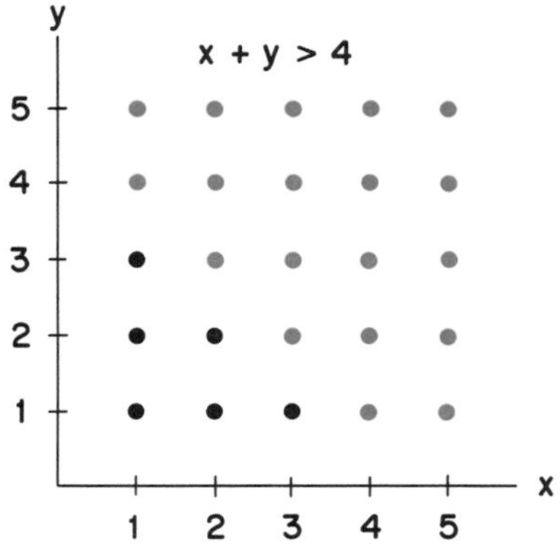

B

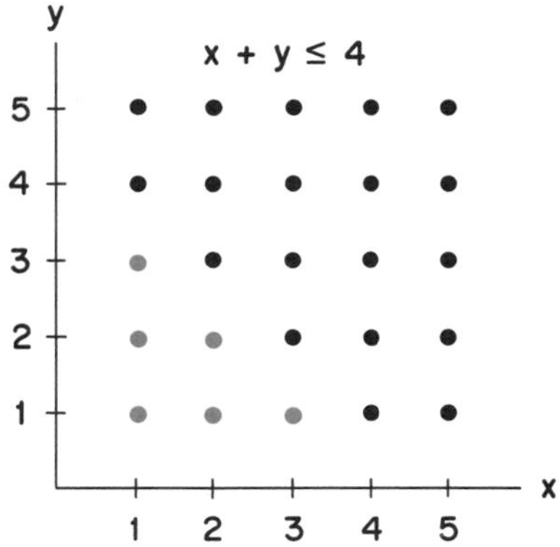

C

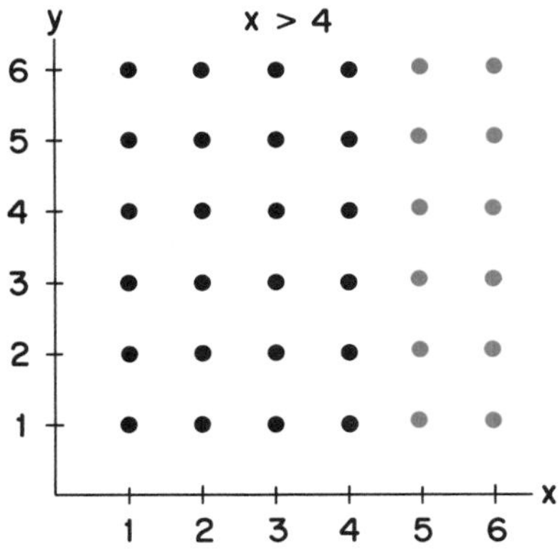

D

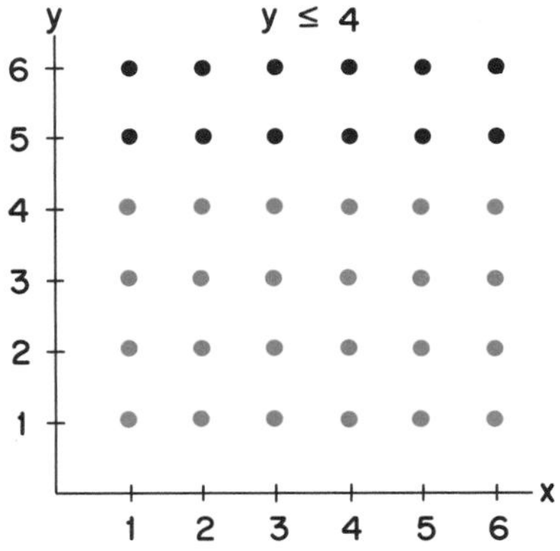

The graphs on page 341 are graphs of the inequalities $x + y > 4$ (Figure A) and $x + y \leq 4$ (Figure B) when $U = \{1, 2, 3, 4, 5\}$.

Observe the graphs of $x > 4$ (Figure C) and $y \leq 4$ (Figure D) when $U = \{1, 2, 3, 4, 5, 6\}$.

EXERCISES

1. Using the set of all real numbers, draw the partial graph of each of the following inequalities:

a. $y > x$ **d.** $x + y < 5$ **g.** $y > 3 - x$ **j.** $2x + y \leq 4$
b. $y < 2x$ **e.** $x - y \leq 3$ **h.** $3x - y < 5$ **k.** $x + 5y < -2$
c. $y > x + 2$ **f.** $x + y \geq 7$ **i.** $2x + 4y > 0$ **l.** $4x - 3y \geq 5$

2. Using the set of all real numbers, draw the partial graph of each of the following solution sets:

a. $\{(x, y) \mid y < x\}$ **b.** $\{(x, y) \mid y > x + 3\}$ **c.** $\{(x, y) \mid x + y \geq 6\}$

3. In each of the following substitute each given element of U for x in the inequality to determine the corresponding values for y. Select, then write, the set listing all the ordered pairs of numbers belonging to the solution set of:

a. $y < x$ when $U = \{0, 1\}$
b. $y > 2x$ when $U = \{1, 2\}$
c. $y < 3x$ when $U = \{1, 2, 3\}$
d. $y > -2x$ when $U = \{-2, -1, 0, 1, 2\}$
e. $y < x + 3$ when $U = \{0, 1, 2, 3\}$
f. $y > x - 1$ when $U = \{1, 2, 3, 4, 5\}$
g. $y < 5 - x$ when $U = \{0, 1, 2, 3, 4\}$
h. $y > -2 - x$ when $U = \{-3, -2, -1\}$
i. $x + y < 2$ when $U = \{-1, 0, 1\}$
j. $x - y > -3$ when $U = \{-3, -2, \ldots, 2, 3\}$
k. $x + y \leq 3$ when $U = \{1, 2, 3, 4\}$
l. $x - y \geq -2$ when $U = \{-2, -1, 0, 1, 2\}$
m. $y \leq -x$ when $U = \{-2, -1, 0, 1, 2\}$
n. $y \geq 3x$ when $U = \{0, 1, 2, 3\}$
o. $2x - y < 4$ when $U = \{1, 2, 3\}$
p. $x + 3y > -1$ when $U = \{0, 1, 2\}$
q. $3x - 2y \not< 0$ when $U = \{0, 1, 2, 3\}$
r. $5x - 3y \not> -3$ when $U = \{-1, 0, 1\}$
s. $4x + 5y < 10$ when $U = \{0, 1, 2, 3, 4, 5\}$
t. $3x + 4y > 2$ when $U = \{-2, -1, 0, 1, 2\}$

4. First draw the graph of $U \times U$, then use heavy or colored dots to show the graph of the solution set of each of the following:

a. $y > x$ when $U = \{0, 1, 2\}$
b. $y < 3x$ when $U = \{1, 2, 3, 4\}$

c. $y > x + 1$ when $U = \{0, 1, 2, 3\}$
d. $y < 5 - x$ when $U = \{1, 2, 3, 4\}$
e. $x + y > -2$ when $U = \{-1, 0, 1\}$
f. $x - y \leq 2$ when $U = \{1, 2, 3, 4, 5\}$
g. $x + y \geq -1$ when $U = \{-2, -1, 0, 1, 2\}$
h. $2x - y < 3$ when $U = \{-3, -2, \ldots, 2, 3\}$
i. $3x - 2y > 0$ when $U = \{0, 1, 2, 3, 4\}$
j. $4x + 3y \not< -1$ when $U = \{-2, -1, 0, 1, 2\}$
k. $4x + y < 6$ when $U = \{0, 1, 2, 3, 4, 5\}$
l. $2x - 3y > 2$ when $U = \{1, 2, 3, 4\}$
m. $x + 2y \not> 1$ when $U = \{-3, -2, \ldots, 2, 3\}$

11–51 DRAWING GRAPHS BY THE SLOPE–INTERCEPT METHOD

Graphs of linear equations are straight lines. Many of these lines slant or incline or slope. We may measure the slope of a line by comparing the vertical change to the horizontal change as we move from one point in the line to another point in the line.

Using the graph, we count from any point in the line a convenient number of units to the right (4), then we count vertically up (3) if the line slants upward (this is the positive direction) or down if the line slants downward (this is the negative direction) until we met the line at a second point. We divide the vertical count (3) by the horizontal count (4).

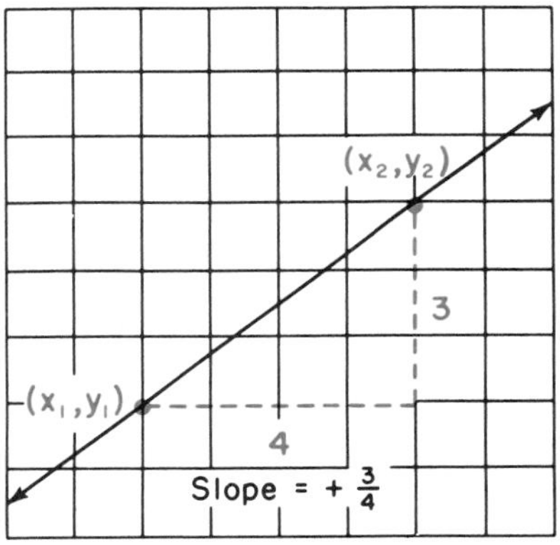

This is the same as dividing the difference between the y-coordinates of two selected points by the difference between the x-coordinates of the same two points. Thus the slope formula is:

$$\text{Slope } (m) = \frac{y_2 - y_1}{x_2 - x_1}$$

Slope is measured by a number, positive if the line slants upward to the right; negative if the line slants downward to the right; zero if the line is horizontal; and no slope if the line is vertical.

The y-form or (slope-intercept form) of an equation in variables x and y is an equivalent equation which has been solved for y. For example, $y = 2x - 5$ is the y-form of the equation $2x - y = 5$.

The y-intercept is the ordinate (y-coordinate) of the point where the line crosses the Y-axis. The y-intercept is positive if this point is above the X-axis, negative if the point is below the X-axis, and zero if the point is at the origin.

Observe that the slope of the line shown at the right is 2 and the y-intercept is -3. The 2 corresponds to the coefficient of x in the y-form of the equation $y = 2x - 3$ and the intercept number -3 corresponds to the constant term (term without x or y) found in the y-form of the equation. Both the correspondence of the slope to the coefficient of x and of the intercept number to the constant term are found to be true in all linear equations arranged in the y-form.

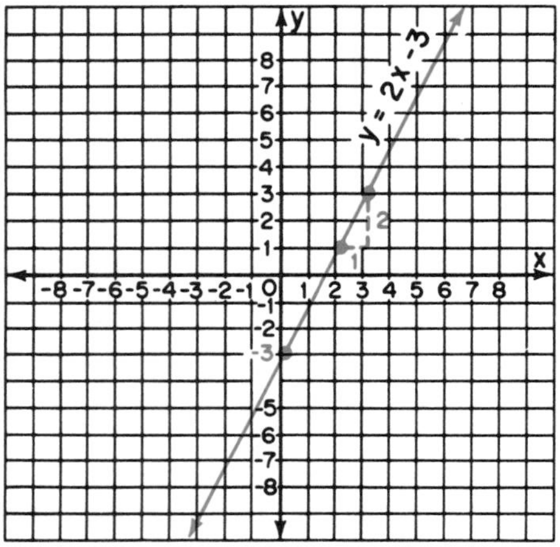

Thus, for any linear equation in the y-form expressed as $y = mx + b$ the coefficient of x, represented by m, indicates the slope and the constant term, represented by b, indicates the y-intercept.

To draw a graph of a linear equation such as $3x + y = 6$ using the slope and the y-intercept, we first express the equation in the form $y = mx + b$. We then use the value of b as the ordinate to locate the y-intercept. We use the slope represented by m to locate a second point, counting to the right a convenient number of units from the y-intercept, then counting the required number of units up or down depending on whether the slope is positive or negative. We draw a line through these two points. We may check whether the graph is correct by substituting for the variables the coordinates of a third point in the line. If these values satisfy the equation, the graph is correct.

$$3x + y = 6$$
$$y = -3x + 6$$

Since the y-intercept number is $+6$ we locate point (0, 6). Since the slope is -3, we count from point (0, 6) 1 unit to the *right* and 3 units *down*, reaching point (1, 3). We then draw a line through points (0, 6) and (1, 3).

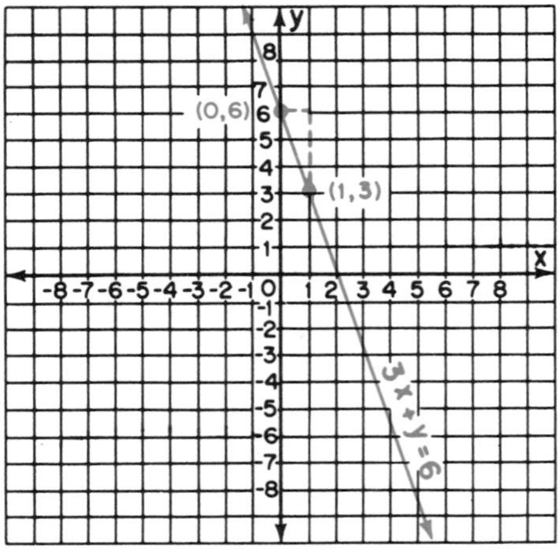

EXERCISES

1. What is the slope of each of the following lines?

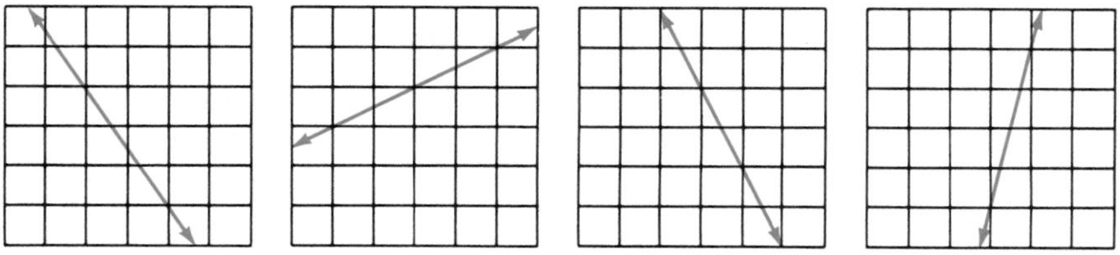

a. **b.** **c.** **d.**

2. Draw a line with a slope of:

a. $+3$ **b.** -2 **c.** $+\frac{2}{3}$ **d.** $-\frac{1}{4}$ **e.** $+\frac{5}{2}$

3. Use the slope formula to determine the slope of a line that passes through two points whose coordinates are:

a. (1, 2) and (4, 8) **c.** (− 3, − 5) and (1, 4)
b. (− 2, 8) and (4, − 4) **d.** (− 5, 7) and (0, 2)

On squared paper for each of the above, plot the two given points, draw the line through the points, then determine the slope by counting, using points other than the given points.

4. What is the slope of a line segment whose endpoints are (− 6, − 3) and (− 1, 7)?

5. Arrange each of the following equations in the y-form:

a. $x + y = 10$ **c.** $3x + y = 7$ **e.** $7x = 2y$ **g.** $5x - 9y = -21$
b. $x - y = 8$ **d.** $x + 4y = -9$ **f.** $3x - 8y = 0$ **h.** $4x + 6y = 18$

6. Directly from the equation without drawing the graph determine the slope and y-intercept number of the graph of each of the following equations:

a. $y = 2x + 6$ **e.** $9x - 5y = 2$ **i.** $2x + 3y = -5$
b. $y = -7x + 3$ **f.** $x = 11 - 4y$ **j.** $x = -4y - 21$
c. $y = -\frac{2}{3}x - 8$ **g.** $3x - 7y = 0$ **k.** $6y - 7x = -10$
d. $x + y = 5$ **h.** $-4x + 8y = 12$ **l.** $10x - 15y = 25$

7. Graph each of the following linear equations by the slope-intercept method:

a. $y = 4x + 3$ **e.** $y = -\frac{3}{4}x + 2$ **i.** $12x + 4y = 16$
b. $y = 3x - 5$ **f.** $y = \frac{1}{2}x - 3$ **j.** $3x - 2y = -12$
c. $y = -2x - 1$ **g.** $x + y = 7$ **k.** $x = 3y + 9$
d. $y = \frac{2}{3}x$ **h.** $2x - y = 1$ **l.** $6x + 5y = 15$

8. What do the graphs of $y = 7x - 2$ and $y = 7x + 8$ have in common?
9. What do the graphs of $y = 3x + 9$ and $y = 5x + 9$ have in common?
10. What is true of all lines having 0 as the **a.** slope? **b.** y-intercept number?

11–52 SYSTEMS OF LINEAR EQUATIONS IN TWO VARIABLES

A pair of equations like $4x - 3y = 9$
$3x + 5y = 14$

forms a *system of linear equations in two variables.* They are sometimes called *simultaneous equations.* Each of these equations has the same two variables. A linear equation in two variables is said to be in *standard form* when written in the form: $ax + by = c$ where x and y are the variables and a, b, and c are rational numbers but a and b cannot both be equal to 0. The equation $3x = 12$ is considered a linear equation in two variables when thought of as $3x+0y=12$.

In section 11–48 we found that the graph of each equation like those above on the real number plane is a straight line. When the graphs of two linear equations in the same two variables are drawn on the same set of axes, we find that only one of the following possibilities exists at any one time:

a. The two lines intersect. See page 397.
b. The two lines are parallel. See page 397.
c. The two lines are one and the same line, they coincide.

When the *two lines intersect,* the point of intersection is obviously on both lines. Consequently the number pair associated with this point of intersection satisfies both equations when we substitute the first component for x and the second component for y in the two equations. Since two straight lines intersect in only one point, this number pair is the only member of the solution set that results from the intersection of the solution sets of the two equations. When the system of linear equations has only one common solution, it is said to be *independent* and *consistent.*

When the *two lines are parallel,* they do not intersect and therefore have no common point. The solution set in this case is the empty set. The system of equations whose graphs are parallel lines is said to be *inconsistent.*

When the *two lines coincide,* they each have exactly the same points. Consequently the solution set consists of an infinite set of number pairs that satisfy both equations. A pair of equations of this type are equivalent equations. This system of linear equations whose graphs coincide is said to be *dependent.*

If we express the two given linear equations in their y-forms (see section 11–51), we can tell without drawing the graphs whether the graphs of these equations form a pair of parallel lines, whether the lines coincide, or whether the lines intersect.

If the y-forms of the two equations indicate different y-intercept numbers but the same coefficients of x, indicating that the slopes of both lines are the same, then the lines are parallel. If the y-forms of the two equations are exactly the same, then the lines coincide.

If the y-forms of the two equations show that the coefficients of x are different, indicating different slopes, then the lines intersect.

EXERCISES

1. Without drawing the graphs, select from the following systems of linear equations those which have graphs which are (1) a pair of parallel lines. (2) A pair of intersecting lines. (3) A pair of lines that coincide.

a. $3x + y = 16$, $x - 3y = 2$ **c.** $6x - 9y = 24$, $4x - 6y = 16$ **e.** $x + 4y = 8$, $3x - 12y = 5$

b. $2x + 7y = 10$, $2x + 7y = -5$ **d.** $x + 2y = 9$, $3x + 6y = 27$ **f.** $5x - y = 6$, $10x - 2y = 14$

2. Which systems of linear equations in problem 1 are dependent? Which are independent and consistent? Which are inconsistent?

To solve a system of two linear equations in two variables means to find the *common solution set* of the two given equations. We shall study three ways of solving systems of linear equations: (1) by graphing, (2) by the addition method, and (3) by the substitution method.

Solving By Graphing

To solve a system of two linear equations by graphing, we draw the graph of each equation on the same axes. See sections 11–48 and 11–51. If the lines intersect, we find the coordinates of the point of intersection. This is the common solution of the two equations. We check by substituting the coordinates of this point of intersection in the two given equations. If the two resulting sentences are true, then this number pair is the required solution set.

Solve graphically: $2x + y = 7$, $4x - 3y = 9$

$2x + y = 7$, $y = 7 - 2x$ $4x - 3y = 9$, $\frac{4x - 9}{3} = y$

x	0	2	5
y	7	3	−3

x	−3	0	4
y	−7	−3	$2\frac{1}{3}$

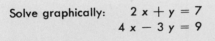

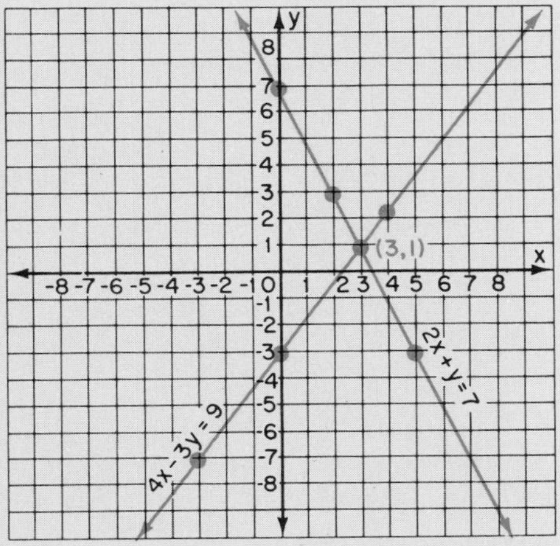

The coordinates of the point of intersection are (3, 1).

Check:

$2x + y = 7$ $4x - 3y = 9$

$(2 \cdot 3) + 1 = 7$ $(4 \cdot 3) - (3 \cdot 1) = 9$

$6 + 1 = 7$ $12 - 3 = 9$

$7 = 7\checkmark$ $9 = 9\checkmark$

Answer, Solution set $\{(3, 1)\}$

EXERCISES

Using the set of all real numbers, solve each of the following systems of equations graphically, then check:

1. $y = 2x$
   $y = x + 3$
2. $y = 11 - x$
   $y = 4x + 1$
3. $y = \frac{1}{4}x + 5$
   $y = \dfrac{3x - 10}{2}$
4. $x + y = 7$
   $x - y = 3$
5. $x + y = 1$
   $3x - 4y = 10$
6. $y = -x$
   $5x + 3y = -6$
7. $x + 2y = -9$
   $7x - 4y = 9$
8. $y = 5$
   $3x + 2y = 1$
9. $x = -4$
   $6x + 5y = -14$
10. $2x + 3y = -6$
    $5x - y = 2$
11. $x = 2y$
    $3x - 7y = 3$
12. $4x - 5y = 25$
    $2x + 7y = 3$
13. $\{(x, y) \mid y = -3x\} \cap \{(x, y) \mid x + y = 4\}$
14. $\{(x, y) \mid 2x + 5y = 3\} \cap \{(x, y) \mid x - 2y = 6\}$

Solving By Addition

To find the common solution set of a system of two linear equations in two variables, we must first reduce the two equations to a single equation in one variable.

When the given equations, arranged in standard form, both have one variable with the same absolute value for its numerical coefficient, we select this variable for elimination. There are two possibilities.

(1) When the numerical coefficients of this variable have opposite signs, we add the left member of the first equation to the left member of the second equation and the right member of the first equation to the right member of the second equation. Here we are using the axiom: when equals are added to equals, the sums are equal.

We solve the resulting equation for the value of the remaining variable. Then we substitute this value in one of the given equations to find the value of the other variable. This resulting number pair is the common solution. We check both given equations to determine whether the found values satisfy them.

Solve and check: $3x + y = 14$
$2x - y = 1$

Solution:

$$\begin{array}{l} 3x + y = 14 \\ \underline{2x - y = 1} \\ 5x \quad\ = 15 \\ \quad\ x = 3 \end{array}$$

$$\begin{aligned} 3x + y &= 14 \\ 3 \cdot 3 + y &= 14 \\ 9 + y &= 14 \\ 9 + (-9) + y &= 14 + (-9) \\ y &= 5 \end{aligned}$$

Check:

$$\begin{aligned} 3x + y &= 14 \\ (3 \cdot 3) + 5 &= 14 \\ 9 + 5 &= 14 \\ 14 &= 14 \checkmark \end{aligned} \qquad \begin{aligned} 2x - y &= 1 \\ (2 \cdot 3) - 5 &= 1 \\ 6 - 5 &= 1 \\ 1 &= 1 \checkmark \end{aligned}$$

Answer, $x = 3$, $y = 5$
or Solution set $\{(3, 5)\}$

(2) When the numerical coefficients of this variable, which was selected for elimination in the two equations, have the same signs, we may choose either one of the following methods:

(a) We may subtract the corresponding members of the given equations by adding the additive inverse of each member of the second equation to the corresponding member of the first equation. An equation with one variable will result. Then we proceed as explained in (1) to get the common solution set.

or (b) We may multiply one of the given equations by -1, using the axiom: when equals are multiplied by the same number, the products are equal. Then we proceed as explained in (1) to get the common solution set.

Solve and check: $2x + 3y = 18$
$4x + 3y = 30$

Solution:

$$\begin{array}{rl} 2x + 3y &= 18 \\ 4x + 3y &= 30 \\ \hline\hline 2x + 3y &= 18 \\ -4x - 3y &= -30 \\ \hline -2x \quad &= -12 \\ x &= 6 \end{array} \qquad \begin{array}{rl} 2x + 3y &= 18 \\ 2 \cdot 6 + 3y &= 18 \\ 12 + 3y &= 18 \\ 12 + (-12) + 3y &= 18 + (-12) \\ 3y &= 6 \\ y &= 2 \end{array}$$

Answer, $x = 6$, $y = 2$ or Solution set $\{(6, 2)\}$

Check:

$$\begin{array}{rl} 2x + 3y &= 18 \\ (2 \cdot 6) + (3 \cdot 2) &= 18 \\ 12 + 6 &= 18 \\ 18 &= 18 \checkmark \end{array}$$

$$\begin{array}{rl} 4x + 3y &= 30 \\ (4 \cdot 6) + (3 \cdot 2) &= 30 \\ 24 + 6 &= 30 \\ 30 &= 30 \checkmark \end{array}$$

When neither variable has a common numerical coefficient, we select a variable for elimination. First we multiply one equation by a number and, if necessary, the second equation by another number so that the numerical coefficient of this variable is the least common multiple of the given numerical coefficients. Then we follow the above procedures (1) or (2) to find the common solution set of the two given linear equations in two variables.

Solve and check: $4x + 5y = 7$
Solution: $7x - 3y = -23$

To eliminate the y variable, we must change both the $5y$ and the $3y$ to a common $15y$. To do this, we multiply both sides of the top equation by 3 and both sides of the bottom equation by 5.

$$\begin{array}{rl} 4x + \ \ 5y &= \ \ 7 \\ 7x - \ \ 3y &= -23 \\ \hline\hline 12x + 15y &= \ \ 21 \\ 35x - 15y &= -115 \\ \hline 47x \quad &= -94 \\ x &= -2 \end{array} \qquad \begin{array}{rl} 4x + 5y &= 7 \\ 4(-2) + 5y &= 7 \\ -8 + 5y &= 7 \\ (-8) + (8) + 5y &= 7 + 8 \\ 5y &= 15 \\ y &= 3 \end{array}$$

Answer, $x = -2$, $y = 3$ or Solution set $\{(-2, 3)\}$

Check:

$$\begin{array}{rl} 4x + 5y &= 7 \\ 4(-2) + (5 \cdot 3) &= 7 \\ -8 + 15 &= 7 \\ 7 &= 7 \checkmark \end{array}$$

$$\begin{array}{rl} 7x - 3y &= -23 \\ 7(-2) - (3 \cdot 3) &= -23 \\ -14 - 9 &= -23 \\ -23 &= -23 \checkmark \end{array}$$

EXERCISES

Solve by the addition method and check:

1. $x + y = 8$
   $x - y = 4$
2. $2x - y = 1$
   $4x + y = -7$
3. $8x - 2y = 24$
   $5x + 2y = 2$
4. $3x + y = -12$
   $x + y = -2$
5. $6x - 7y = 11$
   $9x - 7y = 20$
6. $-5x - 9y = 10$
   $-7x - 9y = 14$
7. $9x + 5y = -1$
   $-2x + 5y = -12$
8. $x - 8y = -8$
   $x + 9y = 26$
9. $10x - 7y = -22$
   $-10x + 3y = 38$
10. $8x + y = 12$
    $7x + 6y = 31$
11. $6x - 8y = 54$
    $12x - 4y = 72$
12. $2x - 3y = -5$
    $5x + 9y = 37$
13. $2x + 5y = 8$
    $7x - 4y = -15$
14. $3x - 2y = 10$
    $4x - 7y = 35$
15. $7x + 6y = 3$
    $9x + 5y = -7$
16. $6x + 4y = 16$
    $10x + 6y = 22$
17. $15x + 8y = 85$
    $9x - 12y = -33$
18. $21x - 10y = -1$
    $14x - 8y = 2$
19. $\{(x, y) \mid x + 3y = 1\} \cap \{(x, y) \mid 3x - y = -17\}$
20. $\{(x, y) \mid 4x - 6y = 14\} \cap \{(x, y) \mid 9x - 10y = 28\}$

Solving By Substitution

We have already found that in order to determine the common solution set of a system of two linear equations in two variables, we must reduce the two given equations with two variables to a single equation with one variable. Thus the substitution method of eliminating one of the variables is most effective when one of the given equations is already arranged so that it expresses one variable in terms of the other or when this arrangement can easily be obtained.

When the system of equations does not include an equation expressing one variable in terms of the other variable, we select the simpler equation and solve it for one variable in terms of the second variable. Using the other given equation, we replace the first variable by the algebraic expression that is equal to it. Then we solve the resulting equation to find the value of the variable. To find the value of the other variable, we substitute the value of the variable, just found, in the rearranged equation. We check by substituting this number pair in the two given equations.

Solve by the substitution method and check: $y = x + 1$, $2x + 7y = 43$

Solution: $y = x + 1$

$$2x + 7y = 43$$
$$2x + 7(x + 1) = 43$$
$$2x + 7x + 7 = 43$$
$$9x + 7 = 43$$
$$9x + 7 + (-7) = 43 + (-7)$$
$$9x = 36$$
$$x = 4$$

$$y = x + 1$$
$$y = 4 + 1$$
$$y = 5$$

Check:

$y = x + 1$	$2x + 7y = 43$
$5 = 4 + 1$	$(2 \cdot 4) + (7 \cdot 5) = 43$
$5 = 5\checkmark$	$8 + 35 = 43$
	$43 = 43\checkmark$

Answer, $x = 4$, $y = 5$ or Solution set $\{(4, 5)\}$

EXERCISES

Solve by the substitution method and check:

1. $y = x$
   $x + y = 6$
2. $x = -5y$
   $2x + 3y = 14$
3. $x = y + 1$
   $5x - 2y = -7$
4. $y = x - 4$
   $3x + 4y = -2$
5. $y = 3x$
   $x - 4y = 11$
6. $y = -2x$
   $6x - 5y = 64$
7. $y = x - 1$
   $3x - 10y = 38$
8. $x + y = 7$
   $2x + 7y = 24$
9. $y = -1$
   $9x - 8y = 71$
10. $2x + y = 1$
    $5x - 3y = 8$
11. $x + 6y = -30$
    $3x - 4y = 20$
12. $2x = 3y$
    $8x - 11y = -2$
13. $\{(x, y) \mid y = -4x\} \cap \{(x, y) \mid 7x - 3y = 5\}$
14. $\{(x, y) \mid x - y = 8\} \cap \{(x, y) \mid 4x + 5y = 5\}$

11–53 PROBLEMS

Problems involving two unknowns may be solved by using a system of two linear equations in two variables. We represent one of the unknowns by one variable and the second unknown by a second variable. We then find related facts in the problem which are translated into two equations in two variables. This system of equations is solved by either method of elimination. The answers are checked directly with the problem.

EXERCISES

Solve each of the following problems by using two variables:

1. The sum of two numbers is 31 and their difference is 7. What are the numbers?
2. Tom has nine more 10¢ stamps than 13¢ stamps. If the value of these stamps is \$2.51, how many stamps of each kind does he have?
3. The difference between two numbers is 27. The larger number is three more than five times the smaller number. Find the numbers.
4. Dolores has 40 coins, some nickels and the rest dimes. The total value of the coins is \$3.10. Find the number of each kind of coin.
5. Tickets to a school play cost 75¢ for reserved seats and 50¢ for regular seats. If 690 tickets were sold and the total receipts amounted to \$403.75, how many of each kind of tickets were sold?
6. What is the cost per pound of each kind of grass seed if 15 pounds of blue-grass seed and 5 pounds of rye-grass seed together cost \$20 but 5 pounds of blue-grass seed and 15 pounds of rye-grass seed together cost \$12?
7. The length of a rectangle is twice its width. What is the length and the width of the rectangle if its perimeter is 78 feet?
8. The number named by a two-digit numeral is six times the number named by the units' digit. If the sum of the numbers named by the digits is 12, what is the number?
9. A woman invested \$16,000; part at 6% annual interest and the rest at 5%. How much did she invest at each rate if the total annual interest is \$910?

ADDITIONAL TOPICS

11–54 RELATIONS AND FUNCTIONS

A set of ordered pairs such as $\{(1, 3), (2, 6), (4, 12), (5, 15)\}$ is a *relation.* Open sentences (equations and inequalities) in two variables like $x = y$, $x > y$, $x < y$, or $x + y = 5$ and the graphs of these sentences each express a relation. In these sentences there is a rule that associates the second number with the first. We obtain the value of one variable, y, (sometimes called the dependent variable) when we assign a number to the other variable, x, (sometimes called the independent variable). In this manner we obtain a set of ordered pairs of numbers which we call a relation. In an ordered pair of numbers the first component is the number written first and the second component is the number written second.

A *function* is a special kind of a relation. It is a relation in which *no two* ordered pairs have the same first component. The second components may be the same but each must be paired with a different first component. If the first component occurs more than once in a set of ordered pairs, the relation *is not* a function.

$A = \{(2, 1), (4, 2), (6, 3), (8, 4)\}$ is a function but $B = \{(0, 1), (0, 2), (1, 2)\}$ is not a function because (0, 1) and (0, 2) have the same first component.

$\{(x, y) \mid y = 2x\}$ is a function but $\{(x, y) \mid y > 2x\}$ is not a function because (1, 5) and (1, 6) are ordered pairs which satisfy $y > 2x$ but they contain the same first component (1).

The set of all the first components of the ordered pairs is called the *domain* of the relation and the set of all the second components is called the *range* of the relation. The domain of the relation $\{(2, 1), (4, 2), (6, 3), (8, 4)\}$ is $\{2, 4, 6, 8\}$ and the range is $\{1, 2, 3, 4\}$.

We can determine whether a relation is a function by testing whether a vertical line intersects the graph of the number pairs. If the vertical line intersects the graph in more than one point, the relation is not a function.

The function is usually designated by letter f but any other letter may be used. The symbol $f(x)$ indicates the value of function f and is read "f of x" or "f at x" or "the value of the function f at x."

In $y = 2x + 3$ and $f(x) = 2x + 3$ both the y and $f(x)$ represent the same value. To find $f(4)$ when $f(x) = 2x + 3$ means to find the value of $2x + 3$ when $x = 4$ just as we find the value of y when $x = 4$ using $y = 2x + 3$. In $f(x) = 2x + 3$ we substitute the 4 for x in $2x + 3$ so that $f(4) = (2 \cdot 4) + 3 = 8 + 3 = 11$; in $y = 2x + 3$ we do the same thing and get $y = 11$. Thus $f(x)$ represents the y value that corresponds to a given x value. Either $(x, f(x))$ or (x, y) may be used to represent any ordered pair of the function. Here the ordered pair of the function is (4, 11).

EXERCISES

1. a. In the relation {(5, 1), (10, 2), (15, 3), (20, 4), (25, 5)} what is the domain of the relation? What is the range of the relation?

b. In the relation {(− 2, − 1), (− 1, 0), (0, 1), (1, 2), (2, 3), (3, 4)} what is the domain of the relation? What is the range of the relation?

c. In the function {(1, 4), (2, 8), (3, 12), (4, 16)} what is the domain of the function? What is the range of the function?

2. Which of the following relations are functions?

a. {(1, 2), (2, 4), (3, 6), (4, 8), (5, 10)}
b. {(1, 3), (2, 3), (3, 3), (4, 3)}
c. {(3, 1), (3, 2), (3, 3), (3, 4)}
d. {(5, 1), (10, 2), (15, 3), (20, 4), (25, 5)}
e. {(1, 1), (1, 2), (2, 2)}
f. {(1, 3), (2, 4), (3, 5), (4, 6)}
g. {(1, 2), (1, 3), (1, 4), (2, 3), (2, 4), (3, 4)}
h. {(5, 4), (6, 5), (7, 6), (8, 7), (9, 8)}
i. {(1, 1), (2, 4), (3, 7), (4, 10), (5, 13), (6, 16)}
j. {(0, − 1), (1, − 1), (1, 0), (2, − 1), (2, 0), (2, 1)}

3. a. If $f(x) = 3x$, find $f(1), f(2), f(0), f(-3), f(-2)$.
b. If $f(x) = -5x$, find $f(0), f(3), f(-1), f(-4), f(6)$.
c. If $f(x) = x + 7$, find $f(4), f(-3), f(0), f(8), f(-1)$.
d. If $f(x) = 3 - 2x$, find $f(0), f(5), f(-2), f(-7), f(4)$.
e. If $f(x) = 4x + 1$, find $f(-1), f(3), f(-9), f(0), f(2)$.

11–55 VARIATION

In a formula or an equation having two variables, any change in one variable will produce a change in the other.

Direct Variation

When a change in one variable produces the same change in the other variable, the second variable *varies directly* as the first. This occurs in sentences where one variable is equal to a constant times another variable. In the formula $i = .06\,p$, i will vary directly as p.

Inverse Variation

When a change in one variable produces an inverse change in the other variable, the second variable *varies inversely* as the first. This occurs in sentences where the product of the two variables equals a constant (in $lw = 6$,

I will vary inversely as *w*) or where one variable equals the quotient of a constant divided by another variable (in $I = \frac{100}{R}$, I will vary inversely as *R*).

Directly as the Square

When a change in one variable produces the square of this change in the other variable, the second variable *varies directly as the square* of the first. This occurs in sentences where one variable is equal to the square of the other variable or to the product of a constant and the square of the other variable. In the formula $A = s^2$, *A* will vary directly as the square of *s*.

EXERCISES

In each of the following examples find what effect a change in one quantity has upon a related quantity:

1. Using the formula $p = 4s$, substitute the values of *s* given in the table below to determine the corresponding values of *p*, then complete the following statements:

s	2	3	4	6	12
p					

a. Doubling the value of *s* (from 3 to 6) will ? the value of *p*.
b. Halving the value of *s* (from 4 to 2) will ? the value of *p*.
c. Trebling the value of *s* (from 2 to 6) will ? the value of *p*.
d. A change in the value of *s* produces the ? change in the value of *p*. Or, *p* varies ? as *s*.

2. Using the formula $a = \frac{360}{n}$, substitute the values of *n* given in the table below to determine the corresponding values of *a*, then complete the following statements:

n	5	10	15	20	40
a					

a. Doubling the value of *n* (from 10 to 20) will ? the value of *a*.
b. Halving the value of *n* (from 10 to 5) will ? the value of *a*.
c. A change in the value of *n* produces an ? change in the value of *a*. Or, *a* varies ? as *n*.

3. Using the formula $A = 6s^2$, substitute the values of s given in the table below to determine the corresponding values of A, then complete the following statements:

s	2	3	4	6	12
A					

a. Doubling (or 2 times) the value of s (from 2 to 4) makes A equal **?** its original value.
b. Halving the value of s (from 6 to 3) makes A equal **?** its original value.
c. Trebling (or 3 times) the value of s (from 4 to 12) makes A equal **?** its original value.
d. A change in the value of s produces the **?** of this change in the value of A. Or, A varies **?** of s.

4. In the formula $A = lw$, if w is doubled and l remains the same, the value of A is **?**.

5. In the formula $r = \frac{d}{t}$, if t is halved and d remains the same, the value of r is **?**.

6. In the formula $A = \pi r^2$, if r is doubled, the value of A is **?**.

Find the required values, using the principles of variation.

7. In the formula $c = \pi d$, $c = 44$ when $d = 14$. Find the value of c when $d = 28$.
8. In the formula $s = 16t^2$, $s = 64$ when $t = 2$. Find the value of s when $t = 8$.
9. In the formula $I = \frac{E}{R}$, $I = 6$ when $R = 20$. Find the value of I when $R = 60$ and E is constant.
10. The area of a circle varies directly as the square of the diameter. Compare the area of a circle having a 18-inch diameter to the area of a circle with a 3-inch diameter.

11–56 TRANSFORMING FORMULAS

Transformation

To transform a formula means to solve for a specific variable in terms of the other related variables of the formula.

To do this, we copy the given formula and solve for the required variable using the principles developed for the solution of numerical equations. See next page left.

Derivation

To derive a formula from two given formulas by eliminating a common variable, we replace the variable to be eliminated in one formula by an equivalent expression found in the second formula. We then transform and simplify as necessary. See below right.

Transform formula $C = np$, solving for p.

Solution:

$$C = np$$
$$\frac{C}{n} = \frac{np}{n}$$
$$\frac{C}{n} = p$$
$$p = \frac{C}{n}$$

Answer, $p = \frac{C}{n}$

Derive a formula for A in terms of π and d, using the formulas $A = \pi r^2$ and $r = \frac{d}{2}$.

$$A = \pi r^2$$
$$A = \pi \times \left(\frac{d}{2}\right)^2$$
$$A = \pi \frac{d^2}{4}$$

or

$$A = \tfrac{1}{4}\pi d^2$$

Answer, $A = \frac{1}{4}\pi d^2$

EXERCISES

1. Transform formula $A = lw$, solving for w.
2. Transform formula $d = rt$, solving for r.
3. Transform formula $i = prt$, solving for t.
4. Transform formula $V = lwh$, solving for w.
5. Transform formula $A = p + i$, solving for i.
6. Transform formula $C = A - L$, solving for A.
7. Transform formula $I = \frac{E}{R}$, solving for E.
8. Transform formula $d = \frac{m}{v}$, solving for v.
9. Transform formula $F = 1.8\,C + 32$, solving for C.
10. Transform formula $\frac{D}{d} = \frac{r}{R}$, solving for R.
11. Using formulas $c = \pi d$ and $d = 2\,r$, derive a formula for c in terms of π and r.
12. Using formulas $A = p + i$ and $i = prt$, derive a formula for A in terms of p, r, and t.
13. Derive a formula for A in terms of p, using the formulas $A = s^2$ and $p = 4\,s$.
14. Using the formulas $W = IE$ and $E = IR$, derive a formula for W in terms of I and R.
15. Derive a formula for P in terms of F and v, using the formulas $P = \frac{W}{t}$, $W = Fd$, and $v = \frac{d}{t}$.

11–57 QUADRATIC EQUATIONS

A *quadratic equation* is an equation which, after its terms have been collected and simplified, contains the second power (but no higher power) of the variable. Every quadratic equation has two roots. In the evaluation of formulas sometimes a quadratic equation develops and its solution is required.

To solve a quadratic equation containing the second power but not the first power of the variable (an incomplete quadratic equation): We divide both sides by the numerical coefficient of the variable. We take the square root of both members of the equation and prefix a $\pm$ sign to the answer. The two answers are numerically equal but opposite in sign.

Solve and check: $6x^2 = 24$.

Solution:

$$6x^2 = 24$$
$$x^2 = 4$$
$$x = \pm 2$$

Check:

$$6 \times (+2)^2 = 24$$
$$6 \times 4 = 24$$
$$24 = 24 \checkmark$$
$$6 \times (-2)^2 = 24$$
$$6 \times 4 = 24$$
$$24 = 24 \checkmark$$

Answer, Solution set $\{+2, -2\}$

If the given equation is a proportion, we first cross multiply, then we solve the resulting equation.

EXERCISES

Solve and check each of the following equations. Whenever necessary, find answer correct to nearest hundredth.

1. $x^2 = 25$
2. $x^2 = 9$
3. $x^2 = 64$
4. $x^2 = 81$
5. $x^2 = 169$
6. $8x^2 = 32$
7. $2x^2 = 98$
8. $5x^2 = 180$
9. $7x^2 = 252$
10. $16x^2 = 400$
11. $x^2 = 19$
12. $x^2 = 7$
13. $x^2 = 30$
14. $x^2 = 48$
15. $x^2 = 54$
16. $5x^2 = 60$
17. $6x^2 = 90$
18. $14x^2 = 84$
19. $9x^2 = 72$
20. $7x^2 = 140$

21. $\frac{x}{3} = \frac{27}{x}$ **22.** $\frac{x}{5} = \frac{20}{x}$ **23.** $\frac{x}{4} = \frac{9}{x}$ **24.** $\frac{x}{125} = \frac{5}{x}$ **25.** $\frac{x}{32} = \frac{8}{x}$

Find the value of:

26. s when $A = 196$, using the formula $A = s^2$.
27. t when $s = 1{,}024$, using the formula $s = 16t^2$.
28. r when $A = 314$ and $\pi = 3.14$, using the formula $A = \pi r^2$.
29. s when $A = 150$, using the formula $A = 6s^2$.

30. r when $A = 616$ and $\pi = \frac{22}{7}$, using the formula $A = 4\pi r^2$.
31. d when $A = 1{,}963.5$, using the formula $A = .7854\, d^2$.
32. I when $W = 720$ and $R = 20$, using the formula $W = I^2R$.
33. r when $V = 6{,}280$, $h = 5$, and $\pi = 3.14$, using the formula $V = \pi r^2 h$.
34. v when $K = 324$ and $m = 8$, using the formula $K = \frac{1}{2}mv^2$.
35. h when $a = 28$ and $b = 21$, using the formula $h^2 = a^2 + b^2$.
36. V when $L = 4{,}225$, $C = .5$, $d = .002$, and $A = 250$, using the formula $L = \frac{1}{2}CdAV^2$.
37. t when $s = 784$ and $g = 32$, using the formula $s = \frac{1}{2}gt^2$.
38. b when $h = 91$ and $b = 35$, using the formula $h^2 = a^2 + b^2$.
39. E when $P = 720$ and $R = 20$, using the formula $P = \frac{E^2}{R}$.
40. V when $F = 60$, $m = 10$, $g = 32$, and $r = 3$, using the formula $F = \frac{mV^2}{gr}$.

CHAPTER REVIEW

1. Which of the following numerals name integers? Which name rational numbers? Which name irrational numbers? Which name real numbers? (11–1) (11–2)

$^{+}.67$ $\quad$ $^{-}\frac{9}{10}$ $\quad$ $^{+}\sqrt{73}$ $\quad$ $^{-}99$ $\quad$ $^{+}6\frac{3}{8}$ $\quad$ $^{-}\frac{36}{12}$

2. Find the absolute value of: **a.** $|^{-}16|$ **b.** $|^{+}10|$ (11–3)
3. Draw the graph of $\{^{-}4, {}^{-}2, 0, {}^{+}1, {}^{+}3\}$ on a number line. (11–4)
4. Which of the following are true? (11–5)

a. $^{+}2 < {}^{+}8$ **b.** $0 > {}^{-}1$ **c.** $^{+}3 \nless {}^{-}4$ **d.** $^{-}5 \ngtr {}^{-}2$ **e.** $^{+}1 < {}^{-}3$

5. If $^{+}300$ feet represents 300 feet above sea level, how can 50 feet below sea level be represented? (11–6)
6. Use the number line as a scale to draw a vector that illustrates each of the following movements. Write the numeral that is represented by each of these vectors. (11–7)

a. From $^{-}3$ to $^{+}6$ **b.** From $^{+}1$ to $^{-}5$

7. **a.** What is the opposite of $^{+}9$? (11–1)
b. What is the negative of $^{-}50$?
c. What is the opposite of the opposite of $^{-}\frac{1}{2}$? (11–8)
d. What is the additive inverse of $^{-}14$?
e. Write symbolically: The negative of negative eight is positive eight.

8. Add on the number line, using vectors:

a. $^{+}3 + {}^{-}7$ **b.** $^{-}4 + {}^{-}2$ **c.** $^{-}1 + {}^{+}8$ **d.** $^{-}6 + {}^{+}4$ (11–9)

9. Add as indicated:

a. $^{-}18 + {}^{+}24$ **b.** $^{+}23 + {}^{-}19$ **c.** $^{-}31 + {}^{-}27$ **d.** $^{+}42 + {}^{-}60$ (11–10)

10. Add as indicated:

a. $^{+}3\frac{1}{2} + {}^{+}7\frac{7}{8}$ **b.** $^{-}\frac{3}{4} + {}^{+}2\frac{2}{3}$ **c.** $^{-}1.8 + {}^{-}.25$ **d.** $^{+}.617 + {}^{-}.62$ **(11–11)**

11. Subtract on the number line, using vectors:

a. $^{-}7 - {}^{+}5$ **b.** $^{-}8 - {}^{-}10$ **c.** $^{+}3 - {}^{-}4$ **d.** $^{+}6 - {}^{+}11$ **(11–12)**

12. Subtract as indicated:

a. $^{-}42 - {}^{+}18$ **b.** $13 - 31$ **c.** $0 - {}^{-}12$ **d.** $^{-}63 - {}^{-}100$ **(11–13)**

13. Subtract as indicated:

a. $^{+}1\frac{1}{5} - {}^{-}2\frac{1}{2}$ **b.** $^{-}\frac{5}{6} - {}^{-}7$ **c.** $8.4 - 9$ **d.** $^{-}.005 - {}^{+}2.41$ **(11–14)**

14. Multiply as indicated:

a. $^{-}15 \times {}^{-}8$ **b.** $^{+}21 \times {}^{-}43$ **c.** $(6)({}^{-}9)({}^{-}5)$ **d.** $({}^{-}2)(3)({}^{-}4)({}^{-}1)$ **(11–15)**

15. Multiply as indicated:

a. $^{-}\frac{3}{8} \times {}^{+}\frac{4}{5}$ **b.** $^{-}3\frac{3}{4} \times {}^{-}2\frac{2}{3}$ **c.** $.64 \times {}^{-}1.87$ **d.** $^{-}80.5 \times {}^{-}.06$ **(11–16)**

16. Divide as indicated:

a. $^{-}30 \div {}^{-}15$ **b.** $^{+}14 \div {}^{-}14$ **c.** $^{-}63 \div 9$ **d.** $^{-}28 \div {}^{-}1$ **(11–17)**

17. Divide as indicated:

a. $^{-}\frac{1}{2} \div {}^{+}\frac{3}{4}$ **b.** $^{-}5\frac{1}{3} \div {}^{-}1\frac{5}{12}$ **c.** $3.496 \div {}^{-}.02$ **d.** $^{-}.1 \div {}^{-}.001$ **(11–18)**

18. Add: **(11–19)**

$\begin{array}{r} +7 \\ \underline{+9} \end{array}$ $\begin{array}{r} -3 \\ \underline{+8} \end{array}$ $\begin{array}{r} -5 \\ \underline{+1} \end{array}$ $\begin{array}{r} -4 \\ \underline{-2} \end{array}$ $\begin{array}{r} +8 \\ \underline{-8} \end{array}$ $(-6) + (+3) + (-9) + (-1)$

19. Subtract: **(11–19)**

$\begin{array}{r} -6 \\ \underline{+7} \end{array}$ $\begin{array}{r} -9 \\ \underline{+4} \end{array}$ $\begin{array}{r} +6 \\ \underline{+8} \end{array}$ $\begin{array}{r} -11 \\ \underline{-5} \end{array}$ $\begin{array}{r} 0 \\ \underline{-6} \end{array}$ $(+7) - (-7)$

20. Multiply: **(11–19)**

$\begin{array}{r} +8 \\ \underline{-9} \end{array}$ $\begin{array}{r} +3 \\ \underline{+3} \end{array}$ $\begin{array}{r} -7 \\ \underline{+4} \end{array}$ $\begin{array}{r} -6 \\ \underline{-6} \end{array}$ $\begin{array}{r} 0 \\ \underline{-8} \end{array}$ $(-3)(+2)(-2)(+3)$

21. Divide: **(11–19)**

$\frac{-15}{-3}$ $\frac{-10}{+5}$ $\frac{+18}{-6}$ $\frac{+24}{+8}$ $\frac{-7}{+7}$ $(-54) \div (-9)$

22. Simplify: **(11–19)**

a. $\dfrac{(-6)(+8)}{(-4) - (+8)}$ **b.** $\dfrac{7(-4-1) - 5(7-5)}{-3(-1+4)}$ **c.** $\dfrac{|2-9| - |-1|^2}{|(-3)(+2)| \times \left|\dfrac{-8}{2}\right|}$

23. Write each of the following as an algebraic expression: (11–21)

a. The sum of angles A, B, and C.
b. The product of the acceleration of gravity (g) and time (t) added to the velocity (V).
c. The difference between the selling price (s) and the margin (m).
d. The annual depreciation (d) divided by the original cost (c).
e. The product of the sum of m and n and the difference between x and y.

24. Write each of the following as an open sentence symbolically: (11–22)

a. Some number x increased by nine is equal to forty-one.
b. Each number y decreased by four is less than sixty.
c. Seven times each number b plus five is greater than twenty-nine.
d. Each number r divided by three is greater than or equal to sixteen.

25. Express as a formula: The average (A) of two numbers m and n equals the sum of the two numbers divided by two. (11–23)

26. Find the value of: (11–24)

a. $d + 5\,y$ when $d = 9$ and $y = -2$.
b. $n + x(n - x)$ when $n = 10$ and $x = 7$.
c. $a^2 + 2\,ab - 3\,b^2$ when $a = 5$ and $b = 3$.
d. $\dfrac{(x-y)^2}{x^2 - y^2}$ when $x = 6$ and $y = -3$.
e. F when $C = 30$, using the formula $F = 1.8\,C + 32$. (11–25)

27. Add: (11–26)

$$\begin{array}{r} +7\,a \\ -8\,a \\ +5\,a \\ \hline \end{array} \qquad \begin{array}{r} 4\,c^2 - 3\,cd - d^2 \\ 5\,c^2 - 7\,cd + 2\,d^2 \\ -6\,c^2 - cd - d^2 \\ \hline \end{array} \qquad (2\,a - b) + (4\,b - a) + (-a - 3\,b)$$

28. Add $9\,c$ to $-5\,b$. (11–26)
29. Simplify: $4\,c - 5\,d - 3 + 8\,d - 9\,d - 7\,c + c - 5$ (11–26)

30. Subtract: (11–27)

$$\begin{array}{r} -3\,m^2 \\ -11\,m^2 \\ \hline \end{array} \qquad \begin{array}{r} 8\,x^3 - 5\,x^2 \qquad + 6 \\ 2\,x^3 - x^2 + 3\,x \qquad \\ \hline \end{array} \qquad (2\,x^2 - 3\,y^2) - (y^2 - 3\,x^2)$$

31. Subtract $-6\,y$ from $4\,x$. (11–27)
32. What must be added to $2\,n^2 - 5$ to get $8\,n$? (11–27)

33. Multiply: (11–28)

$$\begin{array}{r} -9\,a^4b^3y^2 \\ -4\,a^2b^9x^7 \\ \hline \end{array} \qquad (8\,c^3n^2)(-c^4n^3) \qquad (-3\,ac)(4\,bc)(-2\,ab)(3\,a^2bc^3)$$

$$(-3\,x^3)^3 \qquad -4\,d^4x^2(3\,d^4 - 5\,d^3x^2 + dx^5)$$

34. Divide as indicated:

$$\frac{-30m^9x^8y^4}{6m^6x^8y} \qquad (-8c^{12}t) \div (-8c^4) \qquad \frac{56x^8y^4 - 64x^5y^5 + 16x^4y^7}{-8x^4y^3}$$ (11–29)

35. Name each of the following, using negative exponents: $\frac{1}{10^4}$; $\frac{1}{2^5}$ (11–30)

36. Find the numerical value of each of the following: 10^{-5}; 3^{-3} (11–30)

37. Solve and check, using the set of real numbers: (11–33)

a. $n + 17 = 36$ **c.** $-7x = 56$ **e.** $4y + 5 = 33$ **g.** $9n + 27 = 0$

b. $28 = b - 42$ **d.** $\frac{a}{5} = 15$ **f.** $6x - x = 19$ **h.** $\frac{4}{9}x = 36$

38. Find the solution set of: (11–35)

a. $\frac{x}{2} = 2$ when the replacement set is the set of all natural numbers.
b. $4c + 7 = 8$ when the replacement set is the set of all integers.
c. $9n = 27$ when the replacement set is the set of all odd prime numbers.
d. $11b - 5b = 30$ when the replacement set is $\{0, 2, 4, 6, 8, 10\}$.
e. $|x| - 5 = 3$ when the replacement set is the set of all integers greater than -6 but less than 9.

39. Find the value of w when $p = 96$ and $l = 29$, using the formula $p = 2l + 2w$. (11–37)

40. How many miles can an airplane fly on n gallons of fuel if it averages r miles per gallon? (11–38)

41. At what price must a dealer sell a lamp which costs \$28 to make 20% on the selling price? (11–39)

42. On a number line draw the graph of: (11–40)

a. $3x + 7 = 16$ when the replacement set is the set of all real numbers.
b. $-11b = 55$ when the replacement set is the set of all integers.

43. Find the ratio of 72 to 8. Of 6 to 48. Of 18 minutes to 2 hours. (11–41)

44. Solve and check: $\frac{n}{90} = \frac{13}{15}$ (11–42)

45. Find the value of c' when $c = 66$, $d = 21$, and $d' = 35$, using the formula $\frac{c}{c'} = \frac{d}{d'}$. (11–42)

46. Read, or write in words, each of the following: (11–43)

a. $\{n \mid 4n = 12\}$ **b.** $\{x \mid 3x - 2 < 40\}$ **c.** $\{y \mid 5y - 2y > -21\}$

47. Find the solution set of each of the following inequalities when the replacement set is the set of all real numbers: (11–43)

a. $x + 3 \neq 9$ **d.** $\frac{n}{3} \not< 1$ **g.** $-7y > 42$ **j.** $-2x < -10$

b. $y - 4 < 3$ **e.** $3x + 8 \neq 2$ **h.** $4n + 5 \leq -11$ **k.** $-\frac{2}{3}y \not> 4$

c. $6b \not> 12$ **f.** $5y - y < 0$ **i.** $\frac{b}{-2} \geq 1$ **l.** $2a - 6a > -12$

48. Find the solution set of: **(11–43)**

a. $4n - 9 < 7$ when the replacement set is the set of all natural numbers.
b. $5x - 2x > 21$ when the replacement set is the set of one-digit prime numbers.
c. $-6y \leq 6$ when the replacement set is $\{-3, -2, -1, 0, 1, 2, 3\}$
d. $2a + 4 \not> 16$ when the replacement set is the set of all multiples of 2.
e. $\frac{n}{-1} \geq 0$ when the replacement set is the set of all non-positive integers.

49. On a number line draw the graph of each of the following inequalities when the replacement set is the set of all the real numbers: **(11–44—11–45)**

a. $x < 5$ **c.** $-4b \leq 16$ **e.** $-5 \leq n \leq -1$ **g.** $\{x \mid x > -1\}$
b. $x - 2 \geq -2$ **d.** $-3 < x < 6$ **f.** $3x + 7 \not> 1$ **h.** $x + x \neq 8$

50. On a number line draw the graph of: **(11–46)**

a. $x + 6 > 3$ when the replacement set is $\{-4, -3, -2, -1, 0, 1, 2, 3\}$
b. $5x - 9 < 11$ when the replacement set is the set of one-digit odd natural numbers.
c. $-7y \geq 7$ when the replacement set is $\{-1, 0, 1\}$.

51. Write the set listing all the ordered pairs which have 0, 1, 2, 3 as the first component and 0, 1, 2 as the second component. **(11–47)**
52. Read or write in words: $\{(x, y) \mid x - 3y < 2\}$. **(11–47)**
53. Draw the graph of each of the following equations when the replacement set is the set of all real numbers: **(11–48)**

a. $y = x + 3$ **b.** $y = -2x$ **c.** $x + y = 5$ **d.** $3x - 2y = 6$

54. Write the set $U \times U$ when $U = \{1, 2, 3, 4, 5, 6\}$. **(11–48)**
55. Draw the graph of: **(11–48)**

a. $y = -x$ when $U = \{-3, -2, -1, 0, 1, 2, 3\}$
b. $x - y = 1$ when $U = \{1, 2, 3, 4, 5, 6, 7\}$

56. Draw the graph of the formula $i = .04\,p$. **(11–49)**
57. Using the set of all real numbers draw the partial graph of: **(11–50)**

a. $y > -3x$ **b.** $x + y \leq 2$

58. In each of the following substitute each element of U for x in the inequality to determine the corresponding value for y. Select, then write, the set listing all the ordered pairs of numbers belonging to the solution set of: **(11–50)**

a. $y > x$ when $U = \{1, 2, 3, 4\}$
b. $x + y < 1$ when $U = \{-2, -1, 0, 1, 2\}$

59. Draw the graph of $y < 6 - x$ when $U = \{1, 2, 3, 4, 5, 6\}$ **(11–50)**
60. What is the slope of the line that passes through two points whose coordinates are $(2, -5)$ and $(-4, 7)$? **(11–51)**

61. Determine directly from $6x + 3y = 8$ the slope and y-intercept number of its graph. (11–51)

62. Draw the graph of $3x - y = 5$ by the slope-intercept method. (11–51)

63. Solve graphically:

$$x + y = 5$$
$$3x - 2y = 20$$ **(11–52)**

64. Solve by addition method and check:

$$5x - 4y = -22$$
$$7x + 6y = 4$$ **(11–52)**

65. Solve by substitution method and check:

$$y = 6 - x$$
$$4x - 3y = 10$$ (11–39)

66. One number is ten more than four times a second number. The sum of three times the first number and twice the second number is 72. What are the numbers? (11–53)

67. Which of the following relations are functions? (11–54)

a. $\{(4, 1), (4, 2), (4, 3), (4, 4)\}$ **b.** $\{(1, 6), (2, 12), (3, 18), (4, 24), (5, 30)\}$

68. What is the domain of function $\{(1, 3), (2, 5), (3, 7), (4, 9)\}$? What is its range? (11–54)

69. If $f(x) = 3x - 4$, find $f(2)$; $f(-3)$; $f(0)$. (11–54)

70. Which of the following are true? (11–55)

a. In $n = \frac{360}{a}$, n varies inversely as a.

b. In $p = 5s$, p varies directly as s.

c. In $A = \pi r^2$, A varies directly as the square of r.

71. Transform formula $A = bh$, solving for h. (11–56)

72. Derive a formula for V in terms of B and h, using the formulas $V = lwh$ and $B = lw$. (11–56)

73. Solve and check: $6x^2 = 150$. (11–57)

74. Find the value of t when $s = 1{,}296$ and $g = 32$, using the formula $s = \frac{1}{2}gt^2$. (11–57)

REVIEW EXERCISES

1. Add: $62{,}968 + 7{,}809 + 35{,}674 + 21{,}988 + 75{,}347$

2. Subtract: $526{,}043 - 395{,}967$

3. Multiply: $4{,}906 \times 718$

4. Divide: $508\overline{)354{,}584}$

5. Add: $2\frac{7}{8} + 3\frac{3}{4} + 1\frac{9}{16}$

6. Subtract: $8\frac{1}{2} - 6\frac{2}{3}$

7. Multiply: $5\frac{5}{8} \times 1\frac{3}{5}$

8. Divide: $2\frac{1}{4} \div 1\frac{1}{2}$

9. Add: $6.54 + .984 + 26.1$

10. Subtract: $80 - .54$

11. Multiply: $.7854 \times 2.5$

12. Divide: $.9\overline{).0639}$

13. Find 19% of \$426.

14. What per cent of 48 is 42?

15. 5% of what number is 26?

16. How many yards are in 9 miles?

17. What part of a gallon is 1 pint?

18. Change 2 hours 45 minutes to minutes.

19. How many pounds are in 8 short tons?

20. Change $5\frac{1}{2}$ bushels to pecks.

21. Find the square root of 82,011,136. (8–2)
22. a. Change 8.9 kilometers to meters; 650 milligrams to grams. (9–1 and 9–2)
    b. 4.625 liters of water occupies a space of ___ cm^3 and weighs ___ kg. (9–5)
23. The distance from Paris to Madrid is 1,807 kilometers. How many miles is this? (9–16)
24. Is a temperature of 80° C cooler than a temperature of 170° F? (9–17)
25. Express in A.M. or P.M. time: a. 0005 b. 1440 (9–18)
26. If it is midnight in Chicago, what time is it in Los Angeles? In Boston? (9–19)
27. If $U = \{0, 1, 2, \ldots, 11\}$, $A = \{0, 3, 5, 6, 7, 8, 10\}$, $B = \{1, 2, 3, 7, 8, 11\}$, $D = \{3, 8\}$, and $E = \{2, 4, 5, 8, 9\}$,
    a. Which of the following are true: $E \subset B$? $D \subset A$? $B \subset D$? $E \subset A$? (10–1)
    b. Write the resulting set: $A \cap B$; $B \cup E$; $D \cap E$; $\overline{A}$. (10–2)
    c. Draw a Venn diagram to illustrate: $B \cup A$; $A \cap E$; $\overline{B}$. (10–3)
28. a. Is the set of integers closed under the operation of addition? Subtraction? Multiplication? Division? Explain.
    b. Does each integer have an additive inverse? Explain your answer.
    c. Does each non-zero integer have a multiplicative inverse? Explain.
    d. Does the set of integers have an additive identity? If so, what is it?
    e. Does the set of integers have a multiplicative identity? If so, what is it?
    f. Does the commutative property hold for the addition of integers? Subtraction of integers? Multiplication of integers? Division of integers? Explain.
    g. Does the associative property hold for the addition of integers? Subtraction of integers? Multiplication of integers? Division of integers? Explain. (11–10) (11–13)
    h. Does the distributive property of multiplication over addition hold for the set of integers? Explain. (11–15) (11–17)
29. a. Is the set of rational numbers closed under the operation of addition? Subtraction? Multiplication? Division? Explain.
    b. Does each rational number have an additive inverse? Explain.
    c. Does each non-zero rational number have a multiplicative inverse?
    d. Does the set of rational numbers have an additive identity? If so, what is it?
    e. Does the set of rational numbers have a multiplicative identity? If so, what is it?
    f. Does the commutative property hold for the addition of rational numbers? Subtraction of rational numbers? Multiplication of rational numbers? Division of rational numbers? Explain.
    g. Does the associative property hold for the addition of rational numbers? Subtraction of rational numbers? Multiplication of rational numbers? Division of rational numbers? Explain. (11–11) (11–14)
    h. Does the distributive property of multiplication over addition hold for the set of rational numbers? Explain. (11–16) (11–18)

Graphs, Statistics, and Probability

12

12–1 BAR GRAPHS

The *bar graph* is used to compare the size of quantities in statistics.

To construct a bar graph, we draw a horizontal guide line on the bottom of the squared paper and a vertical guide line on the left. We select a convenient scale for the numbers that are being compared, first rounding large numbers. For a vertical bar graph, we place and label the number scale along the vertical guide line; for a horizontal bar graph, we use the horizontal guide line.

We print the items being compared in alternate squares along the second guide line, labeling these items. We mark off for each item the height corresponding to the given number and draw lines to complete bars. All bars should have the same width. We select and print an appropriate title describing the graph.

EXERCISES

1. Use the graph on the right to answer the following questions regarding the heights of some famous dams in the United States.

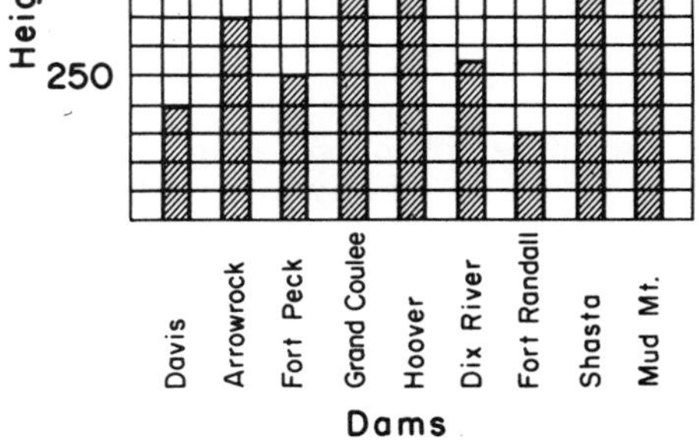

a. How many feet does the side of a small square represent in the vertical scale of the bar graph? $\frac{1}{2}$ of the side of a square?

b. Which dam is the highest?

c. Find the height of each dam:

Davis	Shasta
Grand Coulee	Fort Peck
Fort Randall	Dix River
Arrowrock	Mud Mt.
Hoover	

d. How much higher is the Grand Coulee than the Mud Mountain Dam?

e. How many times as high as the Davis Dam is the Shasta?

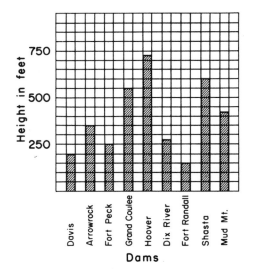

2. Draw a vertical bar graph showing the sales of tickets for the school show by classes. Freshmen, 360 tickets; sophomores, 300 tickets; juniors, 270 tickets; and seniors, 410 tickets.
3. Draw a vertical bar graph picturing the number of accidental deaths by causes during a recent year: motor vehicle, 55,800; falls, 16,900; burns, 6,400; drowning, 8,700; firearms, 2,700; and poison gases, 1,500.
4. Draw a horizontal bar graph representing the stopping distances of good brakes on good pavement.

Speed in m.p.h.	20	30	40	50	60	70
Stopping distance	43 ft.	79 ft.	126 ft.	183 ft.	251 ft.	328 ft.

5. Draw a vertical bar graph showing the areas of the Great Lakes: Superior, 31,820 sq. mi.; Erie, 9,940 sq. mi.; Michigan, 22,400 sq. mi.; Ontario, 7,540 sq. mi.; and Huron, 23,010 sq. mi.

12–2 LINE GRAPHS

The *line graph* is used to show changes and the relationship between quantities.

To construct a line graph, we draw a horizontal guide line on the bottom of the squared paper and a vertical guide line on the left. We select a convenient scale for the related numbers, first rounding large numbers. We place and label this scale along one of the guide lines.

We print and label items below the other guide line, using a separate line for each item. On each of these lines we mark with a dot the location of the value corresponding to the given number. We then draw straight lines to connect successive dots. We select and print an appropriate title describing the graph.

EXERCISES

1. Use the graph on page 367 to answer the following questions regarding the monthly normal precipitation in Cincinnati (solid line) and Phoenix (dotted line).
   a. During what month does the most rain fall in Phoenix? How many inches?
   b. During what month does the least amount of rain fall in Cincinnati? How many inches?
   c. How many inches less rain falls in Phoenix than in Cincinnati during the month of September?
   d. How many times as much rain falls in Cincinnati as in Phoenix during the month of December?

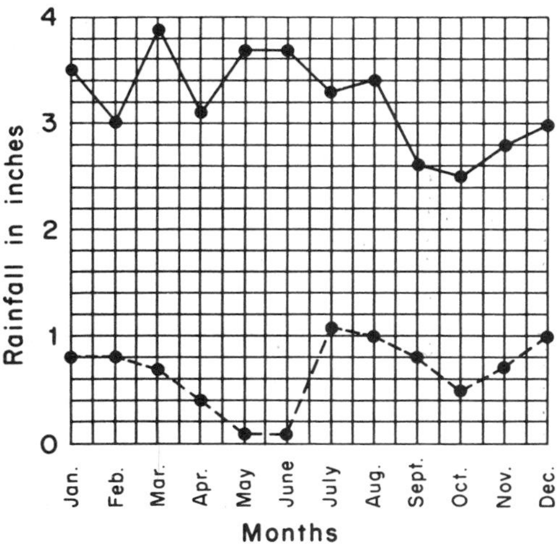

2. Draw a line graph showing that in ten successive progress tests in arithmetic, each containing 30 examples, John solved 14, 17, 19, 20, 25, 24, 28, 27, 29, and 26 examples correctly.

3. Use a solid line to show changes in the retail price of a dozen eggs and a dotted line for the changes in the price of a pound of sliced bacon.

Food	1965	1966	1967	1968	1969	1970	1971	1972	1973	1974
Eggs	52.7¢	59.9¢	49.1¢	52.9¢	62.1¢	61.4¢	52.9¢	52.4¢	78.1¢	78.3¢
Bacon	81.3¢	95.4¢	83.7¢	81.4¢	87.8¢	94.9¢	80.0¢	96.2¢	132.5¢	132.0¢

4. Use a solid line to represent monthly normal temperatures in San Francisco and a dotted line for the temperatures in Baltimore.

CITY	Jan.	Feb.	Mar.	Apr.	May	June	July	Aug.	Sept.	Oct.	Nov.	Dec.
San Francisco	50	52	54	55	57	58	58	59	61	60	56	51
Baltimore	35	36	43	54	67	73	78	76	69	58	47	37

5. Use a solid line to represent the receipts and a dotted line for the expenditures of the United States government, expressed in billions of dollars.

	1966	1967	1968	1969	1970	1971	1972	1973	1974
Receipts	107.0	149.6	153.7	187.8	193.8	188.3	215.3	232.2	264.8
Expenditures	104.7	153.2	172.8	183.1	195.0	210.7	238.3	246.6	268.3

12–3 CIRCLE GRAPHS

The *circle graph* is used to show the relation of the parts to the whole and to each other.

To construct a circle graph, we make a table showing: (a) given facts; (b) fractional part or per cent each quantity is of the whole; (c) the number of degrees representing each fractional part or per cent, obtained by multiplying 360° by the fraction or per cent.

We draw a convenient circle and with a protractor construct successive central angles, using the number of degrees representing each part. We label each sector and select and print an appropriate title describing the graph.

EXERCISES

1. Use the graph on the right to answer the following questions regarding the distribution of marks in a social studies test.
   a. What fractional part of the class received A? B? E? D? C?
   b. How many degrees of the circle represent the number of pupils receiving B? C? A? E?
   c. What is the ratio of the C's to the A's? Of the E's to the B's?
   d. If there are 36 pupils in the class, how many received A? E? C? B? D?

MARKS IN A SOCIAL STUDIES TEST

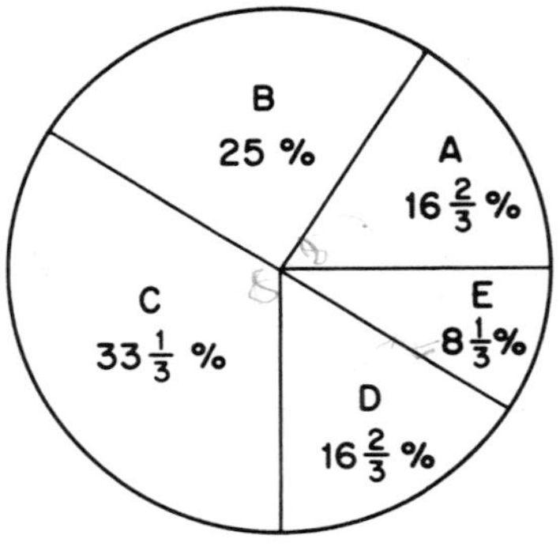

Draw circle graphs showing the following data:

2. A family budget: food, 30%; clothing, 15%; shelter, 25%; operating expenses, 10%; savings, 5%; automobile, 8%; other expenses, 7%.
3. In an eighth grade class, 12 pupils plan to stay home during the summer, 10 are going to the seashore, 4 are going on motor trips, 8 are going camping, and 6 are going to the mountains.
4. The Amerigo Vespucci High School has the following enrollment: 240 freshmen, 216 sophomores, 184 juniors, and 160 seniors.
5. The operating expenses of Thomas and Company during the month of June were as follows: General expense, $600; rent, $800; salaries, $3,840; insurance, $72; depreciation on furniture and equipment, $400; delivery expense, $288.
6. The sources of United States taxes for a recent year were: individual income taxes, $143,000,000,000; corporation income taxes, $42,000,000,000; employment taxes including old age and disability insurance, $66,000,000,000; estate and gift taxes, $5,000,000,000; and excise taxes, $17,000,000,000.

REVIEW EXERCISES

1. Construct a bar graph showing the area of the continents: Africa, 11,500,000 sq. mi.; Antarctica, 5,500,000 sq. mi.; Asia, 16,900,000 sq. mi.; Australia, 2,945,000 sq. mi.; Europe, 3,750,000 sq. mi.; North America, 8,440,000 sq. mi.; South America, 6,800,000 sq. mi.

2. Construct a line graph showing the population of the United States, 1800–1970:

Year	Population	Year	Population
1800	5,308,483	1890	62,947,714
1810	7,239,881	1900	75,994,575
1820	9,638,453	1910	91,972,266
1830	12,866,020	1920	105,710,620
1840	17,069,453	1930	122,775,046
1850	23,191,876	1940	131,669,275
1860	31,443,321	1950	150,697,361
1870	38,558,371	1960	179,323,175
1880	50,155,783	1970	203,184,772

3. Construct a circle graph showing the enrollment at the Township Junior-Senior High School: seventh grade, 440 students; eighth grade, 480 students; ninth grade, 420 students; tenth grade, 380 students; eleventh grade, 320 students; twelfth grade, 360 students.

12–4 FREQUENCY DISTRIBUTION

Data may be arranged in tabular form. For example, we arrange the scores made by a class on a test: 80, 85, 65, 60, 90, 80, 85, 95, 100, 85, 60, 75, 95, 85, 80, 75, 80, 65, 90, 80, 75, 85, 70, 65, 70, 85, 80, 90, 85, 75 by first tallying the scores and then summarizing the results as shown in the following table. The number of times each score occurs is called its *frequency*. Arrangement of data in this form is called a *frequency distribution.*

Score	Tally	Frequency
100	I	1
95	II	2
90	III	3
85	~~IIII~~ II	7
80	~~IIII~~ I	6
75	IIII	4
70	II	2
65	III	3
60	II	2

A frequency distribution can be pictured by a special bar graph called a *histogram* or by a special line graph called a *frequency polygon.*

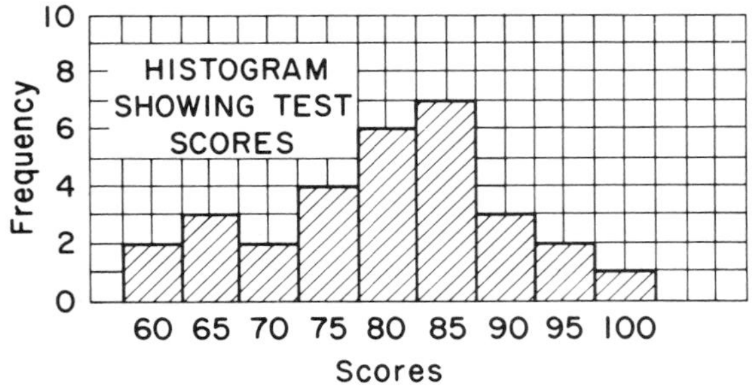

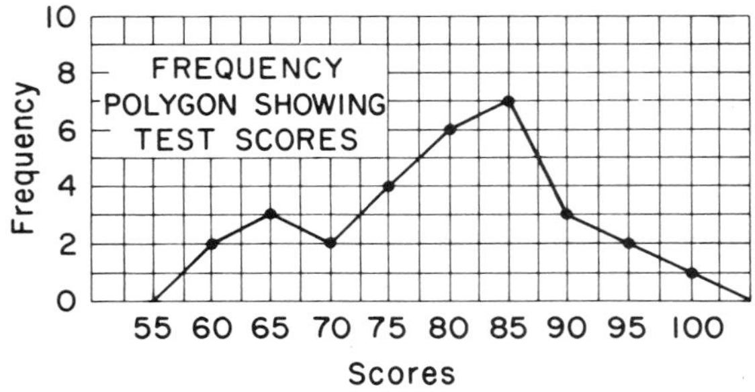

In a histogram each score or group of scores is represented by a bar, its base representing the score and its height representing the frequency of the score. Unlike the conventional bar graph, there is no space left between bars in a histogram.

In a frequency polygon, like the histogram, the horizontal scale indicates the score and the vertical scale indicates the frequency of these scores. The line segments and the axis of the horizontal scale form a polygon.

EXERCISES

1. Make a frequency distribution table for each of the following sets of scores:

a. 7, 9, 5, 2, 8, 9, 6, 7, 5, 8, 7, 9, 10, 6, 4, 8, 6, 7, 9, 8, 10, 5, 7, 9

b. 75, 90, 65, 80, 100, 45, 65, 60, 85, 75, 60, 95, 80, 70, 75, 65, 95, 100, 45, 80

c. 23, 17, 20, 19, 20, 24, 18, 23, 25, 22, 19, 20, 18, 21, 24, 23, 17, 21, 19, 24, 22, 21, 20, 22, 23, 21, 24, 22, 19, 21, 20

d. 39, 47, 43, 45, 43, 40, 38, 45, 44, 46, 48, 47, 45, 42, 43, 40, 44, 45, 47, 50, 44, 36, 47, 45, 42, 46, 41, 49, 46, 42, 41, 48, 47, 45, 42, 44, 43, 45, 42, 46

e. 4, 6, 7, 8, 0, 5, 9, 1, 6, 7, 4, 6, 5, 3, 9, 1, 10, 5, 4, 6, 9, 0, 7, 8, 3, 8, 6, 2, 7, 8, 5, 4, 9, 6, 7, 10, 3, 6, 6, 5, 7, 3, 5, 8, 6

2. Make a histogram for each set of scores given in problem 1.

3. Make a frequency polygon for each set of scores given in problem 1.

12-5 AVERAGES—MEASURES OF CENTRAL TENDENCY

An *average* is a measure of central tendency of data. This data may be scores, measurements, or other numerical facts. Three commonly used averages are the arithmetic mean (or simply mean), the median, and the mode.

The *arithmetic mean* of a set of numbers is determined by dividing the sum of numbers by the number of members in the set. High numbers and low numbers are balanced off to make all numbers the same size. An extremely high number or an extremely low number distorts the picture of central tendency if the arithmetic mean is used.

Find the arithmetic mean of the following set of scores: 80, 96, 85, and 91:

$$\frac{80+96+85+91}{4}$$
$$=\frac{352}{4}$$
$$=88$$

Answer, 88

The *median* of a set of numbers is the middle number when the numbers are arranged in order of size. To find this number we may count from either end, smallest to largest or vice-versa. If the number of members is even, the median is determined by dividing the sum of the two middle numbers by 2. Since the median is a positional average, an extremely high or low number does not affect it.

The *mode* of a set of numbers is the number that occurs most frequently in the set. There may be more than one mode.

A class of 25 pupils made the following scores on a test:
8, 6, 10, 9, 9, 8, 7, 10, 9, 7, 9, 8, 6, 10, 6, 9, 7, 8, 9, 8, 7, 9, 6, 7, and 8.
Find the arithmetic mean, median, and mode.

Score	Tally	Frequency
10	III	3
9	~~IIII~~ II	7
8	~~IIII~~ I	6
7	~~IIII~~	5
6	IIII	4
	Total	25

Arithmetic Mean:

$10 \times 3 = 30$
$9 \times 7 = 63$
$8 \times 6 = 48$
$7 \times 5 = 35$
$6 \times 4 = 24$
sum $= 200$ $\quad \frac{200}{25} = 8$

Arithmetic mean $= 8$

Median or Middle Score
13th score from top or bottom $= 8$

Mode or most frequent score $= 9$

The Greek letter $\sum$, capital sigma, is the symbol that is generally used to indicate in a concise way the sum of a set of numbers.

The symbol $\sum_{i=1}^{6} x_i$ is read "the summation of x sub i for i equal to 1 through 6" and represents the sum "$x_1 + x_2 + x_3 + x_4 + x_5 + x_6$" where each addend (represented by x with a different subscript) is a specific number, score, or measurement.

If $x_1 = 18$, $x_2 = 25$, $x_3 = 16$, $x_4 = 27$, $x_5 = 30$, and $x_6 = 24$,

then $\sum_{i=1}^{6} x_i = x_1 + x_2 + x_3 + x_4 + x_5 + x_6 = 18 + 25 + 16 + 27 + 30 + 24 = 140$

and $\sum_{i=3}^{5} x_i = x_3 + x_4 + x_5 = 16 + 27 + 30 = 73$

Observe that the numeral below the $\sum$ is the first subscript and the numeral above the $\sum$ is the last subscript of the required addends. The sigma notation is used in the following formula for finding the arithmetic mean: $M = \frac{\sum x_i}{n}$ where M represents the arithmetic mean, $\sum x_i$, the sum of the scores, and n, the number of scores.

EXERCISES

1. Find the mean for each of the following sets of scores:

a. 90, 85, 94, 78, 80, 89, 93
b. 18, 16, 19, 20, 15, 12, 18, 17, 14, 17, 16, 19
c. 2.7, 2.6, 2.4, 2.9, 2.4, 2.7
d. 6, 8, 4, 5, 9, 9, 7, 8, 4, 10, 9, 10, 8, 7, 10, 8, 9, 8
e. $3\frac{1}{2}$, $4\frac{3}{8}$, $3\frac{7}{8}$, $4\frac{3}{4}$

2. Find the median for each of the following sets of scores:

a. 5, 6, 8, 7, 8, 10, 9
b. 75, 90, 85, 93, 79, 82
c. 3.6, 4.9, 2.8, 5.7, 6.3
d. $1\frac{1}{2}$, $2\frac{1}{4}$, $1\frac{3}{4}$, $2\frac{3}{8}$, $2\frac{1}{2}$, $2\frac{1}{8}$, $1\frac{7}{8}$, $2\frac{1}{4}$, $2\frac{5}{8}$
e. 15, 17, 13, 10, 19, 14, 18, 13, 16, 18, 19, 17, 14, 15

3. Find the mode for each of the following sets of scores:

a. 3, 9, 8, 7, 6, 7, 5
b. 90, 55, 75, 85, 80, 85, 65, 80, 70
c. 7.4, 8.1, 7.5, 6.9, 7.7, 7.8
d. 16, 19, 25, 14, 18, 32, 29, 33, 27, 21, 19, 25, 30, 24

e. 4, 10, 5, 9, 6, 13, 8, 15, 9, 5, 13, 8, 12, 7, 8, 14, 6, 11, 8, 4, 5

4. Find the mean, median, and mode for each of the following sets of scores:

a. 5, 8, 9, 6, 2, 4, 8, 3, 6, 7, 5, 9, 6
b. 80, 75, 60, 90, 95, 80, 70, 85, 75, 80, 70, 65, 95, 85, 80
c. 11, 10, 13, 12, 15, 16, 14, 12, 13, 10, 14, 16, 13, 15, 11, 13, 16, 10, 13, 15
d. 63, 72, 65, 68, 74, 69, 73, 68, 67, 69, 74, 68, 69, 70, 68, 66, 75, 69, 71
e. 15, 21, 16, 24, 25, 14, 10, 22, 21, 18, 21, 20, 19, 16, 24, 20, 23, 19, 19, 16, 23, 21, 25, 18, 21

5. Tally and arrange each of the following sets of scores in a frequency distribution table, then find the mean, median, and mode:

a. 12, 16, 19, 13, 20, 11, 14, 17, 15, 12, 13, 18, 16, 17, 18, 20, 11, 15, 18, 19, 12, 19, 15, 18, 17, 16, 17, 16, 14, 15, 18, 12
b. 6, 4, 9, 3, 8, 2, 5, 8, 7, 4, 2, 3, 9, 7, 1, 6, 5, 5, 2, 6, 4, 8, 9, 3, 1, 5, 9, 8, 6, 8, 6, 9, 7, 4, 9, 8, 6, 5, 7, 3, 5, 6, 8, 7, 9, 8, 5, 8, 7, 6

6. Find the mean and median of the following set of scores: 65, 69, 77, 800, 75, 73, 68, 71, 66, 70, 72. If the score of 800 were dropped, which measure would be more affected, the mean or the median?

7. If $x_1 = 8$, $x_2 = 15$, $x_3 = 11$, $x_4 = 16$, and $x_5 = 9$, find:

a. $\sum_{i=1}^{5} x_i$ **b.** $\sum_{i=1}^{3} x_i$ **c.** $\sum_{i=1}^{4} x_i$ **d.** $\sum_{i=2}^{5} x_i$ **e.** $\sum_{i=3}^{4} x_i$

8. If $x_1 = 14$, $x_2 = 6$, $x_3 = 5$, $x_4 = 10$, $x_5 = 12$, $x_6 = 2$, $x_7 = 17$, and $x_8 = 9$, find:

a. $\sum_{i=1}^{8} x_i$ **b.** $\sum_{i=1}^{6} x_i$ **c.** $\sum_{i=3}^{7} x_i$ **d.** $\sum_{i=2}^{8} x_i$ **e.** $\sum_{i=4}^{6} x_i$

9. Use the formula $M = \frac{\sum x_i}{n}$ to find the arithmetic mean, M, of the following six scores:

$x_1 = 10$, $x_2 = 9$, $x_3 = 8$, $x_4 = 5$, $x_5 = 4$, and $x_6 = 12$.

10. The individual weights of the seven linemen on the school football team starting with the left end are: 195, 218, 224, 210, 235, 220, and 192 pounds respectively. The individual weights of the four backfield men are: 175; 182, 169, and 180 pounds. Find the arithmetic mean of the weights of the: **a.** linemen **b.** backfield men. **c.** entire team.

stannine

12-6 PERCENTILES AND QUARTILES

We may compare an individual score with all the other scores by giving the score a positional standing or a *rank* from the top. However, unless we know how many scores there are, this comparison is unsatisfactory. For example, Steve ranks ninth in his class. In a class of 12 students, he would have a low rank. In a class of 180 students, he would have a high rank. Thus it is better to make a comparison by using percentile rank.

The *percentile rank* of a score tells us the per cent of all the scores that are below this given score. If the rank of a particular score is the 60th percentile, it means that 60% of all the scores are lower than this score.

$\frac{3}{12} = \frac{1}{4} = 25\%$

$\frac{171}{180} = \frac{19}{20} = 95\%$

For example, if Steve ranked ninth in a class of 12 students, there are 3 students of 12 students or 25% with a lower rank. He would have a percentile rank of 25 or a rank of the 25th percentile. If Steve ranked ninth in a class of 180 students, there are 171 students of 180 students or 95% with a lower rank. He would then have a percentile rank of 95 or a rank of the 95th percentile.

If an arranged set of scores (or other data) is divided into four equal parts, the score at each point of division is called a *quartile*. A percentile also refers to a score.

The *upper quartile* (or third quartile) is the score at the point below which 75% of all the scores fall. It corresponds to the 75th percentile.

The median (or second quartile) corresponds to the 50th percentile.

The *lower quartile* (or first quartile) is the score below which 25% of all the scores fall. It corresponds to the 25th percentile.

EXERCISES

1. Find the percentile rank in each of the following:

a. Marilyn ranked tenth in a class of 100 students.
b. Charlotte ranked sixth in a class of 40 students.
c. Scott ranked third in a class of 50 students.
d. Peter ranked eighteenth in a class of 72 students.
e. Ronald ranked twenty-seventh in a class of 90 students.

2. In a class of 50 students Elaine has a percentile rank of 80. What is her rank in the class? Does she have a higher rank than Ed who ranks fourteenth in this class?

3. If the top 15 students of a graduating class of 120 students are to receive awards, who among the following would get an award? A student with a percentile rank of 75? 90? 85? 80? 89?

4. If all students above the upper quartile are to be exempt from taking the final examination, who among the following would be exempt? A student with a percentile rank of 81? 76? 73? 66? 92?

5. Find the 50th percentile, upper quartile, and lower quartile for the following set of scores: 55, 60, 60, 70, 75, 80, 80, 85, 85, 95, 95, 100.

12-7 MEASURES OF DISPERSION

Measures of dispersion show the spread or scatter or variability of the distribution of scores (or other statistical facts) around the average.

The *range* is the difference between the highest and the lowest scores.

The *deviation from the mean* is the difference between the score and the mean. Positive and negative numbers are used to indicate whether a score is above the mean (+ deviation) or below the mean (− deviation).

The *mean deviation from the mean* is the mean of the set of absolute values of all the deviations from the mean.

Variance is the mean of the set of squares of all the deviations from the mean.

Standard deviation is the principal square root of the variance and is denoted by the Greek letter sigma, σ.

To find the variance of a set of scores, we first find the deviation from the mean for each score and square each of these deviations. We then find the sum of the squares of these deviations and divide this sum by the number of scores in the set. The quotient, thus found, is the variance.

To find the standard deviation, we take the positive square root of the variance.

Find the range, mean deviation from the mean, variance, and standard deviation for the following set of scores: 80, 95, 90, 85, 100.

Scores	Deviation From Mean	Square of Deviation
100	+10	100
95	+5	25
90	0	0
85	−5	25
80	−10	100
450	0	250

Sum of absolute values 30

Range: $100 - 80 = 20$

Mean: $\frac{450}{5} = 90$

Mean Deviation From Mean: $\frac{30}{5} = 6$

Variance: $\frac{250}{5} = 50$

Standard Deviation: $\sqrt{50} = 7.07$

EXERCISES

1. Find the range for each of the following sets of scores:

   a. 82, 53, 75, 94, 67, 46, 85, 62, 90, 79 b. 21, 18, 25, 13, 29, 34, 17, 25, 21, 18, 28

2. Find the mean and mean deviation from the mean for each of the following sets of scores:

a. 100, 85, 70, 90, 80
b. 60, 75, 90, 85, 90, 75, 80, 85
c. 6, 4, 8, 2, 5, 7, 8, 7, 4, 6, 9
d. 20, 15, 16, 25, 17, 22, 18, 15, 20, 22, 17, 21

3. During the semester Peter's test marks in social studies were: 85, 91, 89, 87, 92, and 84 while John's test marks were: 94, 65, 100, 92, 79, and 98. Find the mean and range for each set of marks. Whose marks were more consistent with the mean?

4. Find the range, mean deviation from the mean, variance, and standard deviation for each of the following sets of scores:

a. 5, 6, 7, 8, 9
b. 12, 8, 16, 4, 20
c. 75, 86, 64, 72, 94, 79, 83
d. 48, 56, 67, 69, 70
e. 7, 8, 10, 11, 14, 16, 18
f. 1, 3, 5, 7, 9, 11, 13, 15, 17, 19

12–8 PERMUTATIONS

The set of letters *a*, *b*, and *c* may be arranged in any of 6 different ways: *abc*, *acb*, *bac*, *bca*, *cab*, *cba*. Each of these arrangements is called a *permutation*. Permutations have to do with the order of the things being arranged.

The product of all natural numbers beginning with 1 up to and including a given number is called the *factorial* of the given number.

Three factorial (or factorial three) written as 3! means $1 \times 2 \times 3$ or the product 6.

Six factorial (or factorial six) written as 6! means $1 \times 2 \times 3 \times 4 \times 5 \times 6$ or the product 720.

One factorial (or factorial one) written as 1! is equal to 1.

If from the set of four letters *a*, *b*, *c*, *d* two letters are taken at a time, they may be arranged as: *ab*, *ac*, *ad*, *ba*, *bc*, *bd*, *ca*, *cb*, *cd*, *da*, *db*, *dc*, or in 12 ways. If from the set of three letters *a*, *b*, *c* one letter is taken at a time, they may be arranged in 3 ways: *a*, *b*, *c*.

The symbol ${}_nP_n$ represents the number of permutations (or arrangements) of n things taken n at a time. For this the permutation formula is ${}_nP_n = n!$

The symbol ${}_nP_r$ represents the number of permutations (or arrangements) of n things taken r at a time. For this the permutation formula is:

$${}_nP_r = n(n-1)(n-2)\ldots(n-r+1).$$

Find the number of permutations of 3 things taken 3 at a time:

$${}_nP_n = n!$$
$${}_3P_3 = 3!$$
$${}_3P_3 = 1 \cdot 2 \cdot 3$$
$${}_3P_3 = 6$$

Answer, 6 permutations

Find the number of permutations of 5 things taken 3 at a time:

$${}_nP_r = n(n-1)\ldots(n-r+1)$$
$${}_5P_3 = 5(5-1)\ldots(5-3+1)$$
$${}_5P_3 = 5 \cdot 4 \cdot 3$$
$${}_5P_3 = 60$$

Answer, 60 permutations

EXERCISES

1. Find the number of permutations that can be formed:

a. From 4 things taken 2 at a time.
b. From 10 things taken 10 at a time.
c. From 11 things taken 5 at a time.

2. Find the value of: ${}_7P_2$; ${}_9P_5$; ${}_6P_6$; ${}_{10}P_7$; ${}_8P_3$; ${}_{12}P_4$

3. In how many ways may 9 different books be arranged on a shelf?

4. In how many ways may 5 pupils be seated in 8 seats?

5. Using all the letters each time, in how many ways may the letters of the word *GIRL* be arranged? Write these arrangements of letters.

12–9 COMBINATIONS

A *combination* is a selection of a group of things with no regard to the arrangement or order.

There are 6 permutations of *abc* taken 3 at a time but there is only 1 combination. *abc*, *acb*, *bac*, *bca*, *cab*, and *cba* are all the same combination.

The symbol ${}_nC_r$ or $\binom{n}{r}$ represents the number of combinations of n things taken r at a time. The combination formula is:

$${}_nC_r = \frac{n!}{r!\,(n-r)!}$$

Find the number of combinations of 5 things taken 3 at a time:

$${}_nC_r = \frac{n!}{r!\,(n-r)!}$$

$${}_5C_3 = \frac{5!}{3!2!}$$

$${}_5C_3 = \frac{1 \cdot \cancel{2} \cdot \cancel{3} \cdot \overset{2}{\cancel{4}} \cdot 5}{1 \cdot \cancel{2} \cdot \cancel{3} \cdot 1 \cdot \cancel{2}}$$

$${}_5C_3 = 10$$

Answer, 10 combinations

EXERCISES

1. Find the number of combinations that can be formed:

a. From 8 things taken 4 at a time.
b. From 9 things taken 3 at a time.
c. From 10 things taken 10 at a time.

2. Find the value of: ${}_6C_5$; ${}_4C_4$; ${}_{11}C_6$; $\binom{7}{2}$; $\binom{9}{4}$; ${}_{12}C_8$
3. How many committees of 4 pupils each can be formed from a group of 15 pupils?
4. How many different basketball teams of 5 players each can be selected from a squad of 10 players?
5. How many combinations of 3 letters each can be formed from the word *NUMBER*? Write these combinations of letters.

12-10 PROBABILITY; SAMPLE SPACES; COMPLEMENTARY EVENTS; ODDS

Probability

When the weather bureau predicts that the probability of precipitation (rain or snow) for the next day is 70%, it is informing us that the likelihood for this precipitation to occur is 7 chances out of 10. *Probability* is a numerical measure indicating the chance or likelihood for a particular event to occur. It is usually expressed as a ratio named by a common fraction or a decimal numeral or a per cent. For example, when we toss a coin in the air, it is equally likely that it will land heads up as tails up. Thus the chance that this coin will land heads up is one out of two and the probability ratio is $\frac{1}{2}$. This may be indicated by the notation:

$P(H) = \frac{1}{2}$ which reads "the probability of heads is one half."

The standard deck of 52 playing cards has 13 diamond cards. The probability of drawing at random (without looking or without preference) a diamond card from this full well-shuffled deck is 13 out of 52 or the ratio $\frac{13}{52}$ which is equivalent to $\frac{1}{4}$. The chance to draw a diamond card therefore is 1 out of 4. This may be indicated as $P(D) = \frac{1}{4}$.

In the study of probability, activities like coin-tossing, drawing cards, etc. are used as experiments. Tossing a coin so that it lands heads up is an example of an *event* or an *outcome*, a particular way in which something occurs.

When a coin is tossed, there are 2 outcomes, heads and tails. However, if the desired result is heads, then there is only 1 favorable outcome, heads.

In a bag containing 10 marbles there are 10 possible outcomes when drawing a marble at random. However, if the event is to draw a green marble from a bag of 10 marbles of which 2 are green marbles, there are only 2 favorable outcomes, the two green marbles. The probability of drawing a green marble is $\frac{2}{10}$ which is equivalent to $\frac{1}{5}$ and may be written as $P(\text{green marble}) = \frac{1}{5}$.

Probability is the ratio of favorable outcomes to the total number of possible outcomes. This is indicated by $P = \dfrac{f}{n}$ where P is the probability, f is the number of favorable outcomes, and n is the total number of possible outcomes. If there are no favorable outcomes, the probability is 0. If all are favorable outcomes, the probability is 1.

Scale of Probability

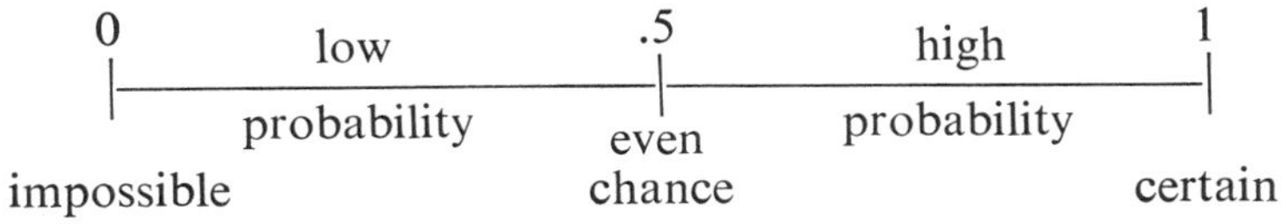

Sample Space

The set of all the possible outcomes of a given experiment is called the *sample space.*

When a coin is tossed, the sample space $\{H, T\}$ indicates the 2 possible outcomes: heads or tails.

When two coins are tossed, the sample space $\{HH, HT, TH, TT\}$ indicates the 4 outcomes: heads heads, heads tails, tails heads, tails tails. Observe that here ordered pairs are used. HH is the abbreviation of (H, H), HT is (H, T), TH is (T, H), and TT is (T, T).

When drawing a marble from a bag containing 3 red marbles and 4 white marbles, the sample space $\{R_1, R_2, R_3, W_1, W_2, W_3, W_4\}$ indicates the 7 possible outcomes since each R represents one of the red marbles and each W represents one of the white marbles. This sample space is sometimes expressed also as $\{R, W\}$.

A *die* is a cube each of whose six faces is marked with a different number of dots, from one through six.

When a die is rolled, the sample space $\{1, 2, 3, 4, 5, 6\}$ indicates the 6 possible outcomes, each representing a different face of the die that may turn up. The plural of die is dice.

An *event* is any subset of the sample space. A *simple event* is a subset containing one outcome of the sample space. When a coin is tossed, $\{T\}$ is a simple event. When a die is rolled, $\{5\}$ is a simple event.

Complementary Events

The event that something happens and the event that this something does not happen are called *complementary events.* If an event can happen in m ways and fail to happen in n ways and each is equally likely, the probability of its happening is $\frac{m}{m+n}$; the probability of its failing to happen is $\frac{n}{m+n}$.

If a bag contains 3 red marbles and 5 marbles of other colors, the probability of drawing at random a red marble is $\frac{3}{8}$. The probability of not drawing a red marble is $\frac{5}{8}$. The sum of their probabilities is 1. If *event A* represents the event of drawing a red marble and *event not A* represents the event of not drawing a red marble, then event A and event not A are complementary events and

$$P(A) + P(\text{not } A) = 1$$
$$\text{or} \quad P(A) = 1 - P(\text{not } A)$$

Odds

When a coin is tossed, the odds of getting a head are 1 to 1 since there is 1 head and 1 tail. When there are 2 green marbles and 8 other marbles in a bag, the odds of selecting a green marble are 2 to 8 or 1 to 4.

The *odds* for an event to happen against its failure to happen are the ratio of its favorable outcomes to unfavorable outcomes.

EXERCISES

1. Experiment — To draw at random a ball from a box that contains 1 red ball and 1 white ball.

a. Write the sample space listing all the possible outcomes (colors) of drawing a ball from the box.
b. List as a subset the favorable outcome for the event of drawing a red ball.
c. What is the probability of drawing a red ball on the first draw?
d. What is the probability of not drawing a red ball on the first draw?
e. What are the odds of drawing a red ball on the first draw? Of not drawing a red ball on the first draw?

2. Experiment — To draw at random a card from a hat containing three cards of the same size numbered 1 to 3 inclusive.

a. Write the sample space listing all the possible outcomes (numerals) of drawing a card from the hat.
b. List as a subset the favorable outcome for the event of drawing a card with numeral 2 on it.
c. What is the probability of drawing the card with numeral 2 on it on the first draw?
d. What is the probability of not drawing the card with numeral 2 on it on the first draw?
e. What are the odds of drawing the card with numeral 2 on it on the first draw? Of not drawing this card on the first draw?

3. Experiment — To pick at random a crayon from a box of ten crayons: 1 blue, 1 white, 1 brown, 1 yellow, 1 orange, 1 green, 1 purple, 1 pink, 1 black, 1 gray.

a. Write the sample space listing all the possible outcomes (colors) of picking a crayon from the box.

b. List as a subset the favorable outcome for the event of picking a yellow crayon from the box.

c. What is the probability of picking a yellow crayon from the box on the first draw?

d. What is the probability of not picking a yellow crayon on the first draw?

e. What are the odds of picking a yellow crayon on the first draw? Of not picking a yellow crayon on the first draw?

4. Experiment — To spin the pointer on a spinner whose circle has eight sectors each of the same size numbered 1 through 8.

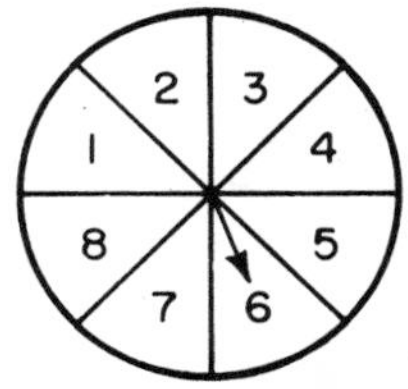

a. Write the sample space listing all the possible outcomes (numerals on sectors) where the pointer may stop.

b. List as a subset the favorable outcome for the event that the pointer stop on the sector marked 6.

c. What is the probability that the pointer stops on the sector marked 6?

d. What is the probability that the pointer does not stop on the sector marked 6?

e. What are the odds that the pointer stops on the sector marked 6? Of not stopping on the sector marked 6?

5. Experiment — To draw at random a slip of paper from twelve identical slips of paper with three slips of paper marked with an *X*.

a. How many outcomes are in the sample space?

b. If the event is to select a slip of paper with an *X* marked on it, how many favorable outcomes are there?

c. What is the probability that a slip with an *X* marked on it will be selected on the first draw?

d. What is the probability that a slip without an *X* marked on it will be selected on the first draw?

e. What are the odds of selecting a slip with an *X* marked on it on the first draw? Of not selecting a slip with an *X* marked on it on the first draw?

6. Experiment — To draw at random a marble from a bag containing 6 blue marbles and 8 yellow marbles.

a. How many outcomes are in the sample space? List them.

b. If the event is to select a blue marble, how many favorable outcomes are there? List them.

c. What is the probability of drawing a blue marble from the bag on the first draw?

d. What is the probability of not drawing a blue marble from the bag on the first draw?

e. What are the odds of drawing a blue marble on the first draw? What are the odds of not drawing a blue marble on the first draw?

7. Experiment — To draw at random two beads from a bowl containing two blue beads and two gold beads, replacing the first bead before drawing the second bead.

a. Write the sample space listing all the outcomes of drawing two beads such as (BG), etc.
b. If the event is to select at random two blue beads, list the favorable outcome.
c. What is the probability of drawing two blue beads on two draws?
d. What is the probability of not drawing two blue beads on two draws?
e. What are the odds of drawing two blue beads on the first two draws? What are the odds of not drawing two blue beads on the first two draws?

8. Experiment — Tossing three coins

a. Write the sample space listing all the outcomes when the three coins are tossed, such as (HHH), (HHT), etc.
b. List the favorable outcomes if the event is to have the three coins fall either all heads or all tails.
c. What is the probability that the three coins will fall either all heads or all tails?
d. What is the probability that the three coins will fall neither all heads nor all tails?
e. What are the odds that the three coins will fall either all heads or all tails? What are the odds that the three coins will fall neither all heads nor all tails?

9. A standard deck of 52 playing cards consists of an ace, king, queen, jack, 10, 9, 8, 7, 6, 5, 4, 3, and 2 in four different suits: clubs, diamonds, hearts, and spades. Club and spade suits are in black; diamond and heart suits are in red. The king, queen, and jack cards are called picture cards.

Experiment — To draw at random a card from a well-shuffled deck:

a. What is the probability that it will be on the first draw:
(1) a black card?
(2) an ace?
(3) a picture card?
(4) a red 2?
(5) a black 7?
(6) a red picture card?
(7) the queen of spades?
(8) any jack?

b. What is the probability that it will not be on the first draw:
(1) the ace of spades?
(2) any king?
(3) a black 9?
(4) a red card?
(5) a picture card?
(6) a black picture card?

c. What are the odds in favor of getting on the first draw:
(1) the king of clubs?
(2) a queen?
(3) a red card?
(4) a picture card?
(5) a black ace?
(6) a red picture card?

d. What are the odds against getting on the first draw:
(1) an ace?
(2) the jack of diamonds?
(3) a red 5?
(4) a black card?
(5) a black picture card?
(6) a picture card?

10. Experiment — Tossing a pair of dice

a. Write the sample space listing all the possible outcomes when a pair of dice is rolled. Observe that each outcome is designated by a number pair such as

(5, 3) in which the first component names the number of dots in the upper face of the first die and the second component names the number of dots in the upper face of the second die. How many possible outcomes are there in the sample space?

b. Write the set of outcomes for the event that the sum of the dots on the two dice is 7. What is the probability that the sum will be 7 when the two dice are rolled? What is the probability that the sum will not be 7? What are the odds that the sum of 7 will occur on the first roll? What are the odds that the sum of 7 will not occur on the first roll?

Write the set of outcomes for each of the following events. Then find the probability for each event to happen and the probability for it not to happen. Also find the odds in favor of the event to happen and the odds against the event to happen.

Event: Sum of the dots on the upper faces of the two dice is:

c. 5 **d.** 11 **e.** 6 **f.** 12 **g.** 9 **h.** 3 **i.** 4 **j.** 2 **k.** 8

11. What is the probability value of:

a. An event having as much chance to happen as it has to fail to happen?
b. An event that is impossible to happen?
c. An event that is absolutely certain to happen?

12. Which has the higher probability:

Drawing at random on the first draw a blue marble from a bag containing 5 blue marbles and 8 green marbles or drawing on the first draw a blue marble from a bag containing 15 blue marbles and 24 white marbles?

13. You have 6 nickels, 9 dimes, and 5 quarters in your pocket. What is the probability of picking a nickel at random from your pocket?

14. A total of 500 raffle tickets were sold at a charity affair. If you hold 15 raffle tickets, what is the probability of your winning the prize?

15. One hundred fifty tickets for the door prize were sold at a charity affair. If you hold 3 of these tickets, what is the probability of your winning the door prize?

16. In a class of 12 boys and 16 girls, a student is to be selected to be the school council member by drawing a name of one of these students. What is the probability that a boy will be selected? What is the probability that a girl will be selected? What are the odds that a boy will be selected? What are the odds that a girl will be selected?

12–11 PROBABILITY OF *A* OR *B*; PROBABILITY OF *A* AND *B*

Probability of *A* or *B*

Mutually exclusive events are events which cannot occur at the same time. A coin falls heads or tails but not both heads and tails at the same time.

When a box contains 4 white marbles, 5 red marbles, and 3 yellow marbles, drawing a white marble or drawing a red marble are mutually exclusive events.

Since there are 12 marbles in the box and 4 are white, the probability of drawing at random a white marble is $\frac{4}{12}$ or $\frac{1}{3}$.

(1) That is: $P(W) = \frac{1}{3}$

Since there are 12 marbles in the box and 5 are red, the probability of drawing at random a red marble is $\frac{5}{12}$.

(2) That is: $P(R) = \frac{5}{12}$

Since there are 12 marbles in the box and the total of white and red marbles is 9, the probability of drawing a white marble or a red marble at random is $\frac{9}{12}$ or $\frac{3}{4}$.

(3) That is: $P(W \text{ or } R) = \frac{3}{4}$

Observe that since $P(W) = \frac{1}{3}$ and $P(R) = \frac{5}{12}$, then $P(W) + P(R) = \frac{1}{3} + \frac{5}{12} = \frac{9}{12} = \frac{3}{4}$, the same answer as that found in (3) for $P(W \text{ or } R)$.

That is: $P(W \text{ or } R) = P(W) + P(R)$

When two events are mutually exclusive, the probability of one event or the other event to happen is equal to the sum of their separate probabilities.

Thus if A and B are mutually exclusive events, then the *probability of A or B* is the sum of the probability of A and the probability of B.

This may be expressed as: $P(A \text{ or } B) = P(A) + P(B)$

or $P(A \cup B) = P(A) + P(B)$

Probability of A and B

Two events are *independent* when the outcome of one event does not affect the outcome of the other.

When we toss two coins, the event of the first coin falling heads up and the event of the second coin falling heads up are independent of each other.

The probability of the first coin falling heads up is $\frac{1}{2}$.

(1) That is: $P(H) = \frac{1}{2}$

The probability of the second coin falling heads up is $\frac{1}{2}$.

(2) That is: $P(H) = \frac{1}{2}$

The sample space for the experiment of tossing two coins may be expressed as: $\{(HH), (HT), (TH), (TT)\}$

or in tabular form:

		Second coin H	Second coin T
First Coin	H	HH	HT
	T	TH	TT

In this sample space the first letter indicates the outcome of the toss of the first coin and the second letter indicates the outcome of the toss of the second coin. Observe there are 4 outcomes: HH, HT, TH, and TT but there is only 1 favorable outcome: HH.

Thus the probability of both coins falling heads up is $\frac{1}{4}$.

(3) That is: $P(H \text{ and } H)$ or $P(HH) = \frac{1}{4}$

We have found that on the first toss $P(H) = \frac{1}{2}$ and on the second toss $P(H) = \frac{1}{2}$. If we multiply these probabilities $(\frac{1}{2} \times \frac{1}{2})$, the product $(\frac{1}{4})$ is the same answer as that found in (3) for $P(H \text{ and } H)$ or $P(HH)$.

The probability that two independent events will both occur is equal to the product of the separate probabilities.

Thus if A and B are independent events, then the *probability of A and B* is the product of the probability of A and the probability of B.

This may be expressed as: $P(A \text{ and } B) = P(A) \cdot P(B)$

or $P(A \cap B) = P(A) \cdot P(B)$

EXERCISES

1. A bag contains 5 white marbles, 6 green marbles, and 9 red marbles. What is the probability of selecting at random from this bag on the first draw:

a. a green marble or a red marble?
b. a white marble or a red marble?
c. a green marble or a white marble?
d. a white, green, or red marble?
e. a blue marble?

2. When you roll a die, what is the probability that on the first roll:

a. a 5 or 6 will turn up?
b. a 1, 2, or 3 will turn up?
c. a 3 or 4 will not turn up?

3. There are nine cards of the same size in a box, each card bearing a different numeral from 1 through 9. What is the probability of selecting at random from this box on the first draw a card marked with:

a. the numeral 4 or 9?
b. a numeral naming a number greater than 1 but less than 8?
c. a numeral naming an odd number?
d. a numeral naming a prime number?
e. a numeral naming a number divisible by 3?

4. If a spinner has twelve sectors, each of the same size, numbered 1 through 12, what is the probability that the pointer will stop on:

a. the sector marked 7 or 11?
b. the sector marked 1, 2 or 3?
c. the sectors marked by numerals naming a number greater than 8?
d. the sectors marked by numerals naming a prime number?
e. the sectors marked by numerals naming an even number?

5. From a well-shuffled deck of 52 playing cards, what is the probability of selecting at random on the first draw:

a. a king of hearts or a queen of spades?
b. a 3 of clubs or 3 of diamonds or 3 of spades?
c. any ace or any jack?
d. any diamond or any club?
e. a picture card or an ace?

6. When a red die and a white die are rolled, what is the probability of getting:

a. a 6 on the white die and a 4 on the red die?
b. a 2 on the white die and a 2 on the red die?

7. One spinner has four sectors, each of the same size, numbered 1 through 4. A second spinner has six sectors, each of the same size, numbered 1 through 6. What is the probability that the pointer on:

a. the first spinner will stop at 3 and on the second spinner at 5?
b. both spinners will stop at 4?

8. One bag contains 6 green balls and 4 yellow balls. A second bag contains 8 green balls and 2 yellow balls. If one ball is selected at random from each bag, what is the probability that:

a. a green ball is selected from the first bag and a yellow ball from the second bag?
b. a yellow ball is selected from the first bag and a green ball from the second bag?
c. both balls are yellow?
d. both balls are green?

9. A bowl contains 9 white marbles and 3 blue marbles. If two marbles are drawn at random from this bowl, replacing the first marble before drawing the second marble, what is the probability that:

a. a white marble is selected on the first draw and a blue marble on the second draw?
b. a blue marble is selected on the first draw and a white marble on the second draw?
c. two blue marbles are drawn?
d. two white marbles are drawn?

10. If two cards are selected at random from a well-shuffled deck of 52 playing cards, replacing the first card before the second card is drawn, what is the probability that:

a. the first card is a 10 and the second card is a jack?
b. the first card is a red card and the second card is a black card?
c. the first card is a black 9 and the second card is a red 9?
d. the first card is a picture card and the second card is an ace?

CHAPTER REVIEW

1. Construct a circle graph showing the enrollment at the Community Junior-Senior High School: seventh grade, 340 students; eighth grade, 300 students; ninth grade, 325 students; tenth grade, 280 students; eleventh grade, 270 students; twelfth grade, 285 students. (12–3)

2. A science class made the following scores in a test: 90, 75, 80, 65, 80, 70, 95, 80, 90, 70, 85, 90, 50, 85, 75, 90, 60, 75, 100, 70, 65, 75, 65, 90, 80.

a. Make a frequency distribution table, a histogram and a frequency polygon. (12–4)

b. Find the mean, median, and mode. (12–5)

3. Find the range, mean deviation from the mean, variance, and standard deviation for the following set of scores: 60, 65, 70, 75, 80, 85, 90. (12–7)

4. Find the median, upper quartile, and lower quartile for the following set of scores: 5, 6, 8, 9, 11, 12, 14, 15. (12–6)

5. If Felipe ranks twenty-first in his class of 75 students, what is his percentile rank? (12–6)

6. Find the value of: ${}_6P_4$ (12–8)

7. Find the value of: ${}_6C_4$ (12–9)

8. Experiment — To draw at random a checker from a box that contains 9 black checkers and 15 red checkers.

a. How many outcomes are in the sample space? List them.

b. If the event is to select a black checker, how many favorable outcomes are there?

c. What is the probability of drawing a black checker from the box on the first draw? Of not drawing a black checker on the first draw?

d. If the event is to select a red checker, how many favorable outcomes are there?

e. What is the probability of drawing a red checker from the box on the first draw? Of not drawing a red checker on the first draw?

f. What are the odds of drawing a black checker on the first draw? What are the odds of drawing a red checker on the first draw? (12–10)

9. From a well-shuffled deck of 52 playing cards, what is the probability of selecting at random on the first draw a card that is:

a. any jack or a red 10? **b.** any spade or any heart? (12–11)

10. One bag contains 8 yellow marbles and 4 blue marbles. A second bag contains 10 yellow marbles and 15 blue marbles. If one marble is selected at random from each bag, what is the probability that:

a. a yellow marble is selected from the first bag and a blue marble from the second bag?

b. a blue marble is selected from the first bag and a yellow marble from the second bag?

c. two yellow marbles are drawn? **d.** two blue marbles are drawn? (12–11)

13 Flow Charts and Logic

13–1 FLOW CHARTS

Flow charts are used to indicate and picture the order of the steps to follow in a program.

Various geometric shapes are used. Ovals indicate the beginning and the end of the program. We use rectangles to enclose instructions and diamonds to enclose questions with a two-choice answer nearby.

For example, the flow chart below indicates the order of the steps in subtracting a common fraction from a larger common fraction.

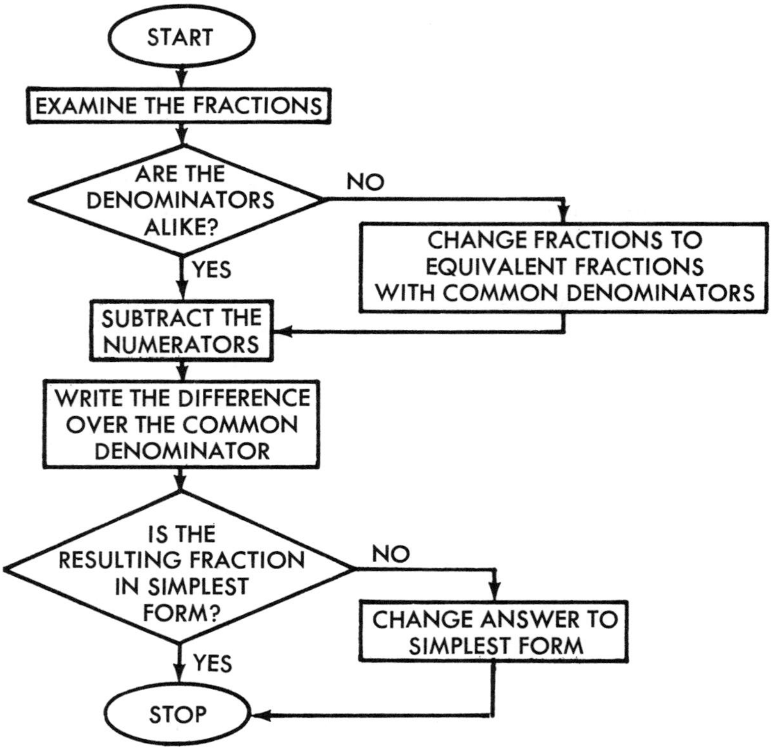

EXERCISES

Construct flow charts indicating the steps in each of the following programs:

1. Adding two common fractions.
2. Rounding a decimal.
3. Adding two mixed decimals.
4. Multiplying two 2-place whole numbers.
5. Dividing one mixed number by another mixed number.

13–2 LOGIC

A compound sentence (or statement) consists of two simple sentences joined by a connective. If the connective is "and," represented by the symbol "$\wedge$," the compound sentence is called a *conjunction*. The number sentence $8 < n$ and $n < 12$ is a conjunction which could be written as $8 < n \wedge n < 12$ or as $8 < n < 12$. Usually in logic the letter p represents the first simple sentence and the letter q the second simple sentence so that $p \wedge q$ represents a conjunction.

If the connective is "or," represented by the symbol "$\vee$," the compound sentence is called a *disjunction*. The number sentence $n > 8$ or $n = 8$ is a disjunction. It could be written as $n > 8 \vee n = 8$ or as $n \geq 8$. In logic $p \vee q$ symbolizes a disjunction.

A conjunction is true if and only if *both* simple sentences are true. If either sentence is false or both sentences are false, the conjunction is false.

A disjunction is true if *either one* of the simple sentences is true or *both* sentences are true. If both sentences are false, the disjunction is false. See the summary in the truth table at the right.

p	q	$p \wedge q$	$p \vee q$
T	T	T	T
T	F	F	T
F	T	F	T
F	F	F	F

The *negation* of a sentence is the opposite of the sentence. The negation of $b = 4$ is $b \neq 4$; the negation of $x \not> 10$ is $x > 10$. If the sentence is true, the negation is false; if the sentence is false, the negation is true. Negation is indicated by the prefix symbol "$\sim$." If p is the sentence, then $\sim p$ (read "not p") is the negation. The negation of a negation is the original sentence.

$$\sim(\sim p) = p$$

p	$\sim p$
T	F
F	T

An *implication* is a conditional sentence (sometimes called a *conditional*) consisting of an *antecedent* (sentence following "if") and a *consequent* (sentence following "then"). The sentences: "If a triangle has 3 equal sides, then it has 3 equal angles.," and "If $4x = 24$, then $x = 6$." are implications. The antededent is sometimes called the *hypothesis* and the consequent is called the *conclusion*. Using p as the antecedent, q for the consequent, and an arrow between them, we express an implication as $p \rightarrow q$ which means "if p, then q" or "p implies q."

If we interchange the antecedent and consequent of a conditional sentence, another implication results which is the *converse* of the first. The converse of $p \rightarrow q$ is $q \rightarrow p$. The expression $p \leftrightarrow q$ symbolizes that an implication and its converse are equivalent since each implies the other.

If we negate both the antecedent and consequent of an implication, the resulting implication is the *inverse* of the first. The inverse of $p \rightarrow q$ is $\sim p \rightarrow \sim q$ which is read "not p implies not q."

If we negate both the antecedent and consequent of an implication and then interchange them, the resulting implication is the *contrapositive* of the first. The contrapositive of $p \rightarrow q$ is $\sim q \rightarrow \sim p$ which is read "not q implies not p."

An example of each of the above implications is:

Implication: If a triangle is equilateral, then it is equiangular.

Converse: If a triangle is equiangular, then it is equilateral.

Inverse: If a triangle is not equilateral, then it is not equiangular.

Contrapositive: If a triangle is not equiangular, then it is not equilateral.

EXERCISES

1. Read, or write in words, each of the following:

a. $6 > 5 \wedge 4 < 7$ **c.** $\sim a$ **e.** $r \leftrightarrow s$

b. $12 < 9 \vee 12 > 9$ **d.** $m \rightarrow n$ **f.** $\sim c \rightarrow \sim d$

2. Write a conjunction for each of the following pairs of sentences and indicate which are true:

a. $2 > 3$; $7 < 10$ **b.** $8 > 6$; $4 < 5$ **c.** $9 > 5$; $1 < 0$ **d.** $3 = 7$; $2 > 5$

e. Express $9 < n \wedge n < 15$ in a more concise form.

3. Write a disjunction for each of the following pairs of sentences and indicate which are true:

a. $4 < 7$; $5 > 6$ **b.** $3 > 10$; $2 = 1$ **c.** $5 < 9$; $3 > 0$ **d.** $6 > 8$; $4 < 5$

e. Express $n < 4 \vee n = 4$ in a more concise form.

4. Write the negation of each of the following:

a. $9y = 27$
b. $8x \neq 16$
c. $2 + 3 > 7$
d. $5 - 2 \nless 4$
e. All students are boys.
f. Every line is straight.

5. a. What do we call a sentence like "If $3x = 18$, then $x = 6$"?
b. What is the antecedent or hypothesis of the above sentence? What is the consequent or conclusion of the above sentence?

6. Write the converse of "If $2a = 10$, then $a = 5$."

7. Which of the following indicates that the implication and its converse are equivalent?

a. $4x = 8 \leftarrow x = 2$
b. $4x = 8 \rightarrow x = 2$
c. $4x = 8 \leftrightarrow x = 2$

8. Write the inverse of "If line *m* is perpendicular to line *n*, then line *n* is perpendicular to line *m*."

9. Write the contrapositive of "If the skies will be clear, the sun will shine."

10. Write the converse, inverse, and contrapositive of each of the following implications:

a. If *b* is greater than *c*, then *c* is less than *b*.
b. If two angles are right angles, then the two angles have the same measure.
c. If lines *a* and *b* are parallel lines, then lines *a* and *b* do not intersect.
d. If $2x + 3 = 11$, then $x = 4$.
e. If I will go to the ball game, then I will miss the television show.

11. If the original implication or statement is true, is the converse always true? Is the inverse always true? Is the contrapositive always true?

12. If the converse is true, is the inverse always true?

REVIEW EXERCISES

1. Add:
89,778 + 6,598 + 975 + 58,969 + 439,676

2. Subtract:
1,200,501
836,492

3. Multiply:
3,079
9,608

4. Divide:
$1,760\overline{)510,400}$

5. Add:
$6\frac{3}{5} + 9\frac{7}{10} + 4\frac{1}{2}$

6. Subtract:
$12 - 5\frac{3}{16}$

7. Multiply:
$3\frac{3}{7} \times 3\frac{1}{2}$

8. Divide:
$10 \div \frac{4}{5}$

9. Add:
$6.35 + .974 + 82.7$

10. Subtract:
$6 - .005$

11. Multiply:
$.132 \times .214$

12. Divide:
$.025\overline{)4.5}$

13. Find $37\frac{1}{2}\%$ of 5,000.

14. 6% of what number is 732?

15. What per cent of 81 is 27?

16. How many inches are in $7\frac{1}{2}$ yards?

17. What part of an hour is 32 minutes?

18. How many ounces are in $4\frac{5}{8}$ pounds?

19. Change 9 km 4 m 7 cm 6 mm to meters.

20. How many centimeters are in 8 inches?

14 Informal Geometry

Geometry is the study of points, lines, planes, and space, of measurement and construction of geometric figures, and of geometric facts and relationships. The word "geometry" means "earth measure." The Egyptians developed many geometric ideas because the flooding of the Nile River destroyed land boundaries which had to be restored. Euclid, a Greek mathematician, in the third century B.C. collected and organized the geometric principles known at that time and wrote a book on geometry.

POINTS, LINES, PLANES, AND SPACE

A geometric *point* is an exact location in space. It has no size nor can it be seen. Any dot we generally use to indicate it is only a representation of a geometric point.

A geometric *line* is a set of points. The pencil or chalk lines we draw are only representations of geometric lines. A line may be extended indefinitely in both directions because it is endless; it has an infinite number of points but no endpoints. A definite part of a line has length but no width or thickness. We cannot see a geometric line.

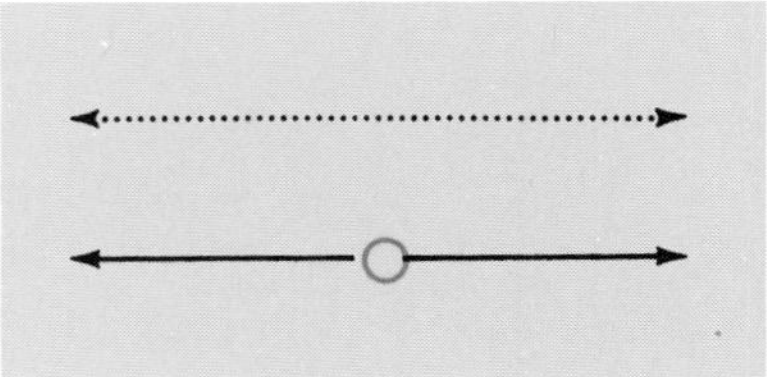

A point separates a line into two *half-lines*. Each half-line extends indefinitely in one direction only and does not include the point that separates the line into two half-lines.

A geometric *plane* or flat surface is a set of points. It is endless in extent and it has no boundaries but it has length and width which can be measured when the plane is limited. A desk top, wall, floor, and sheet of paper are common representations of a limited plane. We cannot see a geometric plane. It extends beyond any line boundaries we use to represent it.

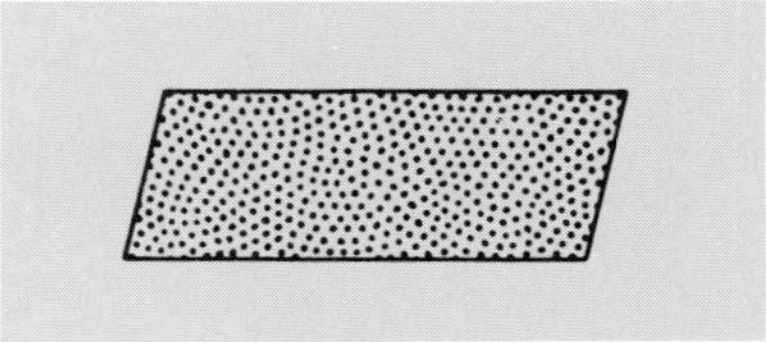

A line separates a plane into two half-planes.

Space is the infinite set of all points. Its length, width, and height are endless. A limited space can be measured. A plane separates space into two half-spaces.

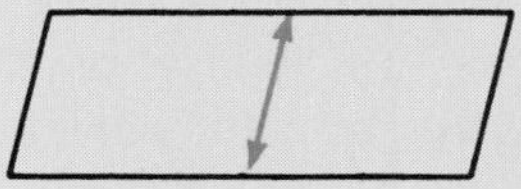

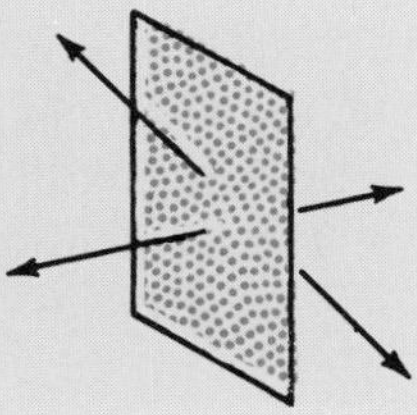

14–1 NAMING POINTS, LINES, AND PLANES

A capital letter is used to label and name a point.

· *B* is "point *B*."

A line is represented as:  The arrowheads are used to show that a line is endless in both directions.

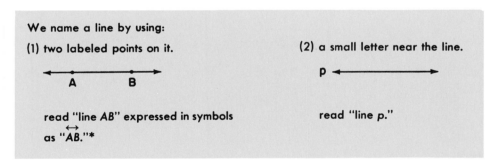

We name a line by using:

(1) two labeled points on it.

read "line *AB*" expressed in symbols as "$\overleftrightarrow{AB}$."*

(2) a small letter near the line.

read "line *p*."

A definite part of a line including both of its endpoints is called a *line segment* or *segment*. It consists of two endpoints and all the points between. We name a line segment by its endpoints.

---

*Also read "line *BA*" and expressed as $\overleftrightarrow{BA}$.

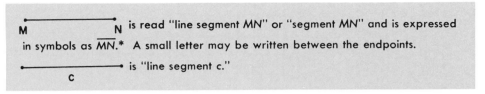

is read "line segment *MN*" or "segment *MN*" and is expressed in symbols as $\overline{MN}$.* A small letter may be written between the endpoints.

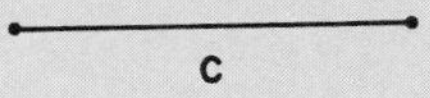

is "line segment c."

A definite part of a line excluding its endpoints is called an *interval.*

A half-line which includes one endpoint is called a *ray*. This endpoint is the one that separates the line into two half-lines. To name a ray we use the letter first which names the endpoint and then the letter which names one other point on the ray.

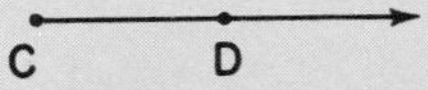

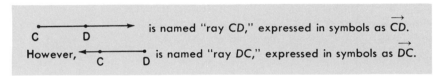

is named "ray *CD*," expressed in symbols as $\overrightarrow{CD}$.

However, 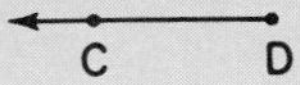

is named "ray *DC*," expressed in symbols as $\overrightarrow{DC}$.

A plane is named by using the letters which name three points not on the same line belonging to it or by two capital letters at opposite corners or by one capital letter as shown.

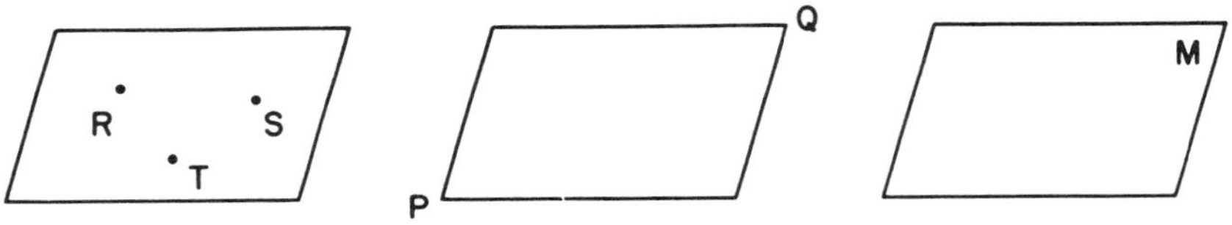

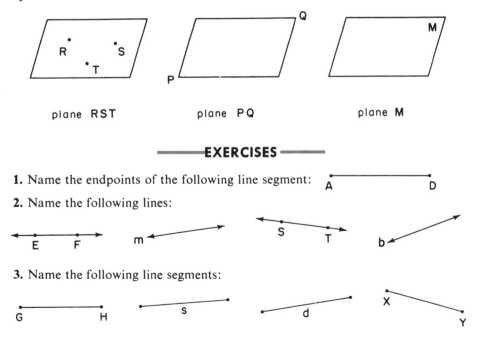

plane RST plane PQ plane M

EXERCISES

1. Name the endpoints of the following line segment: 

2. Name the following lines:

3. Name the following line segments:

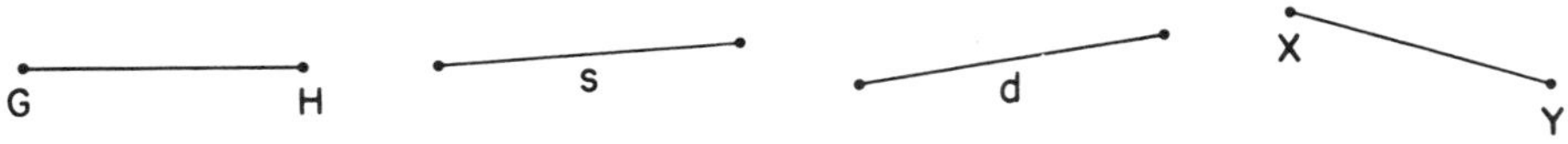

*Also read "line segment *NM*" and expressed as $\overline{NM}$.

4. Express the name of each of the following symbolically:

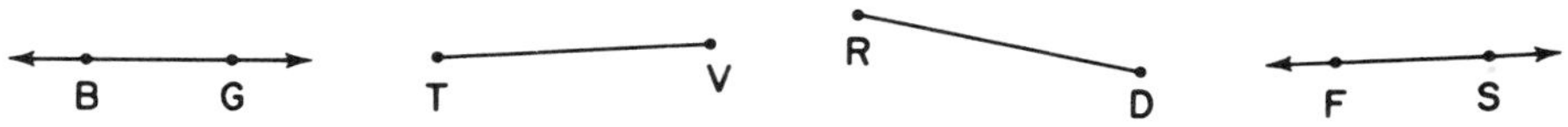

5. Name the following rays and express them symbolically:

6. Name  in three ways. Write the names symbolically.

7. Read, or write in words, each of the following:

$\overleftrightarrow{BC}$, $\overrightarrow{RT}$, $\overline{NO}$, $\overrightarrow{GF}$, $\overleftrightarrow{KL}$, $\overline{EF}$, $\overleftrightarrow{NY}$, $\overrightarrow{PA}$, $\overline{OC}$, $\overleftrightarrow{DR}$

8. Name the point of intersection of $\overleftrightarrow{PQ}$ and $\overleftrightarrow{RS}$.

9. $\overline{AC}$ is divided into 2 parts by point B. Name the two segments.

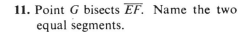

10. Points O and P separate $\overline{MN}$ into 3 parts. Name the three segments.

11. Point G bisects $\overline{EF}$. Name the two equal segments.

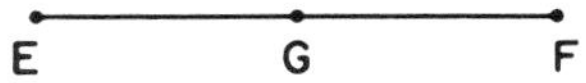

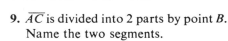

12. Name each of the following planes:

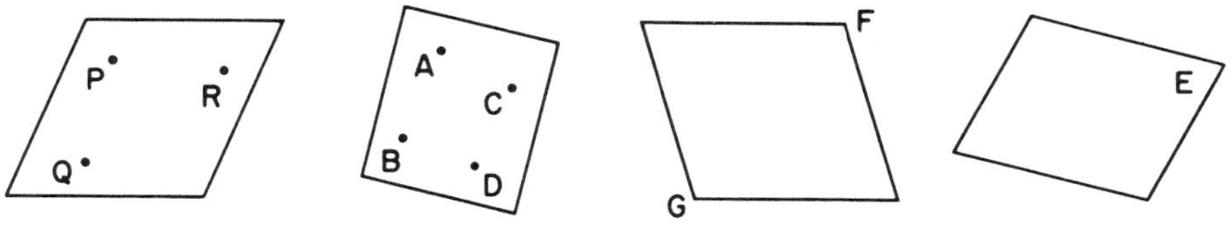

Name the line segments in each of the following figures:

13.

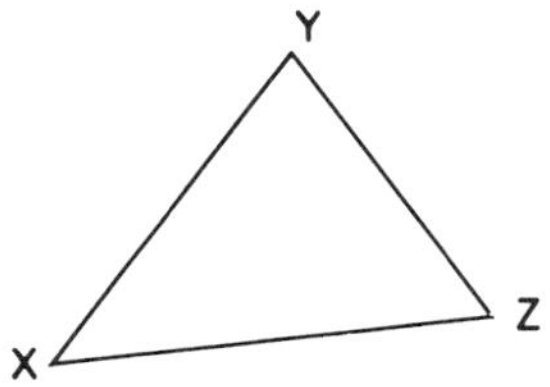

14.

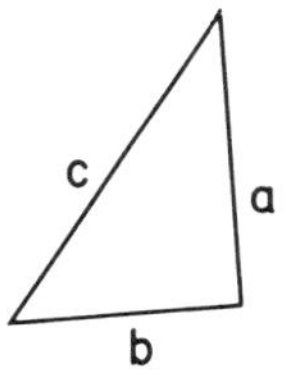

15.

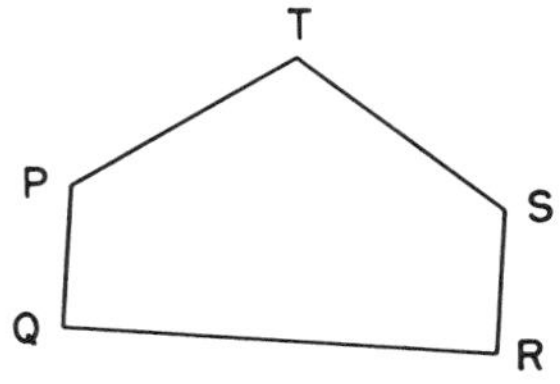

16. $\overline{LM}$ is the sum of what segments?

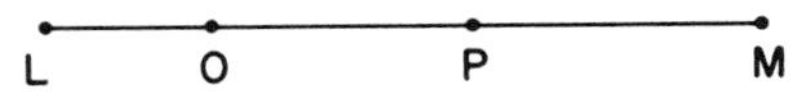

17. $\overline{RS}$ is the difference of what segments?

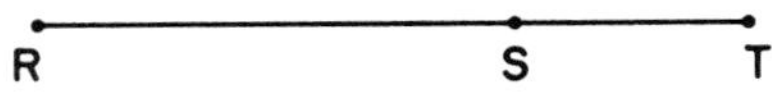

18. Read, or write in words, the following:

$$\overleftrightarrow{GH} \cap \overleftrightarrow{MN}.$$

What does this mean?

Find the following: $\overleftrightarrow{GH} \cap \overleftrightarrow{MN} = \{?\}$

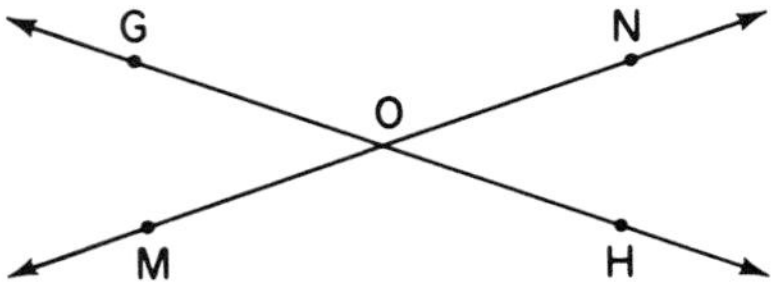

19. Find each of the following:

a. $\overline{EF} \cap \overline{FG} = \{?\}$ **b.** $\overline{FE} \cap \overline{EG} = \{?\}$
c. $\overline{EG} \cap \overline{GF} = \{?\}$

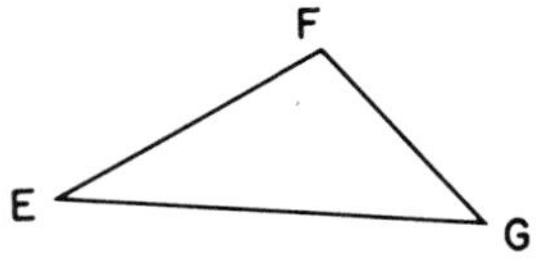

20. Find each of the following:

a. $\overline{BC} \cap \overline{CD} = \{?\}$ **c.** $\overline{BA} \cap \overline{AD} = \{?\}$
b. $\overline{AD} \cap \overline{DC} = \{?\}$ **d.** $\overline{AB} \cap \overline{BC} = \{?\}$

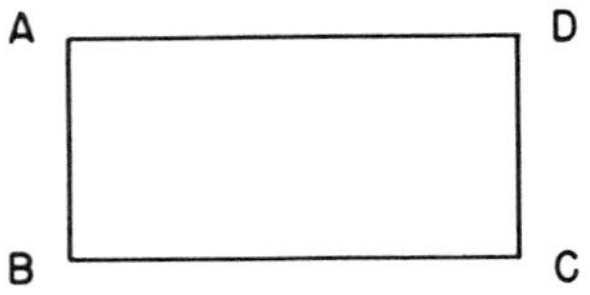

14–2 KINDS OF LINES

Lines may be *straight*, *curved*, or *broken*. Usually a straight line is simply called a line. All these lines are sometimes called curves.

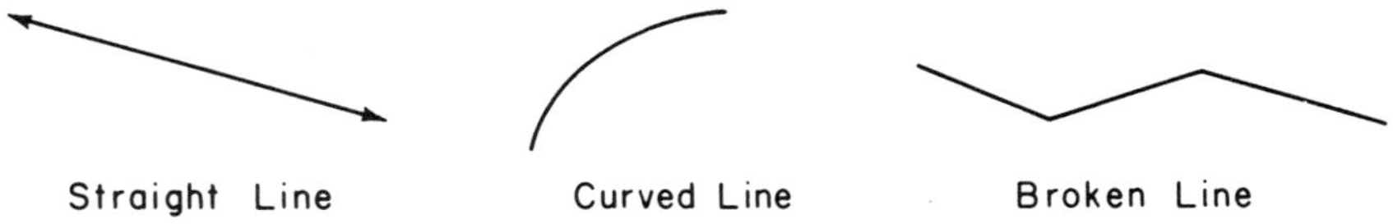

EXERCISES

What kind of line is shown by figure *a*? *b*? *c*? *d*? *e*? *f*? *g*? *h*? *i*? *j*?

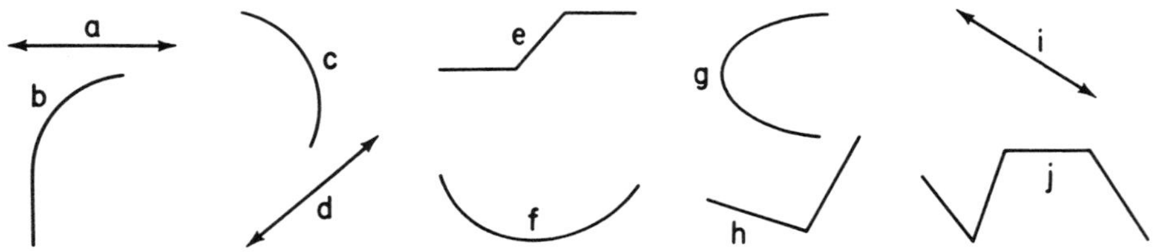

14-3 POSITION OF LINES

Lines may be in *vertical*, *horizontal*, or *slanting* (sometimes called *oblique*) positions.

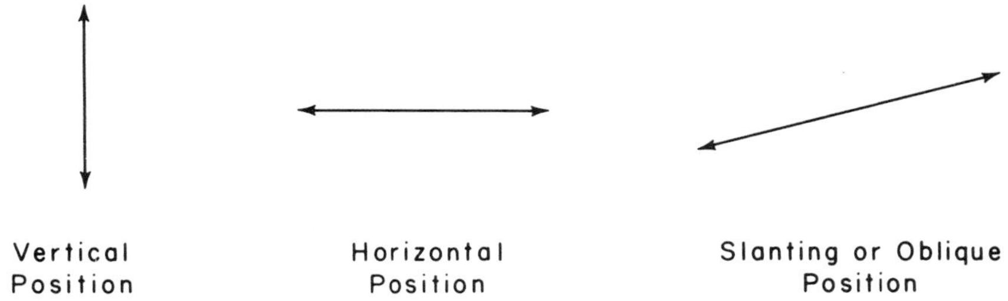

EXERCISES

What position is shown by the line in figure *a*? *b*? *c*? *d*? *e*? *f*? *g*? *h*? *i*? *j*?

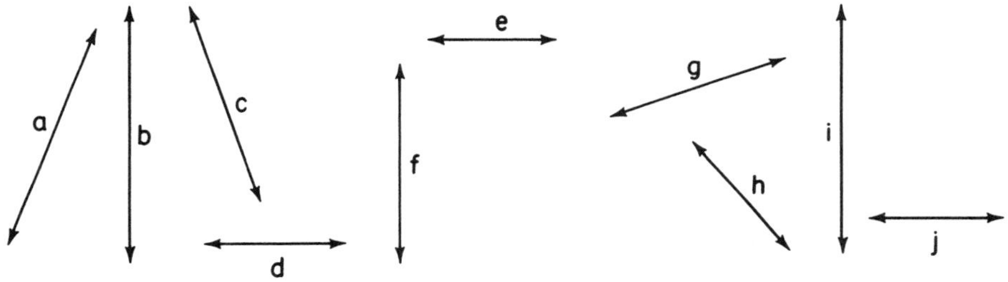

14-4 INTERSECTING, PARALLEL, AND PERPENDICULAR LINES

Lines that meet are *intersecting lines*. Since they have a common point, intersecting lines are sometimes called *concurrent lines*. Two lines in the same plane that do not meet are called *parallel lines*. Two lines not in the same plane that do not meet are called *skew lines*. Two intersecting lines or rays or segments or a line and ray or a line and segment or a ray and segment that form a right angle (see page 401) are said to be *perpendicular* to each other.

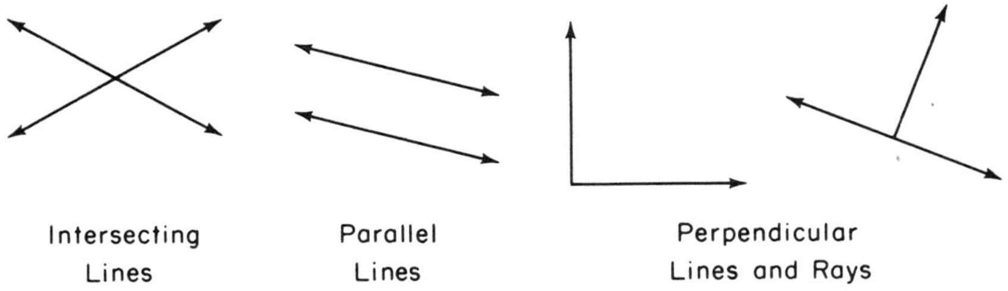

EXERCISES

1. Which of the following are intersecting lines or rays? Which are parallel lines? Which are perpendicular lines or rays?

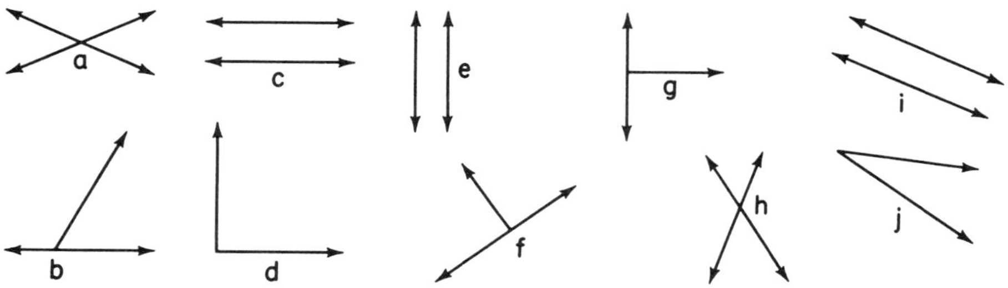

2. Which line is parallel to line *d*?

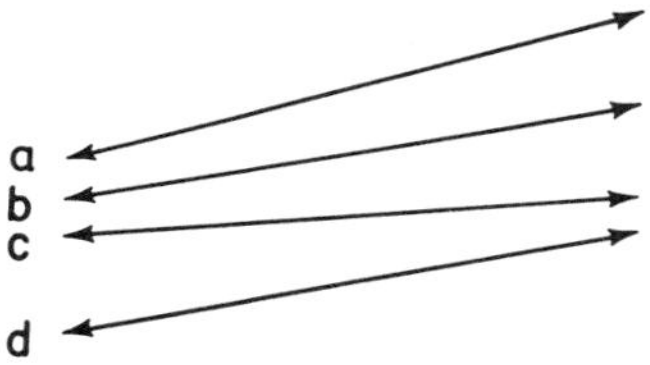

3. a. Which ray is perpendicular to $\overleftrightarrow{CD}$?

b. Which ray is perpendicular to $\overrightarrow{OE}$?

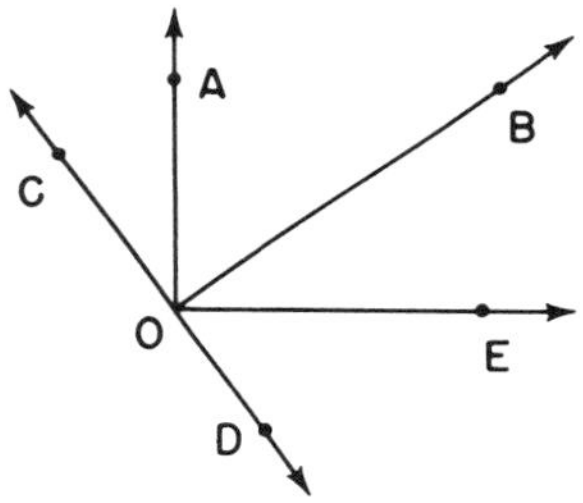

14–5 MORE FACTS ABOUT POINTS, LINES, AND PLANES

(1) Points that lie on the same straight line are called *collinear points.*

(a) Are points *A*, *B*, and *C* collinear?

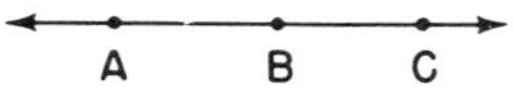

(b) Are points *D*, *E*, and *F* collinear?
Are points *D* and *E* collinear?
Are points *E* and *F* collinear?
Are points *D* and *F* collinear?

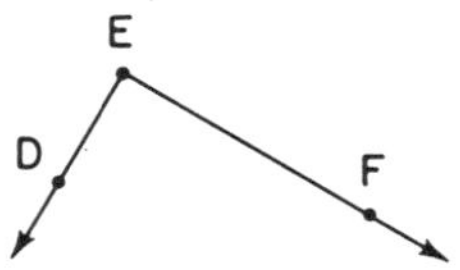

(c) Are points *M*, *O*, and *N* collinear?
Are points *P*, *O*, and *N* collinear?
Are points *Q*, *O*, and *M* collinear?
Are points *P*, *O*, and *Q* collinear?

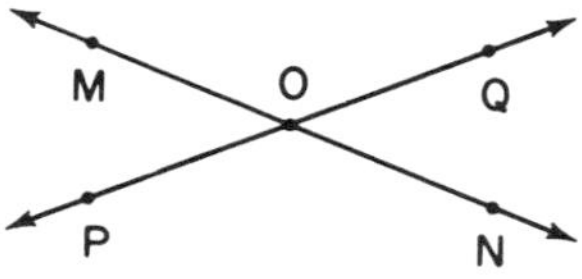

(2) Label a point on your paper as R. Draw a line through point R. Draw a different line through point R. Draw a third line through point R. Can more than one straight line be drawn through a point? How many lines can be drawn through a point? Can we say that *an infinite number of straight lines can be drawn through a point*? Since these lines have a common point, can we call them concurrent lines?

(3) Label two points on your paper as C and D. Draw a straight line through points C and D. Draw another line through points C and D. How many straight lines can be drawn through two points? Can we say that *two points determine a straight line and that one and only one straight line can pass through any two points*? Are any two points collinear?

(4) Draw a pair of intersecting lines. Draw another pair of intersecting lines. At how many points can two straight lines intersect? How many points in common does a pair of intersecting lines have? Can we say that *two straight lines can intersect in only one point*?

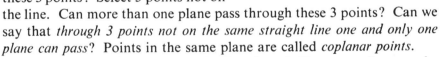

(5) Which of the three kinds of lines is the shortest path between points A and B? Can we say that *the shortest path between two points is along a straight line*?

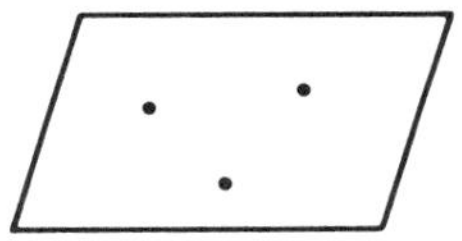

(6) Select 3 points on a straight line. Can more than one plane pass through these 3 points? Select 3 points not on the line. Can more than one plane pass through these 3 points? Can we say that *through 3 points not on the same straight line one and only one plane can pass*? Points in the same plane are called *coplanar points.*

(7) How many points does a geometric plane have? How many lines can be drawn through these points? How many lines does a geometric plane contain? Can we say that *a geometric plane contains an infinite number of points and lines*? Lines in the same plane are sometimes called *coplanar lines.* Can we say that *two intersecting lines determine one and only one plane*?

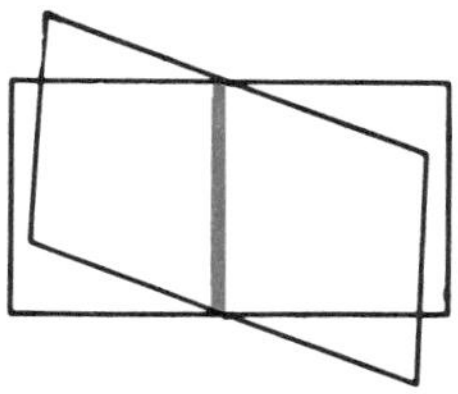

(8) Draw two different planes that intersect as shown. What geometric figure is their intersection? Can we say that *when two different planes intersect, their intersection is a straight line*? Planes that do not intersect are called *parallel planes.* They have no point in common; their intersection is the empty set.

(9) Draw a plane and a line that is not in this plane but which intersects the plane. Can we say that their intersection is one and only one point?

A plane and a line not in this plane are parallel when they have no point in common.

(10) *A line perpendicular to a plane* is a line that is perpendicular to every line in the plane that passes through its foot (the point of intersection). Can we say that *a line that is perpendicular to each of two intersecting lines at their point of intersection is perpendicular to the plane in which these lines lie?*

ANGLES

An *angle* is the figure formed by two different rays having the same endpoint. It is the union of two rays. This common endpoint is called the *vertex* of the angle and the two rays are called the *sides* of the angle. An angle may be considered as the rotation of a ray about a fixed endpoint, the angle being formed as the ray turns from one position to another. An angle is sometimes used to show direction. The symbol "∠" designates the word "angle."

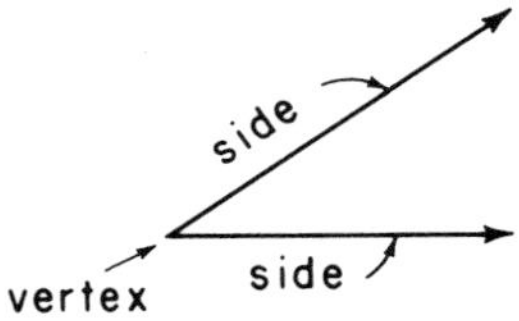

14-6 NAMING ANGLES

Angles are identified or named in the following ways:

(1) By reading the capital letter at the vertex:

Angle C

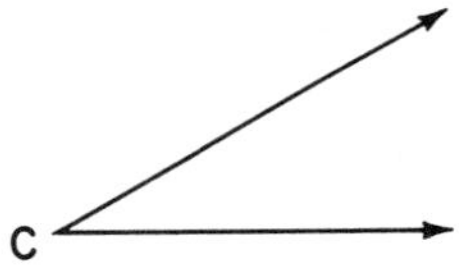

(2) By reading the inside letter or numeral:

Angle a Angle 3

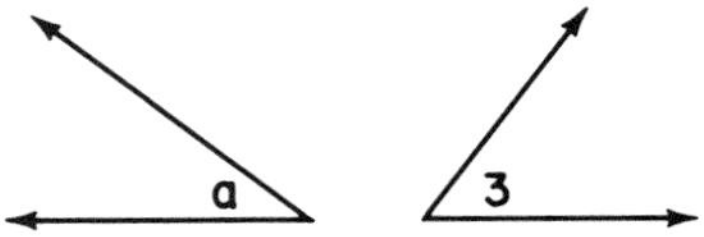

(3) By reading the three letters associated with the vertex and one point on each of the sides. The middle letter always indicates the vertex.

Angle DEF or Angle FED

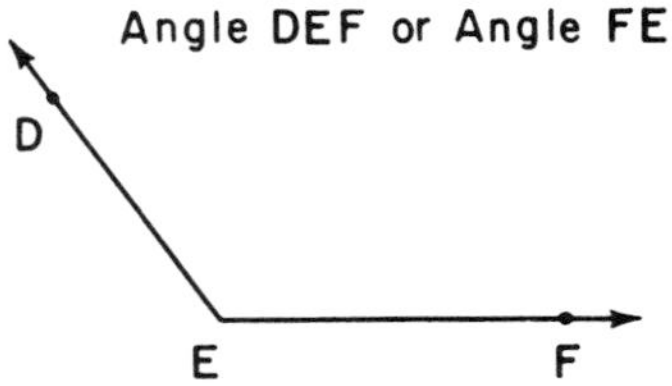

EXERCISES

1. Name the sides and vertex of the following angle:

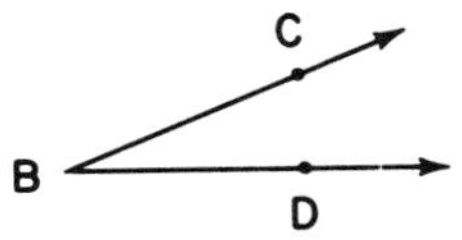

2. What is the symbol used to denote the word "angle"?

3. Name the following angle in three ways:

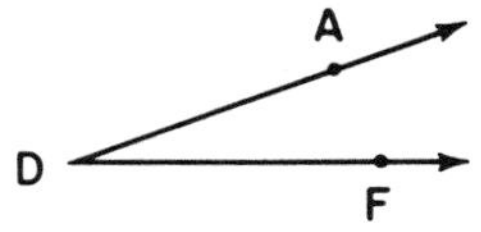

4. Name the following angle in two ways:

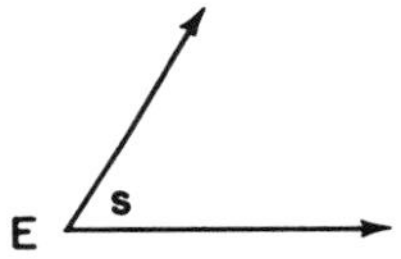

5. Name the following angle in two ways:

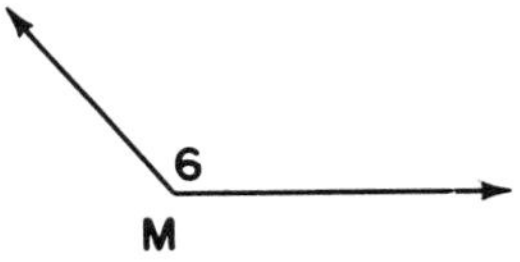

6. Name each of the following angles:

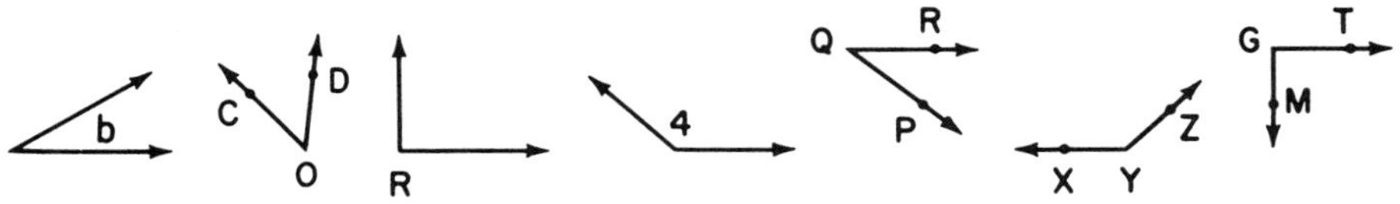

7. Name $\angle 1$, $\angle 2$, and $\angle 3$ in the following figure, using three letters.

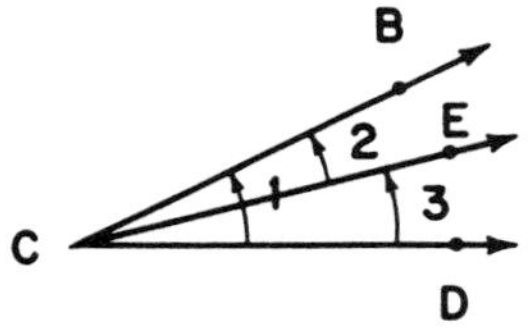

8. Name each angle of the following triangle in four ways.

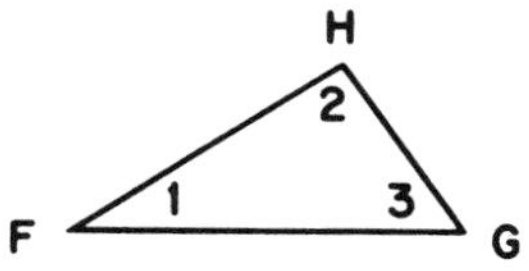

9. Name the four angles formed when $\overleftrightarrow{CD}$ intersects $\overleftrightarrow{EF}$ at point G, using 3 letters for each angle.

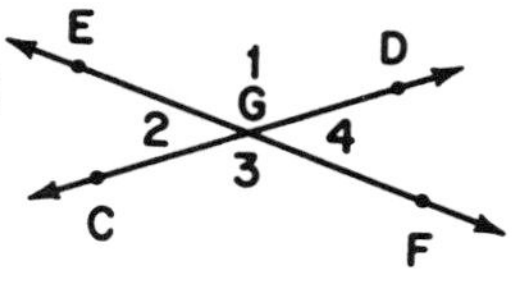

10. In the following triangle, find the angle opposite to:

a. Side DC **b.** Side BD **c.** Side BC

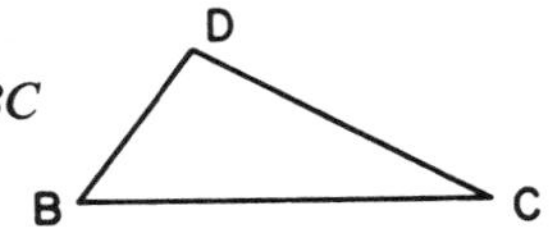

14-7 KINDS OF ANGLES

The *degree*, indicated by the symbol "°", is the unit of measure of angles and arcs. A degree is $\frac{1}{360}$ part of the entire angular measure about a point in a plane. If a circle is divided into 360 equal parts and lines are drawn from the center to these points of division, 360 equal central angles are formed each measuring 1 degree. Each of the corresponding 360 equal arcs also measures 1 degree.

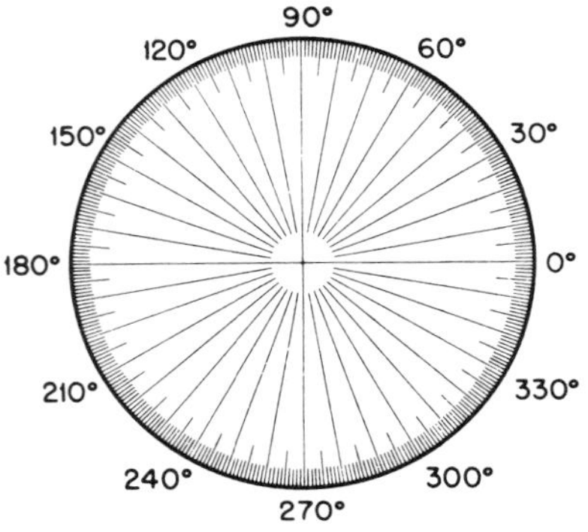

The degree (°) is divided into 60 equal parts called *minutes* (symbol ′). The minute (′) is divided into 60 equal parts called *seconds* (symbol ″).

When a ray turns from one position to another about its fixed endpoint, one complete rotation is equal to 360°.

A *right angle* is one-fourth of a complete rotation; it is an angle whose measure is 90°.

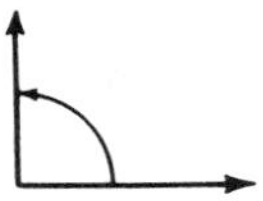
Right

An *acute angle* is an angle whose measure is greater than 0° but less than 90°

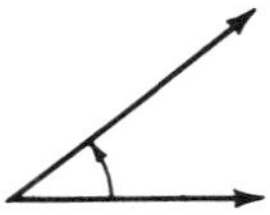
Acute

An *obtuse angle* is an angle whose measure is greater than 90° but less than 180°.

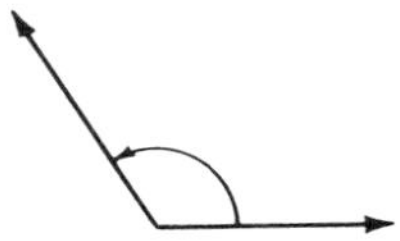
Obtuse

A *straight angle* is one half of a complete rotation; it is an angle whose measure is 180°. The two rays that form a straight angle extend in opposite directions along a straight line that passes through the vertex.

Straight

A *reflex angle* is an angle whose measure is greater than 180° but less than 360°.

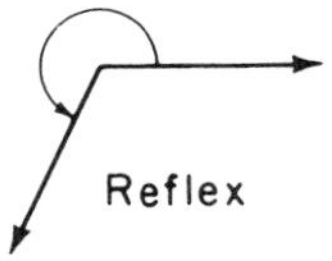
Reflex

EXERCISES

1. What is the unit called that is used to measure angles? What symbol designates it?
2. The measure of an acute angle is greater than ?° and less than ?°.
3. One complete rotation measures ?°.
4. A right angle measures ?° and is ? of a rotation.
5. The measure of an obtuse angle is greater than ?° and less than ?°.
6. A straight angle is ? of a rotation and measures ?°.
7. Which of the following are measures of an acute angle: 90°? 26°? 173°? 89°? 212°?
8. Which of the following are measures of an obtuse angle: 49°? 101°? 305°? 93°? 183°?
9. Indicate by writing corresponding letters which of the following angles are:

   a. acute angles **b.** right angles **c.** obtuse angles **d.** straight angles

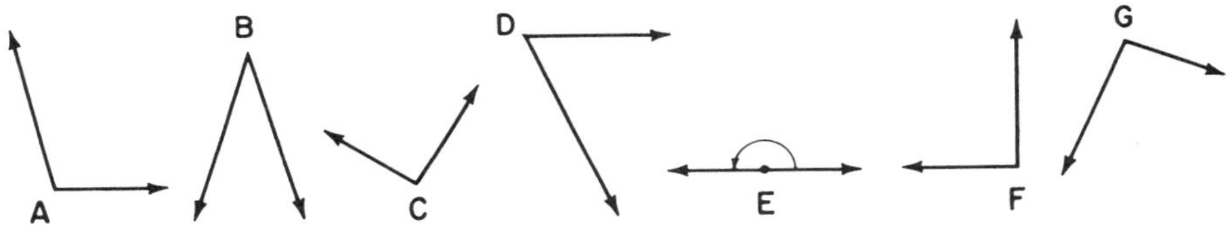

10. What kind of angle is formed by the hands of a clock at 2 o'clock? Describe only the smaller of the two angles formed.
11. Draw five clock faces, then use arrows to indicate the position of the hands at 3 o'clock; at 10 o'clock; at 5 o'clock; at 6 o'clock; at 4 o'clock. What kind of angle is formed by the hands in each case? Describe only the smaller of the two angles formed.
12. **a.** Draw any obtuse angle. **b.** Draw any acute angle.

14-8 GEOMETRIC FIGURES

Simple Closed Plane Figures or Simple Closed Curves

Geometric figures consist of sets of points. Plane geometric figures are figures with all of their points in the same plane. A simple closed plane figure (or curve) begins at a point and returns to this point without crossing itself. It divides a plane into three sets of points, those in the interior, those in the exterior, and those on the figure.

A *polygon* is a simple closed plane figure made up of line segments (called sides). It is the union of three or more line segments. Each pair of intersecting sides meets in a point called a *vertex*. In a polygon the union of two adjacent sides (line segments) forms an angle, since these line segments are parts of rays.

A polygon with all sides of equal length and all angles of equal measure is called a *regular polygon*. It should be noted that the set of points contained in the sides is the figure and not the region enclosed by the figure. A polygon is named by reading the letters at the vertices. A line segment connecting two nonadjacent vertices of a polygon is called a *diagonal*. Some common polygons are: *triangle*, 3 sides; *quadrilateral*, 4 sides; *pentagon*, 5 sides; *hexagon*, 6 sides; *octagon*, 8 sides; *decagon*, 10 sides; and *dodecagon*, 12 sides.

Triangles

When all three sides of a triangle are equal in length, the triangle is called an *equilateral triangle*; when two sides are equal, an *isosceles triangle*; when no sides are equal, a *scalene triangle*. When all three angles of a triangle are equal in size, the triangle is called an *equiangular triangle*; a triangle with a right angle, a *right triangle*; with an obtuse angle, an *obtuse triangle*; and with three acute angles, an *acute triangle*. The *altitude* of a triangle is the perpendicular segment from any vertex of a triangle to the opposite side or extension of that side. The *median* of a triangle is the line segment connecting any vertex of a triangle to the midpoint of the opposite side.

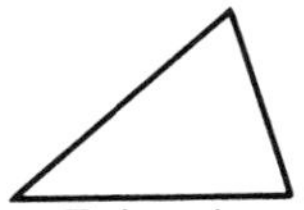
Triangle

In a right triangle the side opposite the right angle is called the *hypotenuse*. The other two sides or legs are the *altitude* and *base* of the triangle. The *base* is generally the side on which the triangle rests. In an isosceles triangle the angle formed by the two equal sides is called the *vertex angle*. It is opposite to the base. The angles opposite the equal sides are called the base angles.

Congruent triangles are triangles which have exactly the same shape and the same size. See page 430.

Similar triangles are triangles which have the same shape but differ in size. See page 431.

Quadrilaterals

The following properties describe special quadrilaterals:

The *rectangle* has two pairs of opposite sides which are equal and parallel and four angles which are right angles.

The *square* has four equal sides with the opposite sides parallel and four angles which are right angles.

The *parallelogram* has two pairs of opposite sides which are parallel and equal.

The *trapezoid* has only one pair of opposite sides that are parallel.

The square is a special rectangle, and the rectangle and square are special parallelograms.

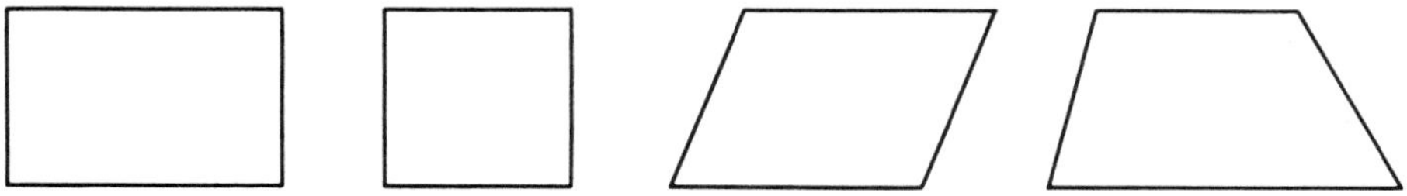

Circles

A *circle* is the set of points in a plane which are equidistant from a fixed point in the plane called the *center*. It is a simple closed curve. The *radius* of a circle is a line segment which has one endpoint at the center of the circle and the other endpoint on the circle. The *diameter* of a circle is a line segment which has both of its endpoints on the circle but passes through the center. A *chord* of a circle is a line segment which has both of its endpoints on the circle. An *arc* is a part of the circle. If the endpoints of an arc are the end-points of a diameter, the arc is a *semi-circle*. An angle whose vertex is at the center of a circle is called a *central angle*. The *circumference* is the distance around the circle. A *tangent to a circle* is a line that has one and only one point in common with the circle.

Concentric circles are circles in the same plane which have the same center but different radii.

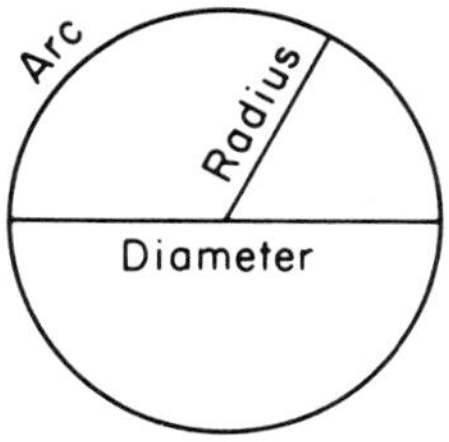

Circle

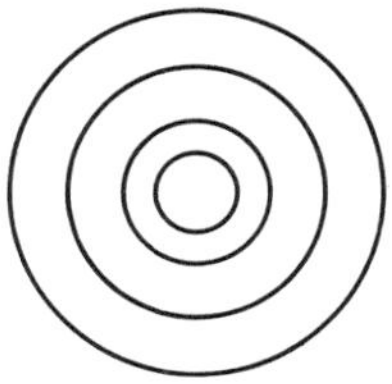

Concentric Circles

Solid or Space Figures

A closed geometric figure consisting of four or more polygons and their interiors, all in different planes, is called a *polyhedron*. The polygons and their interiors are called *faces*. These faces intersect in line segments called *edges*. These edges intersect in points called *vertices*.

Common polyhedra are the rectangular solid (right rectangular prism), the cube, and the pyramid.

(1)

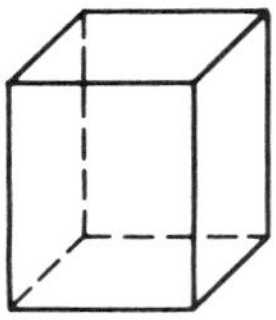

Rectangular Solid

(2)

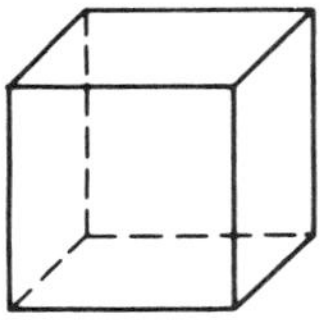

Cube

(3)

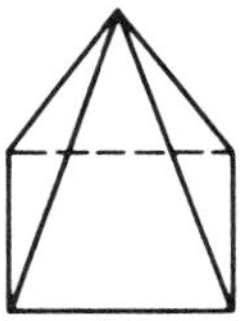

Pyramid

(4)

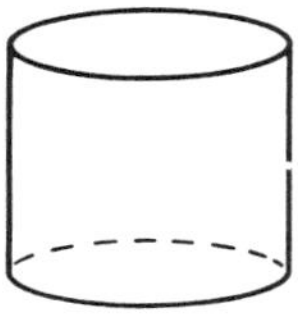

Cylinder

(5)

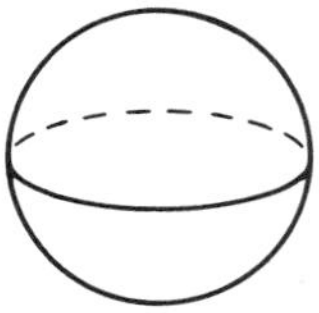

Sphere

(6)

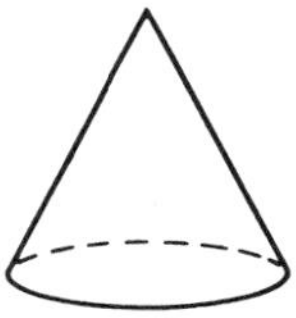

Cone

(1) The *rectangular solid* has six rectangular faces.
(2) The *cube* has six squares for its faces. All the edges are equal in length.
(3) The *pyramid* has any polygon as its base and triangular faces that meet in a common vertex.

Other common solid geometric figures are the cylinder, sphere, and cone.

(4) The *cylinder* has two equal and parallel circles as bases and a lateral curved surface.
(5) The *sphere* has a curved surface on which every point is the same distance from the center within.
(6) The *cone* has a circle for the base and a curved surface that comes to a point called the vertex.

Euler Formula

The Euler formula expresses the relationship of the faces, edges and vertices of a polyhedron. The formula $F + V - E = 2$ tells us that "the number of faces plus the number of vertices minus the number of edges is equal to two."

EXERCISES

1. What is the name of each of the following figures?

a.

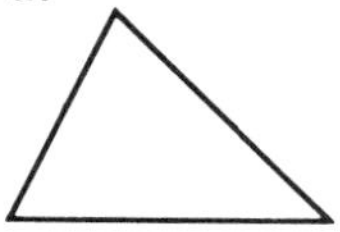

b.

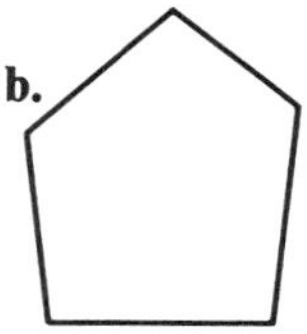

c.

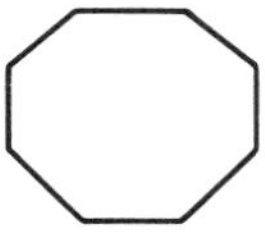

d.

e.

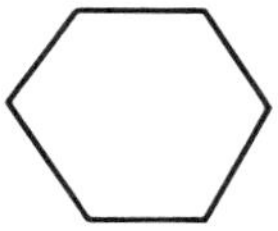

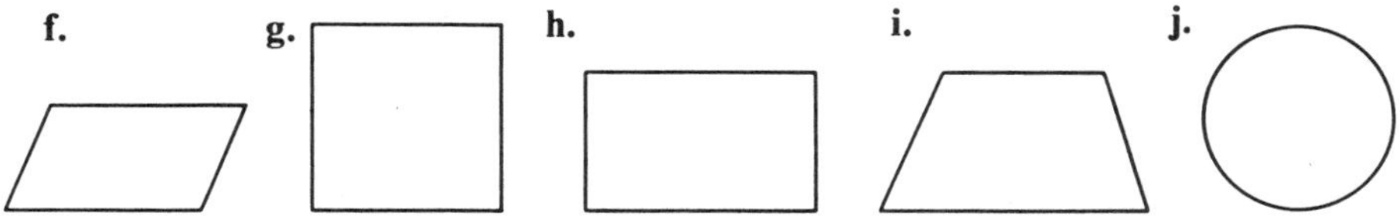

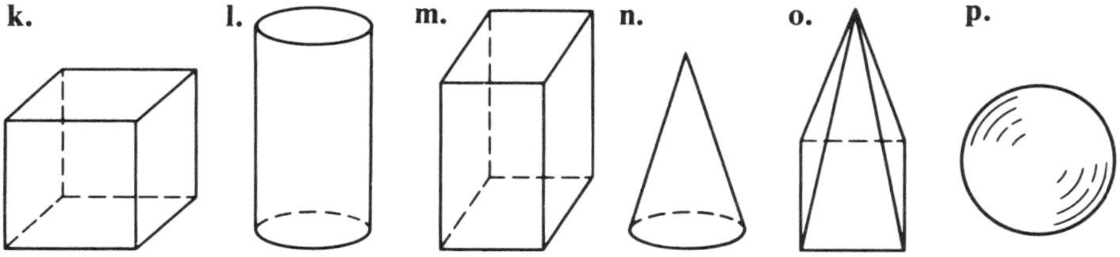

2. What kind of triangle is each of the following?

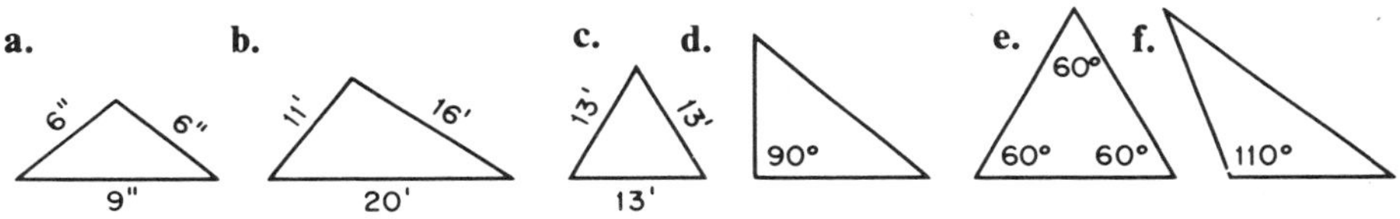

3. Which of the following figures are parallelograms?

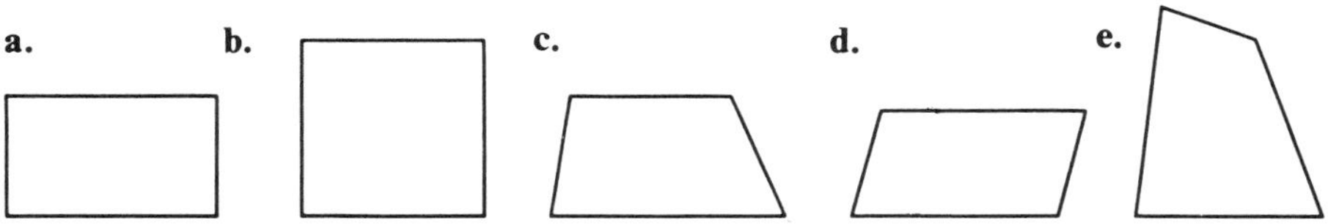

4. How many diagonals can be drawn from any one vertex in each of the following figures?

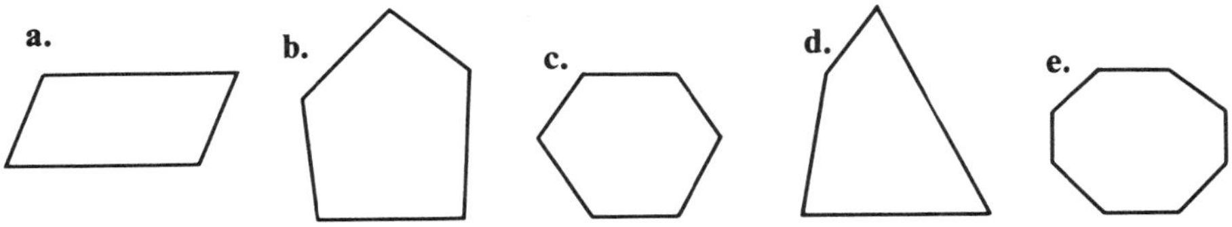

5. **a.** How many faces (F) does a rectangular solid have? How many vertices (V)? How many edges (E)? Does $F + V - E = 2$?

b. How many faces (F) does a pyramid with a square base have? How many vertices (V)? How many edges (E)? Does $F + V - E = 2$?

MEASUREMENT AND CONSTRUCTIONS

14–9 DRAWING AND MEASURING LINE SEGMENTS

A *straightedge* is used to draw line segments. The *ruler,* which is a straightedge with calibrated measurements, is used both to draw and to measure line segments of varying lengths.

Sometimes an instrument called a *compass* is used along with the ruler to measure line segments or to draw line segments of specified lengths. Since the compass is used to draw circles and arcs of circles, the distance between the metal point and the pencil point of the compass corresponds to the radius of the circle or the radius of the arc of the circle that may be drawn.

Generally we measure the length of a line segment by using only the ruler. However when we use both a compass and a ruler to do this, we open the compass so that the metal point and the pencil point fit exactly on the endpoints of the line segment. We then transfer the compass to a ruler to determine the measurement between the metal point and the pencil point.

We reverse this operation to draw a line segment of a given length by using both a compass and a ruler. We first set the compass so that the distance between the metal point and pencil point corresponds to the given measurement on a ruler. We then apply this compass setting to a light pencil working line, using the metal point as one endpoint and the pencil point as the second endpoint.

EXERCISES

1. Using a ruler, draw line segments having the following dimensions:

   a. $2\frac{1}{2}$ in. **b.** $3\frac{3}{4}$ in. **c.** $2\frac{5}{8}$ in. **d.** $3\frac{11}{16}$ in. **e.** $1\frac{15}{16}$ in.

2. Draw a line segment 6 inches long. Using compasses, lay off in succession on this segment lengths of $1\frac{7}{8}$ inches, $2\frac{3}{16}$ inches, and $1\frac{3}{4}$ inches. How long is the remaining segment?

3. Write your estimate, then measure the length of each of the following line segments, using the ruler only for segments in *a, b,* and *c,* and compasses and ruler for segments in *d* and *e.*

   a. ______________________________

   b. ________

   c. ____________________

   d. _________________________

   e. ____________

4. Using the metric scale, draw line segments having the following dimensions:

   a. 12 cm **b.** 105 mm **c.** 9 cm **d.** 86 mm **e.** 35 mm

5. The symbol $m\overline{DC}$ is read "the measure of line segment DC" and represents the length of the segment.

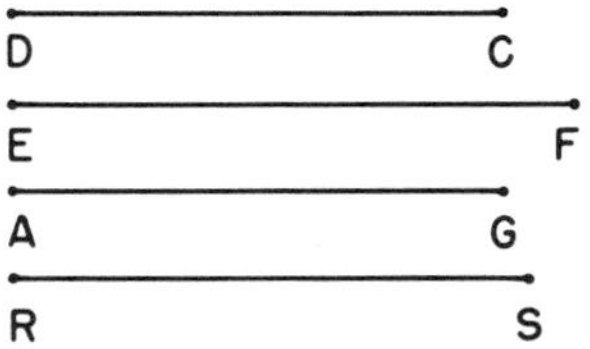

Find $m\overline{DC}$, $m\overline{EF}$, $m\overline{AG}$, $m\overline{RS}$.

Does $m\overline{DC} = m\overline{AG}$?

Does $m\overline{RS} = m\overline{EF}$?

6. Draw any line segment BC. Label a point not on $\overline{BC}$ as M. Draw a line segment MN equal in length to $\overline{BC}$.

7. a. Draw a line segment that is twice as long as $\overline{AG}$ shown in problem 5.

b. Draw a line segment that is three times as long as $\overline{EF}$ shown in problem 5.

14–10 COPYING A LINE SEGMENT

To copy a line segment means to draw a line segment equal in length to the given line segment.

(1) *Using a ruler,* we first measure the given line segment, then we draw another line segment of the same length.

(2) *Using a compass,* we first draw a pencil working line. Then we select a point on this line where the copy of the given line segment is to begin, indicating it by a dot. We open the compass, placing the metal point on one endpoint of the given line segment and the pencil point on the other endpoint. We transfer this fixed compass setting to the working line, placing the metal point on the marked dot.

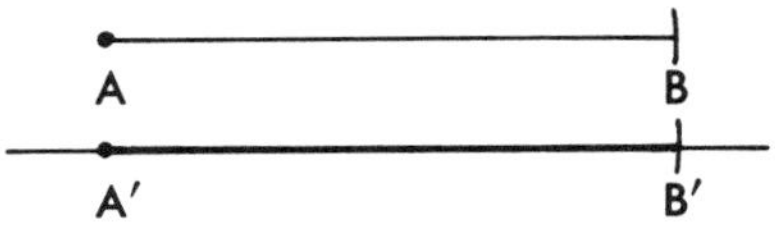

This point forms one endpoint of the copy of the given line segment. We draw an arc, cutting the working line. This intersection forms the second endpoint of the copy of the given line segment. We then draw the segment of the working line between these endpoints heavier to indicate the required line segment.

EXERCISES

1. Make a copy of each of the following line segments, using a ruler:

2. Make a copy of each of the following line segments, using a compass and a straightedge. Check with ruler.

3. Draw any line segment and label the endpoints E and F. Label a point not on $\overline{EF}$ as G. Using a compass and a straightedge, draw a line segment GH equal in length to $\overline{EF}$.

14–11 SCALE

The *scale* shows the relationship between the dimensions of the drawing, plan, or map, and the actual dimensions. A scale like 1 inch = 4 feet may also be written as $\frac{1}{4}$ inch = 1 foot or by the representative fraction $\frac{1}{48}$ or by the ratio 1:48 which shows that each scale inch equals 48 actual inches or 4 feet.

On maps we usually find a scale of miles like:

(1) *To find the actual distance*, we multiply the scale distance by the scale value of a unit (cm in the following):

Scale: 1 cm = 20 km	7.5 × 20 = 150 km
Scale distance 7.5 cm = ? km	*Answer*, 150 km

(2) *To find the scale distance*, we divide the actual distance by the scale value of a unit (inch in the following):

Scale: 1 inch = 8 feet	$8\overline{)18}$ = $2\frac{1}{4}$
? scale distance = 18 feet	*Answer*, $2\frac{1}{4}$ inches

(3) *To find the scale*, we divide the actual distance by the scale distance.

EXERCISES

1. If 1 in. = 30 mi., what distance is represented by: 4 in.? $1\frac{1}{2}$ in.? $\frac{7}{8}$ in.? $2\frac{3}{4}$ in.?
2. If $\frac{1}{8}$ inch = 1 foot, what distance is represented by: 7 in.? $4\frac{1}{2}$ in.? $\frac{3}{4}$ in.? $2\frac{13}{16}$ in.?
3. If the scale is $\frac{1}{48}$, what distance is represented by: 3 in.? $10\frac{1}{2}$ in.? $\frac{5}{8}$ in.? $5\frac{9}{16}$ in.?
4. The scale ratio 1:1,000,000 means 1 scale mm = _ actual mm = _ m = _ km.
5. The scale ratio 1:1,250,000 means 1 scale mm = _ actual mm = _ m = _ km.
6. If the scale is 1:2,000,000, what distance is represented by 1 cm? 25 mm? 3.7 cm? 43 mm?
7. If the scale is 1:1,500,000, what distance is represented by 1 mm? 2 cm? 65 mm? 9.4 cm?
8. If 1 in. = 80 mi., how many in. represent: 560 mi.? 440 mi.? 150 mi.? 65 mi.?
9. If $\frac{1}{4}$ inch = 1 foot, how many inches represent: 32 ft.? 18 ft.? 41 ft.? $25\frac{1}{2}$ ft.?
10. If the scale is $\frac{1}{24}$, how many inches represent: 6 ft.? 21 ft.? $15\frac{1}{2}$ ft.? $29\frac{1}{4}$ ft.?
11. If the scale is 1:2,500,000 how many mm represent: 10 km? 35 km? 60 km? 75 km?
12. If the scale is 1:400,000 how many cm represent: 12 km? 40 km? 56 km? 50 km?
13. If the scale is 1:2,000 how many mm represent: 8 m? 60 m? 42 m? 94 m?
14. Find the scale when: actual length is 350 km and scale length is 7 mm; actual length is 290 mi. and scale length is $3\frac{5}{8}$ in.; actual length is 48 m and scale length is 6 mm.

15. Using the scale 1 in. = 40 mi., draw line segments representing 70 mi.; 105 mi.
16. Using the scale 1:1,000,000 draw line segments representing 25 km; 67 km.
17. Using the scale $\frac{1}{8}$ inch = 1 foot, draw line segments representing 28 feet; 19 feet.
18. What are the actual dimensions of a floor if plans drawn to the scale of $\frac{1}{2}$ inch = 1 foot show scale dimensions of $8\frac{1}{2}$ in. by $9\frac{7}{8}$ in.?

19. Use the scale of miles 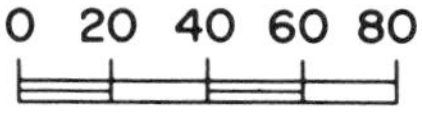

to find the distance represented by each of the following line segments:

a. ____________________ **c.** ______________________________

b. _______________________________ **d.** _______________

e. ____

f. ____

20. Find the scales used to draw the following line segments representing the given distances:

105 mi.
Scale: 1 inch = ?

54 ft.
Scale: 1 inch = ?

108 m
Scale: 1 : ?

360 km
Scale: 1 : ?

21. If the distance from A to B is 102 miles, what is the distance from C to D?

A ________________________________ B C _______________________ D

22. Find the distance from A to B.
From C to A.

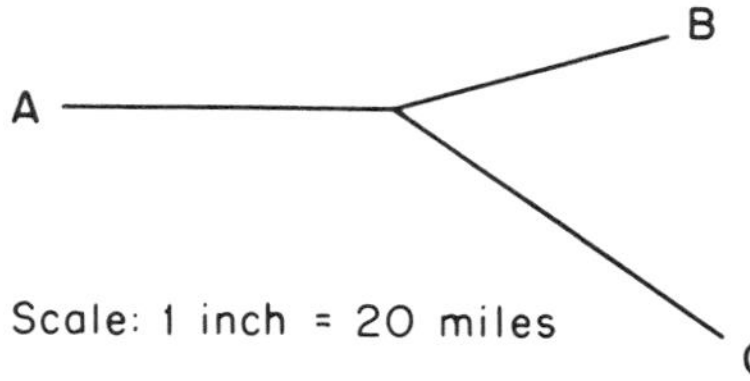

23. Draw a plan of a schoolroom 10 meters long and 9 meters wide, using the scale 1:200.

24. Draw a diagram of a basketball court 75 feet long and 42 feet wide, using the scale $\frac{1}{8}$ inch = 3 feet.

14–12 MEASURING ANGLES

Measuring an angle means to determine how many units of angular measure are contained in it. The protractor is an instrument used to measure an angle.

The size of an angle does not depend on the length of its sides. The symbol "$m\angle ABC$" is read "the measure of angle ABC."

To measure an angle, we place the straight edge of the protractor on one side of the angle with its center mark at the vertex of the angle. We read the number of degrees at the point where the other side of the angle cuts the protractor, using the scale which has its zero on one side of the angle.

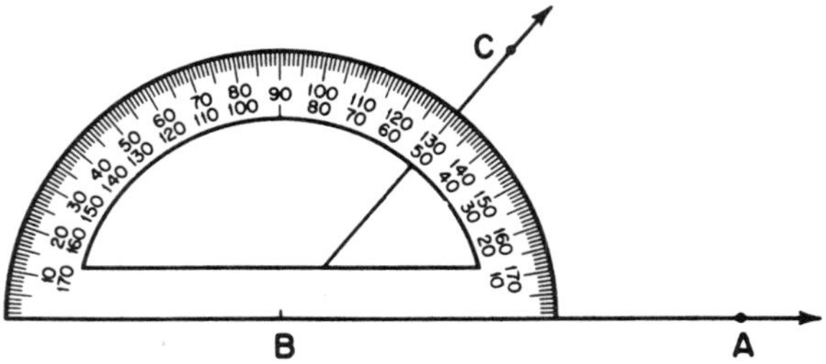

This angle measures 50° written either as $m\angle ABC = 50°$ or briefly as $\angle ABC = 50°$.

EXERCISES

1. Estimate the size of each of the following angles. Then measure each angle with a protractor.

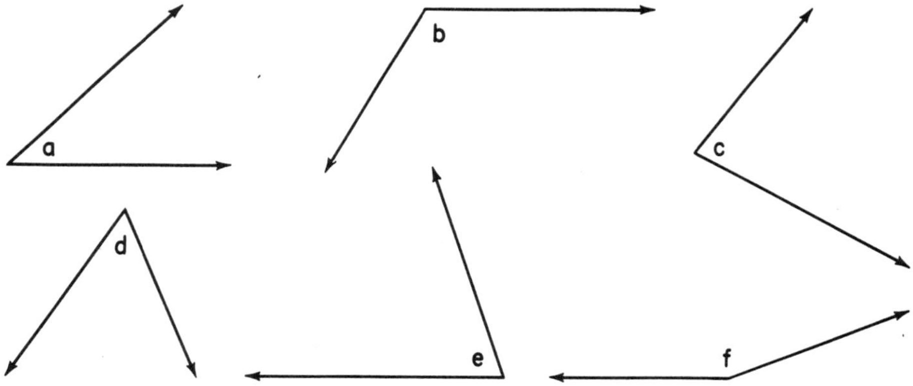

2. Write the amount of error and per cent of error of each estimate you made in Exercise **1.**

3. a. Which of the following angles is smaller?
b. The size of an angle does not depend upon the length of the ?.

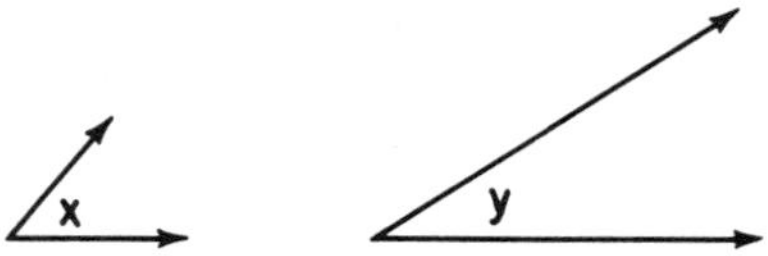

4. How many degrees are in the angle formed by the hands of a clock:

a. at 4 o'clock? **c.** at 3 o'clock?
b. at 2 o'clock? **d.** at 6 o'clock?

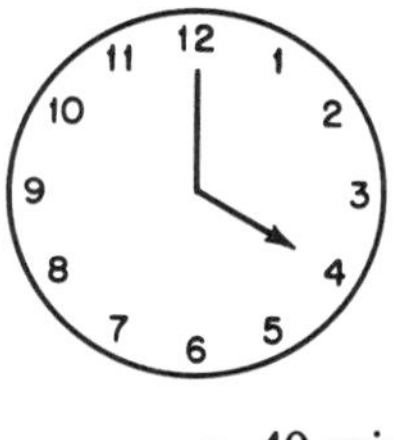

5. Through how many degrees does the minute hand of a clock turn in:

a. 15 minutes? **c.** 10 minutes? **e.** 40 minutes?
b. 30 minutes? **d.** 25 minutes? **f.** 1 hour?

14–13 DRAWING ANGLES

To draw an angle with a given measure, we draw a ray to represent one side of the angle. Sometimes this ray (or line segment which is part of a ray) is already drawn. We place the protractor so that its straight edge falls on this ray, and its center mark is on the endpoint which becomes the vertex of the angle. This vertex may also be any point on the line.

Draw an angle of 30° using *E* as the vertex.

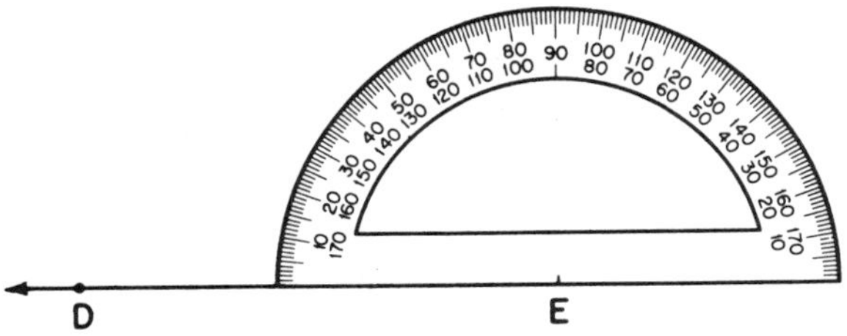

Counting on the scale which has its zero on the ray, we locate the required number of degrees and indicate its position by a dot. We remove the protractor, then we draw a ray from the vertex through this dot.

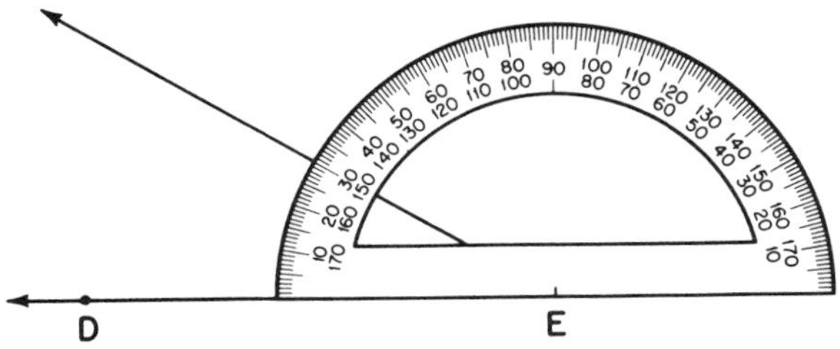

EXERCISES

1. For each of the following first draw a ray with the endpoint either on the left or on the right as required, then with protractor draw the angle of the given measure.

   a. Left endpoint as the vertex:
   40°, 65°, 130°, 155°, 97°, 24°, 109°, 200°, 270°, 345°

   b. Right endpoint as the vertex:
   60°, 75°, 100°, 90°, 84°, 225°, 300°, 340°, 136°, 16°

2. With protractor measure each of the following angles, then draw an angle equal to it.

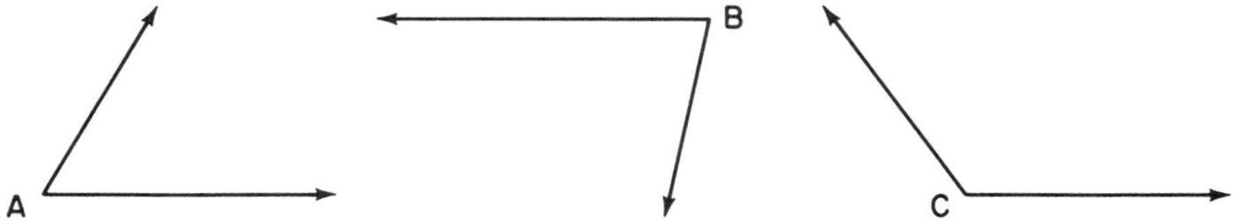

3. With protractor draw: **a.** a right angle **b.** a straight angle

14–14 ANGLES IN NAVIGATION

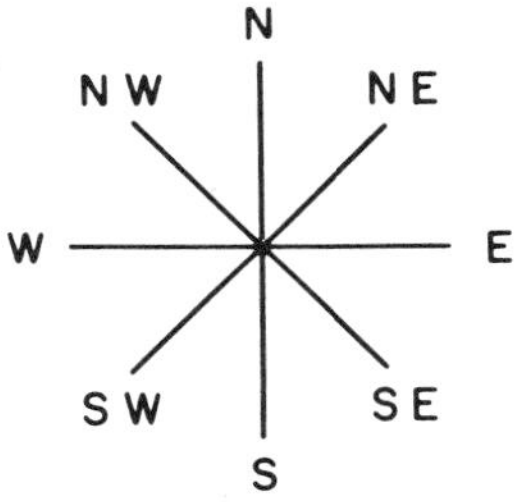

An angle is often used to show the direction (usually called the *course*) in which an airplane flies or the direction from which the wind blows or the direction of one object from another (usually called the *bearing*). Each one of these directions is indicated by an angle measured clockwise from the north direction. This angle is sometimes called an *azimuth*.

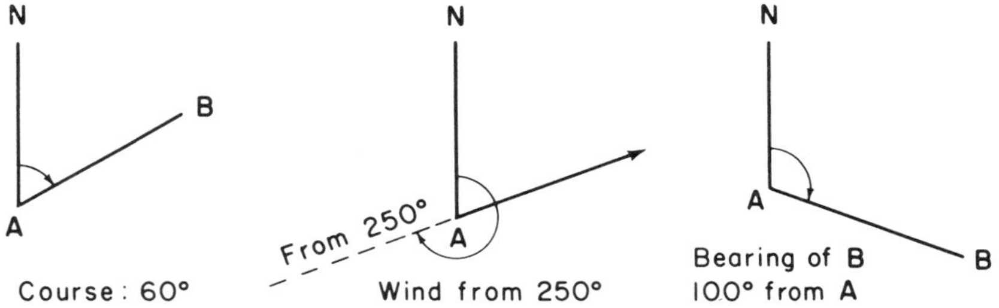

Winds decrease or increase the speed developed by an airplane in flight through the air. *Air speed* is the speed of the airplane in still air. *Ground speed* is the actual speed of the airplane measured by land markings. *Heading* is the direction (expressed by an angle) in which the airplane points. A *vector* is an arrow which represents a speed or force and also indicates its direction. The vector is drawn to scale as one side of the angle which represents the direction measured clockwise from the north.

Position of airplane at the end of 1 hour when flying a heading of 80° at a speed of 200 m.p.h.

Scale 1 inch = 160 m.p.h.

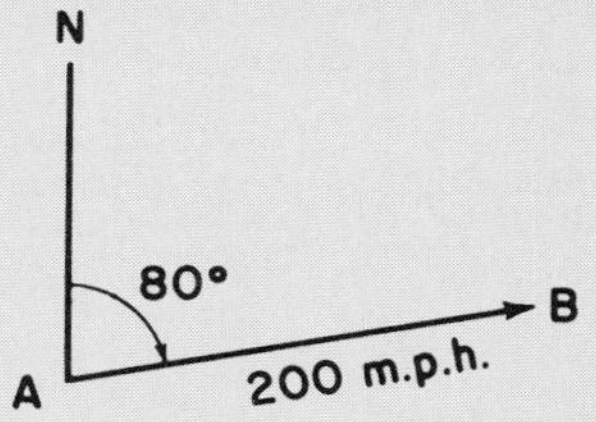

Airplane takes off from point *A*. Reaches point *B* at the end of 1 hour.

EXERCISES

1. If you face north, then turn clockwise to face east, how many degrees do you turn?
2. If you face northeast, then turn clockwise to face northwest, how many degrees do you turn?
3. If you face north, then turn clockwise to face southwest, how many degrees do you turn?
4. First copy north lines, then with protractor draw angles indicating the required direction, measuring clockwise from the north line with point *O* as the vertex:

Due East Due South Southeast Northwest Due West

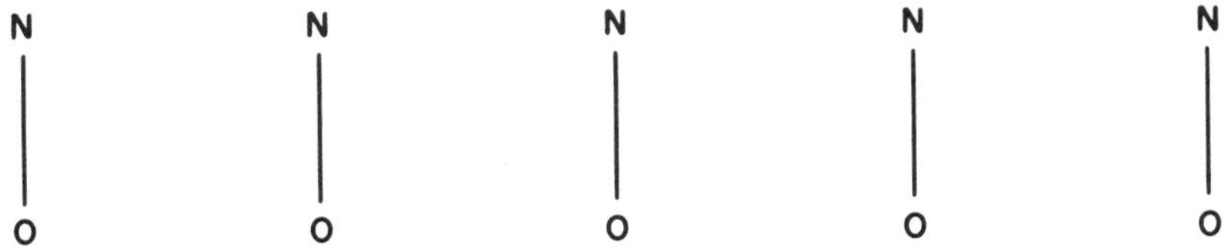

5. Find the course from point *A* to point *B* in each of the following:

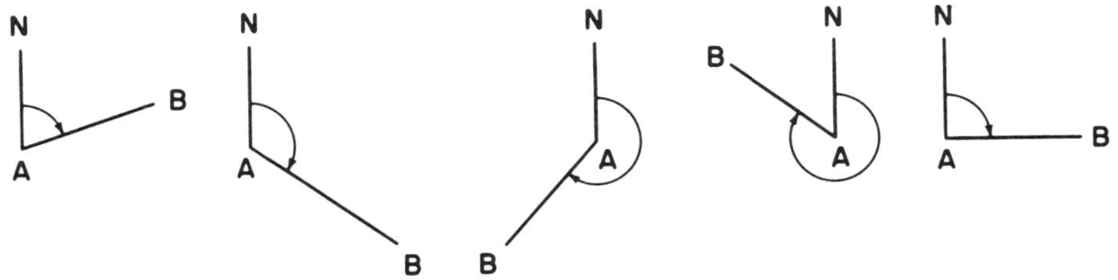

6. Express in degrees each of the following directions from which winds are blowing:

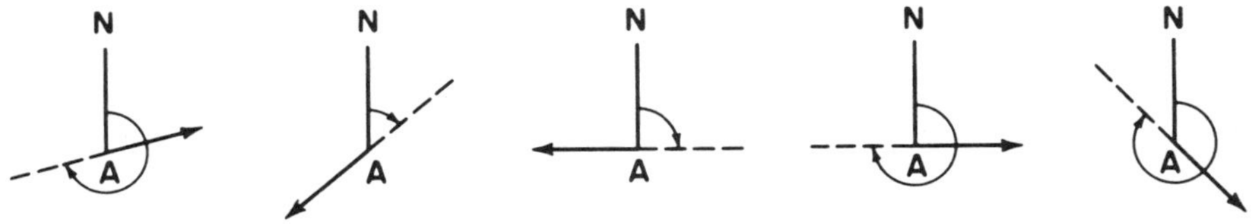

7. Find the bearing of point D from point C in each of the following:

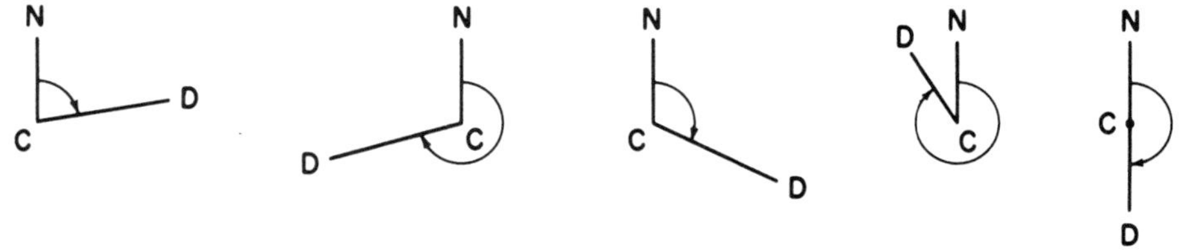

8. a. Draw an angle showing a course of 35°. Of 140°. Of 230°.
b. Draw an angle showing that the wind is blowing from 55°. From 110°. From 345°.

9. Make scale drawings showing the positions of airplanes at the end of one hour flying:
a. Course, 70°; ground speed, 150 m.p.h.; scale 1 in. = 100 mi.
b. Course, 120°; ground speed, 300 m.p.h.; scale 1 in. = 160 mi.
c. Course 200°; ground speed, 475 m.p.h.; scale 1 in. = 200 mi.
d. Heading, 60°; air speed, 180 m.p.h.; scale 1 in. = 120 mi.
e. Heading, 150°; air speed, 410 m.p.h.; scale 1 in. = 160 mi.
f. Heading, 280°; air speed, 570 m.p.h.; scale 1 in. = 240 mi.

10. Draw vectors representing the following wind velocities:
a. 35 m.p.h. wind from 60°, scale 1 in. = 40 m.p.h.
b. 25 m.p.h. wind from 135°, scale 1 in. = 20 m.p.h.
c. 45 m.p.h. wind from 230°, scale 1 in. = 30 m.p.h.

14–15 CONSTRUCTING TRIANGLES

A triangle contains three sides and three angles. A triangle may be constructed when any of the following combinations of three parts are known: (1) Three sides (2) Two sides and an included angle (3) Two angles and an included side. For example:

(1) Three Sides

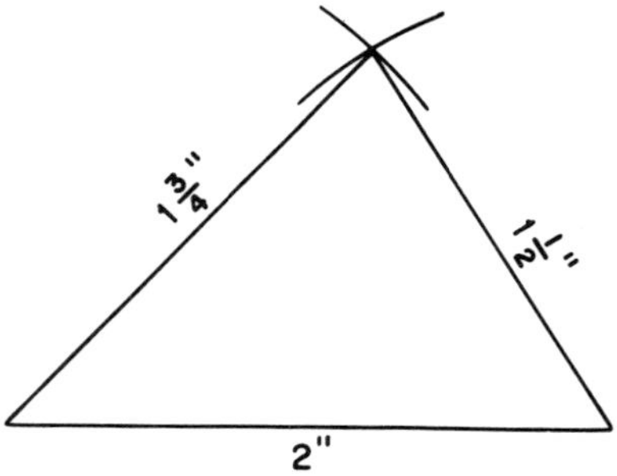

To construct a triangle whose sides measure 2 in., $1\frac{3}{4}$ in., and $1\frac{1}{2}$ in., we first use a compass to lay off a line segment 2 in. long. With one of the endpoints as the center and setting the compass so that the radius is $1\frac{3}{4}$ in., we draw an arc. With the other endpoint as the center and a radius of $1\frac{1}{2}$ in., we draw an arc crossing the first arc. From this point of intersection we draw line segments to the endpoints of the base line to form the required triangle.

EXERCISES

1. Construct triangles having sides that measure:

   a. $2\frac{1}{2}$ in., $1\frac{7}{8}$ in., 2 in. **b.** $2\frac{1}{4}$ in., 3 in., $1\frac{13}{16}$ in. **c.** 42 mm, 28 mm, 31 mm

2. Construct an equilateral triangle whose sides are each 2.6 cm long. Measure the three angles. Are their measures equal?
3. Construct an isosceles triangle whose base is $2\frac{5}{8}$ in. long and each of whose two equal sides is $1\frac{3}{4}$ in. long.
4. Using the scale 1 in. = 160 mi., construct triangles whose sides are:

   a. 240 mi., 180 mi., 260 mi.
   b. 300 mi., 190 mi., 240 mi.
   c. 220 mi., 150 mi., 190 mi.

5. Using the scale 1:4,000,000 construct triangles whose sides are:

   a. 80 km, 100 km, 88 km
   b. 60 km, 52 km, 72 km
   c. 140 km, 120 km, 112 km

6. Construct a triangle with sides equal in length to the following line segments:

7. Construct a scalene triangle. Check whether:

   a. Of any two sides, the side opposite the greater angle is greater.
   b. Of any two angles, the angle opposite the greater side is greater.

(2) Two Sides and an Included Angle

To construct a triangle with sides measuring $1\frac{1}{2}$ in. and $1\frac{1}{8}$ in. and an included angle of 90°, we draw a line segment $1\frac{1}{2}$ in. long. Using the left endpoint as the vertex, we draw an angle of 90°. Along the ray just drawn we measure $1\frac{1}{8}$ in. from the vertex. We then draw a line segment connecting endpoints to form the required triangle.

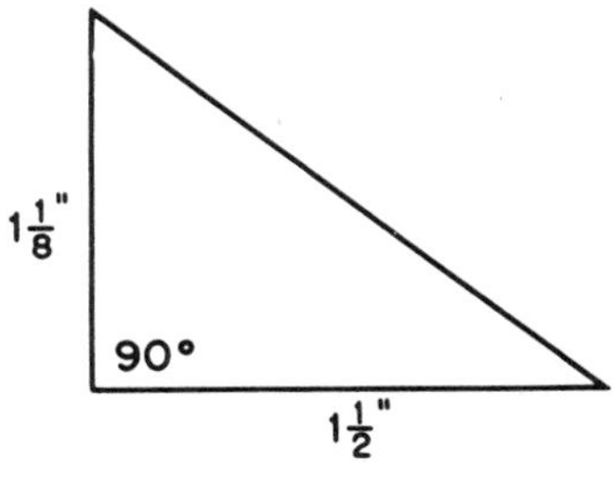

EXERCISES

1. Construct triangles having the following sides and included angles that measure:

   a. 58 mm, 41 mm, 65° **b.** 4 in., $2\frac{7}{8}$ in., 40° **c.** $3\frac{1}{4}$ in., $3\frac{1}{4}$ in., 90°

2. Construct a right triangle having:

   a. A base of 6.8 cm and an altitude of 5.4 cm
   b. A base of $2\frac{11}{16}$ in. and an altitude of $4\frac{1}{4}$ in.

3. Construct an isosceles triangle in which the equal sides each measure $2\frac{5}{8}$ in. and the vertex angle formed by these sides measures 56°. Measure the angles opposite the equal sides. Are their measures equal?

4. Construct a triangle having two sides and an included angle equal to the following two line segments and angle. Also see sections 14–9, 14–10, and 14–19.

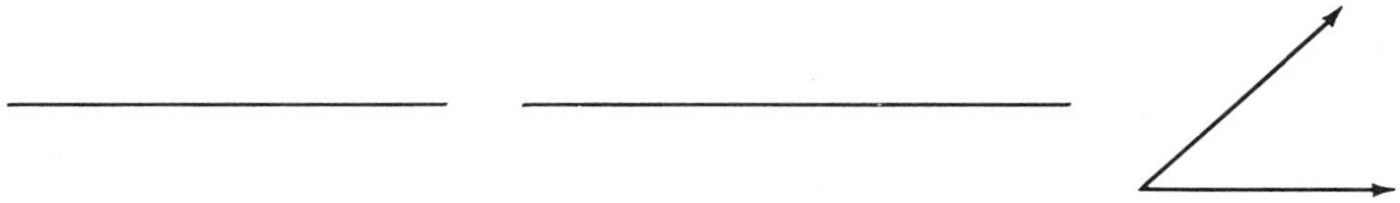

5. Using the scale $\frac{1}{4}$ in. = 1 ft., construct a triangle in which two sides and the included angle measure:

a. 12 ft., 17 ft., and 35° **b.** 21 ft., 18 ft., and 80° **c.** 19 ft., 14 ft., and 60°

(3) Two Angles and an Included Side

To construct a triangle with angles measuring 37° and 27° and the included side measuring $1\frac{7}{8}$ in., we first draw a line segment $1\frac{7}{8}$ in. long. Using the left endpoint as the vertex, we draw an angle of 37°. Using the right endpoint as the vertex, we draw an angle of 27°. We extend the sides until they meet to form the required triangle.

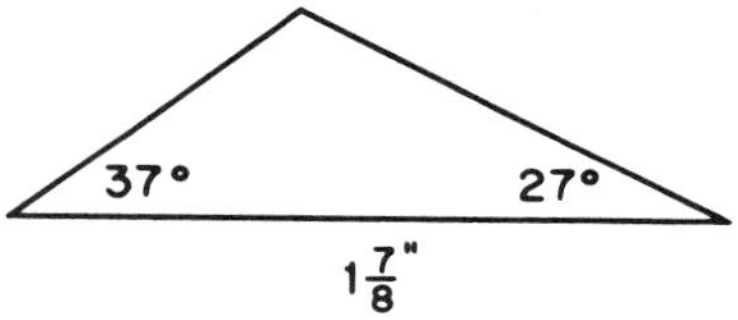

EXERCISES

1. Construct triangles having the following angles and included sides that measure:

a. 80°, 35°, $2\frac{1}{2}$ in. **b.** 110°, 20°, $1\frac{11}{16}$ in. **c.** 45°, 90°, 68 mm

2. Construct a triangle having two equal angles each measuring 40° and the included side measuring $2\frac{3}{4}$ in. Measure the sides opposite the equal angles. Are their measures equal?

3. Construct a triangle having two equal angles each measuring 60° and the included side measuring $3\frac{3}{8}$ in. Measure the third angle. Measure the other two sides. Is the triangle equiangular? Equilateral?

4. Construct a triangle having two angles and an included side equal to the following angles and line segment. Also see sections 14–9, 14–10, and 14–19.

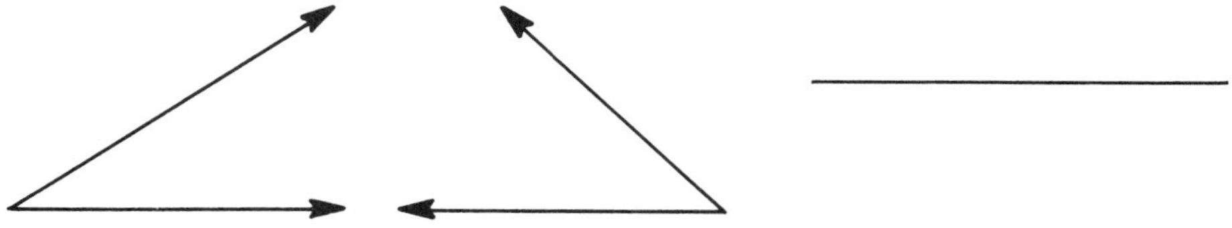

5. Using the scale $\frac{1}{8}$ in. = 1 ft., construct a triangle in which two angles and an included side measure:

a. 45°, 45°, and 20 ft. **b.** 30°, 70°, and 25 ft. **c.** 105°, 40°, and 31 ft.

6. Construct right triangles having the following angles and included sides that measure:

a. 90°, 30°, $2\frac{5}{8}$ in. **b.** 90°, 30°, 7.2 cm **c.** 90°, 30°, $1\frac{15}{16}$ in.

Also select your own measurement for the included side and, using the measures of 90° and 30° for the angles, construct a right triangle. In each of these triangles measure the hypotenuse and the side opposite the 30° angle. Then compare these measurements. In each case is the hypotenuse twice as long as the side opposite the 30° angle?

7. Construct two triangles, each containing two angles measuring 50° and 75°. The included side in the first triangle measures 2 inches and in the second triangle it measures $2\frac{1}{2}$ inches. Are the triangles congruent? Are they similar? See pages 404, 431, and 432.

14–16 CONSTRUCTING A PERPENDICULAR TO A GIVEN LINE AT OR THROUGH A GIVEN POINT ON THE GIVEN LINE

Two lines (or rays or segments) that meet to form right angles are called *perpendicular lines* (or rays or segments). Each line is said to be perpendicular to the other. See page 397. The symbol "⊥" means "is perpendicular to."

(1) *Using a protractor,* we draw a 90° angle with the given point on the line as the vertex. The ray drawn to form the angle is perpendicular to the given line.

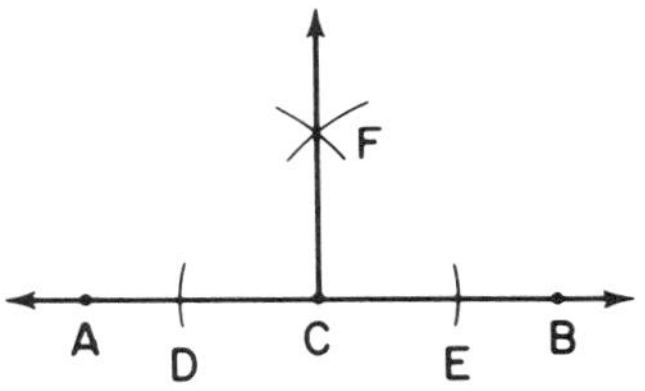

(2) *Using a compass* (see figure): To draw a line (or ray or segment) perpendicular to $\overleftrightarrow{AB}$ at C, we use point C as the center and with any radius we draw an arc cutting $\overleftrightarrow{AB}$ at D and E. With D and E as centers and with a radius greater than $\overline{CD}$, we draw arcs crossing at F. We draw $\overrightarrow{CF}$ which is perpendicular to $\overleftrightarrow{AB}$ at point C. Or we may draw the line FC passing through point C.

EXERCISES

1. Draw any line. Select a point on this line. Construct a line perpendicular to the line you have drawn at your selected point.
2. Construct a rectangle $2\frac{1}{4}$ inches long and $1\frac{13}{16}$ inches wide.
3. Construct a square with each side 46 mm long.
4. Make a plan of a room 42 ft. by 31 ft., using the scale 1 in. = 8 ft.

5. Draw a circle. Draw any diameter of this circle. At the center construct another diameter perpendicular to the first diameter, dividing the circle into four arcs. Are these four arcs equal in length? Use a compass to check.
6. Draw a circle. Select any point on this circle and label it A. Construct a line perpendicular to the radius at point A. A *tangent* to a circle is a line that has one and only one point in common with the circle. Observe that a line that is perpendicular to the radius at a point on the circle is tangent to the circle.

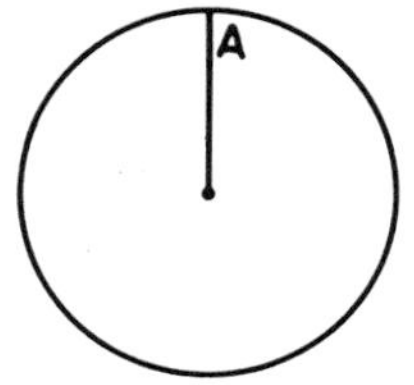

14–17 CONSTRUCTING A PERPENDICULAR TO A GIVEN LINE FROM OR THROUGH A GIVEN POINT NOT ON THE GIVEN LINE

Using a compass (see figure): To draw a ray from point C perpendicular to $\overleftrightarrow{AB}$, we use point C as the center and draw an arc cutting $\overleftrightarrow{AB}$ at D and E. With D and E as centers and a radius of more than one-half the distance from D to E, we draw arcs crossing at F. We draw $\overrightarrow{CF}$ which is perpendicular to $\overleftrightarrow{AB}$. Or we may draw the line FC passing through point C.

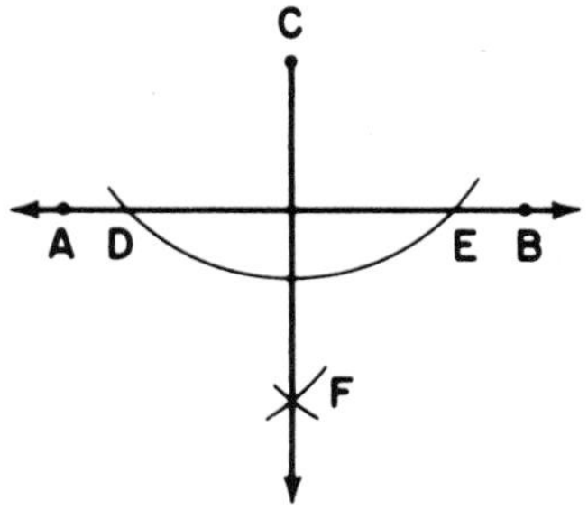

EXERCISES

1. Draw any line. Select a point not on this line. Construct a perpendicular to the line you have drawn from your selected point.
2. Construct an equilateral triangle with each side $3\frac{3}{8}$ inches long. From each vertex construct a perpendicular to the opposite side. Is each angle bisected? Is each side bisected?
3. Draw any acute triangle. From each vertex construct a perpendicular to the opposite side. What do these perpendicular line segments represent in a triangle? Are they concurrent?
4. Draw any right triangle. Construct the altitude to each side. Are they concurrent? If so, what is the common point?
5. Draw a circle. Draw any chord in this circle except the diameter. Construct a perpendicular from the center of the circle to this chord. Check whether the chord is bisected. Extend the perpendicular line segment so that it intersects the circle. Check with a compass whether the arc corresponding to the chord is also bisected. Observe that a radius of a circle that is perpendicular to a chord bisects the chord and its corresponding arc.

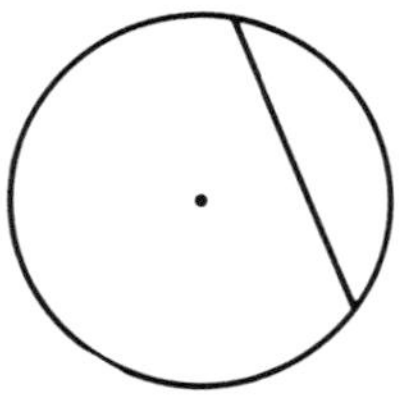

14–18 BISECTING A LINE SEGMENT

To *bisect a line segment* means to divide it into two equal parts. The point on a line segment that separates the line segment into two equal parts is called the *midpoint* of the line segment.

(1) *Using a ruler*, we first measure the line segment, then we mark off half the measurement.

(2) *Using a compass* (see figure): To bisect $\overline{AB}$, we set the compass so that the radius is more than half the length of $\overline{AB}$. With A and B as centers we draw arcs which cross above and below the segment at C and D. Then we draw $\overleftrightarrow{CD}$ bisecting $\overline{AB}$ at E. Observe that $\overleftrightarrow{CD}$ is also perpendicular to $\overline{AB}$. Thus a line like $\overleftrightarrow{CD}$, which both bisects a line segment and is perpendicular to it, is called the *perpendicular bisector* of the given segment.

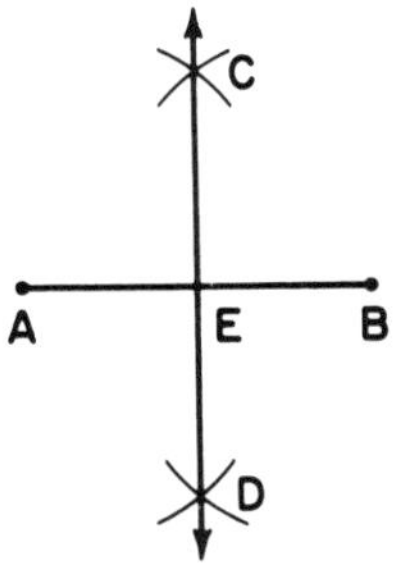

EXERCISES

1. Draw line segments: **a.** $3\frac{3}{4}''$ long. **b.** $2\frac{15}{16}''$ long. **c.** 5.8 cm long. Bisect each line segment, using a compass. Check with a ruler.

2. Copy, then using a compass, bisect each of the following segments. Check each with a ruler.

3. Copy, then using a compass, divide the following line segment into four equal parts. Check with a ruler.

4. Draw any line segment. Bisect it, using a compass. Then use a protractor to check whether each of the four angles formed is a right angle. What name do we give to a line that both bisects and is perpendicular to a line segment?

5. Draw any triangle. Bisect each side by constructing the perpendicular bisector of that side. Are these perpendicular bisectors concurrent? Do they meet in a point equidistant from the vertices of the triangle? Check by measuring. Using this common point as the center and the distance from this point to any vertex of the triangle as the radius, draw a circle through the three vertices of the triangle.

When each side of the triangle is a chord of the circle or each vertex is a point on the circle, we say that the *circle is circumscribed about the triangle* or that the *triangle is inscribed in a circle*.

6. Draw any triangle. Find the midpoint of each side by constructing the perpendicular bisector of that side. Draw the median from each vertex to the midpoint of the opposite side. Are these medians concurrent? Along each median check whether the distance from this common point to the vertex is twice the distance from this common point to the midpoint of the opposite side.

7. Draw a right triangle. Bisect the hypotenuse of the right triangle. Draw the median from the vertex of the right angle to the hypotenuse. Check whether this median is one-half as long as the hypotenuse.

14–19 COPYING A GIVEN ANGLE

To copy a given angle means to construct an angle equal in size to the given angle.

(1) *Using a protractor,* we measure the given angle and draw another angle of the same size.

(2) *Using a compass* (see figure): To construct an angle at point C on $\overleftrightarrow{AB}$ equal to $\angle MNO$, we take point N as center and draw an arc cutting side MN at P and side NO at Q. With the same radius and point C as center, we draw an arc cutting $\overleftrightarrow{AB}$ at D. With a radius equal to PQ and point D as center, we draw an arc crossing the first arc at E. We draw $\overrightarrow{CE}$. $m\angle BCE$ is equal to $m\angle MNO$.

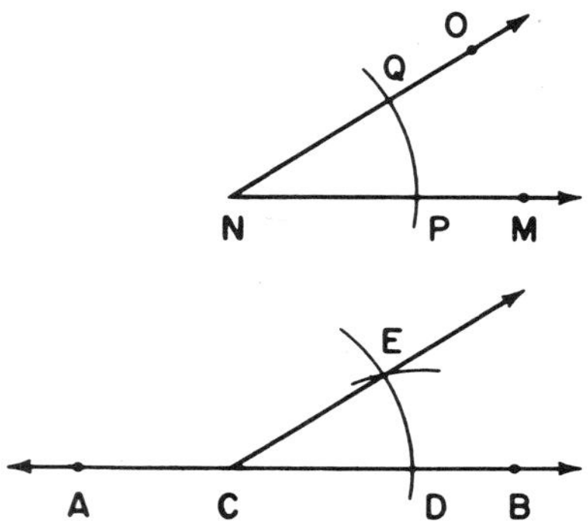

EXERCISES

1. Draw angles having the following measures, using a protractor. For each angle construct with a compass an angle of equal size. Check your copy of the angle with a protractor.

   a. 80° **b.** 130° **c.** 45° **d.** 23° **e.** 157° **f.** 90°

2. Draw an acute angle. Construct with a compass an angle of equal size. Check both angles with a protractor.
3. Draw a right angle, using a protractor. Construct with a compass an angle of equal size. Check with a protractor.
4. Draw any obtuse angle. Construct with a compass an angle of equal size. Check both angles with a protractor.
5. Draw any angle. Construct with a compass an angle having the same measure. Check both angles with a protractor.

14–20 BISECTING AN ANGLE

To *bisect an angle* means to divide it into two equal angles.

(1) *Using a protractor,* we measure the given angle and mark off one-half the measurement. We then draw a ray from the vertex.

(2) *Using a compass* (see figure): To bisect $\angle ABC$ with B as the center and any radius, we draw an arc cutting side AB at D and side BC at E. With D and E as centers and a radius of more than half the distance from D to E, we draw arcs crossing at F. We then draw $\overrightarrow{BF}$ bisecting $\angle ABC$.

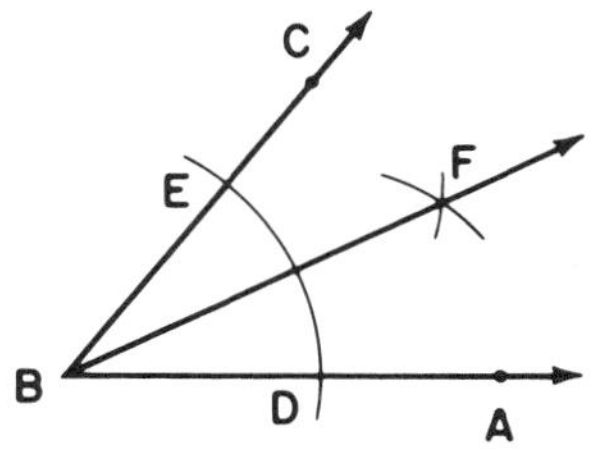

EXERCISES

1. Draw angles having the following measures. Bisect each angle using a compass. Check with a protractor.

   a. 60° **b.** 150° **c.** 48° **d.** 165° **e.** 75° **f.** 121°

2. Draw any angle. Bisect it by using a compass only. Check by measuring your angle and each bisected angle.
3. Copy, then bisect each of the following angles:

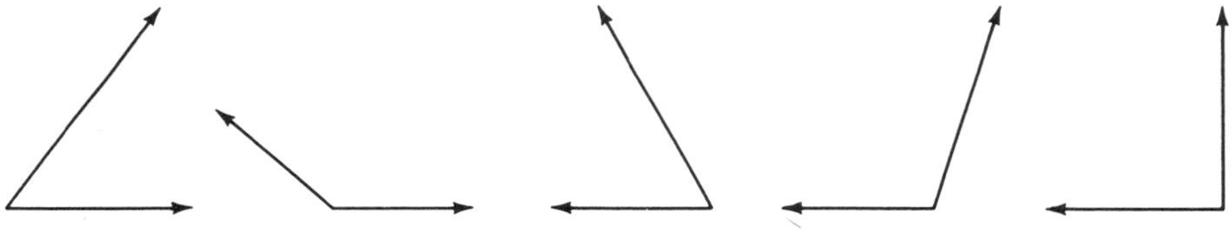

4. Draw any angle. Using a compass only, divide the angle into four equal angles. Check with protractor.
5. Draw any triangle. Bisect each angle. Do the bisectors meet at a common point? Are these bisectors concurrent? From this common point draw a perpendicular to any side. With this common point as center and with the perpendicular distance from the common point to any side as the radius, draw a circle. Observe that each side of the triangle is tangent to the circle.

   When each side of the triangle is tangent to the circle, we say that the *triangle is circumscribed about the circle* or that the *circle is inscribed in the triangle*.
6. Construct with a compass an angle measuring: **a.** 45° **b.** 135°
7. Draw any equilateral triangle. Select an angle and bisect it. Construct from the vertex of this selected angle the altitude (line segment that is perpendicular to the opposite side). Does this perpendicular line bisect this opposite side? From this same vertex also draw the median to the opposite side. In an equilateral triangle, are all three of these lines (angle bisector, altitude, and median) one and the same line? If this is so, will it be true if each of the other angles is selected? Check.
8. Draw any isosceles triangle. Bisect the vertex angle (angle formed by the two equal sides and opposite to the base). Then construct from the vertex of this angle the altitude (line segment that is perpendicular to the base). Does this perpendicular line bisect the base of the isosceles triangle? From this same vertex also draw the median to the base. In an isosceles triangle, are all three of these lines (bisector of the vertex angle, altitude drawn to the base, and the median drawn to the base)

one and the same line?

In an isosceles triangle the angles opposite the equal sides are called *base angles*. Check whether the bisector of a base angle, the altitude and the median both drawn from the vertex of the same base angle are all one and the same line.

9. Draw any scalene triangle and do the following:

a. Select an angle and bisect it.

b. Construct from the vertex of this bisected angle the line segment that is perpendicular (altitude) to the opposite side.

c. Construct the perpendicular bisector of the side opposite to the bisected angle.

d. Draw the median from the vertex of this bisected angle to the opposite side. Are any of these four lines (angle bisector, altitude, perpendicular bisector, and median) the same line? If so, which?

14–21 CONSTRUCTING A LINE PARALLEL TO A GIVEN LINE THROUGH A GIVEN POINT NOT ON THE GIVEN LINE

Lines in the same plane which do not meet are called *parallel lines*. The symbol "$\|$" means "is parallel to."

To construct a line parallel to $\overleftrightarrow{AB}$ through point C (see figure), we draw any line $\overleftrightarrow{DE}$ through C meeting $\overleftrightarrow{AB}$ at F.

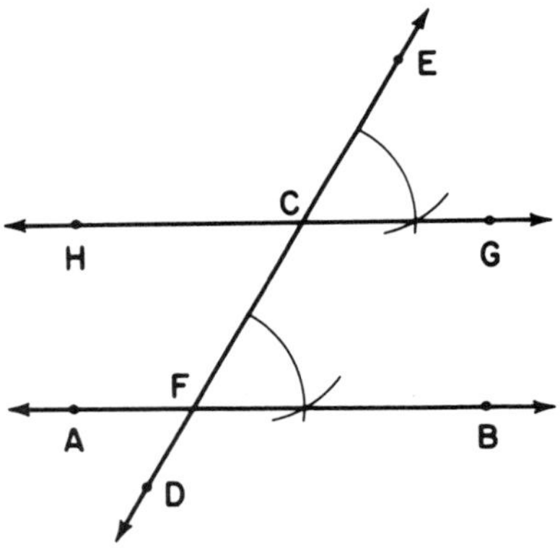

(1) *Using a protractor*, we measure $\angle BFC$ and draw, at point C on $\overleftrightarrow{DE}$, $\angle GCE$ equal to the corresponding $\angle BFC$. Then we extend $\overleftrightarrow{GC}$ through H. $\overleftrightarrow{HG}$ is parallel to $\overleftrightarrow{AB}$.

(2) *Using a compass*, we construct, at point C on $\overleftrightarrow{DE}$, $\angle GCE$ equal to the corresponding $\angle BFC$ by following the procedure explained in the preceding construction (14–19). We then extend GC through H. HG is parallel to AB.

EXERCISES

1. Draw any line. Select a point that is not on this line. Through this point construct a line parallel to the line you have drawn.

2. Construct a parallelogram with a base $3\frac{5}{16}$ inches long, a side $2\frac{7}{8}$ inches long, and an included angle of 55°.

3. Draw any acute triangle. Bisect one of the sides. Draw a line segment from this midpoint, parallel to one of the other sides, until it intersects the third side. Does this line segment bisect the third side? Check by measuring. Also check whether this line segment is one-half as long as the side to which it is parallel. Draw other triangles and check whether the above findings are true no matter which side is used first or which side is used as the parallel side.

14–22 CONSTRUCTING REGULAR POLYGONS AND OTHER POLYGONS

A *regular polygon* is a polygon that is both equilateral (all of its sides are of equal length) and equiangular (all of its angles are of equal size). An *inscribed polygon in a circle* is a polygon whose vertices are points on the circle.

Although there are other ways to construct some of the regular polygons, in general we use a method that is based on the geometric fact that equal central angles of a circle intercept equal arcs and equal chords. A regular polygon constructed in this way is inscribed within a circle.

Therefore *to construct a regular polygon*, we first draw a circle. Then we divide this circle into the same number of equal arcs as there are sides in the required polygon by drawing a corresponding number of equal central angles. We draw line segments (chords) connecting the points of division to form the polygon.

To draw a regular octagon which is a polygon of eight equal sides, we first determine the measure of each of the eight equal central angles by dividing 360° by 8. This measure is 45°. We draw eight central angles each measuring 45°, with its sides (radii) intercepting the circle dividing it into eight equal arcs. We then draw line segments to connect the points of division to form the regular octagon.

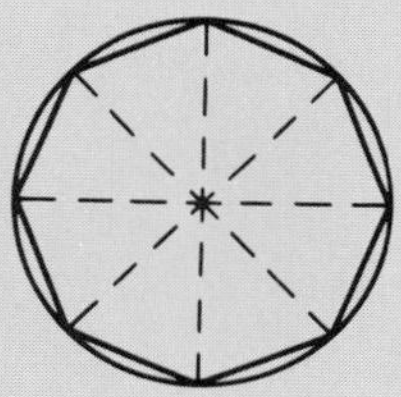

A regular hexagon (6 sides) may be drawn by the alternate way of dividing the circle into six equal arcs by using the radius of the circle as the radius of the arc. An equilateral triangle may be constructed by drawing line segments to connect alternate points of division after the circle is divided into six equal arcs.

A square may be constructed by drawing two diameters of a circle perpendicular to each other to divide the circle into four equal arcs. Line segments are then drawn to connect these points of division.

Rectangles and parallelograms of specific measurement may be constructed provided enough of these measurements are given so that basic constructions may be used.

EXERCISES

1. Construct each of the following regular polygons:

a. Hexagon—6 sides
b. Pentagon—5 sides
c. Square
d. Decagon—10 sides
e. Dodecagon—12 sides
f. Equilateral triangle

2. Construct a rectangle:

a. $2\frac{1}{2}$ inches long and $1\frac{7}{8}$ inches wide **c.** 8.2 cm long and 5.7 cm wide
b. $3\frac{5}{16}$ inches long and $2\frac{3}{4}$ inches wide **d.** $3\frac{11}{16}$ inches long and $2\frac{3}{8}$ inches wide

3. Construct a square whose side measures:

a. $2\frac{1}{2}$ inches **b.** $1\frac{9}{16}$ inches **c.** $\frac{3}{4}$ inch **d.** 75 mm

4. Construct a parallelogram with:

a. A base $2\frac{5}{8}$ inches long, a side $1\frac{13}{16}$ inches long, and an included angle of 45°
b. A base $3\frac{3}{4}$ inches long, a side $2\frac{7}{8}$ inches long, and an included angle of 70°
c. A base 4 inches long, a side $3\frac{11}{16}$ inches long, and an included angle of 60°
d. A base 5 cm long, a side 8.6 cm long, and an included angle of 110°

5. Draw a regular hexagon, each side measuring: **a.** $1\frac{3}{4}$ inches **b.** 44 mm **c.** $3\frac{5}{16}$ inches. In each case draw a circle whose radius has the same measure as each required side.

14-23 SUM OF ANGLES OF POLYGONS

(1) Triangle

Draw any triangle. Measure its three angles. What is the sum of the measures of these three angles? Draw a second triangle. Find the sum of the measures of its three angles. Find the sum of the measures of the three angles of a third triangle.

Do you see that *the sum of the measures of the angles of any triangle is 180°*?

(a) In each of the following, find the measure of the third angle of a triangle when the other two angles measure:

1. 56° and 47° 3. 63° and 84° 5. 4° and 141°
2. 112° and 39° 4. 18° and 25° 6. 83° and 83°

(b) What is the measure of each angle of an equiangular triangle?
(c) If each of the equal angles of an isosceles triangle measures 48°, find the measure of the third angle.
(d) If the vertex angle of an isosceles triangle measures 80°, find the measure of each of the other two angles.
(e) Why can there be only one right or one obtuse angle in a triangle? What kind of angle must each of the other two angles be?
(f) In a right triangle if one of the acute angles measures 51°, what is the measure of the other acute angle?
(g) If two angles of one triangle are equal respectively to two angles of another triangle, why are the third angles equal?

(2) Quadrilateral

Draw a parallelogram and a trapezoid. Measure the four angles in each. What is the sum of the measures of the angles of the parallelogram? Of the trapezoid?

Draw any other quadrilateral. Find the sum of the measures of the angles of the quadrilateral.

Do you see that *the sum of the measures of the angles of any quadrilateral is 360°*?

(a) What is the sum of the measures of the angles of a rectangle? Of a square?

(b) In each of the following, find the measure of the fourth angle of a quadrilateral when the other three angles measure:

1. 56°, 83°, and 108°   2. 121°, 90°, and 90°   3. 115°, 17°, and 154°

(c) The opposite angles of a parallelogram are equal. If one angle measures 68°, find the measures of the other three angles.

(d) Three angles of a trapezoid measure 90°, 90°, and 117°. What is the measure of the fourth angle?

(3) Other Polygons

Draw a pentagon, a hexagon, an octagon, and a decagon. What is the sum of the measures of the angles of each of these geometric figures? Check whether your sum of measures of the angles of each of these figures matches the angular measure determined by substituting the number of sides in the polygon for n in the expression $180(n - 2)$ and performing the required operations. Is this also true for the triangle and the quadrilateral?

Do you see that *the sum of measures of the angles of a polygon of n sides is* $180(n - 2)$?

14–24 PAIRS OF ANGLES

(1) *Complementary angles* are two angles whose sum of measures is 90°.

(a) Which of the following pairs of angles are complementary?

1. $m\angle B = 60°$, $m\angle D = 30°$
2. $m\angle A = 22°$, $m\angle T = 68°$
3. $m\angle 5 = 53°$, $m\angle 7 = 47°$
4. $m\angle r = 21°$, $m\angle s = 59°$
5. $m\angle E = 74°$, $m\angle F = 16°$
6. $m\angle 1 = 100°$, $m\angle 2 = 80°$

(b) Find the measure of the angle that is the complement of each of the following angles:

1. $m\angle P = 27°$
2. $m\angle 3 = 81°$
3. $m\angle t = 55°$
4. $m\angle A = 8°$
5. $m\angle R = 39°$
6. $m\angle B = 64°$

(c) Angle ABC is a right angle.

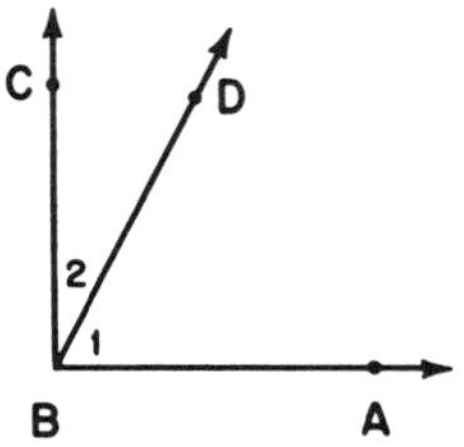

1. If $m\angle 1 = 75°$, find the measure of $\angle 2$.
2. If $m\angle 2 = 24°$, find the measure of $\angle 1$.
3. If $m\angle 1 = 63°$, find the measure of $\angle 2$.
4. If $m\angle 2 = 11°$, find the measure of $\angle 1$.

(d) In the figure $\overrightarrow{FB} \perp \overrightarrow{FD}$ and $\overrightarrow{FC} \perp \overleftrightarrow{AE}$.

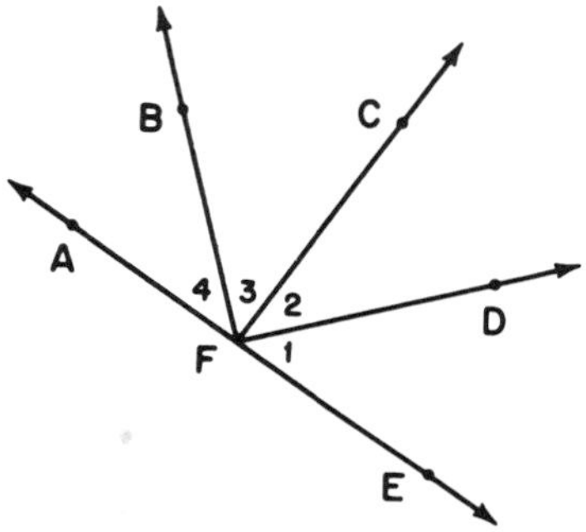

1. If $m\angle 1 = 52°$, find the measure of $\angle 2$. Of $\angle 3$. Of $\angle 4$.
2. If $m\angle 4 = 69°$, find the measure of $\angle 1$. Of $\angle 2$. Of $\angle 3$.
3. If $m\angle 2 = 33°$, find the measure of $\angle 3$. Of $\angle 1$. Of $\angle 4$.
4. If $m\angle 3 = 78°$, find the measure of $\angle 1$. Of $\angle 2$. Of $\angle 4$.

(e) Why is the complement of an acute angle also an acute angle?

(2) *Supplementary angles* are two angles whose sum of measures is 180°.

(a) Which of the following pairs of angles are supplementary?

1. $m\angle B = 125°$, $m\angle R = 55°$
2. $m\angle 1 = 83°$, $m\angle 2 = 107°$
3. $m\angle G = 92°$, $m\angle H = 78°$
4. $m\angle a = 47°$, $m\angle b = 133°$
5. $m\angle M = 51°$, $m\angle N = 39°$
6. $m\angle S = 86°$, $m\angle T = 94°$

(b) Find the measure of the angle that is the supplement of each of the following angles:

1. $m\angle D = 67°$
2. $m\angle b = 102°$
3. $m\angle 3 = 5°$
4. $m\angle T = 90°$
5. $m\angle L = 148°$
6. $m\angle E = 39°$

(c) Measure $\angle CDF$ and $\angle EDF$. Is the sum of their measures 180°? Do you see that *when one straight line meets another, the adjacent angles, which have the same vertex and a common side, are supplementary*?

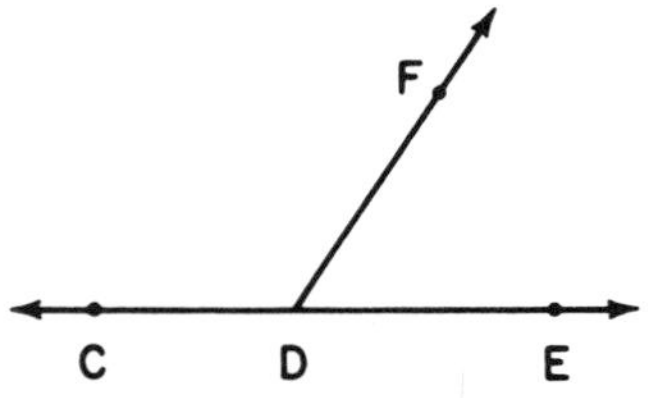

1. Find the measure of $\angle EDF$ when $m\angle CDF = 120°$.
2. What is the measure of $\angle CDF$ when $m\angle EDF = 42°$?

3. What is the measure of $\angle EDF$ when $m\angle CDF = 157°$?
4. Find the measure of $\angle CDF$ when $m\angle EDF = 90°$.

(d) Is the supplement of an obtuse angle also an obtuse angle? Explain your answer.

(3) Opposite or Vertical Angles

Draw two intersecting lines, forming four angles. Measure a pair of angles that are directly opposite to each other. Are their measures equal? Measure the other pair of opposite angles. Are their measures equal?

Do you see that, *when two straight lines intersect, the opposite or vertical angles are equal*?

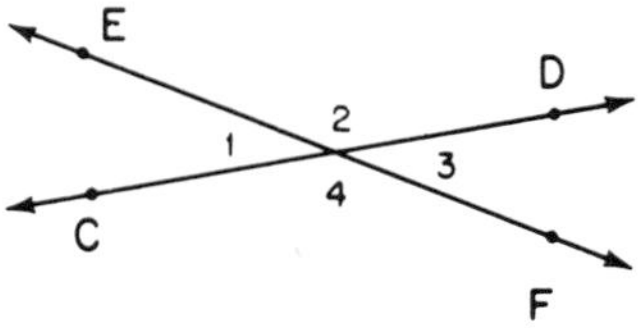

(a) In the drawing at the right, what angle is opposite to $\angle 1$? Does $\angle 1 = \angle 3$? What angle is opposite to $\angle 4$? Does $\angle 4 = \angle 2$?
(b) If $m\angle 4 = 106°$, what is the measure of $\angle 2$? Of $\angle 1$? Of $\angle 3$?
(c) If $m\angle 3 = 59°$, what is the measure of $\angle 1$? Of $\angle 2$? Of $\angle 4$?
(d) If $m\angle 2 = 132°$, what is the measure of $\angle 3$? Of $\angle 4$? Of $\angle 1$?
(e) If $m\angle 1 = 27°$, what is the measure of $\angle 4$? Of $\angle 2$? Of $\angle 3$?

(4) Exterior of a Triangle

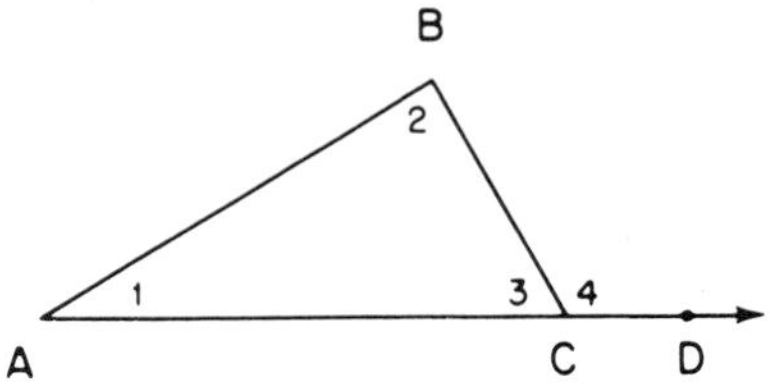

Draw a triangle and extend one side like the drawing at the right. Label the angles as shown in the drawing.

The angle formed by a side of the triangle and the adjacent side extended (see $\angle 4$ in the drawing) is called the *exterior angle*.

Measure the three angles of the triangle and the indicated exterior angle in your drawing.

What is the sum of the measures of $\angle 1$, $\angle 2$, and $\angle 3$? What is the sum of the measures of $\angle 4$ and $\angle 3$?

Why does the measure of $\angle 4$ equal the sum of the measures of $\angle 1$ and $\angle 2$?

Do you see that *the measure of an exterior angle of a triangle is equal to the sum of the measures of the opposite two interior angles*?

(a) If $m\angle 1 = 37°$ and $m\angle 2 = 68°$, find the measure of $\angle 4$. Of $\angle 3$.
(b) If $m\angle 3 = 43°$ and $m\angle 1 = 74°$, find the measure of $\angle 2$. Of $\angle 4$.
(c) If $m\angle 2 = 81°$ and $m\angle 3 = 79°$, find the measure of $\angle 1$. Of $\angle 4$.
(d) If $m\angle 4 = 116°$ and $m\angle 1 = 60°$, find the measure of $\angle 2$. Of $\angle 3$.
(e) If $m\angle 2 = 58°$ and $m\angle 4 = 135°$, find the measure of $\angle 3$. Of $\angle 1$.
(f) Can an exterior angle of a triangle have the same measure as one of the

angles of the triangle? As one of the opposite interior angles of the triangle? Explain your answers.

14–25 PARALLEL LINES AND ANGLE RELATIONSHIPS

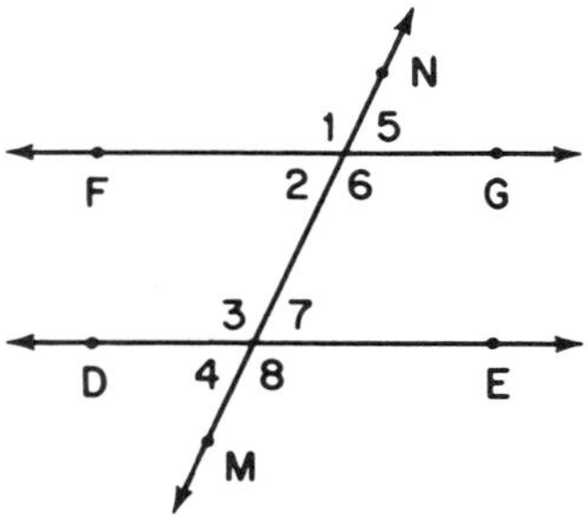

In the figure at the right, $\overleftrightarrow{DE}$ and $\overleftrightarrow{FG}$ are parallel and cut by $\overleftrightarrow{MN}$; $\angle 2$ and $\angle 7$ form a pair of *alternate-interior angles* and $\angle 3$ and $\angle 6$ form another pair of alternate-interior angles. These angles are between the parallel lines and the related angles fall on alternate sides.

$\angle 1$ and $\angle 3$, $\angle 2$ and $\angle 4$, $\angle 5$ and $\angle 7$, and $\angle 6$ and $\angle 8$ are pairs of *corresponding angles.* Each pair of angles is in a corresponding position. Observe that when two parallel lines are cut by a third line (called a transversal), both the corresponding angles are equal and the alternate-interior angles are equal.

Also, when two lines are cut by a transversal making a pair of corresponding angles or a pair of alternate-interior angles equal, the lines are parallel.

EXERCISES

Use the figure below for all problems. Complete each statement as required in problems 1 to 6 inclusive.

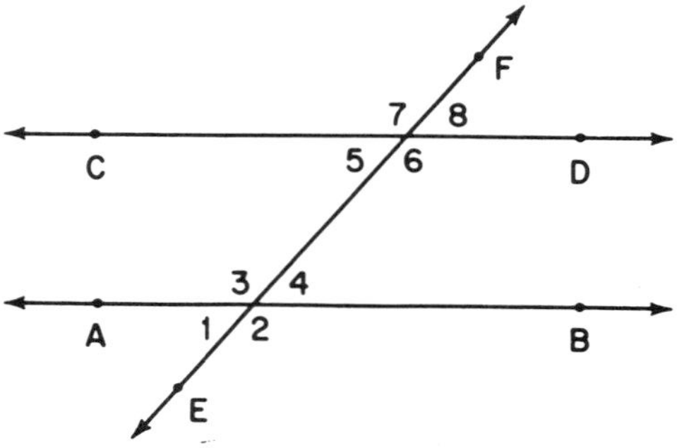

1. If lines *AB* and *CD* are cut by a third line *EF*,
   a. the line *EF* is called a ?.
   b. ∡ 4 and 5 are ? angles.
   c. ∡ 3 and 6 are ? angles.
   d. ∡ 4 and 8 are ? angles.
   e. ∡ 5 and 1 are ? angles.
2. If two parallel lines are cut by a transversal, the alternate-interior angles are ?. Therefore, $\angle 6 = ?$, $\angle 5 = ?$.
3. If two parallel lines are cut by a transversal, the corresponding angles are ?. Therefore, $\angle 2 = ?$, $\angle 7 = ?$, $\angle 1 = ?$, $\angle 8 = ?$.
4. What is the sum of the measures of ∡ 6 and 8? They are a pair of ? angles.
5. What is the sum of the measures of ∡ 1 and 3? They are a pair of ? angles.

6. $\angle 1$ and $\angle 4$ are a pair of ? angles.

7. If line AB is parallel to line CD,

a. Show that $\angle 2 = \angle 7$.
b. Show that $\measuredangle$s 1 and 6 are supplementary angles.
c. Show that $\angle 8 = \angle 1$.
d. Show that $\measuredangle$s 3 and 5 are supplementary angles.
e. Show that $\measuredangle$s 8 and 2 are supplementary angles.

8. If line AB is parallel to line CD, what is the measure of:

a. $\angle 5$ if $m\angle 1 = 75°$?
b. $\angle 6$ if $m\angle 3 = 140°$?
c. $\angle 7$ if $m\angle 4 = 61°$?
d. $\angle 1$ if $m\angle 8 = 57°$?
e. $\angle 8$ if $m\angle 2 = 106°$?
f. $\angle 2$ if $m\angle 7 = 132°$?

9. If two lines are cut by a transversal making a pair of corresponding angles or a pair of alternate-interior angles equal, the lines are ?.

10. Are lines AB and CD parallel when:

a. $m\angle 4 = 39°$ and $m\angle 8 = 39°$?
b. $m\angle 6 = 97°$ and $m\angle 3 = 97°$?
c. $m\angle 4 = 54°$ and $m\angle 6 = 126°$?
d. $m\angle 2 = 135°$ and $m\angle 8 = 50°$?

14–26 CONGRUENT TRIANGLES

Congruent triangles are triangles which have exactly the same shape and the same size. The corresponding sides are equal in length and the corresponding angles are equal in size. The symbol "$\cong$" means "is congruent to."

Two triangles are congruent when any of the following combinations of three parts are known:

(1) Three sides of one triangle are equal to three sides of the second triangle.
(2) Two sides and an included angle of one triangle are equal respectively to two sides and an included angle of the other.
(3) Two angles and an included side of one triangle are equal respectively to two angles and an included side of the other.

EXERCISES

Select in each of the following groups two triangles which are congruent and state the reason why.

1.

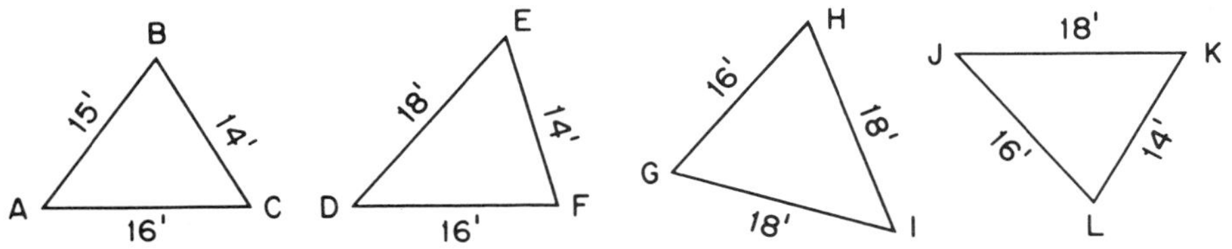

2.

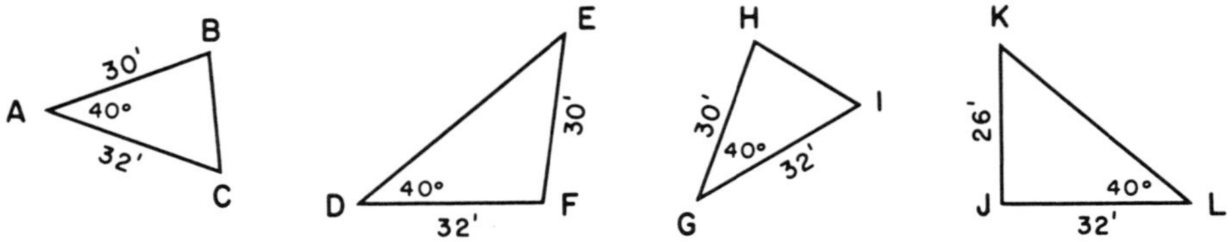

3.

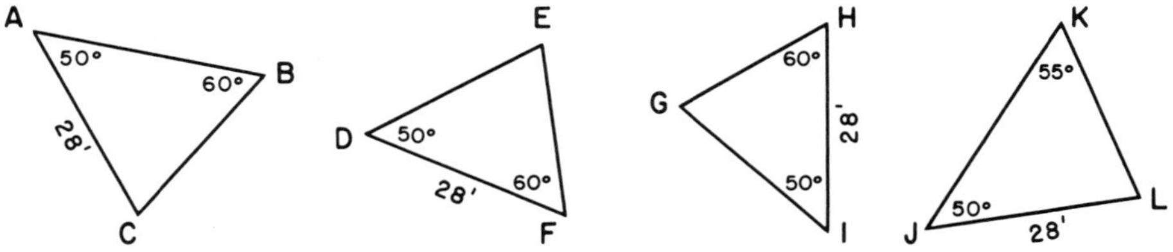

Find the indicated missing parts in the following triangles:

4.

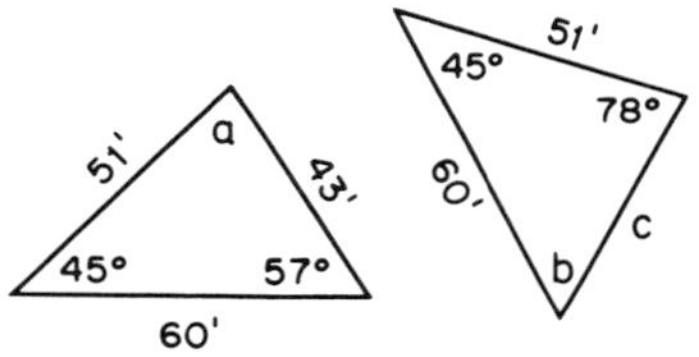

5.

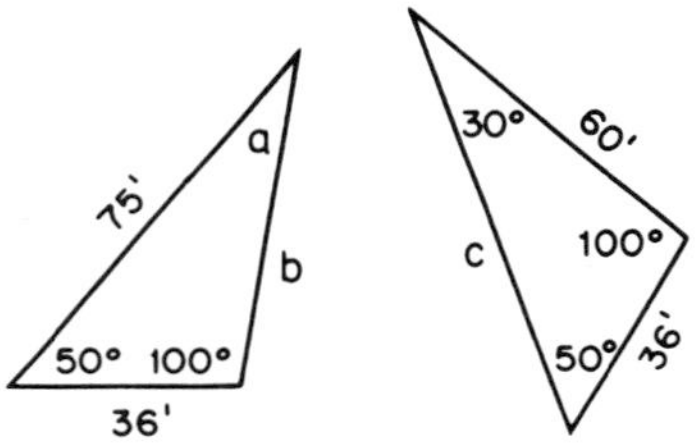

6. In triangle ABC, side AB = side BC and BD bisects $\angle ABC$ making $\angle ABD = \angle DBC$. Prove that triangle ABD and triangle DBC are congruent.

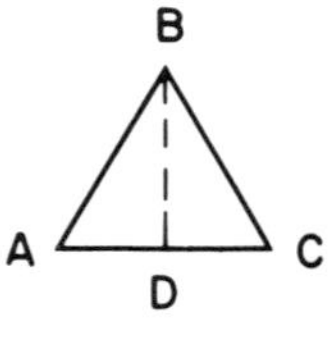

7. In parallelogram $ABCD$, side AB = side DC and side AD = side BC. Prove that the diagonal AC divides the parallelogram into two congruent triangles.

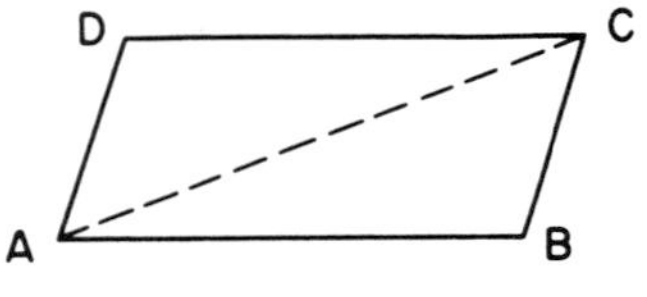

14–27 SIMILAR TRIANGLES

Similar triangles are triangles which have the same shape but differ in size. Two triangles are similar when any of the following conditions are known:

(1) Two angles of one triangle are equal to two angles of the other triangle.

(2) The ratios of the corresponding sides are equal.

(3) Two sides of one triangle are proportional (equal ratios) to two corresponding sides of the other triangle and the included angles are equal.

EXERCISES

Select in each of the following groups two triangles which are similar and state the reason why.

1.

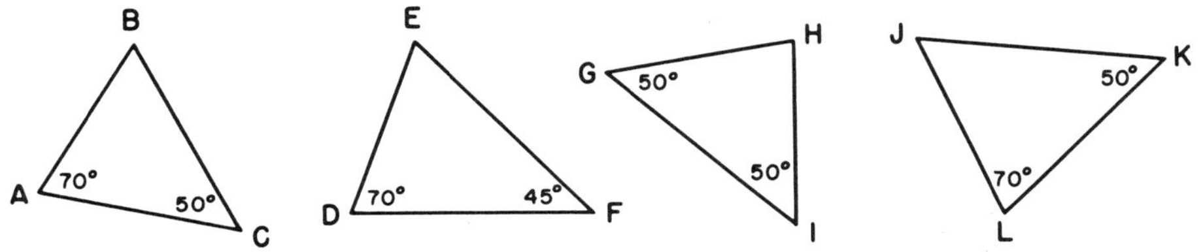

2.

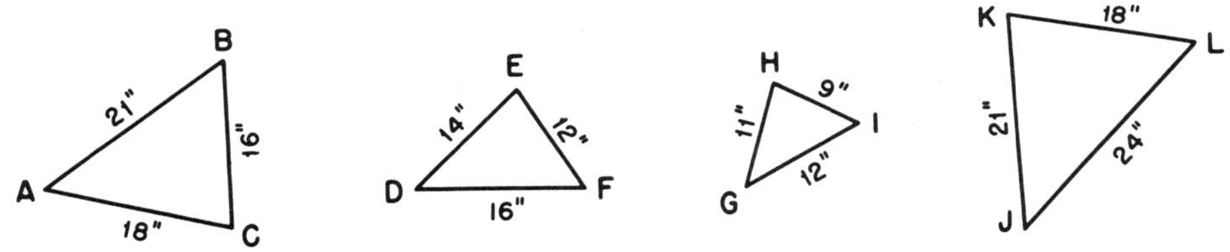

3.

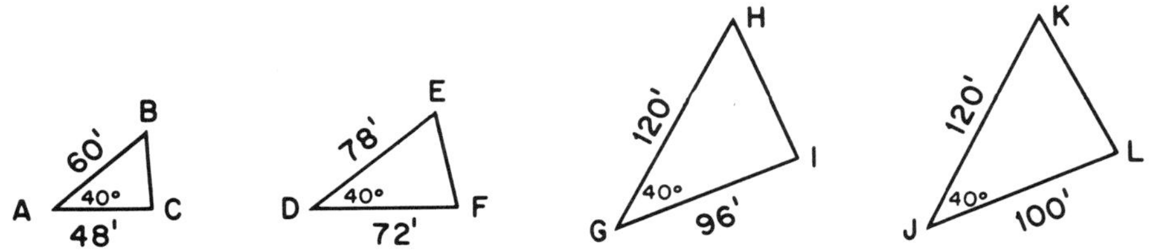

Find the indicated missing parts in the following figures:

4.

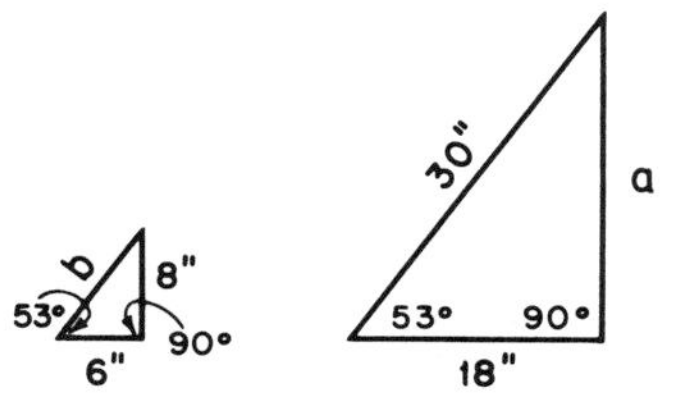

5.

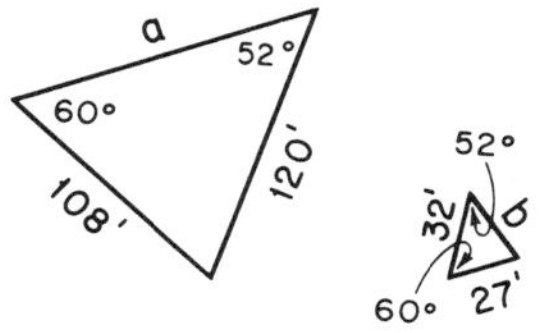

6.

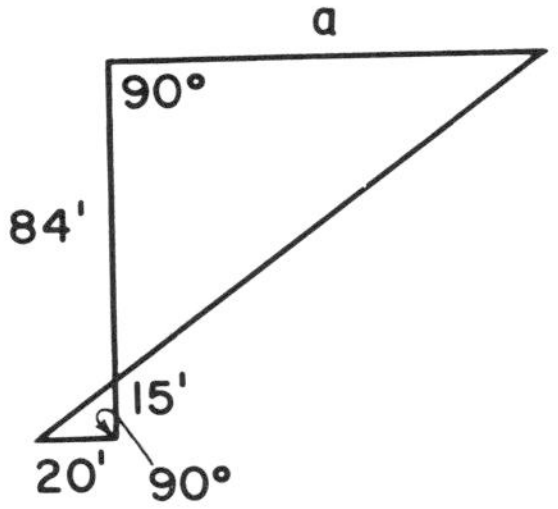

7.

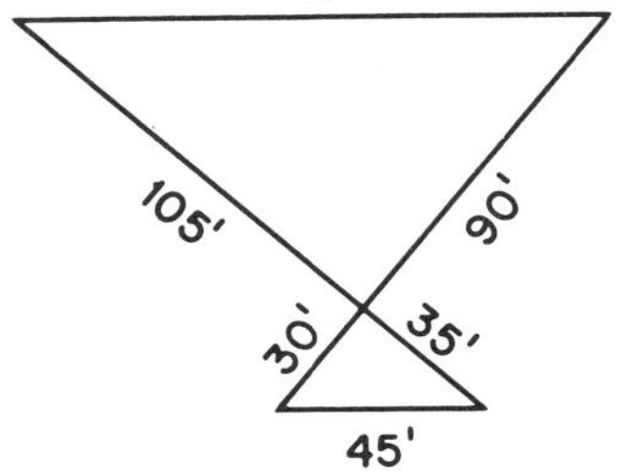

INDIRECT MEASUREMENT

14–28 BY RULE OF PYTHAGORAS

The Rule of Pythagoras, who was a Greek mathematician, is used to find distances by indirect means. It expresses the relationship of the sides of a right triangle. The side opposite the right angle is called the *hypotenuse*. The other two sides or legs are the *altitude* and *base* of the triangle.

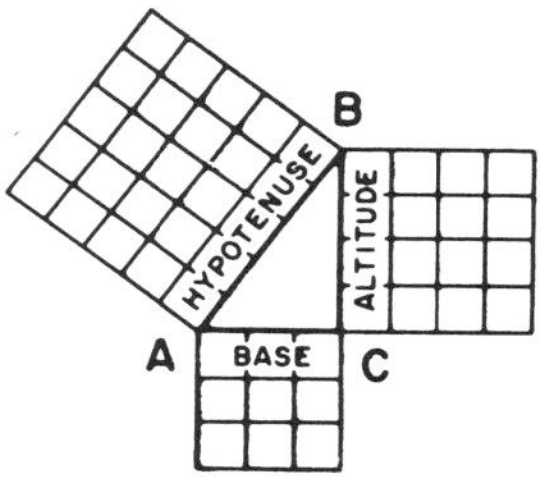

The diagram shows that the area of the square drawn on the hypotenuse is 25 square units and is equal to the sum of the areas of the squares drawn on the altitude and base (16 square units and 9 square units respectively). The rule states: *The square of the hypotenuse is equal to the sum of the squares of the other two sides.* This relationship is expressed by the formula $h^2 = a^2 + b^2$ where h represents the hypotenuse, a the altitude, and b the base.

If any two sides of a right triangle are known, the third side may be determined by the Pythagorean relation expressed in one of the following simplified forms:

$$h = \sqrt{a^2 + b^2} \qquad a = \sqrt{h^2 - b^2} \qquad b = \sqrt{h^2 - a^2}$$

(1) Find the hypotenuse of a right triangle if the altitude is 36 inches and the base is 27 inches.

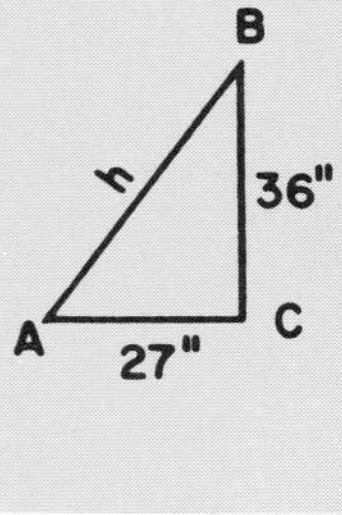

$a = 36$ in.
$b = 27$ in.
$h = ?$

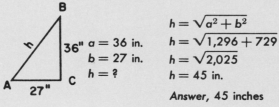

$h = \sqrt{a^2 + b^2}$
$h = \sqrt{1,296 + 729}$
$h = \sqrt{2,025}$
$h = 45$ in.

Answer, 45 inches

(2) Find the altitude of a right triangle if the hypotenuse is 26 feet and the base is 24 feet:

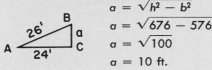

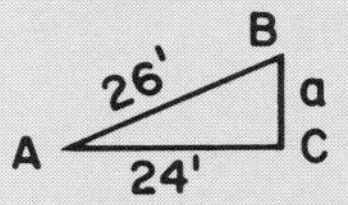

$a = \sqrt{h^2 - b^2}$

$a = \sqrt{676 - 576}$

$a = \sqrt{100}$

$a = 10$ ft.

$h = 26$ ft.

$b = 24$ ft.

$a = ?$

Answer, 10 ft.

(3) Find the base of a right triangle if the hypotenuse is 17 meters and the altitude is 8 meters:

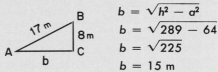

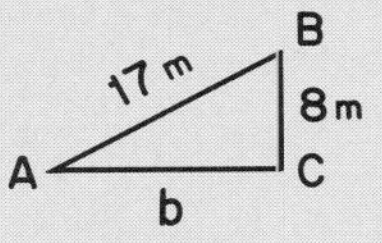

$b = \sqrt{h^2 - a^2}$

$b = \sqrt{289 - 64}$

$b = \sqrt{225}$

$b = 15$ m

$h = 17$ m

$a = 8$ m

$b = ?$

Answer, 15 meters

EXERCISES

1. Find the hypotenuse of each right triangle with the following dimensions:

Altitude	28 in.	192 ft.	108 yd.	75 m	2.1 km
Base	45 in.	56 ft.	315 yd.	180 m	3.2 km

2. Find the altitude of each right triangle with the following dimensions:

Hypotenuse	73 m	136 cm	455 ft.	117 yd.	9.5 mi.
Base	48 m	64 cm	112 ft.	108 yd.	5.7 mi.

3. Find the base of each right triangle with the following dimensions:

Hypotenuse	65 in.	116 yd.	680 ft.	182 m	4.5 km
Altitude	33 in.	84 yd.	104 ft.	168 m	3.6 km

4. How long is a path running diagonally across a rectangular lot 405 feet long and 216 feet wide?

5. If the foot of a ladder is 14 feet from the wall, how high up on the wall does a 50-foot ladder reach?

6. Find the length of the diagonal of a square whose side measures 18 cm.

7. What is the distance from the pitcher's box to second base if the distance from the pitcher's box to home plate is 60 feet and the distance between bases is 90 feet?

8. In the installation of an antenna 16 feet high, how long must a wire be to reach from the top of the antenna to a point on the roof 12 feet from the foot of the antenna?

9. A boat left port, sailing 28 kilometers due west and 45 kilometers due south. How far away from the port was the boat?
10. Mr. Harris plans to brace his 4 ft. by 7 ft. garage door by nailing a board diagonally across. How long a board does he need to reach from corner to corner?
11. What distance did an airplane actually fly when, traveling due east 312 miles from town X to town Y, it drifted off its course in a straight line and landed at town Z, 91 miles due north of town Y?
12. How high is a kite if a girl holds the string 3 ft. from the ground and lets out 200 ft. of string? The distance from a point on the ground directly under the kite to where the girl stands is 56 ft.

14–29 BY SIMILAR TRIANGLES

To measure indirectly a distance or length by similar triangles, we first draw two similar triangles from the given facts. Since the corresponding sides of similar triangles are proportional (have equal ratios), we form and solve a proportion using three given sides and the unknown distance as the fourth side.

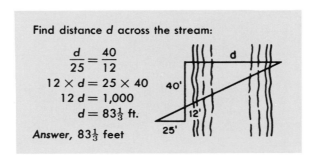

Find distance *d* across the stream:

$$\frac{d}{25} = \frac{40}{12}$$

$$12 \times d = 25 \times 40$$

$$12\,d = 1{,}000$$

$$d = 83\tfrac{1}{3} \text{ ft.}$$

Answer, $83\frac{1}{3}$ feet

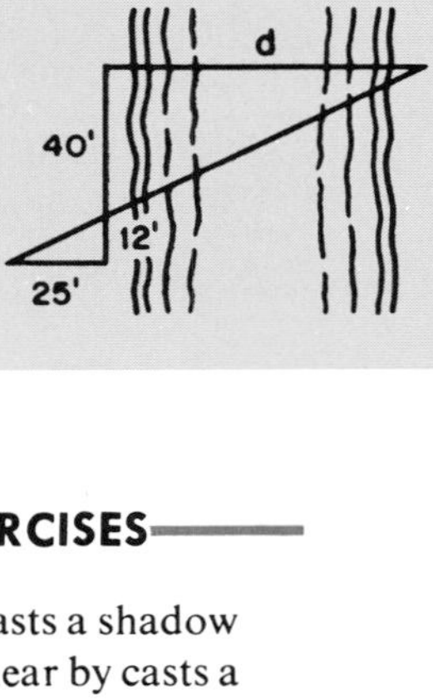

EXERCISES

1. What is the height (BC) of a tree that casts a shadow (AB) of 28 ft. while a 6-foot post (EF) near by casts a shadow (DE) of 4 ft.? Why are triangles ABC and DEF similar?

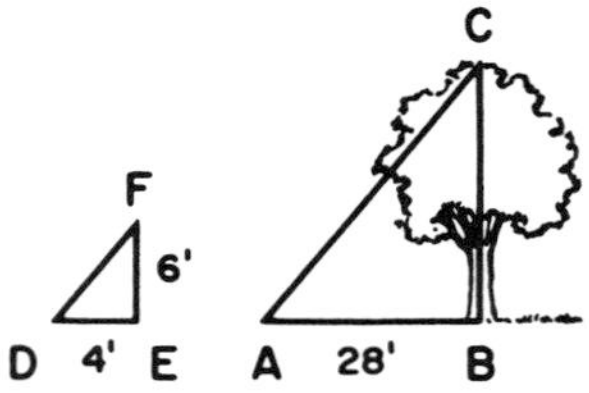

2. Find the height of a flagpole that casts a shadow of 39 ft. at a time when a girl, 5 feet tall, casts a shadow of 3 ft.?

3. A building casts a shadow of 72 ft. At the same time an 8-foot pole casts a shadow of 12 ft. How high is the building?

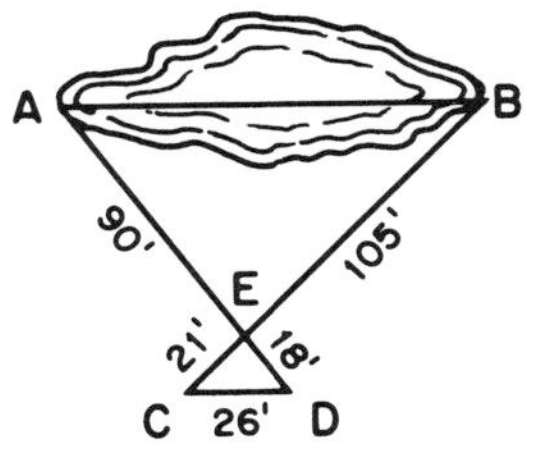

4. Find the length (AB) of the pond shown in the figure at the right. Why are triangles ABE and CDE similar?

5. What is the distance (AB) across the stream shown in the figure at the right? Why are triangles ABC and CDE similar?

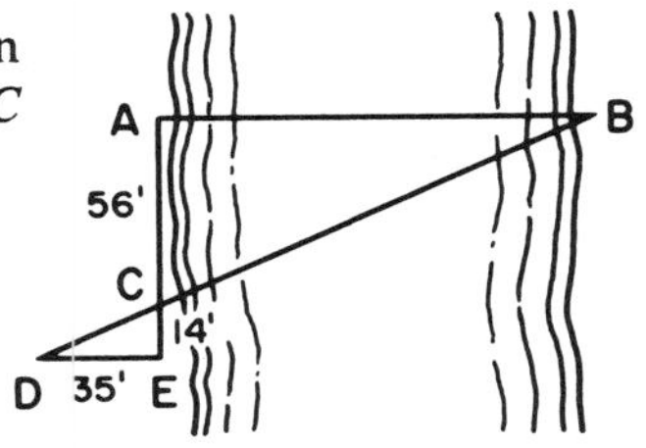

14–30 BY SCALE DRAWING

We determine the required lengths by first finding the scale (see page 410) if it is not known, then measuring the scale lengths and applying the scale to find the actual length.

EXERCISES

1. Determine the required lengths indicated in the following triangles which are drawn to scale:

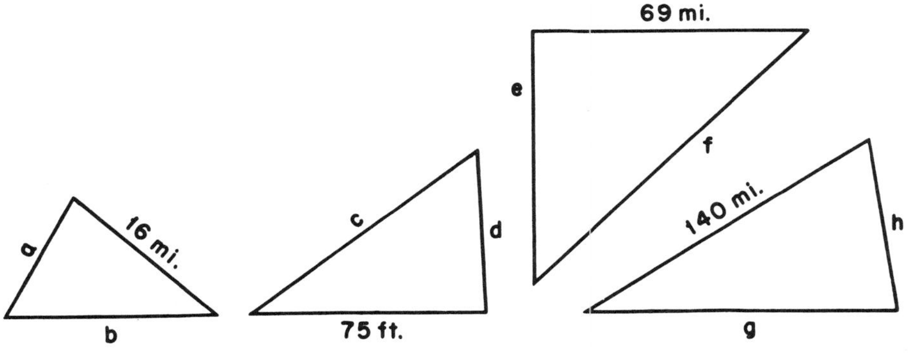

2. An airplane, flying due west (270°) from starting point A at an air speed of 240 m.p.h., was blown off its course by a 30 m.p.h. south wind (180°). Make a drawing, using the scale 1 inch = 80 miles, showing the position of the airplane at the end of one hour. How far was the airplane from its starting point? What course (or track) was it actually flying?

14-31 NUMERICAL TRIGONOMETRY

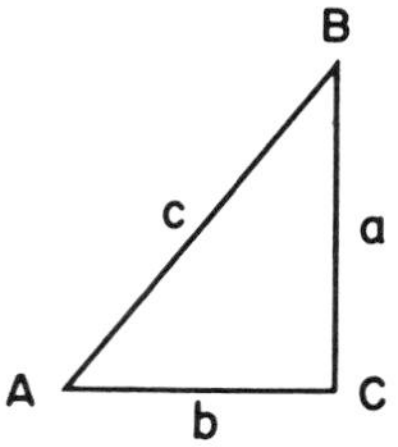

In trigonometry (meaning triangle measure) the relationships between the sides and the angles of the right triangle are used to determine certain parts of the triangle when the other parts are known. The ratios of the sides of the right triangle are related to the acute angles as follows:

The ratio of the side opposite an acute angle to the adjacent side is called the *tangent* of the angle (abbreviated tan). The expression tan A means tangent of angle A. In right triangle ABC, $\tan A = \frac{a}{b}$ and $\tan B = \frac{b}{a}$.

The ratio of the side opposite an acute angle to the hypotenuse is called the *sine* of the angle (abbreviated sin). In the right triangle ABC, $\sin A = \frac{a}{c}$ and $\sin B = \frac{b}{c}$.

The ratio of the adjacent side of an acute angle to the hypotenuse is called the *cosine* of the angle (abbreviated cos). In the right triangle ABC, $\cos A = \frac{b}{c}$ and $\cos B = \frac{a}{c}$.

Angles are measured both vertically and horizontally. An angle measured vertically between the horizontal line and the observer's line of sight to an object is called the *angle of elevation* when the object is above the observer and the *angle of depression* when the object is below the observer.

To solve problems using trigonometric ratios, we draw a right triangle if one is not given and place the given dimensions on it. We select the proper formula and substitute the given values, using the table of trigonometric values when necessary. Then we solve the resulting equation.

Find side a:

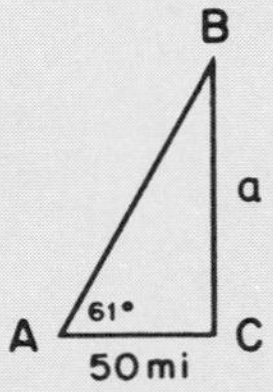

$$\tan A = \frac{a}{b}$$
$$\tan 61° = \frac{a}{50}$$
$$1.8040 = \frac{a}{50}$$
$$a = 90.2 \text{ mi.}$$

$b = 50$ mi.
$A = 61°$
$a = ?$

Answer, 90.2 mi.

Find side b:

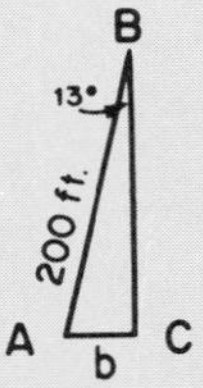

$$\sin B = \frac{b}{c}$$
$$\sin 13° = \frac{b}{200}$$
$$.2250 = \frac{b}{200}$$
$$b = 45 \text{ ft.}$$

$c = 200$ ft.
$B = 13°$
$b = ?$

Answer, 45 ft.

Find side *a*:

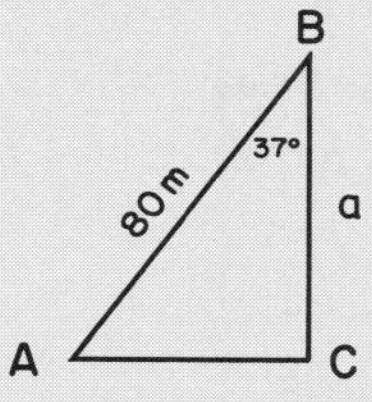

$\cos B = \dfrac{a}{c}$

$\cos 37^\circ = \dfrac{a}{80}$

$.7986 = \dfrac{a}{80}$

$a = 63.9 \text{ m}$

$c = 80 \text{ m}$
$B = 37^\circ$
$a = ?$

Answer, 63.9 m

EXERCISES

Tangent Ratio

When necessary, use the table of trigonometric values on page 441.

1. Find the value of: tan 61°, tan 17°, tan 42° 30′, tan 9° 15′.
2. Find angle A if: $\tan A = .8693$, $\tan A = .5430$, $\tan A = 2.3559$.
3. Find angle B if: $\tan B = 1.9626$, $\tan B = .9574$, $\tan B = .0699$.

Find the indicated parts of the following triangles:

4. Find side a:

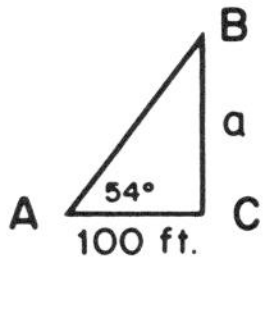

5. Find side b:

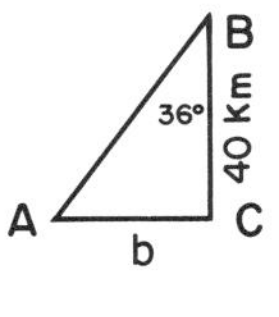

6. Find side b:

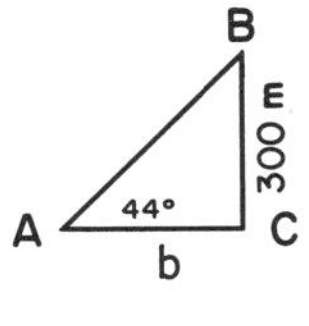

7. Find side a:

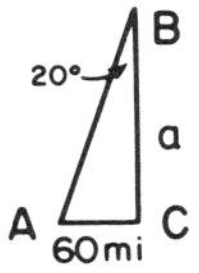

8. Find $\angle A$:

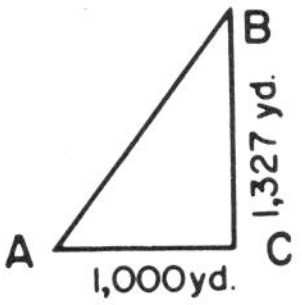

9. Find $\angle B$:

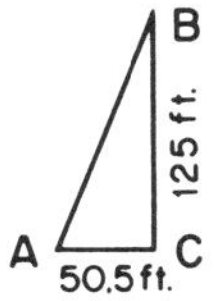

10. The light from a searchlight is observed on a cloud at a horizontal distance of 1,800 feet. What is the height of the cloud if the angle of elevation is 82°?
11. What is the ground distance from an airplane to the airport if at an altitude of 6,000 meters, the angle of depression of the airport is 34°?

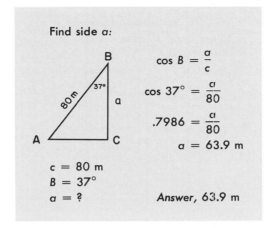

Sine Ratio

When necessary, use the table of trigonometric values on page 441.

12. Find the value of: sin 26°, sin 83°, sin 54° 45′, sin 16° 30′.
13. Find angle A if: sin A = .8290, sin A = .6626, sin A = .3090.
14. Find angle B if: sin B = .5446, sin B = .9288, sin B = .8829.

Find the indicated parts of the following triangles:

15. Find side a:

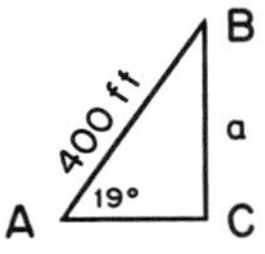

16. Find side b:

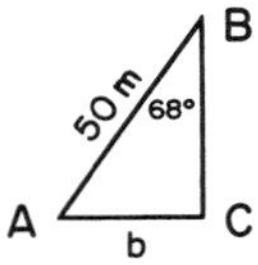

17. Find side b:

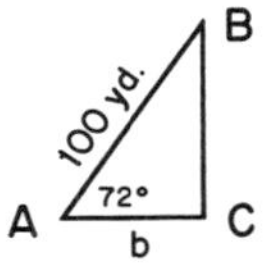

18. Find side a:

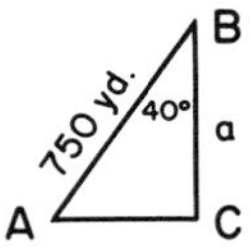

19. Find side c:

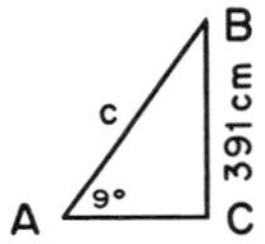

20. Find side c:

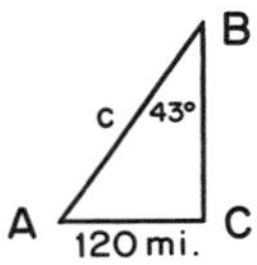

21. Find $\angle A$:

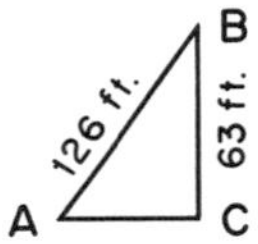

22. Find $\angle B$:

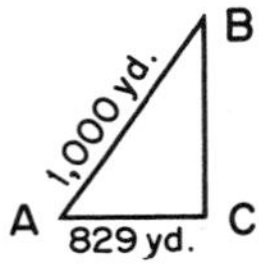

23. How high is a kite if 240 feet of string is let out and the string makes an angle of 52° with the ground?

Cosine Ratio

When necessary, use the table of trigonometric values on page 441.

24. Find the value of: cos 47°, cos 6°, cos 73° 30′, cos 19° 20′.
25. Find angle A if: cos A = .2588, cos A = .8064, cos A = .9994.
26. Find angle B if: cos B = .9744, cos B = .5075, cos B = .0872.

Find the indicated parts of the following triangles, using the cosine ratio:

27. Find side b:

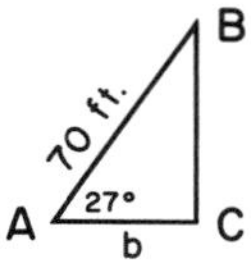

28. Find side a:

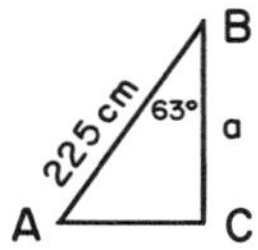

29. Find side a:

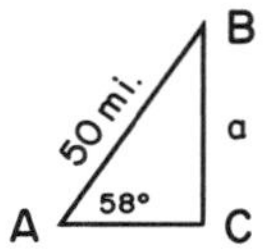

30. Find side b:

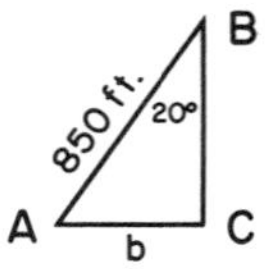

31. Find side c:

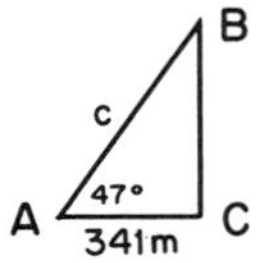

32. Find side c:

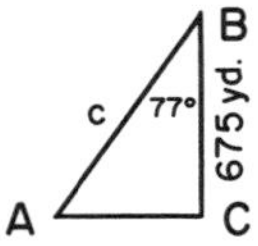

33. Find $\angle A$:

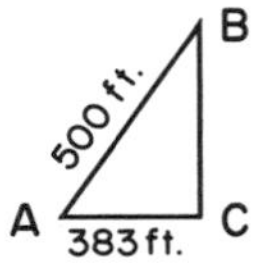

34. Find $B \angle$:

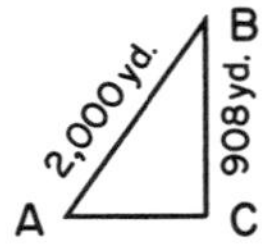

Table of Trigonometric Values

Angle	Sine	Cosine	Tangent	Angle	Sine	Cosine	Tangent
0°	.0000	1.0000	.0000	**46°**	.7193	.6947	1.0355
1°	.0175	.9998	.0175	**47°**	.7314	.6820	1.0724
2°	.0349	.9994	.0349	**48°**	.7431	.6691	1.1106
3°	.0523	.9986	.0524	**49°**	.7547	.6561	1.1504
4°	.0698	.9976	.0699	**50°**	.7660	.6428	1.1918
5°	.0872	.9962	.0875	**51°**	.7771	.6293	1.2349
6°	.1045	.9945	.1051	**52°**	.7880	.6157	1.2799
7°	.1219	.9925	.1228	**53°**	.7986	.6018	1.3270
8°	.1392	.9903	.1405	**54°**	.8090	.5878	1.3764
9°	.1564	.9877	.1584	**55°**	.8192	.5736	1.4281
10°	.1736	.9848	.1763	**56°**	.8290	.5592	1.4826
11°	.1908	.9816	.1944	**57°**	.8387	.5446	1.5399
12°	.2079	.9781	.2126	**58°**	.8480	.5299	1.6003
13°	.2250	.9744	.2309	**59°**	.8572	.5150	1.6643
14°	.2419	.9703	.2493	**60°**	.8660	.5000	1.7321
15°	.2588	.9659	.2679	**61°**	.8746	.4848	1.8040
16°	.2756	.9613	.2867	**62°**	.8829	.4695	1.8807
17°	.2924	.9563	.3057	**63°**	.8910	.4540	1.9626
18°	.3090	.9511	.3249	**64°**	.8988	.4384	2.0503
19°	.3256	.9455	.3443	**65°**	.9063	.4226	2.1445
20°	.3420	.9397	.3640	**66°**	.9135	.4067	2.2460
21°	.3584	.9336	.3839	**67°**	.9205	.3907	2.3559
22°	.3746	.9272	.4040	**68°**	.9272	.3746	2.4751
23°	.3907	.9205	.4245	**69°**	.9336	.3584	2.6051
24°	.4067	.9135	.4452	**70°**	.9397	.3420	2.7475
25°	.4226	.9063	.4663	**71°**	.9455	.3256	2.9042
26°	.4384	.8988	.4877	**72°**	.9511	.3090	3.0777
27°	.4540	.8910	.5095	**73°**	.9563	.2924	3.2709
28°	.4695	.8829	.5317	**74°**	.9613	.2756	3.4874
29°	.4848	.8746	.5543	**75°**	.9659	.2588	3.7321
30°	.5000	.8660	.5774	**76°**	.9703	.2419	4.0108
31°	.5150	.8572	.6009	**77°**	.9744	.2250	4.3315
32°	.5299	.8480	.6249	**78°**	.9781	.2079	4.7046
33°	.5446	.8387	.6494	**79°**	.9816	.1908	5.1446
34°	.5592	.8290	.6745	**80°**	.9848	.1736	5.6713
35°	.5736	.8192	.7002	**81°**	.9877	.1564	6.3138
36°	.5878	.8090	.7265	**82°**	.9903	.1392	7.1154
37°	.6018	.7986	.7536	**83°**	.9925	.1219	8.1443
38°	.6157	.7880	.7813	**84°**	.9945	.1045	9.5144
39°	.6293	.7771	.8098	**85°**	.9962	.0872	11.4301
40°	.6428	.7660	.8391	**86°**	.9976	.0698	14.3007
41°	.6561	.7547	.8693	**87°**	.9986	.0523	19.0811
42°	.6691	.7431	.9004	**88°**	.9994	.0349	28.6363
43°	.6820	.7314	.9325	**89°**	.9998	.0175	57.2900
44°	.6947	.7193	.9657	**90°**	1.0000	.0000	
45°	.7071	.7071	1.0000				

PERIMETER AND CIRCUMFERENCE

The *perimeter* of a polygon is the sum of the lengths of its sides. It is the distance around the polygon.

14–32 PERIMETER OF A RECTANGLE

We see from the figure below that the perimeter of a rectangle is equal to twice its length plus twice its width. Expressed as a formula this relationship is $p = 2l + 2w$ or $p = 2(l + w)$.

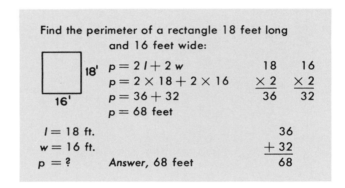

Find the perimeter of a rectangle 18 feet long and 16 feet wide:

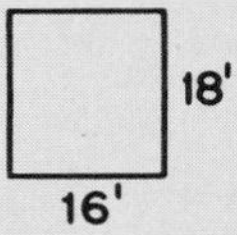

$p = 2l + 2w$
$p = 2 \times 18 + 2 \times 16$
$p = 36 + 32$
$p = 68$ feet

$$\begin{array}{r} 18 \\ \times 2 \\ \hline 36 \end{array} \qquad \begin{array}{r} 16 \\ \times 2 \\ \hline 32 \end{array}$$

$l = 18$ ft.
$w = 16$ ft.
$p = ?$

$$\begin{array}{r} 36 \\ +32 \\ \hline 68 \end{array}$$

Answer, 68 feet

EXERCISES

Find the perimeters of rectangles having the following dimensions:

1.

Length	23 in.	145 ft.	216 yd.	6.5 cm	8.25 m	5.625 km
Width	17 in.	98 ft.	307 yd.	1.2 cm	9.75 m	4.875 km

2.

Length	$5\frac{3}{4}$ ft.	$7\frac{5}{8}$ in.	$4\frac{1}{2}$ yd.	2 ft. 7 in.	6 ft.	9 ft. 11 in.
Width	$3\frac{1}{2}$ ft.	$2\frac{13}{16}$ in.	$3\frac{2}{3}$ yd.	1 ft. 5 in.	7 ft. 4 in.	5 ft. 8 in.

3. How many feet of picture moulding are needed to go around the walls of a room 17 ft. long and 15 ft. wide?
4. If 1 meter is to be deducted for a doorway, how many meters of baseboard are needed for a room 6 meters long and 5 meters wide?
5. At $1.70 a yard, how much will it cost to bind a rug 9 ft. by 12 ft.?
6. A rectangular field is 146 ft. long and 121 ft. wide. How many feet of fencing are required to enclose the field? At $4.20 per yard, how much will it cost?

14–33 PERIMETER OF A SQUARE

Since the 4 sides of a square are of equal length, the perimeter of a square is 4 times the length of its side. Expressed as a formula this is $p = 4s$.

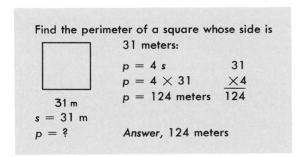

Find the perimeter of a square whose side is 31 meters:

31 m

$s = 31$ m

$p = ?$

$p = 4\,s$

$p = 4 \times 31$

$p = 124$ meters

$$\begin{array}{r} 31 \\ \times 4 \\ \hline 124 \end{array}$$

Answer, 124 meters

EXERCISES

Find the perimeters of squares whose sides measure:

1. **a.** 9 in. **b.** 47 ft. **c.** 6,080 ft. **d.** 1,760 yd.

2. **a.** 0.75 m **b.** 1.25 dm **c.** 28.5 cm **d.** 5.9 km

3. **a.** $\frac{5}{8}$ in. **b.** $32\frac{3}{4}$ ft. **c.** $7\frac{13}{16}$ in. **d.** $29\frac{2}{3}$ yd.

4. **a.** 1 ft. 6 in. **b.** 3 ft. 4 in. **c.** 5 ft. 11 in. **d.** 8 yd. 2 ft.

5. How many yards of linoleum border are needed for a kitchen floor 11 feet by 11 feet?

6. If the distance between bases on a soft-ball diamond is 60 ft., how many yards does a batter run when he hits a home-run?

7. At \$.10 per foot, what would be the cost of weatherstripping for a window measuring 3 ft. 9 in. by 3 ft. 9 in.?

8. How much would it cost, at \$4.80 per yard, to fence a 90-foot square garden?

14–34 PERIMETER OF A TRIANGLE

The perimeter of a triangle is equal to the sum of the lengths of its sides. The formula $p = a + b + c$ is sometimes used. Since the sides of an equilateral triangle are equal, its perimeter is equal to 3 times the length of its side which expressed as a formula is $p = 3s$.

EXERCISES

1. Find the perimeters of triangles with sides measuring:

a. 16 cm, 23 cm, 19 cm

b. 3.5 m, 4.6 m, 2.75 m

c. $9\frac{1}{2}$ yd., $8\frac{3}{4}$ yd., $10\frac{2}{3}$ yd.

d. 1 ft. 7 in., 2 ft., 2 ft. 1 in.

2. Find the perimeters of equilateral triangles with sides measuring:

   a. 26 ft. b. 9.25 km c. $7\frac{9}{16}$ in. d. 2 ft. 5 in.

3. Find the perimeters of isosceles triangles with the following dimensions:

Base	13 in.	74 mm	$4\frac{1}{2}$ ft.	7.5 m	2 ft. 4 in.
Each equal side	17 in.	59 mm	$3\frac{3}{4}$ ft.	4.3 m	1 ft. 11 in.

4. How many meters of fencing are needed to enclose a triangular lot with sides measuring 128 m, 106 m, and 117 m?
5. How many plants are needed to go around the outside of a triangular flower bed if the plants are to be spaced 8 inches apart and the sides of the flower bed measure 18 ft., 14 ft., and 16 ft., respectively?
6. Joan wishes to sew ribbon around her school pennant which measures $29\frac{1}{2}$ in., $29\frac{1}{2}$ in., and 13 in. How many yards of ribbon will she need? At $.45 a yard, how much will the ribbon cost?

14-35 CIRCUMFERENCE OF A CIRCLE

The *circumference* of a circle is the length of the circle or the distance around the circle. To develop the relationship of the circumference, diameter, and radius of a circle, let us do the following:

(1) Draw a circle having a radius of 1 inch. Determine the diameter by measuring it. Is the diameter of a circle twice as long as the radius? Does the formula $d = 2r$ express this relationship when d represents the diameter, and r the radius?

(2) Draw a circle having a diameter of $2\frac{1}{2}$ inches. What is its radius? Is the radius of a circle one-half as long as the diameter? Does the radius equal the diameter divided by two? Does the formula $r = \frac{d}{2}$ express this relationship?

(3) Draw a circle with a radius of $1\frac{1}{2}$ inches. Draw three line segments from the center to points on the circle. Measure these segments. Are all radii of a circle of equal length? Extend each of these line segments until it meets the circle, thus forming a diameter. Measure the diameters. Are all diameters of a circle of equal length?

(4) Measure the diameters and circumferences of three circular objects. Copy the chart (see page 445) and tabulate your measurements. Divide the circumference of each object by its diameter, finding the quotient correct to the nearest hundredth. Find the average of these quotients.

Object	Circumference (c) of object	Diameter (d) of object	Circumference divided by diameter $(c \div d) = \pi$

The circumference is a little more than 3 times as long as the diameter. This constant ratio is represented by the Greek letter π (called pi) which equals 3.14 or $3\frac{1}{7}$ or $\frac{22}{7}$. For greater accuracy 3.1416 is used.

Thus, the circumference of a circle is equal to pi (π) times the diameter.

Expressed as a formula it is: $c = \pi d$.

Since $d = 2r$, we may substitute $2r$ for d in the formula $c = \pi d$ to get $c = \pi 2r$ or better $c = 2\pi r$. Therefore we may say that the circumference of a circle is equal to two times pi (π) times the radius. The diameter of a circle is equal to the circumference divided by pi (π). Expressed as a formula it is $d = \dfrac{c}{\pi}$

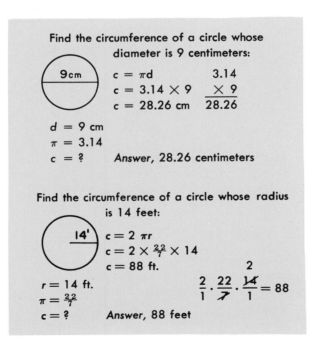

Find the circumference of a circle whose diameter is 9 centimeters:

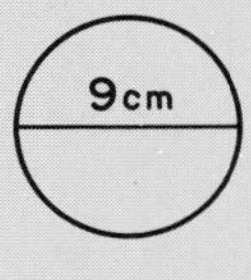

$c = \pi d$

$c = 3.14 \times 9$

$c = 28.26$ cm

$$\begin{array}{r} 3.14 \\ \times\ 9 \\ \hline 28.26 \end{array}$$

$d = 9$ cm

$\pi = 3.14$

$c = ?$

Answer, 28.26 centimeters

Find the circumference of a circle whose radius is 14 feet:

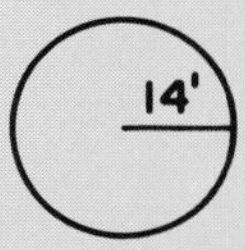

$c = 2\pi r$

$c = 2 \times \frac{22}{7} \times 14$

$c = 88$ ft.

$$\frac{2}{1} \cdot \frac{22}{\cancel{7}} \cdot \frac{\overset{2}{\cancel{14}}}{1} = 88$$

$r = 14$ ft.

$\pi = \frac{22}{7}$

$c = ?$

Answer, 88 feet

EXERCISES

Find the circumference of a circle having a diameter of:

1. a. 8 cm	**b.** 21 ft.	**c.** 37 mm	**d.** 63 mi.
2. a. 2.5 km	**b.** $8\frac{3}{4}$ in.	**c.** $\frac{7}{8}$ mi.	**d.** 1.625 m
3. a. $5\frac{1}{4}$ ft.	**b.** 4 ft. 8 in.	**c.** 3 ft. 6 in.	**d.** 1 ft. 5 in.

Find the circumference of a circle having a radius of:

4. a. 9 mm	**b.** 14 in.	**c.** 85 cm	**d.** 280 ft.
5. a. $7\frac{7}{8}$ in.	**b.** $2\frac{1}{3}$ yd.	**c.** $1\frac{5}{16}$ in.	**d.** 3.75 km
6. a. 8.6 m	**b.** 3 ft. 1 in.	**c.** 1 ft. 3 in.	**d.** 2 ft. 4 in.

7. What distance do you ride in one turn of a Ferris wheel when you sit 18 feet from the center?

8. How many times must you go around a circular track with a diameter of 105 feet in order to run a mile?

9. Over what distance will the tip of the minute hand of a clock move in 1 hour? The minute hand is 12 cm long.

10. How long a metal bar is needed to make a hoop for a barrel with a diameter of 28 inches?

11. What distance does a point on a circular saw travel in one turn if the saw has a diameter of 20 inches?

12. The diameter of the earth is 7,918 miles. Find the circumference of the earth.

13. How many yards of wire fencing are needed to enclose a circular flower bed 16 feet in diameter?

14. The wheels on a truck have a diameter of 35 inches. How many times does each wheel revolve when the truck goes a mile?

15. Find the diameter of a tree whose circumference is 3 ft. 8 in.

AREA

The *area* of the interior (or closed region) of any plane figure is the number of units of square measure it contains. When computing the area of a geometric figure, we express all linear units in the same denomination.

14–36 AREA OF A RECTANGLE

At the right we see that one measure ment indicates the number of square units in a row and the other measurement indicates how many rows there are:

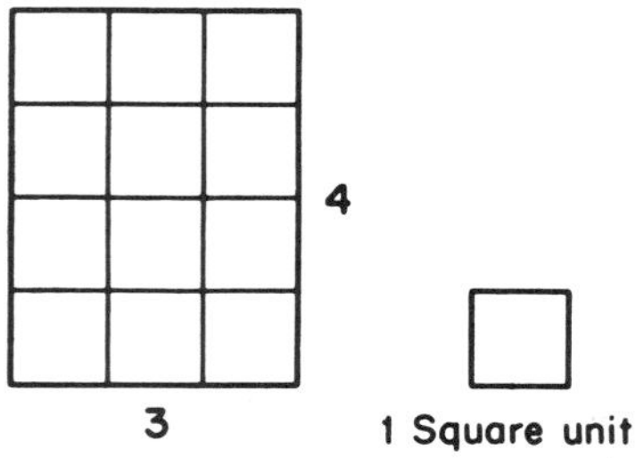

Thus to find the area of the interior of a rectangle we multiply the length by the width. Expressed as a formula this relationship is $A = lw$. Or, the area is equal to the altitude times the base. Formula: $A = ab$.

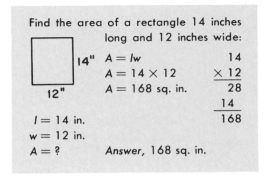

Find the area of a rectangle 14 inches long and 12 inches wide:

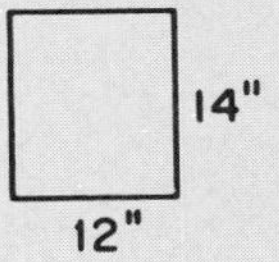

$A = lw$

$A = 14 \times 12$

$A = 168$ sq. in.

$$\begin{array}{r} 14 \\ \times\ 12 \\ \hline 28 \\ 14\ \\ \hline 168 \end{array}$$

$l = 14$ in.
$w = 12$ in.
$A = ?$

Answer, 168 sq. in.

EXERCISES

Find the area of rectangles having the following dimensions:

1.

Length	16 mm	58 ft.	175 yd.	3.4 cm	9.5 m	4.6 km
Width	9 mm	65 ft.	128 yd.	0.8 cm	6.25 m	8.3 km

2.

Length	$7\frac{1}{2}$ in.	$2\frac{3}{4}$ yd.	$6\frac{1}{4}$ in.	5 ft. 9 in.	1 ft. 6 in.	8 ft. 2 in.
Width	5 in.	$8\frac{2}{3}$ yd.	$1\frac{3}{8}$ in.	4 ft.	2 ft. 8 in.	6 ft. 10 in.

3. Find the cost of each of the following:

a. Cementing a sidewalk 15 ft. by 8 ft. at $1.50 per sq. ft.
b. Covering a floor 16 ft. by 14 ft. with broadloom at $9.50 per sq. yd.
c. Sodding a lawn 12 ft. 6 in. by 9 ft. 8 in. at $.20 per sq. ft.

4. What is the picture area on a television screen $18\frac{1}{2}$ in. wide and 14 in. high?
5. The official length of the United States flag is 1.9 times its width. What is the area of a flag 4 ft. wide?
6. How many square feet of floor space remain uncovered if a 9 ft. by 12 ft. rug is placed on a floor 17 ft. long and 14 ft. wide?

14–37 AREA OF A SQUARE

Although we may use the formula $A = lw$ to find the area of the interior of a square, we generally use the formula $A = s^2$ where s is the length of the side of the square. Since the length and width of a square are both equal to the length of the side, s, the formula $A = lw$ becomes $A = s \times s$ or $A = s^2$.

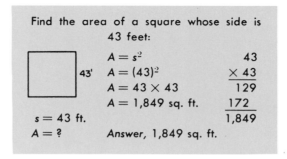

Find the area of a square whose side is 43 feet:

$A = s^2$

$A = (43)^2$

$A = 43 \times 43$

$A = 1{,}849$ sq. ft.

$$\begin{array}{r} 43 \\ \times\ 43 \\ \hline 129 \\ 172 \\ \hline 1{,}849 \end{array}$$

$s = 43$ ft.

$A = ?$

Answer, 1,849 sq. ft.

EXERCISES

Find the areas of squares whose sides measure:

1. **a.** 8 in. **b.** 59 mm **c.** 440 yd. **d.** 5,280 ft.
2. **a.** 9.5 cm **b.** 1.09 m **c.** 3.28 km **d.** 0.62 m
3. **a.** $5\frac{1}{2}$ yd. **b.** $16\frac{1}{2}$ ft. **c.** $5\frac{3}{4}$ ft. **d.** $\frac{7}{8}$ in.
4. **a.** 1 ft. 3 in. **b.** 2 ft. 5 in. **c.** 7 ft. 8 in. **d.** 2 yd. 1 ft.
5. Find the cost of each of the following:
   a. Covering a floor 10 ft. by 10 ft. with linoleum at $7.75 per sq. yd.
   b. Resilvering a mirror 18 in. by 18 in. at $6.90 per sq. ft.
   c. Refinishing a floor 15 ft. 6 in. by 15 ft. 6 in. at $.28 per sq. ft.
6. How many square yards of tarpaulin are needed to cover the infield of a baseball diamond if the distance between bases is 90 feet?
7. How many times as large as a 2-centimeter square is a 10-centimeter square?
8. How many square feet of ground remain when a building 105 ft. square is erected on a lot 180 ft. square?

14-38 AREA OF A PARALLELOGRAM

The *base* of a parallelogram is the side on which it rests. The *altitude* or *height* is the perpendicular segment between the base and its opposite side.

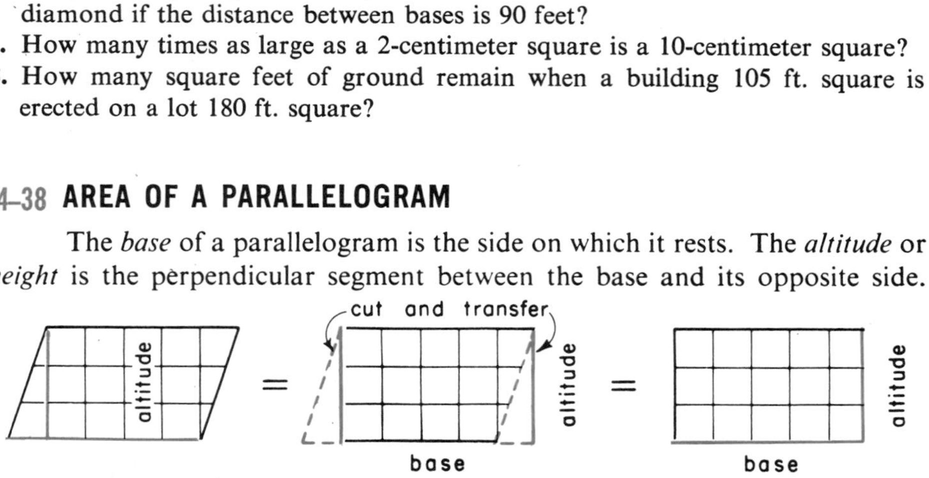

The above diagrams illustrate that the area of the interior of a parallelogram is equal to the product of the altitude and the base, formula $A = ab$, or the product of the base and height, formula $A = bh$.

Find the area of a parallelogram with an altitude of 16 meters and a base of 27 meters:

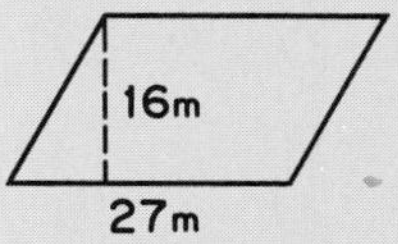

$a = 16$ m
$b = 27$ m
$A = ?$

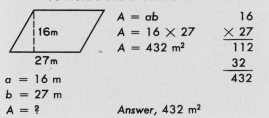

$A = ab$
$A = 16 \times 27$
$A = 432\ m^2$

$$\begin{array}{r} 16 \\ \times\ 27 \\ \hline 112 \\ 32 \\ \hline 432 \end{array}$$

Answer, 432 m^2

EXERCISES

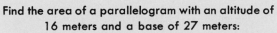

Find the areas of parallelograms having the following dimensions:

1.

Altitude	19 in.	53 ft.	165 mm	3.5 cm	6.8 km	9.5 m
Base	21 in.	37 ft.	86 mm	6 cm	1.4 km	6.25 m

2.

Altitude	$1\frac{7}{8}$ in.	$4\frac{2}{3}$ ft.	$2\frac{1}{2}$ yd.	1 ft. 3 in.	5 yd. 1 ft.	1 yd. 18 in.
Base	$1\frac{3}{4}$ in.	$2\frac{1}{4}$ ft.	$3\frac{3}{4}$ yd.	2 ft.	3 yd. 2 ft.	2 yd. 6 in.

3. What is the area of a lot shaped like a parallelogram with a base of 145 m and an altitude of 124 m?

4. A parking lot has 46 spaces, each a parallelogram with a base of 7 ft. and an altitude of 15 ft. What is the total parking area of the lot?

5. Which parallelogram has the greater area, one with an altitude of 8 ft. 6 in. and a base of 6 ft. 6 in. or one with an altitude of 3 yd. and a base of 2 yd.? How much greater?

6. How many square feet of sod are needed for a lawn shaped like a parallelogram with a base of 32 ft. and an altitude of 15 ft.? At $.16 per sq. ft., what is the cost of the sod?

14–39 AREA OF A TRIANGLE

A diagonal separates a parallelogram into two congruent triangles. The area of the interior of each triangle is equal to one-half the area of the interior of the parallelogram.

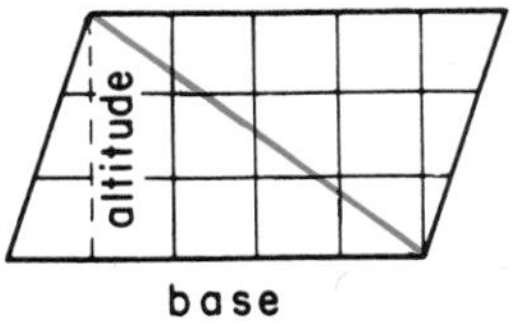

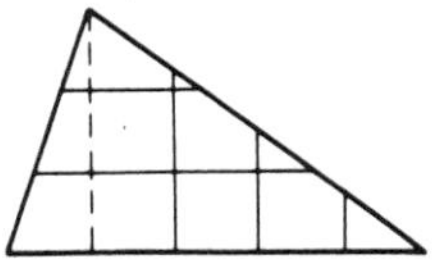

Thus the area of the interior of a triangle is equal to one-half the altitude times the base. Expressed as a formula it is $A = \frac{1}{2}ab$ or $A = \frac{ab}{2}$. Or, the area is equal to one half the base times the height. Formula: $A = \frac{1}{2}bh$.

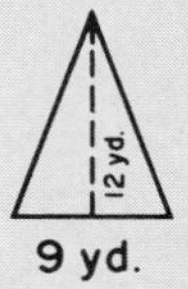

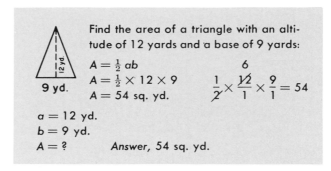

Find the area of a triangle with an altitude of 12 yards and a base of 9 yards:

$A = \frac{1}{2}ab$

$A = \frac{1}{2} \times 12 \times 9$

$A = 54$ sq. yd.

$$\frac{1}{\not{2}} \times \frac{\not{12}^{\,6}}{1} \times \frac{9}{1} = 54$$

$a = 12$ yd.
$b = 9$ yd.
$A = ?$

Answer, 54 sq. yd.

EXERCISES

Find the area of the triangles having the following dimensions:

1.

Altitude	20 ft.	8 mm	53 yd.	$2\frac{1}{2}$ in.	$8\frac{3}{4}$ ft.	7.4 m
Base	32 ft.	13 mm	47 yd.	3 in.	$4\frac{1}{4}$ ft.	5.9 m

2.

Base	$6\frac{2}{3}$ yd.	9.3 km	$1\frac{5}{8}$ in.	4 ft.	1 ft. 4 in.	5 yd. 9 in.
Height	$1\frac{7}{8}$ yd.	4.5 km	$2\frac{3}{4}$ in.	2 ft. 9 in.	1 ft. 11 in.	3 yd. 6 in.

3. What is the area of a triangular plot having a base of 69 m and an altitude of 86 m?

4. How many square feet of felt are used in a pennant with a base of 12 inches and an altitude of 30 inches?

5. A triangular sail has a base of 14 ft. and an altitude of 16 ft. 6 in. How many square feet of surface does each side of the sail expose?

6. Allowing 3 sq. ft. per person, find the number of persons a triangular safety traffic island can hold if it has a base of 16 ft. and an altitude of 4 ft. 6 in.

14–40 AREA OF A TRAPEZOID

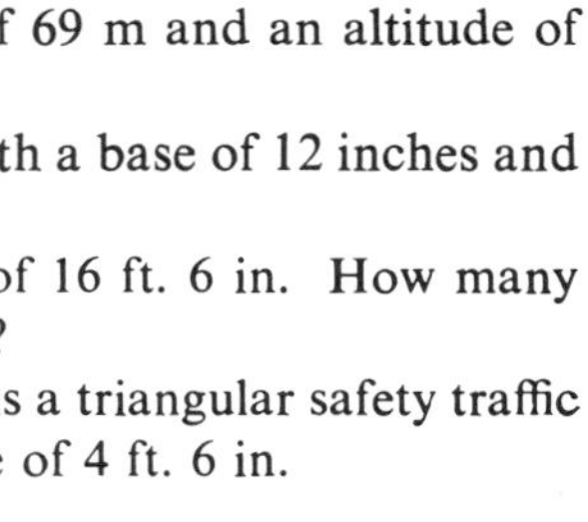

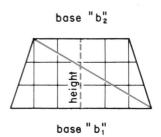

The figure at the right shows that the diagonal separates the trapezoid into two triangles which have a common height but different bases. The area of one triangle is $\frac{1}{2}b_1h$ and of the other is $\frac{1}{2}b_2h$. The area of the interior of the trapezoid is equal to the sum of the areas of the interiors of the two triangles or $\frac{1}{2}b_1h + \frac{1}{2}b_2h$. Using the distributive principle we find that the area of the interior of a trapezoid is equal to the height times the average of the two parallel sides (bases). Expressed as a formula it is $A = h \times \frac{b_1 + b_2}{2}$ or sometimes as $A = \frac{h}{2}(b_1 + b_2)$

Find the area of a trapezoid with bases of 18 feet and 14 feet and a height of 15 feet:

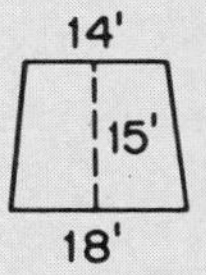

$A = h \times \frac{b_1 + b_2}{2}$

$A = 15 \times \frac{18 + 14}{2}$

$A = 15 \times 16$

$A = 240$ sq. ft.

$$\begin{array}{r} 18 \\ +14 \\ \hline 32 \end{array} \qquad \begin{array}{r} 15 \\ \times 16 \\ \hline 90 \\ 15 \\ \hline 240 \end{array}$$

$$\begin{array}{r} 16 \\ 2\overline{)32} \end{array}$$

$h = 15$ ft.
$b_1 = 18$ ft.
$b_2 = 14$ ft.
$A = ?$

Answer, 240 sq. ft.

EXERCISES

Find the areas of trapezoids having the following dimensions:

1.

Height	9 in.	21 ft.	13 yd.	15 cm	85 mm	37 m
Upper Base	8 in.	17 ft.	46 yd.	19 cm	102 mm	26 m
Lower Base	16 in.	29 ft.	53 yd.	26 cm	97 mm	42 m

2.

Height	$5\frac{3}{4}$ in.	8 in.	6 yd.	8 in.	1 ft.	1 ft. 7 in.
Upper Base	$4\frac{1}{2}$ in.	$2\frac{3}{4}$ in.	$1\frac{2}{3}$ yd.	2 ft.	1 ft. 11 in.	2 ft. 4 in.
Lower Base	$11\frac{1}{2}$ in.	$6\frac{7}{8}$ in.	$3\frac{1}{4}$ yd.	1 ft. 6 in.	2 ft. 9 in.	1 yd. 8 in.

3. Find the area of a section of a tapered airplane wing which has the shape of a trapezoid. The two parallel sides, measuring 4 ft. 3 in. and 5 ft. 9 in., are 16 ft. apart.

4. A porch roof consists of 2 sections each shaped like a trapezoid. The bases of one section are 11 ft. and 13 ft. and of the other 17 ft. and 19 ft. The distance between the bases in both sections is 8 ft. At the spreading rate of 250 sq. ft. per gallon, will a gallon of paint be sufficient to cover the entire roof?

5. Find the cost of seeding a lawn shaped like a trapezoid with bases of 33 ft. and 27 ft. and an altitude of 18 ft. One pound of grass seed covers 135 sq. ft. and costs $2.25.

14–41 AREA OF A CIRCLE

If we divide the interior of a circle into sectors and arrange them in a shape approximating a parallelogram, we see that the area of the interior of a circle equals the area of the interior of the parallelogram whose base is one-half the circumference ($\frac{1}{2}c$) and that the altitude is the radius (r). The formula for this is $A = \frac{1}{2}cr$. Since $c = 2\pi r$, we may substitute $2\pi r$ for c in the formula of $A = \frac{1}{2}cr$ to obtain $A = \frac{1}{2} \cdot 2\pi r \cdot r$ which simplified is $A = \pi r^2$. The area of a circle is equal to pi (π) times the radius squared.

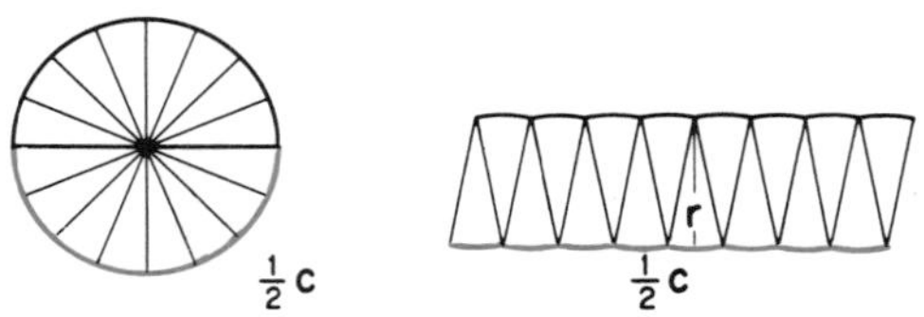

Or by substituting $\frac{d}{2}$ for r in the formula $A = \pi r^2$ we obtain the formula $A = \frac{1}{4}\pi d^2$ which may also be used to find the area of the interior of a circle.

Find the area of a circle having a radius of 8 inches:

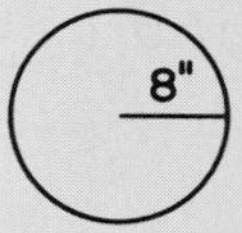

$r = 8$ in.
$\pi = 3.14$
$A = ?$

$A = \pi r^2$
$A = 3.14 \times (8)^2$
$A = 3.14 \times 64$
$A = 200.96$ sq. in.

$$\begin{array}{r} 8 \\ \times 8 \\ \hline 64 \end{array} \qquad \begin{array}{r} 3.14 \\ \times 64 \\ \hline 1256 \\ 1884 \\ \hline 200.96 \end{array}$$

Answer, 200.96 sq. in.

Find the area of a circle having a diameter of 28 millimeters:

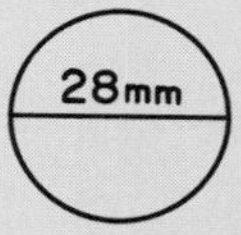

$d = 28$ mm
$\pi = \frac{22}{7}$
$A = ?$

$A = \frac{1}{4}\pi d^2$
$A = \frac{1}{4} \times \frac{22}{7} \times (28)^2$
$A = 616\ \text{mm}^2$

Answer, 616 mm²

EXERCISES

Find the area of a circle having a radius of:

1. a. 6 ft.	**b.** 35 in.	**c.** 78 cm	**d.** 112 mm
2. a. 7.3 km	**b.** $1\frac{3}{4}$ in.	**c.** $\frac{5}{16}$ mi.	**d.** 4.625 m
3. a. $11\frac{3}{8}$ in.	**b.** 2 ft. 11 in.	**c.** 9 yd. 1 ft.	**d.** 8 ft. 9 in.

Find the area of a circle having a diameter of:

4. a. 5 ft.	**b.** 42 in.	**c.** 110 mm	**d.** 56 m
5. a. 8.6 cm	**b.** $4\frac{1}{2}$ ft.	**c.** $3\frac{15}{16}$ in.	**d.** 0.18 km
6. a. $\frac{7}{8}$ in.	**b.** 2 ft. 6 in.	**c.** 8 ft. 2 in.	**d.** 1 yd. 1 ft.

7. Over how many square miles can a program be received if a T.V. station can televise programs for a distance of 42 miles?

8. What is the area of a circular skating rink 50 meters in diameter?
9. The viewing area of a 21-inch circular television screen is how many times as large as the area of a 14-inch circular screen?
10. If the lookout in the "crow's nest" of a ship can see for a distance of 12 miles, over how many square miles can he observe?
11. What is the area of the largest circle that can be cut out of paper 9 in. by 12 in.?
12. If the pressure is the same, how many times as much water will flow through a pipe 24 inches in diameter as through a 2-inch pipe?
13. A circular fountain 10 ft. in diameter has a circular walk 3 ft. wide paved around it. What is the area of the walk? At $1.75 per sq. ft., what was the cost of the walk?
14. How many times as large is the area of the 200-inch mirror of the world's largest telescope compared to the area of the previous record 100-inch telescope?
15. Find the area of the head of the piston if the diameter is $4\frac{3}{8}$ inches.
16. What is the cross-sectional area of the metal of a steel tube with a wall thickness of 0.25 inch and an outside diameter of 3 inches?

14–42 TOTAL AREA OF A RECTANGULAR SOLID AND A CUBE

The total area of the outside surface of a rectangular solid is the area of its six faces. Expressed as a formula it is $A = 2\,lw + 2\,lh + 2\,wh$.

The total area of the outside surface of a cube is the area of its six congruent square faces. Expressed as a formula it is $A = 6\,e^2$ or $A = 6\,s^2$.

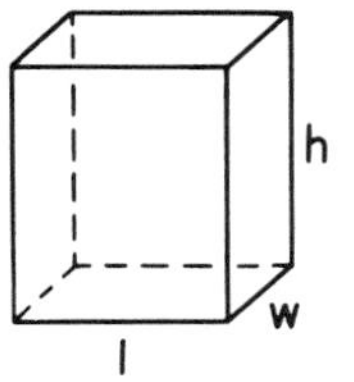

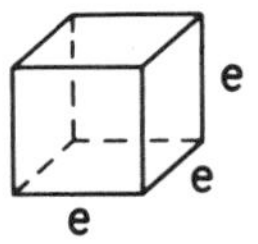

EXERCISES

1. Find the total areas of rectangular solids with the following dimensions:

Length	18 ft.	23 in.	54 mm	9.2 cm	$5\frac{3}{4}$ in.	2 ft. 6 in.
Width	6 ft.	17 in.	25 mm	5.4 cm	4 in.	1 ft. 9 in.
Height	15 ft.	39 in.	40 mm	7.5 cm	$6\frac{1}{2}$ in.	3 ft.

2. Find the total areas of cubes whose sides measure: **a.** 9 m **b.** 25 ft. **c.** 40 mm **d.** 17 in. **e.** $3\frac{1}{2}$ ft. **f.** 5.8 cm **g.** $4\frac{3}{4}$ in. **h.** 2 ft. 8 in.

3. How many square feet of cardboard are needed to make 5 dozen boxes with lids? Each box is to be 12 in. long, 9 in. wide, and 6 in. deep. Each lid is to have an overlapping band 1 inch wide all around. Allow 10% for waste.
4. At $5.40 per sq. yd., find the cost of plastering the walls and ceiling of a room 15 ft. long, 13 ft. wide, and 9 ft. high. Allow 33 sq. ft. for openings.
5. Allowing 82 sq. ft. for windows and doorway, how many gallons of paint are needed to cover the walls and ceiling of a room 17 ft. long, 16 ft. wide, and 10 ft. high with two coats of paint? One gallon will cover 425 sq. ft. in one coat. At $9.25 a gallon, how much will the paint cost?
6. How many double rolls of wallpaper are needed to cover the walls of a room 16 ft. long, 12 ft. wide, and 8 ft. high if each double roll covers 72 sq. ft.? Deduct 52 sq. ft. for openings and allow $\frac{1}{2}$ roll extra paper for matching. Find the cost of the wallpaper at $5.95 per roll.

14–43 LATERAL AREA AND TOTAL AREA OF A RIGHT CIRCULAR CYLINDER

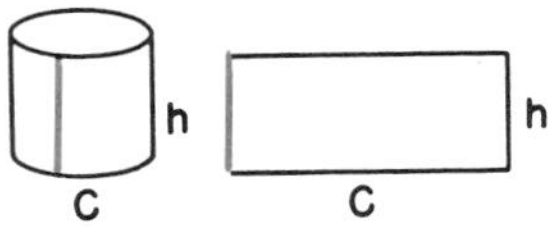

If we cut the label on a can vertically and flatten it out, we see that the lateral area or the area of the curved surface is equal to the area of a rectangle whose dimensions are those of the circumference and the height of the can. The lateral area is equal to the product of the circumference and height or $A = ch$. Since $c = \pi d$, we may substitute πd for c in $A = ch$ which then becomes $A = \pi dh$. Since $c = 2\pi r$, we may substitute $2\pi r$ for c in $A = ch$ which then becomes $A = 2\pi rh$. Thus the lateral area may be computed by using any one of the following formulas:

$$A = ch \quad \text{or} \quad A = \pi dh \quad \text{or} \quad A = 2\pi rh$$

Since the lower and upper bases are congruent circles (area of each is πr^2), the total area of a right circular cylinder equals the lateral area plus the area of the two bases. Any one of the following formulas may be used:

$$A = 2\pi rh + 2\pi r^2 \quad \text{or} \quad A = 2\pi r(r + h) \quad \text{or} \quad A = \pi dh + \tfrac{1}{2}\pi d^2$$

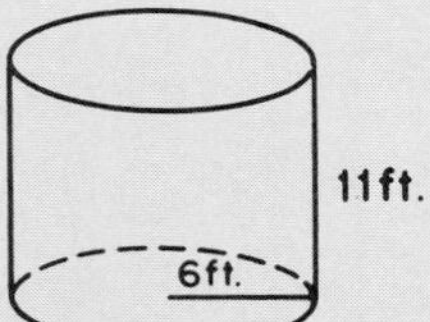

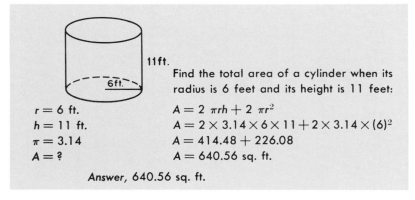

Find the total area of a cylinder when its radius is 6 feet and its height is 11 feet:

$r = 6$ ft.
$h = 11$ ft.
$\pi = 3.14$
$A = ?$

$A = 2\pi rh + 2\pi r^2$
$A = 2 \times 3.14 \times 6 \times 11 + 2 \times 3.14 \times (6)^2$
$A = 414.48 + 226.08$
$A = 640.56$ sq. ft.

Answer, 640.56 sq. ft.

EXERCISES

1. Find the lateral area of cylinders with the following dimensions:

a.

Radius	9 in.	35 mm	16 cm	$4\frac{1}{2}$ in.	$19\frac{1}{4}$ ft.	5 ft. 8 in.
Height	14 in.	40 mm	13 cm	8 in.	$24\frac{2}{3}$ ft.	10 ft.

b.

Diameter	5 m	13 cm	42 ft.	$3\frac{15}{16}$ in.	$10\frac{1}{2}$ ft.	3 ft. 6 in.
Height	6 m	29 cm	25 ft.	4 in.	$15\frac{3}{4}$ ft.	8 ft. 4 in.

2. Find the total areas of cylinders with the following dimensions:

a.

Radius	6 ft.	14 cm	$17\frac{1}{2}$ ft.
Height	8 ft.	30 cm	$27\frac{1}{2}$ ft.

b.

Diameter	4 in.	49 mm	$6\frac{1}{2}$ in.
Height	7 in.	38 mm	$3\frac{1}{8}$ in.

3. A tennis court roller is 18 inches in diameter and 3 feet long. How much surface is rolled in one rotation?
4. How many square feet of paper are needed to make 7,200 labels for cans, each 7 inches high and 4 inches in diameter?
5. How many square feet of asbestos covering are used to enclose the curved surface and the two ends of a water storage tank 14 inches in diameter and 6 feet high?

VOLUME—MEASURE OF SPACE

Space has three dimensions: length, width, and height (or depth or thickness). We live in a three-dimensional world. The volume, also called capacity or cubical contents, is the number of units of cubic measure contained in a given space. When computing the volume of a geometric solid, we express all linear units in the same denomination.

14–44 VOLUME OF A RECTANGULAR SOLID

A one-inch cube contains a volume of 1 cubic inch. The rectangular solid at the right has 4 cubes in each row, 3 rows of cubes and 2 layers of cubes. In one layer there are 4 × 3 or 12 cubes; in two layers there are 4 × 3 × 2 or 24

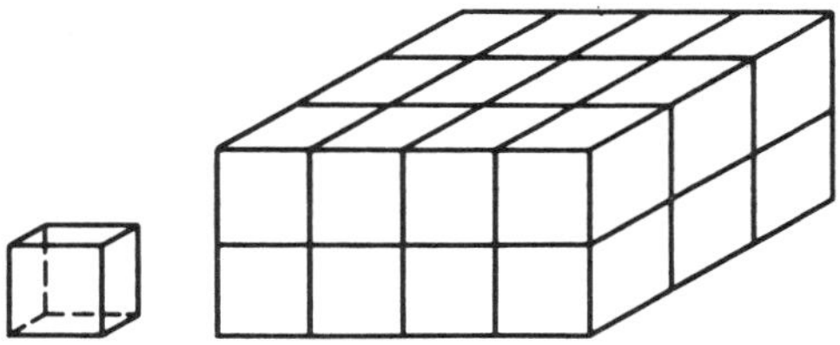

cubes which contain a total volume of 24 cubic inches. The volume of a rectangular solid is equal to the length times the width times the height. Expressed as a formula it is $V = lwh$. Sometimes the formula $V = Bh$ is used where B is the area of the base (lw) of the rectangular solid.

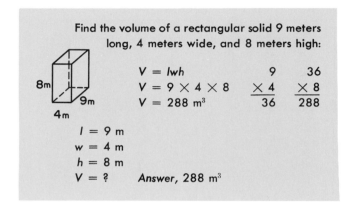

Find the volume of a rectangular solid 9 meters long, 4 meters wide, and 8 meters high:

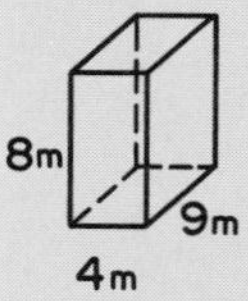

$V = lwh$
$V = 9 \times 4 \times 8$
$V = 288\ m^3$

$$\begin{array}{r} 9 \\ \times 4 \\ \hline 36 \end{array} \qquad \begin{array}{r} 36 \\ \times 8 \\ \hline 288 \end{array}$$

$l = 9$ m
$w = 4$ m
$h = 8$ m
$V = ?$

Answer, 288 m^3

EXERCISES

Find the volume of rectangular solids having the following dimensions:

	Length	15 ft.	27 in.	59 mm	6.5 cm	9.25 m
1.	Width	9 ft.	16 in.	32 mm	4.8 cm	6.2 m
	Height	13 ft.	14 in.	70 mm	5.3 cm	4.75 m

	Length	$4\frac{1}{2}$ in.	$2\frac{1}{4}$ ft.	1 yd.	3 ft. 8 in.	8 ft.
2.	Width	$5\frac{3}{4}$ in.	$1\frac{1}{3}$ ft.	2 ft. 6 in.	4 ft. 2 in.	6 ft. 6 in.
	Height	8 in.	$3\frac{1}{2}$ ft.	2 ft.	5 ft. 6 in.	3 ft. 9 in.

3. A classroom is 10 m long, 9 m wide, and 5 m high. If 30 pupils are assigned to the room, how many cubic meters of air space does this allow for each pupil?

4. How many cubic feet of dirt will fill a box 12 ft. long, 10 in. wide, and 6 in. deep?

5. At $6.80 per cu. yd., how much will it cost to dig a basement 36 ft. long, 14 ft. wide, and 9 ft. deep?

6. A swimming pool 100 ft. long and 32 ft. wide is filled to an average depth of 5 ft. What is the weight of the water if a cu. ft. of water weighs $62\frac{1}{2}$ lb.? How many gallons of water does the swimming pool contain if a cubic foot holds $7\frac{1}{2}$ gallons? At the rate of 400 gallons per minute, how long will it take to fill the pool?

7. How many cartons, 2 ft. by 2 ft. 6 in. by 1 ft., can be stored on a freight car with inside dimensions of 40 ft. by 8 ft. by 8 ft.?

8. What is the weight of a steel bar 16 ft. long, 3 in. wide, and $\frac{3}{4}$ in. thick? A cu. ft. of steel weighs 490 lb.

14–45 VOLUME OF A CUBE

Since the length, width and height of a cube are all equal to the length of the edge of the cube, the formula $V = lwh$ becomes $V = e \times e \times e$ or $V = e^3$. The volume of a cube is equal to the length of its edge cubed. Sometimes the formula $V = s^3$ is used.

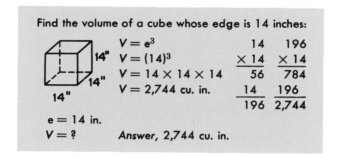

Find the volume of a cube whose edge is 14 inches:

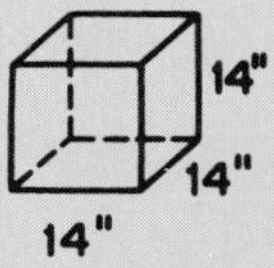

$V = e^3$
$V = (14)^3$
$V = 14 \times 14 \times 14$
$V = 2{,}744$ cu. in.

$$\begin{array}{r} 14 \\ \times 14 \\ \hline 56 \\ 14 \\ \hline 196 \end{array} \qquad \begin{array}{r} 196 \\ \times 14 \\ \hline 784 \\ 196 \\ \hline 2{,}744 \end{array}$$

$e = 14$ in.
$V = ?$

Answer, 2,744 cu. in.

EXERCISES

Find the volumes of cubes whose sides measure:

1. a. 8 in. **b.** 34 ft. **c.** 50 mm **d.** 26 ft.

2. a. .04 m **b.** 39.3 cm **c.** 3.28 m **d.** 21.75 cm

3. a. $16\frac{1}{2}$ ft. **b.** $2\frac{3}{4}$ in. **c.** $1\frac{7}{8}$ in. **d.** $8\frac{2}{3}$ ft.

4. a. 1 ft. 6 in. **b.** 4 ft. 2 in. **c.** 3 yd. 1 ft. **d.** 2 yd. 10 in.

5. Find the weight of a cubical column of sea water, 15 ft. long, wide, and deep. A cubic foot of sea water weighs 64 lb.

6. How many bushels will a storage bin hold if it measures $7\frac{1}{2}$ ft. on each side? $1\frac{1}{4}$ cu. ft. = 1 bushel.

7. What is the weight of a cake of ice, $1\frac{1}{2}$ ft. long, wide, and thick? Ice weighs 57 lb. per cu. ft.

8. If 1 ton of coal occupies 35 cu. ft., how many tons of coal could be stored in a bin measuring $10\frac{1}{2}$ ft. by $10\frac{1}{2}$ ft. by $10\frac{1}{2}$ ft.?

9. Which has a greater volume and how much greater, a group of three 4-centimeter cubes or four 3-centimeter cubes?

10. The volume of a 12-millimeter cube is how many times as large as the volume of a 2-millimeter cube?

14–46 VOLUME OF A RIGHT CIRCULAR CYLINDER

To determine the formula for the volume of a right circular cylinder we apply the principle that the volume is equal to the area of the base of the cylinder times the height ($V = Bh$). The same principle is used in determining the volume of a rectangular solid. Since the area of the base of the cylinder is the area of a circle, the formula $V = Bh$ becomes $V = \pi r^2 h$. Thus the volume of a cylinder is equal to pi (π) times the square of the radius of the base times the height.

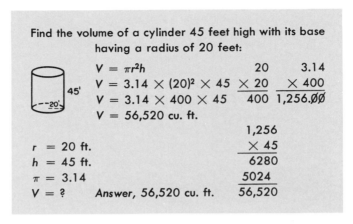

Find the volume of a cylinder 45 feet high with its base having a radius of 20 feet:

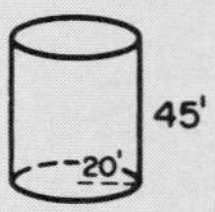

$V = \pi r^2 h$

$V = 3.14 \times (20)^2 \times 45$

$V = 3.14 \times 400 \times 45$

$V = 56{,}520$ cu. ft.

$$\begin{array}{r} 20 \\ \times\ 20 \\ \hline 400 \end{array} \qquad \begin{array}{r} 3.14 \\ \times\ 400 \\ \hline 1{,}256.\not{0}\not{0} \end{array} \qquad \begin{array}{r} 1{,}256 \\ \times\ 45 \\ \hline 6280 \\ 5024 \\ \hline 56{,}520 \end{array}$$

$r = 20$ ft.
$h = 45$ ft.
$\pi = 3.14$
$V = ?$

Answer, 56,520 cu. ft.

When the diameter is known, the formula $V = \frac{1}{4}\pi d^2 h$ may be used.

EXERCISES

Find the volumes of cylinders having the following dimensions:

1.

Radius	6 in.	7 ft.	42 mm	9.8 cm	3.5 m
Height	9 in.	20 ft.	35 mm	6.5 cm	10 m

2.

Radius	$3\frac{1}{2}$ in.	$4\frac{3}{8}$ in.	$5\frac{3}{4}$ ft.	2 ft. 4 in.	1 ft. 3 in.
Height	10 in.	$9\frac{1}{4}$ in.	$4\frac{2}{3}$ ft.	8 ft.	4 ft. 2 in.

3.

Diameter	5 in.	28 ft.	39 mm	4.1 m	2.8 cm
Height	8 in.	25 ft.	56 mm	3.7 m	8.25 cm

4.

Diameter	$3\frac{1}{2}$ in.	$5\frac{1}{4}$ ft.	$6\frac{3}{4}$ in.	1 yd. 6 in.	5 ft. 8 in.
Height	4 in.	$10\frac{2}{3}$ ft.	$2\frac{5}{8}$ in.	10 ft.	2 ft. 10 in.

5. How many cubic yards of dirt must be removed in digging a well 5 ft. in diameter and 54 ft. deep? At $15.40 per cu. yd., how much will the excavation cost?

6. An oil drum is 28 inches in diameter and 3 feet high. How many gallons of oil will it hold?

7. How many gallons will a gasoline tank hold if its diameter is 35 ft. and it is 30 ft. high?

8. How many bushels of silage will a silo hold if it has an inside diameter of 24 ft. and a height of 42 ft.?

9. A water storage tank has a diameter of 20 in. and a height of 6 ft. How many gallons of water will it hold?

14–47 VOLUME OF A SPHERE, RIGHT CIRCULAR CONE, AND PYRAMID

Sphere

The volume of a sphere is equal to $\frac{4}{3}$ times pi (π) times the cube of the radius. Expressed as a formula it is $V = \frac{4}{3}\pi r^3$. Sometimes the formula $V = \frac{\pi d^3}{6}$ is used when the diameter is known.

EXERCISES

1. Find the volumes of spheres having the following radii:

a. 5 in. **b.** 21 cm **c.** $8\frac{3}{4}$ in. **d.** 2 ft. 6 in.

2. Find the volumes of spheres having the following diameters:

a. 18 in. **b.** 49 mm **c.** $2\frac{5}{8}$ in. **d.** 6 ft. 5 in.

Right Circular Cone

The volume of a right circular cone is equal to $\frac{1}{3}$ times pi (π) times the square of the radius of the base times the height. Expressed as a formula it is $V = \frac{1}{3}\pi r^2 h$.

EXERCISES

Find the volumes of right circular cones having the following dimensions:

a. Radius: 6 mm, height: 10 mm

b. Radius: $3\frac{1}{2}$ ft., height: 9 ft.

c. Diameter: 20 ft., height: 35 ft.

d. Diameter: 8 ft. 9 in., height: 15 ft.

Pyramid

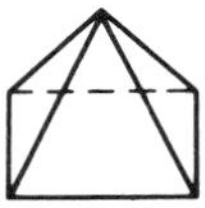

The volume of a pyramid is equal to $\frac{1}{3}$ times the area of the base times the height. Expressed as a formula it is $V = \frac{1}{3}Bh$.

EXERCISES

Find the volumes of square pyramids having the following dimensions:

Side of base	8 mm	200 ft.	7 ft. 6 in.	$5\frac{1}{2}$ in.	6.4 m
Height	13 mm	50 ft.	9 ft.	18 in.	5.1 m

Problems

1. Find the volume of each of the following figures:

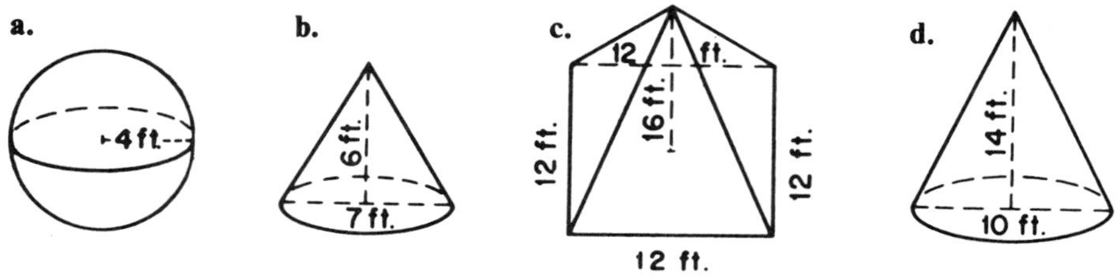

2. What is the weight of a steel ball 7 inches in diameter? A cubic foot of steel weighs 490 lb.
3. How many tons of coal are in a conical pile of coal 21 ft. in diameter and 10 ft. high? 1 ton of coal occupies 35 cu. ft.
4. How many cu. ft. of space are inside a tent in the shape of a square pyramid 16 ft. on each side of the base and 15 ft. high?
5. How many cu. ft. of space does the earth satellite, Vanguard I, occupy if its diameter is 6.4 inches?
6. How many cu. yd. of sand are in a conical pile of sand 14 ft. in diameter and 9 ft. high?

CHAPTER REVIEW

Part 1

1. a. Read, or write in words, each of the following: $\overline{AN}$, $\overrightarrow{FG}$, $\overleftrightarrow{ED}$. (14–1)
   b. Explain the difference between: line and ray; ray and segment; line and segment.
2. Name each of the following: (14–1)

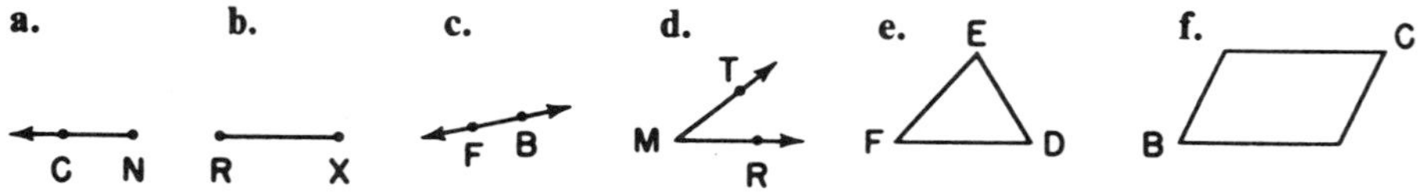

3. a. Find: $\overleftrightarrow{GL} \cap \overleftrightarrow{NT} = \{?\}$ (14–1)

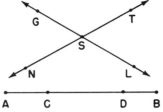

b. Are points G, S, and T collinear?

c. Are points G, S, and L collinear? (14–5)

d. Are $\overleftrightarrow{GL}$ and $\overleftrightarrow{NT}$ concurrent?

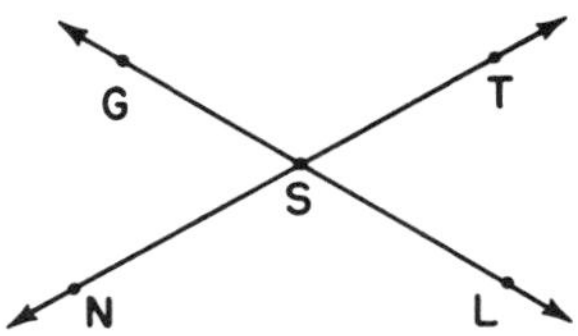

4. Points C and D divide $\overline{AB}$ into three parts. Name the three segments. (14–1)

5. Indicate by writing the corresponding letter which line is: (14–2)

a. A straight line.
b. A curved line.
c. A broken line.

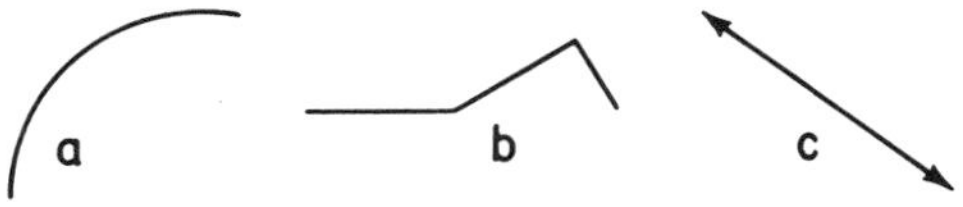

6. Indicate by writing the corresponding letter which line is in: (14–3)

a. A vertical position.
b. A horizontal position.
c. A slanting position.

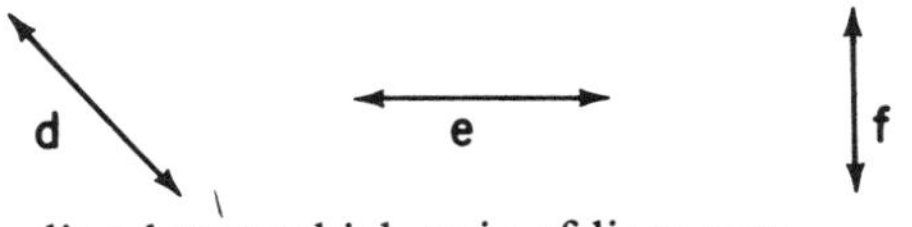

7. Indicate by writing the corresponding letter which pair of lines are: (14–4)

a. Parallel.
b. Perpendicular.

8. If the distance from C to D is 63 miles, what is the distance from X to Y?

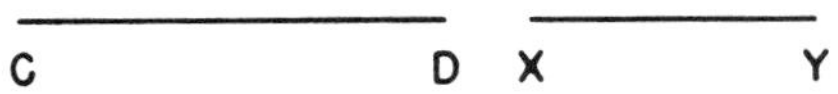

(14–11)

9. a. Can two lines intersect in more than one point?
b. How many points determine a line? Determine a plane? (14–5)
c. Can more than one line pass through 2 points? Can more than one plane pass through 3 points on a line?
d. How many points and lines does a plane contain?
e. What is the intersection of two planes?

10. a. How many diagonals can be drawn from any vertex of a parallelogram?
b. How many faces (F) does a cube have? How many vertices (V)? How many edges (E)? Does $F + V - E = 2$? (14–8)

Part 2

1. With protractor measure the following angle: (14–12)

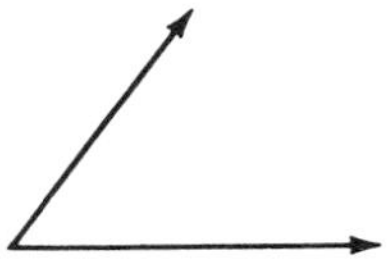

2. a. With protractor draw an angle of 145°. (14–13)
b. Draw a right angle.
c. Draw any acute angle.
d. Draw any obtuse angle. (14–7)

3. Draw a circle with a diameter of $1\frac{3}{8}$ inches. (14–8)
4. Draw a regular hexagon, each side measuring 38 millimeters. (14–22)
5. Construct a triangle with sides measuring $1\frac{1}{2}$ in., $1\frac{3}{4}$ in., and $1\frac{3}{8}$ in. (14–15)
6. Construct a triangle with sides measuring $1\frac{5}{8}$ in., $1\frac{1}{4}$ in., and an included angle of 40°. (14–15)
7. **a.** Draw any line segment. Using compasses, bisect this segment. (14–18)
   b. Draw any angle. Using compasses, bisect this angle. (14–20)
8. Draw any line. Using compasses, construct a perpendicular to this line:
   a. At a point on the line. **b.** From a point not on the line. (14–16, 14–17)
9. Draw with protractor an angle of 65°. Then construct with compasses an angle equal to it. Check with protractor. (14–19)
10. Draw any line. Locate a point outside this line. Through this point construct a line parallel to the first line. (14–21)

Part 3

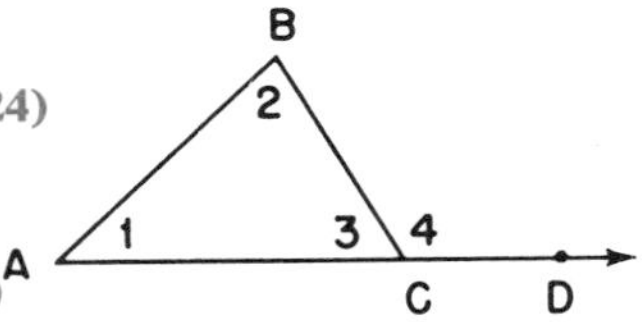

1. What is the complement of an angle of 38°? (14–24)
2. What is the supplement of an angle of 96°?
3. **a.** What is the sum of measures of ∡ 1, 2, and 3 of triangle *ABC*? (14–23)
   b. If $m\angle 1 = 56°$ and $m\angle 4 = 118°$, find the measure of $\angle 3$. Of $\angle 2$. (14–24)

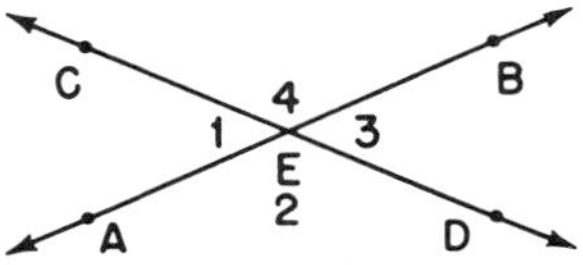

4. $\overleftrightarrow{AB}$ and $\overleftrightarrow{CD}$ intersect at *E*. If $m\angle 4 = 115°$, what is the measure of $\angle 1$? $\angle 2$? $\angle 3$? (14–24)

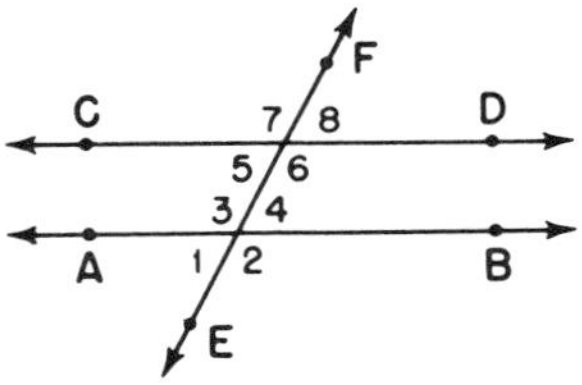

5. Parallel lines *AB* and *CD* are cut by transversal *EF*. If $m\angle 7 = 104°$, what is the measure of $\angle 5$? $\angle 6$? $\angle 4$? $\angle 1$? $\angle 2$? $\angle 8$? $\angle 3$? (14–25)

Part 4

Find the perimeter of:

1. A rectangle 23 centimeters long and 19 centimeters wide. (14–32)
2. A triangle with sides measuring $6\frac{3}{4}$ in., $5\frac{7}{8}$ in., and $7\frac{3}{8}$ in. (14–34)
3. A square whose side measures 34 yards. (14–33)

Find the circumference of:

4. A circle whose diameter is 42 feet. (14–35)
5. A circle whose radius is 57 millimeters. (14–35)
6. How many feet of fencing are required to enclose a rectangular garden 73 ft. long and 59 ft. wide. At $4.30 a yard, how much will the fencing cost? (14–32)
7. How many feet do you ride in one turn of a merry-go-round if you sit 21 feet from the center? (14–35)

Find the area of:

8. A square whose side measures 27 meters. (14–37)
9. A circle whose radius is 49 feet. (14–41)
10. A rectangle 104 centimeters long and 95 centimeters wide. (14–36)
11. A parallelogram with an altitude of 76 ft. and a base of 68 ft. (14–38)
12. A circle whose diameter is 60 meters. (14–41)
13. A triangle whose altitude is 17 inches and base is 26 inches. (14–39)
14. A trapezoid with bases of 91 ft. and 57 ft. and a height of 48 ft. (14–40)
15. A house 42 ft. square and a garage 16 ft. long and 10 ft. wide are built on a lot 108 ft. long and 61 ft. wide. How many sq. ft. of the lot remains? (14–36, 14–37)
16. At $6.00 per sq. ft., how much will it cost to resilver a circular mirror 35 inches in diameter? (14–41)

Find the volume of:

17. A cube whose edge measures 39 millimeters. (14–45)
18. A rectangular solid 17 inches long, 13 inches wide, and 25 inches high. (14–44)
19. A right circular cylinder with a diameter of 28 ft. and a height of 30 ft. (14–46)
20. A sphere whose diameter is 6 centimeters. (14–47)
21. A right circular cylinder with a radius of 4 in. and a height of 9 in. (14–46)
22. A square pyramid 20 ft. on each side and 18 ft. high. (14–47)
23. A right circular cone 14 ft. in diameter and 17 ft. high. (14–47)

Find the total area of:

24. A cube 15 centimeters on a side. (14–42)
25. A rectangular solid 28 m long, 23 m wide, and 14 m high. (14–42)
26. A right circular cylinder 70 inches in diameter and 45 inches high. (14–43)
27. A circular wading pool 40 ft. in diameter is filled to an average depth of 2 ft. How many gallons of water does the pool contain? (14–46)
28. How many bushels will a storage bin hold if it is 15 ft. long, 12 ft. wide, and 8 ft. deep? (14–44)
29. At $1.30 per sq. yd., find the cost of painting the walls and ceiling of a room 18 ft. long, 14 ft. wide, and 10 ft. high? Allow 55 sq. ft. for openings. (14–42)
30. Allowing 2 inches for the seam, how many square feet of tin are needed to make a pipe 14 feet long and 10 inches in diameter? (14–43)

Part 5

1. Find the hypotenuse of a right triangle if the altitude is 105 cm and the base is 56 cm. (14–28)
2. Find the altitude of a right triangle if the hypotenuse is 195 ft. and the base is 48 ft. (14–28)
3. Find the base of a right triangle if the altitude is 270 m and the hypotenuse is 318 m. (14–28)
4. How many feet does a person save by walking diagonally across a field 385 ft. long and 336 ft. wide instead of around two sides? (14–28)

Find the indicated missing parts in the following:

5. Congruent triangles (14–26)

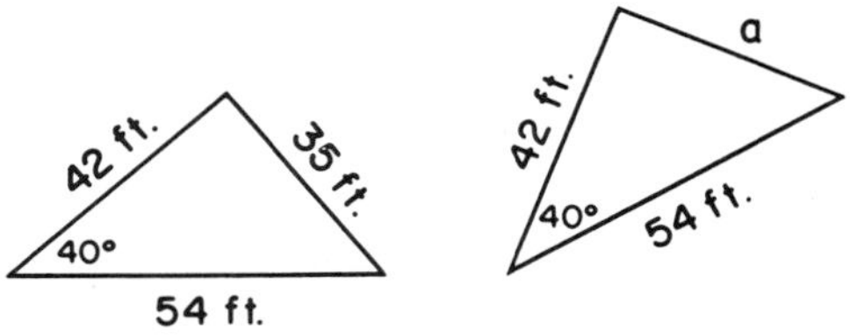

6. Similar triangles (14–27)

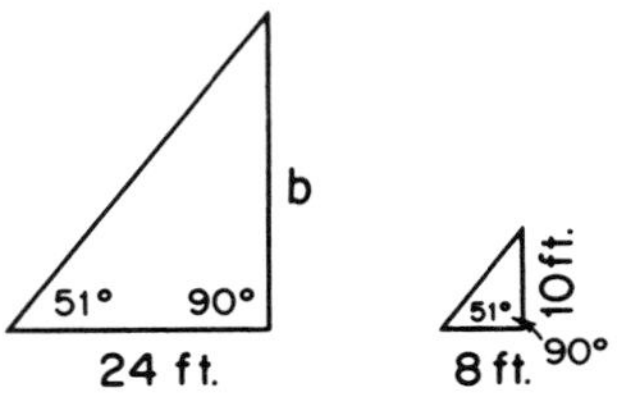

7. Find the height of a telegraph pole that casts a shadow of 10 feet at a time when a boy, 5 feet tall, casts a shadow of 2 feet. (14–29)

8. Find the distance across the stream (MN) shown in the figure below. (14–29)

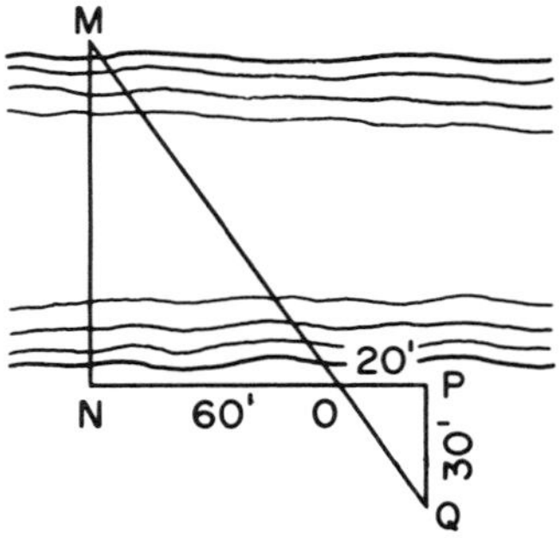

9. Two telegraph poles, 26 ft. and 39 ft. respectively, are 84 ft. apart. Find the length of the wire that is attached to each pole, 2 ft. below the top. (14–28)

10. Find the indicated parts of the following triangles: (14–31)

a. Find side a.

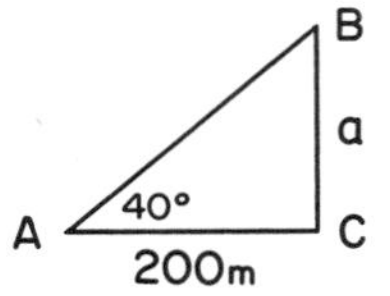

b. Find side b.

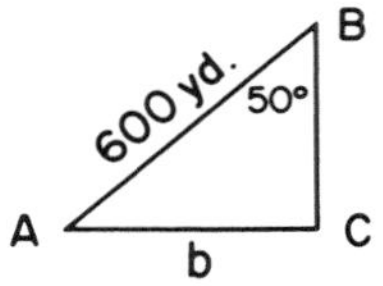

c. Find side b.

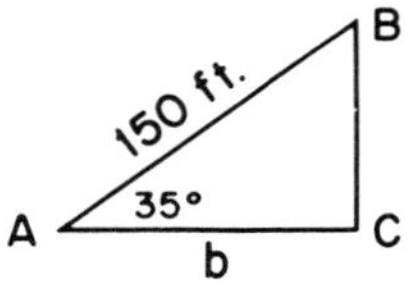

Applications of Mathematics

15

INCOME; TAKE-HOME PAY; INCOME TAX

15–1 COMPUTING INCOME

Income includes earnings (salary or wages) per hour, day, week, month, or year; commissions, profits, fees, tips, interest, dividends, bonuses, and pensions.

EXERCISES

1. Mr. Clark works 40 hours each week. If he receives $6.60 per hour, how much does he earn each week?
2. Find the weekly earnings for each of the following:

No. of hours	32	40	37	35	38	40
Rate per hour	$3.65	$3.96	$6.64	$4.30	$8.75	$13.25

3. If the overtime rate is $1\frac{1}{2}$ times the regular rate, what is the hourly overtime rate corresponding to each of the following regular hour rates:

   a. $3.80 **b.** $5.86 **c.** $7.15 **d.** $4.20 **e.** $6.75 **f.** $9.20 **g.** $10.50
4. Find the weekly earnings (time-and-a-half rate over 40 hr.) for each of the following:

No. of hours	42	45	$50\frac{1}{2}$	43	$47\frac{1}{2}$	48
Rate per hour	$5.80	$3.77	$4.20	$6.90	$12.10	$9.65

5. Find the annual salary of a person earning:

   a. $85 per week **b.** $810 per month **c.** $239.25 per week **d.** $568.75 semi-monthly **e.** $3.40 per hr., 40 hr. week **f.** $465.50 per week **g.** $2,500 monthly
6. If a woman works 40 hours per week, what is her weekly salary, annual salary, and monthly salary at each of the following hourly rates:

   a. $6.90 **b.** $7.85 **c.** $3.68 **d.** $4.05 **e.** $5.17 **f.** $9.46 **g.** $11.50?
7. What is your monthly salary and weekly salary if your annual salary is:

   a. $2,860 **b.** $6,600 **c.** $4,500 **d.** $8,700 **e.** $12,000 **f.** $26,000 **g.** $39,000?

8. Which is a better wage:
   a. \$5,400 per year or \$460 per month? b. \$175 per week or \$8,800 per year?
   c. \$160 per week or \$650 per month? d. \$1,200 per month or \$15,000 per year?
   e. \$13,500 per year or \$260 per week? f. \$215 per week or \$900 per month?
9. Last year Mr. Martin received an 8% increase on his weekly salary of \$290. Two weeks ago his salary was reduced 8%. How does his present salary compare with his original salary before the increase?
10. Miss Caruso receives \$160 per week and 4% commission on sales. Last week her sales totaled \$3,784. What were her total earnings for the week?
11. Mr. Scott receives a salary of \$78 per week and 7% bonus on all sales over the quota of \$875. If his weekly sales are \$2,540, what are his total weekly earnings?
12. A lawyer collected 75% of a debt of \$920. If she charged 15% commission, how much did the lawyer receive?

15–2 WITHHOLDING TAX

Withholding tax is the amount of income tax that employers are required by the federal government to withhold from their employees' earnings. This money is forwarded to the Collector of Internal Revenue's offices, where it is entered on each employee's account. The tax rates may change from year to year.

The employer may use either the percentage method or the income tax withholding tables to compute the amount of income tax to be deducted and withheld from a payment of wages to an employee.

To compute the amount of weekly withholding tax for a married person by the percentage method, first find the difference between the total wage payment and the number of withholding allowances (exemptions) times \$14.40. Then use the following rates:

When this difference is:		Amount of withholding tax:	
Not over \$61		0	
Over—	*But not over—*		*Of excess over—*
\$61	— \$105	15%	— \$61
\$105	— \$223	\$6.60, plus 18%	— \$105
\$223	— \$278	\$27.84, plus 22%	— \$223
\$278	— \$355	\$39.94, plus 25%	— \$278
\$355	— \$432	\$59.19, plus 28%	— \$355
\$432	— \$509	\$80.75, plus 32%	— \$432
\$509		\$105.39, plus 36%	— \$509

To compute the amount of weekly withholding tax for a married person using the income tax withholding tables, locate the row indicating the weekly salary, then select the amount of tax from the column corresponding to the number of withholding allowances claimed.

EXERCISES

1. Use the percentage method to compute the amount of income tax to be withheld each week if a married person's weekly wages are:

a. $72 and 5 allowances are claimed.
b. $90 and 2 allowances are claimed.
c. $165 and 1 allowance is claimed.
d. $280 and 4 allowances are claimed.
e. $525 and 3 allowances are claimed.
f. $58.50 and 2 allowances are claimed.
g. $99.25 and 1 allowance is claimed.
h. $104.50 and 3 allowances are claimed.
i. $230.75 and 4 allowances are claimed.
j. $386.25 and 2 allowances are claimed.

TABLE OF WITHHOLDING TAX—FOR MARRIED PERSONS—WEEKLY PAYROLL PERIODS

And the wages are—		And the number of withholding allowances claimed is—						
At least	But less than	0	1	2	3	4	5	6
		The amount of income tax to be withheld shall be—						
$ 88	$ 90	$ 4.30	$ 2.10	$ 0	$ 0	$ 0	$ 0	$ 0
90	92	4.60	2.40	.20	0	0	0	0
92	94	4.90	2.70	.50	0	0	0	0
94	96	5.20	3.00	.80	0	0	0	0
96	98	5.50	3.30	1.10	0	0	0	0
98	100	5.80	3.60	1.40	0	0	0	0
100	105	6.30	4.10	2.00	0	0	0	0
105	110	7.10	4.90	2.70	.50	0	0	0
110	115	8.00	5.60	3.50	1.30	0	0	0
115	120	8.90	6.40	4.20	2.00	0	0	0
120	125	9.80	7.20	5.00	2.80	.60	0	0
125	130	10.70	8.10	5.70	3.50	1.40	0	0
130	135	11.60	9.00	6.50	4.30	2.10	0	0
135	140	12.50	9.90	7.30	5.00	2.90	.70	0
140	145	13.40	10.80	8.20	5.80	3.60	1.50	0
145	150	14.30	11.70	9.10	6.50	4.40	2.20	.10
150	160	15.70	13.10	10.50	7.90	5.50	3.30	1.20
160	170	17.50	14.90	12.30	9.70	7.10	4.80	2.70
170	180	19.30	16.70	14.10	11.50	8.90	6.30	4.20
180	190	21.10	18.50	15.90	13.30	10.70	8.10	5.70
190	200	22.90	20.30	17.70	15.10	12.50	9.90	7.30
200	210	24.70	22.10	19.50	16.90	14.30	11.70	9.10
210	220	26.50	23.90	21.30	18.70	16.10	13.50	10.90
220	230	28.40	25.70	23.10	20.50	17.90	15.30	12.70
230	240	30.60	27.50	24.90	22.30	19.70	17.10	14.50
240	250	32.80	29.60	26.70	24.10	21.50	18.90	16.30
250	260	35.00	31.80	28.60	25.90	23.30	20.70	18.10
260	270	37.20	34.00	30.80	27.70	25.10	22.50	19.90
270	280	39.40	36.20	33.00	29.80	26.90	24.30	21.70
280	290	41.80	38.40	35.20	32.00	28.90	26.10	23.50
290	300	44.30	40.70	37.40	34.20	31.10	27.90	25.30

2. Use the above table of withholding tax to compute the amount of income tax to be withheld each week if a married person's weekly wages are:

a. $88 and 4 allowances are claimed.
b. $100 and 1 allowance is claimed.
c. $127 and 0 allowances are claimed.
d. $95 and 3 allowances are claimed.

e. $288 and 2 allowances claimed.
f. $96.50 and 5 allowances claimed.
g. $171.40 and 3 allowances claimed.
h. $204.25 and 2 allowances claimed.
i. $262.50 and 1 allowance claimed.
j. $197.75 and 4 allowances claimed.

3. Use the table of withholding tax to compute the amount of income tax withheld for the year if a married person's weekly wages are:

a. $90 and 2 allowances are claimed.
b. $115 and 1 allowance is claimed.
c. $203 and 3 allowances are claimed.
d. $156.50 and 2 allowances are claimed.
e. $246.25 and 4 allowances are claimed.
f. $210.75 and 1 allowance is claimed.

15–3 SOCIAL SECURITY

Social security tax is money deducted from employees' earnings to help pay old age pensions. Employees pay a tax at the rate of 6.13% of the first $22,900 of their annual wages. Employers pay the same amount, matching the amounts paid by the employees. The rate and the amount on which it is applied may change from year to year.

EXERCISES

Find the amount deducted from the employee's weekly wages for social security tax if an employee earns the following each week. Round the answers to the nearest cent.

1. $58
2. $94
3. $410
4. $182
5. $260
6. $350
7. $48.50
8. $63.75
9. $189.25
10. $218.75
11. $96.60
12. $135.10
13. $247.50
14. $184.75
15. $233.30

15–4 TAKE-HOME PAY

Take-home pay is the amount of earnings remaining after the withholding tax and social security tax are deducted from wages.

EXERCISES

In the following problems use the table of withholding tax on page 467 for all wages between $88 and $300; for all other wages use the rates given on page 466. The social security rate is 6.13% on the first $22,900 earned per year.

1. Mr. Smith earns $250 each week and claims 3 withholding allowances. How much income tax is withheld each week? How much is deducted for old age pension? What is his take-home pay?

2. Find the take-home pay if a married person's weekly wages are:

a. $120 with 2 allowances claimed.
b. $86 with 1 allowance claimed.
c. $173 with 4 allowances claimed.
d. $350 with 3 allowances claimed.
e. $104 with 2 allowances claimed.
f. $117.50 with 1 allowance claimed.
g. $96.25 with 3 allowances claimed.
h. $174.80 with 2 allowances claimed.
i. $410.75 with 1 allowance claimed.
j. $241.50 with 4 allowances claimed.

3. Which take-home pay is greater, a weekly wage of $191.50 with 1 withholding allowance claimed or $179.75 with 5 withholding allowances claimed?

15-5 FEDERAL INCOME TAX

An *income tax* is a tax on earnings. Below is a part of the tax table used by a married person who files a joint return. For the rest of this tax table and other tax tables, use the Internal Revenue Service tax publications. Income tax rates are subject to change.

To find the federal income tax owed, select the appropriate table, locate the row indicating the taxable income (actually the adjusted gross income less itemized deductions), then read across to the column headed by the total number of exemptions claimed.

If taxable income is—		And the total number of exemptions claimed on line 7 is—							
Over	But not over	2	3	4	5	6	7	8	9
		Your tax is—							
$7,900	$7,950	$418	$261	$109	$ 0	$ 0	$0	$0	$0
7,950	8,000	427	269	116	0	0	0	0	0
8,000	8,050	435	277	124	0	0	0	0	0
8,050	8,100	444	285	131	0	0	0	0	0
8,100	8,150	452	293	139	0	0	0	0	0
8,150	8,200	461	301	146	0	0	0	0	0
8,200	8,250	469	309	154	6	0	0	0	0
8,250	8,300	476	317	162	14	0	0	0	0
8,300	8,350	484	325	170	21	0	0	0	0
8,350	8,400	491	333	178	29	0	0	0	0
8,400	8,450	499	341	186	36	0	0	0	0
8,450	8,500	506	349	194	44	0	0	0	0
8,500	8,550	514	358	202	51	0	0	0	0
8,550	8,600	521	366	210	59	0	0	0	0
8,600	8,650	529	375	218	66	0	0	0	0
8,650	8,700	536	383	226	74	0	0	0	0
8,700	8,750	544	392	234	81	0	0	0	0
8,750	8,800	553	400	242	89	0	0	0	0
8,800	8,850	561	409	250	96	0	0	0	0
8,850	8,900	570	417	258	104	0	0	0	0
8,900	8,950	578	426	266	111	0	0	0	0
8,950	9,000	587	434	274	119	0	0	0	0
9,000	9,050	595	443	282	127	0	0	0	0
9,050	9,100	604	451	290	135	0	0	0	0
9,100	9,150	612	460	298	143	0	0	0	0
9,150	9,200	621	468	306	151	1	0	0	0
9,200	9,250	629	477	314	159	9	0	0	0
9,250	9,300	638	485	323	167	16	0	0	0
9,300	9,350	646	494	331	175	24	0	0	0
9,350	9,400	655	502	340	183	31	0	0	0
9,400	9,450	663	511	348	191	39	0	0	0
9,450	9,500	672	520	357	199	46	0	0	0
9,500	9,550	680	529	365	207	54	0	0	0
9,550	9,600	689	539	374	215	61	0	0	0
9,600	9,650	697	548	382	223	69	0	0	0
9,650	9,700	706	558	391	231	76	0	0	0
9,700	9,750	714	567	399	239	84	0	0	0
9,750	9,800	723	577	408	247	92	0	0	0
9,800	9,850	731	586	416	255	100	0	0	0
9,850	9,900	740	596	425	263	108	0	0	0
9,900	9,950	748	605	433	271	116	0	0	0
9,950	10,000	757	615	442	279	124	0	0	0

EXERCISES

For these problems use the earnings given as the taxable income in the tax table.

1. How much does a married woman with 3 exemptions, owe for the year if she earns \$8,850 per year? \$170 per week?
2. How much tax does a married man with 5 exemptions, owe for the year if he earns \$9,625 per year? \$185 per week?
3. How much tax does a married person, filing jointly, owe for the year, earning annually:
   a. \$9,150 and claiming 2 exemptions?
   b. \$8,750 and claiming 2 exemptions?
   c. \$8,000 and claiming 3 exemptions?
   d. \$9,600 and claiming 4 exemptions?
   e. \$9,325 and claiming 4 exemptions?
   f. \$7,910 and claiming 2 exemptions?
   g. \$8,475 and claiming 5 exemptions?
   h. \$9,960 and claiming 4 exemptions?
4. How much tax does a married person, filing jointly, owe for the year, earning weekly:
   a. \$163 and claiming 4 exemptions?
   b. \$152 and claiming 3 exemptions?
   c. \$175 and claiming 4 exemptions?
   d. \$190 and claiming 5 exemptions?
   e. \$186 and claiming 4 exemptions?
   f. \$179.50 and claiming 2 exemptions?
   g. \$162.75 and claiming 5 exemptions?
   h. \$191.25 and claiming 2 exemptions?
   i. \$158.90 and claiming 3 exemptions?
   j. \$177.40 and claiming 2 exemptions?
5. The Clark family consists of a husband, a wife, and two young children. If Mr. Clark is the sole wage-earner of the family, how many tax exemptions may he claim? If he earns \$184.50 each week, how much income tax is withheld from his pay each week? How much tax does he owe for the entire year? Does he still owe some taxes or did he overpay? How much?
6. Using the income tax table and table of withholding tax, find the balance of tax due or amount of overpayment for the year when a married person, filing jointly, earns weekly:
   a. \$160 and claims 2 exemptions.
   b. \$189 and claims 4 exemptions.
   c. \$155 and claims 5 exemptions.
   d. \$172 and claims 3 exemptions.
   e. \$191 and claims 4 exemptions.
   f. \$176.50 and claims 4 exemptions.
   g. \$191.75 and claims 2 exemptions.
   h. \$167.25 and claims 3 exemptions.
   i. \$153.40 and claims 5 exemptions.
   j. \$184.60 and claims 2 exemptions.

15–6 WAGE TAX

Many states and cities in the United States have, in addition to the federal income tax, another tax on earnings. This income tax is sometimes called a *wage tax*. It is usually expressed as a fixed rate of per cent of the earnings.

EXERCISES

1. If the wage tax rate is 3% of the wages, how much tax does a man owe for the week if he earns weekly: **a.** \$58? **b.** \$139? **c.** \$225? **d.** \$261.75? **e.** \$309.25?
2. If the wage tax rate is $1\frac{3}{4}$% of the wages, how much tax does a woman owe for the week if she earns weekly: **a.** \$149? **b.** \$74? **c.** \$226? **d.** \$282.50? **e.** \$363.40?

3. Find the take-home pay for each of the following married person's weekly wages, using the table of withholding tax (page 467), a 6.13% deduction for social security tax, and a 2% wage tax:

a. \$155 with 3 allowances claimed.
b. \$270 with 1 allowance claimed.
c. \$91 with 2 allowances claimed.
d. \$225 with 4 allowances claimed.
e. \$118 with 1 allowance claimed.
f. \$87.50 with 2 allowances claimed.
g. \$225.75 with 1 allowance claimed.
h. \$111.10 with 4 allowances claimed.
i. \$135.50 with 1 allowance claimed.
j. \$228.70 with 3 allowances claimed.

CONSUMER SPENDING

15–7 BUYING FOOD

LEE'S FOOD MARKET

See our specials for the week

Groceries

Beverages, 2 bottles for 79¢
Coffee, \$2.39 a pound
Corn, 5 cans for \$1.49
Flour, 5 lb. for 93¢
Peaches, 55¢ a can
Peas, 3 cans for 81¢
Soap, 4 bars for 59¢
Laundry detergent, \$1.98 a pkg.
Soups, 4 cans for 95¢
Sugar, 5 lb. for \$1.49
Tomato juice, 2 cans for \$1.05
Tuna fish, 3 cans for \$2

Meats, Fish, and Dairy

Bacon, \$1.69 a pound
Margarine, 84¢ a pound
Cheese, \$1.96 a pound
Chicken, 60¢ a pound
Eggs, 2 doz. for \$1.79
Flounder, \$1.39 a pound
Haddock, \$1.12 a pound
Turkey, 96¢ a pound
Milk, 2 qt. for \$1.09
Rib roast, \$2.56 a pound
Sirloin, \$2.40 a pound
Smoked ham, \$1.04 a pound

Fruit and Produce

Apples, 39¢ a pound
Bananas, 25¢ a pound
Grapefruit, 3 for 49¢
Grapes, 69¢ a pound
Lemons, 6 for 59¢
Oranges, 99¢ a dozen
Peaches, 49¢ a pound
Pears, 45¢ a pound
Carrots, 2 bunches for 49¢
Celery, 35¢ a stalk
Lettuce, 2 heads for 95¢
Onions, 3 lb. for 45¢
Peppers, 3 for 39¢
Potatoes, 5 lb. for 89¢
String beans, 32¢ a pound
Tomatoes, 48¢ a pound

Bakery

Assorted cakes, 59¢ a half-dozen
Bread, 2 loaves for 91¢
Rolls, 96¢ a dozen
Cookies, \$1.92 a pound

The ¢ symbol indicates cents. For easier computation you may be required to replace this symbol with the dollar mark ($) and a decimal point. For example: 49¢ may be written as $.49.

EXERCISES

When the result of your computation of the cost of a food item contains *any* fractional part of a cent, drop the fraction and add one cent to the cost of the item.

In each of the following problems, use the prices listed by Lee's Food Market.

1. What is the cost of:

a. 1 lb. onions?
b. 1 bottle beverage?
c. 1 doz. eggs?
d. 1 head of lettuce?
e. 1 orange?
f. 1 cake?
g. 1 can corn?
h. 1 bar soap?
i. 1 can tuna fish?
j. 1 can soap?

2. What is the cost of:

a. 2 cans peaches?
b. 3 lb. bananas?
c. 4 lb. flounder?
d. 12 lemons?
e. 5 lb. chicken?
f. 16 cans soup?
g. 6 doz. eggs?
h. 9 lb. rib roast?
i. 12 peppers?
j. 9 cans tuna fish?

3. What is the cost of:

a. $\frac{1}{2}$ doz. rolls?
b. $\frac{1}{4}$ lb. cheese?
c. $\frac{3}{4}$ lb. cookies?
d. $\frac{7}{8}$ lb. smoked ham?
e. 3 oranges?
f. 2 lemons?
g. 4 cakes?
h. 2 lb. onions?
i. 3 bars soap?
j. 2 cans soup?
k. 12 oz. grapes?
l. 7 oz. cookies?
m. 8 oz. string beans?
n. 4 oz. margarine?
o. 10 oz. tomatoes?
p. 14 oz. peaches?
q. 15 oz. pears?
r. 13 oz. haddock?
s. 5 oz. cheese?
t. 14 oz. apples?

4. What is the cost of:

a. 4 cans peas?
b. 3 cans tomato juice?
c. 16 cakes?
d. 5 cans tuna fish?
e. 30 eggs?
f. 15 cans soup?
g. 10 lb. onions?
h. 7 qt. milk?
i. 8 peppers?
j. 32 rolls?
k. $2\frac{1}{2}$ lb. sirloin?
l. $4\frac{1}{4}$ lb. chicken?
m. $1\frac{3}{8}$ lb. string beans?
n. $5\frac{3}{4}$ lb. turkey?
o. $3\frac{11}{16}$ lb. pears?
p. 1 lb. 4 oz. tomatoes?
q. 3 lb. 8 oz. apples?
r. 6 lb. 14 oz. rib roast?
s. 2 lb. 9 oz. bananas?
t. 4 lb. 1 oz. flounder?

5. Find the cost of each of the following orders:

a. 2 doz. eggs, 5 lb. flour, 4 cans soup, 1 pkg. detergent
b. 1 loaf bread, 1 qt. milk, 1 doz. cakes, $\frac{3}{4}$ lb. cheese, 4 lemons
c. 1 lb. coffee, 2 doz. oranges, $1\frac{1}{2}$ lb. margarine, 2 lb. bacon, 8 cans soup
d. 12 oz. margarine, 4 lb. 8 oz. sirloin, 4 grapefruit, 1 can tuna fish, 1 bar soap
e. 6 bottles beverage, 3 heads lettuce, $2\frac{1}{2}$ lb. tomatoes, $2\frac{3}{4}$ lb. apples, 9 rolls
f. 5 lb. potatoes, 2 stalks celery, 5 peppers, 2 cans tomato juice, 6 cans tuna fish
g. 1 bunch carrots, 3 loaves bread, 2 lb. 12 oz. smoked ham, 3 lb. 11 oz. haddock, 5 lb. sugar, 2 lb. coffee, 12 oz. margarine, 4 lb. 5 oz. sirloin
h. 6 bars soap, 10 cans corn, 2 cans peaches, 4 cans tuna fish, 1 doz. oranges, 2 lb. string beans, $1\frac{1}{2}$ lb. flounder, $9\frac{1}{2}$ lb. turkey, $5\frac{1}{4}$ lb. chicken
i. 20 cakes, 15 rolls, 1 lb. 11 oz. tomatoes, 5 lb. potatoes, 2 lb. 14 oz. haddock, 1 doz. eggs, 1 qt. milk, 9 grapefruit, 9 oranges
j. 2 lb. 7 oz. bananas, 3 lb. 4 oz. sirloin, 6 eggs, 1 loaf bread, 3 qt. milk, 1 lb. 5 oz. tomatoes, 2 lb. 6 oz. string beans, 1 pepper, 1 lb. 8 oz. grapes, 3 lb. coffee
k. 4 loaves bread, $1\frac{1}{2}$ doz. cakes, 1 head lettuce, 4 lb. onions, $2\frac{1}{2}$ doz. oranges, $4\frac{3}{4}$ lb. rib roast, 1 lb. 5 oz. pears, 2 bars soap, 3 doz. eggs, 10 cans soup
l. 1 lb. 3 oz. cheese, 2 doz. rolls, 1 lb. 9 oz. haddock, 2 cans peas, 12 cans tuna fish, 8 oz. margarine, 1 can tomato juice, 4 lb. 3 oz. chicken, 3 cans corn, 8 bottles beverage, 3 lemons, 5 grapefruit

6. Find the cost of each of the following weights of food:

a. (1) 2 kg (2) 750 g (3) 4.5 kg (4) 1,400 g when 1 kg of chicken costs $1.40.
b. (1) 1 kg (2) 850 g (3) 200 g (4) 1.2 kg when 500 g of cheese cost $1.30.
c. (1) 900 g (2) 2.1 kg (3) 0.8 kg (4) 1,600 g when 1 kg of onions costs 35¢.
d. (1) 460 g (2) 1 kg (3) 625 g (4) 0.4 kg when 100 g of cookies cost 40¢.
e. (1) 3 kg (2) 700 g (3) 4.9 kg (4) 2,800 g when 1 kg of shad costs $2.
f. (1) 15 kg (2) 2,400 g (3) 3.5 kg (4) 7.4 kg when 5 kg of potatoes cost $1.25.
g. (1) 125 g (2) 680 g (3) 1 kg (4) 0.6 kg when 250 g of salami cost 75¢.
h. (1) 2 kg (2) 800 g (3) 1.7 kg (4) 1,580 g when 1 kg of peas costs 65¢.
i. (1) 3 kg (2) 750 g (3) 2.1 kg (4) 975 g when 500 g of bananas cost 28¢.
j. (1) 1,500 g (2) 4 kg (3) 640 g (4) 0.7 kg when 1 kg of apples costs 75¢.

7. If gasoline costs 22.9¢ per liter, find the cost of 41.8 liters of gasoline.

8. How many 8-packs of 250 ml bottles of beverage should be purchased if you want 4 liters in all? What will the total cost be if an 8-pack costs $1.49?

15–8 BUYING AT A SALE

A person buying an article at a sale expects to pay a reduced price for this article. The regular or full price of an article is generally called the *marked price* or *list price*. The reduced price is sometimes called the *sale price* or *net price*. See section 15–27. The per cent taken off is the *rate of reduction*. To find the rate of reduction, we determine what per cent the reduction is of the regular price.

EXERCISES

1. How much is a basketball reduced in price at a 30% reduction sale if its regular price is \$14.50? What is its sale price?
2. A dress marked \$30 was sold at a reduction of 25%. What was the sale price of the dress?
3. At a clearance sale a discontinued model television set, regularly selling for \$550, can now be purchased at a reduction of 18%. What is its sale price?
4. Find the sale price of each of the following articles:
   a. Football; regular price, \$15.80; reduced 50%.
   b. Bicycle; regular price, \$85; reduced 14%.
   c. Blanket; regular price, \$14.75; reduced 22%.
   d. Rug; regular price, \$375; $\frac{1}{4}$ off.
   e. Camera; regular price, \$51.25; reduced 19%.
5. A department store advertises a 40% reduction on all clothing. Find the reduced price on each of the following items if the original price marked on the tag is:
   a. Coat, \$95
   b. Suit, \$125
   c. Dress, \$44.95
   d. Slacks, \$23.75
   e. Shirt, \$10.95
   f. Hat, \$19.75
   g. Tie, \$4.50
   h. Shoes, \$34.95
   i. Sweater, \$12.99
6. A furniture and appliance store advertises a $33\frac{1}{3}$% reduction on all furniture and a 15% reduction on all appliances. Find the sale price of each of the following articles if the regular price of:
   a. A dinette set is \$645.
   b. A bedroom set is \$980.
   c. A refrigerator is \$430.
   d. A living room set is \$1,095.
   e. An electric blender is \$29.
   f. A studio couch is \$324.85.
   g. An electric iron is \$19.95.
   h. A table is \$113.75.
   i. An electric heater is \$21.98.
   j. A vacuum cleaner is \$89.50.
7. Find the rate of reduction allowed when a typewriter that regularly sells for \$300 was purchased for \$250.
8. Find the rate of reduction allowed on each of the following articles:
   a. Desk; regular price, \$180; sale price, \$150.
   b. Freezer; regular price, \$285; sale price, \$228.
   c. Lawn mower; regular price, \$96.75; sale price, \$64.50.
   d. Automobile; regular price, \$5,000; sale price, \$4,200.
9. Find the regular price of a washing machine that sold for \$200 at a 20% reduction sale.
10. Find the regular price of each of the following:
    a. Tire, \$38 sale price when a 24% reduction was allowed.
    b. Mirror, \$63 sale price when a 10% reduction was allowed.
    c. Office file, \$70 sale price when a $37\frac{1}{2}$% reduction was allowed.
    d. Toaster, \$17.94 sale price when a 40% reduction was allowed.

15-9 UNIT PRICING—WHICH IS THE BETTER BUY?

Generally when buying food or other commodities that are packaged and sold in two different sizes, one must consider whether the larger size will be used without waste. In the problems of this section we shall disregard this consideration.

To determine the "better buy" when two different quantities of the same item are priced at two rates, we first find the *unit price* or the cost per unit of the item at each rate and select the lower unit price as the better buy.

For example: Which is the better buy: 4 grapefruit costing 39¢ or 6 grapefruit costing 54¢?

Grapefruit at the rate of 4 for 39¢ costs $9\frac{3}{4}$¢ each, but at the rate of 6 for 54¢ costs 9¢ each. Therefore, 6 grapefruit for 54¢ is the better buy.

Not always is the larger size of an item a better buy.

For example: The 5.8-oz. tube of brand X toothpaste at a sale sells for 58¢ but the 8-oz. tube sells at the regular price of 96¢. Which is the better buy?

Toothpaste X in the 5.8-oz. tube costs 10¢ an ounce, but in the 8-oz. size costs 12¢ an ounce. Therefore, the 5.8-oz. tube is the better buy.

Observe in the above problem that the unit price is the cost per ounce. In other situations the unit of comparison may be cost per pound, cost per gallon, cost per item, etc. Soon the food stores may be required to label each item with its unit price.

Sometimes in working with unit prices like cost per ounce, we use parts of a cent represented by a numeral like \$.238. Observe that the numeral \$.238 means 23.8 cents or 23.8¢.

EXERCISES

1. Rewrite each of the following, using a dollar mark \$ and a decimal point:

a. 24¢	**c.** 9¢	**e.** 8.6¢	**g.** $3\frac{1}{2}$¢	**i.** 64.5¢
b. 53¢	**d.** 3¢	**f.** 19.7¢	**h.** $20\frac{1}{4}$¢	**j.** 12.48¢

2. Rewrite each of the following, using the ¢ symbol:

a. \$.61	**c.** \$.05	**e.** \$.029	**g.** \$.656	**i.** \$.8305
b. \$.08	**d.** \$.93	**f.** \$.542	**h.** \$.075	**j.** \$.0025

3. Read each of the following in terms of cents:

a. 47.6¢	**b.** 2.1¢	**c.** \$.853	**d.** \$.014	**e.** \$.9875

4. In each of the following exactly the same item is priced at two rates. Which is the better buy in each case and why?

   a. Apples; 2 for 25¢ or 3 for 40¢.
   b. Grapefruit; 5 for 80¢ or 6 for 95¢.
   c. Oranges; 1 doz. for 96¢ or 15 oranges for $1.05.
   d. Soap; 3 bars for 42¢ or 4 bars for 56¢.
   e. Pears; 4 for 39¢ or 9 for 79¢.
   f. Soup; 3 cans for 59¢ or 7 cans for $1.25.
   g. Beverage; 2 bottles for 43¢ or 5 bottles for $1.
   h. Frozen orange juice; 2 cans for 95¢ or 6 cans for $2.75.
   i. Baby food; 4 jars for 93¢ or 10 jars for $2.25.
   j. Shrimp; 2 lb. for $4.98 or 5 lb. for $11.25.

5. Find how much you save on each can when buying in quantity:

   a. 1 dozen cans of corn for $3.60 or 34¢ each.
   b. 3 cans of beans for $1 or 35¢ each.
   c. 2 cans of peas for 57¢ or 29¢ each.
   d. 24 cans of tomato juice for $5 or 25¢ each.
   e. 4 cans of cleanser for $1.25 or 39¢ each.

6. **a.** If a dozen eggs cost 75¢, how much will 20 eggs cost?
   b. If 5 dozen note books cost $14.40, what will 1 book cost?
   c. If 4 tires cost $152, how much will 7 tires cost?
   d. If 8 shirts cost $60, how much will a dozen shirts cost?
   e. If $2\frac{2}{3}$ yd. acrylic fabric cost $8.40, how much will $6\frac{1}{4}$ yd. acrylic fabric cost?

7. Find how much you save by buying the larger size of each of the following items instead of an equivalent quantity in the smaller size:

   a. Spaghetti; 8-oz. box at 26¢, 1-lb. box at 49¢.
   b. Potato chips; 8-oz. bag at 69¢, 1-lb. bag at 95¢.
   c. Peas; 2 cans each $8\frac{1}{2}$ oz., at 49¢, 1-lb. 1-oz. can at 37¢.
   d. Bleach; 1-qt. bottle at 32¢, 1-gal. bottle at 75¢.
   e. Vinegar; 1-gal. bottle at $1.35, 1-pt. bottle at 25¢.
   f. Mayonnaise; 32-oz. jar at $1.33, 1-pt. jar at 79¢.
   g. Beverage; 2-qt. bottle at 79¢, 1-pt. bottle at 25¢.
   h. Egg noodles; 8-oz. pkg. at 33¢, 1-lb. pkg. at 55¢.
   i. Detergent; 1-lb. 6-oz. pkg. at 79¢, 5-lb. 8-oz. pkg. at $3.09.
   j. Paint; 1-gal. can at $12.49, 1-qt. can at $4.59.

8. How much do you save per ounce by buying the larger size instead of the smaller size? Round the answers to the nearest tenth of a cent.

   a. Cookies; 15-oz. pkg. at 87¢, 19-oz. pkg. at $1.05.
   b. Beets; 8-oz. can at 19¢, 1-lb. can at 31¢.
   c. Asparagus; $10\frac{1}{2}$-oz. can at 71¢, $14\frac{1}{2}$-oz. can at 85¢.
   d. Grapefruit juice; 1-pt. 2-oz. can at 30¢, 1-qt. 14-oz. can at 57¢.
   e. Salmon; $3\frac{3}{4}$-oz. can at 69¢, $7\frac{3}{4}$-oz. can at $1.29.

9. Which is the better buy and why?

a. Rolls; 8 for 75¢ or 12 for \$1.
b. Shampoo; $3\frac{1}{2}$-oz tube at 89¢ or 7-oz. tube at \$1.35.
c. Bacon; 1-lb. pkg. at \$1.69 or $\frac{1}{2}$-lb. pkg. at 89¢.
d. Facial tissue; 2 boxes for 87¢ or 5 boxes for \$2.
e. Beverage; 6 bottles, each 1 pt., for \$1.59 or 8 bottles, each 10 oz., for \$1.59.
f. Aspirin; 100 tablets for 95¢ or 300 tablets for \$2.29.
g. Frozen orange juice; 4 cans, each 6 oz., for \$1.99 or 2 cans, each 12 oz., for \$1.89.
h. Detergent; 1-lb. 4-oz. pkg. at 75¢ or 5-lb. 4-oz. pkg. at \$2.98.
i. Mouthwash; 14-oz. bottle at 99¢ or 20-oz. bottle at \$1.29.
j. Salmon; 1-lb. can at \$2.59 or $7\frac{3}{4}$-oz. can at \$1.29.

15–10 COUNT YOUR CHANGE

Usually store clerks, in making change, will add the value of the coins and bills to the amount of the purchase until the sum of the change and the amount of the purchase equals the amount offered in payment.

The symbol @ means "at" and indicates the unit cost. "2 ties @ \$2.50 each" or just "2 ties @ \$2.50" is read "two ties at \$2.50 each."

EXERCISES

1. Find how much change you should get from \$1 if your purchases cost:

a. \$.89 **b.** \$.53 **c.** \$.27 **d.** \$.08 **e.** \$.74

2. Find how much change you should get from \$5 if your purchases cost:

a. \$1.42 **b.** \$4.68 **c.** \$3.06 **d.** \$.87 **e.** \$2.31

3. Find how much change you should get from \$10 if your purchases cost:

a. \$5.26 **b.** \$8.09 **c.** \$2.44 **d.** \$7.92 **e.** \$9.13

4. Find how much change you should get from \$20 if your purchases cost:

a. \$7.64 **b.** \$15.81 **c.** \$13.03 **d.** \$19.18 **e.** \$2.72

5. Find the cost of each of the following:

a. 2 sweaters @ \$14.95 each
b. $1\frac{1}{2}$ lb. butter @ \$1.64 per lb.
c. 3 doz. oranges @ 99¢ per doz.
d. $2\frac{3}{4}$ lb. chicken @ 68¢ per lb.
e. 10 cans soup @ 26¢ each
f. 6 shirts @ \$9.95 each
g. 20 gal. gasoline @ 92.9¢ per gal.
h. 75 sq. yd. broadloom @ \$12.50 per sq. yd.
i. 3.8 m linen @ \$5.60 per m
j. 5 pr. drapes @ \$79.75 per pr.
k. 9 kg potatoes @ 33¢ per kg

6. For each purchase on page 478, determine how much change should be given when the specified amount of money is offered in payment.

	Articles Purchased	Amount Offered in Payment
a.	1 pen @ 89¢	$1.00
b.	1 book @ $3.95	$5.00
c.	2 skirts @ $15.99 each	$35.00
d.	1 pr. shoes @ $34.95, 2 shirts @ $12.95 each	$70.00
e.	1 lb. bacon @ $1.59, 1 lb. tomatoes @ 69¢	$3.00
f.	15 gal. gasoline @ 93.9¢ per gal., 1 qt. oil @ $1.25	$20.00
g.	$\frac{1}{2}$ lb. butter @ 98¢ per lb., 2 doz. eggs @ 79¢ per doz.	$5.00
h.	6 cans soup @ 27¢ each, 5 cans fruit @ 57¢ each	$10.00
i.	4 tires @ $42.67 each	$180.00
j.	2 dresses @ $39.95 each, 1 hat @ $14.75, 3 sweaters @ $11.98 each	$150.00
k.	1 suit @ $109.50, 3 pr. slacks @ $15.95 each, 2 shirts @ $9.95 each	$200.00
l.	4 trees @ $28.50 each, 6 shrubs @ $10.75 each, 1 bag fertilizer @ $17.25	$200.00
m.	1 typewriter @ $189.75, 5 reams paper @ $8.85 each	$250.00

15-11 GAS, ELECTRICITY, WATER, HEATING AND TELEPHONE SERVICES

People generally require all or some of the services discussed here. In order to check bills for the gas, electricity, and water services, one must be able to read a meter measuring the consumption. To read a gas meter or an electric meter or a water meter, we select on each dial of the meter the numeral that was just passed by the dial pointer. To find the consumption during a given period, we subtract the reading at the beginning of the period from the reading at the end of the period. To find the cost, we apply the prevailing rate to the consumption.

The reading of the dials on the gas meter (see page 479) is expressed in *hundred cubic feet*. A reading such as 3692 represents 369,200 cubic feet.

The reading of the dials on the electric meter (see page 479) is expressed in *kilowatt hours*. A reading such as 2713 represents 2,713 kilowatt hours of electrical energy. A kilowatt is 1,000 watts. It is a unit measuring electric power. If an appliance requires the power of 1 kilowatt to be used for 1 hour, then 1 kilowatt hour of electrical energy is furnished. Observe also that if an appliance requires the power of 200 watts but is used for 5 hours, then 1 kilowatt hour of electricity is also used (since $200 \times 5 = 1{,}000$ watts hours $= 1$ kilowatt hour).

The reading of the dials on the water meter is expressed in *tens of gallons*. Although there is a small dial on the meter that measures from 0 to 10 gallons, it is not used as part of the reading. A reading such as 82314 represents 823,140 gallons of water.

EXERCISES

1. What is the reading of each of the following gas meters?

a.

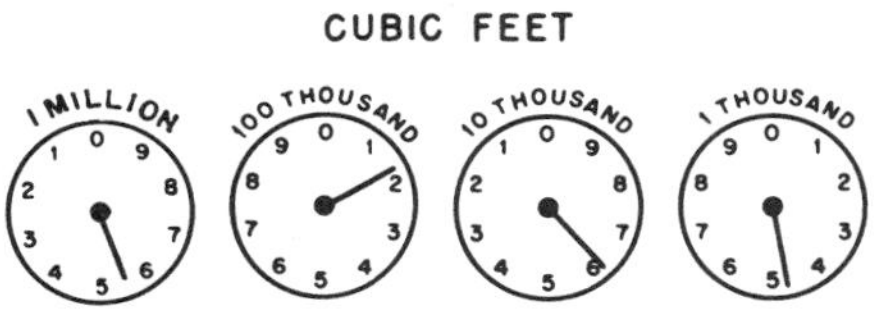

b.

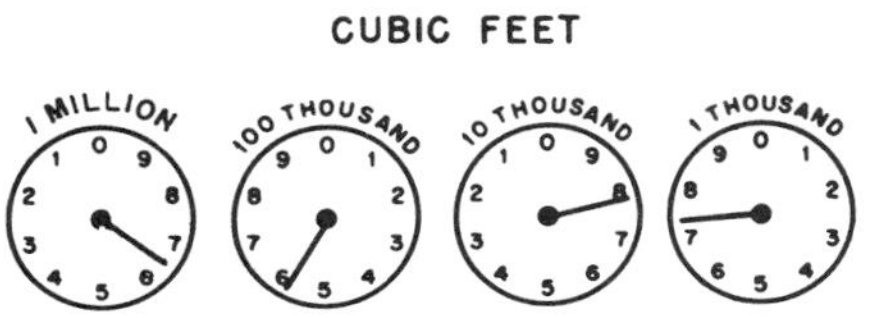

c.

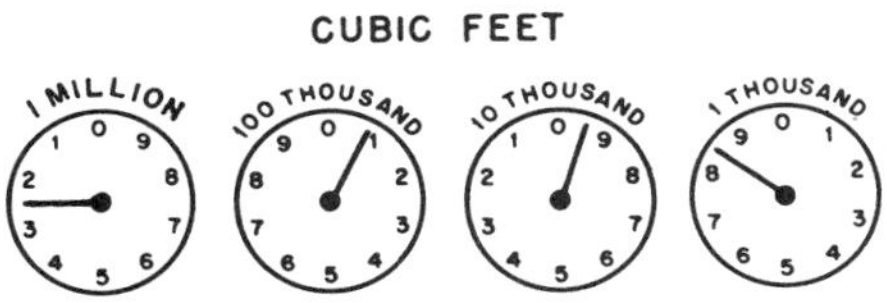

2. What is the reading of each of the following electric meters?

a.

b.

c.

3. Find the cost of the gas consumed during the months of (1) October and (2) November at the rates given below if the meters read on:

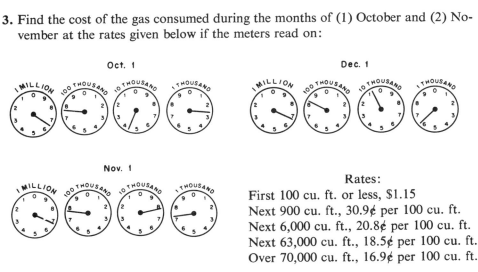

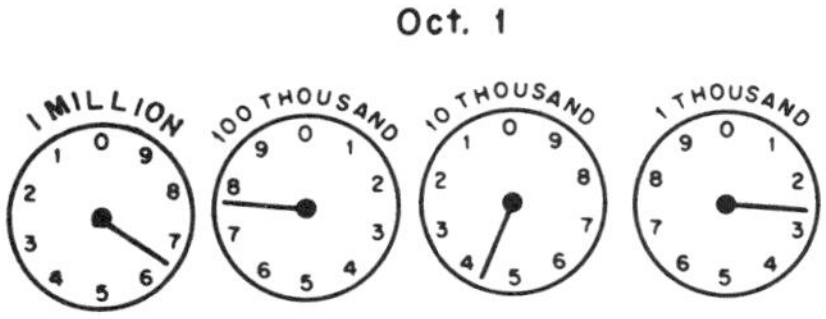

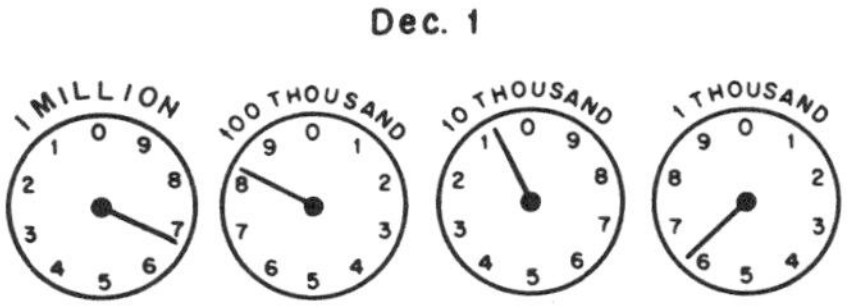

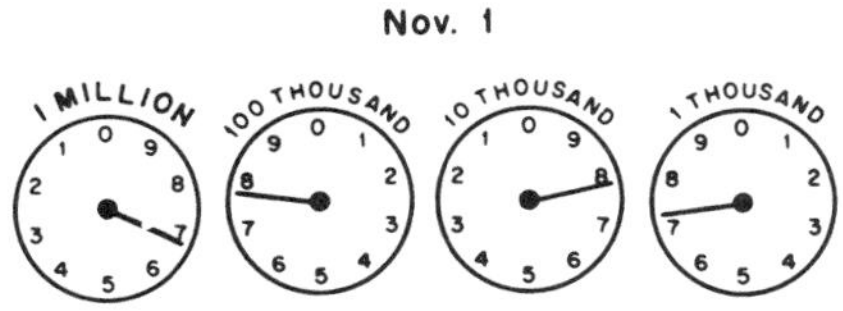

Rates:

First 100 cu. ft. or less, \$1.15
Next 900 cu. ft., 30.9¢ per 100 cu. ft.
Next 6,000 cu. ft., 20.8¢ per 100 cu. ft.
Next 63,000 cu. ft., 18.5¢ per 100 cu. ft.
Over 70,000 cu. ft., 16.9¢ per 100 cu. ft.

4. Given the reading at the beginning and at the end of the period, use the above rates in each of the following to find the cost of the gas consumed during the period.

	Reading At Beginning	At End		Reading At Beginning	At End
a.	4213	4269	**f.**	3738	3832
b.	7794	7821	**g.**	5624	5707
c.	2155	2194	**h.**	9903	0008
d.	0332	0400	**i.**	7295	7519
e.	6529	6604	**j.**	1480	1798

5. Find the cost of the electricity consumed during the months of (1) January and (2) February at the rates given below if the meters read on:

KILOWATT HOURS Jan. 1

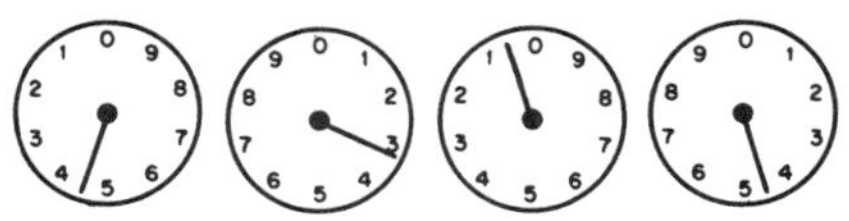

KILOWATT HOURS March 1

KILOWATT HOURS Feb. 1

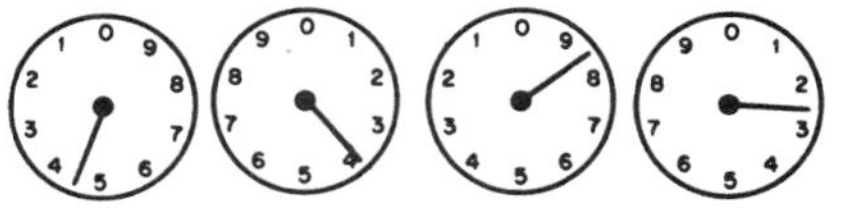

Rates:
First 10 kwh or less, \$3.00
Next 490 kwh @ 3.9¢ each
Over 500 kwh @ 3.1¢ each

6. Given the reading at the beginning and at the end of the period, use the above rates in each of the following to find the cost of the electricity used during the period.

	Reading At Beginning	At End		Reading At Beginning	At End
a.	6895	6932	**f.**	0792	1000
b.	3414	3505	**g.**	5459	5822
c.	1325	1387	**h.**	2616	2687
d.	9962	0068	**i.**	4278	4462
e.	7289	7434	**j.**	8987	9454

7. At an average cost of 4.8¢ per kilowatt hour, find the *hourly* cost of operating each of the following:

a. Dryer, 5,000 watts
b. Heater, 1,500 watts
c. Iron, 1,200 watts
d. Radio, 40 watts
e. TV set, 250 watts
f. 100 watt light bulb
g. Percolator, 600 watts
h. Stereo set, 125 watts
i. Toaster, 800 watts
j. Washer, 1,800 watts

8. In many communities water bills are issued every 3 months. The reading on a certain water meter on March 15th was 34765, on June 15th it was 37024, and on September 15th it was 40009.

a. How many gallons of water were used during the 3-month period ending June 15th? During the 3-month period ending September 15th?

b. Find the cost of the water consumed during each of these periods using the rates:
First 2,500 gallons or less, $5.48
Over 2,500 gallons, $1.28 for every 1,000 gallons

9. Mr. Harris during the year bought 190 gallons of fuel oil at 70.3¢ per gallon, 175 gallons at 69.9¢ each, 210 gallons at 70.1¢ each, 225 gallons at 69.7¢ each, and 195 gallons at 69.6¢ each. How many gallons of fuel oil did he buy? What was the total cost of his fuel oil for the year?

10. The telephone message rate service for 50 calls or less per month costs $5.09 with 4.7¢ additional for each call over 50 calls. The flat rate service which allows unlimited calls costs $7.25 monthly. There is a federal tax of 3% added to the cost of telephone service. If a family expects to average no more than 100 telephone calls per month, which service is cheaper? How much cheaper?

15–12 OWNING AN AUTOMOBILE

EXERCISES

1. a. Mr. Hughes bought a new car for $5,670. After using it for 3 years, he sold it for $2,190. What was the average yearly depreciation of the car?

b. At the end of the first year the speedometer read 9,762 miles, at the end of the second year, 21,299 miles, and at the end of the third year 30,184 miles. How many miles did he drive during the second year? The third year?

c. During the first year his expenses in addition to the average yearly depreciation were: gasoline, 658 gal. at 91.9¢ per gal; oil, 28 qt. at 95¢ per qt.; garage rent, $75 per mo.; insurance, $210 per yr.; license fees, $18 per yr.; miscellaneous expenses, $69.26. How much did it cost Mr. Hughes to run his car for the year? What was the average cost per month? Per week? Per mile?

2. Mrs. Bertman bought a new car for $6,000. She insured it against fire and theft for 80% of its value. If the rate is 75¢ per $100, what is the annual premium?

3. Miss Williams insures her car costing $5,200 against fire and theft for 75% of its value. The rate is 80¢ per $100 for fire and theft insurance. If property damage insurance costs $41.50, $10,000–$20,000 liability insurance costs $69.75, and $100 deductible collision insurance costs $58, what is her total premium?

4. How far can a motorcycle go on a tankful of gasoline if it averages 85 miles on a gallon and the tank holds 2.5 gallons?

5. How long will it take a man, driving at an average speed of 45 m.p.h., to travel from New York to Cincinnati, a distance of 660 miles? If the car averages 15 miles on a gallon of gasoline, how many gallons are required to make the trip?

6. Ms. Burns had the tank of her car filled with gasoline and noted that the speedometer read 14,593. The next time she stopped for gasoline it required 14 gallons to fill the tank and the speedometer read 14,817. How many miles did the car average on a gallon of gasoline?

7. Tom's father took the family on a motor trip. How much did the trip cost if he bought 57 gallons of gasoline at 92¢ per gal. and 2 qt. of oil at 95¢ per qt., and if lodging cost $26.50 per night for 3 nights, meals $97.81, amusements $49.75, and miscellaneous expenses were $74.28?

8. Oil can be bought at 79¢ per qt. in individual cans or $15.84 for a case of 24 cans. How much do you save per quart if you buy oil by the case?
9. At a speed of 40 m.p.h., how many minutes will it take to go a distance of 18 miles?
10. If the driver's reaction time before appling the brakes is one second, how many feet will the car go in one second at each of the following speeds: **a.** 45 m.p.h.? **b.** 40 m.p.h.? **c.** 25 m.p.h.? **d.** 35 m.p.h.? **e.** 60 m.p.h.?
11. The braking distance (distance an automobile travels before coming to a stop after the brakes are applied) on a dry, level road is 30 ft. at 20 m.p.h., 67.5 ft. at 30 m.p.h., 120 ft. at 40 m.p.h., and 187.5 ft. at 50 m.p.h. If the driver's reaction time is $\frac{3}{4}$ second, find the total stopping distance (reaction time distance plus braking distance) at each of the above speeds.

15–13 HOUSEHOLD BILLS

EXERCISES

Determine the amounts of payment due in each of the following bills.

1.

RATES
GENERAL SERVICE
First 200 C.F. or less $2.50
Next 1,800 C.F. @ $3.20 per M.C.F.
Next 8,000 C.F. @ $2.20 per M.C.F.
Next 70,000 C.F. @ $1.90 per M.C.F.
Additional use @ $1.70 per M.C.F.

M.C.F. = 1,000 cu. ft.

LOCAL GAS WORKS CO.
YOUR CITY

Net Payment Period Expires Jan 2, 19___

MR. ROBERT MILLER
249 FOURTH AVE.

READING DATES		METER READINGS		CLASS	HUNDRED CU. FT. USED	AMOUNT
PREVIOUS	PRESENT	PRESENT	PREVIOUS			
Nov 15	Dec 15	404	375	DOM		

UNPAID GAS ACCOUNT______
TOTAL

2.

THE BELL TELEPHONE COMPANY

Account No. 428-6767
MRS. CHARLOTTE JAY Jan. 16, 19___
1025 COUNTY ROAD

	Amount Due
Service and Equipment	17.82
Message Units	
Itemized Charges	2.67
Directory Advertising	
Other Charges and Credits	7.56
Tax — Fed 2 04 State	2.04
Balance from Last Bill	
Total Amount Due	

3.

LOCAL ELECTRIC COMPANY

YOUR CITY

Net Payment Period Expires May 1, 19__

RATES
12 K.W. Hr. or less $1.25
Next 30 K.W. Hr. @ 7¢
Next 48 K.W. Hr. @ 4.25¢
Over 90 K.W. Hr. @ 3.4¢

MR. JOHN KELLY
6318 WEST ST.

READING DATES		METER READINGS		SERVICE CODE	KILOWATT HOURS	AMOUNT OF BILL
PREVIOUS	PRESENT	PRESENT	PREVIOUS			
Mar 6	Apr 5	6143	6068	EL		

4.

Local Department Store

YOUR CITY

MISS ELAINE GOMEZ
1953 WALNUT ST.

Oct. 31, 19__

ITEMS	CHARGES	CREDITS	DATE	LAST AMOUNT IS BALANCE DUE
Balance from Previous Bill				21.79
Dress	30.95		Oct. 3	
Coat	63.75			
Cash		21.79	5	
Shoes	19.50		10	
Chairs	45.75			
Hat	17.50		18	
Chair Returned		15.25	19	
Ties	12.67		26	

15–14 HOUSEHOLD ACCOUNTS

A *cash account* is a record of the money a person receives (receipts) and spends (expenditures). The cash on hand should equal the difference in the amounts of receipts and expenditures. Thus, by writing in the expenditure column the amount of cash on hand at the end of the period, the receipt and expenditure columns will be in balance.

EXERCISES

Make a copy, then balance the following account for the week of April 8. Record the following items, then balance the account for the week of April 15. Receipts: April 18, salary, $325. Expenditures: April 15, rent, $270; food, $22.56; April 17, gasoline, $9.50; telephone, $13.76; April 18, insurance, $25; food, $17.82; April 20, food, $19.14; bank deposit, $20; April 21, movies, $9.

RECEIPTS				EXPENDITURES			
April				April			
8	Cash on hand	99	50	9	Fuel Oil	24	00
11	Salary	325	00	10	Food	27	26
					Electricity	13	68
				11	Shoes	19	90
					Movies	7	00
				12	Food	16	18
				13	Gasoline	4	50
					Bank Deposit	20	00
					Food	19	16
				14	Cash on hand		
	Total				Total		
April 15	Cash on hand						

15–15 INSTALLMENT BUYING; DEFERRED PAYMENT PURCHASE PLANS; ANNUAL PERCENTAGE RATE

Installment buying or *buying on credit* is a purchase plan by which the customer usually pays a certain amount of cash at the time of purchase and then pays in equal installments at regular intervals, usually months, both the balance and the finance charge on this balance. This *finance charge*, also called *carrying charge*, is the interest (see page 496) which usually is applied on the entire balance as being owed for the full time of payment. The carrying charge or finance charge is the difference between the cash price and the total amount paid. However, since in this type of purchase plan part of the original balance owed is being repaid each month, the given percentage rate charged is not the true rate. The true rate charge is much higher since it is based on the lower average balance that is owed during the payment period. Actually the true rate is almost twice the given rate.

In order that people who buy on credit may know what this true rate is so that they may compare finance charge rates, the federal government requires each credit purchase contract or loan contract (see section 15–24) to indicate the *annual percentage rate* which is the true finance charge rate for the year. Annual Percentage Rate Tables are generally used to determine these rates. See page 487.

Many stores use *deferred payment* plans. In these plans a specific monthly payment is made to reduce the balance owed. Sometimes these payments are fixed amounts. Other times these payments decrease as the bracket of the balance owed falls. However a finance charge, usually $1\frac{1}{4}\%$ per month or $1\frac{1}{2}\%$ per month, is applied to the previous monthly unpaid balance and is added to this previous balance. The deferred payment plans also must indicate the annual percentage rate. To compute this we multiply the rate per period by the number of periods per year. Thus the finance charge rate of $1\frac{1}{4}\%$ per month is $12 \times 1\frac{1}{4}\%$ or a 15% annual percentage rate. Credit card accounts are charged a late payment finance charge each month on past due amounts. Here also the annual percentage rates must be indicated.

EXERCISES

1. Find the annual percentage rate if the monthly finance charge rate is:

a. 3% **b.** 2% **c.** $1\frac{3}{4}\%$ **d.** $1\frac{1}{3}\%$ **e.** $2\frac{1}{2}\%$

2. Find the annual percentage rate if the monthly late payment finance charge rate on past due amounts is:

a. 1% **b.** $\frac{2}{3}$ of 1% **c.** $\frac{5}{8}$ of 1% **d.** $\frac{5}{6}$ of 1% **e.** $\frac{3}{4}$ of 1%

3. Mrs. Harris bought a sewing machine for $19 down and $8.50 a month for 10 months. How much did the sewing machine cost on the installment plan? If the cash price was $92.50, how much more did the sewing machine cost on the installment plan?

4. Andrew's father can purchase an air conditioner for the cash price of $240 or $25 down and 12 equal monthly payments of $22.50 each. How much is the carrying charge? How much can be saved by paying cash?

5. Find the carrying charge when each of the following articles are purchased on the installment plan:

	Cash Price	Down Payment	Monthly Payment	Number of Monthly Payments
a. Rug	$89	$15	$7	12
b. Oil Burner	$650	$100	$26.50	24
c. Refrigerator	$249	$49	$12.25	18
d. Camera	$72.50	$13	$6.75	10

6. Find the monthly payment if the amount financed is $385 and the finance charge is $71, both to be paid in 24 equal monthly payments.

7. A washing machine can be purchased for the cash price of $240 or $20 down and both the balance and the finance charge to be paid in 12 equal monthly payments. If 8% is charged on the full balance, how much is paid monthly?

8. Find the monthly payment when each of the following articles are purchased on the installment plan:

	Cash Price	Down Payment	Finance Charge Interest Rate	Number of Monthly Payments
a. Piano	$645	$105	6%	12
b. Sink	$115	$15	8%	12
c. Projector	$96	$24	10%	12
d. TV	$520	$70	9%	12

9. A department store uses the following deferred payment schedule:

Monthly Balance	Monthly Payment	Monthly Balance	Monthly Payment
\$0 to \$10	Full Balance	\$161 to \$175	\$25
\$11 to \$90	\$10	\$176 to \$210	\$30
\$91 to \$120	\$15	\$211 to \$250	\$40
\$121 to \$160	\$20	over \$250	$\frac{1}{5}$ of balance

If finance charges are computed at the monthly rate of $1\frac{1}{4}\%$ on the previous month's balance, find the monthly payment due on each of the following previous month's balances. First compute and add the finance charge to the previous month's balance.

a. \$68 **b.** \$174 **c.** \$145 **d.** \$92.50 **e.** \$234.80

10. Compute the late payment finance charge on each of the following past due amounts at the specified rate:

a. \$65 past due; $1\frac{1}{2}\%$ monthly rate
b. \$140 past due; $1\frac{1}{4}\%$ monthly rate
c. \$28.60 past due; 1% monthly rate
d. \$71.42 past due; $\frac{1}{2}$ of 1% monthly rate
e. \$93.80 past due; $\frac{3}{4}$ of 1% monthly rate
f. \$17.49 past due; $\frac{2}{3}$ of 1% monthly rate

11. To find the annual percentage rate when the amount to be financed, the finance charge for this amount, and the number of monthly payments are known,

(1) Determine the finance charge per \$100 of amount financed by multiplying the given finance charge by 100 and dividing this product by the amount financed.
(2) Then follow down the left hand column of the table to the line indicating the required number of monthly payments, move across this line until you reach the nearest number to the finance charge per \$100.
(3) The rate at the top of this column is the annual percentage rate.

For example,

For 18 monthly payments a finance charge of \$12.29 per \$100 financed is at the annual percentage rate of 15%.

Using the table on page 487, find the annual percentage rate when:

	Total Amount Financed	Finance Charge on Total Amount	Number of Monthly Payments
a.	\$100	\$9.94	15
b.	\$100	\$13.88	20
c.	\$100	\$6.68	9
d.	\$200	\$16.34	12
e.	\$500	\$25.10	6
f.	\$350	\$59.29	24
g.	\$840	\$96.18	18

h.	$1,000	$265.70	36
i.	$1,250	$344.75	42
j.	$2,000	$498.40	30
k.	$1,500	$211.05	21
l.	$600	$34.80	8
m.	$3,000	$583.50	25
n.	$900	$91.80	16
o.	$2,500	$666.25	32
p.	$750	$55.20	10
q.	$5,000	$1,335.00	40
r.	$400	$35.52	12
s.	$1,800	$343.62	27

The following is a condensed section of the Federal Reserve Board tables for computing annual percentage rates for monthly payment plans:

NUMBER OF PAYMENTS	ANNUAL PERCENTAGE RATE										
	14.00%	14.25%	14.50%	14.75%	15.00%	15.25%	15.50%	15.75%	16.00%	17.00%	18.00%
	(FINANCE CHARGE PER $100 OF AMOUNT FINANCED)										
1	1.17	1.19	1.21	1.23	1.25	1.27	1.29	1.31	1.33	1.42	1.50
2	1.75	1.78	1.82	1.85	1.88	1.91	1.94	1.97	2.00	2.13	2.26
3	2.34	2.38	2.43	2.47	2.51	2.55	2.59	2.64	2.68	2.85	3.01
4	2.93	2.99	3.04	3.09	3.14	3.20	3.25	3.30	3.36	3.57	3.78
5	3.53	3.59	3.65	3.72	3.78	3.84	3.91	3.97	4.04	4.29	4.54
6	4.12	4.20	4.27	4.35	4.42	4.49	4.57	4.64	4.72	5.02	5.32
7	4.72	4.81	4.89	4.98	5.06	5.15	5.23	5.32	5.40	5.75	6.09
8	5.32	5.42	5.51	5.61	5.71	5.80	5.90	6.00	6.09	6.48	6.87
9	5.92	6.03	6.14	6.25	6.35	6.46	6.57	6.68	6.78	7.22	7.65
10	6.53	6.65	6.77	6.88	7.00	7.12	7.24	7.36	7.48	7.96	8.43
11	7.14	7.27	7.40	7.53	7.66	7.79	7.92	8.05	8.18	8.70	9.22
12	7.74	7.89	8.03	8.17	8.31	8.45	8.59	8.74	8.88	9.45	10.02
13	8.36	8.51	8.66	8.81	8.97	9.12	9.27	9.43	9.58	10.20	10.81
14	8.97	9.13	9.30	9.46	9.63	9.79	9.96	10.12	10.29	10.95	11.61
15	9.59	9.76	9.94	10.11	10.29	10.47	10.64	10.82	11.00	11.71	12.42
16	10.20	10.39	10.58	10.77	10.95	11.14	11.33	11.52	11.71	12.46	13.22
17	10.82	11.02	11.22	11.42	11.62	11.82	12.02	12.22	12.42	13.23	14.04
18	11.45	11.66	11.87	12.08	12.29	12.50	12.72	12.93	13.14	13.99	14.85
19	12.07	12.30	12.52	12.74	12.97	13.19	13.41	13.64	13.86	14.76	15.67
20	12.70	12.93	13.17	13.41	13.64	13.88	14.11	14.35	14.59	15.54	16.49
21	13.33	13.58	13.82	14.07	14.32	14.57	14.82	15.06	15.31	16.31	17.32
22	13.96	14.22	14.48	14.74	15.00	15.26	15.52	15.78	16.04	17.09	18.15
23	14.59	14.87	15.14	15.41	15.68	15.96	16.23	16.50	16.78	17.88	18.98
24	15.23	15.51	15.80	16.08	16.37	16.65	16.94	17.22	17.51	18.66	19.82
25	15.87	16.17	16.46	16.76	17.06	17.35	17.65	17.95	18.25	19.45	20.66
26	16.51	16.82	17.13	17.44	17.75	18.06	18.37	18.68	18.99	20.24	21.50
27	17.15	17.47	17.80	18.12	18.44	18.76	19.09	19.41	19.74	21.04	22.35
28	17.80	18.13	18.47	18.80	19.14	19.47	19.81	20.15	20.48	21.84	23.20
29	18.45	18.79	19.14	19.49	19.83	20.18	20.53	20.88	21.23	22.64	24.06
30	19.10	19.45	19.81	20.17	20.54	20.90	21.26	21.62	21.99	23.45	24.92
31	19.75	20.12	20.49	20.87	21.24	21.61	21.99	22.37	22.74	24.26	25.78
32	20.40	20.79	21.17	21.56	21.95	22.33	22.72	23.11	23.50	25.07	26.65
33	21.06	21.46	21.85	22.25	22.65	23.06	23.46	23.86	24.26	25.88	27.52
34	21.72	22.13	22.54	22.95	23.37	23.78	24.19	24.61	25.03	26.70	28.39
35	22.38	22.80	23.23	23.65	24.08	24.51	24.94	25.36	25.79	27.52	29.27
36	23.04	23.48	23.92	24.35	24.80	25.24	25.68	26.12	26.57	28.35	30.15
37	23.70	24.16	24.61	25.06	25.51	25.97	26.42	26.88	27.34	29.18	31.03
38	24.37	24.84	25.30	25.77	26.24	26.70	27.17	27.64	28.11	30.01	31.92
39	25.04	25.52	26.00	26.48	26.96	27.44	27.92	28.41	28.89	30.85	32.81
40	25.71	26.20	26.70	27.19	27.69	28.18	28.68	29.18	29.68	31.68	33.71
41	26.39	26.89	27.40	27.91	28.41	28.92	29.44	29.95	30.46	32.52	34.61
42	27.06	27.58	28.10	28.62	29.15	29.67	30.19	30.72	31.25	33.37	35.51
43	27.74	28.27	28.81	29.34	29.88	30.42	30.96	31.50	32.04	34.22	36.42
44	28.42	28.97	29.52	30.07	30.62	31.17	31.72	32.28	32.83	35.07	37.33
45	29.11	29.67	30.23	30.79	31.36	31.92	32.49	33.06	33.63	35.92	38.24

12. The annual percentage rate tables may be used to determine the finance charge and the monthly payment when the annual percentage rate is known.

(1) Using the column headed by the known annual percentage rate, read down this column until you reach the line corresponding to the given number of monthly payments. This is the finance charge per $100 of the amount to be financed.
(2) To determine the finance charge for the given amount financed, multiply the finance charge per $100 by the amount financed, then divide the product by 100.
(3) To find the monthly payment, add the finance charge to the amount financed, then divide this sum by the number of monthly payments.

Using the table on page 487, find the finance charge for each of the following amounts financed at the given annual percentage rates. Also find the monthly payment.

	Amount Financed	Annual Percentage Rate	Number of Monthly Payments
a.	$100	16%	15
b.	$300	15.25%	9
c.	$500	17%	24
d.	$750	15.75%	40
e.	$1,000	14.25%	36
f.	$800	14.50%	6
g.	$1,200	18%	42
h.	$4,000	14.75%	12
i.	$3,600	15.50%	36
j.	$5,500	14.25%	30
k.	$2,300	15%	28
l.	$1,750	14.50%	18
m.	$625	15.25%	10
n.	$1,500	14%	45
o.	$6,000	15.75%	20

15–16 FIRE INSURANCE

Insurance is a plan by which persons share risks so that each person is protected against financial loss. The insured person is charged a sum of money called a *premium.* The insurance company uses the fund created by these premiums to pay the insured persons who suffer losses.

The written contract between the insured person and the insurance company is called the *policy.* The amount of insurance specified in the policy is called the *face of the policy.* The length of time the insurance is in force is called the *term* of the policy.

To find the premium, we multiply the number of $100 of fire insurance by the rate per $100.

The prepaid premium rate on fire insurance for 3 years is 2.7 times the yearly rate. If this premium is paid by yearly installments, then 35% of the total premium is paid each year.

EXERCISES

1. What is the cost of $5,000 fire insurance for 1 year at the rate of $.20 per $100?
2. Find the premium for each of the following 1-year policies:

Face of policy	$2,000	$8,000	$15,000	$6,500	$10,000	$18,000
Yearly rate per $100	$.22	$.16	$.41	$.35	$.18	$.25

3. What is the premium for a 3-year policy if the yearly rate is $.24 per $100 and the face of the policy is $7,500?
4. Find for each of the following 3-yr. policies (a) the total prepaid premium and (b) the yearly premium if paid by yearly installments:

Face of Policy	$20,000	$35,000	$7,500	$10,000	$12,500	$15,000
Yearly rate per $100	$.14	$.18	$.36	$.25	$.22	$.40

5. Find the total premium for each of the following policies:

Value of building	$10,000	$7,500	$8,800	$15,000	$20,000	$17,500
Per cent of value insured	80%	60%	75%	70%	75%	80%
Yearly rate per $100	$.32	$.14	$.30	$.26	$.35	$.28
Term	1 yr.	3 yr.	3 yr.	1 yr.	3 yr.	3 yr.

6. How much does Mrs. Carroll save by taking out a 3-year policy instead of three 1-year policies if she insures her home for $27,500 at the yearly rate of $.22 per $100?
7. Mr. Johnson insured his house worth $18,400 for 75% of its value. He also insured the contents worth $9,000 for 70% of their value. What was his total premium if he bought a 3-year policy for each. The yearly rate on the building was $.12 per $100 and on the contents was $.24 per $100.
8. In a certain community the rate for a 1-year policy on a frame building is $.42 per $100 and on a brick building is $.14 per $100. At these rates what is the premium for $15,000 insurance on a frame building? On a brick building?
9. An owner of a building worth $10,000 took out $6,000 fire insurance. Her policy contains the 80% coinsurance clause. How much would the insurance company pay her if she suffered a fire loss of $4,000?

15–17 LIFE INSURANCE

Life insurance offers financial protection to the dependents of an insured person in event of the insured person's death. The person who receives the money when the insured person dies is called the *beneficiary.*

(1) *Term insurance*—Person is insured for a specified period of time. Premiums are paid only during that time. Beneficiary receives face of policy if insured person dies within the specified time.

(2) *Ordinary life insurance*—Person is insured until death. Premiums are paid until death of the insured person. Beneficiary receives face of policy when insured person dies.

(3) *Limited-payment life insurance*—Person is insured until death. Premiums are paid for a specified period of time. Beneficiary receives face of policy when insured person dies.

(4) *Endowment insurance*—Person is insured for a specified period of time. Premiums are paid only during that time. Beneficiary receives face of policy if insured person dies within the specified time. The insured person or policy holder receives face of policy if alive at the end of the specified time.

To find the premium, we select from the table on page 491 the premium for $1,000 insurance, then we multiply it by the number of $1,000 of insurance.

EXERCISES

1. Average life expectation:

At the age of	Birth	1	5	10	15	20	25
Years male may expect to live	68.9	69.1	65.3	60.5	55.6	51.0	46.5
Years female may expect to live	76.6	76.6	72.7	67.8	62.9	58.1	53.3

At the age of	30	35	40	45	50	55	60
Years male may expect to live	41.8	37.2	32.6	28.1	24.0	20.1	16.5
Years female may expect to live	48.4	43.6	38.9	34.3	29.8	25.5	21.4

a. To what age may a 15-year-old male expect to live?
b. To what age may a 15-year-old female expect to live?
c. How many years longer may a 20-year-old female expect to live than a 20-year-old male?

Annual Life Insurance Premiums per $1,000

Age	5-Year Term		Ordinary Life		20-Payment Life		20-Year Endowment	
	Male	Female	Male	Female	Male	Female	Male	Female
18	3.77	3.74	13.48	12.95	23.97	23.22	45.17	45.17
20	4.01	4.01	14.20	13.62	24.91	24.11	45.30	45.30
22	4.20	4.20	15.00	14.35	25.89	25.05	45.46	45.46
24	4.40	4.40	15.86	15.15	26.94	26.03	45.64	45.64
25	4.50	4.50	16.34	15.59	27.49	26.55	45.75	45.75
26	4.60	4.60	16.83	16.04	28.06	27.09	45.86	45.85
28	4.83	4.83	17.90	17.02	29.26	28.21	46.14	46.10
30	5.09	5.05	19.09	18.11	30.54	29.42	46.48	46.40
35	6.08	5.86	22.71	21.43	34.24	32.88	47.72	47.49
40	8.03	7.44	27.20	25.56	38.49	36.85	49.45	49.00
45	11.10	10.13	32.79	30.73	43.34	41.40	51.81	51.06
50	16.00	14.45	40.07	37.50	49.28	46.94	55.40	54.18

2. Find the annual premium on each of the following policies:
a. Ordinary life for $1,000 issued to a woman at age 25.
b. 20-payment life for $5,000 issued to a woman at age 22.
c. 10-year term for $50,000 issued to a man at age 35.
d. 20-year endowment for $10,000 issued to a man at age 40.
e. 20-payment life for $25,000 issued to a man at age 28.
f. Ordinary life for $30,000 issued to a woman at age 22.
g. 20-year endowment for $15,000 issued to a man at age 26.
h. 10-year term for $20,000 issued to a woman at age 30.
i. Ordinary life for $75,000 issued to a man at age 45.
j. 20-year endowment for $20,000 issued to a woman at age 24.

3. Mr. Harris bought a 20-year endowment life insurance policy for $10,000 at age 22. If he is alive at the age of 42, how much less will he receive than he paid in?

4. At the age of 24 Mr. Wilson bought a 20-payment life policy for $15,000. What is the total amount of premium he pays for the 20 years? Will the beneficiary receive more or less than Mr. Wilson paid in? How much?

5. Approximately how many times as much ordinary life insurance can be bought for a given premium at age 22 as 20-year endowment insurance?
6. Mr. Patton took out at age 25 an ordinary life policy for $20,000. If he died after paying 9 annual premiums, how much more did the beneficiary receive than Mr. Patton paid in?

15–18 SALES TAX

A *sales tax* is a tax on the purchase price of an article.

EXERCISES

1. At the rate of 6% sales tax, what would the tax be on an article that sells for $12.50?
2. Find the sales tax to the nearest cent on each of the following purchases at the given tax rates:

Purchases	$15.00	$36.75	$4.98	$13.60	$59.95	$108.35
Tax Rate	5%	7%	2%	$5\frac{1}{2}$%	4%	$3\frac{1}{2}$%

3. Find the selling price, including tax, of each of the following articles:
   a. A coat selling for $98, tax 2%.
   b. A radio selling for $69, tax 1%.
   c. A football selling for $9.95, tax 3%.
   d. Shoes selling for $17.75, tax 6%.
   e. A lamp selling for $39.50, tax 4%.
   f. A rug selling for $85, tax $1\frac{1}{2}$%.
   g. A dress selling for $12.95, tax 8%.
   h. A TV set selling for $198, tax $2\frac{1}{2}$%.
   i. A chair selling for $39.25, tax 7%.
   j. A hat selling for $14.89, tax $3\frac{1}{2}$%.

15–19 EXCISE TAX

An *excise tax* is a federal tax on certain items. The federal excise tax rate on telephone service is 3% and on fares for air travel is 8%.

EXERCISES

1. Find the amount of tax on local telephone service costing:
   a. $5.80 b. $13.90 c. $15.95 d. $8.20 e. $11.60
2. Find the amount of tax on an air travel fare of:
   a. $72 b. $58.10 c. $141.50 d. $205.80 e. $93.25
3. Find the full amount of the bill including federal excise tax and, where given, state tax on telephone service costing:
   a. $6.65
   b. $9.40
   c. $14.15, 5% state tax
   d. $7.50, 3% state tax
   e. $12.48, 2% state tax
   f. $15.72, 4% state tax
4. Find the total cost of airplane fare including excise tax when the fare excluding excise tax is:
   a. $85 b. $140 c. $97.50 d. $212.75 e. $176.40

15–20 PROPERTY TAX

A tax on buildings and land is called a *real property* or *real estate tax.* The value that tax officers place on property for tax purposes is the *assessed value* of the property.

The tax rate on property may be expressed as: Cents per \$1, mills per \$1, dollars and cents per \$100, dollars and cents per \$1,000, and per cent.

There are 10 mills in 1 cent. A mill is one-thousandth of a dollar (\$.001) or one-tenth of a cent.

To find the amount of taxes on property, we apply the tax rate to the assessed value of the property.

\$.04 per \$1 = 40 mills per \$1 = \$4 per \$100 = \$40 per \$1,000 = 4%

Find the taxes on a house assessed for \$8,500 when the tax rate is \$2.50 per \$100.

$$\$100\overline{)\$8{,}500}\quad = 85$$

```
  $2.50 tax rate per $100
  × 85  number of $100
  1250
 2000
$212.50 taxes
```

Answer, \$212.50 taxes

EXERCISES

1. Write as a decimal part of a dollar: 9 mills; 14 mills; 128 mills; 40 mills
2. Write as cents: 30 mills; 23 mills; 7 mills; 115 mills
3. Write as mills: \$.02; \$.04; \$.006; \$.131

Write each of the following rates as:

4. Cents per \$1: \$8 per \$100; 64.5 mills per \$1; \$39 per \$1,000; 2.9%
5. Mills per \$1: \$13 per \$100; \$47 per \$1,000; 1.7%; \$.91 per \$1
6. Dollars and cents per \$100: \$.09 per \$1; 3.1%; 118 mills per \$1; \$75 per \$1,000
7. Dollars and cents per \$1,000: 4.3%; \$5.85 per \$100; 34.5 mills per \$1; \$.16 per \$1
8. Per Cent: \$3 per \$100; \$93.60 per \$1,000; \$.08 per \$1; 52 mills per \$1
9. A community with an assessed valuation of \$2,500,000 requires \$50,000 as taxes. What should its tax rate be? Express the rate as dollars per \$100.
10. a. If the tax rate is \$6.80 per \$100, how much must you pay for taxes on a house assessed for \$9,900?
    b. How much must you pay for taxes on a building assessed for \$75,000 if the tax rate is \$5.75 per \$100?

Find the amount of taxes on property having the following assessed valuations and tax rates:

11.

Assessed valuation	$4,300	$9,100	$17,500	$11,000	$22,500
Tax rate per $100	$9.60	$3.40	$1.85	$14.10	$5.80

12.

Assessed valuation	$5,600	$10,000	$18,700	$27,800	$13,000
Tax rate per $1	$.15	$.127	$.04½	$.054	$.09

13.

Assessed valuation	$25,300	$22,900	$12,000	$16,800	$39,300
Tax rate per $1	26 mills	74 mills	119 mills	47 mills	61 mills

14.

Assessed valuation	$8,000	$13,500	$7,600	$10,400	$25,000
Tax rate	4%	8.2%	12.9%	9.3%	4.8%

15. A house is assessed for $9,500 and the tax rate is $2.75 per $100. Which will give a greater reduction in taxes, reducing the assessment to $8,200 or reducing the rate to $2.40 per $100? How much greater?
16. A building worth $27,000 is assessed for ⅔ of its value. How much taxes does the owner pay on the property if the tax rate is $7.25 per $100?
17. How much does Lisa's father pay for taxes on his property which is assessed for 40% of its cost of $28,000 if the tax rate is $3.45 per $100?

15–21 BUYING AND RENTING A HOUSE

When a person borrows money to buy a house, the borrower gives the lender a written claim (called a *mortgage*) to the property in case of failure to repay the specified loan or to pay the interest on the loan when due. To find the amount of the mortgage, we subtract the down-payment from the purchase price.

To determine the cost of owning a house, we find the sum of the following expenses: interest (see page 496) on mortgage, cost of repairs, depreciation, cost of insurance on the house, property taxes, and interest lost on the principal paid. Although in the computation of the federal income tax we use the property tax and the interest paid on the mortgage as deductions and the interest received on the invested principal as income, we shall disregard these considerations in problems 7 through 10.

EXERCISES

1. What is your rent for a year if your monthly rent is:
   a. $135? **b.** $490? **c.** $92.50? **d.** $268.25? **e.** $387.50?

2. If you should not spend more than 25% of your income for the rent of a house, what is the highest monthly rent you can afford to pay if you earn:
a. $8,760 per yr.? **b.** $1,000 per mo.? **c.** $217.75 per wk.? **d.** $25,000 per yr.?
3. If you should not buy a house costing more than $2\frac{1}{2}$ times your annual income, what is the highest price you can afford to pay if you earn:
a. $12,000 per yr.? **b.** $825 per mo.? **c.** $96.25 per wk.? **d.** $335 per wk.?
4. A man bought a house for $25,000. He paid $7,500 in cash and gave a mortgage bearing 8% interest for the balance. What is his annual interest? Semi-annual interest?
5. Find the amount of mortgage, annual interest, and semi-annual interest on the mortgage:

Cost of house	$10,800	$26,000	$21,900	$35,000	$40,000
Down-payment	$1,800	$2,600	$5,500	$8,000	$8,000
Annual interest rate	8%	9%	$8\frac{1}{2}$%	$8\frac{1}{4}$%	$7\frac{3}{4}$%

6. A mortgage of $10,000 is to be paid off in 20 years. How much must be paid every 6 months if the mortgage is to be repaid in equal semi-annual payments?
7. What are the costs per month of owning a house if the property tax is $920 per yr., semi-annual interest on the mortgage is $220, repairs average $300 per yr., insurance costs $210 for 3 yr., and interest lost on principal paid amounts to $250 per yr.?
8. Find the monthly carrying charges during the first and second years on a new house that costs $24,000. Down-payment is 20% of the cost, and 8% interest is charged on the balance. House is insured for 80% of cost at 20¢ per $100. House is assessed for 75% of cost; tax rate is $4.75 per $100. Mortgage is to be paid off in 25 years in equal annual installments. Principal paid on house could be invested at 6%.

9. Find the monthly cost of owning a house:

	Cost of House	Down-Payment	Interest Rate on Mortgage	Repairs and Depreciation	Insurance	Assessed Value	Tax Rate per $100	Interest Rate Lost on Principal
a.	$28,500	$6,500	8%	$480	$250	$27,000	$4.95	6%
b.	$16,300	$4,300	$8\frac{1}{2}$%	$372	$126	$15,000	$5.10	$5\frac{1}{2}$%
c.	$37,500	$7,500	$7\frac{1}{2}$%	$540	$324	$19,500	$8.70	8%

10. Is it cheaper to buy or rent the house and how much cheaper per year?

	Cost of House	Down-Payment	Interest Rate on Mortgage	Repairs and Insurance	Assessed Value	Tax Rate per $100	Interest Rate Lost on Principal	Rent per Month
a.	$17,400	$3,400	8%	$308	$15,200	$3.65	6%	$200
b.	$18,500	$6,000	7%	$250	$11,000	$7.25	$5\frac{1}{2}$%	$250
c.	$32,800	$11,400	$8\frac{1}{2}$%	$516	$16,400	$9.20	8%	$500
d.	$20,000	$4,800	$7\frac{1}{2}$%	$350	$12,000	$6.95	9%	$300

MANAGING MONEY

15–22 SIMPLE INTEREST

Just as people pay rent for the use of a house belonging to someone else, people pay interest for the use of money belonging to others. *Interest* is money paid for the use of money. The money borrowed or invested on which interest is paid is called the *principal.* Interest paid on the principal only is called *simple interest.* The interest charged is generally expressed as a per cent of the principal. This per cent is called the *rate of interest.* Unless specified otherwise, the rate of interest is the rate per year. The sum of the principal and the interest is called the *amount.*

(1) To find the interest, we multiply the principal by the rate of interest per year. Then we multiply this product by the time expressed in years. (2 mo. = $\frac{1}{6}$ yr.; 120 da. = $\frac{1}{3}$ yr.; semi-annual = $\frac{1}{2}$ yr.; quarterly = $\frac{1}{4}$ yr.)

The interest (i) equals the principal (p) times the rate of interest per year (r) times the time expressed in years (t). This rule expressed as a formula is: $i = prt$.

Find the interest and the amount due on \$375 borrowed for 4 years at 5%:

\$375 principal	\$375 principal
.05 rate	75 interest
\$18.75 interest for 1 year	\$450 amount
× 4	
\$75.00 interest for 4 years	

Answer, \$75 interest; \$450 amount

Find the interest on \$500 for 2 yr. 8 mo. at 6%:

$p = \$500$	$i = prt$
$r = 6\% = \frac{6}{100}$	$i = 500 \times \frac{6}{100} \times 2\frac{2}{3}$
$t = 2$ yr. 8 mo. $= 2\frac{2}{3}$ yr.	$i = \$80$
$i = ?$	

Answer, \$80 interest

(2) To find the interest by the formula, we write the interest formula, substitute the given quantities, then compute as required. The rate is usually expressed as a decimal but a common fraction may be used.

What is the annual rate of interest if the annual interest on a principal of \$150 is \$9?

$$\frac{\$9}{\$150} = \$9 \div \$150 = \$150\overline{)\$9.00} \quad .06 = 6\%$$

Answer, 6%

(3) To find the amount, we add the interest and the principal.

(4) *Sixty-day or 6% method.*—To find the interest for 60 days at 6%, we move the decimal point in the principal two places to the left.

(5) To find the annual rate of interest, we find what per cent the interest for one year is of the principal.
(6) To find the principal when the annual rate and annual interest are known, we divide the interest by the rate.

EXERCISES

1. Find the interest for 1 year on:

a. \$300 at 7%.
b. \$950 at 6%.
c. \$1,000 at $8\frac{1}{2}$%.
d. \$1,600 at $5\frac{1}{2}$%.

2. Find the interest on:

a. \$500 for 2 yr. at 8%.
b. \$625 for 3 yr. at 9%.
c. \$1,400 for 6 yr. at $7\frac{1}{2}$%.
d. \$2,500 for 8 yr. at $3\frac{3}{4}$%.

3. Find the interest on:

a. \$760 for $3\frac{1}{2}$ yr. at 1%.
b. \$420 for $\frac{2}{3}$ yr. at 11%.
c. \$1,200 for $4\frac{3}{4}$ yr. at $5\frac{1}{2}$%.
d. \$2,700 for $\frac{5}{6}$ yr. at $6\frac{1}{2}$%.

4. Find the interest on:

a. \$800 for 4 mo. at 3%.
b. \$5,200 for 10 mo. at 6%.
c. \$624 for 5 mo. at $7\frac{1}{2}$%.
d. \$1,170 for 8 mo. at $2\frac{1}{2}$%.

5. Find the interest on:

a. \$80 for 1 yr. 9 mo. at 9%.
b. \$4,000 for 5 yr. 6 mo. at 2%.
c. \$2,100 for 2 yr. 11 mo. at $8\frac{1}{2}$%.
d. \$600 for 1 yr. 3 mo. at $7\frac{1}{4}$%.

6. Find the exact interest (1 yr. = 365 da.) on:

a. \$580 for 73 da. at 3%.
b. \$1,740 for 15 da. at 6%.
c. \$3,000 for 45 da. at 8%.
d. \$800 for 146 da. at 5%.

7. Find the interest (1 yr. = 360 da.) on:

a. \$900 for 30 da. at 5%.
b. \$1,600 for 60 da. at 4%.
c. \$480 for 72 da. at 6%.
d. \$7,000 for 90 da. at 9%.

8. Find the semi-annual interest on:

a. \$825 at 10%.
b. \$2,350 at 6%.
c. \$3,900 at $5\frac{1}{2}$%.
d. \$6,400 at 7.9%.

9. Find the quarterly interest on:

a. \$700 at 4%.
b. \$1,250 at 8%.
c. \$6,000 at $1\frac{3}{4}$%.
d. \$10,000 at 9.6%.

10. Use 6%, 60-day method to find the interest on:

a. \$2,850 for 60 da. at 6%.
b. \$349 for 60 da. at 6%.
c. \$1,200 for 30 da. at 6%.
d. \$700 for 30 da. at 3%.

11. What is the amount due when the principal is \$360 and the interest is \$21.60?

12. What is the interest and amount due on a loan of $800 for 3 yr. at 6%?
13. What is the amount due on a loan of $425 for 8 yr. at $7\frac{1}{2}$%?
14. Find the interest and amounts due on the following loans:

   a. $200 for 1 yr. at 6%.
   b. $500 for 7 yr. at 9%.
   c. $1,750 for 2 yr. at $8\frac{1}{2}$%.

15. What is the annual rate of interest if the principal is $280 and the annual interest is $14?
16. What is the annual rate of interest if the principal is $450 and the interest for 5 yr. is $135?
17. Find the annual rates of interest when the interest for:

   a. 1 yr. on $75 is $3.
   b. 4 yr. on $1,200 is $144.
   c. 9 yr. on $860 is $154.80.

18. What is the principal if the amount is $796 and the interest is $31?
19. On what principal is the annual interest $9 when the annual rate of interest is 6%?
20. Find the principal on which the annual interest is:

   a. $23 when the annual rate is 4%.
   b. $90 when the annual rate is 5%.
   c. $37.50 when the annual rate is 3%.

21. Mr. O'Connor owns eight $1,000 bonds, each bearing $3\frac{1}{4}$% interest. How much interest does he receive every 6 months?
22. A woman bought a house for $29,600. She paid 30% down and gave a mortgage bearing 8% interest for the remainder. How much interest does she pay semi-annually?
23. On which investment does Mrs. Johnson receive a higher rate of interest? She receives $157.50 semi-annually on an investment of $7,500 and $344 annually on an investment of $8,600.
24. How much money must be invested at 8% to earn $10,000 per year?
25. How many years does it take $200 invested at 4% simple interest to double itself?

15–23 BANKING

There are two main kinds of banks: savings banks and commercial banks. A third type of banking institution that has grown greatly in recent years is the federal and state savings and loan association, which is very much like a savings bank. Credit unions also render both savings and loan services, but they are usually offered only to particular groups of people, like city employees, factory workers, teachers, etc.

The life of a bank depends on the use of its depositors' money as loans to business people and others at rates of interest higher than the rates of interest that the bank pays to its depositors. The difference between these rates of interest produces money for the bank both to pay its expenses and to earn a profit.

The main service offered by the savings bank is the *savings account*, on which the depositor is paid interest. Savings banks also provide other services such as passbook loans, safe deposit boxes, and money orders. We use a form called a deposit slip to deposit money and a withdrawal slip to withdraw money from our accounts.

The commercial bank provides almost similar services. However it also provides its most important service—the *checking account*, a service not furnished by the savings bank. To deposit money in a checking account, we use a *deposit slip* (see form at the right). To transfer funds from a checking account to persons or business establishments to whom we owe money, we use a form called a *check* (see form below). This method facilitates the payment of bills without any cash being handled directly.

LOCAL TRUST COMPANY

YOUR CITY ______________ 19__
ACCOUNT NO. ______________
NAME ______________
ADDRESS ______________

	DOLLARS	CENTS
CASH →		
CHECKS BY BANK NUMBER ↓		
1		
2		
3		
4		
5		
6		
7		
TOTAL		

No. 294		
1/12 19 83		
To CASH		
FOR ______		
	DOLLARS	CENTS
BAL. BRO'T FOR'D	1000	00
AM'T DEPOSITED	2384	62
TOTAL	1374	62
AM'T THIS CHECK		
BAL. CAR'D FOR'D		

YOUR CITY Hingham 19 83 NO. 294

LOCAL TRUST COMPANY

PAY TO THE ORDER OF CASH $1000.00

One Thousand DOLLARS

David Wright

EXERCISES

Use two deposit slip forms and three check forms if available, otherwise copy them as illustrated, to record the following:

1. On April 2, you have a bank balance of $875.46. Write this balance as the balance brought forward on the first check stub.

2. On April 4, you deposit 13 $10 bills, 6 $5 bills, and a check, numbered 3560–3, drawn on the National Bank for $36.85. Fill in the deposit slip. Write the total amount deposited on the first check stub.
3. On April 9, you send a check to Andrew Jaffe in the amount of $192.87 for equipment purchased. Fill in the first check stub. Carry the balance to the second check stub. Write out the first check.
4. On April 17, you send a check to Addison and Co. in the amount of $403.59 for furniture purchased. Fill in the second check stub. Carry the balance to the third check stub. Write out the second check.
5. On April 23, you deposit 4 $20 bills, 16 $10 bills, 7 $5 bills, 23 $1 bills, 3 half-dollars, 19 quarters, 6 dimes, 3 nickels, a check, numbered 2997–8, drawn on the Bankers Trust Co. for $136.87, a check, numbered 1503–2, drawn on the State Bank for $260.51, and a check, numbered 3117–5, drawn on the First National Bank for $97.18. Fill in the second deposit slip. Write the total amount deposited on the third check stub.
6. On April 24, you send a check to Collins Manufacturing Co. in the amount of $85.07 for merchandise purchased. Fill in the third check stub. Write out the third check.

15–24 BORROWING MONEY

Money may be borrowed from other people, from banks, from loan companies, from savings and loan associations, from credit unions, and on your life insurance policy. When money is borrowed, there is a charge (called interest) for the use of this money. See section 15–22.

Loans which are secured by stocks, bonds, property, etc. are called *collateral loans*. If the loan is not repaid at the specified time, then the lending institution can collect the amount due by selling the collateral. The rate of interest charged on collateral loans is less than the rate of interest charged on unsecured loans.

Some banks will lend money to customers, allowing them to keep the principal until a specified time but requiring payment of interest monthly. This type of loan is called a *demand loan*.

The *discount loan* and the *add-on-interest* loan are the two types of loans most generally used. Annual percentage rates are indicated on loan contracts. See page 484.

When a discount loan is made, the lending company immediately takes off the full interest from the principal (amount of note) and the borrower receives only the difference (called net proceeds). The borrower, however, is required to repay the full principal in a specified number of equal monthly payments. See problem on page 501.

When an add-on-interest loan is made, the borrower receives the full principal. The interest is added on to this principal and the borrower is required to repay the full amount (principal plus interest) in a specified number of equal monthly payments.

For example:

Mr. Harris borrowed \$750 at 6% annual interest and is required to repay his loan in 12 equal monthly payments. What is the actual amount of money that he receives and how much is his monthly payment if he takes a discount loan? If he takes an add-on-interest loan?

Discount Loan

\$750	principal (amount of note)
− 45	interest
\$705	amount received

\$62.50	monthly payment
12)\$750.00	amount to be repaid

Answer, \$705, amount received; \$62.50, monthly payment

Add-on-Interest Loan

\$750	amount received

\$750	principal
+ 45	interest
\$795	amount to be repaid

\$66.25	monthly payment
12)\$795.00	amount to be repaid

Answer, \$750, amount received; \$66.25, monthly payment

Observe in these loans that since the principal is being repaid monthly, the average amount that was borrowed is only about one-half of the original amount of the loan. Therefore the true rate of interest is almost double the stated rate of interest.

When a loan is made to purchase a house, each equal monthly payment to pay off the loan consists of (1) a reduction in principal and (2) the interest on the new reduced balance of principal owed each month instead of the entire amount of the loan, since part of each equal monthly payment goes to pay off the principal.

The repayment of a mortgage loan in this way is sometimes described as *amortizing* the loan. This is different than what we found with discount loans and add-on-interest loans where the borrower pays interest on the full amount of his loan for the entire period of his loan instead of on the balance owed on his loan each month.

Under the terms of many life insurance policies the policyholder is permitted to borrow from the insurance company an amount of money not more than the loan value or the cash surrender value of the policy. The interest charged is simple interest for the period of time the money is borrowed. See section 15–22.

EXERCISES

1. What is the annual percentage rate if the rate of interest charged on a loan:

a. Per month is: 1%? $\frac{1}{2}$%? 0.8%? 3%? $1\frac{3}{4}$%?
b. Per week is: 1%? $\frac{3}{4}$%? $\frac{1}{2}$%? $1\frac{1}{4}$%? 0.2%?
c. Per day is: 1%? $\frac{1}{4}$%? 0.1%? $\frac{3}{4}$%? 0.3%?

2. What is the annual rate of interest you are paying when you:

a. Borrow $1 and pay back $2 at the end of 2 months?

b. Borrow $5 and pay back $5.50 at the end of a week?

c. Borrow $100 and pay back $125 at the end of a week?

d. Borrow 25¢ and pay back 30¢ at the end of 1 day?

e. Borrow $10 and pay back $11 at the end of 1 day?

3. What is the actual amount of money (net proceeds) the borrower receives and how much is the monthly payment in each of the following discount loans?

	Amount of Note	Interest Rate	Period of Loan
a.	$600	6%	12 months
b.	$1,000	7%	18 months
c.	$500	$6\frac{1}{2}$%	24 months
d.	$2,000	$8\frac{1}{4}$%	36 months
e.	$1,800	$7\frac{1}{2}$%	12 months
f.	$3,500	8%	30 months
g.	$5,000	$6\frac{1}{2}$%	42 months
h.	$7,500	6%	48 months
i.	$4,200	$6\frac{3}{4}$%	24 months
j.	$10,000	$7\frac{1}{4}$%	60 months

4. What is the actual amount of money the borrower receives and how much is the monthly payment in each of the following add-on-interest loans?

	Principal	Interest Rate	Period of Loan
a.	$800	6%	12 months
b.	$1,000	8%	30 months
c.	$4,000	$7\frac{1}{2}$%	24 months
d.	$2,500	$6\frac{1}{2}$%	36 months
e.	$1,600	$8\frac{1}{4}$%	18 months
f.	$6,000	7%	24 months
g.	$7,200	$6\frac{3}{4}$%	60 months
h.	$400	9%	42 months
i.	$9,000	$6\frac{1}{4}$%	36 months
j.	$3,600	$6\frac{1}{2}$%	30 months

5. How much interest is due on the following loans on insurance policies?

a. $900 borrowed for 6 months at 6%.

b. $1,700 borrowed for 3 months at $6\frac{1}{2}$%.

c. $2,100 borrowed for 4 months at 5%.

d. $500 borrowed for 45 days at 6%.

e. $600 borrowed for 1 week at 6%.

6. How much interest is due each month on each of the following collateral loans?

a. $2,000 borrowed for 1 year at 8%.

b. $1,400 borrowed for 6 months at $7\frac{1}{2}$%.

c. $7,500 borrowed for 18 months at 7%.

d. $10,000 borrowed for 9 months at $8\frac{1}{4}$%.

e. $25,000 borrowed for 3 months at $7\frac{3}{4}$%.

7. Cost Schedule for Financing a Car

Amount to be Borrowed	Amount of Each Payment		
	24 months	30 months	36 months
$1,000	$45.83	$37.50	$31.94
1,200	55.00	45.00	38.33
1,500	68.75	56.25	47.91
1,800	82.50	67.50	57.50
2,000	91.66	75.00	63.88
2,500	114.58	93.75	79.86
3,000	137.50	112.50	95.83
3,500	160.41	131.25	111.80

a. Find the total payment if $1,200 is borrowed for 24 months. For 30 months. For 36 months. Find the total cost of borrowing $1,200 for 24 months. For 30 months. For 36 months.

b. How much less each month is the payment for a loan of $3,500 for 36 months than it is for 24 months? How many more monthly payments must be made for a 36-month loan than for a 24-month loan? How much more is the total interest charge on the 36-month loan of $3,500 than it is on the 24-month loan?

c. How much less is the total interest charge when you borrow:
(1) $1,800 for 24 months instead of 30 months?
(2) $2,500 for 30 months instead of 36 months?
(3) $2,000 for 24 months instead of 36 months?

d. If your budget will allow a car financing cost of $50 per month maximum and you require a loan of $1,500 to complete the purchase of the car, for what period of time must you borrow the money?

8. Amortizing a Loan

The following schedule illustrates how a mortgage loan for $16,000 at the annual rate of 6% is amortized by equal monthly payments of $177.64. This section shows only the first 6 payments of a 120-month payment schedule.

Payment Number	Payment on Interest	Payment on Principal	Balance of Loan
1	$80.00	$97.64	$15,902.36
2	79.51	98.13	15,804.23
3	79.02	98.62	15,705.61
4	78.53	99.11	15,606.50
5	78.03	99.61	15,506.89
6	77.53	100.11	15,406.78

To find each new monthly balance of the loan, first find the monthly interest on the preceding month's loan balance. Subtract this interest from the monthly payment (in this case $177.64) to find the amount of reduction of principal for the month. Subtract this reduction in principal from the preceding month's loan balance to find the new monthly balance.

For example:

(.06 × $16,000) ÷ 12 = $80 interest for first month
$177.64 − $80 = $97.64 reduction in principal
$16,000.00 − $97.64 = $15,902.36 loan balance after first payment

Check these figures with those in the first row of the schedule on page 503.

(.06 × $15,902.36) ÷ 12 = $79.51 interest for second month
$177.64 − $79.51 = $98.13 next reduction in principal
$15,902.36 − $98.13 = $15.804.23 loan balance after second payment

Check these figures with those in the second row of the schedule on page 503.

a. Determine the interest payment, the payment on principal, and the new monthly loan balance for payments 3, 4, 5, and 6. Check with figures given in the schedule on page 503.
b. Determine the interest payment, the payment on principal, and the new monthly loan balance for each of payments 7 through 12.
c. What was the total amount of interest paid for the first year?
d. How much was the loan reduced at the end of the first year?

15–25 COMPOUND INTEREST

Savings and loan associations and some savings banks advertise that savings left with them earn interest compounded semi-annually or quarterly (every 3 months). Some banks even advertise interest compounded daily on passbook savings or on savings certificates.

Compound interest is interest paid on both the principal and the interest earned previously.

To find how much a given principal will amount to compounded at a given rate for a certain period of time by the use of the table, we select from the table on page 505 how much $1 will amount to based on the given rate and the given number of years (periods) when the interest is compounded annually; and on half the given rate and twice as many periods as the given number of years when the interest is compounded semi-annually; and on one-fourth the given rate and four times as many periods as the given number of years when the interest is compounded quarterly. Then we multiply this amount by the given principal.

To *find the compound interest*, we subtract the principal from the amount.

COMPOUND INTEREST TABLE

Showing How Much $1 Will Amount to at Various Rates

Periods	$\frac{1}{2}$%	1%	$1\frac{1}{4}$%	$1\frac{1}{2}$%	2%	$2\frac{1}{2}$%
1	1.005000	1.010000	1.012500	1.015000	1.020000	1.025000
2	1.010025	1.020100	1.025156	1.030225	1.040400	1.050625
3	1.015075	1.030301	1.037971	1.045678	1.061208	1.076891
4	1.020151	1.040604	1.050945	1.061364	1.082432	1.103813
5	1.025251	1.051010	1.064082	1.077284	1.104081	1.131408
6	1.030378	1.061520	1.077383	1.093443	1.126162	1.159693
7	1.035529	1.072135	1.090851	1.109845	1.148686	1.188686
8	1.040707	1.082857	1.104486	1.126493	1.171659	1.218403
9	1.045911	1.093685	1.118292	1.143390	1.195093	1.248863
10	1.051140	1.104622	1.132271	1.160541	1.218994	1.280085
11	1.056396	1.115668	1.146424	1.177949	1.243374	1.312087
12	1.061678	1.126825	1.160755	1.195618	1.268242	1.344889
13	1.066986	1.138093	1.175264	1.213552	1.293607	1.378511
14	1.072321	1.149474	1.189955	1.231756	1.319479	1.412974
15	1.077683	1.160969	1.204829	1.250232	1.345868	1.448298
16	1.083071	1.172579	1.219890	1.268986	1.372786	1.484506
17	1.088487	1.184304	1.235138	1.288020	1.400241	1.521618
18	1.093929	1.196147	1.250577	1.307341	1.428246	1.559659
19	1.099399	1.208109	1.266210	1.326951	1.456811	1.598650
20	1.104896	1.220190	1.282037	1.346855	1.485947	1.638616

Periods	3%	4%	$4\frac{1}{2}$%	5%	$5\frac{1}{2}$%	6%
1	1.030000	1.040000	1.045000	1.050000	1.055000	1.060000
2	1.060900	1.081600	1.092025	1.102500	1.113025	1.123600
3	1.092727	1.124864	1.141166	1.157625	1.174241	1.191016
4	1.125509	1.169859	1.192519	1.215506	1.238825	1.262477
5	1.159274	1.216653	1.246182	1.276282	1.306960	1.338226
6	1.194052	1.265319	1.302260	1.340096	1.378843	1.418519
7	1.229874	1.315932	1.360862	1.407100	1.454679	1.503630
8	1.266770	1.368569	1.422101	1.477455	1.534687	1.593848
9	1.304773	1.423312	1.486095	1.551328	1.619094	1.689479
10	1.343916	1.480244	1.552969	1.628895	1.708144	1.790848
11	1.384234	1.539454	1.622853	1.710339	1.802092	1.898299
12	1.424561	1.601032	1.695881	1.795856	1.901207	2.012196
13	1.468534	1.665074	1.772196	1.885649	2.005774	2.132928
14	1.512590	1.731676	1.851945	1.979932	2.116091	2.260904
15	1.557967	1.800944	1.935282	2.078928	2.232476	2.396558
16	1.604706	1.872981	2.022370	2.182875	2.355263	2.540352
17	1.652848	1.947901	2.113377	2.292018	2.484802	2.692773
18	1.702433	2.025817	2.208479	2.406619	2.621466	2.854339
19	1.753506	2.106849	2.307860	2.526950	2.765647	3.025600
20	1.806111	2.191123	2.411714	2.653298	2.917757	3.207135

Amount and compound interest on

$500 compounded at 6% at the end of 4 years.

Annually:	Semi-annually:	Quarterly:
Periods: 4 Rate: 6%	Periods: 8 Rate: 3%	Periods: 16 Rate: $1\frac{1}{2}$%
1.262477	1.266770	1.268986
$500	$500	$500
$631.238500	$633.385000	$634.493000
= $631.24	= $633.39	= $634.49
Amount: $631.24	Amount: $633.39	Amount: $634.49
Compound interest: $131.24	Compound interest: $133.39	Compound interest: $134.49

EXERCISES

1. Find the amount when the interest is compounded annually on:

a. $200 for 6 years at 3%.
b. $700 for 11 years at $5\frac{1}{2}$%.
c. $1,500 for 18 years at $4\frac{1}{2}$%.
d. $8,000 for 9 years at 6%.
e. $20,000 for 15 years at 5%.
f. $35,000 for 20 years at $2\frac{1}{2}$%.

2. Find the amount when the interest is compounded semi-annually on:

a. $600 for 1 year at 6%.
b. $450 for 4 years at 5%.
c. $2,000 for 5 years at 3%.
d. $4,800 for 7 years at 4%.
e. $12,000 for 10 years at 6%.
f. $7,500 for 9 years at $2\frac{1}{2}$%.

3. Find the amount when the interest is compounded quarterly on:

a. $100 for 3 years at 6%.
b. $800 for 1 year at 5%.
c. $2,500 for 5 years at 2%.
d. $11,000 for 2 years at 5%.
e. $14,500 for 3 years at 4%.
f. $8,300 for 4 years at 6%.

4. Find the interest earned when the interest is compounded annually on:

a. $400 for 6 years at $5\frac{1}{2}$%.
b. $950 for 11 years at 6%.
c. $1,600 for 17 years at 3%.
d. $2,100 for 13 years at 5%.
e. $8,000 for 4 years at $4\frac{1}{2}$%.
f. $19,000 for 19 years at 6%.

5. Find the interest earned when the interest is compounded semi-annually on:

a. $700 for 7 years at 5%.
b. $860 for 5 years at 4%.
c. $4,000 for 9 years at 6%.
d. $20,000 for 2 years at 3%.
e. $35,000 for 8 years at $2\frac{1}{2}$%.
f. $8,200 for 3 years at 6%.

6. Find the interest earned when the interest is compounded quarterly on:

a. $900 for 2 years at 5%.
b. $3,000 for 5 years at 6%.
c. $1,800 for 4 years at 4%.
d. $17,000 for 1 year at 6%.
e. $9,250 for 3 years at 2%.
f. $40,000 for 5 years at 5%.

7. A bank advertises a 5% interest rate compounded semi-annually on all deposits.
   a. What would a deposit of $1,000 amount to at the end of 6 months? 1 year? 18 months? 2 years? 30 months? 3 years? 42 months? 4 years? 54 months? 5 years? 6 years? 7 years? 8 years? 9 years? 10 years?
   b. What is the interest earned on a $2,500 deposit at the end of 6 months? 1 year? 2 years? 30 months? 3 years? 4 years? 5 years? 66 months? 6 years? 78 months? 7 years? 8 years? 102 months? 9 years? 10 years?

8. A bank advertises a 5% interest rate compounded quarterly on all deposits.
   a. What would a deposit of $2,000 amount to at the end of 3 months? 6 months? 9 months? 1 year? 15 months? 18 months? 21 months? 2 years? 30 months? 3 years? 42 months? 4 years? 51 months? 54 months? 5 years?
   b. What is the interest earned on a $5,000 deposit at the end of 3 months? 6 months? 9 months? 1 year? 15 months? 18 months? 21 months? 2 years? 27 months? 30 months? 33 months? 3 years? 42 months? 4 years? 5 years?

9. At the interest rate of 6% compounded monthly (12 times a year):
   a. What will each of the following deposits amount to at the end of one year?
      (1) $600 (2) $1,400 (3) $8,000 (4) $19,000 (5) $50,000
   b. How much interest is earned at the end of 1 year on each of the following deposits?
      (1) $400 (2) $750 (3) $6,000 (4) $12,500 (5) $30,000
   c. What will each of the following deposits amount to?
      (1) A deposit of $300 at the end of 7 months.
      (2) A deposit of $1,100 at the end of 4 months.
      (3) A deposit of $8,500 at the end of 1 month.
   d. How much interest is earned on:
      (1) A deposit of $900 at the end of 8 months?
      (2) A deposit of $2,150 at the end of 5 months?
      (3) A deposit of $15,000 at the end of 10 months?

10. Will $5,000 compounded annually for 5 years at 6% amount to the same as for 6 years at 5%? If not, the amount of which is greater? How much greater?

11. Find the interest earned on a $25,000 deposit at the end of 1 year at each of the following rates:

    a. 6% compounded semi-annually c. 6% compounded quarterly
    b. 6% compounded annually d. 6% compounded monthly

    Which rate brings the greatest amount of interest? Which rate brings the smallest amount of interest?

15–26 STOCKS, BONDS, MUTUAL FUNDS, SAVINGS CERTIFICATES

A person buying a share of *stock* becomes a part-owner of the business and receives dividends from the profits earned. A person buying a bond is lending money to the business and, therefore, receives interest on the face value of the bond. A *bond* is a written promise of a private corporation or of a local, state, or the national government to pay a given rate of interest at stated times and to repay the face value of the bond at a specified time (*date of maturity*).

The original value of a bond is called *par value* or face value; the value at which it sells at any given time is called *market value*, the value is *above par* when the market value is more than the par value, and *below par* when the market value is less than the par value.

Stock quotations are in terms of dollars. $28\frac{3}{4}$ means \$28.75.

The P-E ratio column heading in the stock quotations refers to an approximate ratio comparing the closing price of a share of each specific stock to the annual earnings. For example, if the P-E ratio of a certain stock is 8, then the price of the stock is approximately 8 times its annual earnings. To find the annual earnings, we divide the closing price by the P-E ratio number.

Bond quotations are in terms of per cent. $97\frac{1}{2}$ means $97\frac{1}{2}\%$.

In the following bond quotation: "Sears $4\frac{3}{4}$s 83 $105\frac{3}{4}$," $4\frac{3}{4}$s represents a $4\frac{3}{4}\%$ rate of interest; 83 is the date of maturity, 1983; and $105\frac{3}{4}$ is the market price, indicating $105\frac{3}{4}\%$ of the face value of the bond.

To find the rate of income or yield of a bond (see column marked Cur Yld in the bond quotations), we find what per cent the annual interest is of the market price of the bond.

Fees are paid to brokers for the service of buying or selling stocks and bonds. These fees are subject to change and at present there is no one set of fixed fees for all brokers.

A mutual fund is an investment company. It obtains money for investment by selling shares of stock of the company. Mutual fund quotations are expressed as a Net Asset Value (NAV) and an offer price. When we purchase shares, we pay the offer price; when we sell shares, we receive the Net Asset Value (NAV) of each share.

Banks and savings and loan associations sell savings certificates with various features. Some certificates are available that pay monthly or quarterly interest and are called *income certificates*. Other certificates which allow all interest earned to accumulate until maturity are called *growth certificates*.

The *E* savings bond is the most popular U.S. government bond. It pays 6% interest compounded quarterly and matures in 5 years.

EXERCISES

1. Find the cost of each of the following purchases, excluding commission:

a. Using the closing prices listed in the stock quotations on page 509:

(1) 100 shares of Skaggs
(2) 100 shares of Squibb
(3) 300 shares of Sou Pac
(4) 50 shares of Sears
(5) 75 shares of Scherg Pl
(6) 30 shares of Stu Wor

Stock Quotations

Year High	Year Low	Stocks and Dividends in Dollars	P-E Ratio	Sales in 100s	High	Low	Close	Net Chg.
$45\frac{1}{4}$	$34\frac{1}{8}$	Safeway 1.80	14	123	$43\frac{7}{8}$	$42\frac{7}{8}$	$43\frac{1}{2}$	$+\frac{3}{4}$
$30\frac{7}{8}$	$25\frac{1}{2}$	S Fe Ind 1.80	6	143	$26\frac{3}{8}$	$25\frac{3}{4}$	$26\frac{1}{4}$	$+\frac{1}{2}$
$67\frac{1}{4}$	$44\frac{1}{2}$	Scherg Pl .80	26	422	$59\frac{3}{4}$	$58\frac{5}{8}$	$59\frac{1}{4}$	$+\frac{1}{2}$
$14\frac{3}{4}$	9	SCM cp .50	3	69	$12\frac{1}{2}$	$12\frac{1}{4}$	$12\frac{1}{2}$	$+\frac{1}{4}$
18	$12\frac{1}{8}$	Scott Pap .68	9	225	17	$16\frac{1}{2}$	17	$+\frac{3}{8}$
$69\frac{1}{4}$	$48\frac{3}{8}$	Sears 1.60a	20	337	$63\frac{3}{4}$	$62\frac{1}{8}$	$63\frac{1}{2}$	$+1\frac{1}{8}$
$48\frac{1}{4}$	$39\frac{1}{4}$	Shell Oil 2.60	4	211	$41\frac{1}{4}$	$40\frac{5}{8}$	$40\frac{3}{4}$	$-\frac{1}{2}$
$17\frac{5}{8}$	$14\frac{3}{8}$	Signal Co .90	2	69	$16\frac{1}{4}$	$15\frac{7}{8}$	$16\frac{1}{8}$	$+\frac{1}{4}$
$13\frac{3}{8}$	$9\frac{1}{2}$	Singer Co .40	....	250	$11\frac{3}{8}$	11	$11\frac{1}{4}$	....
$30\frac{7}{8}$	$21\frac{3}{8}$	Skaggs .80	8	54	$24\frac{5}{8}$	$24\frac{1}{2}$	$24\frac{1}{2}$	$-\frac{1}{8}$
$56\frac{5}{8}$	$43\frac{1}{4}$	Smithkline 2	14	23	$54\frac{1}{4}$	$53\frac{5}{8}$	$54\frac{1}{4}$	$+\frac{1}{4}$
$31\frac{7}{8}$	$26\frac{3}{8}$	Sou Pac 2.24	7	67	$28\frac{3}{8}$	$27\frac{5}{8}$	$28\frac{1}{8}$	$+\frac{7}{8}$
$36\frac{5}{8}$	$25\frac{3}{8}$	Sperry R .76	10	330	$35\frac{1}{4}$	$34\frac{3}{8}$	$35\frac{1}{4}$	$+1\frac{1}{8}$
$41\frac{1}{2}$	$24\frac{5}{8}$	Squibb .84	18	119	$35\frac{7}{8}$	$35\frac{3}{8}$	$35\frac{1}{2}$	$-\frac{1}{4}$
$67\frac{3}{8}$	$52\frac{1}{2}$	Std Brands 2	15	148	$61\frac{3}{4}$	61	$61\frac{3}{4}$	$+\frac{1}{8}$
62	$44\frac{1}{2}$	St Oil Oh 1.36	17	160	$57\frac{3}{8}$	$56\frac{1}{2}$	$57\frac{1}{4}$	$+\frac{3}{4}$
$29\frac{3}{4}$	19	Stu Wor 1.32	15	36	$29\frac{3}{8}$	$28\frac{5}{8}$	$28\frac{5}{8}$	$-\frac{5}{8}$
$191\frac{3}{4}$	153	Supr Oil 1.40	11	23	$172\frac{1}{2}$	171	172	$+2$

b. Using the highest prices listed in the stock quotations:

(1) 100 shares of Scott Pap
(2) 300 shares of Signal Co
(3) 25 shares of Smithkline
(4) 65 shares of St Oil Oh
(5) 140 shares of S Fe Ind
(6) 90 shares of Sears

2. Find the selling price of each of the following sales, excluding commission, using the lowest prices listed in the stock quotations:

a. 100 shares of Std Brands
b. 100 shares of St Oil Oh
c. 400 shares of SCM cp
d. 75 shares of Smithkline
e. 150 shares of Sperry R
f. 80 shares of Sou Pac

3. If the commission is 1.5% of the cost of the stock plus $18, find the total cost, including commission, of each of the following purchases, using the highest quoted prices:

a. 100 shares of SCM cp
b. 200 shares of Scott Pap
c. 50 shares of Sperry R
d. 75 shares of Scherg Pl
e. 60 shares of Squibb
f. 39 shares of Std Brands

4. If the commission is 1.18% of the selling price of the stock plus $28, find the net amount due (selling price of stock less commission) of each of the following sales, using the quoted closing prices:

a. 100 shares of Safeway
b. 400 shares of Singer Co
c. 70 shares of S Fe Ind
d. 25 shares of Supr Oil
e. 64 shares of Shell Oil
f. 180 shares of Signal Co

5. Find the annual earnings per share (to nearest cent) for each of the stocks listed in the stock quotations by using its P-E ratio and closing price.
6. If the commission is 1.3% of the value plus $25 in each of the following purchases and sales, find the amount of profit:
   a. 100 shares of BZ, purchased at 35, sold for 43.
   b. 100 shares of RST, purchased at $26\frac{1}{2}$, sold for $35\frac{1}{4}$.
   c. 50 shares of XYZ, purchased at $43\frac{5}{8}$, sold for $51\frac{1}{2}$.
   d. 35 shares of MPZ, purchased at $64\frac{7}{8}$, sold for $83\frac{3}{4}$.
   e. 80 shares of EIS, purchased at $19\frac{1}{4}$, sold for 30.
7. If the commission is 1.1% of the value plus $30 in each of the following purchases and sales, find the amount of loss:
   a. 100 shares of BBB, purchased at 46, sold for 39.
   b. 100 shares of RLS, purchased at $29\frac{3}{4}$, sold for $23\frac{7}{8}$.
   c. 30 shares of CSJ, purchased at $57\frac{1}{2}$, sold for $48\frac{1}{2}$.
   d. 75 shares of PJJ, purchased at 61, sold for $56\frac{3}{8}$.
   e. 43 shares of EBS, purchased at $17\frac{5}{8}$, sold for $10\frac{1}{4}$.
8. The dividend listed for each stock in the stock quotations is for a year. Determine for each of the following stocks the amount of the quarterly dividend:
   a. S Fe Ind
   b. Shell Oil
   c. Skaggs
   d. Sou Pac
   e. Smithkline
   f. Scott Pap
   g. Stu Wor
   h. Singer Co
   i. Squibb
   j. Scherg Pl
   k. St Oil Oh
   l. Signal Co
9. Find the date of maturity and interest rate for each of the following bonds:
   a. Crane 7s 94
   b. Sa F In $6\frac{1}{4}$s 98
   c. ATT $8\frac{3}{4}$s 2000
   d. Dow $8\frac{1}{2}$s 05
   e. Bell Pa $9\frac{5}{8}$s 14
   f. GM 8.05s 85
10. Find the cost of each of the following purchases, using the closing prices in the bond quotations on page 511 (face value of each bond is $1,000):
    a. One Rey Tob 7s 89
    b. Ten Sears R $7\frac{3}{4}$s 85
    c. Five RCA $9\frac{1}{4}$s 89
    d. Twenty NY Tel 9s 14
    e. Fifteen PGE $9\frac{5}{8}$s 06
    f. Twenty-five PSEG 12s 04
11. Using the highest quoted prices on page 511, find the total cost including commission of:
    a. Five Bell Pa $9\frac{5}{8}$s 14, face value $1,000 each; commission $10 per bond
    b. Ten Exxon $6\frac{1}{2}$s 98, face value $1,000 each; commission $7.50 per bond
    c. Thirty ATT $8\frac{3}{4}$s 2000, face value $1,000 each; commission $5 per bond
12. If the face value of each TWA 11s 86 is $1,000, how much interest will the owner of forty of these bonds receive semi-annually?

Bond Quotations

Bonds	Cur Yld	Vol	High	Low	Close	Net Chg.
Att $8\frac{3}{4}$s 2000	8.7	239	$100\frac{1}{4}$	$99\frac{1}{2}$	$100\frac{1}{8}$	$+\frac{5}{8}$
Atl Rich $7\frac{3}{4}$s 03	8.9	15	87	87	87	-2
Bell Pa $9\frac{5}{8}$s 14	9.2	13	$103\frac{1}{2}$	$103\frac{1}{4}$	$103\frac{1}{2}$	$-\frac{1}{8}$
Crane 7s 94	9.5	17	73	$72\frac{1}{2}$	73	$+\frac{1}{2}$
Dow $8\frac{1}{2}$s 05	8.9	15	$95\frac{1}{4}$	$95\frac{1}{4}$	$95\frac{1}{4}$	$-\frac{1}{4}$
Exxon $6\frac{1}{2}$s 98	7.9	25	$82\frac{1}{2}$	81	82	$-\frac{1}{2}$
Ford $9\frac{1}{4}$s 94	9.0	15	102	102	102	....
GM 8.05s 85	8.0	181	100	$99\frac{1}{2}$	$99\frac{7}{8}$	$+\frac{1}{4}$
NY Tel 9s 14	9.0	35	$99\frac{1}{2}$	$99\frac{3}{8}$	$99\frac{1}{2}$	$+\frac{1}{2}$
PGE $9\frac{5}{8}$s 06	9.6	71	$100\frac{1}{4}$	$100\frac{1}{8}$	$100\frac{1}{4}$	$+\frac{1}{4}$
PSEG 12s 04	10.0	12	$109\frac{7}{8}$	$109\frac{1}{2}$	$109\frac{7}{8}$	$-\frac{1}{8}$
RCA $9\frac{1}{4}$s 90	9.2	4	100	100	100	$+\frac{1}{2}$
Rey Tob 7s 89	8.8	9	80	79	79	-1
Sa F In $6\frac{1}{4}$s 98	cv *	2	$89\frac{1}{2}$	$89\frac{1}{2}$	$89\frac{1}{2}$	$+\frac{1}{2}$
Sears R $7\frac{3}{4}$s 85	8.0	105	$96\frac{3}{4}$	$96\frac{3}{4}$	$96\frac{3}{4}$	$-\frac{1}{8}$
TWA 11s 86	12.0	8	88	88	88	$+1$

*The abbreviation cv indicates convertible bonds.

13. Calculate the current yield, using the closing quoted prices for each of the following $1,000 bonds, then check with the current yield given in the bond quotations:

a. TWA 11s 86
b. Rey Tob 7s 89
c. PSEG 12s 04
d. Exxon $6\frac{1}{2}$s 98
e. Ford $9\frac{1}{4}$s 94
f. Atl Rich $7\frac{3}{4}$s 03
g. Sears R $7\frac{3}{4}$s 85
h. GM 8.05s 85
i. Bell Pa $9\frac{5}{8}$s 06

14. Using the mutual funds quotations at the right:

a. Find the cost of buying each of the following:
(1) 100 shares of Value Line
(2) 100 shares of Fidelity Fd
(3) 60 shares of Mony Fd.
(4) 25 shares of Mass Fd
(5) 80 shares of Istel

b. How much money should you receive when you sell each of the following:
(1) 100 shares of Delaware
(2) 100 shares of Dreyfus Fd
(3) 50 shares of Bullock Fd
(4) 20 shares of Oppenheimer Fd
(5) 75 shares of Wall St Fd.

Mutual Funds

	Net Asset Value (NAV)	Offer Price
Bullock Fd	10.76	11.76
Delaware	8.46	9.25
Dreyfus Fd	9.69	10.62
Fidelity Fd	13.41	14.66
Istel	24.01	24.75
Mass Fd	9.22	10.10
Mony Fd.	8.99	9.83
Oppenheimer Fd	5.68	6.21
Value Line	5.22	5.72
Wall St. Fd	6.47	7.08

15. How much interest do you get monthly when you buy a $25,000 income savings certificate at 5% interest?

16. How much is your 5% growth savings certificate worth at the end of 5 years if it cost:

    a. $50 and the interest is compounded quarterly?
    b. $200 and the interest is compounded semi-annually?
    c. $675 and the interest is compounded semi-annually?
    d. $1,000 and the interest is compounded quarterly?

17. If $25 E bonds cost $18.75 each, $50 E bonds cost $37.50 each, $75 E bonds cost $56.25 each, and $100 E bonds cost $75 each, what is the total cost of eight $25 E bonds, five $50 E bonds, four $75 E bonds, and six $100 E bonds? How much will the owner receive in all at maturity?

APPLICATIONS IN BUSINESS

15–27 DISCOUNT

Discount is the amount that an article is reduced in price. The regular or full price of the article is called the *list price* or *marked price*. The price of the article after the discount is deducted is called the *net price* or *sale price*. The per cent taken off is called the *rate of discount*. This rate is sometimes expressed as a common fraction. Manufacturers and wholesale houses sometimes allow trade discounts.

Often when a bill is paid at the time of purchase or within a specified time thereafter, a reduction called *cash discount* is allowed. The terms of sale are usually stated on the bill. The terms 3/10, *n*/30 mean that a 3% discount is allowed if the bill is paid within 10 days and the full amount is due by the 30th day. The *net amount* is the amount of the bill after the cash discount is deducted.

When two or more discounts are given, they are called *successive discounts*.

(1) To find the discount when the list price and the rate of discount are given, we multiply the list price by the rate of discount.

(2) To find the net price, we subtract the discount from the list price.

Find the discount and net price when the list price is $240 and the rate of discount is 18%:

```
$240 list price          $240.00 list price
 × .18 rate              − 43.20 discount
 1920                    $196.80 net price
 240
$43.20 discount
```

Answer, $43.20 discount; $196.80 net price

(3) To find the rate of discount, we find what per cent the discount is of the list price.

> What is the rate of discount when the list price is $35 and the discount is $14?
>
> $$\frac{\$14}{\$35} = \frac{2}{5} = 40\%$$
>
> *Answer,* 40%

(4) To find the list price when the net price and rate of discount are given, we subtract the given rate from 100%, then divide the answer into the net price.

(5) To find the net amount, we subtract the cash discount from the original amount of the bill. When two or more successive discount rates are given, we do not add the rates but use them one at a time.

EXERCISES

1. Find the discount when the list price is $34.75 and the net price is $26.90.

2. Find the net price when the list price is $9.18 and the discount is $.43.

3. Find the discount when the list price is $56.29 and the rate of discount is 6%.

4. Find the discount and net price when the list price is $349.50 and the rate of discount is 27%.

5. Find the net price when the list price is $1,085 and the rate of discount is 4%.

6. Find the sale price when the regular price is $23.69 and the rate of reduction is 12%.

7. Find the discount and net price:

List price	$84	$9.61	$106.75	$28.50	$695
Rate of discount	10%	3%	25%	40%	$33\frac{1}{3}\%$

8. Find the rate of discount when the list price is $20 and the discount is $4.

9. Find the rate of discount when the list price is $56 and the net price is $48.

10. Find the rate of discount:

List price	$72	$8.50	$28	$120	$24.75
Net price	$54	$7.65	$23.80	$110.40	$21.78

11. On what list price is the discount $4 when the rate of discount is 5%?

12. Find the list price when the net price is $18 and the rate of discount is $33\frac{1}{3}\%$.

13. Find the list price:

Net price	$32	$42.30	$45	$18.75	$27.36
Rate of discount	20%	6%	$37\frac{1}{2}\%$	50%	10%

14. What is the trade discount when the catalogue list price is $12.95 and rate of trade discount is 8%?

15. What is the net price when the catalogue list price is \$76.50 and rate of trade discount is 15%?

16. Find the trade discount and net price:

Catalogue list price	\$112	\$16.75	\$86.40	\$247	\$49.50
Rate of trade discount	35%	10%	$12\frac{1}{2}$%	20%	8%

17. Find the rate of trade discount when the catalogue list price is \$150 and net price is \$125.

18. What is the cash discount when the amount of a bill is \$26 and the rate of cash discount is 3%?

19. What is the net amount when the amount of a bill is \$395.82 and the rate of cash discount is 4%?

20. Find the cash discounts and net amounts:

Amount of bill	\$55	\$142.98	\$106.25	\$318.40	\$569.85
Rate of cash discount	3%	3%	5%	6%	2%

21. Find the cash discount received and net amount paid when the amount of a bill is \$48, terms of payment 2/10, *n*/30, and the bill is paid on the third day after receipt.

22. Find the cash discounts received and net amounts paid:

Amount of bill	\$16	\$239.50	\$89.25	\$348.20	\$169.75
Terms of payment	2/10, *n*/30	5/10, *n*/60	4/15, *n*/60	3/10, *n*/30	4/10, *n*/60
Bill paid on	4th day	15th day	10th day	9th day	12th day

23. Find the cash discounts and net amounts paid:

Amount of bill	\$287	\$104.90	\$643.75	\$986.25	\$1,000
Terms of payment	3/10, *n*/30	2/10, *n*/30	5/15, *n*/60	1/10, *n*/30	4/15, *n*/60
Date of bill	June 6	April 14	May 6	March 6	July 5
Date of payment	June 12	April 29	May 19	March 28	July 19

24. What is the net price when the list price is \$80 and the rates of discount are 20% and 5%?

25. What is the net price when the list price is \$400 and the rates of discount are 25%, 10%, and 5%?

26. What is the net amount when the list price is \$67, rate of trade discount is 30%, and rate of cash discount is 4%?

27. Find the net prices:

List price	$130	$456.40	$247.50	$625	$870
Rates of discount	15%, 8%	20%, 2%	40%, 10%	10%, 2%	30%, 5%

28. Jones and Co. allow all customers 6% discount on cash purchases. Mr. Olsen paid cash when he bought merchandise costing $879. How much discount was he allowed? What was the net price of the merchandise?

29. Mrs. Blake bought a dress marked $58 at a discount of 20%. How much did she pay for it?

30. A furniture store advertised a 25% reduction on all articles. What is the sale price of a living room suite marked $375?

31. Find the sale price of each of the following articles:

a. Hat; regular price, $15; reduced 10%.
b. Table; marked to sell for $125; reduced 20%.
c. Freezer; listed at $290; 15% off.
d. Clock radio; regular price, $49; 25% off.
e. Mirror; regular price, $67.50; reduced 10%.
f. Coat; marked to sell for $79.75; 6% off.
g. Clothes dryer; listed at $199.95; 5% off.
h. Lamp; marked to sell for $49.89; $\frac{1}{3}$ off.
i. Camera; listed at $84.25; 20% off.
j. Living room suite; regular price, $795; reduced 15%.

32. Joe bought a bicycle for $80 less 18% discount. How much did he pay for it?

33. If a dress marked $24 was sold for $19.20, what rate of discount was given?

34. Find the regular price of a washing machine that sold for $202.30 at a 30% reduction sale.

35. What would 40 science workbooks cost at the school discount of 25% if the list price is $2.70 each?

36. A salesman sold a calculator marked $89 with discounts of 15% and 5%. What was the amount of his error if he calculated the net price, using a single discount of 20%?

37. Ms. Evans purchased 6 storm windows marked $24.50 each at a discount of 10%. How much did she pay for them?

38. What rate of discount was given if a piano marked $1,300 was sold at a special sale for $1,040?

39. How much trade discount is allowed if the catalogue lists a heating unit at $369 and the discount sheet shows a 14% discount? What is the net price?

40. If blankets listed for $16.50 each are sold for 15% less, how much did Mrs. Dawson save when she bought 3 blankets?

41. On which is the rate of discount greater and how much greater, a rug reduced from $120 to $90 or one reduced from $80 to $56?

42. If shirts cost $48 a dozen less 10% and 5%, what is the cost of one shirt?

43. The net price of a television set is \$210.21 when a 22% discount is allowed. What is its list price?
44. What is the net price of a gas range listed at \$290 with a trade discount of 15% and an additional cash discount of 3%?
45. Mrs. Chen saved \$9 by buying a digital clock radio at a sale when prices were reduced 18%. What was the regular price of the digital clock radio?

15–28 PROMISSORY NOTES—BANK DISCOUNT

A *promissory note* is a written promise by the borrower to repay the loan. The *date of maturity* is the date when the money is due. The *face of the note* is the amount of money borrowed that is written on the note. When a person borrows money at a bank and gives a note, the bank deducts the interest from the loan in advance. The interest deducted is called *bank discount* and the amount the borrower gets is called *proceeds.* The exact number of days from the date a note is discounted to the date of maturity is called the *term of discount.*

EXERCISES

1. Draw two forms like the following to write notes, dated today, in which you promise to pay:

\$________ ________________ 19___

________________________ *after date* _____ *promise to pay*

to the order of ____________________________

____________________________________ *Dollars*

Value received

No. _____ *Due* __________ ________________

   a. A sum of \$600 to Scott Jaffe 3 months after date with interest at 4%.
   b. A sum of \$1,000 to First National Bank 60 days after date with interest at 6%.

2. Find the date of maturity of each of the following notes:
   a. 30-day note dated April 9.
   b. 90-day note dated June 21.
   c. 4-month note dated March 17.
   d. 60-day note dated November 30.
   e. 45-day note dated August 14.

3. What is the term of discount if the date of discount is January 9 and the date of maturity is February 28?
4. A 30-day note for $800 dated December 11 is discounted at 6% on the same day. Find the bank discount and proceeds.
5. Find the bank discount and proceeds of each of the following notes:

Face of note	$900	$4,000	$5,000	$8,400
Rate of Discount	5%	6%	4%	6%
Term of Discount	4 mo.	90 da.	60 da.	6 mo.

6. A 90-day note for $2,000 dated September 26 with interest at 5% is discounted on November 20 at 6%. Find the bank discount, proceeds, and date of maturity.

15–29 COMMISSION

A salesperson who sells goods for another person is usually paid a sum of money based on the value of the goods sold. This money which is received for the services is called *commission.* When it is expressed as a per cent, it is called the *rate of commission.* The amount that remains after the commission is deducted from the total selling price is called *net proceeds.* When a buyer purchases goods for another person, the buyer also is usually paid a commission for the services based on the cost of the goods. Salespeople and buyers are sometimes called *agents.*

(1) Selling: To find the commission when the amount of sales and rate of commission are given, we multiply the amount of sales by the rate of commission. To find the net proceeds, we subtract the commission from the amount of sales.

Find the commission and net proceeds when the sales are $86.98 and the rate of commission is 7%:

$$\begin{array}{rl} \$86.98 & \text{sales} \\ \times\ .07 & \text{rate} \\ \hline \$6.0886 & \\ = \$6.09 & \text{commission} \end{array} \qquad \begin{array}{rl} \$86.98 & \text{sales} \\ -\ 6.09 & \text{commission} \\ \hline \$80.89 & \text{net proceeds} \end{array}$$

Answer, $6.09 commission; $80.89 net proceeds

Find the rate of commission if the commission is $19.20 on sales of $320:

$$\frac{\$19.20}{\$320} = \$320\overline{)\$19.20}^{\;.06\,=\,6\%} \qquad \underline{19\ 20}$$

Answer, 6% commission

(2) Buying: To find the commission when the cost of goods purchased and rate of commission are given, we multiply the cost of goods purchased by the rate of commission. To find the total cost, we add the commission to the cost of goods purchased.

(3) To find the rate of commission, we find what per cent the commission is of the amount of sales (or cost of goods purchased).

(4) To find the sales when the commission and rate of commission are given, we divide the commission by the rate.

EXERCISES

1. What is the commission if the sales are \$380 and the net proceeds are \$323?
2. What are the net proceeds if the sales are \$509.82 and the commission is \$43.95?
3. What is the commission if the sales are \$721.50 and the rate of commission is 8%?
4. Find the commission:

Sales	\$640	\$96.20	\$867.50	\$175	\$1,300
Rate of commission	10%	25%	9%	8%	5%

5. What are the net proceeds if the sales are \$182.75 and the rate of commission is 20%?
6. Find the commission and net proceeds:

Sales	\$92	\$905.75	\$547.20	\$345	\$2,000
Rate of commission	15%	7%	$16\frac{2}{3}$%	20%	$12\frac{1}{2}$%

7. What is the rate of commission if the commission is \$16 on sales amounting to \$320?

Find the rate of commission:

8.

Sales	\$480	\$525	\$239.50	\$730	\$95.50
Commission	\$72	\$31.50	\$47.90	\$58.40	\$9.55

9.

Sales	\$375	\$304.24	\$681	\$480	\$750
Net proceeds	\$340	\$228.18	\$619.71	\$420	\$625

10. What must the sales be for an 8% rate of commission to bring a commission of \$60?
11. Find the amount of sales when commission is as follows:

Commission	\$46.70	\$50.66	\$82	\$94.50	\$101.35
Rate of commission	10%	17%	$12\frac{1}{2}$%	5%	20%

12. What do the sales amount to if the net proceeds are \$1,680 at the rate of 4% commission?
13. Find the commission on goods purchased:

Cost of goods	\$139	\$928	\$436.75	\$621.90	\$240.80
Rate of commission	6%	5%	8%	10%	$12\frac{1}{2}$%

14. Find the commission and total cost of goods purchased:

Cost of goods	\$594	\$389	\$618.50	\$29.40	\$439.60
Rate of commission	7%	4%	12%	5%	15%

15. At the rate of 4% commission on sales, how much will a salesclerk receive when she sells \$1,358 worth of merchandise?
16. A real-estate agent sold a house for \$19,800. If the agent charges 6% commission, how much money does the owner receive?
17. During the week a salesman sold 20 washing machines at \$254.95 each. At the rate of 2% commission on sales, how much money does the salesman receive?
18. A lawyer collected 70% of a debt of \$1,250 for Mr. Gordon. If the lawyer charges 15% for his services, how much will Mr. Gordon receive?
19. If an agent charges $2\frac{1}{2}$% commission for purchasing goods, how much will he receive for buying 750 boxes of oranges at \$6.90 a box?
20. An agent received \$1,590 commission for selling a house for \$26,500. What rate of commission did she charge?
21. If an agent's commission is 8%, how much will he receive for selling 530 bags of potatoes at \$6.25 a bag?
22. Marilyn receives \$60 per week and 8% commission on sales. If her weekly sales are \$1,859, what are her total earnings?
23. Mr. Morgan is paid \$75 a week and 4% commission on all weekly sales over \$500. How much did he earn during a week when his sales amounted to \$5,928?
24. Charlotte received \$63.36 commission last week. If she was paid at the rate of 6%, what was the amount of her sales?
25. An auctioneer, charging 7% commission, sold goods amounting to \$2,806. How much commission did she receive?
26. How much must a salesman sell each week to earn \$275 weekly if he is paid 5% commission on sales?
27. A merchant, charging 6% commission, sold for a farmer 275 crates of berries at \$7.80 a crate. If freight charges were \$182.50, how much did the farmer receive per crate?
28. An agent received \$63.75 commission for selling 85 bushels of apples at \$7.50 each. What rate of commission did she charge?
29. A salesman sold 148 dozen ties at \$20.50 per dozen. At the rate of $4\frac{1}{2}$% commission, how much money will he receive?

15–30 PROFIT AND LOSS

Merchants are engaged in business to earn a profit. To determine the profit, each merchant must consider the cost of goods, the selling price of the same goods, and the operating expenses. The *cost* is the amount the merchant pays for the goods. The *selling price* is the amount the merchant receives for selling the goods. The *operating expenses* or *overhead* are expenses of running the business. They include wages, rent, heat, light, telephone, taxes, insurance, advertising, repairs, supplies, and delivery costs.

The difference between the selling price and the cost is called the *margin*. The terms *gross profit*, *spread*, and *markup* sometimes are also used. The margin is often erroneously thought of as the profit earned. Actually, the profit earned, called *net profit*, is the amount that remains after both the cost and the operating expenses are deducted from the selling price.

To earn a profit the merchant must sell goods at a price which is greater than the sum of the cost of the goods and the operating expenses. If the selling price is less than the sum of the cost and operating expenses, the goods are sold at a *loss*.

The profit may be expressed as a per cent of either the cost or the selling price. This per cent is called the *rate of profit*. (The loss may also be expressed in a similar way.)

To determine the selling price of the articles they sell, some merchants use the per cent markup (or gross profit) based on the cost while other merchants use the rate based on the selling price. This *per cent markup* is an estimated rate large enough to cover the proportionate amount of overhead expenses to be borne by each article and the amount of profit desired. The per cent markup (or gross profit) is the per cent the margin is of the cost or of the selling price.

To find the selling price: (1) We add the cost, operating expenses, and net profit; or (2) add the cost and margin; or (3) multiply the cost by the per cent markup on the cost, then add the product to the cost.

To find the margin or gross profit or markup: (1) We subtract the cost from the selling price; or (2) multiply the cost by the per cent markup on cost; or (3) add the operating expenses and net profit.

To find the net profit: (1) We subtract the operating expenses from the margin; or (2) add the cost and the operating expenses, and subtract this sum from the selling price.

To find the loss: We add the cost and the operating expenses, and subtract the selling price from this sum.

To find the per cent markup (or gross profit) on cost, we find what per cent the margin is of the cost. When based on selling price, we find what per cent the margin is of the selling price.

To find the rate of net profit (or loss) on the selling price, we find what per cent the net profit (or loss) is of the selling price. When based on the cost, we find what per cent the profit (or loss) is of the cost.

To find the selling price when the cost and per cent markup (or gross profit) on the selling price are given, we subtract the per cent markup from 100%, then divide the answer into the cost.

If the rate of markup on cost is 34%, what is the selling price of an article that costs $25?

```
  $25 cost                    $25.00 cost
× .34 rate                   + 8.50 markup
  100                         $33.50 selling price
  75
$8.50 markup
```

Answer, $33.50 selling price

Find the rate of markup on cost if an article costs $40 and sells for $55:

```
  $55 selling price
− 40 cost
  $15 markup
```

$$\frac{\$15}{\$40} = \frac{3}{8} = 37\tfrac{1}{2}\%$$

Answer, $37\frac{1}{2}$% markup on cost

Find the net profit and rate of net profit on the selling price if an article costs $18.50, operating expenses are $1.50, and it sells for $24:

```
 $18.50 cost                        $24 selling price
+ 1.50 operating expenses          − 20
 $20.00 total cost                   $4 net profit
```

$$\frac{\$4}{\$24} = \frac{1}{6} = 16\tfrac{2}{3}\%$$

Answer, $4 net profit; $16\frac{2}{3}$% net profit on selling price

Find the selling price of an article that costs $17 and the per cent markup on the selling price is 32%:

```
 100% = selling price                 $ 25
− 32% = markup                  .68^)$17.00^
  68% = cost                          13 6
                                       3 40
                                       3 40
```

Answer, $25 selling price

EXERCISES

1. What must the selling price of goods be to allow a net profit of $56 if they cost $287 and the operating expenses are $29?
2. Find the margin or gross profit on an article when the operating expenses are $13.25 and the net profit is $68.50.
3. What is the selling price of a chair if it costs $37.50 and the margin is $18.45?
4. A rug costs $171.90 and sells at $298.50. What is the margin or gross profit?

5. What is the margin on a desk which costs $142 and the rate of markup (margin) on the cost is 35%?
6. A cassette tape recorder costs $62.50. If the rate of markup on cost is 40%, what is the selling price?
7. Find the selling price:

Cost	$87	$169	$29.40	$145.50	$78.60
Markup on cost	25%	32%	30%	40%	$33\frac{1}{3}$%

8. Find the net profit if operating expenses are $26.95 and the margin is $48.75?
9. Find the net profit:

Cost	$34	$91.58	$107	$1.98	$367.39
Operating expenses	$7	$8.45	$9.65	$.27	$26.43
Selling price	$50	$129.98	$145.75	$2.75	$430.75

10. What is the loss if the operating expenses are $21.50 and the margin is $17.35?
11. Find the loss if the cost is $75, operating expenses are $6.95, and the selling price is $78.50.
12. What is the selling price of an article if it costs $60 and the rate of markup on the selling price is 20%?
13. A refrigerator sold for $375. If a 25% markup on the selling price was used, how much did it cost?
14. Find the selling price:

Cost	$140	$54	$22.50	$72.75	$1,200
Markup on selling price	30%	40%	25%	$33\frac{1}{3}$%	15%

15. What is the rate of markup on cost if the margin is $16 and the cost is $48?
16. If the cost is $30, operating expenses are $6.75 and net profit is $5.25, what is the per cent markup on cost?
17. Find the rate of markup (margin or gross profit) on cost:

Cost	$24	$6.50	$95	$.10	$137.50
Selling price	$32	$7.80	$125.40	$.12	$200

18. Find the rate of gross profit on selling price:

Cost	$48	$6.30	$190.99	$.18	$100.20
Selling price	$64	$10.50	$269	$.25	$150.30

19. What is the rate of net profit on the selling price if the net profit is $24 and selling price is $120.

20. Find the rate of net profit on the selling price:

Cost	$25	$18.25	$56.94	$.37	$9.69
Operating expenses	$5	$4.25	$8.25	$.05	$1.71
Selling price	$40	$27	$79.50	$.63	$15

21. If the cost is $18, operating expenses are $3.16, and markup is 30% on cost, what is the rate of net profit on the selling price?

22. What is the rate of loss on the selling price if the loss is $9.50 and the selling price is $190?

23. Find the rate of loss on the selling price if the cost is $49.55, operating expenses are $10.45, and selling price is $50.

24. What is the rate of net profit on the cost if the cost is $52 and net profit is $6.50?

25. Find the rate of net profit on the cost:

Cost	$20	$325	$17.80	$.60	$250
Operating expenses	$9	$28.65	$3.93	$.12	$30.50
Selling price	$36	$392.65	$24.40	$.90	$330

26. What is the rate of loss on the cost if the loss is $8.40 and the cost is $140?

27. Find the rate of loss on the cost if the cost is $36, operating expenses are $7, and selling price is $40.

28. Mr. Harrison's sales last year amounted to $80,000. The cost of the goods sold was $48,000. Operating expenses for the year totaled $12,000. What was the net profit? What per cent of the selling price was the net profit? Find the rate of net profit on the cost. Find the per cent markup on the cost. What per cent of the selling price was the gross profit?

29. A kitchen sink costing $130 was sold for $195. If the overhead expenses were 19% of the selling price, what was the net profit?

30. A radio was bought by a merchant for $45 and marked to sell for $69. It was sold at a discount of 15%. Find the selling price and gross profit.

31. Find the selling price of each of the following articles:

a. Washing machine, costing $220 and 20% markup on cost.
b. Dress, costing $18.40 and 25% markup on cost.
c. Pocket calculator, costing $77 and 30% markup on selling price.
d. Watch, costing $39 and 35% markup on selling price.
e. Sweater, costing $9.52 and 20% markup on selling price.

32. Find the rate of gross profit on the selling price in each of the following sales:

a. Gas range which costs $150 and sells for $250.
b. Television set which costs $215 and sells for $288.
c. Suit which costs $54 and sells for $75.

d. Electric heater which costs $18.60 and sells for $30.
e. Refrigerator which costs $200.60 and sells for $295.

33. A dealer bought softballs at $21 a dozen and sold them for $3 each. What per cent of the selling price was his margin?
34. A merchant purchased 200 lb. of tomatoes, packed in 1-lb. cartons, for $40. She sold 112 lb. at $.49 a lb., 78 lb. at $.39 a lb., and the rest spoiled. What was her gross profit?
35. A dealer purchased 3 dozen footballs for $180. At what price each must he sell them to realize a gross profit of 40% on the cost?
36. A merchant bought 5 dozen pairs of shoes for $720. If she sells them for $18 a pair, what is her gross profit?
37. At what price each must a merchant sell shirts to realize a gross profit of 20% on the selling price if he buys them at $67.20 a dozen?
38. At what price must a dealer sell a rug which costs $208 to realize a profit of 35% on the selling price?
39. A merchant purchased handbags at $95.76 a dozen. What price should each bag be marked so that a 5% discount may be offered and yet a 30% gross profit on the selling price may be realized?
40. A television set was bought by a dealer for $225. She marked it to sell at a margin of 40% of the cost. If it was sold at a discount of 8%, what was the gross profit?
41. A manufacturer sold coats to a retail merchant, making 25% profit on the cost of each coat. If the merchant sold these coats for $45.50 each, making 30% on the cost, how much did each coat cost the manufacturer?
42. A merchant bought 9 dozen ties for $200. He sold 38 ties for $3.50 each, 56 ties for $2.50 each, and the rest for $1.50 each. If his overhead expenses were 20% of the cost, what per cent of the selling price was his net profit? What per cent of the cost was his net profit? Which rate is greater? How much greater? Compare the rate of gross profit on cost with that on selling price.

15–31 BALANCE SHEET—PROFIT AND LOSS STATEMENT

A *balance sheet* tells how much a business is worth. This financial statement shows that the *assets* (value of what is owned) minus the *liabilities* (value of what is owed) is equal to the *capital* or *net worth.*

A *profit and loss statement* tells how much profit is earned or loss is suffered by a business during a certain period of time. This financial statement shows that the *sales* minus the *cost of goods sold* is equal to the *gross profit on sales.* The *net profit* is equal to the *gross profit* minus the *operating expenses.*

EXERCISES

1. Copy the form on page 525, and prepare with these facts a balance sheet as of

December 31, showing the capital of James Harris. Assets: cash, \$4,382.59; accounts receivable, \$12,857.16; notes receivable, \$875; inventory of merchandise, \$25,200; furniture and fixtures, \$3,425; delivery trucks, \$8,200. Liabilities: accounts payable, \$2,195.63; notes payable, \$1,450.

JAMES HARRIS

BALANCE SHEET, DECEMBER 31, 19__

Assets		Liabilities	
Cash	\$	Accounts Payable	\$
Accounts Receivable		Notes Payable	
Notes Receivable		Total Liabilities	
Inventory of Merchandise			
Furniture and Fixtures		James Harris, Capital	
Delivery Trucks			
Total Assets		Total Liabilities and Capital	

2. Copy the form below, then prepare from the following facts a profit and loss statement for the Wilson Company for the year ending December 31. Sales amounted to \$193,575. Inventory on January 1 was \$13,750 and on December 31 was \$29,800. Purchases for the year amounted to \$116,500. The operating expenses were: salaries, \$50,500; rent, \$6,400; delivery expenses, \$3,700; advertising, \$1,900; insurance, \$1,750.

WILSON COMPANY

STATEMENT OF PROFIT AND LOSS
FOR THE YEAR ENDING DECEMBER 31, 19__

Sales		\$
Cost of Goods Sold:		
Inventory of Merchandise, January 1	\$	
Purchases		
Total		
Less Inventory of Merchandise, December 31		
Cost of Goods Sold		\$
Gross Profit on Sales		\$
Less		
Operating Expenses		
Salaries	\$	
Rent		
Delivery Expenses		
Advertising		
Insurance		
Total Expenses		\$
Net Profit for the Year		\$

15–32 PAYROLLS

Factories and some businesses use *time cards* for each employee to record the number of hours he or she works. The payroll clerk uses the time cards to prepare the *payroll.* A *currency break-up sheet* is used to determine the number and kinds of bills and coins needed for each pay envelope. The clerk then prepares a *bank cash memorandum* listing the total number of each kind of coin and bill required.

EXERCISES

Make a copy of each of the following forms. Please do not write in this book.

1. For each of the following time cards compute the number of hours (to the nearest half-hour) worked each day and the total time for the week. Then find the amount of wages due each employee.

TIME CARD

No. 52 NAME Mary Jackson

WEEK ENDING

DAY	IN	OUT	IN	OUT	HOURS
M.	8:59	12:01	12:59	5:02	
T.	8:56	12:00	12:57	5:00	
W.	9:00	12:02	12:58	4:30	
Th.	8:59	12:00	12:57	5:00	
F.	8:58	12:03	1:00	5:03	
S.	9:00	12:01			

RATE $5.80 TOTAL HOURS

WAGES

TIME CARD

No. 128 NAME David Evans

WEEK ENDING

DAY	IN	OUT	IN	OUT	HOURS
M.	8:00	12:01	12:58	5:00	
T.	7:58	12:00	12:59	5:01	
W.	7:59	12:02	1:00	5:02	
Th.	7:57	12:01	12:59	5:00	
F.	8:00	12:02	12:58	5:01	
S.					

RATE $6.47 TOTAL HOURS

WAGES

2. Find the weekly take-home pay (see page 468) of each of the following married persons at the given wage rates:

a. Robert Jones, $242.75 per week; 3 withholding allowances.
b. Janet Bell, $197.50 per week; 2 withholding allowances.
c. Steve Jaffe, 34 hours at $8.10 per hour; 1 withholding allowance.
d. Diane Warren, 39 hours at $5.25 per hour; 3 withholding allowances.
e. Bill Ross, $35\frac{1}{2}$ hours at $4.80 per hour; 2 withholding allowances.

3. Complete the following payroll. Then prepare a currency break-up sheet and bank cash memorandum for payroll as shown on page 527.

PAYROLL FOR WEEK ENDING

NO.	NAME	HOURS WORKED M.	T.	W.	Th.	F.	S.	TOTAL HOURS	HOURLY RATE	WAGES
84	Perez, Maria	8	7	7	7	7	4		$6.25	
85	Elliott, Joseph	8	8	8	8	8	0		4.38	
86	Hooper, Albert	8	7	8	5	8	3		3.51	
87	Browning, George	8	$7\frac{1}{2}$	$7\frac{1}{2}$	8	7	0		7.46	
88	Mitchell, John	$7\frac{1}{2}$	$7\frac{1}{2}$	$7\frac{1}{2}$	7	7	3		5.60	
89	Nelson, Olga	7	8	$7\frac{1}{2}$	8	7	0		5.84	
90	Riley, Irene	7	7	$6\frac{1}{2}$	8	7	3		6.72	
	TOTALS								XXXX	

CURRENCY BREAK-UP SHEET FOR WEEK ENDING ....

NO.	NAME	NET PAY		NOTES $20	$10	$5	$1	COINS 50¢	25¢	10¢	5¢	1¢
	TOTALS											

YOUR CITY BANK
PAY ROLL

......19

FOR ....

NUMBER		AMOUNT
	$20 Notes	
	$10 "	
	$ 5 "	
	$ 1 "	
	50¢ Coins	
	25¢ "	
	10¢ "	
	5¢ "	
	1¢ "	
	TOTAL	

15-33 INVOICES

An *invoice* is a bill that the seller sends to the purchaser listing the terms of purchase and the quantity, description, and unit price of each article purchased. It also includes (1) the extension for each item, the amount obtained by multiplying the quantity of each article by its unit price and written in the column next to the given unit price; and (2) the total amount of the bill, obtained by adding the extensions and written in the last column. A copy of the invoice is usually sent with the order. When the bill is paid, it should be marked paid, dated, and signed by the person receiving the money.

EXERCISES

1. Make a copy of each of the following invoices, then complete them.

CENTRAL PAINT COMPANY
52 Market St.
WASHINGTON, D. C.
June 4, 19__

SOLD TO: R. BAILEY
167 THIRD AVE.
WASHINGTON, D. C.

TERMS: Cash

6 gal.	Paint, outside white	@ $10.85		
3 gal.	Paint, primer white	@ $8.60		
5 gal.	Turpentine	@ $2.75		
2	3-in. brushes	@ $2.50		

MODERN CLOTHING COMPANY
457 Main St.
NEW YORK, N. Y.
December 17, 19__

SOLD TO: MEN'S SPORT SHOP
46 W. CHESTNUT ST.
NEW YORK, N. Y.

TERMS: n/30

4 doz.	Dress shirts	@ $54.00		
2½ doz.	Sport shirts	@ $43.20		
5 doz.	Ties	@ $16.40		
18 pr.	Slacks	@ $11.75		

WILLIAMS LEATHER PRODUCTS
2917 High St.
CHICAGO, ILL.
September 6, 19__

SOLD TO: JOHNSON LUGGAGE CO.
863 NINTH ST.
CHICAGO, ILL.

TERMS: 2/10, n/30

2 doz.	Wallets, men's	@ $49.20		
3 doz.	Handbags	@ $86.45		
1½ doz.	Brief cases	@ $79.00		
	Received Payment September 12, 19__			

KITCHEN SUPPLY COMPANY
917 W. Sixth St.
LOUISVILLE, KY.
Feb. 25, 19__

SOLD TO: WILSON CONSTRUCTION CO.
471 GIRARD AVE.
LOUISVILLE, KY.

TERMS: 3/10, n/30

4	Heat-Well ranges 40″	@ $179.50		
6	Cabinet sinks, 52″	@ $127.50		
2	Dependable refrigerators 6 cu. ft.	@ $282.00		
	Received Payment March 1, 19__			

2. Prepare invoices for each of the following:

   a. Wholesale Food Company sold to Wilson Grocery: 2 cases sardines @ \$16.50, 4 cases canned peas @ \$7.12, 6 cases canned corn @ \$6.75, and 5 cases soap @ \$13.25. Terms were 2/10, *n*/30.

   b. Allied Drug Company sold to Howard Pharmacy: 2 gross prescription bottles @ \$10.50, 4 doz. eye droppers @ 90¢, 3 doz. tubes shaving cream @ \$4.75, 6 doz. tubes tooth paste @ \$5.80, and 2½ doz. packages of hair spray @ \$9.26. Terms were *n*/30.

WHOLESALE FOOD COMPANY
1648 First St.
DALLAS, TEXAS
19__
SOLD TO:
TERMS:

ALLIED DRUG COMPANY
21 East Boulevard
SAN FRANCISCO, CAL.
19__
SOLD TO:
TERMS:

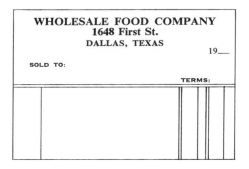

15–34 INVENTORIES AND SALES SUMMARIES

An *inventory* of merchandise is a record of the number of articles of each kind on hand and their value.

A *sales summary* is a record of the goods sold during a specified period of time.

EXERCISES

1. Make a copy of each of the six charts on this and the next page. Then complete the inventories, sales summaries, and invoices. In the inventory of Feb. 15, determine the quantity of each article that should be on hand.

SCHOOL STORE INVENTORY, FEB. 1, 19__

Article	Quantity	Unit Price	Total Value
Assignment books	163	\$.15	
Compasses	38	.19	
Copy books	319	.30	
Erasers	91	.10	
Ball point pens	47	.45	
Loose leaf binders	104	.90	
Loose leaf paper	212	.35	
Pencils	296	.03	
Pencils, automatic	57	.40	
Protractors	83	.25	
Rulers	125	.08	
Tablets	374	.20	
Total	xxxx	xxxx	

SCHOOL STORE SALES SUMMARY FOR WEEK ENDING FEB. 7, 19__

Article	M.	T.	W.	Th.	F.	Total Sold	Unit Price	Total Sales
Assignment books	26	15	9	11	4		\$ .25	
Compasses	8	5	6	3	5		.29	
Copy books	51	28	34	17	24		.49	
Erasers	17	7	8	5	9		.19	
Ball point pens	2	1	3	0	1		.79	
Loose leaf binders	5	12	4	6	3		1.49	
Loose leaf paper	21	18	23	26	35		.69	
Pencils	69	37	31	45	28		.05	
Pencils, automatic	8	4	3	5	9		.39	
Protractors	6	7	4	2	5		.15	
Rulers	39	15	9	12	6		.15	
Tablets	45	34	27	38	29		.35	
Totals							xxxx	

SCHOOL SUPPLY COMPANY
1950 Main St.
OAK LANE, N. Y.

February 5, 19__

SOLD TO: WILSON HIGH SCHOOL
OAK LANE, N. Y.

TERMS: n/30

3 doz.	Compasses	@ $2.28		
15 doz.	Loose leaf paper	@ 4.20		
4 doz.	Erasers	@ 1.20		
18 doz.	Pencils	@ .36		
6 doz.	Rulers	@ .96		

SCHOOL SUPPLY COMPANY
1950 Main St.
OAK LANE, N. Y.

February 11, 19__

SOLD TO: WILSON HIGH SCHOOL
OAK LANE, N. Y.

TERMS: n/30

6 doz.	Assignment books	@ $1.80		
8 doz.	Copy books	@ 3.60		
10 doz.	Tablets	@ 2.40		
9 doz.	Loose leaf paper	@ 4.20		

SCHOOL STORE SALES SUMMARY FOR WEEK ENDING FEB. 14, 19__

Article	M.	T.	W.	Th.	F.	Total Sold	Unit Price	Total Sales
Assignment books	16	13	8	14	6		$ .25	
Compasses	7	9	5	8	4		.29	
Copy books	29	38	33	46	19		.49	
Erasers	11	6	9	12	5		.19	
Ball point pens	2	0	0	1	1		.79	
Loose leaf binders	5	6	3	4	7		1.49	
Loose leaf paper	34	19	22	27	32		.69	
Pencils	41	28	17	35	21		.05	
Pencils, automatic	3	5	6	4	1		.39	
Protractors	6	3	2	9	5		.15	
Rulers	14	5	6	2	4		.15	
Tablets	27	16	18	24	15		.35	
Totals							xxxx	

SCHOOL STORE INVENTORY, FEB. 15, 19__

Article	Quantity	Unit Price	Total Value
Assignment books		$.15	
Compasses		.19	
Copy books		.30	
Erasers		.10	
Ball point pens		.45	
Loose leaf binders		.90	
Loose leaf paper		.35	
Pencils		.03	
Pencils, automatic		.40	
Protractors		.25	
Rulers		.08	
Tablets		.20	
Total	xxxx	xxxx	

2. Find the cost of goods sold and the gross profit earned by the school store during the first two weeks of February.

Inventory, Feb. 1	$______	
Purchases	$______	
Total		$______
Less Inventory Feb. 15		$______
Cost of goods sold		$______

Sales	$______
Less cost of goods sold	$______
Gross profit	$______

15–35 PARCEL POST

Parcel post may be insured as follows:

From $.01 to $15 insurance, the fee is $.50; from $15.01 to $50 insurance, the fee is $.85; from $50.01 to $100 insurance, the fee is $1.10; from $100.01 to $150 insurance, the fee is $1.40; from $150.01 to $200 insurance, the fee is $1.75; from $200.01 to $300 insurance, the fee is $2.25; from $300.01 to $400, the fee is $2.75.

EXERCISES

In the examples on the next page fractions of pounds are computed as full pounds. Third-class rates are used for weights of 1 pound and under. Use the table on page 530 for the exercises.

PARCEL POST RATE TABLE								
Weight 1 pound and not exceed-ing (pounds)	Local	Zones						
		1 and 2	3	4	5	6	7	8
		Up to 150 miles	150 to 300 miles	300 to 600 miles	600 to 1,000 miles	1,000 to 1,400 miles	1,400 to 1,800 miles	Over 1,800 miles
2	$1.15	$1.35	$1.39	$1.56	$1.72	$1.84	$1.98	$2.22
3	1.23	1.45	1.53	1.73	1.86	2.04	2.24	2.61
4	1.29	1.56	1.65	1.82	2.00	2.23	2.50	3.00
5	1.36	1.66	1.77	1.92	2.14	2.43	2.77	3.39
6	1.42	1.71	1.84	2.01	2.28	2.62	3.03	3.78
7	1.47	1.76	1.90	2.11	2.41	2.82	3.29	4.17
8	1.51	1.80	1.97	2.20	2.55	3.02	3.56	4.56
9	1.54	1.85	2.03	2.29	2.69	3.21	3.82	4.95
10	1.57	1.89	2.10	2.39	2.83	3.41	4.08	5.34
11	1.60	1.94	2.17	2.50	3.00	3.65	4.42	5.73
12	1.64	1.98	2.22	2.56	3.09	3.77	4.57	6.12
13	1.67	2.02	2.27	2.63	3.17	3.89	4.72	6.41
14	1.70	2.05	2.32	2.69	3.25	3.99	4.86	6.62
15	1.73	2.09	2.36	2.74	3.33	4.09	4.99	6.80
16	1.76	2.13	2.41	2.80	3.40	4.19	5.11	6.98
17	1.79	2.16	2.45	2.85	3.47	4.28	5.23	7.15
18	1.82	2.20	2.49	2.91	3.54	4.37	5.34	7.31
19	1.86	2.23	2.53	2.96	3.61	4.46	5.45	7.47
20	1.89	2.27	2.58	3.01	3.67	4.54	5.55	7.62
21	1.92	2.30	2.62	3.06	3.74	4.62	5.66	7.76
22	1.95	2.34	2.66	3.14	3.85	4.78	5.80	7.90
23	1.98	2.37	2.72	3.25	3.99	4.96	6.02	8.03
24	2.01	2.44	2.80	3.35	4.12	5.13	6.24	8.16
25	2.04	2.51	2.89	3.46	4.26	5.31	6.46	8.28
26	2.07	2.58	2.97	3.56	4.39	5.48	6.68	8.40
27	2.11	2.65	3.06	3.67	4.53	5.66	6.90	8.52
28	2.14	2.72	3.14	3.77	4.66	5.83	7.12	8.63
29	2.17	2.79	3.23	3.88	4.80	6.01	7.34	8.75
30	2.20	2.86	3.31	3.98	4.93	6.18	7.56	8.85
31	2.68	3.09	3.46	4.09	5.07	6.36	7.78	9.41
32	2.71	3.12	3.49	4.19	5.20	6.53	8.00	9.51
33	2.74	3.16	3.57	4.30	5.34	6.71	8.22	9.61
34	2.77	3.19	3.65	4.40	5.47	6.88	8.44	9.80
35	2.80	3.22	3.74	4.51	5.61	7.06	8.66	10.06

1. Find the cost of sending by parcel post a package weighing:
   - a. 20 lb. to zone 4.
   - b. 11 lb. local.
   - c. 16 lb. to zone 7.
   - d. 29 lb. to zone 2.
   - e. 5 lb. to zone 6.
   - f. 3 lb. 8 oz. to zone 3.
   - g. 1 lb. 14 oz. to zone 1.
   - h. 2 lb. 15 oz. local.
   - i. 8 lb. 7 oz. to zone 8.
   - j. 24 lb. 12 oz. to zone 5.
   - k. 13 lb. local.
   - l. 2 lb. 7 oz. to zone 4.
   - m. 29 lb. 11 oz. to zone 6.
   - n. 19 lb. to zone 8.
   - o. 5 lb. 13 oz. to zone 2.
2. Find the cost of sending by parcel post a package weighing:
   - a. 26 lb. for 200 miles.
   - b. 9 lb. for 1,100 miles.
   - c. 17 lb. for 125 miles.
   - d. 3 lb. for 750 miles.
   - e. 21 lb. for 2,000 miles.
   - f. 1 lb. 8 oz. for 100 miles.
   - g. 12 lb. 1 oz. for 278 miles.
   - h. 23 lb. 14 oz. for 940 miles.
   - i. 4 lb. 15 oz. for 1,575 miles.
   - j. 9 lb. 6 oz. for 1,300 miles.
   - k. 18 lb. for 60 miles.
   - l. 19 lb. 10 oz. for 185 miles.
3. Find the cost of sending by parcel post a package weighing:
   - a. 16 lb. to zone 5, insured for $50.
   - b. 4 lb. to zone 6, insured for $10.
   - c. 11 lb. to zone 3, insured for $100.
   - d. 3 lb. to zone 4, insured for $5.
   - e. 22 lb. to zone 7, insured for $200.
   - f. 9 lb. 11 oz. local, insured for $15.
   - g. 14 lb. 5 oz. to zone 6, insured for $25.
   - h. 12 lb. for 500 miles, insured for $125.
   - i. 2 lb. for 40 miles, insured for $5.
   - j. 24 lb. for 110 miles, insured for $100.
   - k. 5 lb. for 1,350 miles, insured for $10.
   - l. 19 lb. for 2,300 miles, insured for $150.

APPLICATIONS IN INDUSTRY

15–36 LUMBER

The measure of a piece of lumber one foot long, one foot wide, and one inch thick is a unit called a *board foot*. To find the number of board feet in a piece of lumber, multiply the length in feet by the width in feet by the thickness in inches.

EXERCISES

1. How many board feet are there in one piece of lumber 2 in. thick, 4 in. wide, and 18 ft. long?
2. How many board feet are there in six pieces of lumber 3 in. thick, 6 in. wide, and 12 ft. long?
3. Find the number of board feet in each of the following:
   - a. 1 pc. of pine 1 in. by 4 in. and 16 ft. long.
   - b. 4 pc. of fir 1 in. by 10 in. and 12 ft. long.
   - c. 9 pc. of spruce 3 in. by 12 in. and 18 ft. long.
   - d. 15 pc. of hemlock 2 in. by 8 in. and 10 ft. long.
   - e. 20 pc. of pine 4 in. by 6 in. and 14 ft. long.

4. How many 1,000 (M) board feet are in 50 pieces of pine each 3 in. × 4 in. and 12 ft. long?
5. Find the cost of 1 piece of fir 2 in. × 6 in. and 14 ft. long at 68¢ per board foot.
6. What is the cost of 18 pieces of spruce each 4 in. × 12 in. and 16 ft. long at $630 per 1,000 (M) board feet?
7. Find the cost of each of the following:

   a. 10 pc. of fir each 1 in. × 6 in. and 12 ft. long at $650 per M.
   b. 30 pc. of pine each 2 in. × 4 in. and 16 ft. long at $720 per M.
   c. 24 pc. of hemlock each 1 in. × 10 in. and 14 ft. long at $625 per M.
   d. 56 pc. of spruce each 3 in. × 12 in. and 18 ft. long at $640 per M.
   e. 60 pc. of fir each 4 in. × 8 in. and 10 ft. long at $660 per M.

8. How many board feet of flooring 1″ × 3″ cut in 20-ft. lengths are required for a floor 18 ft. by 20 ft.? Allowing 10% for waste, how much will the flooring cost at $680 per M?
9. How many board feet of lumber 1″ × 4″ in 12-ft. lengths are needed for the four walls of a bin each 12 ft. by 15 ft.? What is the cost of the lumber at $620 per M, allowing 15% for waste?
10. Twelve pieces of lumber 1″ × 9″ in 3-ft. lengths are required for repairs. Disregarding waste, how many board feet are needed and what is the cost of the lumber at 66¢ per board foot?

15–37 EXCAVATION, BRICKWORK, AND CONCRETE MIXTURES

EXERCISES

1. How many cubic yards of dirt must be removed to make an excavation 40 ft. long, 32 ft. wide, and 15 ft. deep?
2. At $38.88 per cu. yd., what will it cost to dig a well 4 ft. in diameter and 35 ft. deep?
3. How many bricks are needed to construct a wall 1 ft. thick, 18 ft. high, and 48 ft. long if it takes 20 standard bricks to make 1 sq. ft. of a wall with a thickness of 1 ft.? At $175 per M, what is the cost of the brick?
4. A driveway 15 ft. wide and 54 ft. long is to be paved with concrete 4 in. thick. How many cu. yd. of concrete are required? If a mixture of 650 lb. of cement, 1,300 lb. of sand, and 1,700 lb. of gravel make one cu. yd. of concrete, how many pounds of each are needed for the driveway?
5. How many cu. yd. of concrete are required for a sidewalk 9 ft. wide, 24 ft. long, and 6 in. thick? How much will it cost at $39.50 per cu. yd.?

15–38 LATHING, PLASTERING, PAPERHANGING, PAINTING, AND ROOFING

EXERCISES

1. If a bundle of rock laths covers 32 sq. ft., how many bundles of laths are needed to cover the walls and ceiling of a room 16 ft. long, 12 ft. wide, and 10

ft. high, subtracting 80 sq. ft. for door and window openings? At $2.40 per bundle, how much will the laths cost?

2. At $9.45 per sq. yd., how much will it cost to plaster the walls and ceiling of the room described in Exercise **1**?
3. The walls and ceiling of a room 24 ft. long, 14 ft. wide, and 8 ft. high are to be covered with plasterboard. If 112 sq. ft. are deducted for openings, how many sheets of plasterboard, 4 ft. by 8 ft. each, are required? How much will the plasterboards cost at $2.10 per sheet?
4. A single roll of wall paper covers about 36 sq. ft. How many rolls are needed to cover the walls of a room 16 ft. long, 12 ft. wide, and 9 ft. high if 72 sq. ft. of wall space is deducted for openings and 1 extra roll of paper is allowed for the waste in matching? What is the cost of the wall paper at $4.80 per roll?
5. The walls and ceiling of a room 18 ft. long, 16 ft. wide, and 12 ft. high are to be papered with double rolls each covering 72 sq. ft. If 10% is to be allowed for matching and 96 sq. ft. is to be deducted for openings, how many rolls are needed for the four walls? What is the cost at $7.20 per roll? How many rolls are needed for the ceiling and what is the cost at $3.60 per roll? Find the total cost.
6. A room is 16 ft. long, 15 ft. wide, and 10 ft. high. Deducting 60 sq. ft. for openings, how many gallons of paint are needed to cover the walls and ceiling with two coats if a gallon will cover 400 sq. ft. one coat? At $9.60 per gallon, how much will the paint cost?
7. How many gallons of varnish are needed to cover a floor 19 ft. by 30 ft. two coats if a gallon covers 285 sq. ft. 2 coats?
8. If the pitch (or slope) of a roof is the ratio of the rise to the span, what is the pitch of a roof whose rise is 6 ft. and span is 24 ft.? Find the rise of a roof whose span is 21 ft. and pitch is $\frac{1}{3}$.
9. If three bundles, each containing 26 shingles, cover a square (100 sq. ft.), how many bundles of shingles are needed to cover a roof of 15 ft. by 40 ft.? What is the total cost of the shingles at $6.95 per bundle?
10. How many rolls of roofing are needed to cover the roof of a house 20 ft. by 35 ft. if each roll covers 100 sq. ft. (or 1 square)? At $9.50 per square, what is the cost of the roofing?

15–39 RIM SPEEDS

The *rim speed* of a revolving object (also called *surface speed* or *cutting speed*) is the number of feet per minute that a point on the circumference of the revolving object travels.

EXERCISES

1. What is the cutting speed of a bandsaw with a diameter of 35 inches when it revolves at 650 r.p.m.?
2. Find the surface speed of a pulley with a diameter of 16 inches turning at 300 r.p.m.

3. What is the surface speed of a grinding wheel 21 inches in diameter when it revolves at 800 r.p.m.?
4. A pulley 15 inches in diameter runs at 700 r.p.m. What is the surface speed?
5. Find the cutting speed of a 28-inch bandsaw when it runs at 900 r.p.m.

15–40 GEARS AND PULLEYS

In industry gears and pulleys are used to change the speed at which power is transmitted.

(1) Two meshed gears are related as follows: the number of teeth (T) of the driving gear times its number of revolutions per minute (R) equals the number of teeth (t) of the driven gear times its number of revolutions per minute (r). Formula: $TR = tr$.

(2) Two belted pulleys (wheels mounted so that they turn on axles) are related as follows: the diameter (D) of the driving pulley times its number of revolutions per minute (R) equals the diameter (d) of the driven pulley times its number of revolutions per minute (r). Formula: $DR = dr$.

EXERCISES

1. A 20-inch pulley turning 175 r.p.m. drives a 5-inch pulley. How many revolutions per minute is the 5-inch pulley turning?
2. How many teeth are required on a gear if it is to run at 270 r.p.m. when driven by a gear with 42 teeth running at 180 r.p.m.?
3. What size pulley must be used on a motor running at 1,300 r.p.m. in order to drive a 12-inch pulley at 325 r.p.m.?
4. A 30-inch pulley is belted to a 6-inch pulley. If the 30-inch pulley makes 250 r.p.m. how many revolutions per minute does the smaller pulley make?
5. How many revolutions per minute is a 30-tooth gear running when driven by a 45-tooth gear running at 150 r.p.m.? When driven by an 18-tooth gear running at 210 r.p.m.?

15–41 ELECTRICITY

The *ampere* is the unit that measures the *intensity of current* (I); the *volt* is the unit that measures the *electromotive force* (E); the *ohm* is the unit that measures the *resistance* of the conductor (R); the *watt* is the unit that measures *power* or *rate of work* (W) done by the current. These units are related as follows: The number of *volts* equals the number of *amperes* times the number of *ohms*. Formula: $E = IR$. The number of *watts* equals the number of *amperes* times the number of *volts*. Formula: $W = IE$. A *kilowatt* (K.W.) is 1,000 watts. A *kilowatt hour* (K.W. Hr.) is the unit that measures electrical energy. To find the amount of energy furnished, multiply the power in kilowatts by the time in hours.

EXERCISES

1. How many amperes of current are flowing through a 120-volt circuit with a resistance of 40 ohms?
2. How much power in watts is a toaster consuming when it takes 7.2 amperes at 110 volts?
3. Find the power consumed if a motor takes 10 amperes at 125 volts.
4. At 6 cents per kilowatt hour how much will it cost to burn eight 25-watt lamps for 5 hours each evening for 30 days?
5. How much will it cost at 5 cents per kilowatt hour to operate for 24 hours a motor that uses 15 amperes of current at 120 volts?
6. A 550-watt electric iron is used on a 110-volt circuit. How much current does it take? What is the resistance of the iron?

15–42 LEVERS

A *lever* is a simple machine that is used to move and lift objects. It consists of a board or bar resting on a support called the *fulcrum*. Two forces or weights act on it. One is the lifting force and the other is the weight of the object to be lifted or moved. The distance to the fulcrum from the point where each force or weight is applied is called the lever *arm*.

The common seesaw or teeterboard, on which a heavier person is balanced by a lighter person sitting on the other side of the fulcrum but further away, illustrates the basic property of the lever. It demonstrates that one force or weight (W) times its arm (D) equals the other weight (w) times its arm (d).

As a formula the law of the lever is expressed: $WD = wd$.

EXERCISES

1. Using formula $WD = wd$, find:

   a. W when $D = 8$, $w = 120$, and $d = 6$.
   b. D when $W = 175$, $w = 125$, and $d = 7$.
   c. w when $W = 600$, $D = 4$, and $d = 12$.
   d. d when $W = 1{,}000$, $D = 2\frac{1}{2}$, and $w = 100$.

2. Two boys sit on the opposite side of the fulcrum of a seesaw. Scott weighs 64 pounds and sits 3 feet from the fulcrum. How far from the fulcrum must Steve, who weighs 48 pounds, sit to balance the seesaw?
3. Marilyn sits 4 feet 6 inches from the fulcrum of a teeterboard and balances Charlotte who sits on the opposite side 3 feet 9 inches from the fulcrum. If Marilyn weighs 90 pounds, how much does Charlotte weigh?

4. How much force must be applied at a distance 5 feet 6 inches from the fulcrum to balance a downward force of 160 pounds applied 4 feet 8 inches on the other side of the fulcrum?
5. A 5-foot crowbar is placed under a 100-pound rock. If the fulcrum is 1 foot from the rock, how much force must be exerted to raise the rock?

15–43 READING A MICROMETER

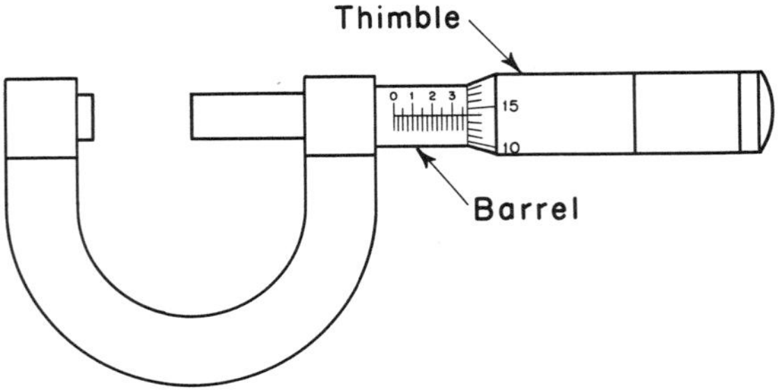

This setting reads .364"

To read a micrometer caliper setting, add the values of the exposed numbers on the barrel (each number = .1″), the uncovered subdivisions on the barrel (each subdivision = .025″), and the number of divisions on the thimble (each division = .001″).

EXERCISES

Read each of the following micrometer settings:

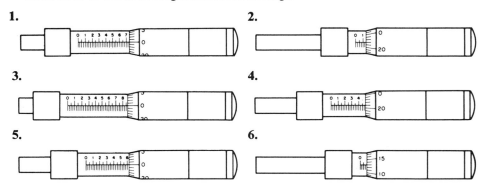

1.

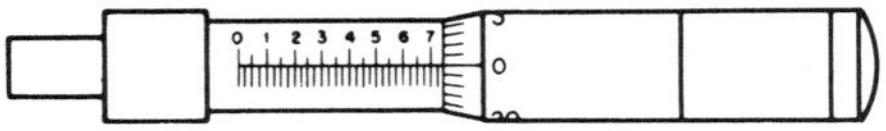

2.

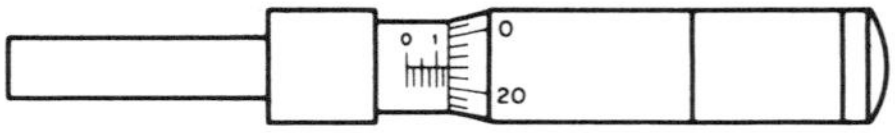

3.

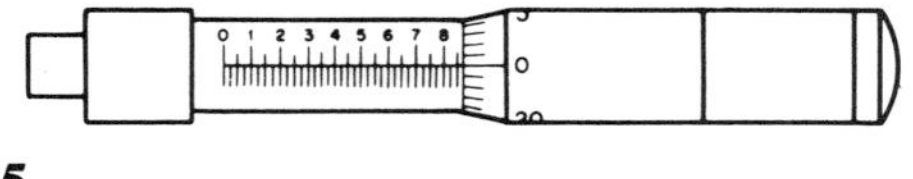

4.

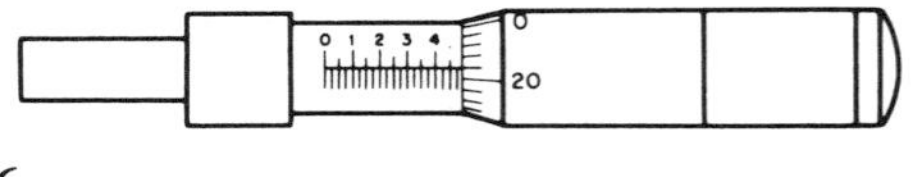

5.

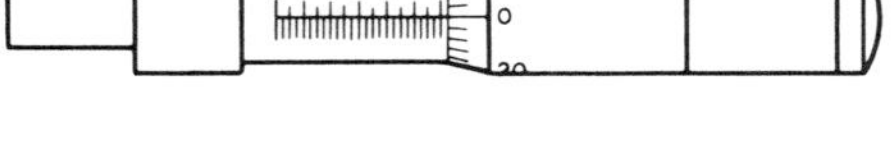

6.

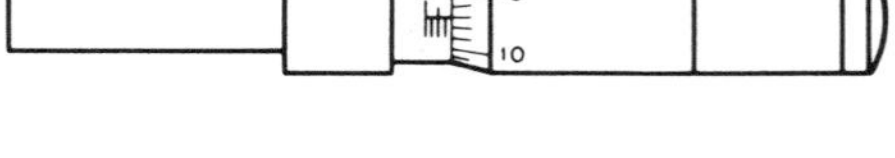

15–44 TOLERANCE

In the manufacture of various parts, a *tolerance* or a difference in the specified size is sometimes allowed and the part is accepted. If a 4-inch part has a tolerance of $\pm \frac{1}{8}''$, it means that any part within the range of $3\frac{7}{8}''$ to $4\frac{1}{8}''$ is acceptable.

EXERCISES

Measure each required part and the corresponding manufactured parts. Which of the manufactured parts are acceptable?

REQUIRED PART	MANUFACTURED PART
1. 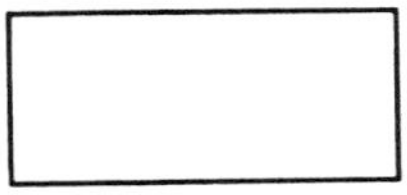 Tolerance $\pm \frac{1}{8}$ inch	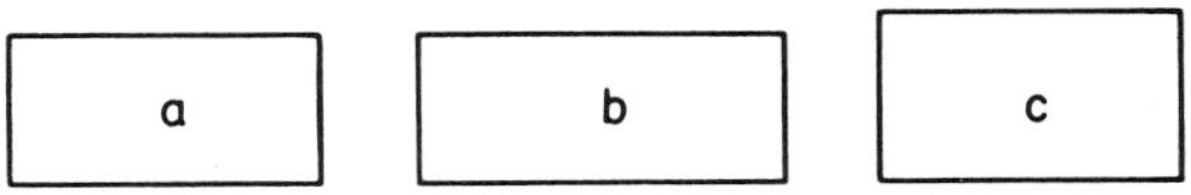
2. 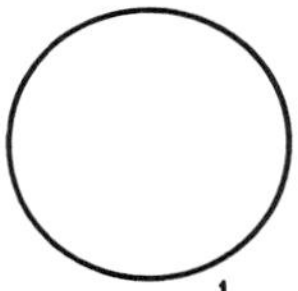 Tolerance $\pm \frac{1}{16}$ inch	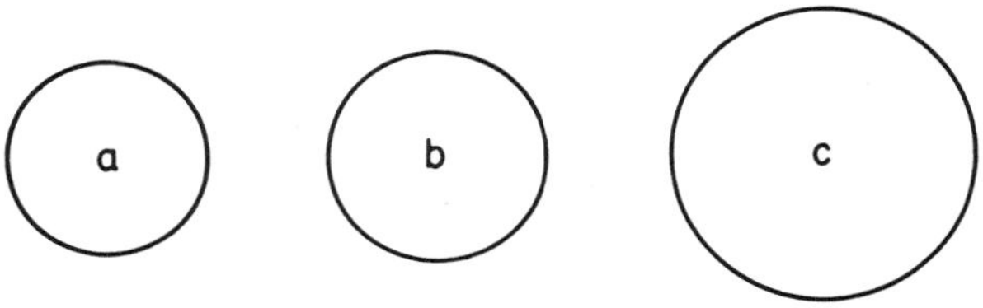
3. 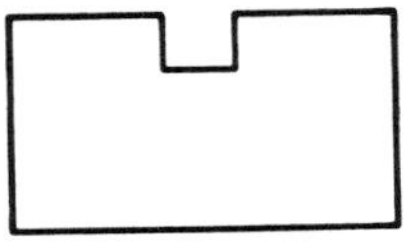 Tolerance $\pm \frac{1}{16}$ inch	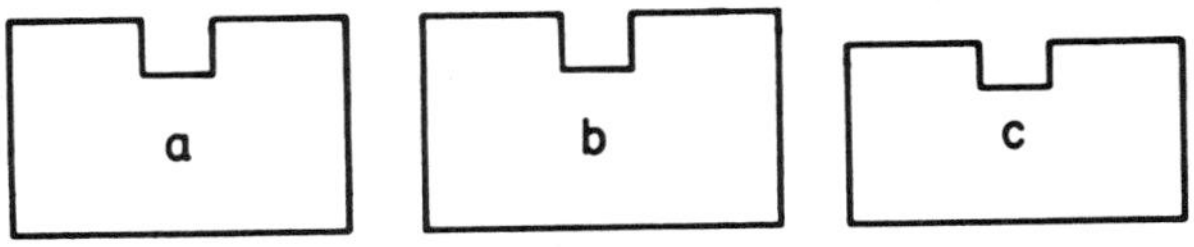

15–45 FINDING MISSING DIMENSIONS

EXERCISES

1. Find the overall length in each of the following figures:

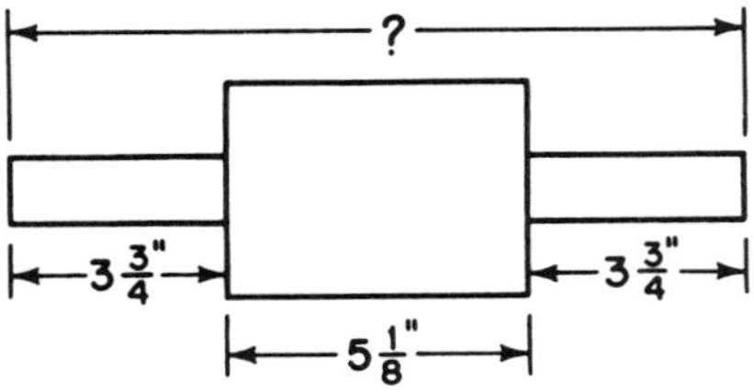

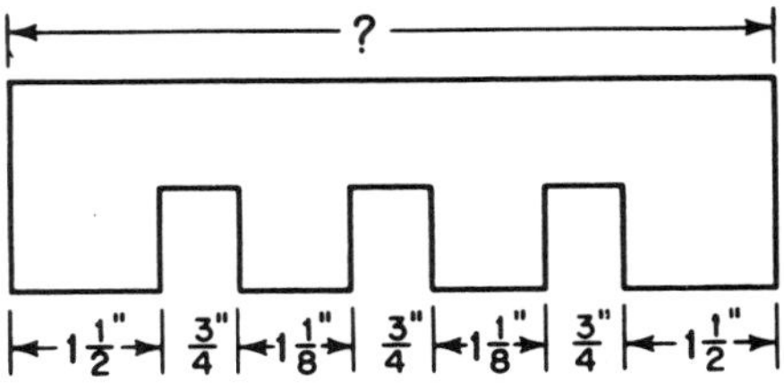

2. Find the missing dimension in each of the following figures:

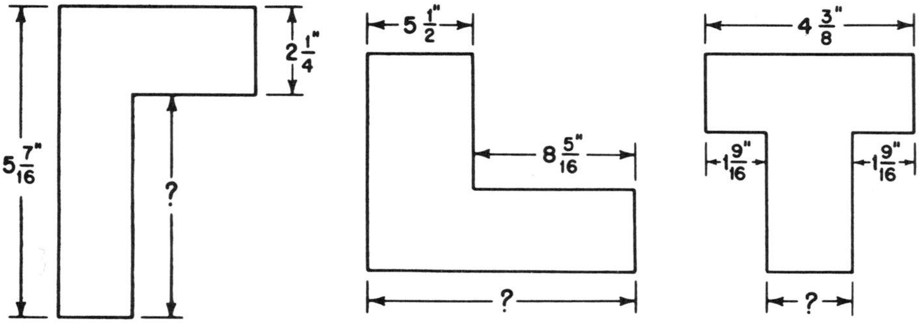

3. Find the missing dimension in each of the following figures:

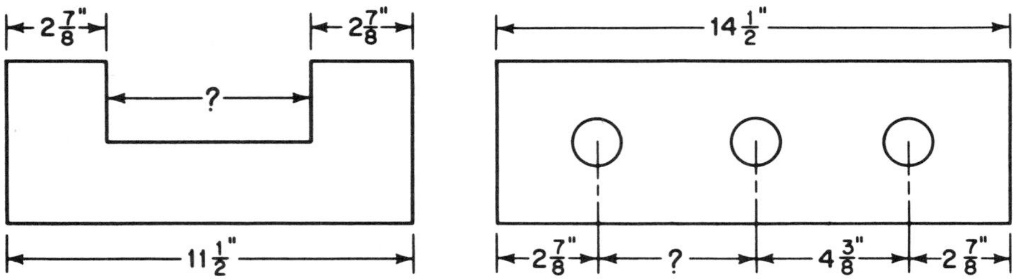

15–46 AREAS OF IRREGULAR FIGURES

EXERCISES

Find the area of each of the following figures:

1.

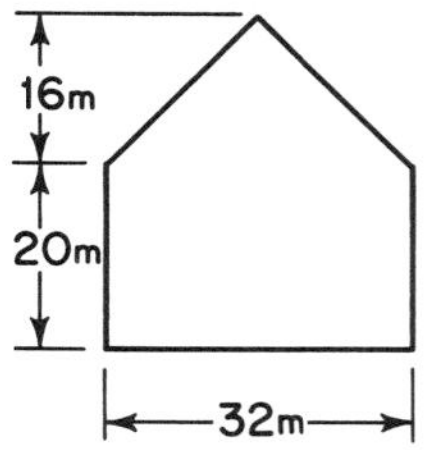

2.

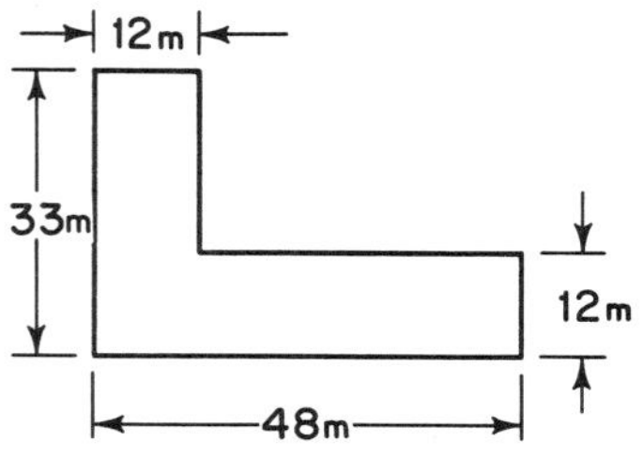

3.

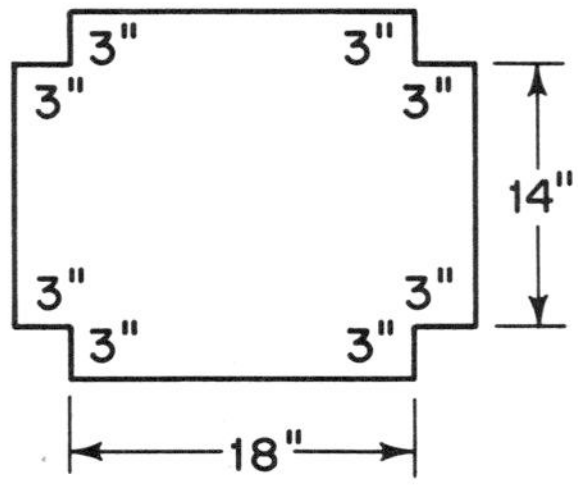

4.

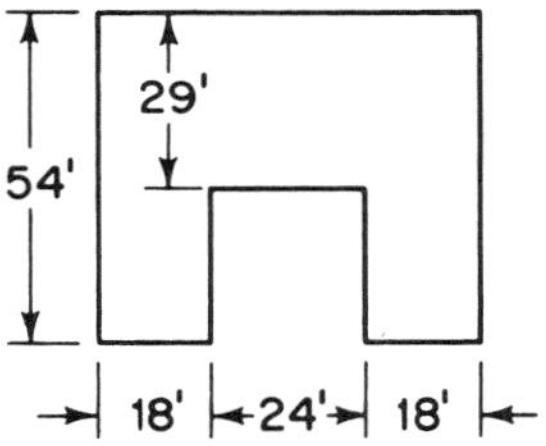

5.

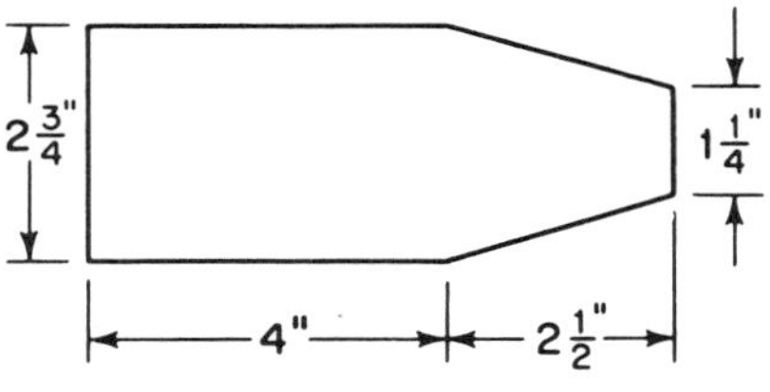

6.

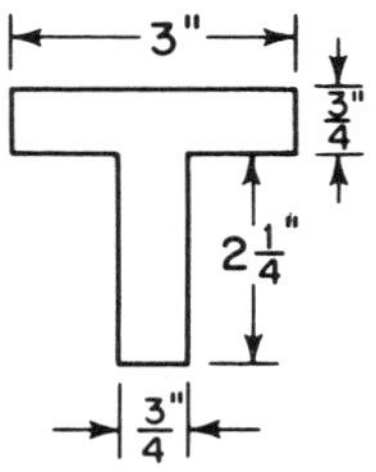

7.

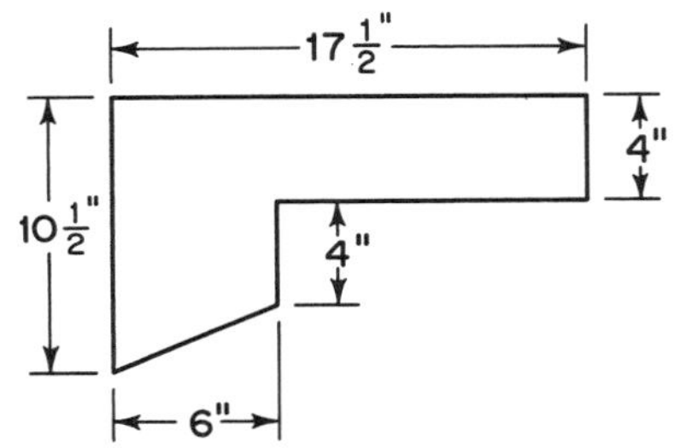

8.

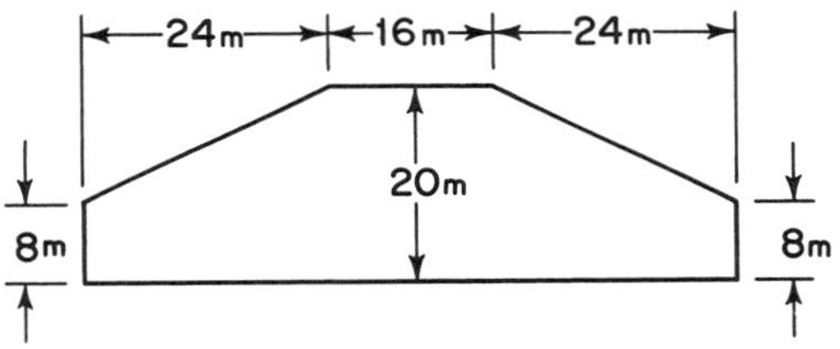

MISCELLANEOUS APPLICATIONS

15–47 LATITUDE AND LONGITUDE—NAUTICAL MILE

Meridians are imaginary semicircles that extend from the North Pole to the South Pole. *Parallels of latitude* are imaginary circles that are parallel to the Equator. The meridian that passes through Greenwich near London, England, is the *prime meridian* from which longitude is calculated. West longitude (W.) extends from this prime meridian (0° longitude) westward half way around the earth to the *International Date Line* (180° longitude); east longitude (E.) extends eastward from the prime meridian to the International Date Line. The *Equator* is 0° latitude; north latitude (N.) is measured north of the Equator; and south latitude (S.) is measured south of the Equator. The *North Pole* is 90° north latitude and the *South Pole* is 90° south latitude.

The position of any point on the earth's surface is determined by the intersection of its meridian of longitude and its parallel of latitude. One minute of arc of latitude equals one nautical mile and one degree equals 60 nautical miles. To find the distance between two points on the same meridian, we find the difference in latitude in minutes of arc, then change to nautical miles. Only along the equator does one minute of arc of longitude also equal one nautical mile.

EXERCISES

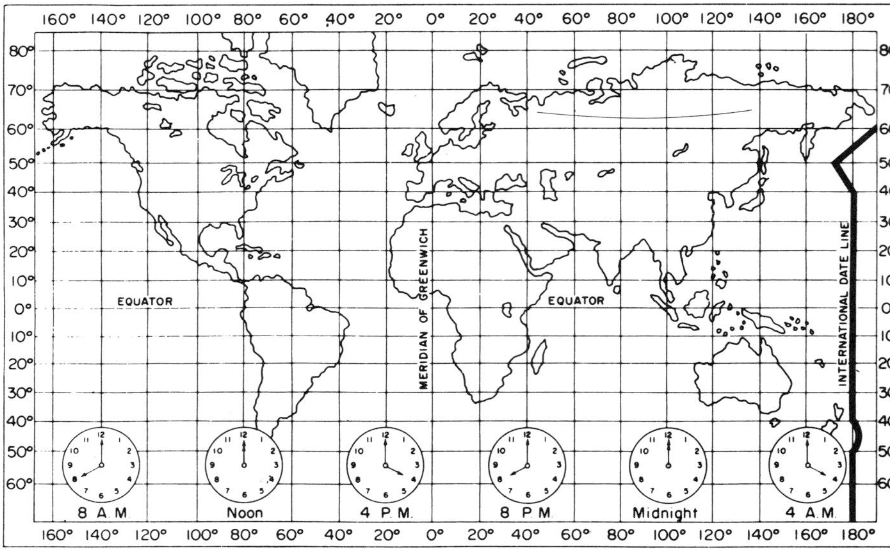

1. Locate on the above map the following points:

   a. 40° N., 60° W. **b.** 0°, 140° W. **c.** 30° S., 80° E. **d.** 25° S., 90° W.

2. Locate on the map the position (approximately) of each of the following cities:

   a. New Orleans, 30° N., 90° W.
   b. Helsinki, 60° N., 25° E.
   c. Philadelphia, 40° N., 75° W.
   d. Melbourne, 38° S., 145° E.
   e. Colon, Panama, 10° N., 80° W.
   f. Buenos Aires, 35° S., 58° W.

3. Find the difference in the following latitudes: **a.** 37° N. and 22° N. **b.** 13° S. and 51° S. **c.** 24° N. and 17° S. **d.** 56° 18′ N. and 19° 45′ N.

4. Find the difference in latitudes of the following cities:

   a. Los Angeles 34° N.; Rio de Janeiro 23° S.
   b. Cape Horn 56° S.; Cape Cod 42° N.
   c. Portland, Me., 43° 40′ N.; Portland, Ore., 45° 31′N.

5. Find the difference in the following longitudes: **a.** 29° W. and 107° W. **b.** 18° E. and 93° E. **c.** 25° W. and 82° E. **d.** 5° 27′ E. and 73° 54′ W.

6. Find the difference in longitudes of the following cities:

   a. Savannah 81° W.; Seattle 122° W.
   b. Honolulu 158° W.; Geneva 6° E.
   c. Sydney 151° E.; Rangoon 96° E.

7. Find how far apart in nautical miles points *A* and *B* are if:

	Location of Point *A*	Location of Point *B*
a.	35° N., 148° W.	21° N., 148° W.
b.	46° 30′ N., 75° 45′ W.	12° 15′ S., 75° 45′ W.

8. How many nautical miles is a ship located 35° 10′ N., 28° 20′ W. from one in distress 33° 40′ N., 28° 20′ W.? Steaming at 18 knots, how long will it take the first ship to reach the second?
9. How many nautical miles apart are St. Louis, 39° N., 90° W. and Guatemala, 15° N., 90° W.? How long should it take a plane at an average speed of 200 knots to fly from one city to the other?
10. How long should it take a plane at an average speed of 500 stat. m.p.h. to fly from Miami, 25° 47′ N., 80° W. to Pittsburgh, 40° 27′ N., 80° W.?

15–48 LONGITUDE AND TIME

The earth rotates about its axis from the west to the east, making one complete revolution each day. Thus, it takes the earth 24 hours to pass through 360° of longitude or 1 hour for 15° of longitude or 4 minutes for each degree of longitude.

The sun or solar time is the same at any given instant for all places located on any one meridian. Thus, to avoid the confusion that would arise if local sun time were used by neighboring towns, the earth is divided into time zones, each about 15° longitude in width. All places within each zone use the sun time of approximately its central meridian. The time in any zone is one hour earlier than the time in the next zone to the east and one hour later than the time in the next zone to the west.

Longitude is sometimes expressed in units of time. Both world time and longitude are calculated from the meridian of Greenwich. Thus, the longitude of any place is equal to the difference between its local sun time and Greenwich time.

To change units measuring longitude to units of time, we divide the units measuring longitude by 15. To change units of time to units of longitude, we multiply the units of time by 15. We simplify the answer when necessary.

EXERCISES

1. Express the following arcs of longitude as units of time:

a. 30°	**e.** 6°	**i.** 15° 30′	**m.** 30″
b. 75°	**f.** 58°	**j.** 4° 15′	**n.** 15′ 45″
c. 135°	**g.** 45′	**k.** 18° 10′	**o.** 90° 45′ 30″
d. 20°	**h.** 20′	**l.** 73° 18′	**p.** 105° 30′ 15″

2. Express the following units of time as arcs of longitude:

a. 3 hr.	**e.** 30 min.	**i.** 1 hr. 30 min.	**m.** 30 sec.
b. 7 hr.	**f.** 45 min.	**j.** 8 hr. 15 min.	**n.** 18 sec.
c. 19 hr.	**g.** 12 min.	**k.** 11 hr. 20 min.	**o.** 45 min. 15 sec.
d. 16 hr.	**h.** 48 min.	**l.** 6 hr. 54 min.	**p.** 4 hr. 30 min. 20 sec.

3. When it is 1200 at Greenwich, what time is it at each of the following longitudes:

   a. 45° E.? b. 90° W.? c. 105° W.? d. 30° 45′ E.? e. 120° 15′ W.?

4. If it is 0800 at 30° W. longitude, what time is it at 75° W. longitude?
5. What time is it at 42° E. longitude if it is 1530 at 28° W. longitude?
6. When it is 1500 at Greenwich, at what longitude is the time:

   a. 1200? b. 0800? c. 2300? d. 1345? e. 1630?

7. If it is 2145 at 15° W. longitude, at what longitude is the time 1715?
8. At what longitude is the time 0730 if it is 1315 at 20° E. longitude?
9. Find the difference between the standard time and the sun time at each of the following cities:

   a. New York City 73° 58′ W.
   b. Salt Lake City 111° 53′ W.
   c. Seattle 122° 19′ W.
   d. Indianapolis 86° 10′ W.

10. If the sun time at Chicago, 87° 37′ W. is 0600, what is the sun time at Baltimore, 76° 37′ W.?
11. What is the difference in standard time at Washington, 77° 04′ W. and Los Angeles, 118° 14′ W.? What is the difference in sun time at these two cities?
12. If the sun time at Oklahoma City, 97° 30′ W. is 1330, what is the sun time at each of the following cities?

    a. Boston, 71° 04′ W.
    b. San Diego, 117° 10′ W.
    c. Louisville, 85° 46′ W.
    d. Denver, 104° 57′ W.

15–49 THE MAGNETIC COMPASS

The direction in which an airplane flies over the earth's surface is called the *course* of the airplane and is expressed as an angle. When it is measured clockwise from the true north (North Pole), it is the *true course*. However, since the compass needle points to the magnetic north, the pilot must correct his true course reading. This correction is called *magnetic variation* or *variation* and the corrected course reading is called the *magnetic course*. The metal parts of the airplane may affect the compass. Thus, the pilot must correct the magnetic course reading. This correction is called *deviation* and the corrected course reading is called the *compass course*.

(1) To find magnetic course from true course, we add west variation but subtract east variation. To find compass course from magnetic course, we add west deviation but subtract east deviation.

(2) To find magnetic course from compass course, we subtract west deviation but add east deviation. To find true course from magnetic course, we subtract west variation but add east variation.

EXERCISES

1. Find the magnetic course:

	True Course	Variation
a.	268°	9° W.
b.	95°	4° E.
c.	157°	0°

2. Find the true course:

	Magnetic Course	Variation
a.	67°	2° W.
b.	301°	10° E.
c.	194°	0°

3. What is the variation when the true course is 110° and magnetic course is 107°?

4. Find the compass course:

	Magnetic Course	Deviation
a.	172°	5° W.
b.	53°	4° E.
c.	298°	0°

5. Find the magnetic course:

	Compass Course	Deviation
a.	42°	6° W.
b.	133°	1° E.
c.	201°	0°

6. What is the deviation when the magnetic course is 89° and the compass course is 93°?

7. Find the compass course:

	True Course	Variation	Deviation
a.	40°	8° W.	1° W.
b.	169°	6° E.	4° E.
c.	324°	11° W.	3° E.
d.	217°	9° E.	6° W.
e.	96°	0°	5° E.

8. Find the true course:

	Compass Course	Deviation	Variation
a.	225°	2° W.	9° W.
b.	61°	3° E.	7° E.
c.	344°	1° W.	5° E.
d.	180°	4° E.	8° W.
e.	15°	0°	3° W.

9. Find the missing numbers:

	True Course	Variation	Magnetic Course	Deviation	Compass Course
a.	28°	5° W.	?	4° E.	?
b.	?	3° E.	?	6° W.	192°
c.	202°	?	199°	?	203°
d.	315°	12° E.	?	5° E.	?
e.	88°	6° E.	?	6° W.	?

10. What compass course should a pilot steer if the true course is 89°, variation is 4° E., and deviation is 7° W.?

11. What is the true course if the compass course is 330°, variation is 5° W., and deviation is 2° W.?

12. The true course between two airports is 127°. If variation is 3° W. and deviation is 5° E., what should the compass course be?

13. A pilot is steering a compass course of 206°. If the variation is 9° E. and the deviation is 5° E., what is his true course?

14. Using the following data, determine a navigator's compass course: true course, 175°; variation, 6° E.; deviation, 4° W.

15. What are the variation and deviation if the true course is 39°, magnetic course is 31°, and the compass course is 36°?

15-50 AVIATION

Some interesting facts:

(1) *Lift-drag ratio* is the ratio of the lift of an airplane to its drag.

(2) *Aspect ratio* is the ratio of the length (span) of an airplane wing to its width (chord).

(3) *Tip speed of a propeller* is the number of feet per second that the tip of the propeller travels.

(4) *Radius of action* is the distance that an airplane can fly on a given amount of fuel in a given direction under known wind conditions and still return to the point where it took off.

(5) *Adiabatic lapse rate*—The temperature of a parcel of air decreases at the rate of $5\frac{1}{2}°$ Fahrenheit for each 1,000 feet it rises following the dry adiabat.

(6) A *Mach number* is the ratio of the speed of an object through a gas to the speed of sound through the gas. The air speed of the airplane is Mach 1 if it is equal to the speed of sound, Mach 2 if it is twice the speed of sound, Mach 0.5 if it is half the speed of sound. Speeds faster than the speed of sound are *supersonic* speeds.

EXERCISES

1. The sectional aeronautical maps of the United States are made at a scale of 1 to 500,000. How many miles do 3 inches on a sectional aeronautical map represent?

2. Using the dry adiabatic lapse rate of $5\frac{1}{2}°$ F. per 1,000 ft., find the temperature at an altitude of 8,000 ft. if the temperature at the surface is 58° F.

3. For a certain air density the true altitude is 3% less than the indicated altitude. What is the true altitude if the indicated altitude is 9,000 ft.?

4. What is the radius of action of an airplane flying at an average ground speed of 235 m.p.h. if it has fuel for 12 hours?

5. The true air speed at a certain altitude and temperature is 8% greater than the indicated air speed. If the indicated air speed is 175 m.p.h., what is the true air speed?

6. The true course from airport *M* to airport *N* is 136°. If the variation is 7° W. and the deviation is 3° E., what compass course should be steered?

7. Find the tip speed of a 7-foot propeller turning at 1,700 revolutions per minute.

8. a. Find the Lift-Drag Ratio if the lift of an airplane is 1,820 lb. and the drag is 140 lb.

b. Find the Aspect Ratio of an airplane having a span of 54 ft. and a wing area of 324 sq. ft.

9. If it is 1200 at Greenwich, what time is it at the following longitudes: 45° W.? 60° E.? 135° W.?

10. How many nautical miles apart are Raleigh, S.C., 35° 47′ N., 78° 30′ W., and Quito, Ecuador, 0° 13′ S., 78° 30′ W.? How many statute miles apart? If it takes an airplane 12 hours to fly between the two cities, what is the average ground speed of the airplane in knots? in statute miles per hour?

11. Make a drawing at the scale 1 inch = 80 miles showing the position, at the end of one hour, of an airplane flying due east at an air speed of 200 m.p.h. but blown off its course by a 45 m.p.h. north wind. How far was the airplane from its starting point? What course (or track) was it actually flying?

12. If the speed of sound is 740 miles per hour:

a. What speed in miles per hour is expressed by:
Mach 1? Mach 1.6? Mach 3? Mach 4.2? Mach 2.67?

b. Express each of the following speeds by a Mach number:
1,480 m.p.h., 370 m.p.h., 2,590 m.p.h., 1,702 m.p.h., 2,000 m.p.h.

15–51 THE WEATHER MAP

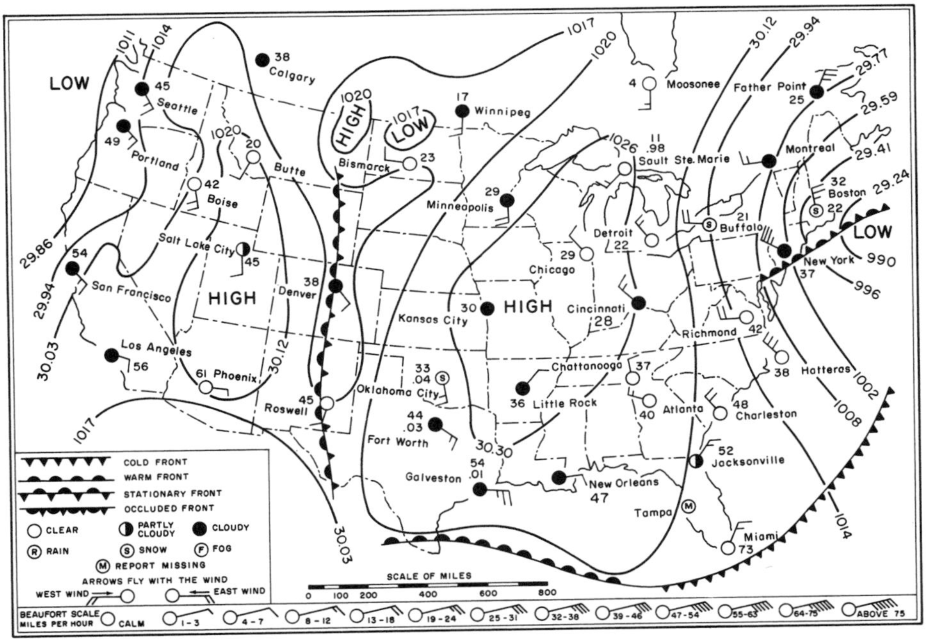

EXERCISES

Use weather map above to find answers for problems 1 through 5.

1. At what given cities is it snowing? Partly cloudy? Clear?

2. If the figures beside the station circle indicate the Fahrenheit temperature and the decimal beneath the temperature indicates the precipitation in inches during the past six hours, what is the temperature and precipitation at Galveston? At Oklahoma City? At Fort Worth? At Salt Lake City? Winnipeg? Detroit? Jacksonville? Phoenix? San Francisco?

3. The black solid lines called *isobars* are lines of equal barometric pressure and are expressed both in *inches* and *millibars*. One inch of pressure equals 33.86 millibars.

   a. Express each of the following pressures in millibars:
   30.30 in., 29.94 in., 29.41 in., 30.12 in.
   b. Express each of the following pressures in inches:
   1020 mb., 1002 mb., 1011 mb., 990 mb.
   c. What is the highest pressure, expressed in millibars, shown on the map? Change pressure to inches and check with map.
   d. What is the lowest pressure, expressed in inches, shown on the map? Change pressure to millibars and check with map.
   e. What is the pressure, expressed both in millibars and inches, at Bismarck? Through what other city does the same isobar pass?

4. Winds move counter-clockwise toward the center of low-pressure areas and clockwise and outward from high-pressure areas.

   a. Check the direction of winds at Detroit, Cincinnati, Little Rock, Oklahoma City, and Minneapolis, cities in a high-pressure area.
   b. Check the direction of winds at Boston, New York, Richmond, and Hatteras, cities in a low-pressure area.

5. Using the Beaufort Scale, give the direction and velocity of winds at:

   a. Boston
   b. Salt Lake City
   c. Kansas City
   d. New Orleans
   e. Seattle
   f. Miami
   g. New York
   h. Minneapolis
   i. Hatteras

Some newspapers include a table showing the existing conditions throughout the country as received in reports to the United States Weather Bureau. Use table on page 547 to find answers for problems 6 through 11.

6. Which city had the highest temperature of the day? Lowest temperature?
7. Where did the most precipitation fall? How much?
8. What was the wind velocity at Buffalo? Norfolk? St. Louis? Indianapolis?
9. If average temperature is determined by finding the average of the given high and low temperatures, what was the average temperature at:

   a. Cleveland?
   b. Chicago?
   c. Altanta?
   d. New York?
   e. Charleston?
   f. New Orleans?
   g. Los Angeles?
   h. El Paso?
   i. San Francisco?
   j. Boston?
   k. Miami?
   l. Seattle?
   m. Detroit?
   n. Cincinnati?
   o. Phoenix?
   p. Philadelphia?

City	Temp. High	Temp. Low	Precipitation	Wind	Weather
Atlanta	46	39	.28	27	Cloudy
Atlantic City	39	37	.45	23	Rain
Bismarck	23	8	T.	12	Snow
Boston	41	29	..	5	Cloudy
Buffalo	45	32	T.	8	Rain
Charleston	56	49	1.00	12	Cloudy
Chicago	36	15	.03	19	Clear
Cincinnati	54	24	.14	19	Clear
Cleveland	53	21	.36	22	Rain
Denver	50	25	..	25	Cloudy
Des Moines	32	10	..	12	Clear
Detroit	41	19	.07	20	Cloudy
Duluth	24	−8	..	28	Clear
El Paso	56	42	..	12	Cloudy
Galveston	68	47	..	20	P. Cloudy
Harrisburg	45	32	.25	3	Rain
Indianapolis	49	22	.02	27	Rain
Jacksonville	71	32	.17	10	Clear
Kansas City	37	22	..	8	Clear
Los Angeles	59	48	..	Calm	Cloudy
Miami	77	65	.14	10	P. Cloudy
Minneapolis	24	−1	..	11	Clear
Nantucket	36	21	..	9	Cloudy
New Orleans	67	49	..	18	Clear
New York	42	35	.04	10	Rain
Norfolk	46	41	.70	23	Cloudy
Philadelphia	42	32	.43	8	Rain
Phoenix	69	40	..	3	Clear
Pittsburgh	55	31	.08	20	Rain
Portland, Me.	42	24	..	5	Cloudy
Portland, Ore.	62	34	..	14	Clear
St. Louis	40	20	.11	14	Clear
Salt Lake City	37	29	..	5	Cloudy
San Antonio	65	39	..	26	Clear
San Francisco	54	49	.01	8	Cloudy
S. Ste. Marie	32	3	.06	17	Snow
Savannah	58	48	.49	12	Clear
Seattle	57	30	..	5	Clear
Washington	44	33	.53	8	Rain
Wilkes-Barre	48	29	.14	8	Rain
Winnipeg	3	−28	..	9	Clear

Temperature in Fahrenheit, precipitation in inches, wind in m.p.h.

10. Which city had the greatest range in temperature? Smallest range?
11. For each of the following cities find how many degrees the average temperature, determined from the table, is above or below the normal average temperature for the date:

	Normal Temperature		Normal Temperature		Normal Temperature
a. Atlanta	48	**i.** Denver	36	**q.** Minneapolis	26
b. Bismarck	21	**j.** Detroit	34	**r.** New Orleans	59
c. Boston	37	**k.** Galveston	60	**s.** New York	39
d. Buffalo	34	**l.** Indianapolis	37	**t.** Philadelphia	41
e. Charleston	55	**m.** Jacksonville	59	**u.** Phoenix	56
f. Chicago	34	**n.** Kansas City	38	**v.** Pittsburgh	38
g. Cincinnati	37	**o.** Los Angeles	59	**w.** St. Louis	40
h. Cleveland	36	**p.** Miami	71	**x.** San Antonio	57

15–52 SPORTS

EXERCISES

Scoring

Find the final scores of the following games:

1. Football

a. Jefferson	14	0	13	0 =
Springdale	7	9	6	7 =

b. Fernwood	0	7	12	13 =
Washington	6	16	0	3 =

2. Basketball

a. Kingston	13	21	17	18 =
Bartram	18	19	16	24 =

b. Central	15	17	14	23 =
Jackson	12	19	18	19 =

3. Baseball

a. North	0	1	1	3	0	2	0	4	1 =
Greenville	2	0	4	0	1	2	1	1	0 =
b. Wagner	1	0	2	0	1	3	0	2	0 =
Dalton	1	1	2	0	0	4	0	0	3 =

4. Golf

Find the score of the first 9 holes, the second 9 holes, and total.

Harrison	4	3	5	4	4	5	3	4	5 =
Santos	4	4	5	4	3	5	4	4	4 =
Harrison	5	3	4	4	4	3	5	4	4 =
Santos	4	5	4	4	5	4	5	3	4 =

5. Bowling—3-game match

Find **a.** which player scored the most points, **b.** which team won more games, and **c.** which team had the greater total match score.

Williams	154	196	205 =		Daniels	163	150	194 =	
Sonlin	148	159	165 =		Watson	188	217	163 =	
Harris	206	178	157 =	___	Rivera	198	182	185 =	___

6. Track

Find the total points scored by each team if 5 points are given for each 1st place, 3 points for each 2nd place, and 1 point for each 3rd place.

	1	2	3	Total points
Team A	5	3	4 =	
Team B	4	3	3 =	
Team C	2	6	1 =	
Team D	3	1	5 =	
Team E	1	2	2 =	

7. Wrestling

Find the total points scored by each team if 5 points are given for a fall, 3 points for a decision, and 2 points for a draw.

	Team X	Team Y
Fall	2	1
Decision	2	3
Draw	1	1
Total points		

8. Cross-country

Find the scores of Teams A, B, C, D, and E by adding the numbers corresponding to the order in which the first 5 members of each team finished. The team with the lowest score (or smallest sum) is the winner. Determine which team won if the runners placed as follows:

1. B **2.** A **3.** D **4.** E **5.** C **6.** A **7.** D **8.** D **9.** E **10.** B **11.** A
12. C **13.** C **14.** A **15.** C **16.** B **17.** D **18.** B **19.** E **20.** E **21.** E **22.** C
23. A **24.** E **25.** B **26.** D **27.** C **28.** A **29.** D **30.** E **31.** B **32.** A

Box Scores

9. Baseball—Find the totals:

Jefferson High

	AB.	R.	H.	P.O.	A.	E.
Jones L.F.	5	1	2	1	0	0
Williams C.F.	4	0	1	2	0	0
Adams 1B.	4	2	3	13	2	1
Harris R.F.	3	0	2	0	1	0
Brunner C.	4	1	0	2	2	0
Wagner S.S.	3	0	1	0	8	0
Carter 3B.	4	0	2	2	1	1
Choy 2B.	4	1	3	4	5	0
Benson P.	4	0	0	0	2	0
Totals						

Fernwood High

	AB.	R.	H.	P.O.	A.	E.
Watson 1B.	5	0	1	9	2	0
Thomas L.F.	4	2	3	4	0	0
Greene 3B.	5	0	1	2	5	2
Kelley C.F.	4	1	2	3	0	0
Gorson S.S.	3	1	0	1	4	1
Hall C.	4	0	1	4	3	0
Barner 2B.	4	2	4	2	3	0
Jenkins R.F.	4	0	2	1	0	0
Morton P.	3	0	1	1	0	0
Totals						

Jefferson High	0	1	0	0	2	0	1	0	1 =
Fernwood High	0	0	3	0	1	0	0	2	x =

Team Standing

10. Find the percent of games won by each team. Express it as a 3-place decimal numeral.

a. Baseball

	Won	Lost	Pct.
North	16	4	
Hudson	15	5	
Central	12	8	
East	10	10	
West	6	14	
South	1	19	

b. Football

	Won	Lost	Pct.
South	10	0	
East	8	2	
Hudson	4	6	
Central	4	6	
North	3	7	
West	1	9	

Averages

11. Baseball—Find the average of each player:

a. Batting

	AB	R	H	Avg.
Bevan	95	17	38	
Riley	96	19	32	
Ward	88	13	29	
Morgan	90	16	27	
Madero	84	18	24	

b. Fielding

	PO	A	E	Avg.
Hatton	108	16	1	
Chapman	60	87	3	
Souza	51	43	2	
Lee	182	34	9	
Jackson	92	26	7	

12. Basketball—

Find the total points scored and average per game to nearest tenth for each player:

	Games	Field Goals	Foul Goals	Total Points	Average
Farley	18	136	74		
Burns	17	129	80		
Monroe	15	123	87		
Schmidt	16	119	68		
Wood	18	108	59		
Jones	12	95	47		

13. Football—

Find the average weight of the line, of the backfield, and of the entire team:

		Wt.
Boland	re	180
Mallon	rt	195
Myers	rg	210
Walsh	c	198
Larsen	lg	205
Webb	lt	192
Ramos	le	183

		Wt.
Stewart	hb	175
Roberts	qb	168
Parker	fb	194
Green	hb	181

TABLE OF SQUARES AND SQUARE ROOTS

No.	Square	Square Root	No.	Square	Square Root	No.	Square	Square Root
1	1	1.000	**34**	1,156	5.831	**67**	4,489	8.185
2	4	1.414	**35**	1,225	5.916	**68**	4,624	8.246
3	9	1.732	**36**	1,296	6.000	**69**	4,761	8.307
4	16	2.000	**37**	1,369	6.083	**70**	4,900	8.367
5	25	2.236	**38**	1,444	6.164	**71**	5,041	8.426
6	36	2.449	**39**	1,521	6.245	**72**	5,184	8.485
7	49	2.646	**40**	1,600	6.325	**73**	5,329	8.544
8	64	2.828	**41**	1,681	6.403	**74**	5,476	8.602
9	81	3.000	**42**	1,764	6.481	**75**	5,625	8.660
10	100	3.162	**43**	1,849	6.557	**76**	5,776	8.718
11	121	3.317	**44**	1,936	6.633	**77**	5,929	8.775
12	144	3.464	**45**	2,025	6.708	**78**	6,084	8.832
13	169	3.606	**46**	2,116	6.782	**79**	6,241	8.888
14	196	3.742	**47**	2,209	6.856	**80**	6,400	8.944
15	225	3.873	**48**	2,304	6.928	**81**	6,561	9.000
16	256	4.000	**49**	2,401	7.000	**82**	6,724	9.055
17	289	4.123	**50**	2,500	7.071	**83**	6,889	9.110
18	324	4.243	**51**	2,601	7.141	**84**	7,056	9.165
19	361	4.359	**52**	2,704	7.211	**85**	7,225	9.220
20	400	4.472	**53**	2,809	7.280	**86**	7,396	9.274
21	441	4.583	**54**	2,916	7.348	**87**	7,569	9.327
22	484	4.690	**55**	3,025	7.416	**88**	7,744	9.381
23	529	4.796	**56**	3,136	7.483	**89**	7,921	9.434
24	576	4.899	**57**	3,249	7.550	**90**	8,100	9.487
25	625	5.000	**58**	3,364	7.616	**91**	8,281	9.539
26	676	5.099	**59**	3,481	7.681	**92**	8,464	9.592
27	729	5.196	**60**	3,600	7.746	**93**	8,649	9.644
28	784	5.292	**61**	3,721	7.810	**94**	8,836	9.695
29	841	5.385	**62**	3,844	7.874	**95**	9,025	9.747
30	900	5.477	**63**	3,969	7.937	**96**	9,216	9.798
31	961	5.568	**64**	4,096	8.000	**97**	9,409	9.849
32	1,024	5.657	**65**	4,225	8.062	**98**	9,604	9.899
33	1,089	5.745	**66**	4,356	8.124	**99**	9,801	9.950

TABLES OF MEASURE

Measure of Length—Metric

Table 1

10 millimeters (mm) = 1 centimeter (cm)
10 centimeters (cm) = 1 decimeter (dm)
10 decimeters (dm) = 1 meter (m)
10 meters (m) = 1 dekameter (dam)
10 dekameters (dam) = 1 hectometer (hm)
10 hectometers (hm) = 1 kilometer (km)

Table 2

1 centimeter (cm) = 10 millimeters (mm)
1 meter (m) = 100 centimeters (cm) = 1,000 millimeters (mm)
1 kilometer (km) = 1,000 meters (m)

Measure of Capacity—Metric

10 milliliters (ml) = 1 centiliter (cl)
10 centiliters (cl) = 1 deciliter (dl)
10 deciliters (dl) = 1 liter
10 liters = 1 dekaliter (dal)
10 dekaliters (dal) = 1 hectoliter (hl)
10 hectoliters (hl) = 1 kiloliter (kl)

Measure of Weight—Metric

10 milligrams (mg) = 1 centigram (cg)
10 centigrams (cg) = 1 decigram (dg)
10 decigrams (dg) = 1 gram (g)
10 grams (g) = 1 dekagram (dag)
10 dekagrams (dag) = 1 hectogram (hg)
10 hectograms (hg) = 1 kilogram (kg)
1,000 kilograms (kg) = 1 metric ton (t)

Metric Measures of Area, Volume, and Miscellaneous Equivalents

See pages 196 and 197.

Measure of Length—Customary

1 foot (ft.) = 12 inches (in.)
1 yard (yd.) = 3 feet (ft.) = 36 inches (in.)
1 rod (rd.) = $16\frac{1}{2}$ feet (ft.) = $5\frac{1}{2}$ yards (yd.)
1 statute mile (stat. mi.) = 5,280 feet (ft.) = 1,760 yards (yd.) = 320 rods (rd.)
1 statute mile (stat. mi.) $\approx$ 0.87 nautical mile (0.8684 naut. mi.)
1 nautical mile (naut. mi.) $\approx$ 6,080 feet (6,080.2 ft.)
$\approx$ 1.15 statute miles (1.1515 stat. mi.)
1 fathom (fath.) $\approx$ 6 feet (ft.)

Measure of Area—Customary

1 square foot (sq. ft.) = 144 square inches (sq. in.)
1 square yard (sq. yd.) = 9 square feet (sq. ft.)
1 square rod (sq. rd.) = 30.25 square yards (sq. yd.)
1 acre = 160 square rods (sq. rd.)
= 4,840 square yards (sq. yd.)
= 43,560 square feet (sq. ft.)
1 square mile (sq. mi.) = 640 acres

Measure of Volume—Customary

1 cubic foot (cu. ft.) = 1,728 cubic inches (cu. in.)
1 cubic yard (cu. yd.) = 27 cubic feet (cu. ft.)

Liquid Measure—Customary

1 pint (pt.) = 16 ounces (oz.)
= 4 gills (gi.)
1 quart (qt.) = 2 pints (pt.)
1 gallon (gal.) = 4 quarts (qt.)

Dry Measure—Customary

1 quart (qt.) = 2 pints (pt.)
1 peck (pk.) = 8 quarts (qt.)
1 bushel (bu.) = 4 pecks (pk.)

Measure of Weight—Avoirdupois—Customary

1 pound (lb.) = 16 ounces (oz.)
1 short ton (sh. tn. or T.) = 2,000 pounds (lb.)
1 long ton (l. ton) = 2,240 pounds (lb.)

Volume, Capacity, and Weight Equivalents—Customary

1 gallon (gal.) ≈ 231 cubic inches (cu. in.)
1 cubic foot (cu. ft.) ≈ $7\frac{1}{2}$ gallons (gal.)
1 bushel (bu.) ≈ $1\frac{1}{4}$ cubic feet (cu. ft.)
≈ 2,150.42 cubic inches (cu. in.)
1 cu. ft. of fresh water weighs $62\frac{1}{2}$ pounds (lb.)
1 cu. ft. of sea water weighs 64 pounds (lb.)

Measure of Time

1 minute (min.) = 60 seconds (sec.)
1 hour (hr.) = 60 minutes (min.)
1 day (da.) = 24 hours (hr.)
1 week (wk.) = 7 days (da.)
1 year (yr.) = 12 months (mo.)
= 52 weeks (wk.)
= 365 days (da.)

Angles and Arcs

1 circle = 360 degrees (°)
1 degree (°) = 60 minutes (′)
1 minute (′) = 60 seconds (″)

Measure of Speed

1 knot = 1 nautical m.p.h.

Metric—Customary Equivalents

1 meter = 39.37 inches
≈ 3.28 feet (3.2808)
≈ 1.09 yards (1.0936)
1 centimeter ≈ .39 or .4 inches (.3937)
1 millimeter ≈ .04 inches (.03937)
1 kilometer ≈ .62 miles (.6214)
1 liter ≈ 1.06 liquid quarts (1.0567)
1 liter ≈ .91 dry quart (.9081)
1 gram ≈ .04 ounce (.0353)
1 kilogram ≈ 2.2 pounds (2.2046)
1 metric ton ≈ 2,204.62 pounds
1 inch = 25.4 millimeters
1 foot ≈ .3 meter (.3048)
1 yard ≈ .91 meter (.9144)
1 mile ≈ 1.61 kilometers (1.6093)
1 liquid quart ≈ .95 liter (.9463)
1 dry quart ≈ 1.1 liters (1.1012)
1 ounce ≈ 28.35 grams (28.3495)
1 pound ≈ .45 kilogram (.4536)
1 short ton ≈ .91 metric ton (.9072)
1 square inch ≈ 6.45 square centimeters (6.4516)
1 cubic inch ≈ 16.39 cubic centimeters (16.3872)

Index